INTERNATIONAL TABLE OF ATOMIC WEIGHTS

Atomic Number	Symbol	Name	Atomic Weight
89	Ac	Actinium	[227]
13	Al	Aluminum	26.9815386
95	Am	Americium	[243]
51	Sb	Antimony	121.760
18	Ar	Argon	39.948
33	As	Arsenic	74.92160
85	At	Astatine	[210]
56	Ba	Barium	137.327
97	Bk	Berkelium	[247]
4	Be	Beryllium	9.012182
83	Bi	Bismuth	208.98040
107	Bh	Bohrium	[272]
5	B	Boron	10.811
35	Br	Bromine	79.904
48	Cd	Cadmium	112.411
55	Cs	Cesium	132.9054519
20	Ca	Calcium	40.078
98	Cf	Californium	[251]
6	C	Carbon	12.0107
58	Ce	Cerium	140.116
17	Cl	Chlorine	35.453
24	Cr	Chromium	51.9961
27	Co	Cobalt	58.933195
29	Cu	Copper	63.546
96	Cm	Curium	[247]
110	Ds	Darmstadtium	[281]
105	Db	Dubnium	[268]
66	Dy	Dysprosium	162.500
99	Es	Einsteinium	[252]
68	Er	Erbium	167.259
63	Eu	Europium	151.964
100	Fm	Fermium	[257]
9	F	Fluorine	18.9984032
87	Fr	Francium	[223]
64	Gd	Gadolinium	157.25
31	Ga	Gallium	69.723
32	Ge	Germanium	72.64
79	Au	Gold	196.966569
72	Hf	Hafnium	178.49
108	Hs	Hassium	[270]
2	He	Helium	4.002602
67	Ho	Holmium	164.93032
1	H	Hydrogen	1.00794
49	In	Indium	114.818
53	I	Iodine	126.90447
77	Ir	Iridium	192.217
6	Fe	Iron	55.845
36	Kr	Krypton	83.798
57	La	Lanthanum	138.90547
103	Lr	Lawrencium	[262]
82	Pb	Lead	207.2
3	Li	Lithium	6.941
71	Lu	Lutetium	174.9668
12	Mg	Magnesium	24.3050
25	Mn	Manganese	54.938045
109	Mt	Meitnerium	[276]
101	Md	Mendelevium	[258]
80	Hg	Mercury	200.59
42	Mo	Molybdenum	95.96
60	Nd	Neodymium	144.242
10	Ne	Neon	20.1797
3	Np	Neptunium	[237]
28	Ni	Nickel	58.6934
41	Nb	Niobium	92.90638
7	N	Nitrogen	14.0067
102	No	Nobelium	[259]
76	Os	Osmium	190.23
8	O	Oxygen	15.9994
46	Pd	Palladium	106.42
15	P	Phosphorus	30.973762
78	Pt	Platinum	195.084
94	Pu	Plutonium	[244]
84	Po	Polonium	[209]
19	K	Potassium	39.0983
59	Pr	Praseodymium	140.90765
61	Pm	Promethium	[145]
91	Pa	Protactinium	231.03588
88	Ra	Radium	[226]
86	Rn	Radon	[222]
75	Re	Rhenium	186.207
45	Rh	Rhodium	102.90550
111	Rg	Roentgenium	[280]
37	Rb	Rubidium	85.4678
44	Ru	Ruthenium	101.07
104	Rf	Rutherfordium	[267]
62	Sm	Samarium	150.36
21	Sc	Scandium	44.955912
106	Sg	Seaborgium	[271]
34	Se	Selenium	78.96
14	Si	Silicon	28.0855
47	Ag	Silver	107.8682
11	Na	Sodium	22.98976928
38	Sr	Strontium	87.62
16	S	Sulfur	32.065
73	Ta	Tantalum	180.94788
43	Tc	Technetium	[98]
52	Te	Tellurium	127.60
65	Tb	Terbium	158.92535
81	Tl	Thallium	204.3833
90	Th	Thorium	232.03806
69	Tm	Thulium	168.93421
50	Sn	Tin	118.710
22	Ti	Titanium	47.867
74	W	Tungsten	183.84
112	Uub	Ununbium	[285]
116	Uuh	Ununhexium	[293]
118	Uuo	Ununoctium	[294]
115	Uup	Ununpentium	[288]
114	Uuq	Ununquadium	[289]
113	Uut	Ununtrium	[284]
92	U	Uranium	238.02891
23	V	Vanadium	50.9415
54	Xe	Xenon	131.293
70	Yb	Ytterbium	173.054
39	Y	Yttrium	88.90585
30	Zn	Zinc	65.38
40	Zr	Zirconium	91.224

Chemistry for Engineering Students

Chemistry for Engineering Students

THIRD EDITION

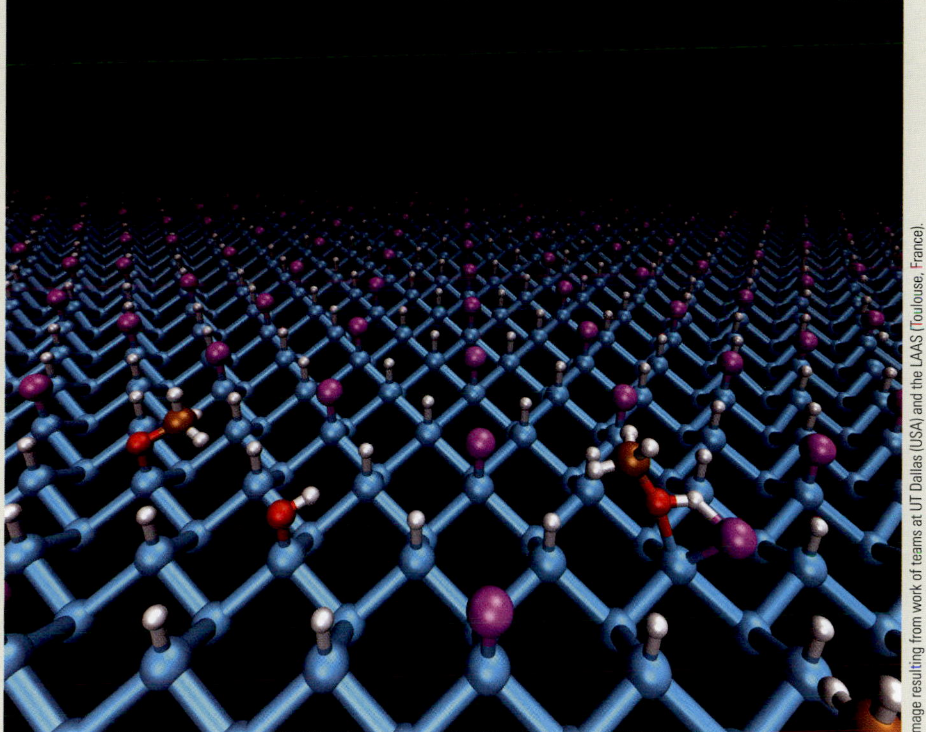

Image resulting from work of teams at UT Dallas (USA) and the LAAS (Toulouse, France).

The cover image shows a portion of a microelectromechanical system, or MEMS, along with a representation of the surface of silicon. MEMS devices such as this one can be used to process or analyze samples of DNA or other biochemical molecules, and provide an excellent example of the interface between chemistry and engineering.

Lawrence S. Brown

Texas A&M University

Thomas A. Holme

Iowa State University

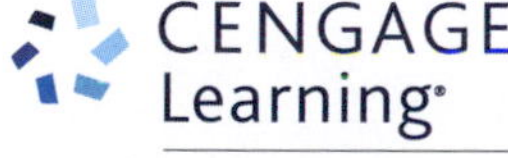

Australia • Brazil • Mexico • Singapore • United Kingdom • United States

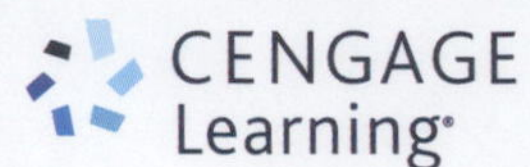

Chemistry for Engineering Students, Third Edition

Lawrence S. Brown, Thomas A. Holme

Product Director: Mary Finch

Senior Product Manager: Maureen Rosner

Content Developer: Alyssa White

Content Coordinator: Brendan Killion

Product Assistant: Karolina Kiwak

Media Developer: Maureen Ross

Brand Manager: Nicole Hamm

Market Development Manager: Janet del Mundo

Senior Content Project Manager: Carol Samet

Art Director: Maria Epes

Manufacturing Planner: Judy Inouye

Rights Acquisitions Specialist: Dean Dauphinais

Production Service: Teresa Christie, MPS North America

Photo Researcher: Jill Reichenbach, Q2A/Bill Smith

Text Researcher: Jill Krupnik, Q2A/Bill Smith

Copy Editor: MPS North America

Illustrator and Compositor: MPS Limited

Text Designer: tani hasegawa

Cover Designer: Bartay Studio

Cover Image: Microfluidic Microchip: David Scharf /Science Source

Cover Image: SNAP Silicon: Image resulting from work of teams at UT Dallas (USA) and the LAAS (Toulouse, France).

Library of Congress Control Number: 2013938466

Student Edition:
ISBN-13: 978-1-285-19902-3
ISBN-10: 1-285-19902-2

Loose-leaf Edition:
ISBN-13: 978-1-305-25667-5
ISBN-10: 1-305-25667-0

Cengage Learning
200 First Stamford Place, 4th Floor
Stamford, CT 06902
USA

Cengage Learning is a leading provider of customized learning solutions with office locations around the globe, including Singapore, the United Kingdom, Australia, Mexico, Brazil, and Japan. Locate your local office at **www.cengage.com/global**

Cengage Learning products are represented in Canada by Nelson Education, Ltd.

To learn more about Cengage Learning Solutions, visit **www.cengage.com**

Purchase any of our products at your local college store or at our preferred online store **www.cengagebrain.com**

Printed in the United States of America
2 3 4 17 16 15 14

About the Authors

Courtesy of Lawrence Brown/photo by Sherry Yennello

Larry Brown has been a faculty member at Texas A&M University since 1988, and in 2013 he was named Presidential Professor for Teaching Excellence. He received his B.S. in 1981 from Rensselaer Polytechnic Institute, and his M.A. in 1983 and Ph.D. in 1986 from Princeton University. During his graduate studies, Larry spent a year working in what was then West Germany. He was a Postdoctoral Fellow at the University of Chicago for two years before moving to Texas A&M. Over the years, he has taught more than 14,000 general chemistry students, most of them engineering majors. Larry's excellence in teaching has been recognized by awards from the Association of Former Students at Texas A&M at both the College of Science and University levels. A version of his class has been broadcast on KAMU-TV, College Station's PBS affiliate and is currently available on iTunesU. From 2001 to 2004, Larry served as a Program Officer for Education and Interdisciplinary Research in the Physics Division of the National Science Foundation. He also coordinates chemistry courses for Texas A&M's engineering program in Doha, Qatar. When not teaching chemistry, he enjoys road bicycling and playing soccer with his daughter Stephanie.

Courtesy of Thomas Holme/photo by Dennis Salisbury

Tom Holme is a Professor of Chemistry at Iowa State University and Director of the ACS Examinations Institute. He received his B.S. in 1983 from Loras College, and his Ph.D. in 1987 from Rice University. He began his teaching career as a Fulbright Scholar in Zambia, Africa and has also lived in Jerusalem, Israel and Suwon, South Korea. He is a fellow of the American Chemical Society and the American Association for the Advancement of Science. His research interests lie in computational chemistry, particularly as applied to understanding processes important for plant growth. He is also active chemical education research and has been involved with the general chemistry for engineers course at both Iowa State University and at the University of Wisconsin–Milwaukee where he was a member of the Chemistry and Biochemistry Department. He has received several grants from the National Science Foundation for work in assessment methods for chemistry, and the "*Focus on Problem Solving*" feature in this textbook grew out of one of these projects. He served as an Associate Editor on the encyclopedia "*Chemistry Foundations and Applications*." He was the lead editor of the laboratory manual for the new AP chemistry curriculum. In 1999 Tom won the ACS's Helen Free Award for Public Outreach for his efforts doing chemical demonstrations on live television in the Milwaukee area.

Brief Contents

Contents

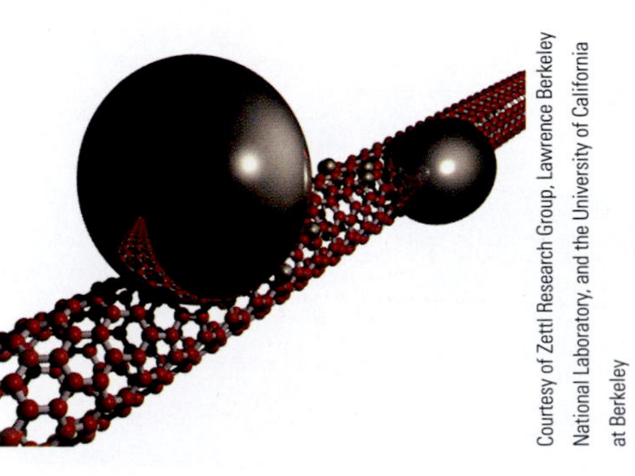

Courtesy of Zettl Research Group, Lawrence Berkeley National Laboratory, and the University of California at Berkeley

© Cengage Learning/Charles D. Winters

SimplyCreativePhotography/Getty Images

Freon-12, CF_2Cl_2

Photo: © Cengage Learning/ Charles D. Winters

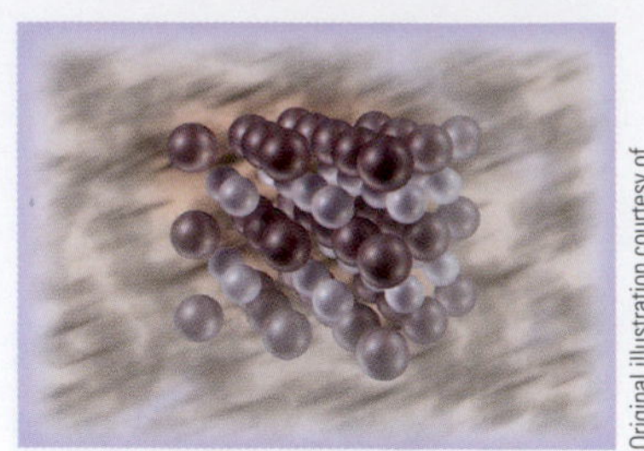
Original illustration courtesy of Raymond Schaak

10 Entropy and the Second Law of Thermodynamics 279

Thomas A. Holme

11 Chemical Kinetics 302

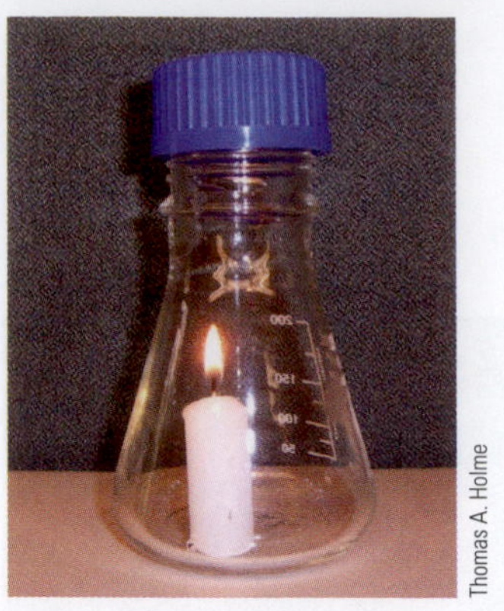

12 Chemical Equilibrium 340

Thomas A. Holme

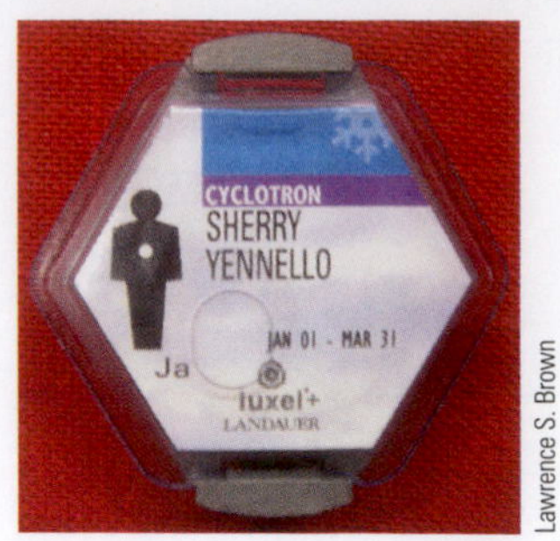
CYCLOTRON
SHERRY
YENNELLO
JAN 01 - MAR 31
luxel+
LANDAUER
Lawrence S. Brown

Appendixes

Preface

The Genesis of This Text

As chemists, we see connections between our subject and virtually everything. So the idea that engineering students should learn chemistry strikes most chemists as self-evident. But chemistry is only one of many sciences with which a practicing engineer must be familiar, and the undergraduate curriculum must find room for many topics. Hence, engineering curricula at more and more universities are shifting from the traditional year long general chemistry sequence to a single semester. And in most cases, these schools are offering a separate one-term course designed specifically for their engineering students. When schools—including our own—originally began offering these courses, there was no text on the market for them, so content from two-semester texts had to be heavily modified to fit the course. Although it is possible to do this, it is far from ideal. It became apparent that a book specifically geared for this shorter course was necessary. ***We have written this book to fill this need.***

Our goal is to instill an appreciation for the role of chemistry in many areas of engineering and technology and of the interplay between chemistry and engineering in a variety of modern technologies. For most engineering students, the chemistry course is primarily a prerequisite for courses involving materials properties. These courses usually take a phenomenological approach to materials rather than emphasizing the chemist's molecular perspective. Thus one aim of this text is to provide knowledge of and appreciation for the chemical principles of structure and bonding that underpin materials science. This does *not* mean that we have written the book as a materials science text, but rather that the text is intended to prepare students for subsequent study in that area.

The book also provides sufficient background in the science of chemistry for a technically educated professional. Engineering, after all, is the creative and practical application of a broad array of scientific principles, so its practitioners should have a broad base in the natural sciences.

Content and Organization

The full scope of the traditional general chemistry course cannot be taught meaningfully in one semester or one or two quarters, so we have had to decide what content to include. There are basically two models used to condense the general chemistry curriculum. The first is to take the approach of an "essentials" book and reduce the depth of coverage and the number of examples but retain nearly all of the traditional topics. The second is to make more difficult and fundamental decisions as to what chemistry topics are proper and relevant to the audience, in this case future engineers. We chose the latter approach and built a 14-chapter book from the ground up to satisfy what we think are the goals of the course:

- Provide a concise but thorough introduction to the science of chemistry.
- Give students a firm foundation in the principles of structure and bonding as a foundation for further study of materials science.

- Show the connection between molecular behavior and observable physical properties.
- Show the connections between chemistry and the other subjects studied by engineering students, especially mathematics and physics.

Taken together, the 14 chapters in this book probably represent somewhat more material than can comfortably fit into a standard semester course. Thus departments or individual instructors will need to make some further choices as to the content that is most suitable for their own students. We suspect that many instructors will not choose to include all of the material on equilibrium in Chapter 12, for example. Similarly, we have included more topics in Chapter 8, on condensed phases, than we expect most faculty will include in their courses.

Topic Coverage

Courtesy of the U.S. Department of Energy's Ames Laboratory

The coverage of topics in this text reflects the fact that chemists constantly use multiple concepts to understand their field, often using more than one model simultaneously. Thus the study of chemistry we present here can be viewed from multiple perspectives: macroscopic, microscopic, and symbolic. The latter two perspectives are emphasized in Chapters 2 and 3 on atoms, molecules, and reactions. In Chapters 4 and 5, we establish more of the connection between microscopic and macroscopic in our treatment of stoichiometry and gases. We return to the microscopic perspective to cover more details of atomic structure and chemical bonding in Chapters 6 through 8. The energetic aspects of chemistry, including important macroscopic consequences, are considered in Chapters 9 and 10, and kinetics and equilibrium are treated in Chapters 11 and 12, respectively. Chapter 13 deals with electrochemistry and corrosion, an important chemistry application for many engineering disciplines. Finally, we conclude with a discussion of nuclear chemistry.

Specific Content Coverage

We know that there are specific topics in general chemistry that are vital to future engineers. We've chosen to treat them in the following ways.

Organic Chemistry: Organic chemistry is important in many areas of engineering, particularly as related to the properties of polymers. Rather than using a single organic chapter, we integrate our organic chemistry coverage over the entire text, focusing on polymers. We introduce organic polymers in Section 2.1 and use polymers and their monomers in many examples in this chapter. Chapter 2 also contains a rich discussion of organic line structures and functional groups and ends with a section on the synthesis, structure, and properties of polyethylene. Chapter 4 opens and ends with discussions of fuels, a topic to which we return in Chapter 9. Chapter 8 contains more on carbon and polymers, and the recycling of polymers provides the context for consideration of the second law of thermodynamics in Chapter 10.

Acid–Base Chemistry: Acid–base reactions represent another important area of chemistry with applications in engineering, and again we have integrated our coverage into appropriate areas of the text. Initially, we define acids and bases in conjunction with the introduction to solutions in Chapter 3. Simple solution stoichiometry is presented in Chapter 4. Finally, a more detailed treatment of acid–base chemistry is presented in the context of equilibria in Chapter 12.

Nuclear Chemistry: A chapter dealing with nuclear chemistry is included for those wishing to teach that topic. Coverage in this chapter includes fundamentals of nuclear reactions, nuclear stability and radioactivity, decay kinetics, and the energetic consequences of nuclear processes.

Mathematics: The math skills of students entering engineering majors generally are stronger than those in the student body at large, and most of the students taking a course of the type for which this book is intended will be concurrently enrolled in an introductory calculus course. In light of this, we include references to the role of calculus where appropriate via our **MathConnections** boxes. These essays expand and review math concepts as they pertain to the particular topic being studied, and appear wherever the links between the topic at hand and mathematics seems especially strong. These boxes are intended to be unobtrusive, so those students taking a precalculus math course will not be adversely affected. The point of including calculus is not to raise the level of material being presented, but rather to show the natural connections between the various subjects students are studying.

Connections between Chemistry and Engineering

Lawrence S. Brown

Because this book is intended for courses designed for engineering majors, we strive to present chemistry in contexts that we feel will appeal to the interests of such students. Links between chemistry and engineering are central to the structure of the text. Each chapter begins and ends with a section called ***INSIGHT INTO*** . . . , which introduces a template or theme showing the interplay between chemistry and engineering. These sections are only the beginning of the connections, and the theme introduced in the initial *Insight* appears regularly throughout that chapter.

We opt for currency in our engineering applications wherever possible, so throughout the book, we discuss recent key innovations in various fields. For example, Chapter 3 includes a discussion of the chemistry and engineering involved in the conversion of biomass to biofuels. In Chapter 7, we describe mesoporous silicon nanoparticles, a front-line research topic that may have important applications in biomedical engineering in the future. Chapter 8 closes with a discussion of the fabrication of micro-electrical-mechanical systems (MEMS).

Approach to Problem Solving

Problem solving is a key part of college chemistry courses and is especially important as a broadly transferable skill for engineering students. Accordingly, this text includes worked problems throughout. All of our Example Problems include a *Strategy* section immediately following the problem statement, in which we emphasize the concepts and relationships that must be considered to work the problem. After the solution, we often include a section called *Analyze Your Answer* that is designed to help students learn to estimate whether or not the answer they have obtained is reasonable. In many examples, we also include *Discussion* sections that help explain the importance of a problem solving concept or point out common pitfalls to be avoided. Finally, each example closes with a *Check Your Understanding* problem or question to help the student to generalize or extend what's been learned in the example problem.

We believe that the general chemistry experience should help engineering students develop improved problem solving skills. Moreover, we feel that those skills should be transferable to other subjects in the engineering curriculum even though chemistry content may not be involved. Accordingly, we include a unique feature at the end of each chapter called **FOCUS ON PROBLEM SOLVING**. In these sections, the questions posed do not require a numerical answer, but rather ask the student to identify the strategy or reasoning to be used in the problem and often require them to identify missing information for the problem. In most cases, it is not possible to arrive at a final numerical answer using the information provided, so students are forced to focus on developing a solution rather than just identifying and executing an algorithm. In this third edition, we have emphasized the use of graphical problem-solving strategies in some of these features. The end-of-chapter exercises include additional problems of this nature so the *Focus on Problem Solving* can be fully incorporated into the course. This feature grew out of an NSF-funded project on assessing problem solving in chemistry classes.

Text Features

We employ a number of features, some of which we referred to earlier, to help students see the utility of chemistry and understand the connections to engineering.

***INSIGHT INTO* Sections** Each chapter is built around a template called *Insights Into* . . . . These themes, which both open and close each chapter, have been chosen to showcase connections between engineering and chemistry. In addition to the chapter opening and closing sections, the template themes are woven throughout the chapter, frequently providing the context for points of discussion or example problems. This special *Insight* icon is used throughout the book to identify places where ideas presented in the chapter opening section are revisited in the narrative.

FOCUS ON PROBLEM SOLVING Sections Engineering faculties unanimously say that freshman engineering students need practice in solving problems. However, it is important to make a distinction here between problems and exercises. Exercises provide a chance to practice a narrow skill, whereas problems require multiple steps and thinking outside the context of the information given. *Focus on Problem Solving* offers students the chance to develop and practice true problem solving skills. These sections, which appear at the end of every chapter, include a mix of quantitative and qualitative questions that focus on the *process* of finding a solution to a problem, not the solution itself. We support these by including additional similar problems in the end-of-chapter material.

MathConnections In our experience, one trait that distinguishes engineering students from other general chemistry students is a higher level of comfort with mathematics. Typically, most students who take a class of the sort for which this book has been written will also be taking a course in calculus. Thus it seems natural to us to point out the mathematical underpinnings of several of the chemistry concepts presented in the text because this should help students forge mental connections between their courses. At the same time, we recognize that a student taking a precalculus math course should not be precluded from taking chemistry. To balance these concerns, we have placed any advanced mathematics into special *MathConnections* sections, which are set off from the body of the text. Our hope is that those students familiar with the mathematics involved will benefit from seeing the origin of things such as integrated rate laws, whereas those students with a less extensive background in math will still be able to read the text and master the chemistry presented.

Example Problems Our examples are designed to illustrate good problem solving practices by first focusing on the reasoning behind the solution before moving into any needed calculations. We emphasize this "think first" approach by beginning with a *Strategy* section, which outlines a plan of attack for the problem. We find that many students are too quick to accept whatever answer their calculator might display. To combat this, we follow most solutions with an *Analyze Your Answer* section, which uses estimation and other strategies to walk students through a double check of their answers. Every example closes with a *Check Your Understanding* exercise to allow students to practice or extend the skill they have just learned. Answers to these additional exercises are included in Appendix J at the end of the book.

End-of-Chapter Features Each chapter concludes with a chapter summary, outlining the main points of the chapter, and a list of key terms, each of which includes the section number where the term first appeared. Definitions for all key terms appear in the Glossary.

Problem Sets Each chapter includes roughly 100 problems and exercises, spanning a wide range of difficulty. Most of these exercises are identified with specific sections to provide the practice that students need to master material from that section. Each chapter also includes a number of *Additional Problems*, which are not tied to any particular section and which may incorporate ideas from multiple sections. *Focus on Problem Solving* exercises follow, as described earlier. The problems for most chapters conclude with *Cumulative Problems*, which ask students to synthesize information from the current chapter with what they've learned from previous chapters to form answers. For the third edition, we have added a number of more challenging problems in several chapters. Answers for all odd-numbered problems appear at the end of the book in Appendix K.

Margin Notes Margin notes in the text point out additional facts, further emphasize points, or point to related discussion either earlier or later in the book. Margin Notes are denoted with an ◂ⓘ icon that is also placed in the narrative and links the margin note with the relevant passage in the text.

New in this Edition

There are several important changes in this third edition of the textbook. As we did for the second edition, we have replaced a number of the "Insight Into. . ." sections to make them more current and to try to include topics that will appeal to a wider range of student interests. Thus, we have introduced two new topics for the chapter-opening insights: Biomass and Biofuel Engineering in Chapter 3 and Trace Analysis in Chapter 6. Both of these themes are more readily connected to engineering applications than those that they replaced. The closing insight sections for Chapters 3, 8, 9, 12, and 13 have also been rewritten to highlight topics with more current relevance. A detailed list of specific changes is given below.

Chapter	Summary of Changes
1	• Updated references to contexts used throughout the book, because several of these contexts have changed. • Improved artwork showing particulate nature of matter and phases. • Cleaned up usage of margin notes throughout. • Added end-of-chapter problems.
2	• Added a section to the text about the newly approved IUPAC atomic masses and explained how the issue of what atomic mass to use in calculations is handled in this book. This change includes a table with the new approved ranges and references to aspects of scientific investigations, such as climate change, that rely on this level of information. • Added end-of-chapter problems that build on connection between isotopes and mass spectroscopy. • Cleaned up usage of margin notes throughout. • Cleaned up artwork for several figures.

Chapter	Summary of Changes
3	Major Change • Changed the context for the entire chapter from explosions to biofuels. ○ Includes new opening insight ○ Includes changes in several example problems ○ Includes changes at several points in the text where references to context are made ○ Includes changes for several figures (3.1, 3.2, 3.10) ○ Includes changes to end-of-chapter problems related to the context theme Other changes • Changed Closing Insight section from explosive and green chemistry to carbon sequestration. • Cleaned up usage of margin notes throughout.
4	• Cleaned up usage of margin notes throughout. • Replaced "Focus on Problem Solving" with a new style of problem for this feature that includes graphical reasoning. • Replaced several end-of-chapter problems and added several end-of-chapter problems.
5	• Cleaned up usage of margin notes throughout. • Added several end-of-chapter problems.
6	Major Change • Changed the context for the entire chapter from light bulbs to trace analysis. ○ Includes new opening insight ○ Includes changes in several example problems ○ Includes changes at several points in the text where references to context are made. ○ Includes changes for several figures (6.1, 6.2, 6.10, which replaces what was 6.16) ○ Includes changes to end-of-chapter problems related to the context theme Other changes • Correction made in Math Connections feature. • Cleaned up usage of margin notes throughout. • Corrected artwork in Figure 6.20 to reflect accurate values of atomic radii. • Changed wording in the introductory paragraph of the closing insight so it is not dependent on the now-removed context.
7	• Cleaned up usage of margin notes throughout. • Cleaned-up artwork for several figures. • Added several end-of-chapter problems. • Corrected quantitative information on atomic radii in Figure 7.1. • Removed mentions of sp^3d and sp^3d^2 hybridization to reflect current best understanding of their non-utility in describing bonding.
8	• Cleaned up usage of margin notes throughout. • Cleaned-up artwork for several figures. ○ Includes replacing Figures 8.1 and 8.24 to get the science depiction to be more accurate • Added several end-of-chapter problems. • Replaced closing insight section on the invention of new materials with one on micro-electrical-mechanical systems (with references to new cover art included)
9	• Cleaned up usage of margin notes throughout. • Added several end-of-chapter problems. • Replaced closing insight section on batteries with one on power distribution and the electrical grid with specific references to computer science demands. ○ Some of the more important concepts related to batteries were moved to the section in Chapter 13 on batteries. • Updated artwork that had dated information. • Updated several end-of-chapter problems to reflect content changes in the chapter.

Chapter	Summary of Changes
10	• Cleaned up usage of margin notes throughout. • Cleaned up artwork for several figures. • Reworked Carnot cycle description of entropy. • Replaced the final two paragraphs of the closing insight section to include the role of plastics in marine environments—includes adding a new figure in addition to text content. • Fixed factual errors in one end-of-chapter problem and added new end-of-chapter problems.
11	• Cleaned up usage of margin notes throughout. • Cleaned up artwork for several figures, including updates on artwork that has data from specific years. • Replaced "Focus on Problem Solving" with a new style of problem for this feature that includes graphical reasoning.
12	• Cleaned up usage of margin notes throughout. • Cleaned up numerical information related to equilibrium constant for a few reactions in the text and appendixes. • Added several end-of-chapter problems and cleaned up sub-headings used in the end-of-chapter problems. • Replaced closing insight section on borates and boric acid with one on bendable concrete.

13	• Cleaned up usage of margin notes throughout.
	• Added material about cathodic protection to Section 13-3.
	• Revised information about batteries, changing examples, in Section 13-5.
	  ○ Includes moving materials on primary vs. secondary batteries from Chapter 9 to here
	  ○ Includes moving information about energy density of batteries from Chapter 9 to here
	• Replaced closing insight on corrosion prevention with a new closing insight on lithium-ion batteries and their use in aerospace engineering.
	• Replaced end-of-chapter problems with new ones appropriate for the new materials added.

14	• Cleaned up usage of margin notes throughout. • Added a several paragraph description within Section 14-6 about the Fukushima reactor accident and the release of radioactivity from it. • Added a large number of additional end-of-chapter problems.

Supporting Material

Please visit www.cengage.com/chemistry/brownholme/chemengineer3e for more information about student and instructor resources for this text.

Acknowledgments

We are very excited to see this book move forward in this third edition, and we are grateful for the help and support we have enjoyed from a large and talented team of professionals. There are many people without whom we never could have done this. But foremost among them are our families, to whom this book is again dedicated.

The origin of this text can be traced back many years, and as we move into the third edition we would like once more to thank a few people who were instrumental in getting this project started. Jennifer Laugier first brought the two of us together to work on a book for engineering students. Jay Campbell's work as developmental editor for the first edition was tremendous, and without his efforts the book may never have been published. When Jay became involved, the project had been languishing for some time, and the subsequent gains in momentum were clearly not coincidental. The editorial leadership team at that time, consisting of Michelle Julet, David Harris, and Lisa Lockwood, was also crucial in seeing this project come to fruition. The decision to launch a book in a market segment that has not really existed was clearly not an easy one, and we appreciate the confidence that everyone at what was then Brooks/Cole placed in us.

In the development of this third edition, our Cengage Learning team includes a mix of the familiar and the new. We would like to thank our product manager, Lisa Lockwood, whose continued support is always appreciated. Our new Content Developer for this edition is Alyssa White, and both Lisa and Alyssa contributed greatly to discussions on where to focus our efforts in this revision. Alyssa has guided us through the entire revision process, and showed wonderful flexibility when we had trouble keeping up with the original production schedule. Media Editor Maureen Ross is coordinating work on the new MindTap electronic version of the text. Assistant Editor Brendan Killion is overseeing revisions to ancillary materials. Carol Samet at Cengage and Teresa Christie at MPS Limited have overseen all aspects of the actual production process. Richard Camp as copy editor has helped us to be much more consistent in our handling of a number of style issues, and Jill Reichenbach provided photo research. David Shinn has helped us check the page proofs, providing many valuable comments that have improved the accuracy of the text. The book in your hands truly reflects the best efforts of many hard-working professionals, and we are grateful to all of them for their roles in this project.

It has been nearly eight years since the first edition was published, and throughout that time we have received useful feedback from numerous students and colleagues. Much of that feedback is informal, including e-mail from students or faculty members pointing out errors they have found or letting us know about sections they really liked. Although there is no way to list all of the people who have contributed in this way, we do sincerely thank you all.

Faculty members from a wide variety of institutions also provided more formal comments on the text at various stages of its development. We thank the following reviewers for their contributions to the current revision.

Darrel Axtell, *Saint Martin's University*

Simon Bolt, *University of Houston*

Patricia Muisener, *University of South Florida*

Diana Phillips, *Kettering University*

Steve Rathbone, *Blinn College*

We also thank the following reviewers for their contributions to the development of the earlier editions of the book.

Paul A. DiMilla, *Northeastern University*

Walter England, *University of Wisconsin–Milwaukee*

Mary Hadley, *Minnesota State University, Mankato*

Andy Jorgensen, *University of Toledo*

Karen Knaus, *University of Colorado–Denver*

Pamela Wolff, *Carleton University*

Grigoriy Yablonsky, *Saint Louis University*

Robert Angelici, *Iowa State University*

Allen Apblett, *Oklahoma State University*

Jeffrey R. Appling, *Clemson University*

Rosemary Bartoszek-Loza, *The Ohio State University*
Danny Bedgood, *Charles Sturt University*
James D. Carr, *University of Nebraska*
Victoria Castells, *University of Miami*
Paul Charlesworth, *Michigan Technological University*
Richard Chung, *San Jose State University*
Charles Cornett, *University of Wisconsin—Platteville*
Robert Cozzens, *George Mason University*
Ronald Evilia, *University of New Orleans*
John Falconer, *University of Colorado*
Sandra Greer, *University of Maryland*
Benjamin S. Hsaio, *State University of New York at Stony Brook*
Gerald Korenowski, *Rensselaer Polytechnic Institute*
Yinfa Ma, *University of Missouri—Rolla*
Gerald Ray Miller, *University of Maryland*
Linda Mona, *Montgomery College*
Michael Mueller, *Rose-Hulman Institute of Technology*
Kristen Murphy, *University of Wisconsin—Milwaukee*
Thomas J. Murphy, *University of Maryland*
Richard Nafshun, *Oregon State University*
Scott Oliver, *State University of New York at Binghamton*
The late Robert Paine, *Rochester Institute of Technology*
Steve Rathbone, *Blinn College*
Jesse Reinstein, *University of Wisconsin—Platteville*
Don Seo, *Arizona State University*
Mike Shaw, *Southern Illinois University—Edwardsville*
Joyce Solochek, *Milwaukee School of Engineering*
Jack Tossell, *University of Maryland*
Peter T Wolczanski, *Cornell University*

Larry Brown
Tom Holme
March, 2013

Student Introduction

Chemistry and Engineering

As you begin this chemistry course, odds are that you may be wondering "Why do I have to take chemistry anyway? I'll never really need to know any of this to be an engineer." So we'd like to begin by offering just a few examples of the many links between our chosen field of chemistry and the various branches of engineering. The most obvious examples, of course, might come from chemical engineering. Many chemical engineers are involved with the design or optimization of processes in the chemical industry, so it is clear that they would be dealing with concepts from chemistry on a daily basis. Similarly, civil or environmental engineers working on environmental protection or remediation might spend a lot of time thinking about chemical reactions taking place in the water supply or the air. But what about other engineering fields?

Much of modern electrical engineering relies on solid-state devices whose properties can be tailored by carefully controlling their chemical compositions. And although most electrical engineers do not regularly make their own chips, an understanding of how those chips operate on an atomic scale is certainly helpful. As the push for ever smaller circuit components continues, the ties between chemistry and electrical engineering will grow tighter. From organic light-emitting diodes (OLEDs) to single molecule transistors, new developments will continue to move out of the chemistry lab and into working devices at an impressive pace.

Some applications of chemistry in engineering are much less obvious. At 1483 feet, the Petronas Towers in Kuala Lumpur, Malaysia, were the tallest buildings in the world when they were completed in 1998. Steel was in short supply in Malaysia, so the towers' architects decided to build the structures out of something the country had an abundance of and local engineers were familiar with: concrete. But the impressive height of the towers required exceptionally strong concrete. The engineers eventually settled on a material that has come to be known as high-strength concrete, in which chemical reactions between silica fume and Portland cement produce a stronger material, more resistant to compression. This example illustrates the relevance of chemistry even to very traditional fields of engineering, and we will discuss some aspects of the chemistry of concrete in Chapter 12, including the development of novel bendable concrete.

About This Text

Both of us have taught general chemistry for many years, and we are familiar with the difficulties that students may encounter with the subject. Perhaps more importantly, for the past several years, we've each been teaching engineering students in the type of one-semester course for which this text is designed. The approach to subjects presented in this text draws from both levels of experience.

We've worked hard to make this text as readable and student friendly as possible. One feature that makes this book different from any other text you could have used for this course is that we incorporate connections between chemistry and engineering as a fundamental component of each chapter. You will notice that each chapter begins and ends with a section called ***INSIGHT INTO***. . . . These sections are only

the beginning of the connections, and the theme introduced in the initial insight appears regularly throughout that chapter. This special icon identifies material that is closely related to the theme of the chapter opening *Insight* section. We've heard many students complain that they don't see what chemistry has to do with their chosen fields, and we hope that this approach might help you to see some of the connections.

Engineering students tend to take a fairly standard set of courses during their first year of college, so it's likely that you might be taking calculus and physics courses along with chemistry. We've tried to point out places where strong connections between these subjects exist, and at the same time to do this in a way that does not disadvantage a student who might be taking a precalculus math class. Thus we may refer to similarities between equations you see here and those you might find in a physics text, but we do not presume that you are already familiar with those equations. In the case of math, we use special sections called **MathConnections** to discuss the use of math, and especially calculus, in chemistry. If you are familiar with calculus or are taking it concurrently with this class, these sections will help you to see how some of the equations used in chemistry emerge from calculus. But if you are not yet taking calculus, you can simply skip over these sections and still be able to work with the needed equations.

Although our primary intent is to help you learn chemistry, we also believe that this text and the course for which you are using it can help you to develop a broad set of skills that you will use throughout your studies and your career. Foremost among them is problem solving. Much of the work done by practicing engineers can be characterized as solving problems. The problems you will confront in your chemistry class clearly will be different from those you will see in engineering, physics, or math. But taken together, all of these subjects will help you formulate a consistent approach that can be used to attack virtually any problem. Many of our students tend to "jump right in" and start writing equations when facing a problem. But it is usually a better idea to think about a plan of attack before doing that, especially if the problem is difficult or unfamiliar. Thus all of our worked examples include a *Strategy* section in which we outline the path to a solution before starting to calculate anything. The *Solution* section then puts that strategy into action. For most numerical examples, we follow the solution with a section we call *Analyze Your Answer*, in which we use estimation or comparison to known values to confirm that our result makes sense. We've seen many students who believe that whatever their calculator shows must be the right answer, even when it should be easy to see that a mistake has been made. Many examples also include a *Discussion* section in which we might talk about common pitfalls that you should avoid or how the problem we've just done relates to other ideas we've already explored. Finally, each example problem closes with a *Check Your Understanding* question or problem, which gives you a chance to practice the skills illustrated in the example or to extend them slightly. Answers to these *Check Your Understanding* questions appear in Appendix J.

While we are thinking about the example problems, a few words about rounding and significant figures are in order. In solving the example problems, we have used atomic weights with the full number of significant figures shown in the periodic table inside the back cover. We have also used as many significant figures as available for constants such as the speed of light or the universal gas constant. Where intermediate results are shown in the text, we have tried to write them with the appropriate number of significant figures. But when those same intermediate results are used in a subsequent calculation, we have *not* rounded the values. Instead, we retain the full calculator result. Only the final answer has actually been rounded. If you follow this same procedure, you should be able to duplicate our answers. (The same process has been used to generate the answers to numerical problems appearing in Appendix K.) For problems that involve finding the slope or intercept of a line, the values shown have been obtained by linear regression using the algorithms built into either a spreadsheet or a graphing calculator.

A unique feature of this text is the inclusion of a *Focus on Problem Solving* question at the end of each chapter. These questions are designed to force you to think about the *process* of solving the problem rather than just getting an answer. In many cases, these problems do not include sufficient information to allow you to reach a final solution. Although we know from experience that many beginning engineering students might find this frustrating, we feel it is a good approximation to the kind of problems that a working engineer might confront. Seldom would a client sit down and provide every piece of information that you need to solve the problem at hand.

One of the most common questions we hear from students is "How should I study for chemistry?" Sadly, that question is most often asked after the student has done poorly on one or more exams. Because different people learn best in different ways, there isn't a single magic formula to ensure that everyone does well in chemistry. But there are some common strategies and practices that we can recommend. First and foremost, we suggest that you avoid getting behind in *any* of your classes. Learning takes time, and very few people can master three chapters of chemistry (or physics, or math, or engineering) the night before a big exam. Getting behind in one class inevitably leads to letting things slide in others, so you should strive to keep up from the outset. Most professors urge students to read the relevant textbook material before it is presented in class. We agree that this is the best approach, because even a general familiarity with the ideas being presented will help you to get a lot more out of your class time.

In studying for exams, you should try to make a realistic assessment of what you do and don't understand. Although it can be discomforting to focus on the problems that you don't seem to be able to get right, spending more time studying things that you have already mastered will probably have less impact on your grade. Engineering students tend to focus much of their attention on numerical problems. Although such calculations are likely to be very important in your chemistry class, we also encourage you to try to master the chemical concepts behind them. Odds are that your professor will test you on qualitative or conceptual material, too.

Finally, we note that this textbook is information rich. It includes many of the topics that normally appear in a full year college chemistry course, but it is designed for a course that takes only one semester. To manage the task of paring down the volume of materials, we've left out some topics and shortened the discussion of others. Having the Internet available means that you can always find more information if what you have read sparks your interest.

We are excited that this book has made it into your hands. We hope you enjoy your semester of learning chemistry and that this book is a positive part of your experience.

Larry Brown
Tom Holme
March, 2013

1

Introduction to Chemistry

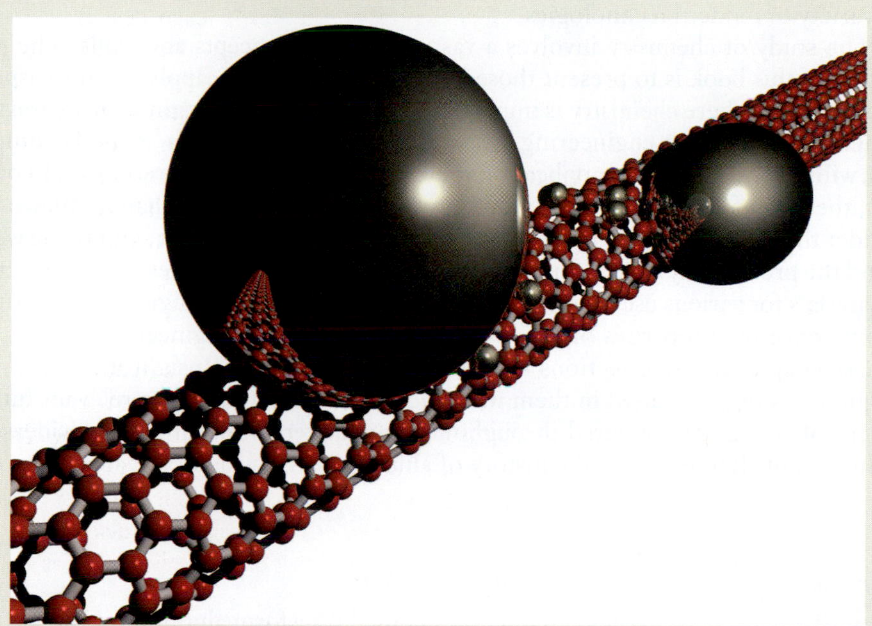

Scientists from Lawrence Berkeley National Laboratory and the University of California at Berkeley developed this nanoscale "conveyor belt." Individual metal atoms are transported along a carbon nanotube from one metal droplet to another. This research offers a possible means for the atomic-scale construction of optical, electronic, and mechanical devices. *Courtesy of Zettl Research Group, Lawrence Berkeley National Laboratory, and the University of California at Berkeley*

CHAPTER OUTLINE

In the not too distant future, engineers may design and assemble miniature mechanical or electronic devices, gears, and other parts fabricated on an atomic scale. Their decisions will be guided by knowledge of the sizes and properties of the atoms of different elements. Such devices might be built up atom by atom: each atom would be specified based on relevant design criteria and maneuvered into position using techniques such as the "conveyor belt" shown above. These nanomachines will be held together not by screws or rivets but by the forces of attraction between the different atoms—by chemical bonds. Clearly, these futuristic engineers will need to understand atoms and the forces that bind them together. In other words, they will need to understand chemistry.

Nanoscience deals with objects whose sizes are similar to those of atoms and molecules. Try a web search for "nanoscience" or "molecular machines" to learn more.

At least for now, though, this atomic-level engineering remains in the future. But what about today's practicing engineers? How do their decisions depend on knowledge of chemistry? And from your own perspective as an engineering student, why are you required to take chemistry?

The Accreditation Board for Engineering and Technology, or ABET, is a professional organization that oversees engineering education. According to ABET's definition, "Engineering is the profession in which a knowledge of the mathematical and natural sciences gained by study, experience, and practice is applied with judgment to develop ways to utilize, economically, the materials and forces of nature for the benefit of mankind." So as one of the sciences, chemistry is clearly included in the realm of knowledge at the disposal of an engineer. Yet engineering students do not always recognize the role of chemistry in their chosen profession. One of the main goals of this textbook is to instill an appreciation of the role of chemistry in many areas of engineering and technology and in the interplay between chemistry and engineering in a variety of modern technologies.

The study of chemistry involves a vast number of concepts and skills. The philosophy of this book is to present those basic ideas and also to apply them to aspects of engineering where chemistry is important. Each chapter will begin with an example of chemistry related to engineering. Some of these examples, such as the burning of fuels, will involve fairly clear applications of chemical principles and reactions. In other cases, the role of chemistry may be less immediately apparent. In Chapter 10, we will consider the recycling of plastics and examine some factors that limit both the feasibility and the profitability of recycling. Other themes will involve the design and selection of materials for various uses and the ways that even very small changes in composition can influence the properties of alloys that are often used in engineering designs. All of these chapter-opening sections have titles that begin with "Insight into . . . ," and the questions that are raised in them will guide our exploration of the relevant fundamentals of chemistry presented throughout that chapter. Our first case considers the production of aluminum and the history of aluminum as a structural material.

Chapter Objectives

After mastering this chapter, you should be able to

- describe how chemistry and engineering helped transform aluminum from a precious metal into an inexpensive structural material.
- explain the usefulness of the macroscopic, microscopic, and symbolic perspectives in understanding chemical systems.
- draw pictures to illustrate simple chemical phenomena (like the differences among solids, liquids, and gases) on the molecular scale.
- explain the difference between inductive and deductive reasoning in your own words.
- use appropriate techniques to convert measurements from one unit to another.
- express the results of calculations using the correct number of significant figures.

INSIGHT INTO

1-1 Aluminum

If you are thirsty, you might ask yourself several questions about what to drink. But you probably wouldn't ask, "Where did the can that holds this soda come from, and why is it made of aluminum?" The aluminum can has become so common that it's easy to take for granted. What makes aluminum an attractive material for this type of application, and how did it become such a familiar part of life?

Each year, about 130 billion aluminum cans are produced in the United States.

You probably can identify a few properties of aluminum that make it suitable for use in a soda can. Compared with most other metals, aluminum is light but fairly strong. So a typical aluminum can is much lighter than a comparable tin or steel can. This means that the can does not add much weight compared to the soda itself, so the cans are easier to handle and cheaper to ship. A soda can made of lead certainly would be less convenient. The fact that aluminum does not readily undergo chemical reactions that might degrade it as the cans are transported and stored is also important. But although all of those features of the aluminum can are nice, they wouldn't be of much practical use if aluminum were not readily available or if the can cost more than the soda itself.

The widespread availability of aluminum results from an impressive collaboration between the basic science of chemistry and the applied sciences of engineering. In the 19th century, aluminum was a rare and precious material. In Europe, Napoleon was emperor of a sizable portion of the continent, and he would impress guests by using extravagant aluminum tableware. In the United States, when architects wanted a suitably impressive material for the capstone at the top of the Washington Monument, a tribute to the "father of our country," they chose aluminum. Weighing in at 100 ounces, the capstone of the monument was the largest single piece of pure aluminum ever cast at that time. Yet today, sheets of aluminum weighing more than 100 pounds are regularly found in many metal shops. Why was aluminum so expensive then, and what changed to make it so affordable now?

Initial discussion of this question can be framed in terms of **Figure 1.1**, which looks rather broadly at the interactions of human society with the earth. Society, represented by the globe, has needs for goods and materials. Currently, and for the foreseeable future, the raw materials needed to make these goods must somehow be extracted from the earth. When the goods are used up, the leftovers become waste that must be disposed of, completing the cycle by returning the exhausted materials to the ecosystem. Ultimately, the role of engineering in this cycle is to maximize the efficiency with which materials are extracted and minimize the amount of waste that is returned.

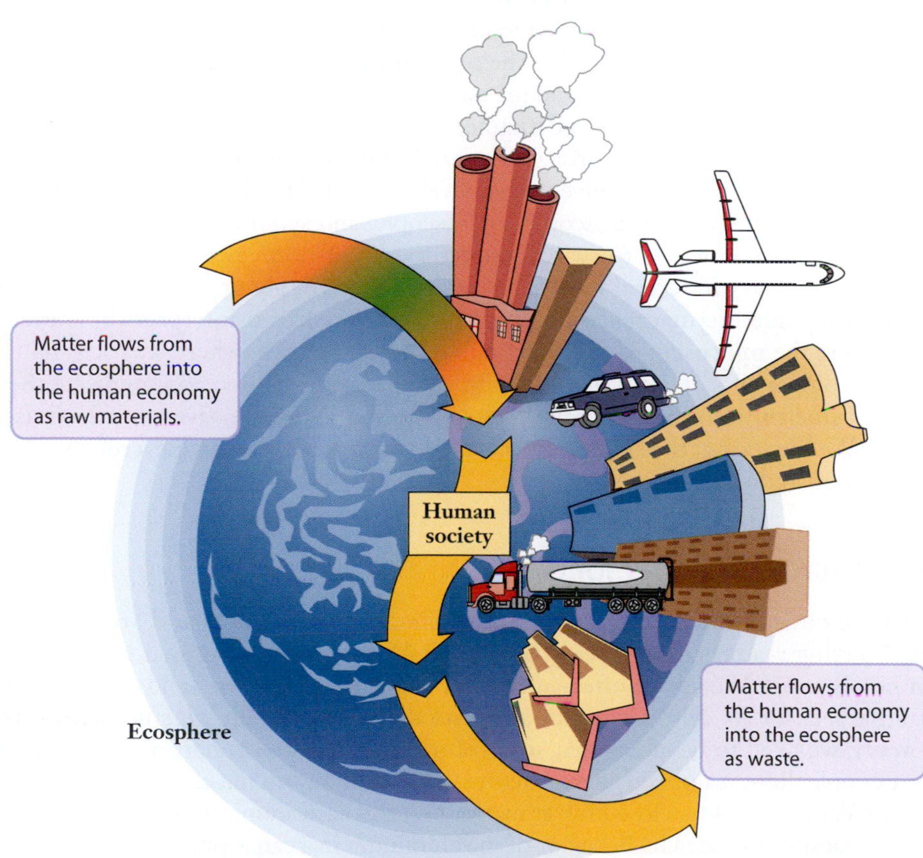
Matter flows from the ecosphere into the human economy as raw materials.
Human society
Ecosphere
Matter flows from the human economy into the ecosphere as waste.

Figure 1.1 ■ The interactions of human society with the earth can be thought of largely in terms of the conversion of matter from raw materials into waste. Much of engineering consists of efforts to optimize the processes used in these conversions. And as the science of matter, chemistry is an important element of the knowledge exploited in engineering those processes.

The aluminum in bauxite is typically found in one of three minerals: gibbsite, bohmite, and diaspore.

Let's think about aluminum in this context. Pure aluminum is never found in nature. Instead, the metal occurs in an ore, called *bauxite*, that is composed of both useless rock and aluminum in combination with oxygen. So before aluminum can be used in our soda can, it must first be extracted, or "won," from its ore and purified. Because aluminum combines very readily with oxygen, this presents some serious challenges. Some of these challenges are chemical and will be revisited in Chapter 13 of this text. Some of the early steps, however, can be solved by clever applications of physical properties, and we will consider a few of them as we investigate introductory material in this chapter. When confronted with a complex mixture of materials, such as an ore, how does a chemist think about separating the mixture?

To look into this type of question, we should adopt the approach that is commonly taken in science. The term **scientific method** has various possible definitions, and we'll look at this concept further in Section 1-4. But at this point, we will consider it an approach to understanding that begins with the observation of nature, continues to hypothesis or model building in response to that observation, and ultimately includes further experiments that either bolster or refute the hypothesis. In this definition, the hypothesis is an educated guess at ways to explain nature. In this chapter, we will see how this method relates to chemistry in general and also to issues relating to materials like aluminum and their use in society.

1-2 The Study of Chemistry

Chemistry has been called the "central science" because it is important to so many other fields of scientific study. So, even if you have never taken a chemistry course, chances are good that you have seen some chemistry before. This text and the course in which you are using it are designed to help you connect pieces of information you have already picked up, increase your understanding of chemical concepts, and give you a more coherent and systematic picture of chemistry. The ultimate goal of introductory college chemistry courses is to help you appreciate the chemical viewpoint and the way it can help you to understand the natural world. This type of perspective of the world is what enables chemists and engineers to devise strategies for refining metals from their ores, as well as to approach the many other applied problems we'll explore.

This coherent picture involves three levels of understanding or perspectives on the nature of chemistry: macroscopic, microscopic, and symbolic. By the end of this course, you should be able to switch among these perspectives to look at problems involving chemistry in several ways. The things we can see about substances and their reactions provide the **macroscopic perspective**. We need to interpret these events considering the **microscopic (or "particulate") perspective**, where we focus on the smallest components of the system. Finally, we need to be able to communicate these concepts efficiently, so chemists have devised a **symbolic perspective** that allows us to do that. We can look at these three aspects of chemistry first, to provide a reference for framing our studies at the outset.

The Macroscopic Perspective

When we observe chemical reactions in the laboratory or in the world around us, we are observing matter at the macroscopic level. **Matter** is anything that has mass and can be observed. We are so often in contact with matter that we tend to accept our intuitive feel for its existence as an adequate definition. When we study chemistry, however, we need to be aware that some of what we observe in nature is not matter. For example, light is not considered matter, because it has no mass.

When we take a close look at matter—in this case aluminum—we can see that various questions arise. The behavior of the aluminum in a can is predictable. If the can is tossed into the air, little will happen except that the can will fall to the earth under

Figure 1.2 ■ All of the common kitchen items shown here are made of aluminum. The metal's light weight, corrosion resistance, and low cost make it a likely choice for many consumer products.

the force of gravity. Aluminum cans and other consumer goods like those shown in **Figure 1.2** do not decompose in the air or undergo other chemical reactions. If the aluminum from a soda can is ground into a fine powder and tossed into the air, however, it may ignite—chemically combining with the oxygen in air. It is now believed that the Hindenburg airship burned primarily because it was covered with a paint containing aluminum powder and not because it was filled with hydrogen gas. (You can easily find a summary of the evidence by doing a web search.)

One of the most common ways to observe matter is to allow it to change in some way. Two types of changes can be distinguished: physical changes and chemical changes. The substances involved in a physical change do not lose their chemical identities. **Physical properties** are variables that we can measure without changing the identity of the substance being observed. Mass and density are familiar physical properties. Mass is measured by comparing the object given and some standard, using a balance. **Density** is a ratio of mass to volume. (This property is sometimes called **mass density**). To determine density, both mass and volume must be measured. But these values can be obtained without changing the material, so density is a physical property. Familiar examples of physical properties also include color, viscosity, hardness, and temperature. Some other physical properties, which will be defined later, include heat capacity, boiling point, melting point, and volatility.

Chemical properties are associated with the types of chemical changes that a substance undergoes. For example, some materials burn readily, whereas others do not. Burning in oxygen is a chemical reaction called **combustion**. Corrosion—the degradation of metals in the presence of air and moisture—is another commonly observed chemical change. Treating a metal with some other material, such as paint, can often prevent the damage caused by corrosion. Thus an important chemical property of paint is its ability to prevent corrosion. Chemical properties can be determined only by observing how a substance changes its identity in chemical reactions.

We will discuss corrosion and its prevention in detail in Chapter 13.

Both chemical and physical properties of aluminum are important to its utility. A structural material is useful only if it can be formed into desired shapes, which requires it to be malleable. **Malleability** is a measure of a material's ability to be rolled or hammered into thin sheets, and metals are valuable in part because of their malleability. It is a physical property because the substance remains intact—it is still the same metal, just in a different shape. An aluminum can is formed during its manufacturing process, but its shape can be changed, as you have perhaps done many times when you crushed a can to put it into a recycling bin. Similarly, the chemical

Aluminum is generally found second, behind gold, in rankings of metal malleability.

properties of aluminum are important. Pure aluminum would be very likely to react with the acids in many popular soft drinks. So aluminum cans are coated inside with a thin layer of polymer—a plastic—to keep the metal from reacting with the contents. This demonstrates how knowing chemical properties can allow product designers to account for and avoid potentially harmful reactions.

When we observe chemical reactions macroscopically, we encounter three common states, or **phases**, of matter: solids, liquids, and gases. At the macroscopic level, **solids** are hard and do not change their shapes easily. When a solid is placed in a container, it retains its own shape rather than assuming that of the container. Even a powdered solid demonstrates this trait because the individual particles still retain their shape, even though the collection of them may take on the shape of the container.

Two other states of matter are plasmas and Bose-Einstein condensates. But these do not exist at ordinary temperatures.

Liquids can be distinguished from solids macroscopically because, unlike solids, liquids adapt to the shape of the container in which they are held. They may not fill the entire volume, but the portion they do occupy has its shape defined by the container. Finally, **gases** can be distinguished macroscopically from both liquids and solids primarily because a gas expands to occupy the entire volume of its container. Although many gases are colorless and thus invisible, the observation that a gas fills the available volume is a common experience; when we walk through a large room, we are not concerned that we will hit a pocket with no air.

The aluminum that we encounter daily is a solid, but during the refining process, the metal must become molten, or liquid. Handling the molten metal, pouring it into containers, and separating impurities provide both chemical and engineering challenges for those who design aluminum production plants.

Often, chemical and physical properties are difficult to distinguish at the macroscopic level. We can assert that boiling water is a physical change, but if you do nothing more than observe that the water in a boiling pot disappears, how do you know if it has undergone a chemical or physical change? To answer this type of question, we need to consider the particles that make up the water, or whatever we observe, and consider what is happening at the microscopic level.

The Microscopic or Particulate Perspective

The most fundamental tenet of chemistry is that all matter is composed of atoms and molecules. This is why chemists tend to think of everything as "a chemical" of one sort or another. In many cases, the matter we encounter is a complex mixture of chemicals, and we refer to each individual component as a chemical substance. We will define these terms much more extensively as our study of chemistry develops, but we'll use basic definitions here. All matter comprises a limited number of "building blocks," called **elements**. Often, the elements are associated with the periodic table of elements, shown inside the back cover of this textbook and probably hanging in the room where your chemistry class meets. **Atoms** are unimaginably small particles that cannot be made any smaller and still behave like a chemical system. When we study matter at levels smaller than an atom, we move into nuclear or elementary particle physics. But atoms are the smallest particles that can exist and retain the chemical identity of whatever element they happen to be. **Molecules** are groups of atoms held together so that they form a unit whose identity is distinguishably different from the atoms alone. Ultimately, we will see how forces known as "chemical bonds" are responsible for holding the atoms together in these molecules.

The word atom *comes from the Greek word "atomos" meaning indivisible.*

The particulate perspective provides a more detailed look at the distinction between chemical and physical changes. Because atoms and molecules are far too small to observe directly or to photograph, typically we will use simplified, schematic drawings to depict them in this book. Often, atoms and molecules will be drawn as spheres to depict them and consider their changes.

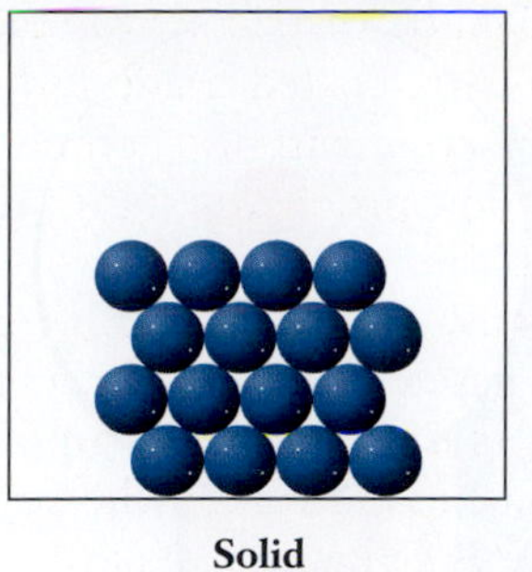

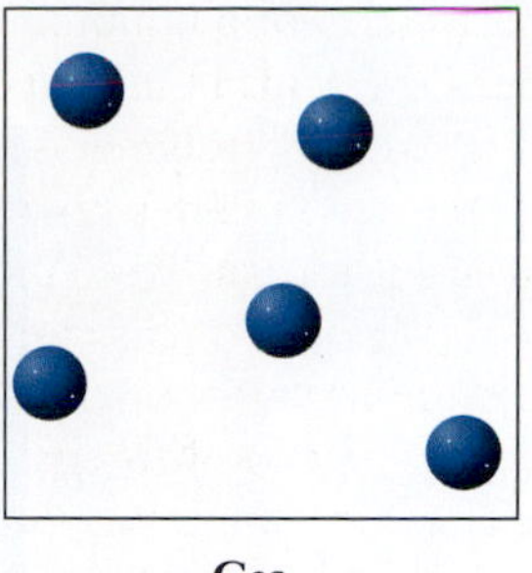

Solid Liquid Gas

Figure 1.3 ■ Particulate-level views of the solid, liquid, and gas phases of matter. In a solid, the molecules maintain a regular ordered structure, so a sample maintains its size and shape. In a liquid, the molecules remain close to one another, but the ordered array breaks down. At the macroscopic level, this allows the liquid to flow and take on the shape of its container. In the gas phase, the molecules are very widely separated, and move independently of one another. This allows the gas to fill the available volume of the container.

To correctly depict the relative densities of a gas and a liquid, much more space would need to be shown between particles in a gas than can be shown in a drawing like Figure 1.3.

If we consider solids, liquids, and gases, how do they differ at the particulate level? **Figure 1.3** provides a very simple but useful illustration. Note that the atoms in a solid are packed closely together, and it is depicted as maintaining its shape—here as a block or chunk. The liquid phase also has its constituent particles closely packed, but they are shown filling the bottom of the container rather than maintaining their shape. Finally, the gas is shown with much larger distances between the particles, and the particles themselves move freely through the entire volume of the container. These pictures have been inferred from experiments that have been conducted over many years. Many solids, for example, have well-ordered structures, called crystals, so a particulate representation of solids usually includes this sense of order.

How can we distinguish between a chemical change and a physical change in this perspective? The difference is much easier to denote at this level, though often it is no more obvious to observe. If a process is a physical change, the atoms or molecules themselves do not change at all. To look at this idea, we turn to a familiar molecule—water. Many people who have never studied chemistry can tell you that the chemical formula of water is "H two O." We depict this molecule using different sized spheres; the slightly larger sphere represents oxygen and the smaller spheres represent hydrogen. In **Figure 1.4**, we see that when water boils, the composition of the individual molecules is

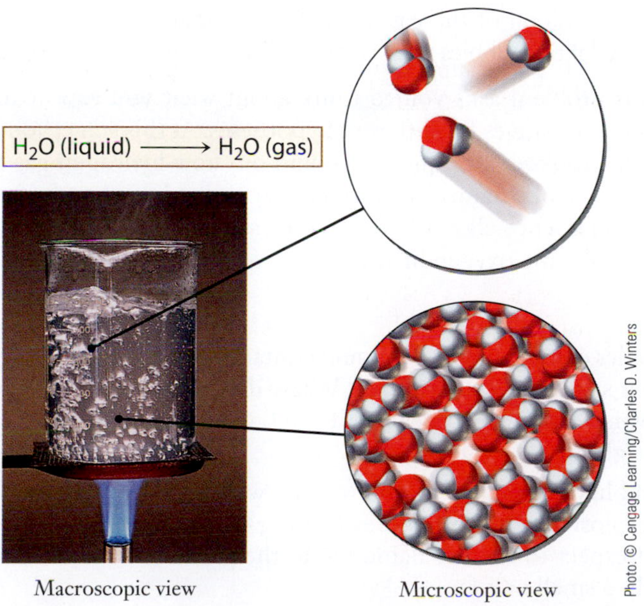

Figure 1.4 ■ The boiling of water is a physical change, in which liquid water is converted into a gas. Both the liquid and gas phases are made up of water molecules; each molecule contains two hydrogen atoms and one oxygen atom. The particulate-scale insets in this figure emphasize that fact and also show that the separation between water molecules is much larger in the gas than in the liquid.

Figure 1.5 ■ If a suitable electric current is passed through liquid water, a chemical change known as electrolysis occurs. In this process, water molecules are converted into molecules of hydrogen and oxygen gases, as shown in the particulate scale insets in the figure.

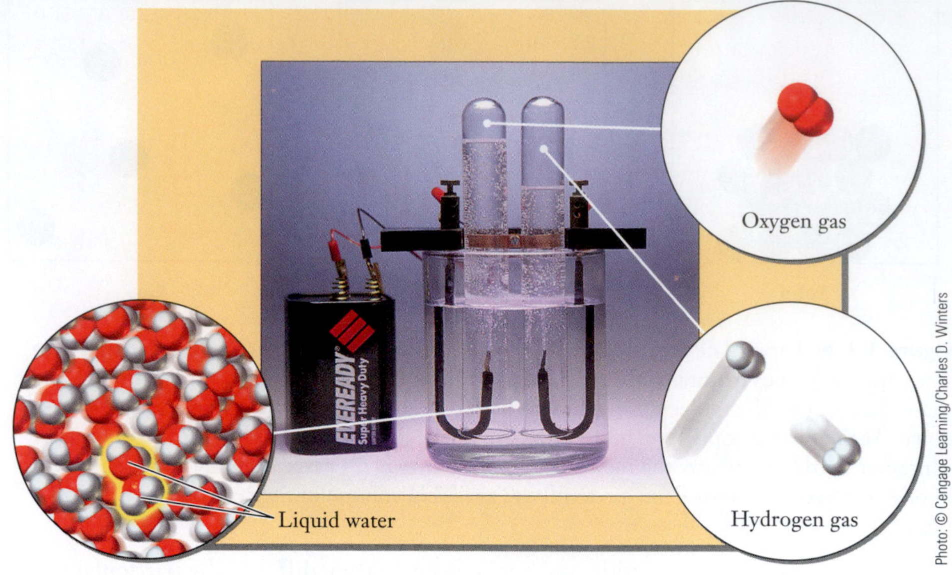

Photo: © Cengage Learning/Charles D. Winters

the same in the liquid phase and the gas phase. The molecular makeup of water has not been altered, and this fact is characteristic of a physical change.

Contrast this with **Figure 1.5**, which depicts a process called electrolysis at the particulate level; electrolysis occurs when water is exposed to an electric current. Notice that the molecules themselves change in this depiction, as water molecules are converted into hydrogen and oxygen molecules. Here, then, we have a chemical change.

If we observe these two reactions macroscopically, what would we see and how would we know the difference? In both cases, we would see bubbles forming, only in one case the bubbles will contain water vapor (gas) and in the other they contain hydrogen or oxygen. Despite this similarity, we can make observations at the macroscopic level to distinguish between these two possibilities. Example Problem 1.1 poses an experiment that could be set up to make such an observation.

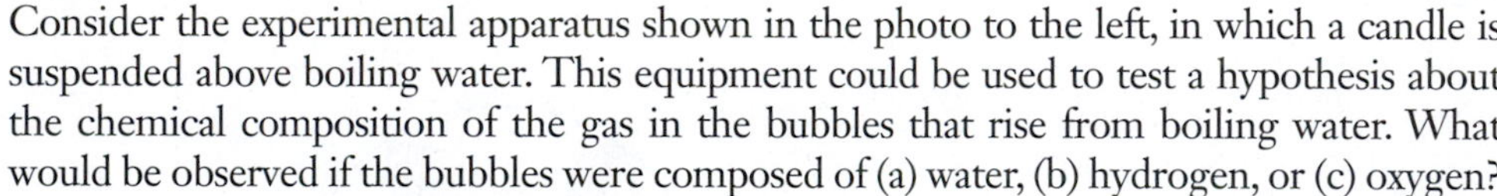

EXAMPLE PROBLEM 1.1

Thomas Holme and Keith Krumnow

Consider the experimental apparatus shown in the photo to the left, in which a candle is suspended above boiling water. This equipment could be used to test a hypothesis about the chemical composition of the gas in the bubbles that rise from boiling water. What would be observed if the bubbles were composed of (a) water, (b) hydrogen, or (c) oxygen?

Strategy This problem asks you to think about what you expect to observe in an experiment and alternatives for different hypotheses. At this stage, you may need to do a little research to answer this question—find out how hydrogen gas behaves chemically in the presence of a flame. We also have to remember some basic facts about fire that we've seen in science classes before. To be sustained, fire requires both a fuel and an oxidizer—usually the oxygen in air.

Solution

(a) If the bubbles coming out of the liquid contain water, we would expect the flame to diminish in size or be extinguished. Water does not sustain the chemical reaction of combustion (as oxygen does), so if the bubbles are water, the flame should not burn as brightly.

(b) You should have been able to find (on the web, for example) that hydrogen tends to burn explosively. If the bubbles coming out of the water were hydrogen gas, we would expect to see the flame ignite the gas with some sort of an explosion. (Hopefully, a small one.)

(c) If the bubbles were oxygen, the flame should burn more brightly. The amount of fuel would remain the same, but the bubbles would increase the amount of oxygen present and make the reaction more intense.

Check Your Understanding Work with students in your class or with your instructor to construct this apparatus and see whether or not your observations confirm any of these hypotheses. Draw a picture showing a particulate-level explanation for what you observe.

Symbolic Representation

The third way that chemists perceive their subject is to use symbols to represent the atoms, molecules, and reactions that make up the science. We will wait to introduce this perspective in detail in the next two chapters, but here we point out that you certainly have encountered chemical symbols in your previous studies. The "H two O" molecule we have noted is never depicted as we have done here in the quotation marks. Rather, you have seen the symbolic representation of water, H_2O. In Chapter 2, we will look at chemical formulas in more detail, and in Chapter 3, we will see how we use them to describe reactions using chemical equations. For now, we simply note that this symbolic level of understanding is very important because it provides a way to discuss some of the most abstract parts of chemistry. We need to think about atoms and molecules, and the symbolic representation provides a convenient way to keep track of these particles we'll never actually see. These symbols will be one of the key ways that we interact with ideas at the particulate level.

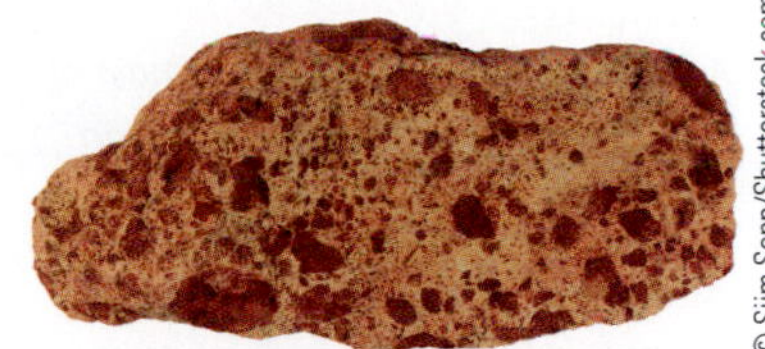

A sample of bauxite.

How can we use these representations to help us think about aluminum ore or aluminum metal? The macroscopic representation is the most familiar, especially to the engineer. From a practical perspective, the clear differences between unrefined ore and usable aluminum metal are apparent immediately. The principal ore from which aluminum is refined is called bauxite, and bauxite looks pretty much like ordinary rock. There's no mistaking that it is different from aluminum metal. At the molecular level, we might focus on the aluminum oxide (also called alumina) in the ore and compare it to aluminum metal, as shown in **Figure 1.6.** This type of drawing emphasizes the fact that the ore is made up of two different types of atoms, whereas only one type of atom is present in the metal. (Note that metals normally contain small amounts of impurities, sometimes introduced intentionally to provide specific, desirable properties. But in this case, we have simplified the illustration by eliminating any impurities.) Finally, Figure 1.6 also shows the symbolic representation for aluminum oxide—its chemical formula. This formula is slightly more complicated than that of water, and we'll look at this type of symbolism more closely in Chapter 2.

A block of aluminum.

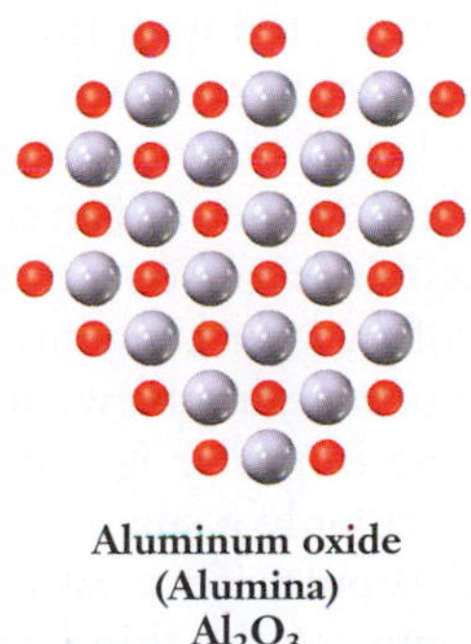

Aluminum oxide
(Alumina)
Al_2O_3

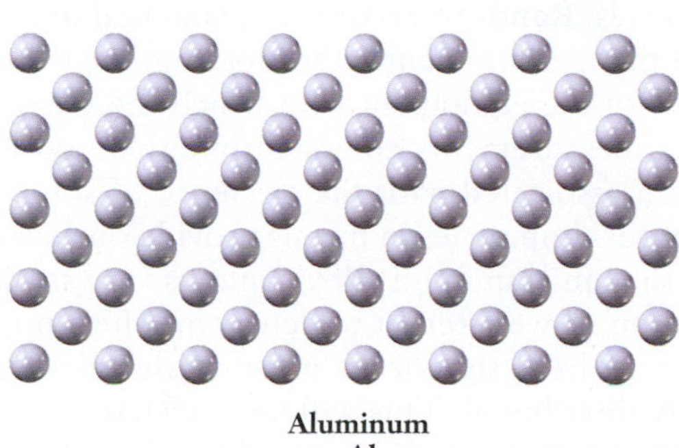

Aluminum
Al

Figure 1.6 ■ A particulate-level representation of aluminum oxide (left) and pure aluminum (right). The gray spheres represent aluminum atoms, and the red spheres represent oxygen.

1-3 The Science of Chemistry: Observations and Models

Chemistry is an empirical science. In other words, scientists who study chemistry do so by measuring properties of chemical substances and observing chemical reactions. Once observations have been made, models are created to help organize and explain the data. This structure of observations and models provides the backdrop for the science that we'll explore throughout this book. Although scientists and engineers both rely on their knowledge, intellect, and creativity, there are differences in the fruits of their efforts. Scientists often strive to create models for understanding nature, whereas engineers work to exploit or constrain nature to achieve some specific purpose. Both fields must begin with the observation of nature.

Observations in Science

Observations in chemistry are made in a wide variety of ways for a wide variety of reasons. In some cases, the observations are made because materials with certain properties are needed. For example, containers that hold liquids such as soft drinks need to be strong enough to hold the liquid but light in weight so they don't significantly increase the cost of transporting the product. Before aluminum cans were widely used, steel cans were the containers that society demanded. But steel is relatively heavy, so there was an incentive to find a different packaging material. Scientists and engineers worked together to make observations that confirmed the desirability of aluminum for this use.

A single aluminum can has a mass of roughly 14 grams, or about half an ounce.

Observations of nature involve some level of uncertainty in most cases. As an analogy, consider the attendance at a football game. We may be able to count with complete accuracy how many people use tickets to attend a football game, which gives the paid attendance. But that number is not really how many people are there, because it does not include members of the press, vendors, and coaching staffs, among others. This example points out one characteristic of making observations: we must be careful to define what we intend to observe. Carefully defining the measurement to be made, however, does not eliminate all sources of uncertainty. Virtually any scientific measurement must be made more than once to be valid.

Because we cannot observe nature with complete certainty, we need to establish the types of uncertainty we encounter in making observations. To do this, two terms are used: accuracy and precision. Although these words may be synonymous in everyday speech, each has its own meaning in scientific or engineering usage. **Accuracy** indicates how close the observed value is to the "true" value. **Precision** is the spread in values obtained from the measurement. A precise observation has several measurements close in value. **Figure 1.7** illustrates the concepts of accuracy and precision in measurement and shows how the two terms differ.

Error in measurement is unavoidable. Again, characteristics of error fall into categories. **Random error** is fundamental to any measurement. Random error may make the measurement either too high or too low and is associated with the limitations of the equipment with which the measurement is made. **Systematic error** makes measurements consistently either too high or too low. This type of error is often associated with the existence of some unknown bias in the measurement apparatus. Impurities in metals provide one example of possible error sources. Suppose that an aluminum alloy contains very small amounts of another element, such as silicon. If we attempt to determine the amount of silicon in a sample, we might expect to have significant error in the measurement if the impurity is not uniformly distributed. This will be especially true if we measure only a small sample. Systematic errors are also possible. Consider what would happen if we estimated the density of aluminum using soda cans and did not account for the thin layer of

(a) Poor precision and poor accuracy

(b) Good precision and poor accuracy

(c) Good precision and good accuracy

Figure 1.7 ■ The dartboards illustrate the concepts of precision and accuracy and show the difference between the two terms. The bull's-eye on the target represents the true value of a quantity, and each dart represents a measurement of that quantity. The darts in the left-hand panel are widely scattered and far away from the target; this is neither precise nor accurate. In the center panel, the darts are closely bunched but are not close to the target; this is good precision but poor accuracy. In the right-hand panel, we see good precision and good accuracy because the darts are closely clustered and close to the target.

polymer that is applied to the can. Because the density of the polymer coating is likely to be different from that of aluminum, the measurement would be systematically incorrect.

Interpreting Observations

Not all experiments provide direct information about the questions they ask. In many cases, we must infer answers from the data that are obtained. Two types of reasoning that are useful are inductive and deductive reasoning.

Inductive reasoning begins with a series of specific observations and attempts to generalize to a larger, more universal conclusion. We have asserted that all gases expand to occupy the full volumes of their containers. This universal conclusion was first drawn by inductive reasoning based on observations of many different gases under many different conditions. Because we have not observed every possible gas in every possible container under every possible set of conditions, it could be argued that some counterexample could be found eventually.

Deductive reasoning takes two or more statements or assertions and combines them so that a clear and irrefutable conclusion can be drawn. This process can be summarized as an "If A and B, then C" approach that is studied in detail in courses in formal logic.

To look at these types of reasoning, consider the developments that led to the affordable refining of aluminum. Even in the 19th century, aluminum ore was available. Metallurgists and other scientists were aware that the combination of aluminum with oxygen in ores is very stable. Observations allowed them to deduce that the chemical bonding forces in the alumina must be strong and would be difficult to overcome. Initial schemes for refining the metal involved heating and adding materials, such as carbon, that would react with oxygen that might be liberated. This method was already in use for refining iron, so inductive reasoning suggested that it might be generalized to other metals, including aluminum. This was a reasonable hypothesis, but it turned out that the methods used in refining iron didn't work for aluminum. As new discoveries were made about electrochemistry, people realized that these discoveries could lead to a viable scheme for refining aluminum. This is deductive reasoning: an electric current can be used to drive a difficult chemical reaction, and if purifying aluminum involves a difficult chemical reaction, then an electric current might be able to drive the purification of aluminum. Further experiments showed that this was correct, but it was still necessary to make it energetically and economically efficient. Eventually, Charles Hall devised a method that reduced the energy input needed for electrochemical refining of aluminum ore. Later, when this approach was coupled with the relatively low cost of hydroelectricity, the availability of aluminum increased greatly. Ultimately, both deductive and inductive reasoning played vital roles in this technological advance.

Models in Science

The number of observations made in the history of science is tremendous. To organize this vast amount of information, scientists create models and theories to make sense of a range of observations. The words "model" and "theory" are sometimes used interchangeably, although some scientists feel that the distinctions between the two are important. Usually, the term **model** refers to a largely empirical description, such as the fact that gas pressure is proportional to temperature. The word **theory**, on the other hand, most often refers to an explanation that is grounded in some more fundamental principle or assumption about the behavior of a system. In the connection between gas pressure and temperature, for example, the kinetic theory of gases uses arguments from physics to explain why the molecular nature of gases leads to the observed proportionality. Models are important for a number of reasons. First, they allow us to summarize a large number of observations concisely. Second, they allow us to *predict* behavior in circumstances that we haven't previously encountered. Third, they represent examples of creative thinking and problem solving. Finally, constructing and refining models can lead us ultimately to a more fundamental understanding of a problem.

We'll explore gases and the kinetic theory of gases in Chapter 5.

Models usually take time to develop. Faced with some initial observations, a creative scientist will try to explain why some pattern exists. The explanation may be qualitative, or it may include a mathematical component to allow quantitative testing or predictions. This insight is considered tentative until additional observations support the hypothesis. The way that science develops, then, involves a cyclic process. Data are obtained, a hypothesis is advanced, and more data are gathered to support or refute the hypothesis. A refuted explanation can be discarded or modified to fit the new data and then tested again. Eventually, this process of proposing explanations and testing the implications of those proposals leads to a theory that explains some facet of nature. Models or theories are frequently dynamic, evolving as new information is obtained. Scientists always accept the possibility that a theory can be overturned by new observations that invalidate their explanations of nature. The theories and models we will discuss in this text, however, are supported by a substantial number of experimental measurements and generally have been in place for many years. Although we hear phrases like "it's just a theory" in casual conversation or even in political discussions of public policy, scientists only use the word to describe a very well-established model.

A few theories become so sufficiently refined, well tested, and widely accepted that they come to be known as **laws**. Because some of these laws have been accepted for hundreds of years, they may appear self-evident. Without serious questioning, for example, we accept the fact that mass should be conserved. Yet the law of conservation of mass did originate in experimental observations of chemical and physical changes carried out by early scientists. As with any theory, a careful scientist is at least implicitly open to the possibility that a principle viewed as a law could be refuted by some unexpected experiment. But, typically, if we encounter a law, we can expect it to summarize nature reliably enough so that we need not be concerned that it will be "repealed."

Mass is conserved in ordinary chemical reactions. In nuclear reactions, which we will discuss in Chapter 14, mass and energy can be interconverted.

The process just described is commonly called the scientific method, as we defined it in the "Insight" section of this chapter. The word "method" implies a more structured approach than actually exists in most scientific advancement. Many of the advances of science happen coincidentally, as products of serendipity. The stops and starts that are characteristic of scientific development, however, are guided by the process of hypothesis formation and observation of nature. Skepticism is a key component of this process. Explanations are accepted only after they have held up to the scrutiny of experimental observations.

The models that were important in discovering affordable ways to refine aluminum were largely associated with the nature of chemical bonding. We'll revisit this topic in Chapters 6 and 7 and see how the models of chemical bonding were developed.

1-4 Numbers and Measurements in Chemistry

We can observe the world in a variety of ways. For example, we may see a professional basketball player and comment that he is tall, a reasonable statement. If we consider that same person on the basketball court, however, we may want to know *how* tall he is. In this case, the answer to the question might be "six-ten." To a basketball fan in the United States, this answer makes sense because that person ascribes additional meaning to the answer—specifically, that the numbers are in feet and inches: 6 feet 10 inches. A fan from another country might not find any meaning in the two numbers, six and ten, because they wouldn't make sense in metric units. After all, no one is six meters tall.

Units can have consequences far more important than confusion over height. NASA attributed the loss of a Mars probe in 1999 to errors associated with the confusion of different units used in the design.

The example of the basketball player offers a good illustration of the difference between qualitative and quantitative information and also points to the need to be specific in the way we communicate information. Science and engineering are regarded as quantitative professions, and usually this reputation is correct. But scientists also look at the world around us in general ways, akin to the assessment that the player is "tall." Such general or qualitative observations can be crucial in establishing a systematic understanding of nature. As that understanding deepens, though, we usually invoke quantitative or numerical measurements and models. When discussing chemistry, qualitatively or quantitatively, it is always important to communicate our observations and results as clearly as possible. In quantitative observations, this usually means making sure that we carefully define the terms we use. As we move through this course, you will encounter many examples of everyday words that take on much more specific meanings when used in a scientific context. Similarly, when we talk about numerical measurements, we need to be very careful about the ways we use numbers and units.

Units

The possible misunderstanding between basketball fans from different countries about the height of a player provides an analogy to an unfortunate reality in studying science and engineering. Science has grown over the centuries, with contributions from a number of cultures and individuals. The legacy of centuries of development is the existence of a large number of units for virtually every basic measurement in science. Energy, for example, has long been important for human civilization, and accordingly there are many units of energy. Distance and mass have also been measured for literally millennia.

The internationalization of science and engineering led to the establishment of a standard system that provides the needed flexibility to handle a wide array of observations. In this International System of Units (*Système International d'Unités*, or SI), carefully defined *units* are combined with a set of *prefixes* that designate powers of ten. This allows us to report and understand quantities of any size, as illustrated in **Figure 1.8**.

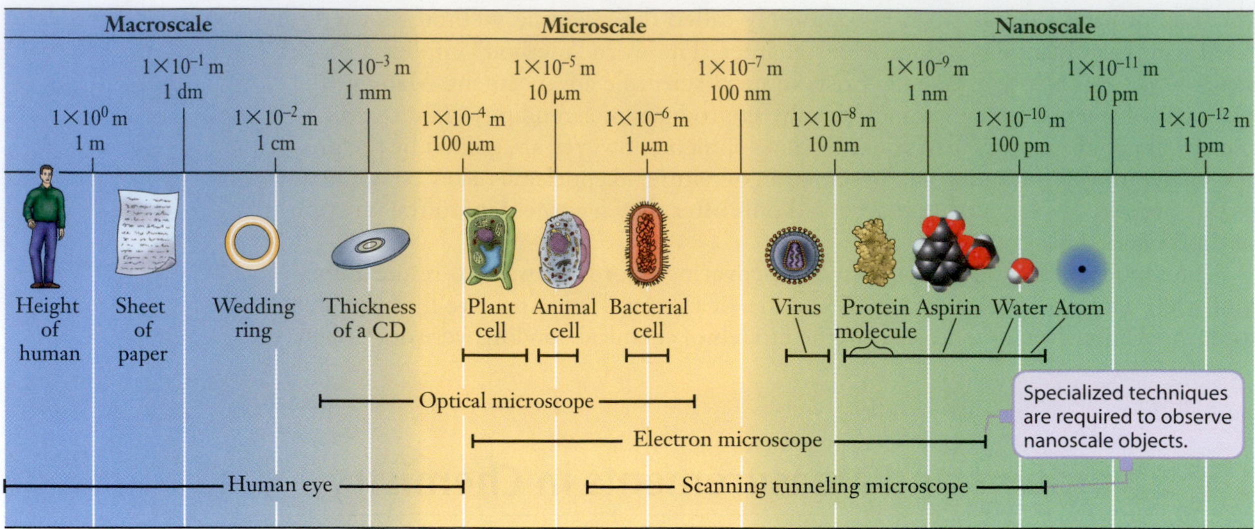

Figure 1.8 ■ The typical dimensions of the objects shown extend over many orders of magnitude and help point out the usefulness of the prefixes in the SI system.

When observations are reported in this system, the base unit designates the type of quantity measured. For example, we can immediately recognize any quantity reported in meters (m) as a distance. **Table 1.1** shows the base units for a number of the quantities that chemists may want to measure. These base units, however, are not always conveniently matched to the size of the quantity to be measured. When we consider the sizes of single atoms or molecules, we will see that they are on the order of 0.0000000001 meters. Dealing with so many decimal places invites confusion, so usually it is preferable to choose a unit that is better matched to the scale of the quantity. The standard way to do this is to use prefixes that alter the "size" of any given base unit. A familiar example is the concept of the "kilo" unit. A kilometer is 1000 meters, and a kilobyte, familiar in computer technology, is roughly 1000 bytes. The prefix *kilo-* implies that there are 1000 of whatever the base unit is. Prefixes such as this exist for numbers large and small, and these prefixes are provided in **Table 1.2**. The distance of 0.0000000001 meters could be reported as either 0.1 nm or 100 pm. Measuring time provides another interesting example of units. Laser based experiments can measure

Because powers of 2 are important in computer science, a kilobyte is actually defined as 1024 bytes ($1024 = 2^{10}$).

Table ■ 1.1

Base quantities of the SI system of units

Property	Unit, with abbreviation
Mass	kilogram, kg
Time	second, s
Distance	meter, m
Electric current	ampere, A
Temperature	kelvin, K
Number of particles	mole, mol
Light intensity	candela, cd

The kilogram is considered the base unit for mass, but the names for mass units are derived by affixing prefixes to the gram. A kilogram is 1000 grams.

Table ■ 1.2

Prefixes used in the SI system

Factor	Name	Symbol	Factor	Name	Symbol
10^{24}	yotta	Y	10^{-1}	deci	d
10^{21}	zetta	Z	10^{-2}	centi	c
10^{18}	exa	E	10^{-3}	milli	m
10^{15}	peta	P	10^{-6}	micro	μ
10^{12}	tera	T	10^{-9}	nano	n
10^{9}	giga	G	10^{-12}	pico	p
10^{6}	mega	M	10^{-15}	femto	f
10^{3}	kilo	k	10^{-18}	atto	a
10^{2}	hecto	h	10^{-21}	zepto	z
10^{1}	deka	da	10^{-24}	yocto	y

the progress of chemical reactions on a timescale of 10^{-15} s. Therefore, scientific papers that report such experiments use femtoseconds, fs, to report time.

Not every quantity can be measured directly in terms of just the seven base units shown in Table 1.1, of course. Some units comprise combinations of these base units and therefore are termed *derived units*. The SI unit for energy, for example, is the joule (J), and 1 J is defined as 1 kg m^2 s^{-2}.

In principle, any quantity could be expressed in terms of appropriate combinations of SI base units. But in practice, many other units are so firmly entrenched that they remain in common use. One simple example of this is the measurement of times that are longer than seconds. Minutes, days, and years are used in these circumstances, rather than kiloseconds, and so on. In chemistry labs—and on soda containers—volumes are most often reported in liters (L) or milliliters (mL) rather than the SI unit of cubic meters (m^3). Chemists also use a wide variety of units to describe concentration, which measures how much of a particular substance is present in a mixture. Metals often contain minor impurities, and in some cases the units used are simply percentages; other units used include **parts per million (ppm)** and **parts per billion (ppb)**. The ppm unit tells how many particles of a particular substance are present for every 1 million particles in the sample. In ppb, the sample size is 1 billion particles. Impurity concentrations on the order of even a few ppm may cause problems in some applications, or they may be added intentionally to impart some desirable property. Later in this chapter, we will take up the issue of converting a measurement from one unit to another, as is frequently required in scientific and engineering calculations.

In Chapter 6, we will discuss trace analysis techniques that can be used to measure the composition of materials down to the ppm or ppb level.

Although many of these SI units have found their way gradually into everyday use, the units for temperature may be the least familiar. You are probably used to seeing temperatures in either degrees Fahrenheit or degrees Celsius, but generally, even in "fully metric" countries, the weather report does not use the Kelvin temperature scale. In general, **temperature scales** arise from the choice of two standard reference points that can be used to calibrate temperature with the use of a thermometer. The familiar Fahrenheit scale originally chose body temperature as one reference and set it at 100°F. (Accuracy of measurement was clearly not a priority when this temperature scale was proposed.) The second reference point was the coldest temperature that could be achieved by adding salt to ice water, a practice

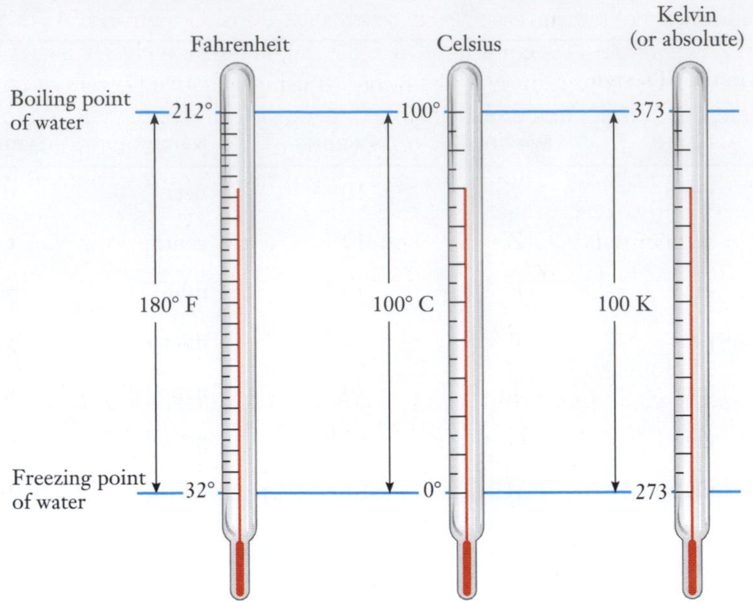

Figure 1.9 ■ The Fahrenheit, Celsius, and Kelvin temperature scales are compared. The freezing point of water can be expressed as 32°F, 0°C, or 273 K. The boiling point of water is 212°F, 100°C, or 373 K.

that lowers the melting point of ice. This established 0°F, and the temperature range between the two points was divided into 100 equal units. The scale is now defined by setting the freezing point of water at 32°F and the boiling point of water at 212°F.

The Celsius scale was developed in a similar way, but with the freezing point of pure water set at 0°C and the boiling point of water at 100°C. **Figure 1.9** shows the relationship between these scales. Conversions between the two scales are given by the following expressions:

Converting between temperature scales differs from other unit conversions because temperature is the only quantity for which the meaning of zero depends on what units we are using.

$$°F = (1.8 \times °C) + 32 \qquad (1.1)$$

$$°C = (°F - 32)/1.8 \qquad (1.2)$$

Scientific uses of temperature require yet another temperature scale. The choice of the kelvin as the standard reflects mathematical convenience more than familiarity. The Kelvin scale is similar to the Celsius scale but draws its utility from the fact that the lowest temperature theoretically possible is zero kelvin. It violates the laws of nature to go below 0 K, as we will see in Chapter 10. The mathematical importance of this definition is that we are assured we will not divide by zero when we use a formula that has temperature in the denominator of an expression. Conversions between Celsius degrees and kelvins are common in science and are also more straightforward:

$$K = °C + 273.15 \qquad (1.3)$$

$$°C = K - 273.15 \qquad (1.4)$$

Engineers in some disciplines use the Rankine temperature scale (°R), which is an absolute scale whose degrees are the same size as those of the Fahrenheit scale.

Numbers and Significant Figures

We often encounter very small and very large numbers in chemistry problems. For example, pesticide production in the world exceeds millions of tons, whereas pesticide residues that may harm animals or humans can have masses as small as nanograms. For either type of number, **scientific notation** is useful. Numbers written using scientific notation factor out all powers of ten and write them separately. Thus the number 54,000 is written as 5.4×10^4. This notation is equivalent to $5.4 \times 10{,}000$, which clearly is 54,000. Small numbers can also be written in scientific notation using negative powers of ten because 10^{-x} is identical to $1/10^x$. The number 0.000042 is 4.2×10^{-5} in scientific notation.

When numbers are derived from observations of nature, we need to report them with the correct number of significant figures. **Significant figures** are used to indicate the amount of information that is reliable when discussing a measurement. "Pure" numbers can be manipulated in a mathematical sense without accounting for how much information is reliable. When we divide the integer 5 by the integer 8, for example, the answer is *exactly* 0.625, and we would not round this to 0.6. When numbers associated with an observation are used, however, we must be more careful about reporting digits. To understand why, consider whether to accept a wager about some measurement.

For example, an almanac lists the population of Canada as 34,482,779. Suppose a study concluded that 24% of the people who live in Canada speak French. Based on this information alone, the wager is that 8,275,587 Canadians speak French. Should we accept the bet? What if the wager was that roughly 8.0 million Canadians speak French? Is this a better bet? This scenario shows the importance of significant figures. The figure of 8,275,587 is not believable because we don't really know if 24% is *exactly* 24.000000%. It could just as well be 23.95% and the correct answer would then be 8,258,626. Both answers might qualify as "roughly 8 million," an answer with a more reasonable number of significant figures.

This type of reasoning has been formalized into rules for significant figures (digits) in numbers reported from scientific observations. When a measurement is reported numerically, generally we consider that each digit given is known accurately, with one important exception. There are special rules for zero such that it is sometimes significant and sometimes not. When a zero establishes the place for the decimal in the number, it is not significant. Thus the measurement 51,300 m has two zeros that are not significant, as does the measurement 0.043 g. A zero is significant when it is the final digit *after* a decimal place or when it is *between* other significant digits. Thus the zeros in both 4.30 mL and 304.2 kg are significant. When numbers are written properly in scientific notation, all of the digits reported are significant. Example Problem 1.2 provides some additional practice in determining the number of significant figures in measurements.

Scientific notation provides an advantage here because any digits shown in the number are significant. There are no zeros needed for placing the decimal.

EXAMPLE PROBLEM 1.2

An alloy contains 2.05% of some impurity. How many significant figures are reported in this value?

Strategy Use the general rule for significant figures: digits reported are significant unless they are zeros whose sole purpose is to position the decimal place.

Solution In this case, all digits reported are significant, so there are three significant figures in the number.

Check Your Understanding How many significant figures are reported in each of the following measurements? (a) 0.000403 s, (b) 200,000 g

We also need to account for significant figures in assessing values that we obtain from calculations. The general principle is that a calculated value should be reported with a number of significant figures that is consistent with the data used in the calculation. Our earlier story about the number of French-speaking Canadians provides some insight into this issue as well. The wager that roughly 8.0 million such people live in Canada was more attractive. The reason: 8.0 million has the same number of significant figures as the value 24%, so it provides a similar quality of information. Three key rules will be required to determine the number of significant figures in the results of calculations.

Rule 1: For multiplication and division, the number of significant figures in a result must be the same as the number of significant figures in the factor with the fewest significant figures. When 0.24 kg is multiplied by 4621 m, the result on a calculator reads 1109.04, but if significant figures are correctly reported, the result will be 1100, or 1.1×10^3 kg m. The value 0.24 kg has only two significant figures, and so the result should also have just two significant figures.

Rule 2: For addition and subtraction, the rules for significant figures center on the position of the first doubtful digit rather than on the number of significant digits. The result (a sum or difference) should be rounded so that the last digit retained is the first uncertain digit. If the numbers added or subtracted are in scientific notation with the same power of ten, this means that the result must have the same number of digits to the right of the decimal point as in the measurement that has the fewest digits to the right of the decimal point. The number 0.3 m added to 4.882 m yields 5.182 on a calculator, but it should be reported as 5.2 m. The first uncertain digit in the values added is the "3" in 0.3 m, so the result should be rounded to one decimal place. In this case, the last allowed digit was "rounded up" because the first nonsignificant digit, 8, was greater than 5. There are two possible conventions for rounding in such cases. In this text, we will round down for numbers 4 and smaller and round up for numbers 5 and larger. Example Problem 1.3 provides some practice in working with significant figures.

EXAMPLE PROBLEM 1.3

Report the result for the indicated arithmetic operations using the correct number of significant figures. Assume that all values are measurements.
(a) 4.30×0.31 (b) $4.033 + 88.1$ (c) $5.6/1.732 \times 10^4$

Strategy Check the rules for significant figures as needed and determine which rules apply. Carry out the calculation and express the result in the correct number of significant figures.

Solution

(a) $4.30 \times 0.31 = 1.3$ (b) $4.033 + 88.1 = 92.1$ (c) $5.6/(1.732 \times 10^4) = 3.2 \times 10^{-4}$

Check Your Understanding Determine the value of the following expressions using the correct number of significant figures: (a) 7.10 m + 9.003 m, (b) 0.004 g $\times$ 1.13 g

The rules above apply to any numbers that result from most of the measurements that we might make. But in the special case of countable objects, we must also consider one additional rule.

Rule 3: When we count discrete objects, the result has no ambiguity. Such measurements use exact numbers, so effectively they have infinite significant figures. Thus if we need to use information such as four quarts in a gallon or two hydrogen atoms in a water molecule, there is no limitation on significant figures. Note that this rule also applies when we work with the various prefixes in the SI system. There are *exactly* 100 centimeters in a meter, so the factor of 100 would never limit the number of significant figures in a calculation.

1-5 Problem Solving in Chemistry and Engineering

Calculations play a major role in the practice of chemistry and in its application to real-world issues and problems. And engineering designs routinely rely on a tremendous number of calculations. The types of exercises we will introduce in this text, though focused on chemistry, will provide practice with techniques that can also be used in engineering applications. To a chemist, the questions associated with aluminum ore require looking into the nature of chemical bonding and how to overcome the stability of strong bonds between aluminum and oxygen. To an engineer, the problems to be addressed in refining the ore might focus on how to deliver enough electricity when and where it is needed. Both disciplines, however, need careful use of quantitative or numerical reasoning.

Using Ratios

We encounter and use ratios regularly. We discuss the speed of our cars in miles per hour and buy our fruit at \$1.09 a pound. Most of us use such ratios intuitively. If we look at the way we do this, we can deduce a formal set of rules to apply to chemical observations and situations. Let's say your grocery bill indicates that you paid \$4.45 for a 5.0-pound bag of apples. How much did you pay per pound? Although you may not think about it, you obtain the answer by creating an appropriate ratio and completing some simple arithmetic.

$$\text{Price} = \frac{\$4.45}{5.0 \text{ pounds}} = \$0.89 \text{ per pound}$$

The notation "per" means for each of the specified units. Price per pound is the cost of *one* pound. Miles per hour tell the distance traveled in *one* hour. This is useful because one is a convenient number in multiplication. We obtain this type of information by creating and simplifying a ratio.

With the information we have about the bag of apples, we have another option to create a ratio. We could obtain the number of pounds of apples per dollar.

$$\frac{5.0 \text{ pounds}}{\$4.45} = 1.1 \text{ pounds of apples per dollar}$$

For this example, such a manipulation may not be particularly useful—the store probably doesn't sell 1.1 pound bags of apples. Still, it provides useful insight. When we form the ratio and carry out the indicated arithmetic, the result tells us how much of the numerator is equivalent to one unit of the denominator. So, in general, given two equivalent quantities A = B, we can write either of two ratios:

A/B (tells how much A is in one unit of B)

and

B/A (tells how much B is in one unit of A)

Example Problem 1.4 shows how this type of manipulation fits into a problem solving strategy with non-chemistry examples before we look at it in a chemistry context.

EXAMPLE PROBLEM 1.4

Shrimp are usually labeled with a "count" that indicates the average number of shrimp per pound. The bigger the shrimp, the smaller the count. Suppose that your supermarket is offering 20-count shrimp for \$5.99 per pound. How much should you expect to pay for one dozen shrimp?

Strategy We can determine the average cost per shrimp and then multiply by the number of shrimp needed to find the total price.

Solution

$$\frac{\$5.99}{20 \text{ shrimp}} \times 12 \text{ shrimp} = \$3.59$$

Analyze Your Answer When we calculate a numerical result, it is generally a good idea to pause and see if the answer makes sense. Such a check can often catch errors that might arise from pushing the wrong key on a calculator or even errors in the way we set up the calculation itself. In this case, we found that the cost of 12 shrimp was \$3.59, which is a little more than half of the \$5.99 price for 20 shrimp. Because 12 is a bit more than half of 20, this seems sensible.

Check Your Understanding A farmer purchases pesticide in 5-gallon drums that cost \$23.00 each. If 65 gallons of the pesticide are applied to a field, what was the total cost? How might the situation be more complicated if the amount needed were not a multiple of 5 gallons?

For some problems in chemistry, the units of measurement can be used to determine how to write the appropriate ratio. In the previous example, by considering the units in an algebraic sense, the "shrimp" (with the 20) in the denominator are "canceled" by the same unit (with the 12) in the numerator:

$$\frac{\$5.99}{20 \cancel{\text{shrimp}}} \times 12 \cancel{\text{shrimp}} = \$3.59$$

This type of reasoning is called **dimensional analysis,** or the **factor-label method** for calculation. In some cases, the dimensions can provide clues about what ratio to define. We will point out this type of reasoning in problems where it may be used.

Ratios in Chemistry Calculations

We use ratios in a variety of common calculations in chemistry. One that you will undoubtedly also encounter in engineering is the need to convert between units of different sizes. Example Problem 1.5 introduces this type of manipulation.

EXAMPLE PROBLEM 1.5

Visible light is commonly described in terms of its wavelength, which is usually given in units of nanometers. In subsequent calculations, this measurement often needs to be expressed in units of meters. If we are considering orange light of wavelength 615 nm, what is its wavelength in meters?

Strategy We need to establish a ratio that relates nanometers and meters. Then we can use that ratio to determine the answer.

Solution

$$1 \text{ m} = 1 \times 10^9 \text{ nm}$$

We can write this as a ratio. Because we want to convert *from* nm *to* m, we'll need m in the numerator and nm in the denominator:

$$\frac{1 \text{ m}}{10^9 \text{ nm}}$$

Then, we just complete the calculation:

$$615 \text{ nm} \times \frac{1 \text{ m}}{10^9 \text{ nm}} = 6.15 \times 10^{-7} \text{ m}$$

Analyze Your Answer Writing units on all quantities involved helps ensure that we use the correct ratio. Even if we aren't familiar with the typical wavelengths of light, we can still check that the answer makes sense. Because nanometers are much smaller than meters, the fact that we got a much smaller number for our answer seems correct. Note that conversions between SI units are between exact numbers when considering significant figures.

Check Your Understanding Large amounts of electrical power, measured in watts (W), are used to produce aluminum from ore. If a production plant uses 4.3 GW in a certain period of time, how many W are used?

Another common type of problem that uses ratios occurs when two quantities are related to each other either by some chemical or physical property. One such property is mass density. Mass density is defined as the mass of a substance per unit volume. This definition is itself a ratio that allows us to convert between mass and volume, as shown in Example Problem 1.6.

EXAMPLE PROBLEM 1.6

The mass density of water at 25°C is 0.997 g per mL. A child's swimming pool holds 346 L of water at this temperature. What mass of water is in the pool?

Strategy We want to use the density as a ratio, but first we must express the volume in appropriate units. We can convert the volume of the pool from liters to milliliters and then use the density as given.

Solution

Because 1 L = 1000 mL, 1000 mL/1 L provides a conversion from L to mL.

The given density is 0.997 g/1 mL, and this provides the connection between volume and mass:

$$346 \text{ L} \times \frac{1000 \text{ mL}}{1 \text{ L}} \times \frac{0.997 \text{ g}}{1 \text{ mL}} = 3.45 \times 10^5 \text{ g}$$

Analyze Your Answer It may be difficult to have an intuitive feel for the size of an answer. But 346 L would be nearly 100 gallons of water, which would be fairly heavy. A pound is equivalent to 454 g, so the answer we found (3.45×10^5 g) is more than 700 pounds, and this is at least plausible. If we had inverted the ratio used to convert from mL to L, for example, we would have gotten an answer of just 0.345 g, or less than a gram. Clearly, that would be too small for even a very tiny pool, so we would have had a chance to find and correct the error.

Discussion Here we have written our solution as a single two-step process in which we first converted the volume from liters to milliliters and then used the density to find the mass. Some students feel more comfortable breaking these steps down into separate calculations; we could have found the volume explicitly as 346,000 mL and then multiplied that by the density. As long as the operations are carried out correctly, the result will be the same. If you do choose to break calculations down into smaller pieces, though, be careful not to round your answers until you reach the final result.

Also note that the order of our steps is not really important. We could have converted the density to 997 g/L and then multiplied by the volume.

Check Your Understanding A liquid pesticide has a density of 1.67 g per mL. What is the mass of liquid in a full 20.0-L container?

Calculations in chemistry use ratios very often. Example Problem 1.6 also shows how they are used in combinations. Learning how to carry out these sequential manipulations will be a key part of solving many of the problems we encounter in chemistry. One guide that often helps to determine how to construct the appropriate ratio is to note the units used. For example, the equality between 1000 mL and 1 L leads to two ways to express the ratio, 1000 mL/1 L or 1 L/1000 mL. Which ratio did we need? Because we started with liters, we required the former ratio, with 1 L in the denominator. We will often include discussions about the ratios we form in the "strategy" part of the example problems.

Mass density is more than a convenient way to define a ratio for calculations. It is also an important characteristic of materials. Among the reasons that aluminum is a popular structural material is that its relatively low density is often important. When a design calls for a strong but lightweight material, aluminum is a good candidate. We'll look at this aspect further in the closing insight in Section 1-6 when we look at various materials used in bicycle frames.

Mass density also is important in determining buoyancy because less dense objects float on more dense ones.

Conceptual Chemistry Problems

Although numerical calculations will always be an important component of chemistry, they are only part of the field. Sometimes, to ensure that the concepts involved in chemistry are understood, we'll also work problems that focus on the particulate representation and other concepts. Often, the strategies employed in this type of question are different from those we have just described.

When we try to visualize a chemical concept using drawings, we will start from the conventions that were introduced for the particulate perspective of chemistry in Section 1-2. In this type of problem, we could be asked, for example, to draw a diagram that depicts what happens to the molecules when steam condenses into liquid water. Example Problem 1.7 illustrates this type of thinking.

EXAMPLE PROBLEM 1.7

Dry ice is solid carbon dioxide; it is called "dry" because it goes directly from a solid to a gas without becoming a liquid under ordinary conditions. Draw a picture that shows what the carbon dioxide molecules might look like as a solid and as a gas.

Strategy To answer conceptual questions such as this, we must depict the differences between the two phases. Our drawing can be schematic, but it must imply the differences clearly. In this case, we know that a solid will have closely packed molecules, whereas a gas will exhibit significant space between particles.

Solution Our picture shows several things. First, we have included some information about the chemical composition of the molecules by using two different colors for the spheres that represent the atoms. Second, the solid is indicated at the bottom and is distinguishable because the molecules are shown closely packed together in an orderly array. Finally, the gas is depicted by only a few molecules, and they are widely spaced.

Discussion Obviously, the solutions to conceptual questions like this are less exact than numerical answers. So your drawing might look somewhat different yet still be "correct." Neither the gas nor the solid needs to look exactly like the picture we have drawn, but the drawing itself should impart the essential concepts we need to understand when thinking about the particulate perspective of this process.

Check Your Understanding Draw a picture that shows a molecular scale view of steam condensing into liquid water.

We will continue to use conceptual problems throughout the text to help provide practice in thinking about the behavior of molecules and atoms.

Visualization in Chemistry

One area where conceptual understanding in chemistry is different from that in most engineering pursuits is the way we visualize systems. Chemistry provides multiple simultaneous ways to view problems, including the completely abstract perspective of atoms and molecules that will never be observed directly. Methods to visualize this level of chemistry provide an important tool in the way that chemistry is taught and learned. We can explore this idea by thinking about the refining of aluminum.

The most common aluminum ore is bauxite, which consists of aluminum oxides and rock. The first step in producing aluminum metal is separating the aluminum oxides from the remainder of the rock. As shown in **Figure 1.10**, this actually entails several steps; the first is digestion of the ore. Process engineers design "digesters" in which crushed ore, caustic soda, and lime are mixed at high temperatures to create a slurry. If we look at **Figure 1.11**, we can visualize this process at the microscopic level.

The chemical name for caustic soda is sodium hydroxide, and its formula is NaOH. Lime is calcium oxide, CaO.

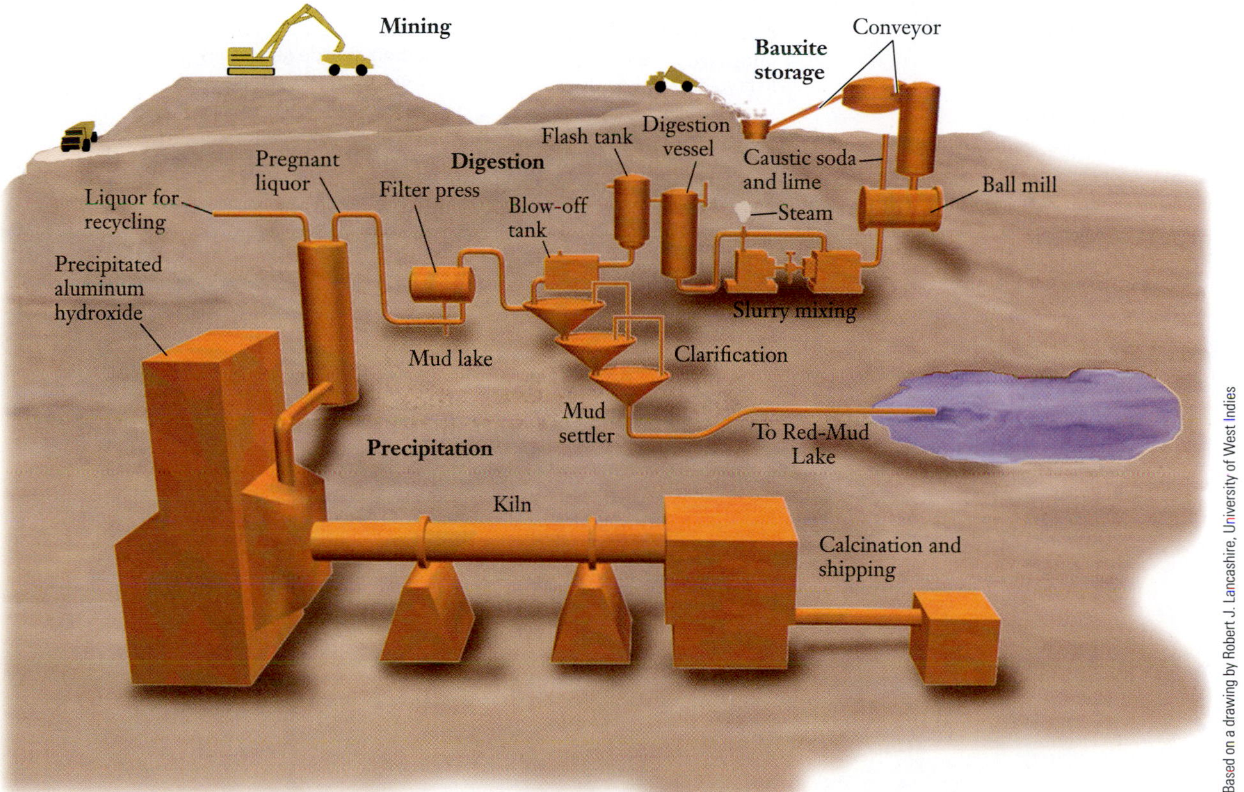

Figure 1.10 ■ Several steps in the processing of bauxite are represented in this diagram.

Figure 1.11 ■ Bauxite is treated with caustic soda and lime in a process called *digestion*. The alumina in the ore reacts and dissolves. This step separates the aluminum from the rest of the ore, which contains a variety of other minerals.

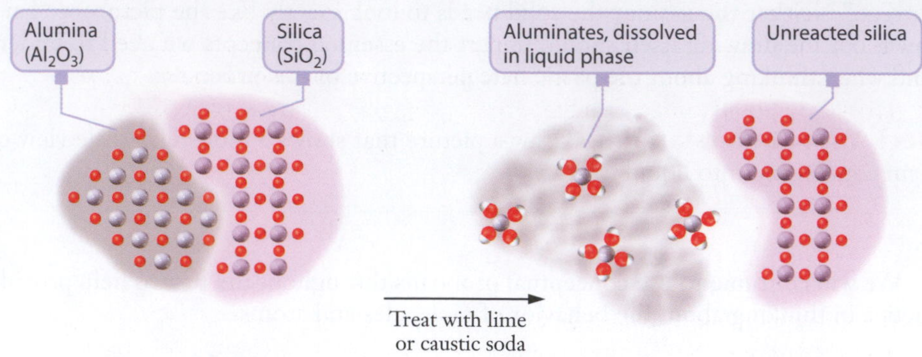

The molecular level pictures show alumina as atoms of aluminum and oxygen, and the rock is shown as silicon and oxygen (or silica). This is a simplified depiction based on the fact that much of the earth's crust contains a large percentage of these elements. (Bauxite ore also contains significant amounts of iron oxide and other minerals, but these behave much like silica in the digestion process.) Using this level of visualization to consider the reactions in the digester, we see that alumina reacts with the caustic soda and lime, but silica does not. In this way, we can look at a large-scale industrial process while thinking in the microscopic perspective.

Note that the conceptual drawings in Figures 1.11 and 1.12 are schematic. They do not attempt to impart molecular detail but rather aim to provide a sense of the particles involved in the chemical process.

The next step in aluminum refining is smelting. At this stage, a chemical reaction is induced where the aluminates, now dissolved in a material called cryolite (Na_3AlF_6), lose oxygen atoms to a carbon rod, forming relatively pure aluminum and carbon dioxide. This portion of the process is shown in **Figure 1.12**. Although this molecular-level visualization again shows how the chemical reaction rearranges atoms at the particulate level, it is worth remembering that this process is carried out on a massive scale: the U.S. aluminum industry produces about 2.6 million metric tons of aluminum per year.

A metric ton is equal to 1000 kg, or roughly 2200 pounds.

Figure 1.12 ■ These particulate level illustrations provide a simplified view of the atomic-scale process involved in smelting aluminum.

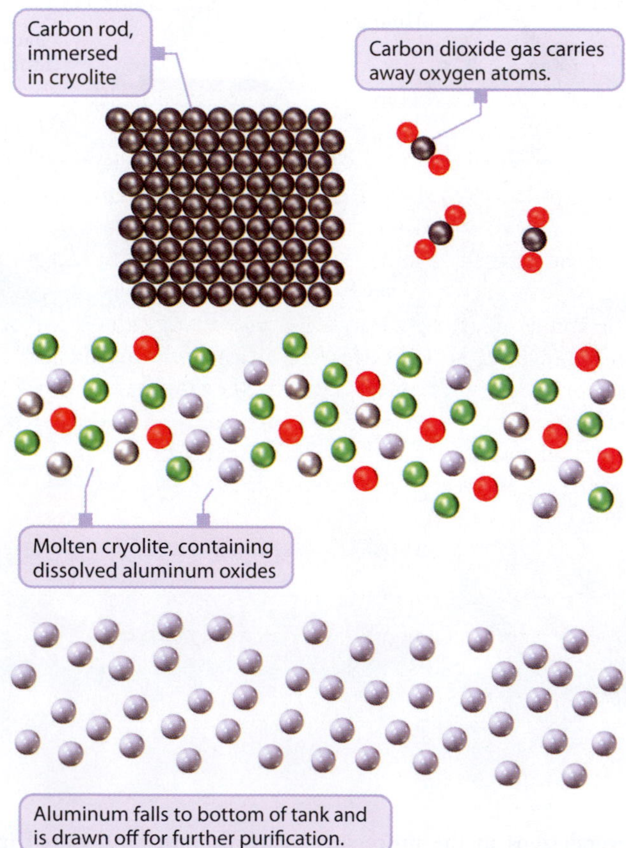

The visualization techniques introduced in this section will help us to develop a particulate level understanding of many concepts we encounter throughout the text. At this point, let's take some time to look at one application where the properties of aluminum metal make it a popular material choice for engineering a consumer product—a bicycle.

1-6 Material Selection and Bicycle Frames

For the average rider who doesn't wish to spend a great deal of money on a bike, the frame is likely to be some alloy of steel. Using other materials for the frame can improve certain aspects of performance but also increases the cost of the bicycle. The consumer's choices are confronted first by the engineer who designs the frame.

Among the properties that an engineer must consider in choosing a material for a frame are strength, density (which affects the weight of the frame), and stiffness. Although we are already familiar with density, talking about the strength and stiffness of materials requires some additional vocabulary. Stiffness is related to a property called the **elastic modulus** of the material. The elastic modulus measures the amount of stretch or compression a material experiences when it is stressed. Something with a high modulus will stretch very little even when experiencing large forces, and a bicycle frame composed from such a material would seem stiff. We will see this concept again when we discuss chemical bonding because the ability of a material to stretch is related to how closely and strongly the atoms in it are connected to each other.

The strength of a material is formally measured by a property called the yield strength. The **yield strength** measures the amount of force required to produce a specified deformation of the material. A stronger material can withstand greater forces before it deforms and so has higher yield strength. Finally, because there are benefits from having a lighter frame, materials with low densities are also desirable.

Measurements such as elastic modulus or yield strength depend on temperature. For most engineering designs, however, this dependence is more critical in plastics than in metals.

The three most common metals for bicycle frames are aluminum, steel, and titanium. **Table 1.3** shows typical values for the elastic modulus, yield strength, and density of these materials.

Looking at this table, we can see that the physical properties of these materials afford trade-offs in types of performance in the frame. Although aluminum is lightweight, it is neither as strong nor as stiff as steel or titanium. Steel is the strongest, stiffest, and heaviest of the materials. It's important to note the range of yield strengths in this table because the details of manipulating the strength of aluminum or steel depend on both the chemical composition and the manner in which the material is treated in fabricating the frame.

Of course, the engineer has more than the choice of material available in making a design for a bicycle frame. The size and connectivity of the tubing can also be adjusted, for example. Aluminum frames generally feature tubing with a much larger diameter than that in steel or titanium frames. These larger diameter tubes give the frame itself a much stiffer feel—so much so that many cyclists find aluminum frames too stiff to give a comfortable ride. Many modern frames also use tubing that is not round but rather is oval to withstand specific types of stress commonly encountered in cycling. Some very expensive

Table 1.3

Elastic modulus, yield strength, and density of some materials used in bicycle frames

Material	Elastic Modulus (psi)	Yield Strength Range (psi)	Density (g/cm^3)
Aluminum	10.0×10^6	$5.0 \times 10^3 - 6.0 \times 10^4$	2.699
Steel	30.0×10^6	$4.5 \times 10^4 - 1.6 \times 10^5$	7.87
Titanium	16.0×10^6	$4.0 \times 10^4 - 1.2 \times 10^5$	4.507

frames for competitive cycling are made from more exotic materials. New aluminum alloys containing trace amounts of scandium, for example, can achieve lower frame weights without the large tubing diameters that cause the jarring ride for which aluminum is usually known. Carbon fiber composites, made by combining strong fibers of carbon with plastics or epoxies, offer another option for lightweight frames. Unlike aluminum or titanium, the properties of such composites will not be uniform in all directions. This is because the carbon fibers themselves can be oriented in a particular direction, and this influences the material's strength. The forces experienced by a bike frame tend to be oriented in particular directions, so the frame can be engineered to be stronger in the directions where it is most likely to be stressed. Varying the amount of carbon fiber in the composite offers a further ability to tune the strength and weight of the material. Because of their high strength-to-weight ratios and the fact that they can be molded easily into exotic shapes that reduce air resistance, carbon composites are frequently used in the extremely aerodynamic bikes preferred for high-speed time trial racing.

Chemistry and chemists have played an essential role in helping engineers develop and exploit these new materials. And the intersection between chemistry and engineering illustrated in this chapter for bicycles and soda cans is also apparent in many other familiar products. Throughout this text, we'll look at places where the chemistry we learn has an impact on engineering designs. At this point, we know enough to say that if a lightweight bicycle frame is your desire, you may want to take advantage of aluminum and its low mass density. In Chapter 2, we take the next step and look at atoms and molecules in more detail and simultaneously introduce polymers—another important class of engineering materials.

FOCUS ON PROBLEM SOLVING

Engineering students often wonder whether their college chemistry courses will prove useful later in their careers. Through the opening and closing "Insight" sections of the chapters of this book, we'll make connections between the content of the chemistry course and places where that content is relevant in engineering applications. But this type of connection isn't the only place where things learned in chemistry courses help future engineers.

Engineers solve problems. And the exposure to various problem-solving techniques in courses like this will help you to develop and diversify your skills. We will emphasize this connection after each chapter with a special section like this one. In these "Focus on Problem Solving" sections, we will look at chemistry problems related to the chapter just presented. The difference will be that the "correct" answer will not be a number but rather the identification of an appropriate strategy for approaching the problem.

Question Describe how you would determine which has a greater mass, a sphere of iron shot with a radius of 4.00 mm or a cube of nickel with an edge length of 4.00 mm. Which formulas are needed, and what other information would have to be looked up?

Strategy Like many real-world problems, this question does not have enough information given to provide an answer. To devise a strategy, we have to think about what it takes to compare two materials—in this case, iron and nickel. Moreover, we are asked to compare masses of the two materials, so that provides a clue. The dimensional information given, along with formulas that are familiar or can be looked up, provides information about volume. Mass and volume were related in this chapter by density, and density values for common materials can be found easily.

Solution To answer this problem, first we need to look up the densities of both iron and nickel. Then, we need to use the data provided to determine the volume of each sample. For a cube, we can calculate the volume by taking the given measure for the side (s) and cubing it (s^3). The volume of a sphere is given by $(4/3)\pi r^3$. Multiplying the density by volume provides the mass, so with the data we looked up for density and the calculated volumes we could answer the question.

SUMMARY

Chemistry is the science of matter, and since all engineering designs involve matter, the links between chemistry and engineering are many. We began to explore the role of materials in engineering by considering aluminum. By applying some very simple chemical concepts, we can begin to understand the transformation of aluminum from a precious metal to a common and inexpensive material.

One common trait of an experienced chemist is the ability to consider a given situation from a number of perspectives. Both the physical and chemical properties of substances can be considered at the macroscopic or microscopic (particulate) level depending on the nature of the question or problem being considered. In addition, chemists often use symbolic representations to describe what is happening in chemical systems. Becoming comfortable with these different perspectives can give students an edge in understanding many chemistry problems.

Chemistry is an empirical science. It relies on experimental observations to develop an understanding of matter. The path from observation toward an understanding of the universe typically involves several steps, relying on inductive or deductive reasoning, or both. By applying reasoning skills to our observations, we construct models for understanding chemical phenomena. Then these models are refined and adapted over time. The ideas that we will explore in this text involve many models or theories that have been developed through the scientific method.

Many of the observations of nature that are used to develop theories and models need to be quantitative; that is, they must assess what is being observed with some level of numerical detail. The need for numerical observations throughout the development of chemistry (and other sciences) has given rise to systematic ways to communicate this information. A number alone is not sufficient to impart all the meaning of a measurement; experimental observations are expected to include units of measurement. An important skill in the study of both chemistry and engineering is the ability to manipulate numerical information, including the units attached to that information. The use of ratios to convert between a measurement in one unit and desired information in another related unit represents a core skill for problem solving in chemistry and engineering. The method of dimensional analysis, sometimes called the factor-label method, provides one common way to carry out these transformations.

We will build on these fundamental ideas as we proceed in the study of chemistry. The ability to look at problems from several perspectives, extract and manipulate numerical information, and ultimately gain a broad understanding of the chemical principles that underlie the behavior of the universe will provide an interesting challenge as we survey the connections between chemistry and engineering in this book.

KEY TERMS

accuracy (1-3)
atoms (1-2)
chemical properties (1-2)
combustion (1-2)
deductive reasoning (1-3)
density (1-2)
dimensional analysis (1-5)
elastic modulus (1-6)
elements (1-2)
factor-label method (1-5)
gases (1-2)
inductive reasoning (1-3)
laws (1-3)
liquids (1-2)
macroscopic perspective (1-2)
malleability (1-2)
mass density (1-2)
matter (1-2)
microscopic perspective (1-2)
molecules (1-2)
particulate perspective (1-2)
parts per billion (ppb) (1-4)
parts per million (ppm) (1-4)
phases of matter (1-2)
physical properties (1-2)
precision (1-3)
random error (1-3)
scientific method (1-1)
scientific models (1-3)
scientific notation (1-4)
significant figures (1-4)
solids (1-2)
symbolic perspective (1-2)
systematic error (1-3)
temperature scales (1-4)
units (1-4)
yield strength (1-6)

PROBLEMS AND EXERCISES

INSIGHT INTO Aluminum

1.1 Use the web to determine the mass of a steel beverage can from the 1970s and the mass of a modern aluminum can. How much more would a 12-pack of soda weigh in steel cans?

1.2 Which properties of aluminum might concern you if you had to use the aluminum tableware that Napoleon employed to impress his guests?

1.3 Where does the scientific method start? What is the first step?

1.4 Use the web to determine the amount of aluminum used in the United States in a single year. What is the primary use for this material?

1.5 Use the web to find current prices offered for aluminum for recycling. Is there variation in the price based on where in the United States the aluminum is returned?

1.6 Use the web to determine the differences in the amounts of aluminum recycled in states where there are deposits on aluminum cans versus states where recycling is voluntary. What is the most reliable way to estimate this value? What uncertainty is there in the estimate?

The Study of Chemistry

1.7 When we make observations in the laboratory, which perspective of chemistry are we normally using?

1.8 Which of the following items are matter and which are not? **(a)** a flashlight, **(b)** sunlight, **(c)** an echo, **(d)** air at sea level, and **(e)** air at the top of Mount Everest

1.9 Which macroscopic characteristics differentiate solids, liquids, and gases? (List as many as possible.)

1.10 How can a liquid be distinguished from a fine powder? What type of experiment or observation might be undertaken?

1.11 Some farmers use ammonia, NH_3, as a fertilizer. This ammonia is stored in liquid form. Use the particulate perspective to show the transition from liquid ammonia to gaseous ammonia.

1.12 Do the terms *element* and *atom* mean the same thing? If not, how do they differ?

1.13 Label each of the following as either a physical process or a chemical process. **(a)** rusting of an iron bridge, **(b)** melting of ice, **(c)** burning of a wooden stick, **(d)** digestion of a baked potato, **(e)** dissolving of sugar in water

1.14 Why do physical properties play a role in chemistry if they do not involve any chemical changes?

1.15 Physical properties may change because of a chemical change. For example, the color of an egg white changes from clear to white because of a chemical change when it is cooked. Think of another common situation when a chemical change also leads to a physical change.

1.16 Which part of the following descriptions of a compound or element refers to its physical properties and which to its chemical properties?

(a) Calcium carbonate is a white solid with a density of 2.71 g/cm^3. It reacts readily with an acid to produce gaseous carbon dioxide.

(b) Gray powdered zinc metal reacts with purple iodine to give a white compound.

1.17 Use a molecular level description to explain why gases are less dense than liquids or solids.

1.18 All molecules attract each other to some extent, and the attraction decreases as the distance between particles increases. Based on this idea, which state of matter would you expect has the strongest interactions between particles: solids, liquids, or gases?

Observations and Models

1.19 We used the example of attendance at a football game to emphasize the nature of observations. Describe another example where deciding how to count subjects of interest could affect the observation.

1.20 Complete the following statement. Data that have a small random error but otherwise fall in a narrow range are **(a)** accurate, **(b)** precise, or **(c)** neither.

1.21 Complete the following statement: Data that have a large systematic error can still be **(a)** accurate, **(b)** precise, or **(c)** neither.

1.22 Two golfers are practicing shots around a putting green. Each golfer takes 20 shots. Golfer 1 has 7 shots within 1 meter of the hole, and the other 13 shots are scattered around the green. Golfer 2 has 17 shots that go into a small sand trap near the green and 3 just on the green near the trap. Which golfer is more precise? Which is more accurate?

1.23 Use your own words to explain the difference between deductive and inductive reasoning.

1.24 Suppose that you are waiting at a corner for a bus. Three different routes pass this particular corner. You see buses pass by from the two routes that you are not interested in taking. When you say to yourself, "My bus must be next," what type of reasoning (deductive or inductive) are you using? Explain your answer.

1.25 When a scientist looks at an experiment and then predicts the results of other related experiments, which type of reasoning is she using? Explain your answer.

1.26 What is the difference between a hypothesis and a question?

1.27 Should the words *theory* and *model* be used interchangeably in the context of science? Defend your answer using information found in a web search.

1.28 What is a law of nature? Are all scientific laws examples of laws of nature?

Numbers and Measurements

1.29 Describe a miscommunication that can arise because units are not included as part of the information.

1.30 What is the difference between a qualitative and a quantitative measurement?

1.31 Identify which of the following units are base units in the SI system: grams, meters, joules, liters, amperes.

1.32 What is a "derived" unit?

1.33 Rank the following prefixes in order of increasing size of the number they represent: centi-, giga-, nano-, and kilo-.

1.34 The largest computers now include disk storage space measured in petabytes. How many bytes are in a petabyte? (Recall that in computer terminology, the prefix is only "close" to the value it designates in the metric system.)

1.35 Historically, some unit differences reflected the belief that the quantity measured was different when it was later revealed to be a single entity. Use the web to look up the origins of the energy units *erg* and *calorie*, and describe how they represent an example of this type of historical development.

1.36 Use the web to determine how the Btu was initially established. For the engineering applications where this unit is still used today, why is it a sensible unit?

1.37 How many micrograms are equal to one gram?

1.38 Convert the value 0.120 ppb into ppm.

1.39 How was the Fahrenheit temperature scale calibrated? Describe how this calibration process reflects the measurement errors that were evident when the temperature scale was devised.

1.40 Superconductors are materials that have no resistance to the flow of electricity, and they hold great promise in many engineering applications. But to date, superconductivity has only been observed under cryogenic conditions. The highest temperature at which superconductivity has been observed is 138 K. Convert this temperature to both °C and °F.

1.41 Express each of the following temperatures in kelvins. **(a)** −10.°C, **(b)** 0.00°C, **(c)** 280.°C, **(d)** 1.4×10^3°C

1.42 Express **(a)** 275°C in K, **(b)** 25.55 K in °C, **(c)** −47.0°C in °F, and **(d)** 100.0°F in K.

1.43 Express each of the following numbers in scientific notation. **(a)** 62.13, **(b)** 0.000414, **(c)** 0.0000051, **(d)** 871,000,000, **(e)** 9100

1.44 How many significant figures are there in each of the following? **(a)** 0.136 m, **(b)** 0.0001050 g, **(c)** 2.700×10^{-3} nm, **(d)** 6×10^{-4} L, **(e)** 56003 cm^3

1.45 How many significant figures are present in these measured quantities? **(a)** 1374 kg, **(b)** 0.00348 s, **(c)** 5.619 mm, **(d)** 2.475×10^{-3} cm, **(e)** 33.1 mL

1.46 Perform these calculations and express the result with the proper number of significant figures.

(a) (4.850 g − 2.34 g)/1.3 mL

(b) $V = \pi r^3$, where $r = 4.112$ cm

(c) $(4.66 \times 10^{-3}) \times 4.666$

(d) 0.003400/65.2

1.47 Calculate the following to the correct number of significant figures. Assume that all these numbers are measurements.

(a) $x = 17.2 + 65.18 - 2.4$

(b) $x = 13.0217/17.10$

(c) $x = (0.0061020)(2.0092)(1200.00)$

(d) $x = 0.0034 + \dfrac{\sqrt{(0.0034)^2 + 4(1.000)(6.3 \times 10^{-4})}}{(2)(1.000)}$

1.48 In an attempt to determine the velocity of a person on a bicycle, an observer uses a stopwatch and finds the length of time it takes to cover 25 "squares" on a sidewalk. The bicycle takes 4.82 seconds to travel this far. A measurement of one of the squares shows that it is 1.13 m long. What velocity, in m/s, should the observer report?

1.49 A student finds that the mass of an object is 4.131 g and its volume is 7.1 mL. What density should be reported in g/mL?

1.50 Measurements indicate that 23.6% of the residents of a city with a population of 531,314 are college graduates. Considering significant figures, how many college graduates are estimated to reside in this city?

1.51 A student weighs 10 quarters and finds that their total mass is 56.63 grams. What should she report as the average mass of a quarter based on her data?

1.52 A rock is placed on a balance and its mass is determined as 12.1 g. When the rock is then placed in a graduated cylinder that originally contains 11.3 mL of water, the new volume is roughly 17 mL. How should the density of the rock be reported?

Problem Solving in Chemistry and Engineering

1.53 A package of eight apples has a mass of 1.00 kg. What is the average mass of one apple in grams?

1.54 If a 1.00-kg bag containing eight apples costs $1.48, how much does one apple cost? What mass of apples costs $1.00?

1.55 A person measures 173 cm in height. What is this height in meters? feet and inches?

1.56 The distance between two atoms in a molecule is 148 pm. What is this distance in meters?

1.57 Carry out the following unit conversions. **(a)** 3.47×10^{-6} g to μg, **(b)** 2.73×10^{-4} L to mL, **(c)** 725 ns to s, **(d)** 1.3 m to km

1.58 Carry out each of the following conversions. **(a)** 25.5 m to km, **(b)** 36.3 km to m, **(c)** 487 kg to g, **(d)** 1.32 L to mL, **(e)** 55.9 dL to L, **(f)** 6251 L to cm^3

1.59 Convert 22.3 mL to **(a)** liters, **(b)** cubic inches, and **(c)** quarts.

1.60 If a vehicle is traveling 92 m/s, what is its velocity in miles per hour? (0.62 miles = 1.00 km)

1.61 A load of asphalt weights 254 lb. and occupies a volume of 220.0 L. What is the density of this asphalt in g/L?

1.62 One square mile contains exactly 640 acres. How many square meters are in one acre?

1.63 A sample of crude oil has a density of 0.87 g/mL. What volume in liters does a 3.6-kg sample of this oil occupy?

1.64 Mercury has a density of 13.6 g/mL. What is the mass of 4.72 L of mercury?

1.65 The area of the 48 contiguous states is 3.02×10^6 mi^2. Assume that these states are completely flat (no mountains and no valleys). What volume of water, in liters, would cover these states with a rainfall of two inches?

1.66 The dimensions of aluminum foil in a box for sale in supermarkets are $66\frac{2}{3}$ yards by 12 inches. The mass of the foil is 0.83 kg. If its density is 2.70 g/cm^3, then what is the thickness of the foil in inches?

1.67 Titanium is used in airplane bodies because it is strong and light. It has a density of 4.55 g/cm^3. If a cylinder of titanium is 7.75 cm long and has a mass of 153.2 g, calculate the diameter of the cylinder. ($V = \pi r^2 h$, where V is the volume of the cylinder, r is its radius, and h is the height.)

1.68 Wire is often sold in pound spools according to the wire gauge number. That number refers to the diameter of the wire. How many meters are in a 10-lb. spool of 12-gauge aluminum wire? A 12-gauge wire has a diameter of 0.0808 in. and aluminum has a density of 2.70 g/cm^3. ($V = \pi r^2 l$)

1.69 An industrial engineer is designing a process to manufacture bullets. The mass of each bullet must be within 0.25% of 150 grains. What range of bullet masses, in mg, will meet this tolerance? 1 grain = 64.79891 mg.

1.70 An engineer is working with archaeologists to create a realistic Roman village in a museum. The plan for a balance in a marketplace calls for 100 granite stones, each weighing 10 denarium. (The denarium was a Roman unit of mass: 1 denarium = 3.396 g.) The manufacturing process for making the stones will remove 20% of the material. If the granite to be used has a density of 2.75 g/cm^3, what is the minimum volume of granite that the engineer should order?

1.71 Draw a molecular scale picture to show how a crystal differs from a liquid.

1.72 Draw a molecular scale picture that distinguishes between alumina and silica. Is your picture structurally accurate or schematic?

1.73 On average, Earth's crust contains about 8.1% aluminum by mass. If a standard 12-ounce soft drink can contains approximately 15.0 g of aluminum, how many cans could be made from one ton of the Earth's crust?

1.74 As computer processor speeds increase, it is necessary for engineers to increase the number of circuit elements packed into a given area. Individual circuit elements are often connected using very small copper "wires" deposited directly onto the surface of the chip. In some current generation processors, these copper interconnects are about 65 nm wide. What mass of copper would be in a 1-mm length of such an interconnect, assuming a square cross section? The density of copper is 8.96 g/cm^3.

1.75 The "Western Stone" in Jerusalem is one of the largest stone building blocks ever to have been used. It has a mass of 517 metric tons and measures 13.6 m long, 3.00 m high and 3.30 m wide. What is the density of this rock in g/cm^3? (1 metric ton = 1000 kg)

1.76 A load of bauxite has a density of 3.15 g/cm^3. If the mass of the load is 115 metric tons, how many dump trucks, each with a capacity of 12 cubic yards, will be needed to haul the whole load?

INSIGHT INTO Material Selection and Bicycle Frames

1.77 Suppose that a new material has been devised with an elastic modulus of 22.0×10^6 psi for a bicycle frame. Is this bike frame likely to be more or less stiff than an aluminum frame?

1.78 Rank aluminum, steel, and titanium in order of increasing stiffness.

1.79 Compare the strengths of aluminum, steel, and titanium. If high strength were needed for a particular design, would aluminum be a good choice?

1.80 Aluminum is not as strong as steel. What other factor should be considered when comparing the desirability of aluminum versus steel if strength is important for a design?

1.81 Use the web to research the differences in the design of steel-framed bicycles versus aluminum-framed bicycles. Write a short paragraph that details the similarities and differences you discover.

1.82 Use the web to research the elastic modulus and yield strength of carbon fiber composites. How do these materials compare to aluminum, steel, and titanium?

1.83 Use the web to research the relative cost of aluminum, steel, and titanium frames for bicycles. Speculate about how much of the relative cost is due to the costs of the materials themselves.

FOCUS ON PROBLEM SOLVING EXERCISES

1.84 A student was given two metal cubes that looked similar. One was 1.05 cm on an edge and had a mass of 14.32 grams; the other was 2.66 cm on a side and had a mass of 215.3 grams. How can the student determine if these two cubes of metal are the same material using only the data given?

1.85 Battery acid has a density of 1.285 g/mL and contains 38.0% sulfuric acid by mass. Describe how you would determine the mass of pure sulfuric acid in a car battery, noting which item(s) you would have to measure or look up.

1.86 Unfermented grape juice used to make wine is called a "must." The sugar content of the must determines whether the wine will be dry or sweet. The sugar content is found by measuring the density of the must. If the density is lower than 1.070 g/mL, then sugar syrup is added until the density reaches 1.075 g/mL. Suppose that you have a sample taken from a must whose mass is 47.28 g and whose volume is 44.60 mL. Describe how you would determine whether or not sugar syrup needs to be added and if so, how would you estimate how much sugar syrup to add?

1.87 A solution of ethanol in water has a volume of 54.2 mL and a mass of 49.6 g. What information would you need to look up and how would you determine the percentage of ethanol in this solution?

1.88 Legend has it that Archimedes, a famous scientist of Ancient Greece, was once commanded by the king to determine if a crown he received was pure gold or a gold–silver alloy. He was not allowed, however, to damage the crown (by slicing off a piece, for example). If you were assigned this same task, what would you need to know about both gold and silver, and how would you make a measurement that would tell you if the crown was pure gold?

1.89 Imagine that you place a cork measuring 1.30 cm $\times$ 4.50 cm $\times$ 3.00 cm in a pan of water. On top of this cork, you place a small cube of lead measuring 1.15 cm on a side. Describe how you would determine if the combination of the cork and lead cube will still float in the water. Note any information you would need to look up to answer the question.

1.90 A calibrated flask was filled to the 25.00-mL mark with ethyl alcohol and was found to have a mass of 19.7325 g. In a second experiment, 25.0920 g of metal beads were put into the container and the flask was again filled to the 25.00-mL mark. The total mass of the metal plus the alcohol was 43.0725 g. Describe how to determine the density of the metal sample.

2 Atoms and Molecules

CHAPTER OUTLINE

The ability to be formed or molded into various shapes is one of the properties that make polymers a material of choice for many consumer products, including the small plastic boxes coming off of this production line.

Atoms and molecules are the building blocks of chemistry. You've probably been hearing this since middle school, so the existence of atoms is not something that you are likely to question or challenge. Chances are that you rarely think about atoms or molecules, however, when you come across items in your day-to-day life. When chemists want to understand some aspect of the world around them, though, they focus their attention at the level of atoms and molecules. So an important part of studying chemistry is learning how to interpret nature by thinking about what atoms and molecules are doing. Certainly, it can take some time to get comfortable with this type of thinking. Individual atoms and molecules are difficult to observe, so often we must infer what they are doing from indirect evidence. Chemistry has matured tremendously during the past century by building an increasingly thorough understanding of atoms and molecules. We will introduce some of the basic concepts of atoms and molecules here and then refine these ideas as we progress through the text.

A variety of techniques known as scanning probe microscopy can produce images of single atoms under some conditions.

Chapter Objectives

After mastering this chapter, you should be able to

- name at least three common polymers and give examples of their uses.
- define the terms *atom*, *molecule*, *isotope*, *ion*, *compound*, *polymer*, *monomer*, and *functional group* in your own words.

- describe the nuclear model for the atom and identify the numbers of protons, electrons, and neutrons in a particular isotope from its chemical symbol.
- calculate the atomic weight of an element from the masses and abundances of its isotopes.
- explain the difference between a molecular formula and an empirical formula.
- determine the number of atoms in a molecule from its chemical formula.
- describe the arrangement of elements in the periodic table and explain the usefulness of the table.
- obtain a correct chemical formula from a line drawing of an organic molecule.
- use standard chemical nomenclature to deduce the names of simple inorganic compounds from their formulas or vice versa.
- describe different forms of polyethylene and how their properties and applications are related to their molecular structures.

2-1 Polymers

As human civilization and technology have progressed, historical eras have been closely associated with materials from which important objects have been made. The Stone Age gave way to the Bronze Age, which in turn was followed by the Iron Age. These labels were chosen much later, through the lens of history, and it may be dangerous to try to characterize our own time period. But it isn't hard to imagine that future archaeologists or historians might label the late 20th century and early 21st century as the Polymer Age. As you go about your life, the many plastics and synthetic fibers you encounter are examples of what chemists call **polymers**. The properties and applications of these polymers are so diverse that you may not even recognize that they have anything in common (see **Figure 2.1**). Hard and durable plastics are routinely used as structural materials for things like computer cases and casual furniture. Softer, flexible plastics give us sandwich bags and Saran® wrap. Other polymers make up the nylon and rayon that are found in our carpets and clothing. Still more polymer materials, such as the filling in many bulletproof vests, offer incredible combinations of light weight and high strength. The diversity of polymer properties is truly impressive. And yet if we take the chemist's approach and turn our attention to the atomic and molecular level, we will see that all of these polymers have much in common.

a

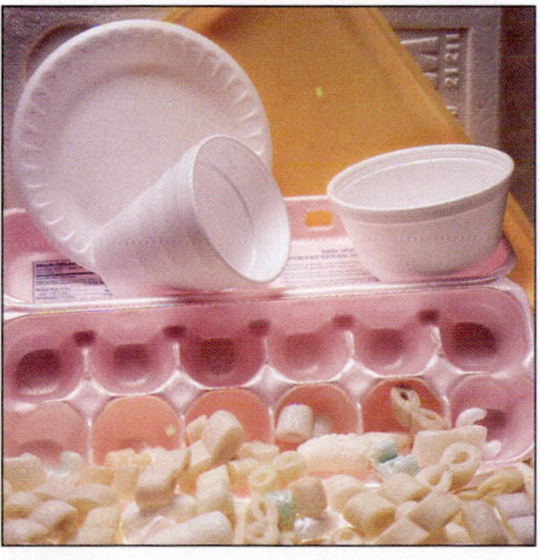

b

c

Figure 2.1 ■ Polymers are the materials of choice for a host of everyday objects. All of the items shown in the left-hand photo are made from high-density polyethylene. The objects in the center are made from polystyrene, and those in the right-hand photo are made from poly(vinyl chloride).

Traced back to its Greek origins, the word *polymer* literally means "many parts." That definition offers us a clue as to what these seemingly disparate substances all have in common. All polymers are made up of very large molecules. These large molecules are made up of many smaller molecules, linked end to end. A typical polymer molecule might contain hundreds or even thousands of these smaller constituent molecules, which are called **monomers**. If we look deeper into the composition of the polymer, we will find that these monomers are themselves made up of assemblies of atoms. But because polymer molecules are so large, it is often helpful to think of them as chains of monomers rather than collections of atoms.

Because these molecules are so large, polymers are also called macromolecules.

The observable macroscopic properties of any particular polymer depend on the identity of its constituent monomers, the number of monomers present, and the way that the monomers are connected to one another. To illustrate how dramatically the properties of a polymer can depend on its atomic composition, let's consider three common items and look at the polymers from which they are made. The plastic bottles in which your milk, juice, or shampoo come are usually made of a plastic called polyethylene. Polyethylene molecules are composed entirely of just two elements, carbon and hydrogen. The carbon atoms are linked together in a long chain that is called the **polymer backbone** of the polymer molecule, and there are two hydrogen atoms attached to each carbon. The molecular model at the top of **Figure 2.2** shows a portion of a polyethylene molecule. If one of the two hydrogen atoms on every other carbon is replaced with a chlorine atom, as shown in the middle of Figure 2.2, we will have poly(vinyl chloride). This polymer is commonly referred to as PVC. Plastic pipe made from PVC has been widely used in plumbing for many years, so you have probably

Other uses of polyethylene are discussed in Section 2-7 at the end of this chapter.

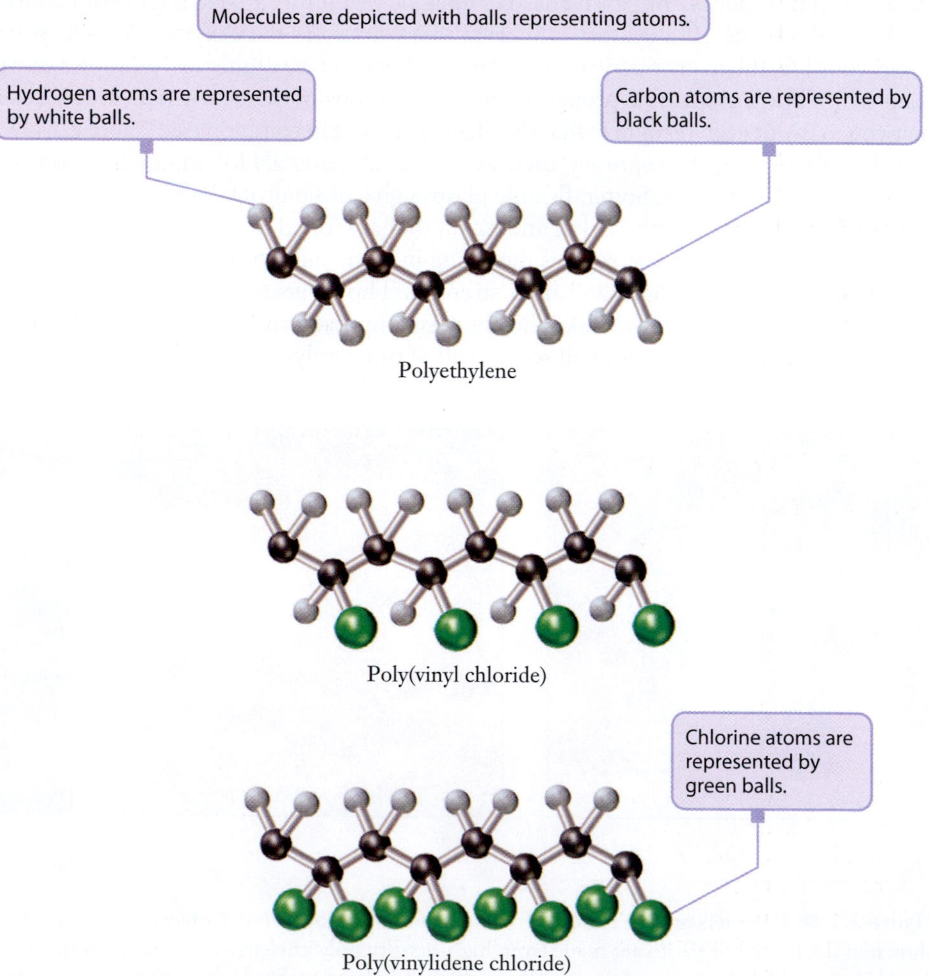
Molecules are depicted with balls representing atoms.
Hydrogen atoms are represented by white balls.
Carbon atoms are represented by black balls.
Polyethylene
Poly(vinyl chloride)
Chlorine atoms are represented by green balls.
Poly(vinylidene chloride)

Figure 2.2 ■ Models showing how atoms are arranged in molecules of polyethylene, poly(vinyl chloride), and poly(vinylidene chloride).

seen PVC pipe in either your house or a hardware store. As you might guess, the PVC used for pipes is much harder and stronger than the polyethylene in soda bottles. Yet the chemical composition and structures of these two materials are very similar. Suppose that we replace the second hydrogen on every other carbon atom with chlorine, too. Then we will have poly(vinylidene chloride), which is the "plastic wrap" used in almost every kitchen to cover leftovers.

These three common examples show how strongly the physical properties of a polymer are influenced by its chemical composition. To begin to explore the world of polymers systematically, we will need to know a little more about the atoms that are the simple building blocks of these giant molecules.

2-2 Atomic Structure and Mass

The comparison of polyethylene, poly(vinyl chloride), and poly(vinylidene chloride) demonstrates that the identity of the atoms in a molecule can have a tremendous impact on that molecule's properties. Let's begin our exploration by examining the structure of atoms, so that we can address the question of how a chlorine atom differs from a hydrogen atom. To do that, we'll need to zoom in one more level, to the realm of subatomic particles.

Fundamental Concepts of the Atom

Our current model of the structure of atoms has been accepted for nearly a century, but it took great creativity and many ingenious experiments to develop. The atom is composed of a small, compact core called the **nucleus** surrounded by a disperse cloud of **electrons**. The nucleus is composed of two types of particles: **protons** and **neutrons**. The electrons occupy so much space compared to the nucleus that it is impossible to show it to scale in an illustration. Consider **Figure 2.3(a)**, which is similar to pictures you've seen before in high school chemistry or physical science books. The figure shows the relative positions of the protons, neutrons, and electrons. But if the protons and neutrons were actually the size shown, then the electrons would be hundreds of meters away. Another misunderstanding promoted by this type of illustration is the picture of electrons following regular orbits around the nucleus. A better model of atomic structure views the electrons as clouds of negative charge that surround the nucleus, as opposed to particles that orbit around it in an orderly way (**Figure 2.3(b)**).

We will examine the structure of atoms in much greater detail in Chapter 6.

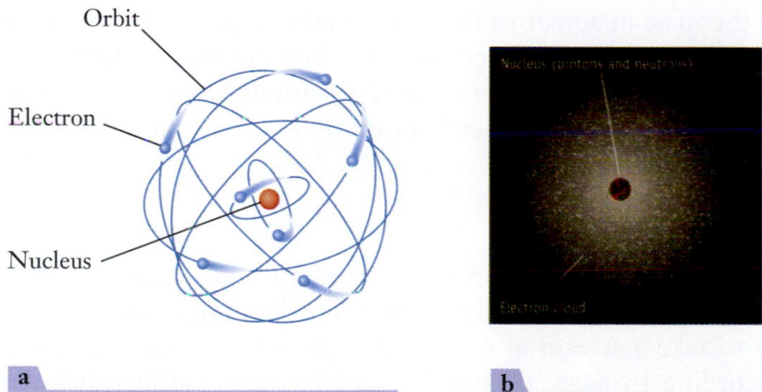

Figure 2.3 ■ Atoms have often been depicted as resembling a solar system: the nucleus is at the center, and the electrons orbit around it, as seen here in **(a)**. Although such pictures do help to emphasize the way that protons, neutrons, and electrons are distributed in the atom, they cannot illustrate accurately the currently accepted model of atomic structure. Instead, we depict the electrons as clouds of negative charge surrounding the nucleus, as shown in **(b)**. In such pictures, the density of the small dots is related to the probability of finding an electron at a particular location.

Protons and neutrons are themselves made up of even smaller particles, known as quarks.

We generally depict the charges in units of the electron charge, so that the charge of an electron is written as 1− and that of a proton is written as 1+.

The atomic mass unit is also referred to as a dalton and is sometimes abbreviated as u.

Now we turn our attention to the numbers of protons, neutrons, and electrons in the atom. Electric charge provides an important constraint on these numbers. Protons are positively charged, electrons are negatively charged, and neutrons are neutral. Atoms themselves are also electrically neutral, so the numbers of protons and electrons present must be such that their charges will cancel each other. You may know from physics that the SI unit of charge is the coulomb (C). Experiments have shown that the electrical charges on a proton and an electron are equal and opposite. Every electron carries a charge of -1.602×10^{-19} C, whereas every proton carries a charge of $+1.602 \times 10^{-19}$ C. So for an atom to remain neutral, the numbers of electrons and protons must be equal. Because neutrons have no charge, the number of neutrons present is not restricted by the requirement for electrical neutrality. For most elements, the number of neutrons can vary from one atom to another, as we'll see.

Atomic Number and Mass Number

The number of protons in a particular atom, referred to as the **atomic number**, identifies the element. Carbon atoms make up the backbone of nearly all polymers, so we will consider them first. The atomic number of carbon is six, which tells us that a neutral carbon atom has six protons. Electrical neutrality requires that a carbon atom also must have six electrons. The great majority of carbon atoms—roughly 99%—also contain six neutrons. But some carbon atoms contain seven or even eight neutrons. Atoms of the same element that have different numbers of neutrons are called **isotopes.** Protons and electrons govern nearly all of the important chemical properties of atoms, so generally isotopes cannot be separated chemically. But the existence and even the relative abundance of isotopes can be proven by careful examinations of the mass of atoms.

Protons and neutrons have similar masses; each is nearly 2000 times more massive than the electron. So the mass of any atom is concentrated in its nucleus. Individual atoms are so small and light that reporting their masses in conventional units such as kilograms or grams is not convenient. Instead, we use a unit that is appropriate to the atomic scale—the **atomic mass unit**, or **amu**:

$$1 \text{ amu} = 1.6605 \times 10^{-24} \text{ g}$$

Both the neutron and the proton have masses very close to 1 amu. The mass of a neutron is 1.009 amu, and that of a proton is 1.007 amu. The mass of an electron, in contrast, is just 0.00055 amu. So for many practical purposes, we can determine the mass of an atom simply by counting the number of protons and neutrons. That number will be the mass in amu, to a fairly reasonable approximation. Because of this, the combined total of protons and neutrons is called the **mass number** of the atom. Because isotopes are atoms of the same element with different numbers of neutrons, they will have the same atomic number but different mass numbers.

Isotopes

How do we know that these isotopes exist? Modern instruments called mass spectrometers provide direct experimental evidence. The first important function of a mass spectrometer is to take a stream of microscopic particles—atoms or molecules—and "sort them" according to mass. (**Figure 2.4** explains how the instrument does this.) Once the particles have been separated by mass, the second key function of the mass spectrometer is to measure accurately the number of particles with a given mass. The data are usually presented as a "mass spectrum." Any time we refer to a spectrum, we will be noting a measurement that is made over a range of values of some variable. In this case, that variable is mass, so the mass spectrum is really just a plot showing the number of particles detected as a function of mass. When a peak is seen at a particular mass, it means that the sample analyzed has some component with that mass.

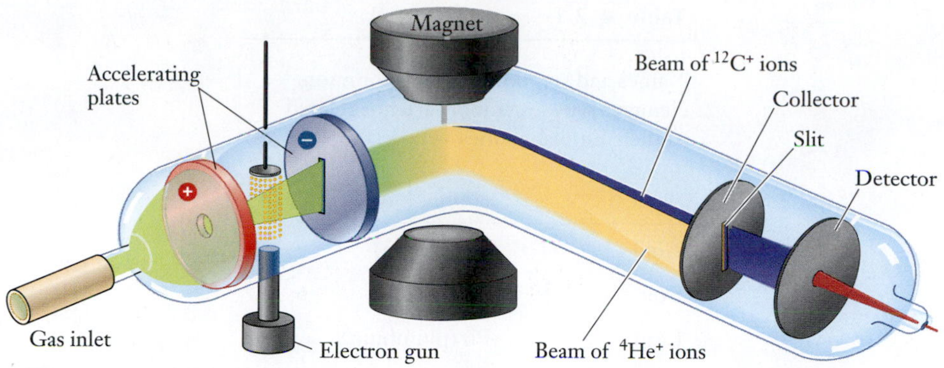

Figure 2.4 ■ The schematic diagram shown here illustrates the key principles in the functioning of a mass spectrometer. A stream of gas to be analyzed enters at the left, and an electron gun causes some of the atoms to lose an electron, forming charged particles called *ions*. These ions are then accelerated to the right by an electric field, so that a beam of ions passes into a magnetic field. The magnetic field deflects the ions, and the extent of that deflection depends on the charge-to-mass ratio of each ion. For a given charge, lighter particles are deflected more severely than heavier ones. So if the sample contained both $^{4}He^+$ and $^{12}C^+$ ions, as shown here, the helium ions would be deflected much more than the carbon ions. This allows a slit to select ions of a particular charge-to-mass ratio, which then strike a detector. The current at this detector produces a signal that is proportional to the number of ions found with the desired charge-to-mass ratio, and this in turn is related to the amount of the parent gas molecule that entered the spectrometer.

Figure 2.5 shows such a mass spectrum for a sample of carbon. Looking at the graph, we immediately see a large peak centered at mass 12. That represents the isotope called carbon-12, whose nucleus contains six protons and six neutrons. This isotope is actually used to define the amu: an atom of carbon-12 has a mass of *exactly* 12 amu. But if we look at the mass spectrum closely, we also see a much smaller peak centered near mass 13. This tells us that there is a small amount of a second isotope, carbon-13, with seven neutrons. Comparing the relative sizes of the two peaks, we could determine that the carbon-12 isotope accounts for roughly 99% of the carbon atoms. More accurate measurement gives a value of 98.93%, with just 1.07% of carbon-13. It is also possible to determine that the exact mass of the carbon-13 isotope is 13.003355 amu. The percentages describing the relative amounts of each isotope are referred to as **isotopic abundances**.

A number of radioactive isotopes of carbon are also known. The most common of these is ^{14}C. Its abundance is measured in the carbon dating of archaeological objects.

Atomic Symbols

All the information about the structure of the atom, which we have just discussed, can be written in scientific shorthand, using atomic symbols. The general atomic symbol

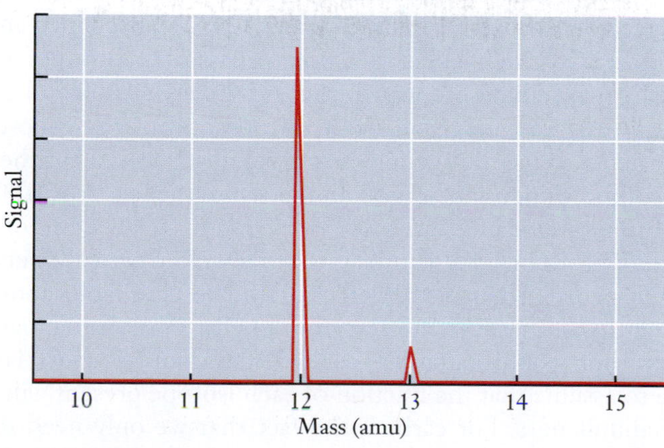

Figure 2.5 ■ A sketch of the mass spectrum of elemental carbon is shown. The large peak is due to ^{12}C, and the smaller peak to the right is ^{13}C. The size of the ^{13}C peak here is somewhat exaggerated; it would actually be just 1/99th the size of the ^{12}C peak.

Table ■ 2.1

Names and symbols of some common elements whose symbols are not related to their English names

Name	Symbol (name origin)
Gold	Au (aurum)
Iron	Fe (ferrum)
Lead	Pb (plumbum)
Mercury	Hg (hydrargyrum)
Silver	Ag (argentum)
Sodium	Na (natrium)

can be written as

$$^{A}_{Z}\mathrm{E}$$

Here E represents the atomic symbol for the element in question, the superscript A is the mass number, and the subscript Z is the atomic number. The symbol for carbon-12, for example, is $^{12}_{6}\mathrm{C}$.

Many atomic symbols are fairly obviously derived from the name of the element, such as the use of C for carbon in our example. For other elements, the symbol is based on the Latin name. The symbol for iron, for example, is Fe, derived from the Latin name *ferrum.* An atom of iron with 26 protons and 30 neutrons is represented as $^{56}_{26}\mathrm{Fe}$. A listing of some common elements whose symbols are not based on their English names is provided in **Table 2.1**. A full list of elements and their symbols can be found in Appendix A at the back of this book.

The term "ferrous metals" refers to iron or alloys such as steel that contain significant amounts of iron.

Atomic Masses and Weights

When you look at an entry in the periodic table, you see some of the information we've just defined, such as the atomic symbol and atomic number. Most periodic tables include additional information as well. Almost always, the atomic weight is given. This number provides the average mass in amu of an atom of the element. If you look up carbon in the periodic table inside the back cover of this book, you will find the box shown in **Figure 2.6**. The atomic weight appears under the symbol: 12.011. But we have already said that the mass of an atom of carbon-12 is exactly 12 amu, and that of carbon-13 is 13.003355 amu. So the value of 12.011 does not seem to be the mass of *any* individual atom of carbon. Then how are atomic weights defined and determined?

We use "atomic mass" to refer to the mass of one individual atom, and "atomic weight" to refer to the average mass of a large number of atoms of a particular element.

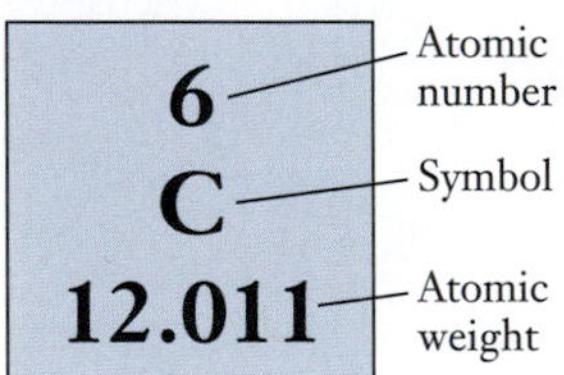

Figure 2.6 ■ Entry for carbon from a periodic table. The atomic number (6) and the atomic weight (12.011) are shown, along with the symbol for the element (C). Some tables may display additional information, and the exact layout may vary from one table to another. But once you are familiar with the table itself, usually it is easy to interpret whatever data are shown.

The atomic weight is defined as the *average* mass of an atom of a particular element. Carbon has two stable isotopes with masses of 12.000000 and 13.003355 amu, respectively. So why is the average mass 12.011 and not something closer to 12.5? The answer is that when we take the average mass, we must account for the relative abundance of each isotope. Suppose that we could measure the mass of a 100-atom sample. Based on the isotopic abundances, we would expect to have 99 atoms of carbon-12 and only a single atom of carbon-13. In any sample that we can actually weigh, the number of atoms will be far greater than 100. Even using the best available laboratory balances, the smallest quantity of matter that can be weighed is about a nanogram, or 10^{-9} g. A nanogram of carbon would contain more than 10^{13} atoms. For such large numbers of atoms, it is safe to assume that the fraction of each isotope present will be determined by the natural abundances. For carbon, the fact that we only need to consider two stable isotopes makes the calculation fairly simple. We can multiply the mass by the fractional abundance to weight each isotope's contribution to the atomic weight:

$$\text{Carbon-12:} \quad 12.000000 \times 0.9893 = 11.87$$

$$\text{Carbon-13:} \quad 13.003355 \times 0.0107 = 0.139$$

$$\text{Weighted average mass} = 11.87 + 0.139 = 12.01$$

The value of 12.011 found in the periodic table is obtained using additional significant figures on the isotopic abundance numbers. Example 2.1 explores this type of calculation further.

EXAMPLE PROBLEM 2.1

The chlorine present in PVC has two stable isotopes. ^{35}Cl with a mass of 34.97 amu makes up 75.77% of the natural chlorine found. The other isotope is ^{37}Cl, whose mass is 36.95 amu. What is the atomic weight of chlorine?

Strategy To determine the atomic weight, we must calculate the average mass weighted by the fractional abundance of each chlorine isotope. Because there are only two stable isotopes, their abundances must add up to 100%. So we can calculate the abundance of ^{37}Cl from the given abundance of ^{35}Cl.

Solution First, we calculate the abundance of the chlorine-37 isotope:

$$\text{Abundance of } ^{37}\text{Cl} = 100\% - 75.77\% = 24.23\%$$

Now we can calculate the contribution of each isotope to the atomic weight:

$$^{35}\text{Cl:} \quad 34.97 \times 0.7577 = 26.50$$

$$^{37}\text{Cl:} \quad 36.95 \times 0.2423 = 8.953$$

$$\text{Weighted average mass} = 26.50 + 8.953 = 35.45$$

So the atomic weight of chlorine is 35.45 amu.

Analyze Your Answer Based on the relative percentages, we should be able to decide if this answer makes sense. The individual isotopes have masses of roughly 35 and 37, so a 50/50 ratio would lead to an average mass of about 36. But the actual abundance of the ^{35}Cl isotope is greater than that of ^{37}Cl, so the average mass should be closer to 35. Thus our answer of 35.45 seems reasonable. And of course we can check the answer by consulting a periodic table.

Discussion Some elements have several stable isotopes, but we can always do the same type of calculation by accounting for the mass and fractional abundance of each isotope.

Check Your Understanding There are three naturally occurring isotopes of the element silicon, which is widely used in producing computer chips. Given the masses and abundances below, calculate the atomic weight of silicon.

Isotope	Abundance	Mass
^{28}Si	92.2%	27.977 amu
^{29}Si	4.67%	28.977 amu
^{30}Si	3.10%	29.974 amu

Table ■ 2.2

Range of observed atomic weight values for selected elements

Element	Atomic Weight Range	Conventional Atomic Weight
Hydrogen	[1.00784; 1.00811]	1.008
Lithium	[6.938; 6.997]	6.94
Boron	[10.806; 10.821]	10.81
Carbon	[12.0096; 12.0116]	12.011
Nitrogen	[14.00643; 14.00728]	14.007
Oxygen	[15.99903; 15.99977]	15.999
Silicon	[28.084; 28.086]	28.085
Sulfur	[32.059; 32.076]	32.06
Chlorine	[35.446; 35.457]	35.45
Thallium	[204.382; 204.385]	204.38

Until recently, the isotopic abundances and therefore the atomic weights of all the elements were regarded as constants of nature. But careful measurement using modern techniques has made clear that the isotopic abundances of a number of elements can vary somewhat depending on the source of the sample. For carbon, atomic weights ranging from 12.009662 to 12.011505 have been observed for various samples. The International Union of Pure and Applied Chemistry (IUPAC) is responsible for establishing the officially recognized values for atomic weights. The most recently issued table of atomic weights acknowledges the variation in atomic weight for different samples, and so lists ranges of values for ten elements, as shown in **Table 2.2.** The numbers in the square brackets are derived from the minimum and maximum atomic weights that have been observed for that element. You can see that in most cases these ranges are fairly narrow. IUPAC also defines a set of what it now calls "conventional" atomic weights for those ten elements. In this textbook, we have chosen to list these conventional values in the periodic table, and we will use these values for all calculations requiring atomic weights.

Small variations in atomic weights due to different isotopic abundances play an important role in the analysis of ice core samples used in the study of climate change.

2-3 Ions

In developing our ideas about the composition of atoms in the previous section, we used the fact that atoms are electrically neutral to conclude that the numbers of protons and electrons must be equal. When the number of protons and the number of electrons do not match, the result is a species with a net charge, called an **ion**. Such species are no longer atoms, and their behavior is markedly different from that of atoms.

The operation of the mass spectrometer, illustrated in Figure 2.4, depends on the instrument's ability to convert an atom into an ion. The separation of the particles by mass is generally based on the behavior of charged particles in a magnetic field, and the detector that counts the particles typically detects only ions, not neutral atoms. Ions can also play important roles in many chemical processes, including several that are important in the large-scale production of polymers.

When an ion is derived from a single atom, it is called a **monatomic ion**. When groups of atoms carry a charge, they are called **polyatomic ions**. Monatomic or polyatomic ions may carry either negative or positive charges. Negatively charged ions are called **anions**, and they contain more electrons than protons. Similarly, an ion with more protons than electrons has a positive charge and is called a **cation.**

The word cation *is pronounced as "kat-ion," not "kay-shun."*

Table ■ 2.3

Examples of monatomic ions

Cation Name	Symbol	Anion Name	Symbol
Sodium ion	Na^+	Fluoride ion	F^-
Lithium ion	Li^+	Chloride ion	Cl^-
Potassium ion	K^+	Bromide ion	Br^-
Magnesium ion	Mg^{2+}	Sulfide ion	S^{2-}
Aluminum ion	Al^{3+}	Nitride ion	N^{3-}

We write symbols for ions analogously to those for atoms, adding the charge on the species as a superscript to the right of the atomic symbol. For monatomic ions, the number of protons still specifies the element whose symbol we use. **Table 2.3** provides some examples of monatomic ions. Notice that the monatomic anions have names ending in *-ide*, whereas cations simply have "ion" added to the name of the element.

The behavior and interaction of electrical charges are important topics in physics, but they also provide a basis for thinking about many aspects of chemistry. For our current interests, we will point out just two fundamental ideas about electric charge. First, *opposite charges attract each other and like charges repel one another.* And second, *electric charge is conserved.* These two ideas have important implications for the formation of ions in chemical processes. First of all, because charge is conserved, we can say that if a neutral atom or molecule is to be converted into an ion, then some oppositely charged particle—most likely an electron or another ion—must be produced at the same time. Moreover, because opposite charges attract one another, some energy input is always needed to convert a neutral atom or molecule into a pair of oppositely charged particles.

Mathematical Description

The statement that "opposites attract and likes repel" can be quantified mathematically. **Coulomb's law**, which you may recall from a physics class, describes the interaction of charged particles. The attraction of opposite charges and the repulsion of like charges are both described mathematically by one simple equation:

$$F = \frac{q_1 q_2}{4\pi\varepsilon_0 r^2} \tag{2.1}$$

Here q_1 and q_2 are the charges, ε_0 is a constant called the permittivity of a vacuum, and r is the distance between the charges. F is the force the objects exert on one another as a result of their charges. Looking at this expression, when both charges have the same sign—either positive or negative—the resultant value for the force is a positive number. When the charges are opposite, the value is negative. This is consistent with the usual sign conventions used throughout chemistry and physics for force and energy; a negative value of F from Equation 2.1 indicates an attractive force and a positive value a repulsive one.

Now consider the effect of varying the distance, r, between two ions. If two positively charged particles are initially very far apart (effectively infinite distance), the r^2 term in the denominator of Equation 2.1 will be very large. This in turn means that the force F will be very small, and so the particles will not interact with each other significantly. As the two like charges are brought closer together, the r^2 term in the denominator shrinks and so the (positive) force grows larger: the particles repel each other. If we somehow force the particles closer together, the repulsive force will continue to grow. The distance dependence of the coulombic force is illustrated in **Figure 2.7**.

Figure 2.7 ■ The figure shows how the coulombic force (Equation 2.1) varies with the distance *r* between two particles with opposite or like charges. When the charges have the same signs, the particles will repel one another, so the value of the force is positive. If the charges have opposite signs, the particles will attract one another and the value of the force will be negative.

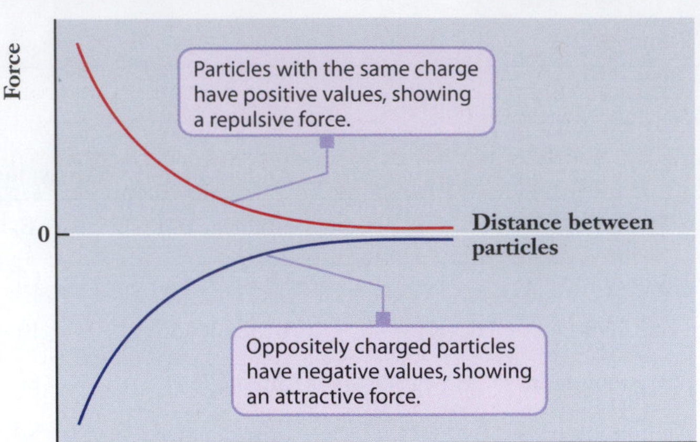

Force
0
Particles with the same charge have positive values, showing a repulsive force.
Distance between particles
Oppositely charged particles have negative values, showing an attractive force.

Ions and Their Properties

Many monatomic cations and anions exist. These ions can exist in the gas phase, and many are important in atmospheric chemistry. But we encounter ions most frequently when dealing with the chemistry of substances dissolved in water. For example, sodium atoms *lose* an electron relatively easily to form the sodium cation, Na^+. Because it still has 11 protons, this ion retains the symbol of sodium, yet it does not behave at all like an atom of sodium. Consider an order of french fries. You may have heard news stories about the high amount of sodium in an order of fries, and concerns have been raised about the possible health effects of too much sodium in our diets. This statement could be confusing because here the word "sodium" does not refer to sodium metal. In fact, if we place sodium metal on freshly made french fries, the metal will burst into flame! The sodium we hear about in stories on diet and health is actually sodium ion, which is added to the fries when salt is sprinkled on. Too much salt might still be a health concern, but we certainly don't worry about the salt igniting. There is a big difference between ions and atoms, at least in this case.

In contrast to sodium, chlorine readily *gains* an extra electron, forming the chloride ion Cl^-. Again, there is a noticeable difference between the ion and the atom of chlorine. The table salt we discussed above is sodium chloride, which contains chloride anions. Just like sodium, these chloride ions are present in french fries or any other salted foods. Chlorine atoms, on the other hand, combine in pairs to form a yellowish-green gas, Cl_2, which irritates the lungs and can be toxic. The behavior of the ion is clearly much different from that of the neutral atom or molecule.

Chlorine gas released in a January 2005 train accident in South Carolina led to eight deaths and forced many residents from their homes for days.

Polymers do not generally contain ions. But ions are important in the chemical reactions used to produce many common plastics. As a result of their electric charge, ions are often much more reactive than neutral atoms or molecules. So small amounts of ions are often used to initiate or sustain the chemical reactions that join monomers together to form polymers.

2-4 Compounds and Chemical Bonds

A basic picture of atoms is a good starting point for understanding the properties of polymers. But to begin to see how the observable properties of a polymer might be related to its atomic and molecular makeup, we will need to consider the connections between atoms. Which atoms are actually attached to one another? And what different types of connections—or **chemical bonds**—are involved? Once again we'll begin by trying to establish some vocabulary that will help us understand compounds and the chemical bonds that hold them together.

Chemical Formulas

A **chemical compound** is a pure substance made up of atoms of two or more elements joined together by chemical bonds. In any compound, the atoms combine in fixed whole number ratios. In any such combination, the resultant substance behaves differently from the atoms alone. In many compounds, atoms combine to form discrete particles called **molecules**. Molecules can be broken down into their constituent atoms, but the resulting collection of atoms no longer behaves like the original molecule. Other materials are composed of vast arrays or extended structures of atoms or ions but do not form discrete molecules. Alloys, metals, and ionic solids fall into this category of chemical compounds. We've seen how we can use atomic symbols as shorthand notation to designate atoms. That same idea can be extended to describe the composition of either molecules or extended compounds in a simple symbolic representation.

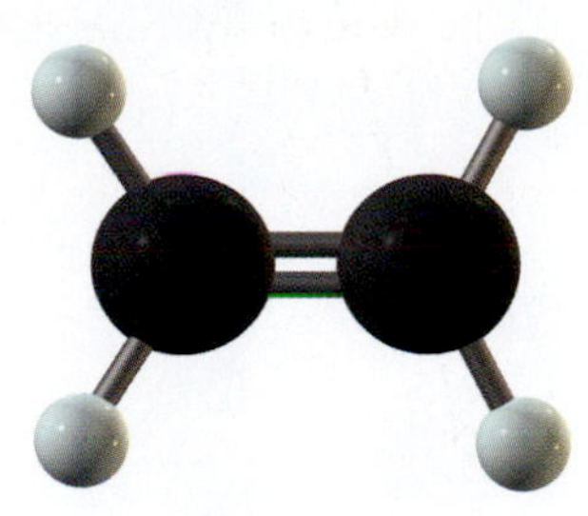

Ethylene, C_2H_4

A **chemical formula** describes a compound in terms of its constituent elements. We will actually encounter two distinct types of chemical formulas: molecular formulas and empirical formulas. The **molecular formula** of a compound is a kind of parts list that describes the atomic composition of a molecule efficiently. The molecular formula of the ethylene monomer from which polyethylene is produced is C_2H_4; this tells us that there are two carbon atoms and four hydrogen atoms per molecule. The **empirical formula** tells us only the relative ratio between the numbers of atoms of the different elements present. Let's consider ethylene again. The ratio of carbon atoms to hydrogen is 1:2. So the empirical formula is CH_2. When dealing with an empirical formula, it is important to realize that it does not tell how large or small an individual molecule of the compound might be; only the relative numbers of atoms of each element are given. We often emphasize this fact by writing a subscript "n" on the entire formula. For ethylene, this would give us $(CH_2)_n$, which means that each molecule must contain some integral number of CH_2 units.

In Section 3-5, we will learn how to find the empirical formula for a compound from experimental data.

Empirical formulas, or a minor variation on them, are especially common when dealing with polymer molecules. Because polymer molecules are so large, the exact number of monomer units in a molecule is generally not very important. And in fact, the exact length of the polymer chains is often not the same for all molecules in a given sample. Instead, there is usually some range of chain lengths that will exist, depending on how the polymer was actually produced. As long as the chains are all within some reasonable range of lengths, the macroscopic properties of the polymer are not affected substantially. So polymer formulas are most often written like empirical formulas. The repeating unit contributed by each monomer molecule is written in parentheses or brackets, and a subscript n is used to emphasize that a large number of these units will be found in any individual molecule. For polyethylene, we would write the formula as $—[CH_2CH_2]_n—$. Here the dashes are added to stress that these units are attached end to end to build up the long chain of the polymer. For the most common forms of polyethylene, the number of monomer units (i.e., the value of n) is in the tens of thousands. We could write similar formulas for the other polymers mentioned in the opening section on pages 30 and 31:

In the closing Insight section for this chapter, we will look more closely at the way large changes in chain length influence the properties of polyethylene.

Poly(vinyl chloride):	$—[CH_2CHCl]_n—$
Poly(vinylidene chloride):	$—[CH_2CCl_2]_n—$

There are four rules that allow us to write most formulas that we will need in this textbook.

1. Indicate the types of atoms in the substance by their atomic symbols.
2. The number of each atom in the compound is indicated by a subscript to the right of the atomic symbol. For example, the molecular formula of ethylene, C_2H_4, tells us that each molecule contains two carbon atoms and four hydrogen atoms.
3. Groups of atoms can be designated by using parentheses. Subscripts outside these parentheses mean that *all* atoms enclosed in the parentheses are multiplied by the value indicated in the subscript.

4. Water molecules associated with certain compounds called **hydrates** are indicated separately from the rest of the compound.

Example Problem 2.2 shows how to interpret chemical formulas by inverting some of these rules.

EXAMPLE PROBLEM 2.2

We cannot generally produce a polymer by simply mixing a large sample of the desired monomers. Instead, additional compounds called *initiators* or *catalysts* are almost always needed to start a polymerization. One polymerization catalyst is diethylaluminum chloride, $Al(C_2H_5)_2Cl$. How many of each type of atom are in a molecule of this compound?

Strategy The subscripts in a formula indicate how many atoms of each type are in the molecule. The parentheses designate a group of atoms, and the subscript associated with the parentheses multiplies each atom in the group.

Solution In each molecule of $Al(C_2H_5)_2Cl$, there is one aluminum atom, one chlorine atom, and two groups of C_2H_5. Each of the C_2H_5 groups contains two carbon atoms and five hydrogen atoms. We multiply those numbers by two because there are two C_2H_5 groups present; so we have four carbon atoms and ten hydrogen atoms.

Discussion The number of atoms present might be easier to see if we wrote this formula as $AlC_4H_{10}Cl$. Right now you might feel that would be simpler. But when we write it as $Al(C_2H_5)_2Cl$, we are actually conveying some additional information about the way the atoms are connected to one another. Specifically, we are showing that the carbon and hydrogen atoms are arranged as two C_2H_5 groups and that each of these groups is attached to the aluminum atom. Later, we will learn that such a C_2H_5 unit is called an *ethyl group*.

Check Your Understanding A compound with the rather imposing name of 2,2′-azo-bis-isobutyrylnitrile is used to initiate the growth of some polymers, including poly(vinyl chloride). If the molecular formula is $C_8H_{12}N_4$, how many of each type of atom are in a molecule of the compound? What is the empirical formula of this compound?

Chemical Bonding

Atoms combine to make compounds by forming chemical bonds. Several different types of chemical bonds are possible, and once we learn to recognize them, these types of bonds will help us to understand some of the chemical properties of many substances.

All chemical bonds share two characteristics. First, all bonds involve exchange or sharing of electrons. We will return to this concept often in this text as we investigate chemical reactions and properties of molecules. Second, this exchange or sharing of electrons results in lower energy for the compound relative to the separate atoms. A chemical bond will not form, or will have only a fleeting existence, unless it lowers the overall energy of the collection of atoms involved.

Chemical bonds can be divided into three broad categories: ionic, covalent, and metallic. Some compounds are composed of collections of oppositely charged ions that form an extended array called a **lattice**. The bonding in these compounds is called **ionic bonding**. To form the ions that make up the compound, one substance loses an electron to become a cation, while another gains an electron to become an anion. We can view this as the transfer of an electron from one species to another. **Figure 2.8** shows this concept for one ionic compound, NaCl.

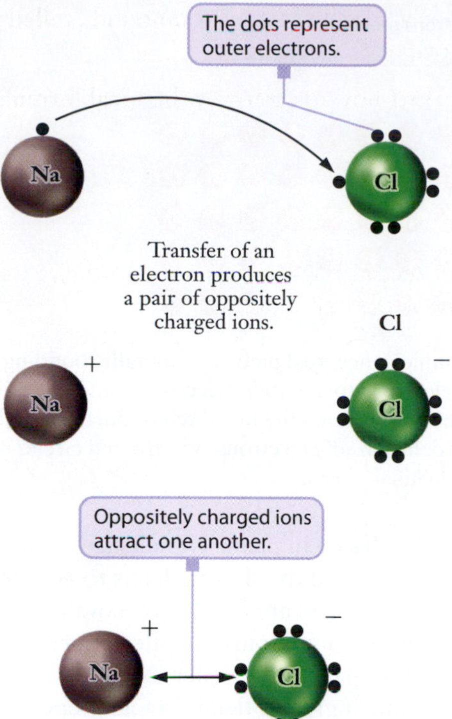
The dots represent outer electrons.
Na
Cl
Transfer of an electron produces a pair of oppositely charged ions.
Cl
+
Na
−
Cl
Oppositely charged ions attract one another.
+
Na
−
Cl

Figure 2.8 ■ A conceptual illustration showing the transfer of one electron from a sodium atom to a chlorine atom, forming a pair of ions (Na^+ and Cl^-). Once electron transfer takes place, coulombic force draws the ions together.

Ionic compounds form extended systems or lattices of alternating positive and negative charges, such as that shown in **Figure 2.9**. Although the formula NaCl correctly indicates that sodium and chlorine are present in a 1:1 ratio, we cannot really identify an individual "molecule" of NaCl. To emphasize this distinction, we sometimes refer to a **formula unit**, rather than a molecule, when talking about ionic compounds. The formula unit is the smallest whole number ratio of atoms in an ionic compound.

Lattices are often depicted as having shapes that are essentially cubic, but there are actually 17 different geometric shapes, not all of which are cubic.

Metals represent another type of extended system, but here the chemical bonding is totally different. In metals, the atoms are once again arranged in a lattice, but positively and negatively charged species do not alternate. Instead, the nuclei and some fraction of their electrons comprise a positively charged "core" localized at these lattice points, and other electrons move more or less freely throughout the whole array. This is called **metallic bonding**. Metallic bonding leads to electrical conductivity because electrons can move easily through the bulk material. **Figure 2.10** shows a schematic illustration of the concept of metallic bonding.

When electrons are shared between pairs of atoms rather than donated from one atom to another or mobile across an entire lattice, we have **covalent bonds**. In covalent

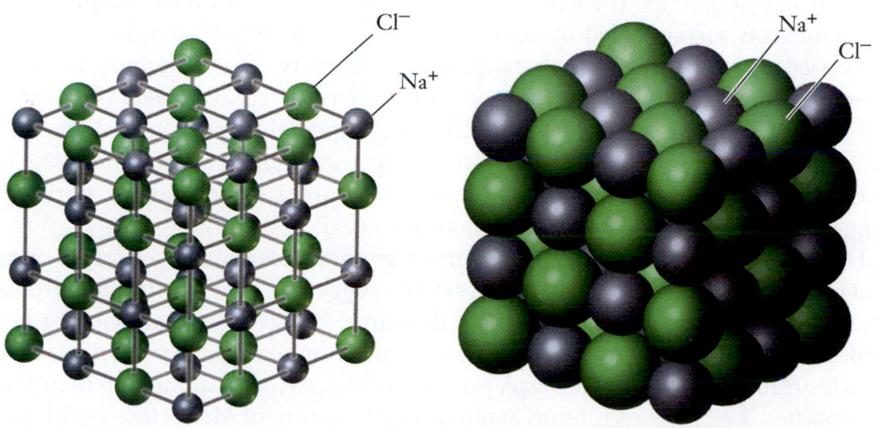
Cl−
Na+
Na+
Cl−

Figure 2.9 ■ Two different representations of the NaCl crystal structure are shown. In each case, the green spheres represent chloride anions, and the gray spheres denote sodium cations. The view on the left emphasizes the positions of the ions, and that on the right better illustrates their relative sizes. In a macroscopic salt crystal, additional ions would simply extend this structure, repeating the same alternating pattern.

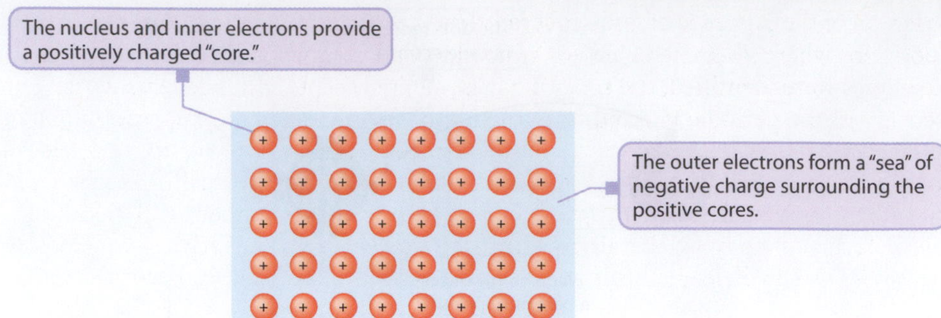
The nucleus and inner electrons provide a positively charged "core."
The outer electrons form a "sea" of negative charge surrounding the positive cores.

Figure 2.10 ■ In this simple conceptual picture of metallic bonding, each metal atom contributes one or more electrons to a mobile "electron sea." The ability of the electrons to move freely through this "sea" allows the metal to conduct electricity. Here the blue area depicts those mobile (or "delocalized") electrons, and the red circles represent the positively charged "cores" of the individual atoms.

Water, H_2O

Carbon dioxide, CO_2

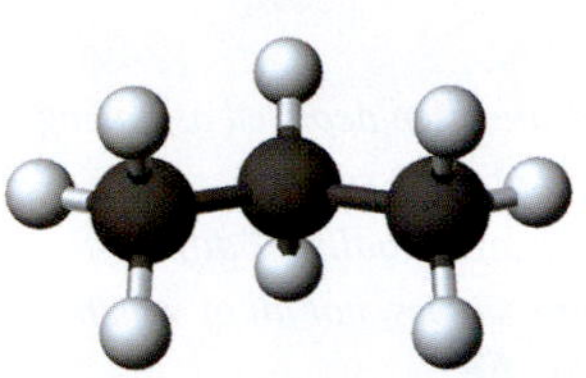

Propane, C_3H_8

bonds, electrons are usually shared in pairs. Two electrons (and sometimes four or six) are located between two nuclei and the sharing leads to an attraction between the nuclei. The long chains in all polymers are formed by covalent bonds in which electrons are shared between adjacent carbon atoms. Smaller, more familiar molecules such as water, carbon dioxide, and propane are simpler examples. All three types of chemical bonds will be discussed in much greater detail in Chapters 7 and 8.

Polymer molecules are built up by the successive addition of monomers to form characteristic long-chain backbones. The bonds that hold the monomers to one another, as well as the bonds between atoms within each monomer unit, are covalent bonds. But ionic bonding is important in many compounds that are used to initiate or sustain the reactions needed to grow a polymer.

2-5 The Periodic Table

One of the most recognizable tools of chemistry is the periodic table. It is prominently displayed in practically every chemistry classroom, and many chemists own T-shirts, neckties, or coffee mugs on which the periodic table appears as a professional badge of honor. Why do chemists hold this table in such high regard? Once you are familiar with it, the periodic table summarizes a wealth of information about the behavior of elements, organizing them simultaneously in ascending order of atomic number and in groups according to chemical behavior. An experienced chemist—or even a good chemistry student—can get a rough idea of an element's properties simply from where that element sits in the periodic table.

Today we can purchase or download artistic renderings of the periodic table, color-coded to display properties of particular interest for different applications. But like many developments in science, the emergence of what we now accept as the periodic table was accompanied by some degree of controversy. A number of scientists had devised various schemes for arranging the elements. These attempts to organize the understanding of the elements were not well received, however. One proposal by John Newlands in 1866 would have grouped elements similarly to musical octaves. This idea was literally laughed at during a scientific meeting where one critic sarcastically asked whether a listing by alphabetical order had been tried, noting, "any arrangement would present occasional coincidences."

One of the difficulties that persisted in the development of the periodic table was the controversy over assigning atomic weights. Without a unique scale for atomic weight, scientists drew different conclusions about how to arrange the elements "in order."

Despite the skepticism of the 19th-century scientific community, efforts to organize the elements persisted. Numerous observations suggested a regularity, or **periodicity**, in the behavior of the elements known at that time. By 1869, Russian scientist Dmitri Mendeleev had published his first periodic table and enumerated the **periodic law**: when properly arranged, the elements display a regular and periodic variation in their chemical properties. The most significant and impressive feature of Mendeleev's work was his prediction of the existence of undiscovered elements. He left holes in his proposed table at positions where no known element seemed to fit. Later, when the elements to fill in these holes were identified, the scientific community accepted Mendeleev's work. The discovery of the periodic law and construction of the periodic table represents one of the most significant creative insights in the history of chemistry. Prior to Mendeleev's time, chemists had to learn the properties of each element individually. As more and more elements were discovered, that task became increasingly daunting. The periodic table helped the study of chemistry to expand quickly by providing a simple, visual means to organize the elements in terms of their chemical and physical properties.

Periods and Groups

The modern periodic table simultaneously arranges elements in two important ways: the horizontal rows of the table, called **periods**, and the vertical columns, called **groups**. The term "period" is used for the rows because many important properties of the elements vary systematically as we move across a row. **Figure 2.11** shows a plot of the density of elements, all in their solid state, as a function of atomic number. From the graph, it is clear that density varies according to a fairly regular pattern that goes through a series of minima and maxima. Different colors are used for the data points in this graph to show how the variation in density is correlated with position in the periodic table. Each color represents a period (row) in the table. Because the elements in the periodic table are arranged in order of increasing atomic number, moving across each segment of the graph corresponds to moving from left to right across the corresponding row of the periodic table. You can see readily that as we move across a row in this way, the density of the elements is initially small, increases until passing through a maximum, and then decreases again. **Figure 2.12** shows the same data, with the density represented by the shading of each element's box. This representation clearly shows how the density of the elements varies regularly across each row of the table. The rows in the table are numbered 1 through 7 sequentially from top to bottom.

Although the properties of the elements can vary widely across a period, each column collects elements that have similar chemical properties. Most elements can combine with hydrogen to form compounds. The graph in **Figure 2.13** shows the number of hydrogen atoms with which an atom of each element will combine, and the regular variation in the plot clearly shows that this is a periodic property. Elements in a group (column) combine with the same number of hydrogen atoms. Fluorine, chlorine, and bromine each combine with one atom of hydrogen, for example, and all fall in the same group.

These types of chemical similarities were among the evidence that led to the development of the periodic table, so some of the groups predate the general acceptance

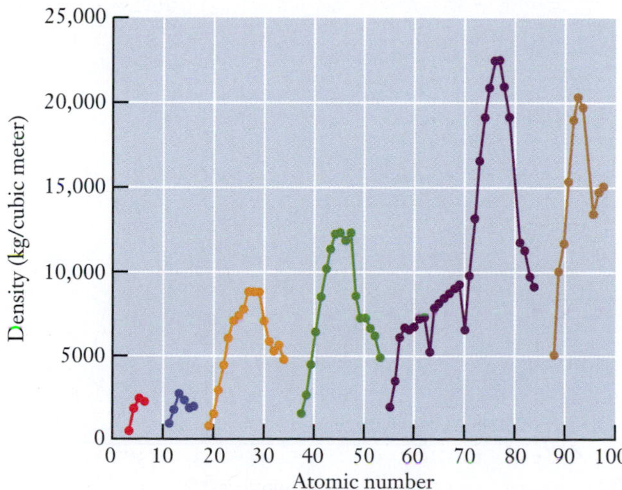

Figure 2.11 ■ The density of elements in their solid states is plotted as a function of atomic number. Here density is in units of kg/m^3. The various colors represent the different periods (rows) in the periodic table. Notice how the same general pattern repeats as we move across each row: the density is low at the left-hand edge of the row (Group 1), increases through a maximum, and then decreases as we continue to move from left to right across the table. (Only those elements that exist as solids under ordinary conditions are shown.)

Figure 2.12 ■ The data from Figure 2.11 are presented here in a different form. The shading of the boxes in the periodic table represents the density of each element; darker shading indicates higher density. The general trends in density are apparent as you look across a row or down a column.

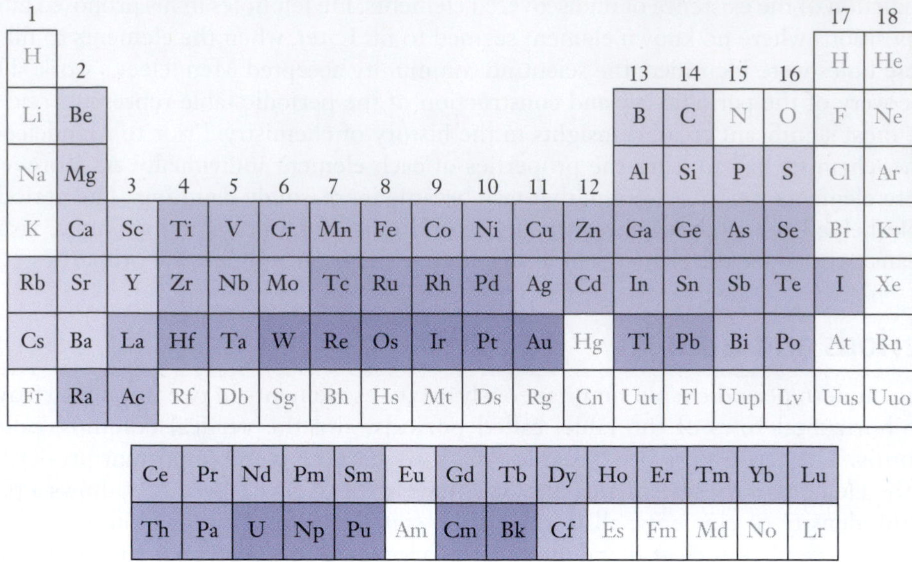

of the table. These groups of elements were assigned names and those names have remained with them. Thus the elements in the far left-hand column (Li, Na, K, Rb, and Cs) are known collectively as **alkali metals**. Similarly, Be, Mg, Ca, Sr, and Ba are called **alkaline earth metals**, and F, Cl, Br, and I are referred to as **halogens**. He, Ne, Ar, Kr, and Xe were discovered much later than most of the other elements, and they have been named **rare gases** or **noble gases**. Other groups are named, but their names are less commonly used and won't be mentioned here.

1 2 3 4 5 6 7 8 9 10 11 12 13 14 15 16 17 18
H He
Li Be B C N O F Ne
Na Mg Al Si P S Cl Ar
K Ca Sc Ti V Cr Mn Fe Co Ni Cu Zn Ga Ge As Se Br Kr
Rb Sr Y Zr Nb Mo Tc Ru Rh Pd Ag Cd In Sn Sb Te I Xe
Cs Ba La Hf Ta W Re Os Ir Pt Au Hg Tl Pb Bi Po At Rn
Fr Ra Ac Rf Db Sg Bh Hs Mt Ds Rg Cn Uut Fl Uup Lv Uus Uuo
Ce Pr Nd Pm Sm Eu Gd Tb Dy Ho Er Tm Yb Lu
Th Pa U Np Pu Am Cm Bk Cf Es Fm Md No Lr

The noble gas elements were once called inert gases because they were considered completely unreactive. Several noble gas compounds are now known, so the term inert *is no longer used.*

There are also names for different regions of the table. Elements in the two groups on the left side of the table and the six groups on the right side are collectively referred to as **representative elements**, or **main group elements**. Elements that separate these two parts of the representative groups in the main body of the periodic table are called **transition metals**. Iron is an example of a transition metal. The elements that appear below the rest of the periodic table are called **lanthanides** (named after the element lanthanum, $Z = 57$) and **actinides** (named after the element actinium, $Z = 89$).

In addition to these names, several numbering systems have been used to designate groups. Current convention dictates numbering from left to right starting with 1 and proceeding to 18. Thus, for example, the group containing C, Si, Ge, Sn, and Pb is referred to as Group 14. We'll use these 1–18 numbers in this textbook, but you may also encounter older labeling schemes elsewhere.

Figure 2.13 ■ The graph shows the number of hydrogen atoms with which an individual atom of various elements will combine. The periodicity of this chemical property is evident from the cyclic graph.

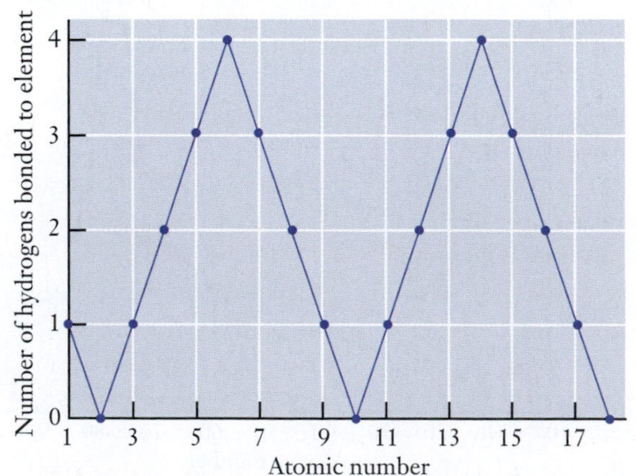

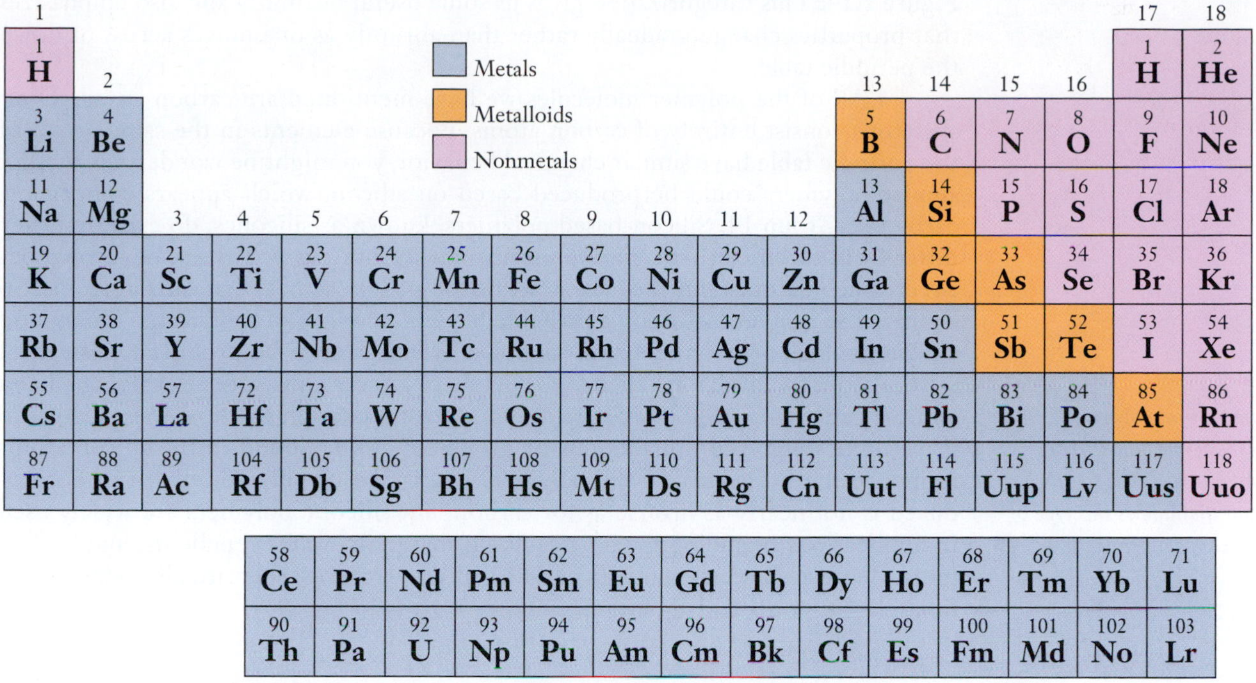

Metals

Metalloids

Nonmetals

1	2	3	4	5	6	7	8	9	10	11	12	13	14	15	16	17	18
1 H																1 H	2 He
3 Li	4 Be											5 B	6 C	7 N	8 O	9 F	10 Ne
11 Na	12 Mg											13 Al	14 Si	15 P	16 S	17 Cl	18 Ar
19 K	20 Ca	21 Sc	22 Ti	23 V	24 Cr	25 Mn	26 Fe	27 Co	28 Ni	29 Cu	30 Zn	31 Ga	32 Ge	33 As	34 Se	35 Br	36 Kr
37 Rb	38 Sr	39 Y	40 Zr	41 Nb	42 Mo	43 Tc	44 Ru	45 Rh	46 Pd	47 Ag	48 Cd	49 In	50 Sn	51 Sb	52 Te	53 I	54 Xe
55 Cs	56 Ba	57 La	72 Hf	73 Ta	74 W	75 Re	76 Os	77 Ir	78 Pt	79 Au	80 Hg	81 Tl	82 Pb	83 Bi	84 Po	85 At	86 Rn
87 Fr	88 Ra	89 Ac	104 Rf	105 Db	106 Sg	107 Bh	108 Hs	109 Mt	110 Ds	111 Rg	112 Cn	113 Uut	114 Fl	115 Uup	116 Lv	117 Uus	118 Uuo

58 Ce	59 Pr	60 Nd	61 Pm	62 Sm	63 Eu	64 Gd	65 Tb	66 Dy	67 Ho	68 Er	69 Tm	70 Yb	71 Lu
90 Th	91 Pa	92 U	93 Np	94 Pu	95 Am	96 Cm	97 Bk	98 Cf	99 Es	100 Fm	101 Md	102 No	103 Lr

Figure 2.14 ■ The colors in this periodic table identify each element as a metal, nonmetal, or metalloid. Notice how the metals are grouped toward the left and bottom of the table, and nonmetals are toward the upper right.

Metals, Nonmetals, and Metalloids

Another way to classify an element is as a **metal**, **nonmetal**, or **metalloid**. Once again, the periodic table conveniently arranges elements so that one can place a given element easily into one of these categories.

As we move across the periodic table, atomic properties vary gradually rather than abruptly. So the boundaries between these categories are ambiguous in some ways.

Most of the elements are metals. Their general location in the periodic table is toward the left and bottom, as seen in the coloring of the periodic table in **Figure 2.14**. Metals share a number of similarities in chemical and physical properties. Physically, metals are shiny, malleable, and ductile (meaning they can be pulled into wires). They also conduct electricity, so wires are always made from metals. Chemical properties can also be used to distinguish metals. Metallic elements tend to form cations in most of their compounds, for example.

Nonmetals occupy the upper right-hand portion of the periodic table. There are fewer nonmetals than metals. But when we consider the relative importance of elements, nonmetals hold their own because of their role in the chemistry of living things. Most of the molecules that make up the human body consist predominantly or exclusively of the nonmetallic elements carbon, hydrogen, oxygen, nitrogen, sulfur, and phosphorus. As our examples so far might lead you to guess, polymers also consist almost exclusively of nonmetallic elements. In contrast to metals, nonmetals are not shiny, malleable, or ductile nor are they good conductors of electricity. These physical properties provide one means by which we can distinguish metals from nonmetals. Chemically, nonmetals tend to form anions rather than cations.

Whether an element is a metal or nonmetal may seem simple to determine based on the physical properties cited above. However, some elements cannot be classified easily as either metal or nonmetal. The question whether or not a substance conducts electricity, for example, does not always have a simple yes or no answer. Lacking a reliable means of drawing a clean boundary between the two categories, scientists have generally chosen to refer to intermediate cases as metalloids or **semimetals**. In the periodic table, metalloids are clustered along a diagonal path, as shown in

Figure 2.14. This categorization gives us some useful flexibility and also emphasizes that properties change gradually rather than abruptly as one moves across or down the periodic table.

All of the polymer molecules we have mentioned are carbon based. Their skeletons consist entirely of carbon atoms. Because elements in the same group of the periodic table have similar chemical behavior, you might be wondering whether similar polymers could be produced based on silicon, which appears right below carbon in Group 14. Silicon-based polymers, known as silicones, do exist, but they differ from carbon polymers in important ways. Covalent bonds can be formed between silicon atoms, but they are not as strong as those between carbon atoms. So chains of silicon atoms become unstable beyond a length of around ten atoms, and silicon analogs of polymers such as polyethylene cannot be produced. Instead of pure silicon, the backbone chains in silicone polymers consist of alternating silicon and oxygen atoms. The Si—O bond is strong enough to allow these chains to grow quite long. Additional atoms or groups of atoms bound to the silicon atoms influence the properties of the polymer. The range of polymers that can be produced is not nearly as diverse as for carbon, but silicone polymers are widely used in applications including greases, caulking materials, water repellents, and surfactants. Because silicones can withstand high temperatures, they are also widely used in kitchen utensils and cookware.

Proteins are natural polymers whose backbones include carbon, oxygen, and nitrogen. The specific arrangement that allows this type of backbone is called a peptide bond.

2-6 Inorganic and Organic Chemistry

Just as engineering can be broken down into various specialties, chemistry can be viewed as a collection of subfields as well. Two of the most fundamental areas of chemistry are **organic chemistry** and **inorganic chemistry**. These names arise from the fact that at one time organic chemistry would have been defined as the chemistry of living things. A more modern definition is that organic chemistry is the study of compounds of the element carbon. As we've already seen, this includes naturally occurring biological molecules and also nearly all synthetic polymers. Inorganic chemistry is the study of all other elements and their compounds. At first glance, this may strike you as a bizarre distinction because carbon is just one element in a periodic table that contains more than 100 others. But the chemistry of carbon is so rich, diverse, and important that organic chemists actually outnumber inorganic chemists. Because this text is intended as a brief overview of important chemical principles, we will focus most of our attention on the behavior of molecules in general rather than delving too deeply into the details of either subfield. In this section, we will describe briefly some important similarities and differences between organic and inorganic chemistry and introduce terminology and notation that we will need as we progress through the text.

The division of chemistry into subdisciplines is a historic artifact. Much modern research in chemistry takes place at the interface of two or more subdisciplines.

Inorganic Chemistry—Main Groups and Transition Metals

Many inorganic compounds exist as relatively small molecules whose atoms are joined together through covalent bonds. One such compound is silicon tetrachloride, $SiCl_4$, which has important uses in the production of semiconductors. **Figure 2.15** shows some visual representations of $SiCl_4$ that we might use to illustrate how its atoms are actually arranged into molecules. Four chlorine atoms surround a central silicon atom, and each chlorine shares one pair of electrons with the silicon.

We will study chemical bonding and molecular shapes in detail in Chapter 7.

Silicon and chlorine are both main group elements, found in Groups 14 and 17 of the periodic table, respectively. As mentioned in the previous section, elements from the same group tend to display similar chemical properties. Thus once we know that $SiCl_4$ exists, we might expect that other pairs of elements from the same groups might

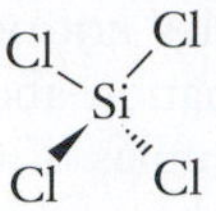

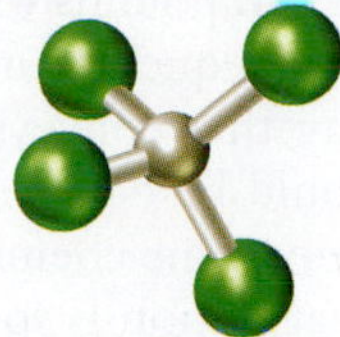

Figure 2.15 ■ This figure presents three depictions of $SiCl_4$. In the drawing at the left, each atom is represented by its symbol and the lines between the symbols depict chemical bonds. In the center panel is a "ball and stick" model, where each atom is a ball, and the bonds are shown as sticks connecting the balls. In the right-hand panel is a "space filling" model, where atoms are shown as balls that overlap one another strongly. Each type of model is commonly used, and each has its strengths and weaknesses.

The solid and dashed triangles in the structure on the left indicate that one of the chlorine atoms would be in front of the plane of the page and one would be behind that plane.

form similar compounds. And this prediction is correct: compounds such as $SnCl_4$ and CF_4 do exist and have structures and bonds analogous to those in Figure 2.15 for $SiCl_4$.

Other compounds of the main group elements form extended ionic structures, such as that of NaCl in Figure 2.9. But despite the difference in the types of chemical bonds employed, we can still readily predict that similar compounds should exist for other pairs of elements from the same groups. From the periodic table, we see that sodium is in Group 1 and chlorine is in Group 17. So we can expect that other pairs of elements from these columns of the table will form ionic solids, too. Again, our prediction is accurate; compounds such as LiCl, NaF, and KBr have structures analogous to that of NaCl. The reason for the existence of these similar compounds is simple. All of the metals in Group 1 form cations with a 1+ charge, and all of the elements in Group 17 form anions with a 1− charge. Any of these cations can combine with any of the anions in a 1:1 ratio to form neutral compounds.

The chemistry of transition metals is somewhat more complicated than that of the main group elements, though, because most transition metals can form multiple cations with different charges. Iron commonly forms two different monatomic cations: Fe^{2+} and Fe^{3+}. As a result of this, iron can form a more diverse set of compounds than Group 1 metals. It can combine with chlorine to form either $FeCl_2$ or $FeCl_3$, and these two compounds have significantly different physical properties (**Figure 2.16**). Largely because they can form multiple cations, the chemistry of transition metals does not vary as sharply from group to group. Regardless of their positions in the periodic table, for example, most transition metals can form cations with a 2+ charge. Thus predictions based simply on group number are not as reliable here as they are for the representative elements. When considering transition metals and their compounds, we must rely more heavily on knowledge of the specific chemistry of each element.

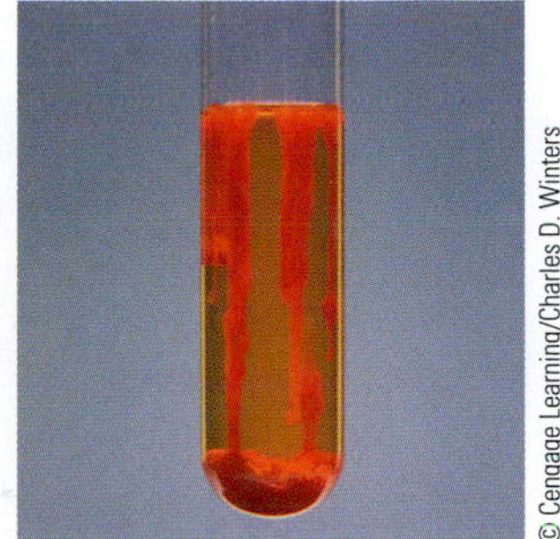

Iron(III) chloride, $FeCl_3$
(Here forming as the solid at the bottom of the test tube)
Orange-brown color
Density 2.90 g cm^{-3}
Melts at 306°C

Iron(II) chloride, $FeCl_2$
Greenish-yellow color
Density 3.16 g cm^{-3}
Melts at 670°C

Figure 2.16 ■ Transition metals typically form more than one type of cation, giving them a very diverse chemistry. Iron, for example, forms cations with both 2+ and 3+ charges, and this allows it to form two different ionic compounds with chlorine. $FeCl_2$ and $FeCl_3$ have different appearances and properties.

Organic Chemistry

All organic compounds feature carbon skeletons, similar to those we have already seen in our introduction to polymers. Other elements frequently found in organic compounds include hydrogen, oxygen, and nitrogen. Despite this rather short list of elements, more than 18 million organic compounds exist. This vast number of compounds arises from some unusual aspects of the chemistry of carbon itself. Most importantly, carbon atoms readily attach to one another to form chains, and these chains can grow quite long. Many of the polymer molecules we have been discussing in this chapter contain thousands of carbon atoms. Furthermore, some of these long chains are straight, whereas others are branched at one or more places. And finally, any pair of carbon atoms can bond to one another in three different ways, by sharing either one, two, or three pairs of electrons. When taken all together, these factors allow carbon to form a vast array of compounds.

The diversity of organic compounds presents some challenges. It is not uncommon for several different organic compounds to have the same molecular formula, for example, but to display different properties depending on exactly how the atoms are arranged into molecules. (Different compounds with the same molecular formula are called *isomers*.) So organic chemists frequently must depict molecules not only by their formulas but also in some way that conveys important information about the arrangement of the atoms. This could be done using structural formulas of the sort we saw in Figure 2.15. But because organic chemistry often deals with very large compounds and complex structures, that option is somewhat unwieldy. A shorthand notation known as a **line structure** has emerged as the most common method for describing organic compounds simply and unambiguously. The line structure is a modified version of the structural formula. As in any structural formula, lines are used to depict bonds between atoms. But in a line drawing, many of the elemental symbols are omitted. By definition, an organic compound is based on carbon atoms. So to reduce clutter in a line drawing, the 'C' symbols for carbon atoms are not written. Furthermore, because organic compounds almost always contain many hydrogen atoms, the 'H' symbol for any hydrogen atom that is attached directly to a carbon atom is also not written. Symbols are written for any elements other than carbon and hydrogen, as well as for any hydrogen atoms that are not directly attached to carbon. We can illustrate the relationship between a structural formula and a line drawing in Example Problem 2.3.

EXAMPLE PROBLEM 2.3

Poly(methyl methacrylate) is widely known as Plexiglas®. The structural formula for the monomer, methyl methacrylate, is shown below. Write the corresponding line structure for methyl methacrylate:

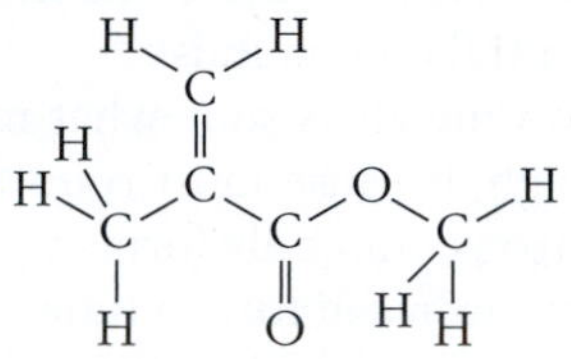

Strategy We convert the structure into a line drawing by removing the symbols for all carbon atoms and for hydrogen atoms attached directly to carbons. Bonds to or between carbon atoms remain, so carbon atom positions become either intersections between lines or ends of lines. Bonds between carbon and hydrogen atoms are omitted.

Solution First, we will remove the symbols and bonds for all of the hydrogen atoms because they are all bound directly to carbon:

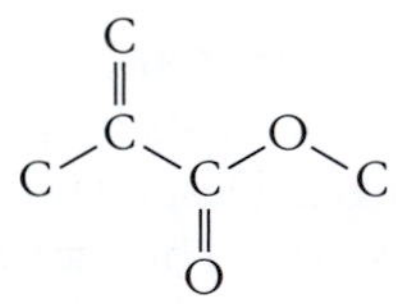

Next, we remove the symbols for the carbon atoms, leaving intact the lines that depict the remaining bonds. This gives us the final line structure:

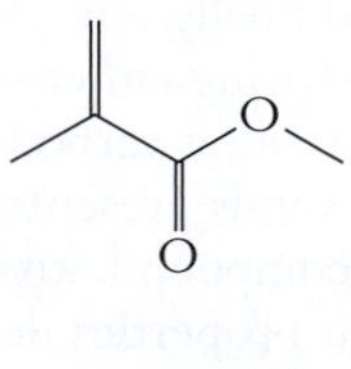

Discussion The line structure is much more compact than the original structural formula. An experienced chemist quickly recognizes where the atoms whose symbols are not shown need to be.

Check Your Understanding The structural formula for styrene, which is the monomer for the common plastic polystyrene, is shown below. Convert this to a line drawing.

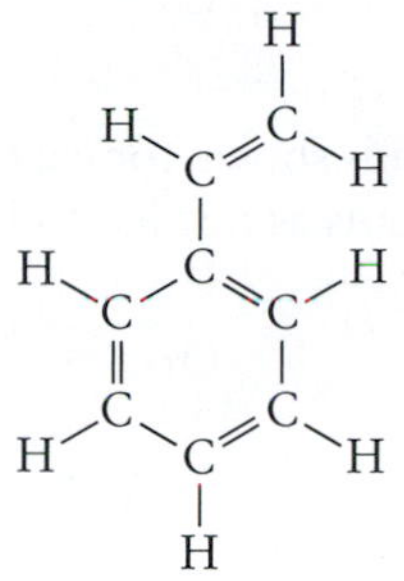

We will use these line structures throughout the rest of this textbook, and you may also encounter them in other places, such as the information sheets that accompany prescription drugs. In many instances, it will be necessary to interpret the line drawing to determine the molecular formula, so we should develop a way to do that systematically. In addition to the rules we used before to transform a structural diagram into a line structure, we will need to introduce two important generalizations about chemical bonding.

1. A hydrogen atom in an organic molecule can form only one covalent bond to one other atom.
2. Every carbon atom in an organic molecule always forms exactly four covalent bonds.

In combination, these two facts allow us to fill in all of the carbon and hydrogen atoms that are not explicitly written in a line structure. First, we place a carbon anywhere that we see either an intersection between lines or the end of a line. Then we add hydrogen atoms as needed to bring each carbon's number of bonds up to four. Example Problem 2.4 illustrates this process.

EXAMPLE PROBLEM 2.4

A temperature-resistant plastic called poly(phenylene oxide) is a key component of resins such as GE's Noryl®, which is widely used in computer cases and automobile dashboards. The line structure below represents 2,6-dimethylphenol, which is the monomer from which poly(phenylene oxide) is made. What is the molecular formula for 2,6-dimethylphenol?

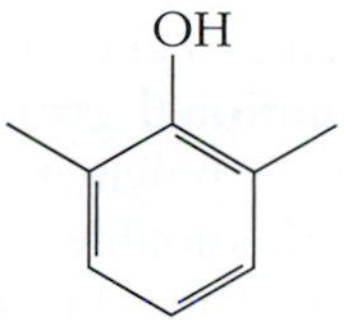

Strategy First, we can pencil in carbon atoms at the appropriate positions. Then we will add hydrogen atoms as needed. Once all of the atoms have been identified, it will be easy to count them to produce the needed formula.

Solution We place a carbon atom at the end of a line or the intersection between lines:

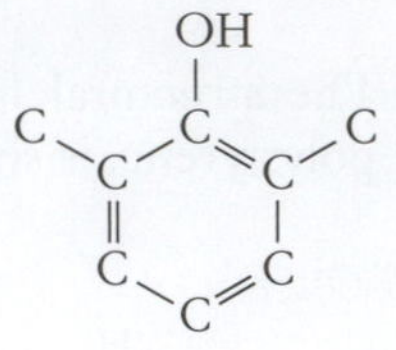

Next, we count the number of bonds shown for each carbon. If that number is three or less, we add as many hydrogen atoms as needed to bring it up to four:

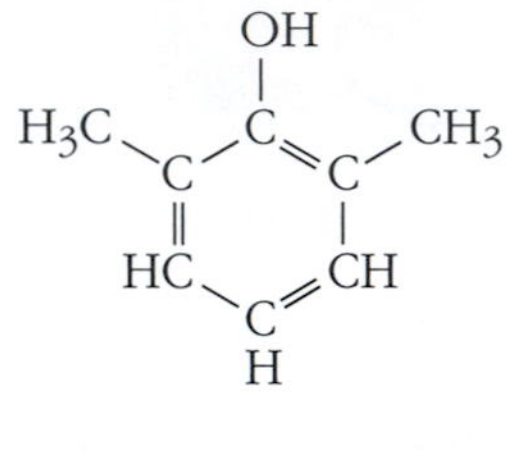

Now all of the atoms are shown explicitly. Counting, we arrive at the molecular formula as $C_8H_{10}O$.

Discussion The double lines in the ring in this structure represent double bonds, in which two pairs of electrons are shared between two atoms. Notice that in locating the carbon atoms, we treated the double lines the same as we did the single lines: each intersection represents a carbon atom, no matter how many lines meet.

Check Your Understanding Once, poly(vinylpyrrolidone) was used in the manufacture of hairsprays, and it is still used in the glue that holds the layers together in plywood. The line structure for the vinylpyrrolidone monomer is shown below. Find the corresponding molecular formula.

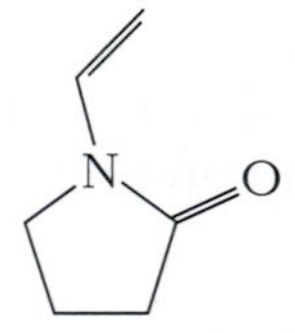

Functional Groups

Given the vast number of organic compounds, the need for some systematic way to understand their chemistry should be apparent. One of the most important concepts for an organic chemist is the idea that certain arrangements of atoms tend to display similar chemical properties whenever they appear together. Such an arrangement of atoms is called a **functional group**. One of the simplest functional groups, and one that is central to many polymerization reactions, is a pair of carbon atoms joined by a double bond. If the double bond is converted to a single bond, then each carbon atom can form a new bond to another atom. Thus the characteristic reaction of a carbon-carbon double bond is **addition**, in which new atoms or groups of atoms are attached to a molecule. Line structures make it very easy to identify any C=C groups in a molecule, and thus to locate positions at which addition reactions might be feasible.

The role of addition reactions in producing polymers will be examined in Section 2-8.

Table ■ 2.4

Some common functional groups

Functional Group	Class of Compounds	Example
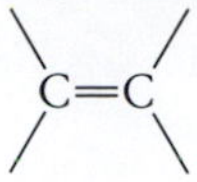 C=C (with four single bonds)	Alkenes	Ethylene
—C≡C—	Alkynes	Acetylene
—X (X = F, Cl, Br, I)	Organic halides	Methyl chloride
—OH	Alcohols, phenols	Ethanol, phenol
C—O—C	Ethers	Diethyl ether
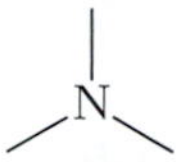 N (with three single bonds)	Amines	Methylamine
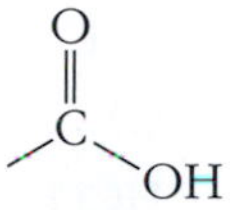 C(=O)OH	Carboxylic acids	Acetic acid
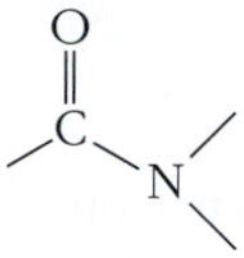 C(=O)N	Amides	Acetanilide
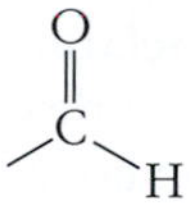 C(=O)H	Aldehydes	Formaldehyde
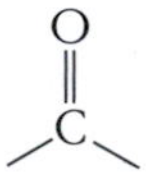 C=O	Ketones	Methyl ethyl ketone

The simplest organic compounds are **hydrocarbons**, molecules that contain only carbon and hydrogen atoms. We can imagine the formation of more complicated molecules by replacing one or more of a hydrocarbon's hydrogen atoms with a functional group. Compounds in which a hydrogen atom is replaced by an —OH functional group, for example, are collectively referred to as alcohols. The presence of the —OH group conveys certain properties to this class of molecules, including the ability to mix with water to a much greater extent than the corresponding hydrocarbons. Often, the notion of functional groups influences the way in which we choose to write chemical formulas. If the chemical formula for an alcohol is written so that the —OH group is emphasized, then it will be easier to recognize that this group is present. So the formula for the simplest alcohol, methanol, is most often written as CH_3OH rather than as CH_4O. Similarly, ethanol is generally written as C_2H_5OH rather than C_2H_6O. Other common functional groups are listed in **Table 2.4**.

Not all organic compounds that contain —OH groups are alcohols. Carboxylic acids contain a —COOH functional group, for example.

2-7 Chemical Nomenclature

Although only a limited number of elements exist, the number of compounds that may be formed from those elements is virtually boundless. Given the vast number of molecules that can be made, we require a systematic means of assigning names to chemical

Table ■ 2.5

Common cations

Sodium ion	Na^+	Potassium ion	K^+
Magnesium ion	Mg^{2+}	Calcium ion	Ca^{2+}
Iron(II) ion	Fe^{2+}	Copper(I) ion	Cu^+
Iron(III) ion	Fe^{3+}	Copper(II) ion	Cu^{2+}
Silver ion	Ag^+	Zinc ion	Zn^{2+}
Ammonium ion	NH_4^+	Hydronium ion	H_3O^+

Table ■ 2.6

Common anions

Halides	F^-, Cl^-, Br^-, I^-	Sulfate	SO_4^{2-}
Nitrate	NO_3^-	Hydroxide	OH^-
Phosphate	PO_4^{3-}	Cyanide	CN^-
Carbonate	CO_3^{2-}	Hydrogen carbonate	HCO_3^-

compounds. This system should be sufficiently well defined that a person who knows the rules can draw the structure of any compound, given its systematic name. This naming process for molecules is often referred to as **chemical nomenclature**. We will establish some of the basic premises of this system now and supplement this initial set of rules as needed when we encounter new situations and types of compounds later.

Binary Systems

Compounds that contain only two elements are called **binary compounds**. Fe_2O_3, for example, is a binary compound. Many such compounds exist and they can be conveniently grouped according to their bonding tendencies. Thus we encounter slightly different rules when we name binary molecules held together by covalent bonds than when we assign names to ionic compounds. This means it is important that you learn to recognize from its formula whether a compound is ionic or covalent. This will become easier as you get more comfortable working with chemical formulas. A good way to start is to begin to recognize the elements as metals or nonmetals. When two nonmetals combine, they usually form a covalent compound. But when metals and nonmetals combine with one another, they frequently form ionic compounds. It is also handy to learn to recognize common polyatomic ions, such as those listed in **Tables 2.5** and **2.6**. The presence of these ions is a sign that a compound is ionic.

Table ■ 2.7

Greek prefixes for the first ten numbers

Number	Prefix
One	Mono-
Two	Di-
Three	Tri-
Four	Tetra-
Five	Penta-
Six	Hexa-
Seven	Hepta-
Eight	Octa-
Nine	Nona-
Ten	Deca-

Naming Covalent Compounds

In some cases, a given pair of elements can form compounds in a number of different ways. Nitrogen and oxygen, for example, form NO, N_2O, NO_2, N_2O_3, N_2O_4, and N_2O_5, all of which are stable enough to observe. So it is critical that our naming system distinguishes these different molecules. To accomplish this, the nomenclature system uses a prefix to specify the number of each element present. The first ten of these prefixes, which arise from the Greek roots for the numbers, are listed in **Table 2.7**.

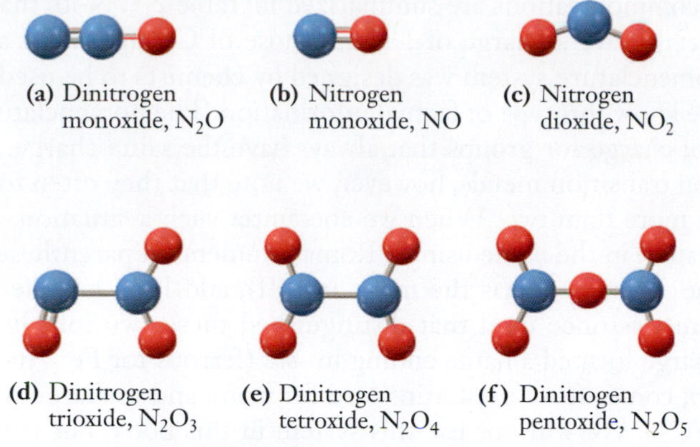

Nitrogen forms a number of binary compounds with oxygen.

In a binary compound, the element that appears first in the formula also appears first in the name. The first element retains its full name, whereas the second element is described by replacing the ending from its name with the suffix *-ide*. *Both elements will be preceded by a number-designating prefix* ***except*** *that when there is only one atom of the first element, it does not carry the prefix* mono-. An example of this procedure is provided in Example Problem 2.5.

EXAMPLE PROBLEM 2.5

What are the systematic names of the following covalent compounds? (a) N_2O_5, (b) PCl_3, (c) P_4O_6

Strategy The first element listed retains its full name and adds a prefix when more than one atom of it is in the compound. The second element will retain only the root of its name, followed by *-ide*, and it too takes a prefix to indicate the number of atoms.

Solution (a) N_2O_5: dinitrogen pentoxide, (b) PCl_3: phosphorus trichloride (*remember*: this is not called *mono*phosphorus trichloride), (c) P_4O_6: tetraphosphorus hexoxide. (The *a* in *hexa-* is dropped here to simplify pronunciation.)

Check Your Understanding What are the names of the following compounds? (a) CS_2, (b) SF_6, (c) Cl_2O_7

Naming Ionic Compounds

The iron chlorides shown earlier in Figure 2.16 are examples of binary ionic compounds. Because ionic compounds must be neutral, the positive and negative charges of the ions must balance each other and only one formula unit is possible. Therefore, once one of the charges is indicated in the name, the entire formula is known. The rules of nomenclature dictate that the positively charged species, the cation, be specified with enough information about its charge to indicate the complete formula. Unfortunately for novices in chemistry, some groups of the periodic table have only one possible cationic charge, whereas other groups (such as Group 8, which contains iron) have more than one possibility.

The most common cations are summarized in Table 2.5. Note that the cations of all Group 1 metals have a charge of 1+ and those of Group 2 have a charge of 2+. Because the nomenclature system was designed by chemists to be used by chemists, it assumes that we know this type of factual information. The nomenclature system gives no indication of charge for groups that always have the same charge. When we look at the charges in transition metals, however, we note that they often form two cations and some even more than two. When we encounter such a situation, we indicate the charge of the cation in the name using a Roman numeral in parentheses following the elemental name. Thus Fe^{2+} has the name iron(II), and Fe^{3+} has the name iron(III). An older system was once used that distinguished these two ions by using a suffix. The smaller charge formed a name ending in *-ous* (ferrous for Fe^{2+} in the case of iron whose root, *ferr*, comes from the Latin word for iron), and the larger charge ended in *-ic* (ferric for Fe^{3+}). We will not use this system in this book, but you may see these older names on some laboratory chemicals.

We've already seen some examples of elements that form monatomic anions. The most common are halogens. There are also several polyatomic anions that we will encounter often in this book. Those listed in Table 2.6 are the most common anions that we will see throughout the text.

Generally, the bonds between the atoms within a polyatomic ion are covalent.

The naming convention for monatomic ions is familiar from the discussion of covalent molecules above: the name of a monatomic anion is the root of the element name with the suffix *-ide* added. Thus Cl^- is chloride, Br^- is bromide, and so on. We can now name $FeCl_2$ and $FeCl_3$ as iron(II) chloride and iron(III) chloride, respectively.

Many compounds contain polyatomic anions, including those shown in Table 2.6. Most often, the names of these polyatomic ions are memorized rather than being obtained by a systematic nomenclature rule. There is, however, a system for polyatomic anions that contain oxygen and one other element, **oxyanions**. The base name of the oxyanion is provided by the element that is not oxygen. If there are two possible groupings of the element with oxygen, the one with more oxygen atoms uses the suffix *-ate* and the one with fewer oxygens uses the suffix *-ite*. When there are four possible oxyanions, we add a prefix *per-* to the *-ate* suffix for the most oxygens and a prefix *hypo-* to the *-ite* suffix for the least oxygens. Chlorine is the classic example of an element that forms four oxyanions, whose names are provided in **Table 2.8**.

Table 2.8

Oxyanions of chlorine

ClO^-	Hypochlorite
ClO_2^-	Chlorite
ClO_3^-	Chlorate
ClO_4^-	Perchlorate

Once we know how to name both of the ions, an ionic compound is named simply by combining the two names. The cation is listed first in the formula unit and in the name. Example Problem 2.6 provides some examples of the way to determine the name of an ionic compound.

EXAMPLE PROBLEM 2.6

Determine the name of the following ionic compounds: (a) Fe_2O_3, (b) Na_2O, (c) $Ca(NO_3)_2$

Strategy We must determine the names of the constituent ions first. The anions will provide a hint about the cation charges if we need it.

Solution

(a) Fe_2O_3: As noted in Table 2.6, oxygen is always a 2− ion in these compounds, so there is a total charge of 6− on the three oxide ions in the formula unit. Therefore, the two iron ions must have a total charge of 6+, requiring 3+ from each iron. So the name is iron(III) oxide.

(b) Na_2O: Sodium from Group 1 always has a 1+ charge and oxygen always 2−. Therefore, the name is sodium oxide. No Roman numeral is needed for sodium because it has only one common ionic charge.

(c) $Ca(NO_3)_2$: Calcium is in Group 2, so the calcium ion always carries a charge of 2+. NO_3^- is a common polyatomic anion called nitrate. The name is calcium nitrate.

Check Your Understanding Name each of the following ionic compounds: (a) $CuSO_4$, (b) Ag_3PO_4, (c) V_2O_5

Occasionally throughout this text, we will encounter new classes of chemical compounds that will need more rules to determine their names. We will introduce these nomenclature systems when necessary.

INSIGHT INTO

2-8 Polyethylene

We close this chapter by taking a closer look at polyethylene, which is probably the most common polymer in the world. The U.S. domestic production of all forms of polyethylene in 2011 totaled more than 37 billion pounds. And in its various forms, polyethylene can be found in items ranging from plastic grocery bags to children's toys, automobile gas tanks, and even bulletproof vests. Let's start at the beginning, with a look at the way polyethylene is produced.

In Section 2-4, we pointed out that polyethylene is built up from a monomer called ethylene and that the molecular formula for ethylene is C_2H_4. We also noted that the formula for polyethylene itself is often written as $—[CH_2CH_2]_n—$. To begin to see how the monomers can be combined to form the polymer, we need to look at the structural formula for ethylene, which is shown below.

From our discussion of line structures, we know that each carbon atom in a molecule always forms a total of four covalent bonds. For the carbon atoms in ethylene, two of those bonds are to hydrogen atoms, and the other two make up a double bond between the two carbon atoms. To link the monomers together and form the polymer, then, we will need to convert this double bond into a single bond, leaving each carbon with one bond available for linking to the next monomer in the chain. The polymerization reaction begins with the addition of a small amount of an initiator molecule that easily breaks down, producing highly reactive fragments called **free radicals**. (The free radical is denoted as "R·" below, where the "·" represents an unshared electron.) One of these free radicals attaches itself to a single ethylene molecule, opening its double bond and leaving one end unbonded:

Free radicals are reactive wherever they form. Physiological processes including aging have been tied to the presence of free radicals, so their importance is wide ranging.

$$\mathrm{R\cdot} \curvearrowright \mathrm{H_2C{=}CH_2} \longrightarrow \mathrm{R{-}CH_2{-}CH_2\cdot}$$

The unbonded end of the ethylene now takes on the role of the free radical, attacking a second ethylene monomer and attaching it to the growing chain. As long as the number of available ethylene monomers remains large, the polymer can continue to grow in this fashion.

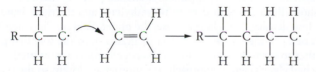

$$\mathrm{R{-}CH_2{-}CH_2\cdot} \curvearrowright \mathrm{H_2C{=}CH_2} \longrightarrow \mathrm{R{-}CH_2{-}CH_2{-}CH_2{-}CH_2\cdot}$$

Linking together thousands of monomers in this way would generate a single polyethylene molecule, whose structural formula would look like that below, only extending over a much greater chain length:

$$
\begin{array}{ccccccccccccccccccccc}
 & H & & H & & H & & H & & H & & H & & H & & H & & H & & H & \\
 & | & & | & & | & & | & & | & & | & & | & & | & & | & & | & \\
- & C & - & C & - & C & - & C & - & C & - & C & - & C & - & C & - & C & - & C & - \\
 & | & & | & & | & & | & & | & & | & & | & & | & & | & & | & \\
 & H & & H & & H & & H & & H & & H & & H & & H & & H & & H &
\end{array}
$$

Eventually, the free radical end of the growing chain will meet up with another free radical, either from an initiator molecule or from another growing chain. When this happens, the chain will stop growing. When the monomer units grow end to end, as shown above, the result is known as linear polyethylene because all of the carbon atoms lie along a more or less straight backbone.

It is also possible to grow polyethylene under conditions that lead to branched chains, in which some of the hydrogen atoms along the backbone are themselves replaced by polyethylene-like chains. The contrasting forms of the linear and branched molecules are shown in **Figure 2.17**.

Although linear polyethylene is conceptually simpler, branched polyethylene is actually cheaper and easier to produce. Perhaps surprisingly, the linear and

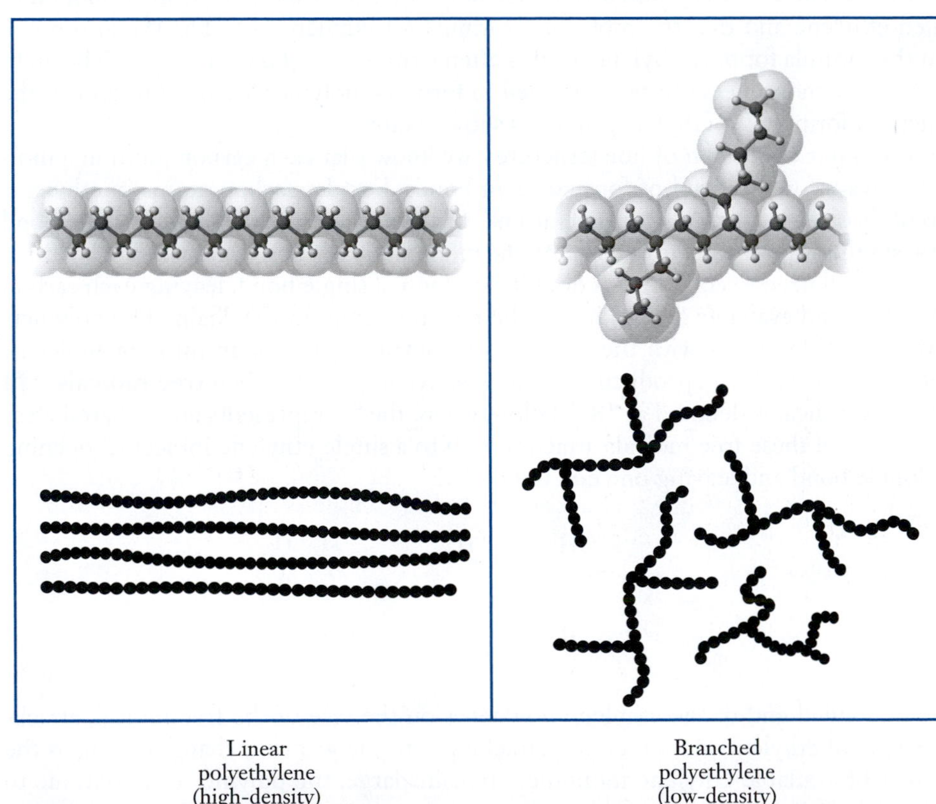

Figure 2.17 ■ Differences between linear and branched polyethylene are illustrated. The left-hand panel shows linear, or high-density, polyethylene, and the right-hand panel shows the branched, or low-density, form. In each case, the upper diagram illustrates the molecular structure for part of a polymer chain. The lower diagrams show the way that polymer chains would pack together to form the solid plastic. (Hydrogen atoms are omitted in these drawings for clarity.) Branched chains cannot approach one another as closely, so the resulting material has a much lower density.

branched forms of polyethylene are actually polymers with significantly different macroscopic properties. When arranged side by side, the long, linear polyethylene molecules can pack together very tightly, producing a relatively dense plastic. So linear polyethylene is also known as **high-density polyethylene (HDPE)**. It is a strong and hard material, used in bottles, kitchenware, and as a structural plastic in many children's toys. Because their chains cannot be stretched out straight, branched polyethylene molecules cannot be packed together as closely as those of linear polyethylene. This looser packing of the molecules produces a plastic with a much lower density, so branched polyethylene is usually referred to as **low-density polyethylene (LDPE)**. LDPE is commonly used in applications where relatively little strength is needed, such as plastic films, sandwich bags, and squeeze bottles.

Recent advances in polymerization technology have made it possible to grow extremely long linear polyethylene chains, stretching to hundreds of thousands of monomer units. Because the individual molecules are relatively large and heavy, this has been called **ultra-high molecular weight polyethylene (UHMWPE)**. The very long chains are so strong that this material is replacing Kevlar® (another polymer) as the standard filling for bulletproof vests. UHMWPE can also be formed into large sheets, and these have been used as ice substitutes for skating rinks.

This quick glimpse at some of the many forms and uses for polyethylene shows how the observable properties of polymers are closely linked to the chemical structure of the individual molecules. It also illustrates how the interplay between chemistry and engineering can lead to polymers that are selectively produced to meet certain design specifications.

FOCUS ON PROBLEM SOLVING

Question Boron is widely used in the production of enamels and glasses. Naturally occurring boron has an atomic weight of 10.811 amu. If the only isotopes present are ^{10}B and ^{11}B, describe how you would determine their relative abundances. Include in your description any information that you would need to look up.

Strategy This problem is the inverse of those in this chapter in which we used isotopic abundances to find atomic weights. Here we must work from the atomic weight to find the isotopic abundances. Since there are two isotopes, we have two unknowns. So we will need to write two equations that relate the percentages or fractions of ^{10}B and ^{11}B. We would also need to know the mass of each isotope, but presumably we could look those up.

Solution The abundances of the two isotopes must add to 100%. This gives us a first equation:

$$(\text{Fraction of } ^{10}\text{B}) + (\text{fraction of } ^{11}\text{B}) = 1.00$$

The fact that the atomic weight is 10.811 provides another equation:

$$(\text{Fraction of } ^{10}\text{B}) \times (\text{mass of } ^{10}\text{B}) + (\text{fraction of } ^{11}\text{B}) \times (\text{mass of } ^{11}\text{B}) = 10.811 \text{ amu}$$

Assuming that we are able to find values for the masses of the two isotopes, we now have two equations in two unknowns, so the problem can be solved. (We could get a rough approximation by assuming that ^{10}B has a mass of 10 amu and ^{11}B has a mass of 11 amu, but that assumption would not lead to a very accurate result.)

SUMMARY

The widely taught description of an atom as a massive, positively charged nucleus surrounded by lighter, fast-moving electrons is based on the results of many ingenious experiments. Several details of this model are important in building our knowledge of chemistry. The number of protons in a nucleus determines the chemical identity of an atom, and must be equal to the number of electrons so that the atom is electrically neutral. If the numbers of protons and electrons are not the same, the result is a charged particle, which is called an ion. Because of their electric charge, ions behave quite differently from neutral atoms. For example, the sodium in our diet is invariably sodium ions, not sodium atoms.

Electric charge plays a central role in determining the structure of both atoms and molecules. The attraction of oppositely charged particles and the repulsion of like charged particles, as described by Coulomb's law, is central to our understanding of chemical bonding. The various types of chemical bonds, including covalent, ionic, and metallic, can all be understood in terms of the interactions between negative and positive charges.

Once we establish that atoms will bond together to form chemical compounds, we are faced with the burden of summarizing a vast number of compounds. Many classification schemes have been developed over the years to assist us in organizing data on a wide range of chemical substances. The periodic table is the most common and important device for such a purpose. It succinctly summarizes many properties of the elements, especially their chemical tendencies. The periodic table also helps us learn chemistry by providing a template for trends. Thus we can remember that metallic elements are found toward the left and bottom of the table while nonmetals are found toward the upper right. Without the periodic table, remembering which elements are in each category would be much more challenging.

Other categorizations also help organize our study of chemistry. In some cases, broad categories such as organic versus inorganic chemistry are helpful. For example, chemical nomenclature, the system we use to name compounds, is different for these two branches of chemistry. The way that we impart information about molecules symbolically also varies. In organic chemistry, because carbon is involved in all of the molecules, we use the fact that carbon always forms four chemical bonds to simplify the depiction of molecules as line structures. For inorganic chemistry, in which we often encounter binary chemical compounds, we can devise a fairly broad system of nomenclature based on a relatively small number of rules. Learning this system is an important step in chemical communication.

KEY TERMS

actinides (2-5)
addition reaction (2-6)
alkali metals (2-5)
alkaline earths (2-5)
anion (2-3)
atomic mass unit (2-2)
atomic number (2-2)
binary compound (2-7)
cation (2-3)
chemical bond (2-4)
chemical compound (2-4)
chemical formula (2-4)
chemical nomenclature (2-7)
Coulomb's law (2-3)
covalent bonds (2-4)
electrons (2-2)
empirical formula (2-4)
formula unit (2-4)
free radicals (2-8)
functional group (2-6)
groups (2-5)
halogens (2-5)
high-density polyethylene (HDPE) (2-8)
hydrates (2-4)
hydrocarbons (2-6)
inorganic chemistry (2-6)
ion (2-3)
ionic bonding (2-4)
isotopes (2-2)
isotopic abundance (2-2)
lanthanides (2-5)
lattice (2-4)
line structure (2-6)
low-density polyethylene (LDPE) (2-8)
main group elements (2-5)
mass number (2-2)
metal (2-5)
metallic bonding (2-4)
metalloid (2-5)
molecular formula (2-4)
molecule (2-4)
monomer (2-1)
neutrons (2-2)
noble gases (2-5)
nonmetal (2-5)
nucleus (2-2)
organic chemistry (2-6)
oxyanions (2-7)
periodic law (2-5)
periodicity (2-5)
periods (2-5)
polyatomic ions (2-7)
polymer backbone (2-1)
protons (2-2)
representative elements (2-5)
semimetals (2-5)
transition metals (2-5)
ultra-high molecular weight polyethylene (UHMWPE) (2-8)

PROBLEMS AND EXERCISES

INSIGHT INTO Polymers

2.1 Define the terms *polymer* and *monomer* in your own words.

2.2 How do polymers compare to their respective monomers?

2.3 Look around you and identify several objects that you think are probably made from polymers.

2.4 Which one element forms the backbone of nearly all common polymers? Which other elements are also found in common household polymers?

2.5 The fact that a polymer's physical properties depend on its atomic composition is very important in making these materials so useful. Why do you think this would be so?

2.6 Use the web to research the amount of PVC polymer produced annually in the United States. What are the three most common uses of this polymer?

2.7 Use the web to research the amount of polyethylene produced annually in the United States. What are the three most common uses of this polymer?

Atomic Structure and Mass

2.8 In a typical illustration of the atom such as Figure 2.3, which features lead to misunderstandings about the structure of atoms? Which ones give important insight?

2.9 Why is the number of protons called the atomic number?

2.10 Which isotope in each pair contains more neutrons? **(a)** chlorine-35 or sulfur-33, **(b)** fluorine-19 or neon-19, **(c)** copper-63 or zinc-65, **(d)** iodine-126 or tellurium-127

2.11 Define the term *isotope*.

2.12 Write the complete atomic symbol for each of the following isotopes. **(a)** carbon-13, **(b)** phosphorus-31, **(c)** sodium-23, **(d)** boron-10

2.13 How many electrons, protons, and neutrons are there in each of the following atoms? **(a)** magnesium-24, ^{24}Mg, **(b)** tin-119, ^{119}Sn, **(c)** thorium-232, ^{232}Th, **(d)** carbon-13, ^{13}C, **(e)** copper-63, ^{63}Cu, **(f)** bismuth-205, ^{205}Bi

2.14 Consider the following nuclear symbols. How many protons, neutrons, and electrons does each element have? What elements do R, T, and X represent?

(a) $^{30}_{14}R$ **(b)** $^{89}_{39}T$ **(c)** $^{133}_{55}X$

2.15 Mercury is 16.716 times more massive than carbon-12. What is the atomic weight of mercury? Remember to express your answer with the correct number of significant figures.

2.16 How can an element have an atomic weight that is not an integer?

2.17 Explain the concept of a "weighted" average in your own words.

2.18 The element gallium, used in gallium arsenide semiconductors, has an atomic weight of 69.72 amu. There are only two isotopes of gallium, ^{69}Ga with a mass of 68.9257 amu and ^{71}Ga with a mass of 70.9249 amu. What are the isotopic abundances of gallium?

© Cengage Learning/Charles D. Winters

Gallium melts just above room temperature.

2.19 The atomic weight of copper is 63.55 amu. There are only two isotopes of copper, ^{63}Cu with a mass of 62.93 amu and ^{65}Cu with a mass of 64.93 amu. What is the percentage abundance of each of these two isotopes?

2.20 The following table presents the abundances and masses of the isotopes of zinc. What is the atomic weight of zinc?

Isotope	Abundance	Mass
^{64}Zn	48.6%	63.9291 amu
^{66}Zn	27.9%	65.9260 amu
^{67}Zn	4.10%	66.9271 amu
^{68}Zn	18.8%	67.9249 amu
^{70}Zn	0.60%	69.9253 amu

2.21 Naturally occurring uranium consists of two isotopes, whose masses and abundances are shown below.

Isotope	Abundance	Mass
^{235}U	0.720%	235.044 amu
^{238}U	99.275%	238.051 amu

Only ^{235}U can be used as fuel in a nuclear reactor, so uranium for use in the nuclear industry must be enriched in this isotope. If a sample of enriched uranium has an atomic weight of 235.684 amu, what percentage of ^{235}U is present?

2.22 The table on the next page provides the identity of the two naturally occurring isotopes for four elements and the atomic weights for those elements. (In each case, the two isotopes differ in mass number by two.) Which element has the mass spectrum shown? Explain your answer.

Isotopes	Atomic weight (amu)
^{121}Sb, ^{123}Sb	121.8
^{185}Re, ^{187}Re	186.2
^{191}Ir, ^{193}Ir	192.2
^{203}Tl, ^{205}Tl	204.4

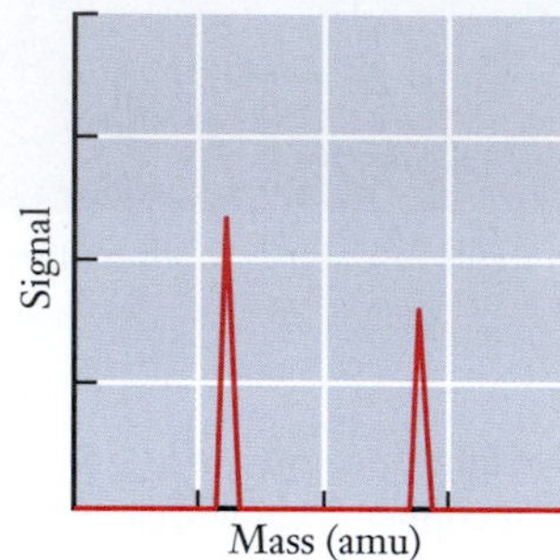

2.23 Chlorine has only two isotopes, one with mass 35 and the other with mass 37. One is present at roughly 75% abundance, and the atomic weight of chlorine on a periodic table is 35.45. Which must be the correct mass spectrum for chlorine?

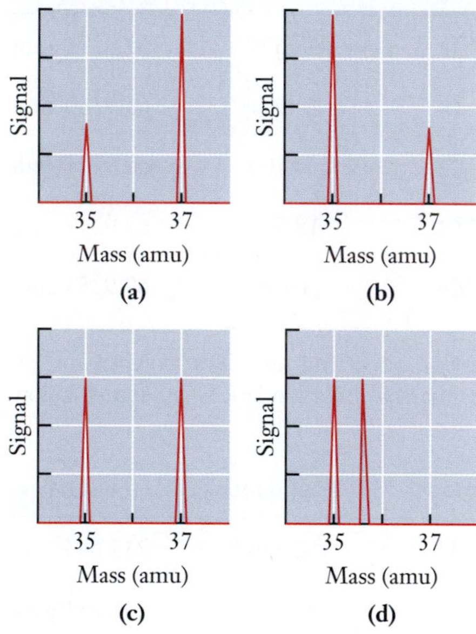

Ions

2.24 Define the term *ion* in your own words.

2.25 What is the difference between cations and anions?

2.26 Provide the symbol of the following monatomic ions, given the number of protons and electrons in each. **(a)** 8 protons, 10 electrons, **(b)** 20 protons, 18 electrons, **(c)** 53 protons, 54 electrons, **(d)** 26 protons, 24 electrons

2.27 How many protons and electrons are in each of the following ions? **(a)** Na^+, **(b)** Al^{3+}, **(c)** S^{2-}, **(d)** Br^-

2.28 Identify each of the following species as an anion, a cation, or a molecule. **(a)** CO_3^{2-}, **(b)** CO_2, **(c)** NH_4^+, **(d)** N^{3-}, **(e)** CH_3COO^-

2.29 Write the atomic symbol for the element whose ion has a 2– charge, has 20 more neutrons than electrons, and has a mass number of 126.

2.30 In what region of the periodic table are you likely to find elements that form more than one stable ion?

2.31 Give the symbol, including the correct charge, for each of the following ions. **(a)** barium ion, **(b)** titanium (IV) ion, **(c)** phosphate ion, **(d)** hydrogen carbonate ion, **(e)** sulfide ion, **(f)** perchlorate ion, **(g)** cobalt (II) ion, **(h)** sulfate ion

2.32 An engineer is designing a water-softening unit based on ion exchange. Use the web to learn what ions typically are "exchanged" in such a system. Given that the ion exchanger cannot build up a large positive charge, what can you conclude about the relative numbers of the various ions involved?

2.33 Use the web to find a catalyst for a polymerization reaction that uses an ion. What are the apparent advantages of using this catalyst for creating the polymer?

2.34 Using Coulomb's law, explain how the difference between attractive and repulsive interactions between ions is expressed mathematically.

Compounds and Chemical Bonds

2.35 How many atoms of each element are represented in the formula $Ba(OH)_2$?

2.36 Which of the following formulas contains the most hydrogen atoms? C_2H_6, $(NH_4)_2CO_3$, H_2SO_4, or $Fe(OH)_3$.

2.37 In general, how are electrons involved in chemical bonding?

2.38 What is the difference between an ionic bond and a covalent bond?

2.39 When talking about the formula of an ionic compound, why do we typically refer to a formula unit rather than a molecule?

2.40 Which formula below is correct for an ionic compound? What is incorrect about each of the others? **(a)** $AlO_{3/2}$, **(b)** Ca_2Br_4, **(c)** $Mg(PO_4)_{3/2}$, **(d)** $BaCO_3$

2.41 Explain the differences between ionic and metallic bonding.

2.42 Conduction of electricity usually involves the movement of electrons. Based on the concept of metallic bonding, explain why metals are good conductors of electricity.

2.43 Describe how a covalently bonded molecule is different from compounds that are either ionic or metallic.

2.44 Explain the difference between a molecular formula and an empirical formula.

2.45 Why are empirical formulas preferred for describing polymer molecules?

2.46 The molecular formula for the ethylene monomer is C_2H_4. What is its empirical formula?

2.47 Polybutadiene is a synthetic elastomer, or rubber. The corresponding monomer is butadiene, which has the molecular formula C_4H_6. What is the empirical formula of butadiene?

The Periodic Table

2.48 What distinguished the work of Mendeleev that caused scientists to accept his concept of the periodic table when others before him were not believed?

2.49 How does the periodic table help to make the study of chemistry more systematic?

2.50 What is a period in the periodic table? From what does it derive its name?

2.51 How do binary compounds with hydrogen illustrate the concept of periodicity?

2.52 Of the following elements, which two would you expect to exhibit the greatest similarity in physical and chemical properties? Cl, P, S, Se, Ti. Explain your choice.

2.53 Name the group to which each of the following elements belongs. **(a)** K, **(b)** Mg, **(c)** Ar, **(d)** Br

2.54 What are some of the physical properties that distinguish metals from nonmetals?

2.55 Identify the area of the periodic table in which you would expect to find each of the following types of elements: **(a)** a metal, **(b)** a nonmetal, **(c)** a metalloid

2.56 Why are nonmetals important even though they account for only a very small fraction of the elements in the periodic table?

2.57 What is a metalloid?

2.58 A materials engineer has filed for a patent for a new alloy to be used in golf club heads. The composition by mass ranges from 25 to 31% manganese, 6.3 to 7.8% aluminum, 0.65 to 0.85% carbon, and 5.5 to 9.0% chromium, with the remainder being iron. What are the maximum and minimum percentages of iron possible in this alloy? Use Figure 2.12 to make a prediction about how the density of this alloy would compare with that of iron; justify your prediction.

2.59 Classify the following elements as metals, metalloids, or nonmetals. **(a)** Si, **(b)** Zn, **(c)** B, **(d)** N, **(e)** K, **(f)** S

2.60 A materials engineer wants to make a new material by taking pure silicon and replacing some fraction of the silicon atoms with other atoms that have similar chemical properties. Based on the periodic table, what elements probably should be tried first?

Inorganic and Organic Chemistry

2.61 The chemistry of main group elements is generally simpler than that of transition metals. Why is this so?

2.62 Calcium and fluorine combine to produce ionic calcium fluoride, CaF_2. Use the periodic table to predict at least two other compounds that you would expect to have structures similar to that of CaF_2.

2.63 What is meant by the phrase *organic chemistry*?

2.64 Based on what you've learned in this chapter, would you classify the chemistry of polymers as organic or inorganic? Why?

2.65 What is a *functional group*? How does the concept of the functional group help to make the study of organic chemistry more systematic?

2.66 The molecule shown below is responsible for the smell of popcorn. Write the correct molecular formula for this compound.

N
O

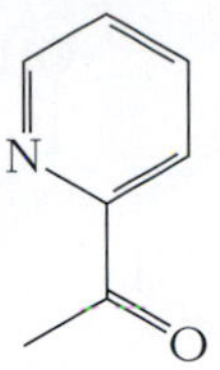

2.67 Not all polymers are formed by simply linking identical monomers together. Polyurethane, for example, can be formed by reacting the two compounds shown below with one another. Write molecular and empirical formulas for each of these two substances.

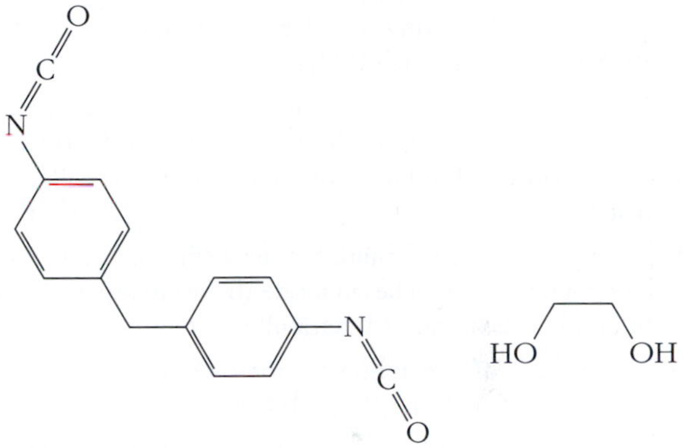

O
C
N
N
C
O
HO
OH

2.68 The compound shown below forms an amorphous solid (a glass) at room temperature and has been used as a medium for storing information holographically. Write the correct molecular formula for this molecule.

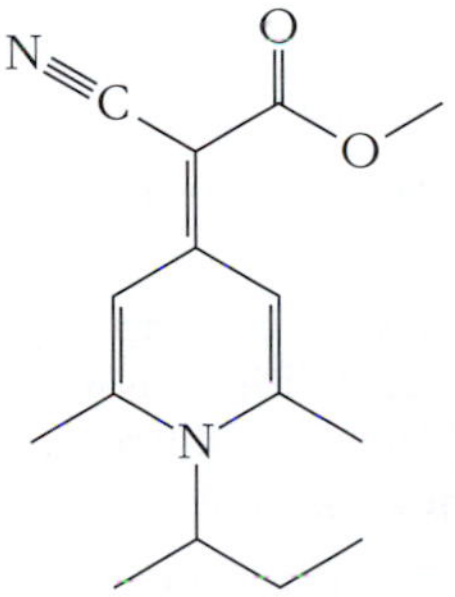

N
C
O
O
N

2.69 The accompanying figure shows the structure gamma-aminobutanoic acid, or GABA. This molecule is a neurotransmitter. Some of the effects of alcohol consumption are due to the interaction between ethanol and GABA. Write the correct molecular formula for this compound.

H_2N
OH
O

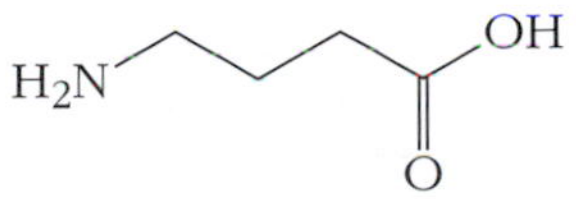

2.70 The figure below shows the structure of the adrenaline molecule. Write the correct molecular formula for this substance.

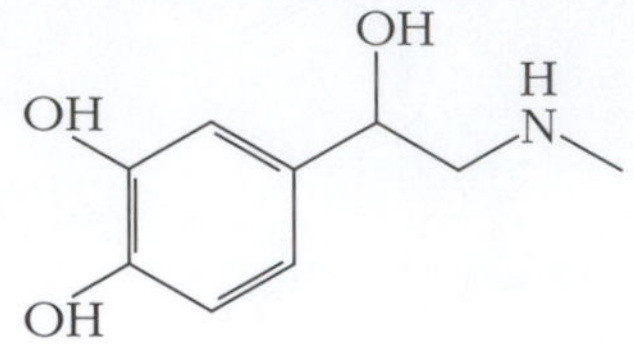

Chemical Nomenclature

2.71 Why are there different rules for naming covalent and ionic binary compounds?

2.72 Which binary combinations of elements are most likely to give ionic substances?

2.73 Name the following covalent compounds: **(a)** N_2O_5, **(b)** S_2Cl_2, **(c)** NBr_3, **(d)** P_4O_{10}

2.74 Give the formula for each of the following compounds: **(a)** sulfur dichloride, **(b)** dinitrogen pentaoxide, **(c)** silicon tetrachloride, **(d)** diboron trioxide (commonly called boric oxide)

2.75 Write the molecular formula for each of the following covalent compounds: **(a)** sulfur hexafluoride, **(b)** bromine pentafluoride, **(c)** disulfur dichloride, **(d)** tetrasulfur tetranitride

2.76 Name each of the following ionic compounds. **(a)** K_2S, **(b)** $CoSO_4$, **(c)** $(NH_4)_3PO_4$, **(d)** $Ca(ClO)_2$

2.77 Name each of the following compounds. **(a)** $MgCl_2$, **(b)** $Fe(NO_3)_2$, **(c)** Na_2SO_4, **(d)** $Ca(OH)_2$, **(e)** $FeSO_4$

2.78 Give the formula for each of the following ionic compounds. **(a)** ammonium carbonate, **(b)** calcium iodide, **(c)** copper (II) bromide, **(d)** aluminum phosphate, **(e)** silver (I) acetate

2.79 Name the following compounds. **(a)** PCl_5, **(b)** Na_2SO_4, **(c)** Ca_3N_2, **(d)** $Fe(NO_3)_3$, **(e)** SO_2, **(f)** Br_2O_5

INSIGHT INTO Polyethylene

2.80 What is a free radical? How are free radicals important in the formation of polyethylene?

2.81 How do molecules of low-density polyethylene and high-density polyethylene differ? How do these molecular scale differences explain the differences in the macroscopic properties of these materials?

2.82 Why do you think an initiator molecule is needed to induce the polymerization of ethylene?

2.83 Use the web to determine the amount of low-density polyethylene and high-density polyethylene produced annually in the United States. Which uses predominate in the applications of these two materials?

FOCUS ON PROBLEM SOLVING EXERCISES

2.84 Describe how you can identify the isotope, X, in this puzzle. The nucleus contains one more neutron than proton, and the mass number is nine times larger than the charge on the ion X^{3+}.

2.85 Many transition metals produce more than one ion. For example, iron has ions with charges of 2+ and 3+ that are both common. How could you use the compounds of a transition metal with oxygen to determine the charge of the metal ion? Use iron as your example.

2.86 Naturally occurring europium has an average atomic weight of 151.965 amu. If the only isotopes of europium present are ^{151}Eu and ^{153}Eu, describe how you would determine the relative abundance of the two isotopes. Include in your description any information that would need to be looked up.

2.87 Strontium has four stable isotopes. Strontium-84 has a very low natural abundance, but ^{86}Sr, ^{87}Sr, and ^{88}Sr are all reasonably abundant. Knowing that the atomic weight of strontium is 87.62, which of the more abundant isotopes predominates?

2.88 A candy manufacturer makes chocolate-covered cherries. Although all of the products look roughly the same, 3% of them are missing the cherry. The mass of the candy with a cherry is 18.5 g; those missing the cherry weigh only 6.4 g. **(a)** How would you compute the average mass of a box of 100 of these chocolate covered cherries from this manufacturer? **(b)** How is this question analogous to the determination of atomic weights?

2.89 Two common oxides of iron are FeO and Fe_2O_3. Based on this information, tell how you would predict two common compounds of iron and chlorine.

Cumulative Problems

2.90 Use a molecular level description to distinguish between LDPE and HDPE.

2.91 Engineers who design bicycle frames are familiar with the information in Table 1.3, which lists the densities of aluminum (2.699 g/cm^3), steel (7.87 g/cm^3), and titanium (4.507 g/cm^3). How does this information compare with Figure 2.12, and what would it suggest for changes in this figure if more shades were used for the density color-coding? (Iron is the principal component of steel.)

2.92 Use the web to look up the density of different forms of steel, such as stainless steel or magnetic steel, and discuss whether or not the differences in the densities follow what might be predicted by looking at the periodic properties of elements.

2.93 LDPE has a density in the range of 0.915–0.935 g/cm^3, and HDPE has a density in the range of 0.940–0.965 g/cm^3. You receive a small disk, 2.0 cm high with a 6.0-cm diameter, from a manufacturer of polyethylene, but its label is missing. You measure the mass of the disk and find that it is 53.8 g. Is the material HDPE or LDPE?

3

Molecules, Moles, and Chemical Equations

SimplyCreativePhotography/Getty Images

Ethanol production plants like this one are common throughout the midwestern United States. Such facilities are typically built near sources of corn, sugar beets, or other crops in order to minimize transportation costs.

CHAPTER OUTLINE

Today, roughly 20 million chemical compounds are known. More are being discovered or created as you read this. Although chemical nomenclature provides a systematic way to facilitate discussion of this tremendous diversity of substances, that is just a small step toward understanding chemistry. Much of the science of chemistry is aimed at understanding the reactions that produce or consume chemical substances. You may not realize it, but chemical reactions are central to many of the technologies that you use. Chemical reactions in batteries produce electricity for your iPhone® and laptop computer, and the burning of hydrocarbons powers your car. Other reactions, such as those designed to synthesize the light-filtering compounds used in LCD computer displays, play less obvious roles in technology. Over the course of this text we will explore many aspects of chemical reactions, and we will also look at a number of examples to show how chemical reactions can be put to good use through clever engineering. One promising technology for the future of engineering lies in utilizing natural reactions that have occurred on Earth for eons—using biomass to generate fuels.

Chapter Objectives

After mastering this chapter, you should be able to

- describe the chemical processes used in biomass production and conversion to biofuels.
- explain balancing a chemical equation as an application of the law of conservation of mass.
- list at least three quantities that must be conserved in chemical reactions.
- write balanced chemical equations for simple reactions, given either an unbalanced equation or a verbal description.
- explain the concept of a mole in your own words.
- interpret chemical equations in terms of both moles and molecules.
- interconvert between mass, number of molecules, and number of moles.
- determine a chemical formula from elemental analysis (i.e., from % composition).
- define the concentration of a solution and calculate the molarity of solutions from appropriate data.
- calculate the molarity of solutions prepared by dilution or calculate the quantities needed to carry out a dilution to prepare a solution of a specified concentration.
- distinguish between electrolytes and nonelectrolytes and explain how their solutions differ.
- describe the species expected to be present (ions, molecules, etc.) in various simple solutions.
- recognize common strong acids and bases.
- write molecular and ionic equations for acid–base neutralization reactions.

INSIGHT INTO

3-1 Biomass and Biofuel Engineering

It may be hard to imagine that significant energy resources for the future may depend on chemistry that has existed since the dawn of life on Earth. But that's one way to view the increasing role of biomass in the way we think about energy. The term **biomass** usually refers to biological material from plants. The potential of biomass arises from the fact that vast amounts of energy from the sun impinge on the earth every second. Despite the growing importance of various solar technologies, the majority of that energy goes unused. And the most prominent mechanism for chemical capture of solar irradiance is photosynthesis in plants. Plants use sunlight to create sugars during growth, and in that process we can think of energy as being stored in the sugar molecules. But common experience tells us we can't use those sugars directly for powering vehicles. So if we want to understand the potential role of biomass in meeting the world's energy needs, we will need to consider both the growth of plants and the engineering challenges of turning plant biomass into usable fuels.

Like nearly all living things, plants are made up mostly of water. If we ignore that component, though, the majority of the mass of plants is due to carbon. And that carbon is obtained from carbon dioxide in the air. Thus the production of biomass by plants is primarily a process where molecules that contain a single carbon atom, CO_2, are transformed into carbohydrate molecules containing five or six carbon atoms. Plants obtain the energy needed to complete this task by harvesting sunlight. The net result of phtosynthesis is that carbon dioxide and water are transformed into sugars and oxygen. There are numerous steps in this process, but we might ask whether or not there are simple ways to represent these results. In order to discuss chemical reactions, we will have to be able to depict those reactions succinctly in a

Carbohydrate molecules contain carbon, hydrogen, and oxygen in ratios that often appear to be simple integers of C and H_2O, which is the historical source of the name.

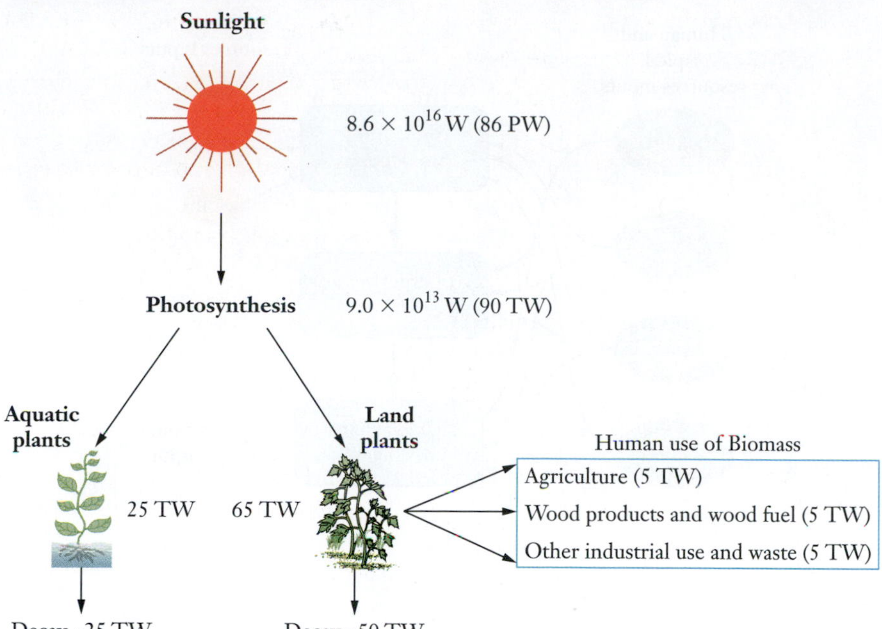

Figure 3.1 ■ Light impinging on Earth from the sun supplies 86 PW of power, of which 90 TW is taken in during photosynthesis in land- and water-based plants. Current human usage of biomass accounts for roughly 15 TW.

way that is easily understood. To consider the potential role of biomass as a source of energy, we will also need to look at these reactions quantitatively. How much biomass is produced annually, and how does it funnel into the human economy? **Figure 3.1** provides some numbers about the scale of this worldwide process. The influx of solar energy is roughly 86 petawatts (PW). The biomass produced by photosynthesis has the potential to supply 90 terrawatts (1 TW = 10^{12} W) of power, and current human usage of biomass accounts for about 15 TW.

1 W = 1 J/s. And we saw in Table 1.2 that peta- is the prefix corresponding to 10^{15}.

The chemical basis of plant biomass production is the photosynthetic formation of sugars. That doesn't mean, however, that commercial-scale production of biomass can rely on plants and sunlight alone. Modern agricultural methods can significantly improve the amount of biomass produced through appropriate use of fertilizers and pesticides on crops. How do chemists and agricultural engineers know what chemicals are needed for these tasks? How are the amounts of the chemicals that are needed for industrial-scale production produced? Once again, to start understanding these questions we need to learn more about chemical reactions.

Another major hurdle for the long-term utility of biomass as an energy solution lies in the conversion of biomass from plants into fuels that can be used for energy production in society. If we stop and think about the nature of fuels, both the ability to refine and transport them to locations where they can be used represent vital aspects of any economically viable fuel. It's certainly possible to burn wood, the longest-standing use of a biomass fuel, but modern technology usually requires fuels with both a higher energy density and a cleaner use cycle than the soot-laden burning of wood. Therefore, considerable research effort is underway to explore ways to convert biomass into fuels whose molecules are similar to those obtained from traditional fossil fuel mining and refining. Such fuels derived from biomass are collectively referred to as **biofuels**. Just what are the chemical choices available for the conversion of biomass and how do we understand them? Again, we will need to consider a variety of chemical reactions to start to answer these questions.

Energy density takes into account both the energy that can be produced from a fuel and the fuel's mass. We'll explore that further in Chapter 9.

Finally, we also have to recognize that the infrastructure for using fossil fuel–based energy has been established for over a century. Any new energy-harvesting concept therefore must be able to compete economically with that existing infrastructure to be

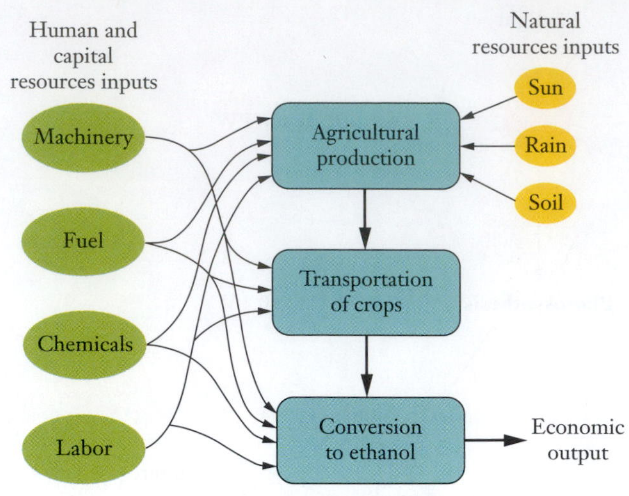

Figure 3.2 ■ This schematic diagram shows the important considerations in an economic assessment of biofuels. To offer a viable alternative to fossil fuels, the overall cost of biofuels must be competitive. Factors that need to be accounted for include the costs of producing the crops that supply the biomass (land, irrigation, fertilizer, labor, etc.), cost for the conversion from crops to usable biofuel, and transportation costs.

a viable alternative. Biomass-based energy is no exception to this fundamental rule of modern economics. **Figure 3.2** depicts some of the important economic factors in the overall process of biofuel production. It is clear from this figure that there are a large number of inputs into the process of producing a biofuel. Engineering challenges to make these processes more efficient are vital steps to the long-term prospects of meaningful biofuel usage. The connection between the chemical reactions needed to grow biomass and convert it to biofuels and the engineering solutions required to have that process reliable at the large scales needed provide important questions for us to consider as we start looking at chemical reactions in more detail.

3-2 Chemical Formulas and Equations

Although the possible routes from biomass to biofuels vary considerably, all involve chemical reactions. A realistic description of the chemistry involved in biomass conversion would require us to consider many individual reactions. To discuss the science behind these processes, we must be able to describe a chemical reaction concisely. In Chapter 2, we've already seen how chemical formulas provide a concise way to represent chemical compounds, and we will now describe how chemical equations build on this to accomplish the same goal for chemical reactions.

Writing Chemical Equations

Chemical equations are designed to represent the transformation of one or more chemical species into new substances. To ensure that their meanings are clear, we follow a set of conventions when writing chemical equations. Each chemical equation has two sides, and we usually envision the reaction as proceeding from left to right. The original materials are called the **reactants** and they appear on the left-hand side of the equation. The compounds that are formed from the reaction are called **products** and appear on the right-hand side of the equation. An arrow is used to represent the changes that occur during the reaction. Thus we can write a completely generic chemical equation:

$$\text{Reactants} \rightarrow \text{Products}$$

This would usually be read as "reactants go to products" or "reactants give products."

We use chemical formulas to identify the specific reactants and products. The physical states of the compounds are often designated; (s) indicates a solid, (ℓ) a liquid, (g) a

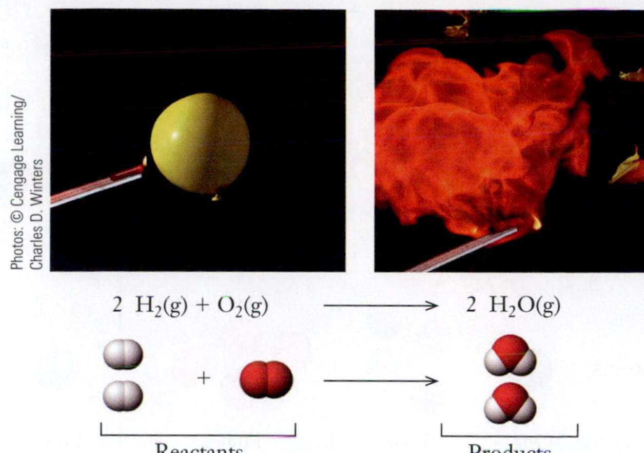

Figure 3.3 ■ A balloon filled with a mixture of H_2 and O_2 explodes in a classroom demonstration. The gas mixture in the balloon is stable until ignited by the flame from a candle. Once ignited, the balloon explodes in a ball of flame as the reaction takes place. The accompanying microscopic-scale illustration shows the molecular species present before and after reaction.

gas, and (aq) a substance dissolved in water. We can write the reaction of hydrogen with oxygen, which is often done as a lecture demonstration, as a simple example:

$$2\ H_2(g) + O_2(g) \rightarrow 2\ H_2O(g)$$

When a substance dissolves in water, the result is called an aqueous solution, and the "aq" label is short for aqueous.

All substances involved in this equation are gases, as indicated by the "(g)" notation. One characteristic of this reaction is that it will only occur at high temperatures. As seen in the photos in **Figure 3.3**, the mixture of H_2 and O_2 is stable unless ignited. If our written equations are going to describe fully the chemical processes they represent, then we need ways to indicate such conditions in addition to identifying the chemicals involved. Generally, this is done by placing a symbol over the arrow. Here we will introduce two such symbols: one for heat and one for light. A reaction that requires heat (or high temperatures) is indicated with a Δ (the Greek letter delta) above the arrow, and one that is initiated by light energy is designated with an $h\nu$. We show examples of these symbols below:

The symbol $h\nu$ is used because it represents the energy of the light, as we will learn in Chapter 6.

$$2\ H_2(g) + O_2(g) \xrightarrow{\Delta} 2\ H_2O(g)$$

$$H_2(g) + Cl_2(g) \xrightarrow{h\nu} 2\ HCl(g)$$

The first equation is a more complete description of the reaction between hydrogen and oxygen. In the second equation, light energy is used to initiate the reaction of hydrogen with chlorine, producing hydrogen chloride. Such a light-induced process is called a **photochemical reaction**.

Balancing Chemical Equations

If you look carefully at the reactions we've been writing, you'll notice more than just the formulas and the states of the substances in the equation. Numerical information about the relative amounts of the substances involved is also given. For example, we included the number "2" in front of both the $H_2(g)$ and $H_2O(g)$ in the equation for the reaction between hydrogen and oxygen. Why are these numbers there, and what do they mean? If you've studied chemistry before in high school, you probably remember that these coefficients are often needed to *balance* a chemical equation.

The underlying premise of the chemical equation is that it is a written representation of a chemical reaction. So any reasonable representation must be consistent with all of our observations of the actual reaction. One of the most fundamental laws of nature is **the law of conservation of matter**: matter is neither created nor destroyed. If we specifically exclude nuclear reactions from our consideration, this law can be phrased more specifically. Atoms of one element are neither created nor destroyed in a chemical reaction. A chemical reaction simply rearranges the atoms present into new compounds. In its written representation of nature,

If we also consider nuclear reactions, we must acknowledge that matter and energy can be interconverted. Nuclear chemistry is discussed in Chapter 14.

Figure 3.4 ■ The chemical reaction for the burning of methane (CH_4) in oxygen illustrates the concept of atom balance for chemical equations. The equation is shown first in symbolic form, with each compound represented by its chemical formula. The individual molecules are pictured in the center row and then are broken down into their constituent atoms. The number of atoms of each element is the same on the left- and right-hand sides.

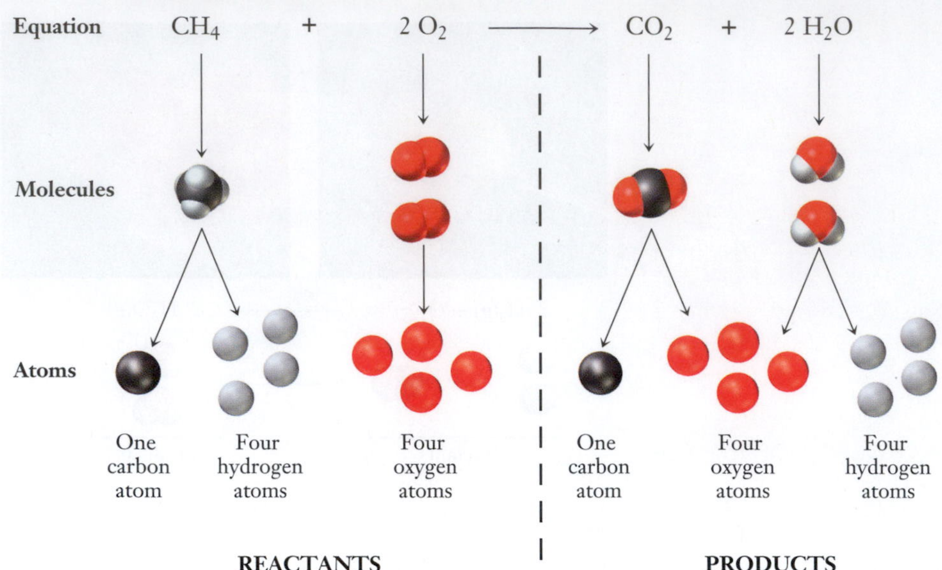

therefore, the chemical equation must not "create or destroy" atoms. To uphold this condition, we must have the same number of atoms of each element on both sides of the chemical equation (see **Figure 3.4**). An equation that does not meet this condition cannot accurately represent the observed chemical reaction and is said to be *unbalanced*.

In many cases, the most efficient way to balance the equations is "by inspection," which really means by trial and error. Even when this exploratory method is used, however, some systematic strategies can help make the balancing process easier. In all cases, the way we obtain balance is to introduce numbers in front of appropriate formulas in the equation. Chemists use the term **stoichiometry** to refer to the various quantitative relationships between the amounts of reactants and products in a chemical reaction. So the numbers used to balance a chemical equation are called **stoichiometric coefficients**. The stoichiometric coefficient multiplies the number of atoms of each element in the formula unit of the compound that it precedes. Example Problem 3.1 shows one strategy for balancing a common class of reactions, the combustion of hydrocarbon fuels.

EXAMPLE PROBLEM 3.1

Propane, C_3H_8, is used as a fuel in many applications, including gas barbecue grills. Because of its widespread use, extensive research is underway to develop ways to produce propane from biomass. When propane burns, it combines with oxygen, O_2, to form carbon dioxide and water. Write a balanced chemical equation describing this reaction.

Strategy This problem requires two steps. First, we must read the problem and determine which substances are reactants and which are products. This will allow us to write an unbalanced "skeleton" equation. Then, we must proceed to balance the equation, making sure that the same number of atoms of each element appears on both sides.

Solution

Step 1: The problem notes that propane burns by combining with oxygen. The reactants are therefore C_3H_8 and O_2. We also are told that the products are CO_2 and H_2O. So we have enough information to write the unbalanced equation. (Here we will write "blanks" in front of each formula to emphasize that we still need to determine the coefficients.)

$$\_\_\_C_3H_8 + \_\_\_O_2 \longrightarrow \_\_\_CO_2 + \_\_\_H_2O$$

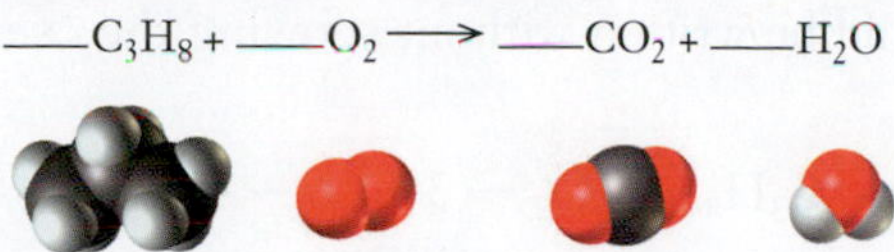

Step 2: Balancing an equation like this is aided by making some observations. In this case, both carbon and hydrogen appear in only one place on each side of the equation. Our first steps will be to balance these two elements, because there will be no other way to adjust them. Let's begin with carbon. (This choice is arbitrary.) To achieve the required balance, we can use stoichiometric coefficients of one for C_3H_8 and three for CO_2, giving three carbon atoms on each side of the equation. At this point, we have

$$1\ C_3H_8 + \_\_\_O_2 \longrightarrow 3\ CO_2 + \_\_\_H_2O$$

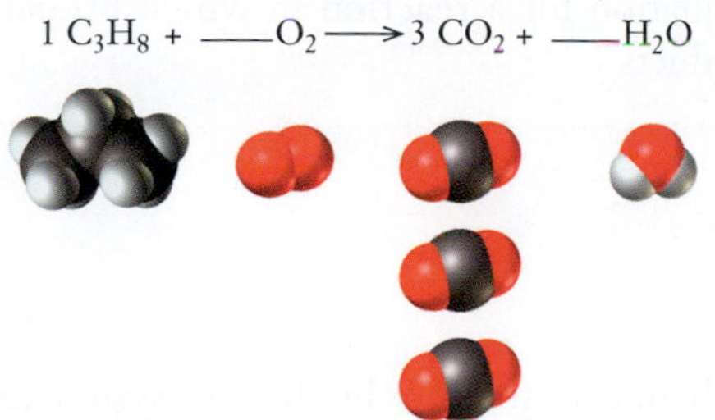

(The coefficient of one in front of the propane would normally be omitted, but it is written explicitly here for emphasis.) Next, we will balance hydrogen. The propane already has its coefficient set from our work on carbon, so it dictates that there are eight H atoms to be accounted for by water on the product side of the equation. We need to insert a coefficient of four to achieve this balance:

$$1\ C_3H_8 + \_\_\_O_2 \longrightarrow 3\ CO_2 + 4\ H_2O$$

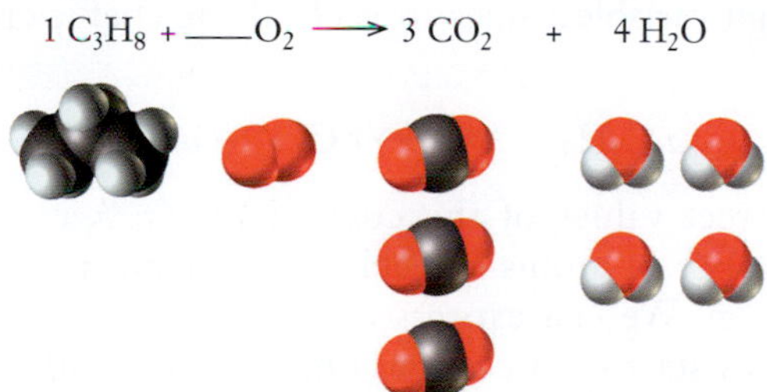

This leaves us with only oxygen to balance. We have saved this for last because it appears in three compounds rather than two and also because it appears as an element rather than a compound on the left side of the equation, which makes it easy to adjust the number of oxygen atoms without upsetting the balance of the other elements. Because there are ten O atoms on the product side of the equation, we need ten on the reactant side. We can easily achieve this by using $10/2 = 5$ for the stoichiometric coefficient of the O_2 reactant. This gives us the following balanced chemical equation:

$$1\ C_3H_8 + 5\ O_2 \longrightarrow 3\ CO_2 + 4\ H_2O$$

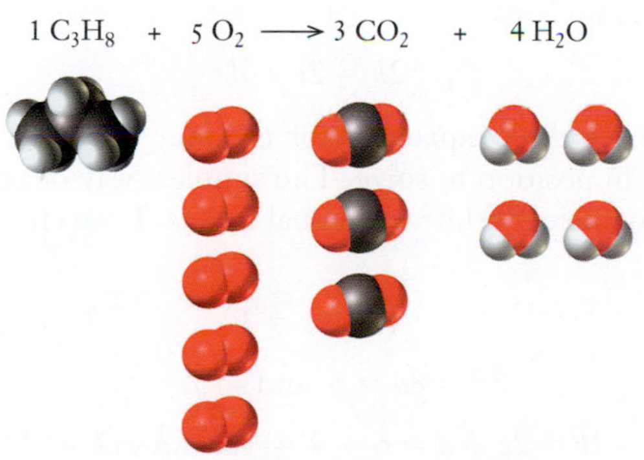

Normally, this would be written without showing the coefficient of one in front of the propane:

$$C_3H_8 + 5\ O_2 \rightarrow 3\ CO_2 + 4\ H_2O$$

Analyze Your Answer Our final equation has three carbon atoms, eight hydrogen atoms, and ten oxygen atoms on each side, so it is balanced.

Check Your Understanding The reaction we wrote above describes the complete combustion of propane. Under many conditions, though, hydrocarbons such as propane may not burn cleanly or completely, and other products may be formed. One such possible reaction is the formation of formaldehyde (CH_2O) and water. Write a balanced chemical equation for a reaction in which propane and oxygen gas form CH_2O and H_2O as products.

MathConnections

Usually, we balance chemical equations by the part systematic, part trial-and-error approach presented in Example 3.1. But it is fairly simple to transform that process into a more mathematical form, and doing so may help to emphasize the connection between balancing equations and the laws of conservation of matter. Let's illustrate the mathematical approach by reconsidering the equation we balanced in Example Problem 3.1:

$$\_\_\_\_C_3H_8 + \_\_\_\_O_2 \rightarrow \_\_\_\_CO_2 + \_\_\_\_H_2O$$

It will help to assign variable names to each of the coefficients in the balanced equation:

$$a\ C_3H_8 + b\ O_2 \rightarrow c\ CO_2 + d\ H_2O$$

When the correct values of the coefficients (a, b, c, d) are inserted, we know that the number of atoms of each element must be the same on both sides of the equation. We can express that idea as an equation for each element involved. Let's start with carbon. On the left-hand side, we have three carbon atoms for each molecule of C_3H_8, for a total of $3a$ carbon atoms. On the right side, we have c carbon atoms. So we can set these two terms equal to one another:

$$3a = c$$

Next, we can write similar equations for the other elements. Looking at hydrogen, we get

$$8a = 2d$$

And for the oxygen, we have

$$2b = 2c + d$$

At this point, we have three equations for the four unknown coefficients, so we don't appear to be in position to solve. The simplest way out of this predicament is just to set one coefficient arbitrarily equal to one. If we choose to set $a = 1$, we get the following:

$$c = 3a = 3$$

$$2d = 8a = 8, \text{ and so } d = 4$$

$$2b = 2c + d = 6 + 4 = 10, \text{ and so } b = 5$$

Inserting these into the original equation gives us a balanced equation:

$$C_3H_8 + 5\ O_2 \rightarrow 3\ CO_2 + 4\ H_2O$$

Because all of the resulting coefficient values are integers, we have found the smallest whole number coefficients that balance the equation, and our result is identical to that found in Example 3.1. If any fractional coefficients were obtained, we would simply multiply through by a constant to get whole numbers. This approach is quite general, and although it is not particularly popular with chemists, it has been used to write computer algorithms for balancing chemical equations.

For most people, the best way to learn how to balance chemical equations is just to practice until you gain confidence. The rules encountered for balancing equations are usually lists of steps that are prohibited rather than maps to be followed. For any reaction, there is more than one way to obtain the balanced equation, though some methods may be dramatically more difficult than others. A look at some common difficulties encountered by students new to balancing chemical equations may prove useful in developing successful strategies.

One method that might appear to provide a route to the desired balanced chemical equation is to change the formulas of the molecules involved. Consider, for example, formaldehyde (CH_2O) that might be formed in the burning of biomass or the incomplete combustion of a hydrocarbon like propane. As long as there is still oxygen available, any formaldehyde that is formed along the way can also burn. If CH_2O burns incompletely, it will form carbon monoxide (CO) and water:

$$CH_2O + O_2 \rightarrow CO + H_2O$$

Counting the number of atoms on each side of the equation reveals that we need one more oxygen on the product side. Many students find it tempting to write

$$CH_2O + O_2 \rightarrow CO_2 + H_2O$$

This is undeniably a balanced chemical equation. In making this change, however, the entire meaning of the equation has been altered. Although this is a balanced equation, it does not depict the same reaction that we set out to describe. Carbon monoxide is no longer shown as a product, so this is *not* a valid way to balance the equation. Similarly, we can't write the oxygen as O rather than O_2, because the formaldehyde is reacting with oxygen *molecules*, not oxygen *atoms*. We may not change the chemicals involved in the equation, only their stoichiometric coefficients. So let's take another look at our correct but unbalanced equation:

$$CH_2O + O_2 \rightarrow CO + H_2O$$

We still have two O atoms on the right and three on the left. One possible way to balance this equation is to multiply the O_2 reactant by 1/2, so that there will be two oxygen atoms on each side:

$$CH_2O + \frac{1}{2}O_2 \rightarrow CO + H_2O$$

Generally, it is preferable to avoid using fractional coefficients because they could be interpreted as implying the existence of fractional molecules. But we can easily eliminate any fractions by multiplying the entire equation by a factor that removes them. In this case multiplying by 2 will do the trick, converting the 1/2 to 1. This brings us to a final form of the balanced equation:

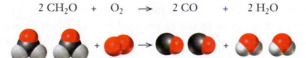

$$2\ CH_2O + O_2 \rightarrow 2\ CO + 2\ H_2O$$

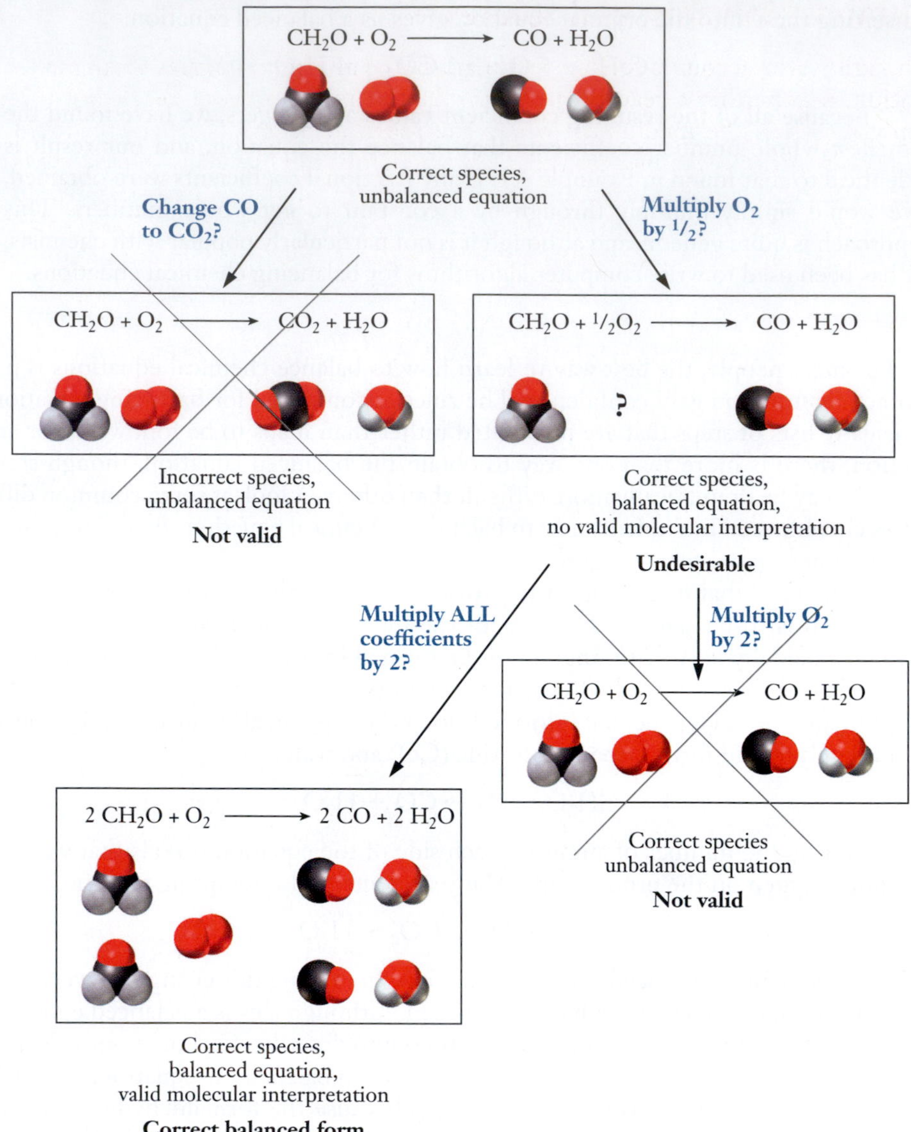

Figure 3.5 ■ The molecular drawings here illustrate some common errors in balancing chemical equations.

Note that we must multiply *all* of the stoichiometric coefficients by 2, *not* just the coefficient 1/2 in front of the oxygen. Otherwise, the balance that we have worked to establish would be destroyed. Some of these ideas are depicted pictorially in **Figure 3.5**.

3-3 Aqueous Solutions and Net Ionic Equations

For a chemical reaction to occur, the reactants involved must be able to come into contact with one another. In the gas or liquid phase, the molecules that make up the reactants move readily, allowing this contact to occur. In the solid phase, though, such motion is uncommon and reactions occur only very slowly if at all. One way to enable the needed contact between reactants that are solids under normal conditions is to dissolve them. Water is the most common medium for producing such solutions. Reactions that occur in water are said to take place in **aqueous solution**. To describe this important group of chemical reactions, first we need to define some key terms.

Solutions, Solvents, and Solutes

Although water accounts for the vast majority of all the molecules in an aqueous solution, it is neither a reactant nor a product in many aqueous reactions. Water simply serves as the medium in which the reaction occurs. The entire liquid is called a **solution**, meaning that it is a homogeneous mixture of two or more substances. In any solution, the component present in greater amounts—usually much greater amounts—is called the **solvent**. In an aqueous solution, water is the solvent. Minor components of the solution are called **solutes**. The key feature of a solution is that the solutes *dissolve* in the solvent. When a substance dissolves, its particles are dispersed in the solvent, and typically many solvent molecules will surround any individual molecule or ion of solute. Although water is the most common solvent, it is not the only one. Not all solutions are even liquids. The same ideas can be applied to homogeneous mixtures of gases or solids. Air is a good example of a gaseous solution, and alloys like brass can be described as solid solutions.

If we want to describe a pure substance, all we need to specify is its identity. If you pick up a bottle in the chemistry lab that is labeled "distilled water," for example, you would know exactly what is in the container. But if the same container says only that it contains a solution of salt (NaCl) dissolved in water, you can't be as sure what the contents are. Specifically, you would probably want to know something about the relative amounts of salt and water that were mixed to prepare the solution. So we need to specify at least one other key piece of information: the **concentration**. If there are many solute particles present, the solution is said to be *concentrated.* When there are few solute particles present, the solution is called *dilute.* **Figure 3.6** shows two $CuSO_4$ solutions with different concentrations, and you can easily see the difference in the color of the those solutions. In Section 3-5, we will discuss various units that can be used to report solution concentrations.

Some solutes can be dissolved in water to produce very concentrated solutions, whereas other substances do not dissolve to any measurable extent. So we can characterize various compounds in terms of their **solubility** in water. Those compounds that dissolve readily are said to be **soluble**, whereas those that do not dissolve significantly

Photos: Lawrence S. Brown

Figure 3.6 ■ In this sequence of photos, one of the authors prepares aqueous solutions of $CuSO_4$. In the upper left panel, solid $CuSO_4$—the solute—is transferred to a flask. In the upper right panel, water—the solvent—is added. The flask is shaken to speed the dissolution process (lower left). The final photo shows two $CuSO_4$ solutions of different concentrations. The solution on the left has the higher concentration, as seen from its darker color.

Table ■ 3.1

Solubility guidelines for ionic compounds in water at room temperature

Usually Soluble	Exceptions
Group 1 cations (Li^+, Na^+, K^+, Rb^+, Cs^+), ammonium (NH_4^+)	No common exceptions
Nitrates (NO_3^-), nitrites (NO_2^-)	Moderately soluble: $AgNO_2$
Chlorides, bromides, iodides (Cl^-, Br^-, I^-)	Insoluble: $AgCl$, Hg_2Cl_2, $PbCl_2$, $AgBr$, Hg_2Br_2, $PbBr_2$, AgI, Hg_2I_2, and PbI_2
Fluorides (F^-)	Insoluble: MgF_2, CaF_2, SrF_2, BaF_2, PbF_2
Sulfates (SO_4^{2-})	Insoluble: $BaSO_4$, $PbSO_4$, $HgSO_4$ Moderately soluble: $CaSO_4$, $SrSO_4$, Ag_2SO_4
Chlorates (ClO_3^-), perchlorates (ClO_4^-)	No common exceptions
Acetates (CH_3COO^-)	Moderately soluble: $AgCH_3COO$
Usually Insoluble	**Exceptions**
Phosphates (PO_4^{3-})	Soluble: $(NH_4)_3PO_4$, Na_3PO_4, K_3PO_4
Carbonates (CO_3^{2-})	Soluble: $(NH_4)_2CO_3$, Na_2CO_3, K_2CO_3
Hydroxides (OH^-)	Soluble: $LiOH$, $NaOH$, KOH, $Ba(OH)_2$ Moderately soluble: $Ca(OH)_2$, $Sr(OH)_2$
Sulfides (S^{2-})	Soluble: $(NH_4)_2S$, Na_2S, K_2S, MgS, CaS

In Chapter 12, we will see that solubility is not an "either/or" process, but that there is continuous variability. These rules summarize the main quantitative tendencies we will learn then.

are called **insoluble.** Chemists often use a series of *solubility rules* to predict whether or not a particular compound is likely to dissolve in water. The most common rules are summarized in **Table 3.1**, and Example Problem 3.2 illustrates their use.

EXAMPLE PROBLEM 3.2

Which of the following compounds would you predict to be soluble in water at room temperature? (a) $KClO_3$, (b) $CaCO_3$, (c) $BaSO_4$, (d) $KMnO_4$

Strategy Solubility guidelines for common ions are given in Table 3.1. So we will identify the ions in each compound and consult the table as needed to determine the solubilities.

Solution

(a) $KClO_3$ is potassium chlorate. From the solubility guidelines in Table 3.1, we see that compounds containing K^+ and ClO_3^- tend to be soluble and that no common exceptions are mentioned. So we predict that $KClO_3$ should be soluble.

(b) $CaCO_3$ is calcium carbonate. Again consulting the table, we see that carbonates are generally insoluble and that $CaCO_3$ is not listed among the exceptions. Thus $CaCO_3$ should be insoluble.

(c) $BaSO_4$ is barium sulfate. Although most sulfates are soluble, $BaSO_4$ is listed in the table as an exception to that rule. Therefore we expect $BaSO_4$ to be insoluble.

(d) $KMnO_4$ is potassium permanganate. Although the permanganate ion (MnO_4^-) is not listed in Table 3.1, the table does tell us that all compounds of K^+ are soluble. So we would predict that $KMnO_4$ should be soluble.

Discussion Here we simply consulted the table to check the solubility of each compound. Chemists generally gain familiarity with many of these solubility rules and learn to recognize soluble and insoluble salts without consulting such a table. You should check with your instructor to see whether you are expected to memorize these rules and exceptions.

Check Your Understanding Which of the following compounds would be soluble in water at room temperature? (a) NH_4Cl, (b) KOH, (c) $Ca(CH_3COO)_2$, (d) $Ba_3(PO_4)_2$

Although these rules imply that solubility is a simple yes or no question, the reality is more complicated than that. Saying that a compound is soluble does not mean that we could dissolve limitless amounts of it in a small beaker of water. If we keep adding more of the solute, eventually we will observe that the added material does not dissolve. When this occurs we have established a *saturated* solution. The concentration at which a given solution will become saturated depends on the identities of the solute and solvent, so it is useful to have a quantitative measure of solubility. The units used to report this number can vary, but one common choice is mass of solute per 100 g of solvent. For example, the solubility of table salt in water at room temperature is 35.7 g NaCl/100 g H_2O. This corresponds to about one-third of a cup of salt in a cup of water. Notice that solubility gives us a ratio, which we could use in the same way as we did the mass density in Example Problem 1.6 on page 21.

One final observation to make about solutions is that many ionic compounds will dissociate into individual ions when they dissolve in water. Thus the salt solution above actually contains Na^+ and Cl^- ions, not NaCl molecules. The availability of freely moving charges allows these solutions to conduct electricity. Any substance that dissolves in water to produce an aqueous solution that conducts electricity is called an **electrolyte**. Substances whose solutions do not conduct electricity are called **nonelectrolytes**. We can divide electrolytes further into two groups. **Strong electrolytes** dissociate or ionize completely, so that only individual ions are present in the solution, with virtually no intact molecules. In contrast, **weak electrolytes** dissociate or ionize only partially; their solutions contain both intact molecules and individual ions in measurable quantities. **Figure 3.7** shows the differences between these classes of solutes.

Chemical Equations for Aqueous Reactions

We can describe the process by which a compound dissolves in water by a chemical equation. When a covalently bonded material such as sugar dissolves in water, the molecules remain intact:

$$C_6H_{12}O_6(s) \rightarrow C_6H_{12}O_6(aq)$$

By contrast, when an ionic solid dissolves in water, it breaks into its constituent ions. This is called a **dissociation reaction**. The dissolving of sodium chloride discussed above is a common example of this process:

$$NaCl(s) \rightarrow Na^+(aq) + Cl^-(aq)$$

In both of these chemical equations, the water molecules are not shown explicitly, although their presence is indicated by the "(aq)" labels on the product side. This omission reflects the general tendency of a solvent to dissolve a solute but not to react chemically with it. When water appears in a chemical equation, the reaction involves water as either a reactant or product.

In addition to these fairly simple reactions by which we describe compounds dissolving in water, many important reactions take place in water. The chemical equations we write to describe these reactions can be written in any of three forms; the choice of

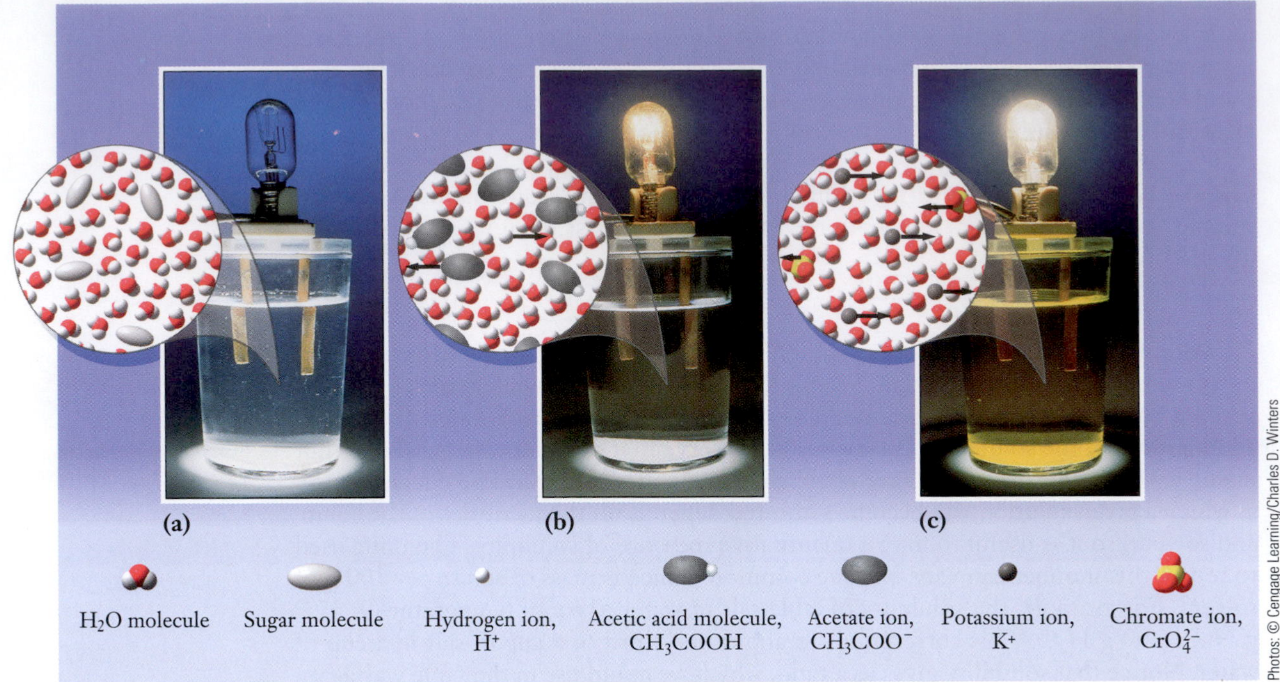

Figure 3.7 ■ The photos show a classroom demonstration in which a pair of copper rods is dipped into different solutions. If the solution conducts electricity, a circuit is closed and the lightbulb lights. **(a)** Ordinary sugar ($C_{12}H_{22}O_{11}$) is a nonelectrolyte, so a solution of sugar dissolved in water does not conduct electricity. **(b)** Acetic acid (CH_3COOH) is a weak electrolyte. An acetic acid solution contains low concentrations of ions and conducts electricity well enough to light the bulb dimly. **(c)** Potassium chromate (K_2CrO_4) is a strong electrolyte. A potassium chromate solution contains higher concentrations of ions and conducts electricity well enough to light the bulb brightly. The illustrations accompanying each photo emphasize the solute species in each solution.

equation is based mainly on the context in which the equation is used. We'll illustrate these three types of equations for the synthesis of ammonium nitrate from ammonia and nitric acid. This reaction is important in the production of nitrogen-containing fertilizer, and such fertilizers are important contributors to increased biomass production from agriculture.

Ammonium nitrate is often the preferred nitrogen fertilizer for surface application and has the additional benefit of not altering the acidity (or pH) of the soil very much when applied. Unfortunately, ammonium nitrate can also be used to generate explosives and has considerable explosive potential itself. This potential has been exploited in the use of ammonium nitrate in bombings, including the 1995 attack on the Murrah Federal Building in Oklahoma City. This nefarious use of ammonium nitrate has led to more restrictive regulations for its purchase than for other nitrogen-containing fertilizers.

In the commercial preparation of ammonium nitrate, pure ammonia (NH_3) in the gas phase is combined with concentrated aqueous nitric acid (HNO_3). This produces a highly concentrated aqueous solution of ammonium nitrate, which is then dried and formed into small pellets (called prills). The **molecular equation** shows the complete formula of each compound involved. Both intact and dissociated compounds are shown, followed by the (aq) to designate an aqueous solution when appropriate:

$$HNO_3(aq) + NH_3(g) \rightarrow NH_4NO_3(aq)$$

Frequently, the compounds involved in aqueous chemistry are strong electrolytes like the HNO_3 and NH_4NO_3 in this example. These substances dissociate or ionize when they dissolve in water, so the actual solution does not contain intact molecules of these compounds. The presence of ions is made clear if we write the **total ionic equation** rather than the molecular equation. This form emphasizes what is actually present in the reacting mixture by writing strong electrolytes as separated ions in the solution. The total ionic equation for our example reaction from above is thus

$$H^+(aq) + NO_3^-(aq) + NH_3(g) \rightarrow NH_4^+(aq) + NO_3^-(aq)$$

The H^+ ion is rather special because it is present in all aqueous solutions of acids. Strictly speaking, an H^+ ion would be nothing more than a proton. But as we will see in more detail later, such a bare proton in aqueous solution would actually be tightly surrounded by water molecules. Thus the H^+ ion is frequently written as H_3O^+ to remind us of this fact. If we write the ion in this way, our total ionic equation becomes

$$H_3O^+(aq) + NO_3^-(aq) + NH_3(g) \rightarrow NH_4^+(aq) + NO_3^-(aq) + H_2O(\ell)$$

Note that now we must include a molecule of water on the right-hand side to balance the atoms that we added when we switched our representation of the H^+ ion from H^+ to H_3O^+.

Careful inspection of this equation shows that one ion, $NO_3^-(aq)$, appears on both sides. Thus it is not an active participant in the reaction. Ions that are uninvolved in the chemistry are referred to as **spectator ions**. In many instances, we are not particularly concerned with these spectator ions and need not include them in the chemical equation. ◀ℹ When the spectator ions are left out, the result is a **net ionic equation**. Using the same reaction as an example, we would have either

The process of removing the spectator ions from the equation is similar to what you would do if you encountered an algebraic equation that had the same term on both sides. ℹ

$$H^+(aq) + NH_3(g) \rightarrow NH_4^+(aq)$$

or

$$H_3O^+(aq) + NH_3(g) \rightarrow NH_4^+(aq) + H_2O(\ell)$$

This depends on whether we choose to write H^+ or H_3O^+ to represent the cation formed from the nitric acid. Because we have removed anions from both sides of the equation, we now have a net charge of 1+ on each side. This is one way to recognize that you are looking at a net ionic equation: there is often a nonzero net charge on each side of the equation. If the equation is properly balanced, though, the net charge on each side must be the same. This rule reflects the fact that electrical charge must be conserved.

At this point, you are probably wondering, "Which of these equations is the correct one?" The answer is that all of these forms provide valid descriptions of the reaction, so none is intrinsically "better" than the others. In this particular case, we would probably not choose the net ionic equation, because the identity of the anion (NO_3^-) is important if our aim is to produce ammonium nitrate for use as a fertilizer. But in many other cases, we might have little or no interest in the identity of a spectator ion, so the simplicity of the net ionic equation might make it attractive. The total ionic equation is more cumbersome and is used less often.

Acid–Base Reactions

Two especially important categories of aqueous solutions are acids and bases. Examples of them are easy to find in our everyday lives as well as in the chemical industry. For our present purposes, we will define an **acid** as any substance that dissolves in water to produce H^+ (or H_3O^+) ions, and a **base** as any substance that dissolves in water to produce OH^- ions. **Table 3.2** lists some common acids and bases. Like other solutes, acids and bases can be either strong or weak electrolytes. Strong acids or bases dissociate or ionize completely in water, so that the resulting solution contains

Table ■ 3.2

Strong and weak acids and bases

Strong Acids		Strong Bases	
HCl	Hydrochloric acid	$LiOH$	Lithium hydroxide
HNO_3	Nitric acid	$NaOH$	Sodium hydroxide
H_2SO_4	Sulfuric acid	KOH	Potassium hydroxide
$HClO_4$	Perchloric acid	$Ca(OH)_2$	Calcium hydroxide
HBr	Hydrobromic acid	$Ba(OH)_2$	Barium hydroxide
HI	Hydriodic acid	$Sr(OH)_2$	Strontium hydroxide
Weak Acids		**Weak Bases**	
H_3PO_4	Phosphoric acid	NH_3	Ammonia
HF	Hydrofluoric acid	CH_3NH_2	Methylamine
CH_3COOH	Acetic acid		
HCN	Hydrocyanic acid		

Note: *All* common strong acids and bases are shown, but only representative examples of weak acids and bases are listed.

essentially no intact solute molecules. We can write the following chemical equations for the dissolution of HCl and NaOH:

$$\text{Strong acid: } HCl(g) + H_2O(\ell) \rightarrow H_3O^+(aq) + Cl^-(aq)$$

$$\text{Strong base: } NaOH(s) \rightarrow Na^+(aq) + OH^-(aq)$$

Unlike H^+ ions, hydroxide ions are not represented as having combined with water from the solvent, so water doesn't appear in the equation for a strong base.

We will revisit the concepts of acids and bases in Chapter 12 and provide additional detail at that time.

The only common strong acids and bases are those shown in Table 3.2.

Weak acids or bases undergo only partial ionization, so that their solutions contain intact molecules as well as ions. When we write equations for the dissolution of weak electrolytes like these, we use a "two-way" arrow that emphasizes that the reaction does not proceed completely from left to right. For common weak acids, it is relatively easy to write the needed ionization equation. Many weak acids contain the —COOH functional group, and the H atom from that group tends to be lost in solution. Acetic acid is a good example:

$$\text{Weak acid: } CH_3COOH(aq) + H_2O(\ell) \rightleftarrows H_3O^+(aq) + CH_3COO^-(aq)$$

For weak bases, the situation is slightly less obvious. Unlike the strong bases listed in Table 3.2, most weak bases do not contain —OH groups. So it may not be obvious that dissolving them will produce hydroxide ions in solution (i.e., that they are bases at all). The most common weak base is ammonia (NH_3), which reacts in water according to the following equation:

$$\text{Weak base: } NH_3(aq) + H_2O(\ell) \rightleftarrows NH_4^+(aq) + OH^-(aq)$$

Again we use the two-way arrow to show that intact NH_3 molecules as well as NH_4^+ and OH^- ions will be present in the solution. Many other weak bases are amines. They can be thought of as derivatives of ammonia in which one or more of the H atoms have been replaced by methyl groups or longer hydrocarbon chains.

Because many acidic and basic solutions occur in nature, observations about acids and bases date back hundreds of years. One of the most important observations is that a solution cannot be both acidic and basic at the same time. Mixing an acid and a base leads to a reaction known as **neutralization**, in which the resulting solution is neither acidic nor basic. To understand the origins of this neutralization, we might start with the definitions of acid and base. Acid solutions contain H_3O^+ ions, and bases contain OH^- ions. So for a solution to be both an acid and a base simultaneously, it would need to contain both of these species. Looking at those ions, it should be easy to see why this is not feasible. Hydronium ions and hydroxide ions combine readily to form water:

$$H_3O^+(aq) + OH^-(aq) \rightarrow 2\ H_2O(\ell)$$

This reaction will take place whenever an acid and a base are combined and will always prevent a solution from being both acidic and basic simultaneously. Example Problem 3.3 looks at writing equations for a neutralization reaction.

EXAMPLE PROBLEM 3.3

When aqueous solutions of acetic acid and potassium hydroxide are combined, a neutralization reaction will occur. Write molecular, total ionic, and net ionic equations for this process.

Strategy This is the reaction of a weak acid with a strong base. The reaction between an acid and a base always produces water as a product, along with an ionic compound formed from the remaining ions. (This second product is often called a **salt**.) We can begin by using this idea to write the molecular equation. To generate the ionic equations, then, we account for the dissociation or ionization of strong electrolytes into their constituent ions.

Solution A hydrogen ion from the acid and the hydroxide ion from the base will produce water. The remaining ions will be an acetate anion (CH_3COO^- from the acid) and a potassium cation (K^+, from the base). Combining these gives us potassium acetate, KCH_3COO. This lets us write the molecular equation:

$$CH_3COOH + KOH \rightarrow H_2O + KCH_3COO$$

To produce the total ionic equation, we must determine which of these species to write as ions. According to Table 3.2, acetic acid is a weak acid. This means that it is only partially ionized in solution, and so we write it as an intact molecule. KOH, on the other hand, is a strong base, and so it will dissociate completely. So we write this as a pair of ions, K^+ and OH^-. According to Table 3.1, both potassium ions and acetate ions tend to produce soluble compounds. So, KCH_3COO will be soluble and hence will dissociate into its constituent ions. Putting all of this together lets us write the total ionic equation:

$$CH_3COOH(aq) + K^+(aq) + OH^-(aq) \rightarrow H_2O(\ell) + K^+(aq) + CH_3COO^-(aq)$$

Looking at this equation, we see potassium ions on both sides. Thus potassium is a spectator ion and can be deleted to give the net ionic equation:

$$CH_3COOH(aq) + OH^-(aq) \rightarrow H_2O(\ell) + CH_3COO^-(aq)$$

Check Your Understanding Write molecular, total ionic, and net ionic equations for the reaction between hydrochloric acid and calcium hydroxide.

A number of commercial products are available for cleaning up acid spills, and they all rely on neutralization. Typically, a mixture of compounds such as calcium carbonate, magnesium oxide, and sodium carbonate is employed, often together with a dye that changes color when the acid spill has been neutralized.

Neutralization is only one example of aqueous reaction chemistry. Another common type of reaction is the **precipitation reaction**, in which a solid (called the precipitate) is

Figure 3.8 ■ The photos show a demonstration in which clear and colorless solutions of KI and $Pb(NO_3)_2$ are mixed and react to form a precipitate of PbI_2. (The precipitate is bright yellow.) The accompanying illustration shows a molecular scale representation of the individual reactant solutions and the final product.

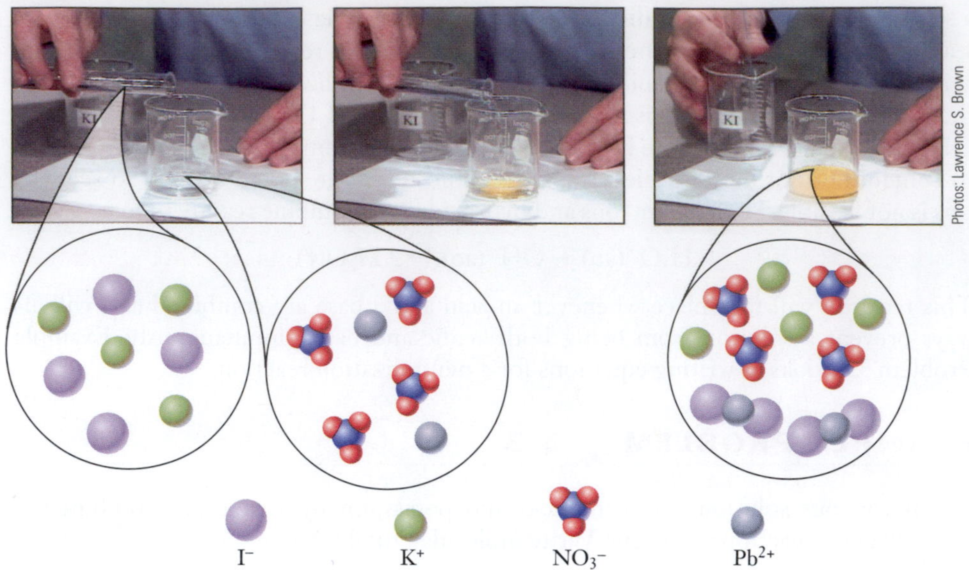

formed and drops out of solution as shown in **Figure 3.8**. Example Problem 3.4 provides an example of the three forms of chemical equations for precipitation reactions.

EXAMPLE PROBLEM 3.4

When aqueous sodium carbonate is added to barium chloride, the solution turns cloudy white with solid barium carbonate. Write the molecular, ionic, and net ionic equations for this reaction.

Strategy First, we will use the nomenclature rules that we presented in Chapter 2 to write formulas for all of the compounds involved. Then, we construct a balanced chemical equation using complete formula units. Next we must identify any strong electrolytes that are present, and we write them as dissociated ions to give the total ionic equation. Finally, to obtain the net ionic equation, we must eliminate any spectator ions from both sides of the equation.

Solution Formulas for the substances listed are Na_2CO_3, $BaCl_2$, and $BaCO_3$. Although $BaCO_3$ is the only product listed in the problem statement, the sodium and chlorine atoms must also appear somewhere on the right-hand side of our equation. So NaCl will also be produced as a product. We write reactants on the left and products on the right to set up the molecular equation and then balance:

$$Na_2CO_3(aq) + BaCl_2(aq) \rightarrow BaCO_3(s) + 2\ NaCl(aq)$$

Next, we recognize that the three substances denoted as "(aq)" in that equation will exist as dissociated ions rather than intact molecules. This leads us to the total ionic equation:

$$2\ Na^+(aq) + CO_3^{2-}(aq) + Ba^{2+}(aq) + 2\ Cl^-(aq) \rightarrow BaCO_3(s) + 2\ Na^+(aq) + 2\ Cl^-(aq)$$

Finally, we can cancel the sodium and chloride ions because they appear in equal numbers on both sides of the equation. This leads us to the net ionic equation:

$$CO_3^{2-}(aq) + Ba^{2+}(aq) \rightarrow BaCO_3(s)$$

Check Your Understanding When sodium sulfate, Na_2SO_4, reacts with lead nitrate, $Pb(NO_3)_2$, the products are solid lead sulfate and aqueous sodium nitrate. Write the molecular, ionic, and net ionic equations for this reaction.

Note that for a precipitation reaction, the net ionic equation does not have a net charge on either the reactant or product side.

3-4 Interpreting Equations and the Mole

The chemical equations that we have been learning to write are symbolic descriptions of chemical reactions. But to interpret these equations, we must think about them from another point of view, in terms of the actual substances and processes they represent. As often happens in chemistry, we can do this at either the microscopic or the macroscopic level. The microscopic interpretation visualizes reactions between individual molecules, and that interpretation is the one we have used thus far. The macroscopic interpretation pictures reactions between bulk quantities. Neither perspective is inherently better than the other; we simply use the point of view that is best suited to whatever circumstances we are considering. To make the connection between the two worlds, we need to look at the way chemists count the molecules in a large sample.

Interpreting Chemical Equations

In describing how to write and balance a chemical equation, we adopted a way of looking at reactions that seems natural. We spoke of one molecule reacting with another molecule to form some new compound that we called the product. Real chemical reactions rarely involve just one or two molecules, however, so we will need to think of chemical equations in terms of larger quantities, too. Let's return to our first example, the explosive reaction between hydrogen and oxygen to produce water:

$$2\ H_2(g) + O_2(g) \rightarrow 2\ H_2O(g)$$

We have read this chemical equation as saying that "two H_2 molecules react with one O_2 molecule to form two H_2O molecules." But suppose we carried out the same reaction starting with 20 H_2 molecules rather than two. The balanced equation says that we need one oxygen molecule for every two H_2 molecules. So in this case, we could say that 20 H_2 molecules react with 10 O_2 molecules to form 20 H_2O molecules.

Two key features of these statements must be emphasized. First, both statements are correct because they use the ratios established by the stoichiometric coefficients. There are twice as many hydrogen molecules as there are oxygen molecules in each case. Second, both statements refer to *numbers of particles*. When we interpret a chemical reaction, we must remember that the stoichiometric coefficients *always* refer to numbers of particles. In the example we just considered, the numbers were 2 and 1 or 20 and 10. We could just as easily have said 2,000,000 and 1,000,000 or 91,200,000 and 45,600,000. Provided we use numbers that satisfy the correct stoichiometric ratio, we will always have a reasonable description of an actual chemical reaction.

Avogadro's Number and the Mole

When we begin to consider chemical reactions with macroscopic amounts of materials, the numbers of molecules involved become staggeringly large. But people generally don't like to work with huge numbers, because it is very difficult to develop an intuitive feeling for them. So inevitably, when confronted with the fact that any real sample contains an almost preposterous number of molecules, scientists invented a way to make the numbers manageable. The simple way to do this is just to count the molecules in terms of some large quantity rather than individually. The particular quantity that chemists have chosen to count by is called the **mole**. A mole is defined as the number of atoms in *exactly* 12 grams of ^{12}C. This number is also referred to as **Avogadro's number**, and its value is 6.022×10^{23}. ◄ⓘ When chemists want to count molecules, they count by moles, so that the resulting numbers will be easier to work with.

Because of its fundamental importance, Avogadro's number has been measured extremely carefully. The internationally accepted value is actually $6.02214129 \times 10^{23}$. ⓘ

There are several reasons why this number is useful in chemistry. First, it is a large number that gives a useful unit for counting particles as small as atoms. In this sense, the mole is to a chemist what a dozen is to a baker. Second, a mole is the same number of particles no matter what substance we are talking about. So a mole of H_2 contains the same number of particles as a mole of mannose ($C_6H_{12}O_6$), a simple sugar found in plants; each has 6.022×10^{23} molecules. The masses or volumes of a mole will vary from one substance to another, as shown in **Figure 3.9**, but the number of particles present is the same in each case.

You may be thinking that the value of Avogadro's number seems like a rather peculiar choice for a counting unit. Why not just make it an even 10^{23}, for example, and simplify some calculations? The answer is that the chosen value of Avogadro's number produces a very convenient connection between the mass of an atom or molecule and the mass of a mole of those atoms or molecules. Our definition of the mole said that a mole of ^{12}C has a mass of exactly 12 g. Earlier, we said that an atom of ^{12}C has a mass of exactly 12 amu. This is not merely a coincidence, but rather a very deliberate decision. Once we have defined the amu and the gram independently, Avogadro's number is fixed as the number of amu in one gram. It may help to envision a hypothetical experiment: suppose that we had a balance that could accurately measure the mass of individual atoms and a way of loading atoms onto that balance one at a time. We could then determine Avogadro's number by loading carbon-12 atoms onto the balance, one by one, until the mass read exactly 12 grams. Of course, no such experiment is possible, so Avogadro's number has been measured by more indirect methods.

The mass of 6.022×10^{23} atoms of any element is the **molar mass** of that element, and its value in grams per mole is given in most periodic tables. The molar mass of silicon, for example, is greater than that of carbon because an atom of silicon is heavier than an atom of carbon. And the molar mass of each element takes into account the natural isotopic abundances. So the molar mass of carbon is 12.011 g/mol, reflecting the weighted average between the masses of ^{12}C and ^{13}C in the same way as we discussed in Chapter 2.

The mole is the key to the macroscopic interpretation of a chemical equation. Consider the same equation we have been discussing:

$$2\,H_2(g) + O_2(g) \rightarrow 2\,H_2O(g)$$

If we want to read this in terms of moles, we can say "2 moles of H_2 and 1 mole of O_2 react to form 2 moles of H_2O." Because each mole contains the same number of molecules, the 2:1 **mole ratio** between the reactants is the same as the 2:1 ratio for the numbers of molecules. Chemical equations and their stoichiometric coefficients always provide ratios of numbers of particles, *not masses.*

Figure 3.9 ■ One mole samples of various elements are shown. Back row (left to right): bromine, aluminum, mercury, and copper. Front row (left to right): sulfur, zinc, and iron.

Determining Molar Mass

Because balanced chemical equations are always expressed in terms of numbers of particles, it will be convenient to have some easy way to determine the number of particles in a given sample of some substance. But there is no simple laboratory instrument that can measure the number of moles in a sample directly. Instead, we usually determine the number of moles indirectly from the mass of a sample.

The link between the mass of a sample and the number of moles present is the molar mass of the substance in question. To determine the molar mass of a compound, we can exploit the idea of conservation of mass. Consider 1 mole of water as an example. We know that 1 mole of the compound must contain Avogadro's number of H_2O molecules. Furthermore, we know that each of those molecules must contain one O atom and two H atoms. Avogadro's number of oxygen atoms is 1 mole, and we know from the periodic table that 1 mole of O atoms has a mass of 16.0 g. Because each molecule contains two hydrogen atoms, the entire 1-mole sample will contain 2 moles of H atoms. Again consulting the periodic table, we see that 1 mole of H atoms has a mass of 1.0 g, so 2 moles must have a mass of 2.0 g. The masses of the O and H atoms must be the same as the mass of the mole of H_2O, so we can simply add them to get 18.0 g as the mass of 1 mole of H_2O. In other words, the molar mass of H_2O is 18.0 g/mol. We can do the same thing for any compound; the sum of the molar masses of all of the atoms is the molar mass of the compound. In Example Problem 3.5, we determine the molar masses of several compounds.

For simplicity, we use whole numbers for the molar masses of hydrogen and oxygen here. In general, you should use molar masses with at least as many significant figures as the actual data involved in the problem you are working.

EXAMPLE PROBLEM 3.5

Determine the molar mass of each of the following compounds, all of which are used as fertilizers for the production of biomass. (a) calcium sulfate, $CaSO_4$, (b) urea, $CO(NH_2)_2$, (c) carnallite, $H_{12}Cl_3KMgO_6$.

Strategy We must determine the mass contributed by each element and then add them up to calculate the molar mass. When parentheses appear in the formula, each atom inside the parentheses must be multiplied by its own subscript *and* by the subscript appearing after the right-hand parenthesis.

Solution

(a) $CaSO_4$:	1 mole Ca:	1 × 40.078 g/mol	=	40.078 g/mol
	1 mole S:	1 × 32.06 g/mol	=	32.06 g/mol
	4 moles O:	4 × 15.999 g/mol	=	63.996 g/mol
		Molar mass	=	136.13 g/mol
(b) $CO(NH_2)_2$:	1 mole C:	1 × 12.011 g/mol	=	12.011 g/mol
	1 mole O:	1 × 15.999 g/mol	=	15.999 g/mol
	4 moles H:	4 × 1.008 g/mol	=	4.032 g/mol
	2 moles N:	2 × 14.007 g/mol	=	28.014 g/mol
		Molar mass	=	60.056 g/mol
(c) $H_{12}Cl_3KMgO_6$:	12 moles H:	12 × 1.008 g/mol	=	12.096 g/mol
	3 moles Cl:	3 × 35.45 g/mol	=	106.35 g/mol
	1 mole K:	1 × 39.0983 g/mol	=	39.0983 g/mol
	1 mole Mg:	1 × 24.3050 g/mol	=	24.3050 g/mol
	6 moles O:	6 × 15.999 g/mol	=	95.994 g/mol
		Molar mass	=	277.84 g/mol

Analyze Your Answer Looking at these three compounds, we see that urea has a smaller molar mass than calcium sulfate even though it contains more atoms per molecule. Does this make sense? The key factor is that calcium and sulfur are both significantly heavier elements than those found in urea, so it is not surprising that urea is the least massive of these three molecules. Carnallite has the largest molar mass because each molecule contains several relatively heavy atoms.

Discussion The number of significant figures to which the molar masses of the elements are known varies from one element to another, as you can see if you examine the periodic table inside the back cover of this book. So the appropriate number of significant figures in the molar masses here varies according to the elements in each molecule.

Check Your Understanding What are the molar masses of the following compounds that can be obtained from the refinement of biomass? (a) xylaric acid, $C_5H_8O_7$ (b) diphenolic acid, $C_4H_7(C_6H_5O)_2COOH$ (c) aspartic acid, $C_2H_5N(COOH)_2$

3-5 Calculations Using Moles and Molar Masses

We've seen the importance of the concept of the mole when dealing with chemical reactions involving macroscopic quantities of material. But generally it is not possible to measure the number of moles in a sample directly, because that would imply that we can count molecules. Molar masses provide a crucial connection and allow us to convert from masses, which can be measured easily, to numbers of moles. The mass and the number of moles are really just two different ways of expressing the same information—the amount of a substance present. The molar mass functions much like a unit conversion between them, allowing us to go from mass to moles or from moles to mass. Example Problem 3.6 illustrates such a use of molar mass.

EXAMPLE PROBLEM 3.6

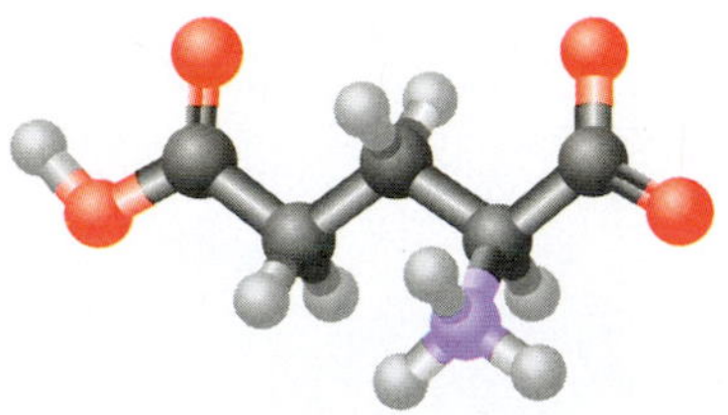
Glutamic acid, $C_5H_9NO_4$

A 245.3-g sample of glutamic acid, $C_5H_9NO_4$, is recovered from an experiment using fermentation to convert biomass. How many moles of $C_5H_9NO_4$ are in this sample? How many molecules is this?

Strategy We are asked to convert from mass to moles. So we must determine the molar mass of the substance and then use it to carry out the conversion. Once we know the number of moles, we can easily find the number of molecules because we know that 1 mole contains Avogadro's number of molecules.

Solution First, we will calculate the molar mass of glutamic acid.

$$\begin{array}{llllll} C_5H_9NO_4: & 5\text{ moles C:} & 5 \times 12.011 & \text{g/mol} & = & 60.055\ \text{g/mol} \\ & 9\text{ moles H:} & 9 \times 1.008 & \text{g/mol} & = & 9.072\ \text{g/mol} \\ & 1\text{ mole N:} & 1 \times 14.007 & \text{g/mol} & = & 14.007\ \text{g/mol} \\ & 4\text{ moles O:} & 4 \times 15.999 & \text{g/mol} & = & \underline{63.996\ \text{g/mol}} \\ & & & \text{Molar mass} & = & 147.130\ \text{g/mol} \end{array}$$

Now we can use the molar mass to convert from mass to moles:

$$245.3\ \text{g}\ C_5H_9NO_4 \times \frac{1\ \text{mol}\ C_5H_9NO_4}{147.130\ \text{g}\ C_5H_9NO_4} = 1.667\ \text{mol}\ C_5H_9NO_4$$

Finally, we can convert from moles to molecules using Avogadro's number:

$$1.667 \text{ mol } C_5H_9NO_4 \times \frac{6.022 \times 10^{23} \text{ molecules } C_5H_9NO_4}{1 \text{ mol } C_5H_9NO_4}$$

$$= 1.004 \times 10^{24} \text{ molecules } C_5H_9NO_4$$

Analyze Your Answer Admittedly, it is difficult to have an intuitive feeling for the number of molecules in any sample. In assessing our answer, we can begin by noting that it is a very large number. This makes sense because we are calculating the number of molecules in a macroscopic amount of glutamic acid. We could also check a little more closely by looking at the intermediate step in which we found the number of moles. Because the molar mass of glutamic acid is slightly less than 150 g/mol, we should have somewhat less than 2 moles in the 245-g sample. This is consistent with the value we found.

A common student error in this type of calculation is to divide by Avogadro's number rather than multiply. This leads to a very small numerical answer. A number of molecules less than one is not possible, of course, and is a sure sign of an error in the calculation.

Discussion The mass of this sample corresponds to a little more than half a pound, so the sample would probably be roughly the size of a tennis ball. Note the enormous number of molecules in this fairly small amount of material.

Check Your Understanding Although ethanol is perhaps the most widely known biofuel, considerable work is being done to produce butanol, C_4H_9OH, from biomass as well. How many moles are present in 1.00 gallon of pure butanol, which has a density of 0.810 g/mL?

In some cases, we might calculate the number of moles of a substance needed in a particular reaction. Then, if we want to carry out that reaction, we would need to know what mass of sample to prepare. We can easily invert the process above to convert from the number of moles to mass, as shown in Example Problem 3.7.

EXAMPLE PROBLEM 3.7

One goal of biomass conversion is the development of feedstocks, chemicals that can be easily converted to raw materials for chemical processes like the synthesis of polymers. One important feedstock molecule that can be derived from biomass is 2-propanol, C_3H_7OH. If 423 moles of the compound are needed for a reaction, how many pounds of C_3H_7OH should be used?

Strategy We are asked to convert from a number of moles to a mass. We know that the link between these two quantities is the molar mass of the compound, so we can start by calculating that. Once we know the molar mass, we can use it to find the required mass of 2-propanol. Finally, we can convert that result from grams to pounds.

Solution We will need to calculate the molar mass of the compound.

C_3H_7OH:	3 moles C:	3 × 12.011 g/mol	=	36.033 g/mol
	8 moles H:	8 × 1.008 g/mol	=	8.064 g/mol
	1 moles O:	1 × 15.999 g/mol	=	15.999 g/mol
		Molar mass	=	60.096 g/mol

Now we can use the molar mass and the required number of moles to find the appropriate mass:

$$423 \text{ mol } C_3H_7OH \times \frac{60.096 \text{ g } C_3H_7OH}{1 \text{ mol } C_3H_7OH} = 2.54 \times 10^4 \text{ g } C_3H_7OH$$

Finally, we can convert this from grams to pounds:

$$2.54 \times 10^4 \text{ g } C_3H_7OH \times \frac{1 \text{ lb.}}{453.59 \text{ g}} = 56.0 \text{ lb. } C_3H_7OH \text{ needed}$$

Analyze Your Answer We can assess the reasonableness of this answer by comparing it to Example Problem 3.6. In this problem, the mass of 2-propanol is about 100 times larger than the glutamic acid mass in the previous problem. The number of moles of 2-propanol is about 250 times larger. Why the difference? The molar mass of 2-propanol is less than half that of glutamic acid, so we should expect the ratio of moles to be larger than the ratio of masses.

Check Your Understanding Approximately 80 billion moles of ammonium nitrate (NH_4NO_3) were produced globally in 2008. How many tons of ammonium nitrate is this?

Elemental Analysis: Determining Empirical and Molecular Formulas

When a new molecule is synthesized, an **elemental analysis** is routinely performed to help verify its identity. This test, which measures the mass percentage of each element in the compound, is also frequently done as part of the process of identifying any substance whose composition is unknown. The mass percentages describe the compound's composition, and so they must be related to its chemical formula. But the data obtained from elemental analysis describe the composition in terms of the *mass* of each element, whereas the formula describes the composition in terms of the *number of atoms* of each element. So these are two different representations of very similar information, and the molar masses of the elements provide a connection between them. The process of obtaining the empirical formula of a compound from its percent composition by mass is illustrated in Example Problem 3.8.

EXAMPLE PROBLEM 3.8

Replacing fossil fuels with biomass is important for many uses beyond energy production. Compounds derived from biomass may be useful as starting materials for many chemical processes. Nitroanaline had been observed in experiments on biomass from pine needles and can be used as a precursor for pharmaceuticals. It contains 52.17% carbon, 4.38% hydrogen, 20.28% nitrogen, and 23.17% oxygen by mass. Determine the empirical formula of nitroanaline.

Strategy The empirical formula is based on the mole ratios among the elements in the compound, and the data given are in terms of mass. Molar mass provides the link between mass and moles, as usual. We can begin by choosing a convenient mass of the compound—usually 100 g—and use the percentages given to find the mass of each element in that sample. Then we will convert those masses into numbers of moles of each element. The ratios between those numbers of moles must be the same for *any* sample of the compound. Finally, we will need to convert those ratios into whole numbers to write the empirical formula.

Solution Consider a 100-g sample of nitroaniline. From the percentages given, that sample will contain 52.17 g C, 4.38 g H, 20.28 g N, and 23.17 g O. We will convert each of these masses into moles:

$$52.17 \text{ g C} \times \frac{1 \text{ mol C}}{12.011 \text{ g C}} = 4.344 \text{ mol C in 100 g nitroaniline}$$

$$4.38 \text{ g H} \times \frac{1 \text{ mol H}}{1.008 \text{ g H}} = 4.35 \text{ mol H in 100 g nitroaniline}$$

$$20.28 \text{ g N} \times \frac{1 \text{ mol N}}{14.007 \text{ g N}} = 1.448 \text{ mol N in 100 g nitroaniline}$$

$$23.17 \text{ g O} \times \frac{1 \text{ mol O}}{15.999 \text{ g O}} = 1.448 \text{ mol O in 100 g nitroaniline}$$

Those numbers tell us the ratios of C:H:N:O in the compound. We might be tempted to write a formula of $C_{4.344}H_{4.35}N_{1.448}O_{1.448}$. But we know that the correct empirical formula requires whole numbers of each element. In this case, it may be easy to see the correct ratios, but we need to approach such a problem systematically. The usual strategy is to divide all of the numbers of moles we obtained above by the smallest number of moles. This ensures that the smallest resulting number will always be one. In this case, we will divide all four numbers of moles by 1.448 because it is the smallest value of the four:

$$\frac{4.344 \text{ mol C}}{1.448} = 3.00$$

$$\frac{4.35 \text{ mol H}}{1.448} = 3.00$$

$$\frac{1.448 \text{ mol N}}{1.448} = 1$$

$$\frac{1.448 \text{ mol O}}{1.448} = 1$$

The result is a small whole number ratio: 3 moles C:3 moles of H:1 mole N:1 mole O. So the empirical formula is C_3H_3NO. The mass percentage data allow us to determine only the empirical formula, and not the molecular formula, because we do not have any information about how large or small a molecule of the compound might be.

Analyze Your Answer We know that the coefficients in a chemical formula must be whole numbers. The fact that all four of the coefficients we found are so close to integers means that our proposed formula is plausible.

Discussion In working with this type of problem, it is a good idea to use accurate molar masses for the elements and to carry as many significant figures as possible throughout the calculation. This ensures that no large rounding errors occur and helps make it easy to decide whether the final coefficients obtained are whole numbers.

The steps we used above will produce whole numbers for all of the elements only if at least one of the subscripts in the empirical formula is a one. Otherwise, the result for one or more elements will still not be an integer. In nearly all such cases, though, such nonintegral results will be readily identified as small rational fractions (e.g., 1.5 or 2.33). We can obtain integral values by multiplying *all* of the coefficients by an appropriate integer. This is demonstrated in the "Check Your Understanding" exercise that follows.

Check Your Understanding A compound called cryptomeridiol is among the active ingredients found in herbs used in traditional Chinese medicine. It contains 74.95% C, 11.74% H, and 13.31% O. Determine the empirical formula of this compound.

If we also know the molar mass of the compound, the process used in the previous example can easily be extended to allow the determination of the molecular formula.

The connection between mass and moles can also be applied in other ways. Materials scientists and engineers often describe the composition of alloys in terms of either weight percentage (wt %) or mole percentage (mol %). Converting between these two units involves the use of molar mass, as shown in Example Problem 3.9.

An alloy is a solution in which both the solute and solvent are solids. Materials engineers can choose or design alloys to have useful properties for a particular application.

EXAMPLE PROBLEM 3.9

Alloys of palladium and nickel are often used in the manufacture of electronic connectors. One such alloy contains 70.8 mol % Pd and 29.2 mol % Ni. Express the composition of this alloy in weight percentage.

Strategy In order to get started, we need to choose an amount of the alloy. Although any amount would work, 1 mole might be convenient. Once we have specified a number of moles, we can use molar masses to determine the mass of each metal and the total mass. And from those we can calculate the weight percentages.

Solution The molar masses of palladium and nickel are 106.42 g/mol and 58.69 g/mol, respectively. From the given mole percentages, we know that 1 mole of alloy would contain 0.708 mole of Pd and 0.292 mole of Ni. We can use these values to determine the mass of each component in 1 mole of alloy:

$$m_{Pd} = 0.708 \text{ mol Pd} \times \frac{106.42 \text{ g Pd}}{1 \text{ mol Pd}} = 75.4 \text{ g Pd}$$

$$m_{Ni} = 0.292 \text{ mol Ni} \times \frac{58.6934 \text{ g Ni}}{1 \text{ mol Ni}} = 17.1 \text{ g Ni}$$

So the total mass of 1 mole of alloy is just

$$75.4 \text{ g Pd} + 17.1 \text{ g Ni} = 92.5 \text{ g alloy}$$

Finally, calculate the percentage by mass (weight) of each metal:

$$\text{wt \% (Pd)} = \frac{75.4 \text{ g Pd}}{92.5 \text{ g alloy}} \times 100\% = 81.5 \text{ wt \% Pd}$$

$$\text{wt \% (Ni)} = \frac{17.1 \text{ g Ni}}{92.5 \text{ g alloy}} \times 100\% = 18.5 \text{ wt \% Ni}$$

Analyze Your Answer It makes sense that the weight percentage of the palladium is higher than the mole percentage because palladium has a higher molar mass than nickel.

Discussion There is nothing special about the assumption of a mole in this problem. It was chosen because it was likely to provide numbers for the masses that were a "comfortable" size to think about. If the problem were worked in the other direction (from wt % to mol %), we might choose 100 g as the amount of material to use in the calculation.

An additional unit sometimes used for alloy composition is atomic percentage (at %). The conversion between at % and wt % involves the same steps as the mol % and wt % conversion in this problem.

Check Your Understanding Eighteen-carat gold typically contains 75% gold, 16% silver, and 9% copper by weight. Express this composition in mol percentages.

Molarity

Molar mass is useful for chemical calculations because it provides a connection between a quantity that is easy to measure (mass) and one that is conceptually important (moles). Another readily measured quantity is volume. When we work with aqueous solutions, we often use volume in calculations rather than mass. It should not be surprising, therefore, that we would want to define quantities that will help us relate a volume measurement to the number of moles.

Many different ways to express the concentration of a solute in a solvent have been developed. To define any unit of concentration, we must know both the amount of solute and the quantity of solvent in the solution. The most commonly used concentration unit in chemistry is **molarity** or **molar concentration**. Molarity, represented by the symbol M, is defined as the number of moles of solute per liter of solution:

Molarity is another ratio between otherwise unrelated variables—similar to mass density. It can be used in calculations in the same way as density.

$$\text{Molarity (M)} = \frac{\text{moles of solute}}{\text{liter of solution}} \tag{3.1}$$

The definition of molarity points to the ways in which it can be used. It provides a relationship among three things: molar concentration, moles of solute, and liters of solution. If we know any two of these quantities, we can determine the third one. We can measure the volume of a solution in the laboratory. Multiplying that volume (expressed in liters) by molarity readily gives moles.

The same relationship also allows us to find the number of moles of solute present if we know both the volume and molarity of any solution. If we use n to represent the number of moles, V for volume, and M for molarity, we can simply rearrange the definition of molarity:

$$n = M \times V \tag{3.2}$$

Example Problem 3.10 illustrates the calculation of molarity.

EXAMPLE PROBLEM 3.10

Many ways to produce biofuels by degrading biomass have been tried, with some more successful than others. One method that has not worked is to treat bacterial biomass with aqueous solutions of sodium hypochlorite (NaClO) to try to digest the complex biomolecules present. In an attempt to research this process further, a solution is prepared by dissolving 45.0 g of NaClO in enough water to produce exactly 750 mL of solution. What is the molarity of the solution?

Strategy To obtain molarity we need two quantities—moles of solute and liters of solution. Neither of these quantities is given directly in the question, but both can be obtained readily from information given. We must use the molar mass of NaClO to convert from mass to moles and convert the volume from milliliters to liters. Then we can use the definition of molarity to obtain the molarity of the solution.

Solution First, we compute the moles of solute:

$$45.0 \text{ g NaClO} \times \frac{1.00 \text{ mol NaClO}}{74.44 \text{ g NaClO}} = 0.605 \text{ mol NaClO}$$

Then, convert the solution volume from mL to L:

$$750 \text{ mL} \times \frac{1.00 \text{ L}}{100 \text{ mL}} = 0.750 \text{ L}$$

Finally, calculate the molarity:

$$\text{Molarity} = \frac{\text{moles of solute}}{\text{litres of solution}} = \frac{0.605 \text{ mol}}{0.750 \text{ L}} = 0.806 \text{ M NaClO}$$

Analyze Your Answer Our answer is somewhat smaller than, but fairly close to, 1 M. Does this make sense? The amount of NaClO was about 2/3 of a mole and the volume of the solution was 3/4 of a liter. The ratio of 2/3 to 3/4 is 8/9, close to but a little bit smaller than 1, so our result seems reasonable.

Discussion As usual, this problem could also be set up as a single calculation including the same conversions. Whichever approach you choose, you should always check to be sure that your units work out correctly. Here we have units of mol/L for our answer, which is appropriate for molarity. Confirming this helps avoid careless errors.

Check Your Understanding An important biomass research area is the possible conversion of the corn stover (the leaves and stalks that remain after harvesting) in addition to the kernels. One promising method uses dilute sulfuric acid for part of the process. How many moles of H_2SO_4 would be present in 45.0 L of 0.150 M H_2SO_4?

Dilution

One of the most common procedures encountered in any science laboratory is **dilution**, the process in which solvent is added to a solution to decrease the concentration of the solute. In dilution, the amount of solute does not change. *The number of moles of solute is the same before and after dilution.* Because we already know that the number of moles of solute equals the product of molarity and volume, we can write the following equation, where the subscripts denote the initial (i) and final (f) values of the quantities involved:

This dilution formula can be used when there is only one solute present. If a reaction is taking place, the problem should be approached as a stoichiometry problem, which we will take up in Chapter 4.

$$M_i \times V_i = M_f \times V_f \quad (3.3)$$

The simplicity of equations such as this is alluring. One of the most common errors made by students in chemistry is to use this relationship in situations where it is not valid. *The only place to use this relationship is in the process of dilution or concentration of a solution.* Example Problem 3.11 illustrates a common laboratory situation where this equation is valid and useful.

EXAMPLE PROBLEM 3.11

A chemist requires 1.5 M hydrochloric acid, HCl, for a series of reactions. The only solution available is 6.0 M HCl. What volume of 6.0 M HCl must be diluted to obtain 5.0 L of 1.5 M HCl?

Strategy First, we need to recognize that this is the dilution of the concentrated solution to prepare the desired one. The underlying concept we must use is that the number of moles of HCl will be the same before and after dilution, and this means we can use Equation 3.3. We know the desired final molarity and final volume, as well as the initial molarity. So we can solve for the needed initial volume.

Solution

Initial concentration of HCl: $M_i = 6.0$ M
Final concentration of HCl: $M_f = 1.5$ M
Final volume of solution: $V_f = 5.0$ L

The unknown is the initial volume, V_i. Rearranging Equation 3.3,

$$V_i = \frac{M_f \times V_f}{M_i}$$

Inserting the known quantities on the right-hand side,

$$V_i = \frac{1.5\text{ M} \times 5.0\text{ L}}{6.0\text{ M}} = 1.3\text{ L}$$

To obtain the desired quantity of diluted HCl, the chemist should begin with 1.3 L of the concentrated solution and add enough water to bring the volume up to 5.0 L.

Analyze Your Answer Once we are certain that we are dealing with a dilution problem, one way to make sure the answer makes sense is to think about which solution is concentrated and which is dilute. You will always need smaller volumes of the concentrated solution. Did we get it right this time? It seems likely that we did: we were asked to calculate how much of the concentrated solution we would need, so the volume we calculate should be the smaller of the two volumes involved in the problem.

Check Your Understanding If 2.70 mL of 12.0 M NaOH is diluted to a volume of 150.0 mL, what is the final concentration?

INSIGHT INTO

3-6 Carbon Sequestration

One thing that you may have noticed as we worked through examples in this chapter is that the reaction products for combustion of biofuels, or fossil fuels for that matter, are generally carbon dioxide and water. Although doing laboratory-scale work on reactions such as these produces negligible amounts of products, the human economy burns vast amounts of fuels every year. Indeed, over the past 150 years, the concentration of CO_2 in the atmosphere has clearly been increasing. Because CO_2 is a greenhouse gas, there is considerable evidence that this change in concentration is a key contributor to climate change. While the potential consequences of climate change are less certain than the fact that the planet has warmed up, there are calls from many corners to consider ways to combat the continued rise in CO_2 concentration in the atmosphere. One place where chemistry and engineering may play a role in these actions is **carbon sequestration**, the process of removing carbon from the atmosphere or from gases (such as flue gases) that are entering the atmosphere.

A greenhouse gas absorbs infrared radiation and thereby may trap energy that might otherwise escape the atmosphere.

Although the concept of carbon sequestration is fairly simple, the global scale of the problem poses serious engineering challenges. As seen in **Figure 3.10**, the amount

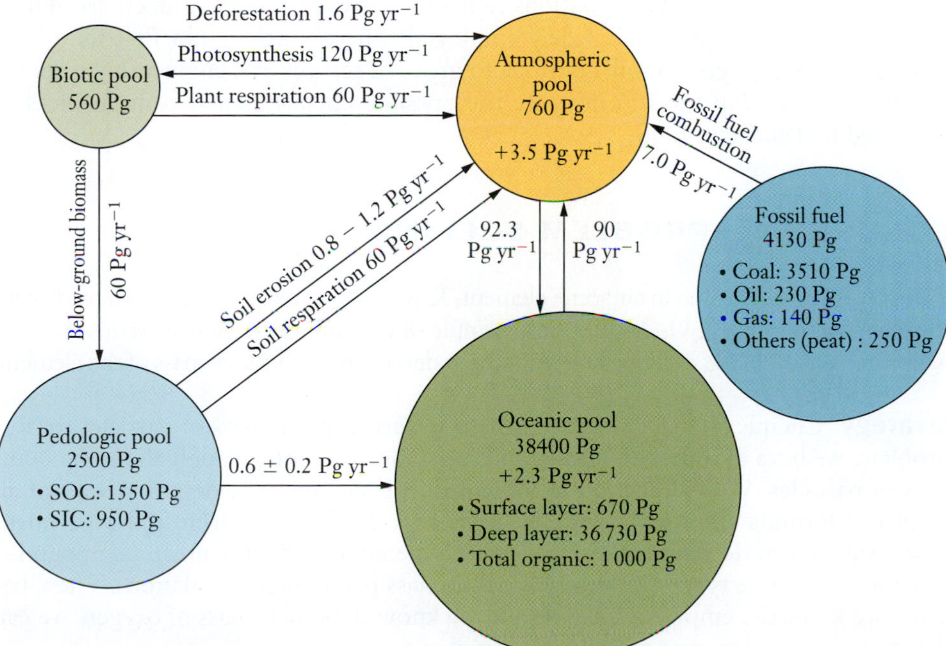

Figure 3.10 ■ The flow of carbon among various reservoirs is shown schematically. The largest flows of carbon are due to natural processes like photosynthesis. Although the flow of carbon into the atmosphere due to burning of fossil fuels is small compared to the overall atmospheric carbon content, it is still sufficient to upset the balance among the carbon reservoirs. Carbon sequestration offers one way to try to address this imbalance.

We saw in Table 1.2 that peta- is the prefix corresponding to 10^{15}.

of carbon present on Earth and the amount that moves between the various reservoirs in any given year are both large enough to be measured in petagrams (Pg). Looking at this figure, many of the annual flows (depicted by the arrows) are generally in balance. Carbon flows between the oceans and the atmosphere in both directions, for example, with slightly larger amounts moving towards the oceans. Similarly, photosynthesis fixes roughly 120 Pg yr^{-1}, but respiration from plants and from soils offsets that amount. Thus the combustion of fossil fuels, despite contributing only a relatively small amount of 7.0 Pg yr^{-1}, is enough to disturb the balance and lead to a steady increase in the amount of carbon in the atmospheric pool.

There are actually many ways that carbon sequestration might be accomplished. Biomass production, because it would potentially increase the amount of carbon captured via photosynthesis, is sometimes considered as one biotic method for the reduction of net carbon flow into the atmosphere. From an engineering perspective, proposals for abiotic methods to remove CO_2 likely require more creative ideas and designs in order to become practical on a large scale. One broad method for trapping carbon, called injection, is based on the idea that CO_2 could somehow be pumped into an environment where chemical reactions will trap it. In ocean injection, CO_2 is released deep underwater where it may react or, because of the very high pressures at the ocean bottom, form stable pools of liquid CO_2. Research into ocean injection has been taking place for decades, but significant challenges related to ensuring that unintended consequences of the process do not harm ocean life remain.

Another form of injection is called geological injection. The basic idea is to recover CO_2 from industrial processes and pump it deep into rock formations where it can react to form stable carbonates. An example reaction is the formation of the mineral magnesite (primarily $MgCO_3$) from the mineral olivine (primarily Mg_2SiO_4):

$$Mg_2SiO_4 + 2\ CO_2(g) \rightarrow 2\ MgCO_3 + SiO_2$$

One potential advantage for the development of the needed engineering for such work is that CO_2 has been used for enhanced oil recovery (EOR) from geological formations for many years. In the case of EOR, the CO_2, or other gases, are pumped into a reservoir where recovery of oil has been decreasing to effectively push the oil into the wellbore to bring it to the surface. While this technology accounts for roughly 60% of the EOR in the United States, it is still at a scale that is much smaller than would be needed to use geological injection as a major component of carbon sequestration strategies in the future. So, many questions remain for chemists and engineers to address before this technology can be widely employed. There's little doubt that the global economy's need for energy will continue to grow, so science and engineering reserach will need to develop ways to mitigate the possible environmental consequences of increased fuel consumption.

FOCUS ON PROBLEM SOLVING

Question When heated in air, some element, X, reacts with oxygen to form an oxide with the chemical formula X_2O_3. If a 0.5386-g sample of this unknown element yields an oxide with a mass of 0.7111 g, describe how you would determine the atomic mass of the element.

Strategy Chemical formulas are expressed in terms of numbers of particles. In this problem, we have to start with masses and somehow infer information about the numbers of particles. We did a similar problem in this chapter: the determination of an empirical formula from percentage by mass data. In this case, from the experiment described we can determine the mass of each element in the unknown compound—providing the same type of data we had from mass percentage calculations. Then, because we know the empirical formula and we know the molar mass of oxygen, we can determine the molar mass of the unknown element.

Solution First, we would determine the mass composition. The mass of element X in the oxide sample must be assumed to be the 0.5386 g that was used to generate the oxide. Subtracting this value from the 0.7111 g of the oxide sample then provides the mass of the oxygen. The mass of oxygen can be converted to moles of oxygen using the molar mass (which could be looked up if we don't know it). Then, we could use the mole ratio from the empirical formula given in the problem to assert that whatever value we obtain, the number of moles of X would be two-thirds of this result. (This follows from the fact that there are two X atoms for every three oxygen atoms.) Then, we would have a value for the number of moles of X and the mass of X in the sample—taking the ratio provides the molar mass, from which we could use a periodic table to look up the identity of element X.

SUMMARY

Chemical reactions are the most important events in chemical systems. So the ability to describe reactions succinctly is vital to our study of chemistry. Such descriptions rely on the balanced chemical equation, a written representation of the reaction accurately reflecting the fact that matter is neither created nor destroyed.

One requirement for a reaction to occur is that the chemicals involved must be able to mix and interact with one another. Thus the physical state of the reactants can be important. In particular, an aqueous solution is often used as the medium for a reaction. To describe reactions in the aqueous phase effectively, we often need to specify the molar concentration of the solutions used. We also have options about how to write the chemical reaction itself—as a molecular equation, for example, or a net ionic equation. Several classes of chemical reactions, such as precipitation reactions or acid–base neutralizations, are sufficiently common that additional definitions have been established to improve our ability to communicate about them.

Ultimately, numerical information is often desirable, so chemists have devised a way to count the enormous numbers of particles involved in reactions on laboratory or industrial scales. The unit chemists use to count atoms or molecules is called the mole, and it is specified by Avogadro's number, 6.022×10^{23}. The molar mass of an element can be obtained readily from the periodic table, and that of a compound can be found by adding the molar masses of its constituent atoms. Because we can determine the molar mass of virtually any chemical substance, we can convert between a mass (that can be readily measured) and the number of particles in moles (that can be used in chemical calculations.) Molar concentrations provide a similar ability to convert between a measurable quantity (in this case volume of solution) and moles.

There are several types of calculations that can be undertaken once we recognize the concept that a mole is simply a way to count particles. We can determine the empirical formula of a compound based on the mass percentage analysis of its elements. We can also determine the molar concentration of solutions prepared by dissolving a known mass of the solute or by diluting a concentrated solution.

KEY TERMS

acid (3-3)
aqueous solution (3-3)
Avogadro's number (3-4)
biofuel (3-1)
biomass (3-1)
carbon sequestration (3-6)
base (3-3)
chemical equation (3-2)
concentration (3-3)
dilution (3-5)
dissociation reaction (3-3)
electrolyte (3-3)
elemental analysis (3-5)
insoluble (3-3)
law of conservation of matter (3-2)
molar concentration (3-5)
molar mass (3-4)
molarity (3-5)
mole (3-4)
mole ratio (3-4)
molecular equation (3-3)
net ionic equation (3-3)
neutralization (3-3)
nonelectrolyte (3-3)
photochemical reaction (3-2)
precipitation reaction (3-3)
products (3-2)
reactants (3-2)
salt (3-3)
solubility (3-3)
soluble (3-3)
solute (3-3)
solution (3-3)
solvent (3-3)
spectator ions (3-3)
stoichiometric coefficients (3-2)
stoichiometry (3-2)
strong electrolyte (3-3)
total ionic equation (3-3)
weak electrolyte (3-3)

PROBLEMS AND EXERCISES

INSIGHT INTO Biomass and Biofuel Engineering

3.1 Based on Figure 3.1, determine (a) the percentage of sunlight that is captured by photosynthesis, (b) the percentage of photosynthesis that occurs in land plants and the percentage that occurs in aquatic plants, and (c) the percentage of the energy captured by all plants that is eventually released when the plants decay.

3.2 What are the chemical products of photosynthesis? Which of these chemicals are used to generate ethanol from biomass?

3.3 Jan Baptist von Helmont is credited with repudiating the idea that trees gain their mass from the soil. In an experiment carried out in 1648, he grew a willow tree in a pot with the amount of soil kept constant. After five years the tree had gained roughly 74 kg. Although von Helmont concluded the weight gain was from water rather than soil, we now know it was from CO_2 and H_2O taken in through photosynthesis. Use $C_6H_{12}O_6$ as a representative formula for the carbohydrates produced, and write a simplified equation for the overall process of photosynthesis.

3.4 A newspaper article states that biomass has actually been used as an energy source throughout human history. Do you agree or disagree with this statement? Defend your answer.

3.5 Figure 3.2 shows the production of ethanol as the "economic output" of biofuels. Use the web to research whether this is the only commercially viable product of the process.

Chemical Formulas and Equations

3.6 Define the terms *reactants* and *products* as they refer to chemical reactions.

3.7 Which symbols are used to indicate solids, liquids, gases, and aqueous solutions in chemical equations?

3.8 How is the addition of heat symbolized in a chemical equation? The addition of light energy?

3.9 What law of nature underpins the concept of a balanced chemical equation?

3.10 Define the term *stoichiometric coefficient.*

3.11 Balance these equations.

(a) $Al(s) + O_2(g) \rightarrow Al_2O_3(s)$

(b) $N_2(g) + H_2(g) \rightarrow NH_3(g)$

(c) $C_6H_6(\ell) + O_2(g) \rightarrow H_2O(\ell) + CO_2(g)$

3.12 Balance the following equations.

(a) $CaC_2(s) + H_2O(\ell) \rightarrow Ca(OH)_2(s) + C_2H_2(g)$

(b) $(NH_4)_2Cr_2O_7(s) \rightarrow Cr_2O_3(s) + N_2(g) + H_2O(g)$

(c) $CH_3NH_2(g) + O_2(g) \rightarrow CO_2(g) + N_2(g) + H_2O(g)$

3.13 An explosive whose chemical formula is $C_3H_6N_6O_6$ produces water, carbon dioxide, and nitrogen gas when detonated in oxygen. Write the chemical equation for the detonation reaction of this explosive.

3.14 A number of compounds are used in cement, and reactions among them occur when water is added. In one, CaO reacts with Al_2O_3 and water to form $Ca_3Al_2(OH)_{12}$. Write a balanced chemical equation for this process.

3.15 Ethanol, C_2H_5OH, is found in gasoline blends used in many parts of North America. Write a balanced chemical equation for the combustion of C_2H_5OH to form CO_2 and H_2O.

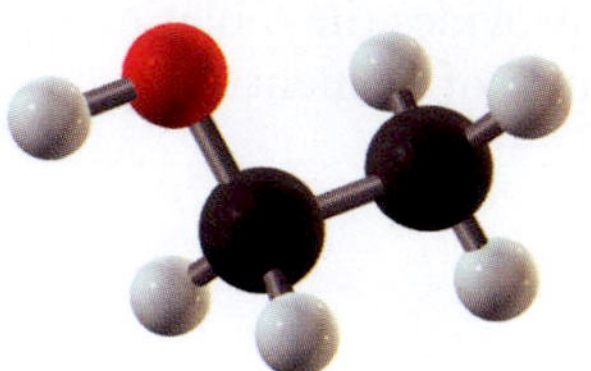

Ethanol, C_2H_5OH

3.16 Balance the following equations.

(a) reaction to produce "superphosphate" fertilizer:

$$Ca_3(PO_4)_2(s) + H_2SO_4(aq) \rightarrow Ca(H_2PO_4)_2(aq) + CaSO_4(s)$$

(b) reaction to produce diborane, B_2H_6:

$$NaBH_4(s) + H_2SO_4(aq) \rightarrow B_2H_6(g) + H_2(g) + Na_2SO_4(aq)$$

(c) reaction to produce tungsten metal from tungsten(VI) oxide:

$$WO_3(s) + H_2(g) \rightarrow W(s) + H_2O(\ell)$$

(d) decomposition of ammonium dichromate:

$$(NH_4)_2Cr_2O_7(s) \rightarrow N_2(g) + H_2O(\ell) + Cr_2O_3(s)$$

3.17 Write balanced chemical equations for the following reactions.

(a) production of ammonia, $NH_3(g)$, by combining $N_2(g)$ and $H_2(g)$

(b) production of methanol, $CH_3OH(\ell)$, by combining $H_2(g)$ and $CO(g)$

(c) production of sulfuric acid by combining sulfur, oxygen, and water

3.18 Diborane and related compounds were proposed as rocket fuels in the 1950s. A representative reaction for this class of molecules is that of B_2H_6 and O_2 to form B_2O_3 and H_2O. Write the balanced chemical equation for this process.

3.19 Silicon nitride, Si_3N_4, is used as a reinforcing fiber in construction materials. It can be synthesized from silicon tetrachloride and ammonia. The other product is ammonium chloride. Write the balanced chemical reaction for this process.

3.20 The following pictures show a molecular-scale view of a chemical reaction between the compounds AB_2 and B_2. (A atoms are shown in blue and B atoms in white). The box on the left represents the reactants at the instant of mixing, and the box on the right shows what is left once the reaction has gone to completion. Write a balanced chemical equation for this reaction. As usual, your equation should use the smallest possible whole number coefficients for all substances.

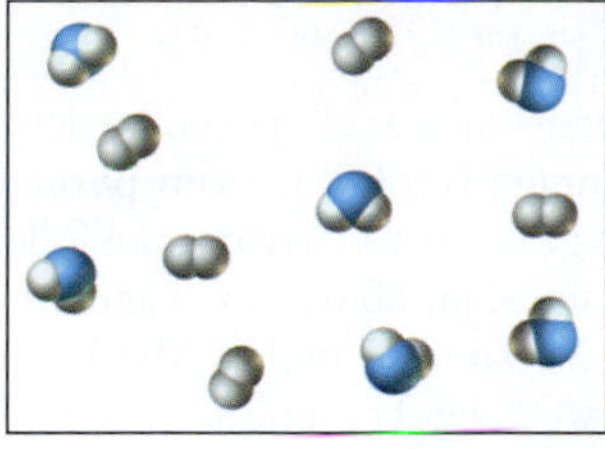
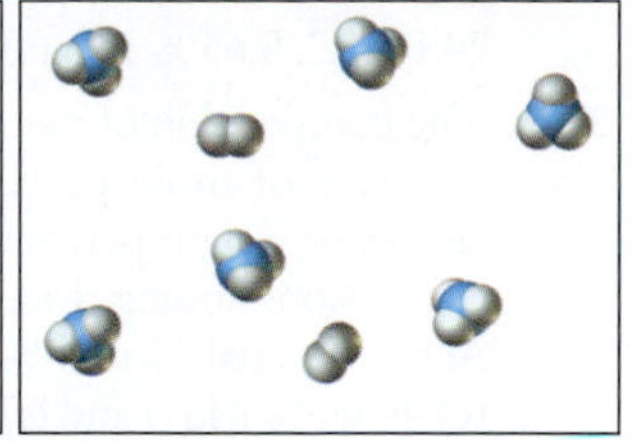

Aqueous Solutions and Net Ionic Equations

3.21 Define the terms *solution*, *solute*, and *solvent*.

3.22 When a substance dissolves, does a chemical reaction always occur? If not, what does happen?

3.23 What is a concentrated solution? A dilute solution?

3.24 Define the term *saturated solution*.

3.25 What is an electrolyte? How can you differentiate experimentally between a weak electrolyte and a strong electrolyte? Give an example of each.

3.26 Classify the following compounds as electrolytes or nonelectrolytes: **(a)** potassium chloride, KCl, **(b)** hydrogen peroxide, H_2O_2, **(c)** methane, CH_4, **(d)** barium nitrate, $Ba(NO_3)_2$

3.27 The following compounds are water-soluble. What ions are produced by each compound in aqueous solution? **(a)** KOH, **(b)** K_2SO_4, **(c)** $LiNO_3$, **(d)** $(NH_4)_2SO_4$

3.28 Decide whether each of the following is water-soluble. If soluble, tell what ions are produced. **(a)** Na_2CO_3, **(b)** $CuSO_4$, **(c)** NiS, **(d)** $BaBr_2$

3.29 The solubility of NaCl in water is 35.7 g NaCl/100 g H_2O. Suppose that you have 500.0 g of NaCl. What is the minimum volume of water you would need to dissolve it all? (Assume that the density of water is 1.0 g/mL.)

3.30 A packaging engineer is working on a new design for cold packs. A burst chamber in the pack will contain 14.0 mL of water, which will be released and used to dissolve ammonium nitrate. **(a)** If the solubility of NH_4NO_3 is 190 g/100 g H_2O, what mass of this compound should be used with this amount of water? **(b)** The design specifications indicate that these chemicals can comprise no more than 35% of the mass of the product. What would the product weigh if it just meets these design specs?

3.31 Classify each of these as an acid or a base. Which are strong and which are weak? What ions are produced when each is dissolved in water? **(a)** KOH, **(b)** $Mg(OH)_2$, **(c)** HClO, **(d)** HBr, **(e)** LiOH, **(f)** H_2SO_3

3.32 Define the term *spectator ion*.

3.33 What is the difference between a total ionic equation and a net ionic equation?

3.34 Balance the following equations and then write the net ionic equation.

(a) $(NH_4)_2CO_3(aq) + Cu(NO_3)_2(aq) \rightarrow CuCO_3(s) + NH_4NO_3(aq)$

(b) $Pb(OH)_2(s) + HCl(aq) \rightarrow PbCl_2(s) + H_2O(\ell)$

(c) $BaCO_3(s) + HCl(aq) \rightarrow BaCl_2(aq) + H_2O(\ell) + CO_2(g)$

(d) $CH_3CO_2H(aq) + Ni(OH)_2(s) \rightarrow Ni(CH_3CO_2)_2(aq) + H_2O(\ell)$

3.35 Balance the following equations, and then write the net ionic equation.

(a) $Zn(s) + HCl(aq) \rightarrow H_2(g) + ZnCl_2(aq)$

(b) $Mg(OH)_2(s) + HCl(aq) \rightarrow MgCl_2(aq) + H_2O(\ell)$

(c) $HNO_3(aq) + CaCO_3(s) \rightarrow Ca(NO_3)_2(aq) + H_2O(\ell) + CO_2(g)$

(d) $(NH_4)_2S(aq) + FeCl_3(aq) \rightarrow NH_4Cl(aq) + Fe_2S_3(s)$

3.36 In principle, it may be possible to engineer the trapping of carbon dioxide by a precipitation reaction with $Ca(OH)_2$ to form calcium carbonate and water. Write the molecular, total ionic, and net ionic equations for this reaction.

Interpreting Equations and the Mole

3.37 Explain the concept of the mole in your own words.

3.38 How many entities are present in each of the following?

(a) 3.21 mol argon atoms

(b) 1.3×10^{-12} mol marbles

(c) 3×10^6 mol water molecules

(d) 7.63×10^{-18} mol basketballs

3.39 If a typical grain of sand occupies a volume of 1.3×10^{-4} cm^3, what is the volume (in cm^3) of 1 mole of sand (ignoring the space between grains)? What is the volume in liters?

3.40 Estimate the size of a particle 1 mole of which would fill your bedroom or dorm room.

3.41 Calculate the molar mass of each of the following compounds. **(a)** Fe_2O_3, iron(III) oxide, **(b)** BCl_3, boron trichloride, **(c)** $C_6H_8O_6$, ascorbic acid (vitamin C)

3.42 Calculate the molar masses (in grams per mole) of each of the following. **(a)** cane sugar, $C_{12}H_{22}O_{11}$, **(b)** laughing gas, N_2O, and **(c)** vitamin A, $C_{20}H_{30}O$.

3.43 Calculate the molar mass of each of these compounds and the mass percentage of each element. **(a)** PbS, lead(II) sulfide, galena, **(b)** C_2H_6, ethane, a hydrocarbon fuel, **(c)** CH_3COOH, acetic acid, an important ingredient in vinegar, **(d)** NH_4NO_3, ammonium nitrate, a fertilizer

3.44 When an automobile air bag deploys, solid sodium azide decomposes rapidly to produce nitrogen gas, and this gas inflates the bag. What is the molar mass of sodium azide (NaN_3)?

3.45 Calculate the molar mass of the following compounds. **(a)** magnesium phosphate, **(b)** sodium sulfide, **(c)** dinitrogen tetroxide

3.46 Determine the molar mass of these ceramic materials. **(a)** HfN, **(b)** ThO_2, **(c)** $BaTiO_3$

Calculations Using Moles and Molar Masses

3.47 A chemist needs exactly 2 moles of NaOH to make a solution. What mass of NaOH must be used?

3.48 What mass of ozone (O_3) gives 4.5 moles of the substance?

Ozone, O_3

3.49 Calculate the mass in grams of each the following. **(a)** 2.5 mol of aluminum, **(b)** 1.25×10^{-3} mol of iron, **(c)** 0.015 mol of calcium, **(d)** 653 mol of neon

3.50 Calculate the mass in grams of 13.5 mol of **(a)** vinyl chloride, C_2H_3Cl, the starting material for a plastic, **(b)** capsaicin, $C_{18}H_{27}NO_3$, the substance that makes red chili peppers "hot," and **(c)** stearic acid, $C_{18}H_{36}O_2$, used in soaps.

3.51 How many moles are present in the given quantities of explosives? **(a)** 358.1g trinitrotoluene (TNT), $C_7H_5N_3O_6$, **(b)** 82.6 g nitromethane, CH_3NO_2, **(c)** 1.68 kg RDX, $C_3H_6N_6O_6$

3.52 A test of an automobile engine's exhaust revealed that 3.7 g of NO_2 was emitted in 10 minutes of operation. How many moles of NO_2 would this engine release if it were used for a 45-minute commute, assuming that this measured number is representative of the emission under all circumstances?

3.53 Modern instruments can measure a mass as small as 5 nanograms. If one observed 5.0 ng of CO_2, how many molecules were measured?

3.54 How many H atoms are present in 7.52 g of propane, C_3H_8?

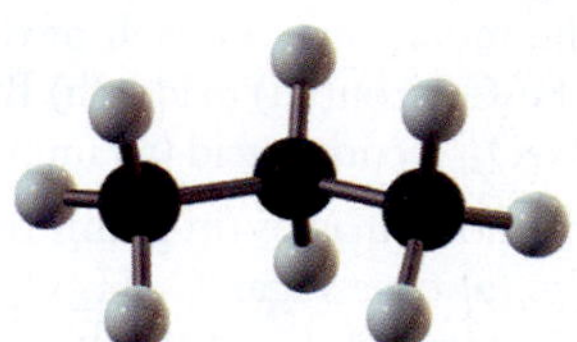

Propane, C_3H_8

3.55 How many O atoms are present in 214 g of mannose ($C_6H_{12}O_6$)

3.56 A sample of $H_2C_2O_4 \cdot 2H_2O$ of mass 3.35 g is heated to drive off the waters of hydration (designated separately in the chemical formula). What mass of $H_2C_2O_4$ remains once the water has been removed by heating?

3.57 An average person inhales roughly 2.5 g of O_2 in a minute. How many molecules of oxygen are inhaled in **(a)** 1 minute, **(b)** 1 hour, **(c)** 1 day by an average person?

3.58 A large family of boron-hydrogen compounds has the general formula B_xH_y. One member of this family contains 88.5% B; the remainder is hydrogen. What is its empirical formula?

3.59 Mandelic acid is an organic acid composed of carbon (63.15%), hydrogen (5.30%), and oxygen (31.55%). Its molar mass is 152.14 g/mol. Determine the empirical and molecular formulas of the acid.

3.60 Determine the simplest formulas of the following compounds. **(a)** the food enhancer monosodium glutamate (MSG), which has the composition 35.51% C, 4.77% H, 37.85% O, 8.29% N, and 13.60% Na, **(b)** zircon, a diamond-like mineral, which has the composition 34.91% O, 15.32% Si, and 49.76% Zr, **(c)** nicotine, which has the composition 74.0% C, 8.65% H, and 17.4% N

3.61 The composition of materials such as alloys can be described in terms of mole percentage (mol %), atom percentage (at %), or weight percentage (wt %). Carry out the following conversions among these units. **(a)** 60 wt % Cu and 40 wt % Al to at %, **(b)** 25 mol % NiO and 75 mol % MgO to wt %, **(c)** 40 wt % MgO and 60 wt % FeO to mol %

3.62 Copper can have improved wear resistance if alloyed with ceramic alumina, Al_2O_3. If a copper alloy has 8.5 wt % Al_2O_3, what is its composition in mol %?

3.63 Calculate the molarity of each of the following solutions.

(a) 1.45 mol HCl in 250. mL of solution

(b) 14.3 mol NaOH in 3.4 L of solution

(c) 0.341 mol KCl in 100.0 mL of solution

(d) 250 mol $NaNO_3$ in 350 L of solution

3.64 What is the molarity of each ion present in aqueous solutions prepared by dissolving 20.00g of the following compounds in water to make 4.50 L of solution? **(a)** cobalt(III) chloride, **(b)** nickel(III) sulfate, **(c)** sodium permanganate, **(d)** iron(II) bromide

3.65 How many moles of solute are present in each of these solutions?

(a) 48.0 mL of 3.4 M H_2SO_4

(b) 1.43 mL of 5.8 M KNO_3

(c) 3.21 L of 0.034 M NH_3

(d) 1.9×10^3 L of 1.4×10^{-5} M NaF

3.66 How many grams of solute are present in each of these solutions?

(a) 37.2 mL of 0.471 M HBr

(b) 113.0 L of 1.43 M Na_2CO_3

(c) 21.2 mL of 6.8 M CH_3COOH

(d) 1.3×10^{-4} L of 1.03 M H_2SO_3

3.67 Determine the final molarity for the following dilutions.

(a) 24.5 mL of 3.0 M solution diluted to 100.0 mL

(b) 15.3 mL of 4.22 M solution diluted to 1.00 L

(c) 1.45 mL of 0.034 M solution diluted to 10.0 mL

(d) 2.35 L of 12.5 M solution diluted to 100.0 L

3.68 Determine the initial volume (in mL) needed to generate each of the following desired solutions by dilution.

(a) 10.0 L of 0.45 M solution from 3.0 M solution

(b) 1.50 L of 1.00 M solution from 11.7 M solution

(c) 100.0 mL of 0.025 M solution from 1.15 M solution

(d) 50.0 mL of 3.3×10^{-4} M solution from 0.25 M solution

3.69 Commercially available concentrated sulfuric acid is 18.0 M H_2SO_4. Calculate the volume of concentrated sulfuric acid required to prepare 2.50 L of 1.50 M H_2SO_4 solution.

3.70 Magnesium is lighter than other structural metals, so it is increasingly important in the design of more efficient vehicles. Mg^{2+} ions are present in seawater, and the metal is often prepared by "harvesting" these ions and converting them to Mg metal. The average magnesium content of the oceans is about 1270 g Mg^{2+} per ton of seawater, and the density of seawater is about 1.03 g/mL. What is the molarity of Mg^{2+} ions in seawater? The design for a concept car calls for 103 kilograms of magnesium per vehicle. How many gallons of seawater would be required to supply enough magnesium to build one of these cars?

INSIGHT INTO Carbon Sequestration

3.71 Carbon dioxide is just one of many greenhouse gases in the atmosphere. What property makes a gas a greenhouse gas?

3.72 Biomass production and use is a small component in the increase of CO_2 in the atmosphere. (a) What is the main source of CO_2 increase in the atmosphere? (b) In what way is this source similar to the use of biomass fuels?

3.73 What is meant by the term *carbon reservoir*? What are the two largest carbon reservoirs for our planet?

3.74 What is meant by the term *carbon sequestration?*

3.75 Explain how the burning of fossil fuels, which contributes only 7.0 Pg yr^{-1} to the atmospheric store of carbon, can upset the balance of carbon in the atmosphere if photosynthesis fixes 120 Pg yr^{-1}.

3.76 Are most proposals to carry out carbon sequestration based on chemical or physical properties of CO_2? Defend your answer.

Additional Problems

3.77 Nitric acid (HNO_3) can be produced by the reaction of nitrogen dioxide (NO_2) and water. Nitric oxide (NO) is also formed as a product. Write a balanced chemical equation for this reaction.

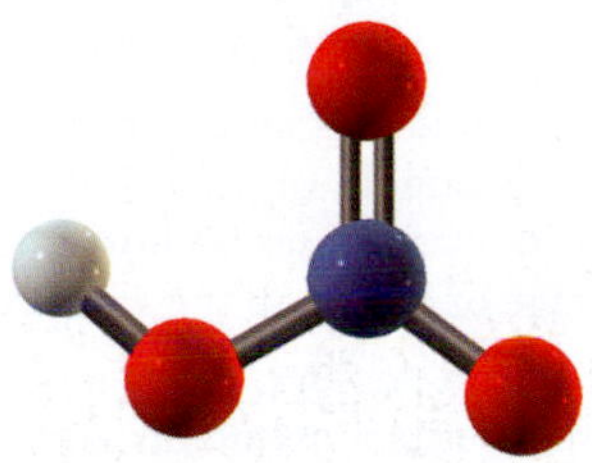

Nitric acid, HNO_3

3.78 One step in the enrichment of uranium for use in nuclear power plants involves the reaction of UO_2 with hydrofluoric acid (HF) solution. The products are solid UF_4 and water. Write a balanced chemical equation for this reaction.

3.79 Pyridine has the molecular formula C_5H_5N. When pyridine reacts with O_2, the products are CO_2, H_2O, and N_2. Write a balanced equation for this reaction.

3.80 Pyrrole has the molecular formula C_4H_5N. When pyrrole reacts with O_2, the products are CO_2, H_2O, and N_2. Write a balanced equation for this reaction.

3.81 Hydrogen cyanide (HCN) is extremely toxic, but it is used in the production of several important plastics. In the most common method for producing HCN, ammonia (NH_3) and methane (CH_4) react with oxygen (O_2) to give HCN and water. Write a balanced chemical equation for this reaction.

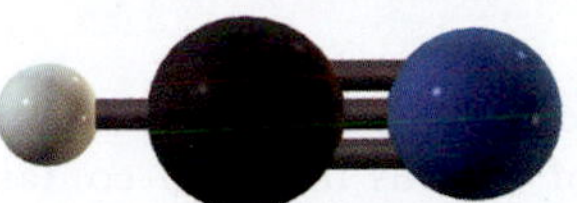

Hydrogen cyanide, HCN

3.82 Many chemical reactions take place in the catalytic converter of a car. In one of these reactions, nitric oxide (NO) reacts with ammonia (NH_3) to give nitrogen (N_2) and water. Write a balanced equation for this reaction.

3.83 Adipic acid is used in the production of nylon, so it is manufactured in large quantities. The most common method for the preparation of adipic acid is the reaction of cyclohexane with oxygen. Balance the skeleton equation shown below.

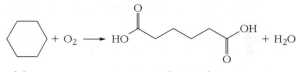

cyclohexane $+ O_2 \longrightarrow$ HO–C(=O)–$(CH_2)_4$–C(=O)–OH $+ H_2O$ (adipic acid)

3.84 Calcium carbonate (limestone, $CaCO_3$) dissolves in hydrochloric acid, producing water and carbon dioxide. An *unbalanced* net ionic equation for this reaction is given below. Balance it.

$$CaCO_3(s) + H_3O^+(aq) \rightarrow H_2O(\ell) + CO_2(g) + Ca^{2+}(aq)$$

3.85 Answer each of the following questions. Note that none of them should require difficult calculations.

(a) How many molecules are present in 1 mole of octane (C_8H_{18})?

(b) How many moles of fluorine atoms are present in 4 moles of C_2F_6?

(c) What is the approximate mass (to the nearest gram) of 3×10^{23} carbon atoms?

3.86 Consider two samples of liquid: 1 mole of water (H_2O) and 1 mole of ethanol (C_2H_5OH). Answer each of the following questions. Note that none of these should require significant calculations.

(a) Which sample contains more atoms?

(b) Which sample contains more molecules?

(c) Which sample has a larger mass?

3.87 Cumene is a hydrocarbon, meaning that it contains only carbon and hydrogen. If this compound is 89.94% C by mass and its molar mass is 120.2 g/mol, what is its molecular formula?

3.88 Methyl cyanoacrylate is the chemical name for the substance sold as Super Glue, and it has the chemical formula

$C_5H_5NO_2$. Calculate the number of molecules of this substance in a 1.0-ounce tube of Super Glue, assuming that the glue is 80% methyl cyanoacrylate by mass.

3.89 A low-grade form of iron ore is called taconite, and the iron in the ore is in the form Fe_3O_4. If a 2.0-ton sample of taconite pellets yields 1075 pounds of iron when it is refined, what is the mass percentage of Fe_3O_4 in taconite?

3.90 The characteristic odor of decaying flesh is due to the presence of various nitrogen-containing compounds. One such compound, called putrescine, was analyzed and found to contain 54.49% carbon, 13.72% hydrogen, and 31.78% nitrogen by mass. If the molar mass of putrescine is known to be between 85 and 105, what is its molecular formula?

3.91 Iron–platinum alloys may be useful as high-density recording materials because of their magnetic properties. These alloys have been made with a wide range of composition, from 34.0 at % Pt to 81.8 at % Pt. Express this range in mol %.

3.92 Some aluminum–lithium alloys display the property of superplasticity, meaning they can undergo tensile deformation by large amounts (1000 times or more) without breaking. If such an alloy has 4 wt % Li, what is its composition in mol %? Explain the relative magnitudes of the mole percentage and weight percentage based on the molar masses of aluminum and lithium.

3.93 Which (if any) of the following compounds are electrolytes? **(a)** glucose, $C_6H_{12}O_6$, **(b)** ethanol, C_2H_5OH, **(c)** magnesium sulfide, MgS, **(d)** sulfur hexafluoride, SF_6

3.94 Classify the following compounds as acids or bases, weak or strong. **(a)** perchloric acid, **(b)** cesium hydroxide, **(c)** carbonic acid, H_2CO_3, **(d)** ethylamine, $C_2H_5NH_2$

3.95 What is the mass in grams of solute in 250. mL of a 0.0125 M solution of $KMnO_4$?

3.96 What volume of 0.123 M NaOH in milliliters contains 25.0 g NaOH?

3.97 Nitric acid is often sold and transported as a concentrated 16 M aqueous solution. How many gallons of such a solution would be needed to contain the entire 1.65×10^7 pounds of HNO_3 produced in the United States in 2008?

3.98 Twenty-five mL of a 0.388 M solution of Na_2SO_4 is mixed with 35.3 mL of 0.229 M Na_2SO_4. What is the molarity of the resulting solution? Assume that the volumes are additive.

3.99 As computer processor speeds increase, it is necessary for engineers to increase the number of circuit elements packed into a given area. Individual circuit elements are often connected using very small copper "wires" deposited directly onto the surface of the chip. In current-generation processors, these copper interconnects are about 22 nm wide. How many copper atoms would be in a 1-mm length of such an interconnect, assuming a square cross section. (The density of copper is 8.96 g/cm^3.)

3.100 As chip speeds increase, the width of the interconnects described in Problem 3.99 must be reduced. A hypothetical limit to this process would be reached if the interconnect was just one copper atom wide. Use the density of copper (8.96 g/cm^3) to estimate the diameter of a copper atom. (Optical interconnects are being developed, and are likely to replace copper in this application within a few years.)

3.101 Materials engineers often create new alloys in an effort to improve the properties of an existing material. ZnO based semiconductors show promise in applications like light-emitting diodes, but their performance can be enhanced by the addition of small amounts of cadmium. One material that has been studied can be represented by the formula $Zn_{0.843}Cd_{0.157}O$. (These materials are solid solutions, and so they can have variable compositions. The noninteger coefficients do not imply fractional atoms.) Express the composition of this alloy in terms of **(a)** at %, **(b)** mol %, and **(c)** wt %.

FOCUS ON PROBLEM SOLVING EXERCISES

3.102 The protein that carries oxygen in the blood is called hemoglobin. It is 0.335% Fe by mass. Given that a molecule of hemoglobin contains four iron atoms, describe how you can calculate the molar mass of hemoglobin. Do you need to look up anything to do it?

3.103 The chlorophyll molecule responsible for photosynthesis in plants contains 2.72% Mg by mass. There is only one Mg atom per chlorophyll molecule. How can you determine the molar mass of chlorophyll based on this information?

3.104 In one experiment, the burning of 0.614 g of sulfur produced 1.246 g of sulfur dioxide as its only product. In a second experiment run to verify the outcome of the first, a sample was burned and 0.813 g of sulfur dioxide was obtained, but the student running the experiment forgot to determine the initial mass of sulfur burned. Describe how the data already obtained could be used to calculate the initial mass of the sulfur burned in the second experiment.

3.105 $MgCl_2$ is often found as an impurity in table salt (NaCl). If a 0.05200-g sample of table salt is found to contain 61.10% Cl by mass, describe how you could determine the percentage of $MgCl_2$ in the sample.

3.106 Under the Clean Air Act, companies can purchase the right to emit certain pollutants from other companies that emit less than they are allowed by law. Suppose that the cost to purchase a pollution credit for 1 ton of sulfur dioxide is $300. Describe how you would determine how many SO_2 molecules could be emitted per dollar at this price.

3.107 The average person exhales 1.0 kg of carbon dioxide in a day. Describe how you would estimate the number of CO_2 molecules exhaled per breath for this average person.

3.108 The simplest approximate chemical formula for the human body could be written as $C_{728}H_{4850}O_{1970}N_{104}Ca_{24}P_{16}K_4S_4Na_3Cl_2Mg$. Based on this formula, describe how you would rank by mass the ten most abundant elements in the human body.

3.109 For the oxides of iron, FeO, Fe_2O_3, and Fe_3O_4, describe how you would determine which has the greatest percentage by mass of oxygen. Would you need to look up any information to solve this problem?

Cumulative Problems

3.110 Consider common sugars such as glucose ($C_6H_{12}O_6$) and sucrose ($C_{12}H_{22}O_{11}$). What type of chemical bonding would you expect to find in these chemicals?

3.111 For the compounds that commonly undergo precipitation reactions, what type of chemical bonding is expected?

3.112 If you have 32.6 g of sodium carbonate that is dissolved to give 2.10 L of solution, what is the molarity of the solution? What is the molarity of the sodium ions?

3.113 If you have 21.1 g of iron(II) nitrate that is dissolved to give 1.54 L of solution, what is the molarity of the solution? What is the molarity of the nitrate ions?

3.114 What type of reasoning were we using when we developed the equation for dilution, $M_i \times V_i = M_f \times V_f$?

3.115 Most periodic tables provide molar masses with four or five significant figures for the elements. How accurately would you have to measure the mass of a sample of roughly 100 g to make a calculation of the number of moles of the chemical to have its significant figures limited by the molar mass calculation rather than the mass measurement?

4 Stoichiometry

CHAPTER OUTLINE

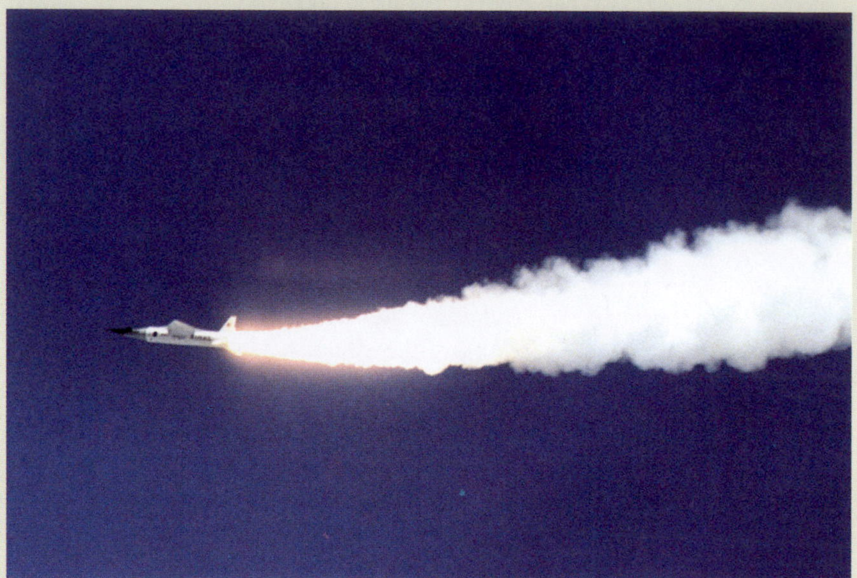

Chemical reactions initiated in the Pegasus rocket accelerate NASA's X-43A aircraft to hypersonic speeds. In a test flight, the X-43A approached Mach 10, ten times the speed of sound. *NASA Dryden Flight Research Center*

A tremendous number of chemical compounds exist in nature and undergo myriad reactions. By building and exploiting a systematic understanding of reactivity, chemists also have produced an impressive array of man-made compounds. The polymers that we encountered in Chapter 2 are handy examples, as are most modern pharmaceuticals. Other practical applications of chemistry involve the development of new routes to synthesize existing natural compounds. As you might imagine, making any of these syntheses commercially feasible requires detailed and *quantitative* understanding of the reactions involved. The economics of any chemical process obviously depend on the amounts of each reactant needed to produce a given amount of product. For processes carried out on an industrial scale, even very small changes in efficiency can have enormous impact on profitability. The quantitative relationships between the amounts of reactants and products in a chemical reaction are referred to as stoichiometry. In this chapter, we'll use a variety of chemical reactions associated with various fuels to illustrate important stoichiometric concepts.

Chapter Objectives

After mastering this chapter, you should be able to

- describe the chemical composition of gasoline.
- write balanced chemical equations for the combustion of fuels.

- calculate the amount of product expected from a chemical reaction, given the amounts of reactants used.
- calculate the amounts of reactants needed in a chemical reaction to produce a specified amount of product.
- identify a limiting reactant and calculate the amount of product formed from a nonstoichiometric mixture of reactants.
- calculate the percentage yield of a chemical reaction.
- identify at least two common additives in gasoline and explain why they are used.

4-1 Gasoline and Other Fuels

Gasoline is the most widely used fuel in our society: daily consumption in the United States averages more than 350 million gallons. To explore the chemistry of gasoline, we must begin by examining its composition. The fuel we know as gasoline is actually a rather complex mixture, typically containing more than 100 different chemical compounds. The exact composition of gasoline varies somewhat depending on factors including the grade of gas, geographic location, and time of year. But the predominant compounds are **hydrocarbons**, molecules containing only carbon and hydrogen atoms. Most of the hydrocarbon molecules in gasoline are **alkanes**, compounds whose carbon atoms are linked together by single bonds. Most of the alkanes in gasoline contain between 6 and 11 carbon atoms.

Table 4.1 lists the names and formulas of several small alkanes, along with molecular models showing their structures. The general formula for any alkane is C_nH_{2n+2}, where n is an integer.

For alkanes with four or more carbon atoms, there are actually multiple possible structures having the same formula. The structures shown in Table 4.1 are known as straight-chain forms because all of the carbon atoms are linked from end to end in a single chain. Other forms involve branched chains, similar to what we saw in Section 2-8 for polyethylene. Two or more structures with the same chemical formula are called **isomers**. For C_5H_{12}, three possible isomers exist. In addition to the straight-chain form shown in Table 4.1, we could also have the two branched forms shown below. As the number of carbons in the chain increases, the number of possible isomers increases rapidly; for $C_{10}H_{22}$, 75 isomers are possible.

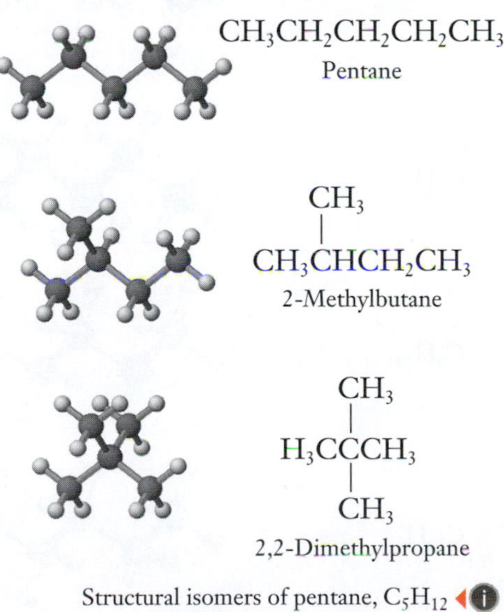

Structural isomers of pentane, C_5H_{12}

The names 2-methylbutane and 2,2-dimethylpropane are examples of the systematic IUPAC nomenclature used for organic compounds. The numbers indicate the positions at which various functional groups are found.

Table ■ 4.1

The first ten alkanes

Compound	Formula	Structure
Methane	CH_4	
Ethane	C_2H_6	
Propane	C_3H_8	
Butane	C_4H_{10}	
Pentane	C_5H_{12}	
Hexane	C_6H_{14}	
Heptane	C_7H_{16}	
Octane	C_8H_{18}	
Nonane	C_9H_{20}	
Decane	$C_{10}H_{22}$	

When gasoline is burned in an engine, all of these various compounds undergo combustion simultaneously, reacting with oxygen from the air. As you might imagine, an accurate description of the combustion of such a complex mixture of compounds would be quite a challenge. So for our present purposes, it will be helpful to use a simplifying model. The simplest possible model is one in which a single compound is used to represent the gasoline mixture, and the most common choice for that compound is octane, C_8H_{18}. If we start by assuming that gasoline can be represented reasonably by octane, then it is fairly easy to write and balance a chemical equation for its combustion. If we further assume complete combustion, then our equation should show octane and oxygen as reactants, and carbon dioxide and water as products.

Gasoline actually contains several different isomers of octane, and the equations we write would apply to any of those substances.

$$2\ C_8H_{18} + 25\ O_2 \rightarrow 16\ CO_2 + 18\ H_2O$$

Because we made two significant assumptions, this equation is a highly idealized model for the combustion of gasoline. The use of octane to represent all of the hydrocarbons in gasoline is mainly for simplicity. If we chose to, it would not be very difficult to write similar combustion equations for each hydrocarbon that is actually present. But the assumption of complete combustion is more drastic. You probably know that typical automobile exhaust contains a number of compounds besides carbon dioxide and water vapor. Most states require periodic emissions testing to measure the levels of carbon monoxide and hydrocarbons in a car's exhaust, and some states or local areas also require additional tests for other types of compounds. Because these compounds don't appear among the products in our equation above, their presence implies that our simple model does not show the full picture. What additional factors could we consider to get a more complete description of engine chemistry?

We know that gasoline itself contains a wide array of hydrocarbons. So the presence of hydrocarbons in the exhaust most likely indicates that some of the gas has made it through the engine and into the exhaust stream without burning. This could mean that some hydrocarbon molecules have simply not reacted at all before leaving the engine. Or it could mean that some larger molecules have broken down into smaller fragments and that some of those fragments have made it into the exhaust. What about the presence of carbon monoxide? We wrote our original equation based on the assumption of complete combustion. This means that all carbon atoms are converted to CO_2. If insufficient oxygen is present in the engine cylinders, then combustion will be incomplete, and CO is produced instead of CO_2.

Additional products of incomplete combustion are also possible. These include formaldehyde (CH_2O) and soot (particles of carbon).

In **Figures 4.1** and **4.2** we have written two chemical equations with the same reactants but different products. How do the molecules know which reaction to

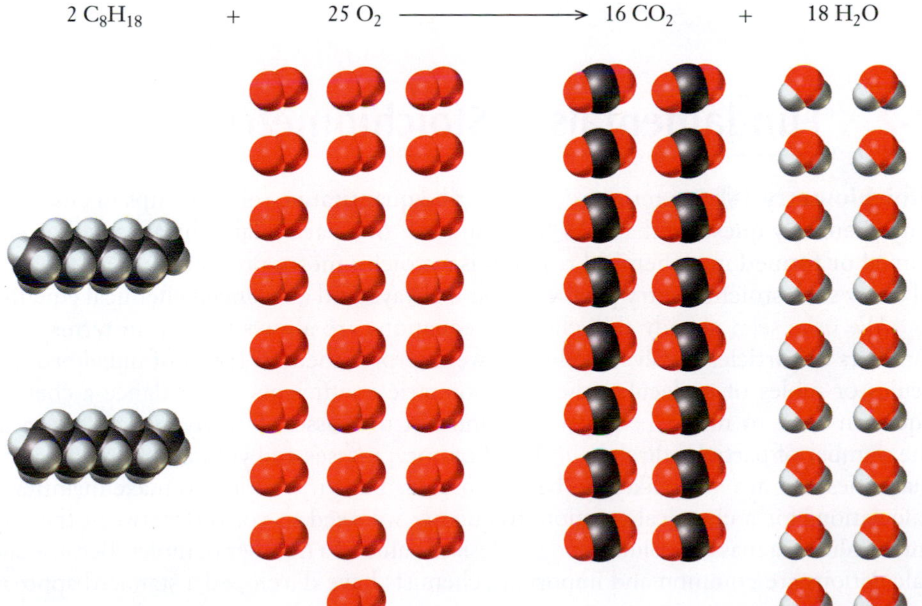

Figure 4.1 ■ This molecular-scale picture of the combustion of octane to give carbon dioxide and water shows the relative number of molecules of each compound involved.

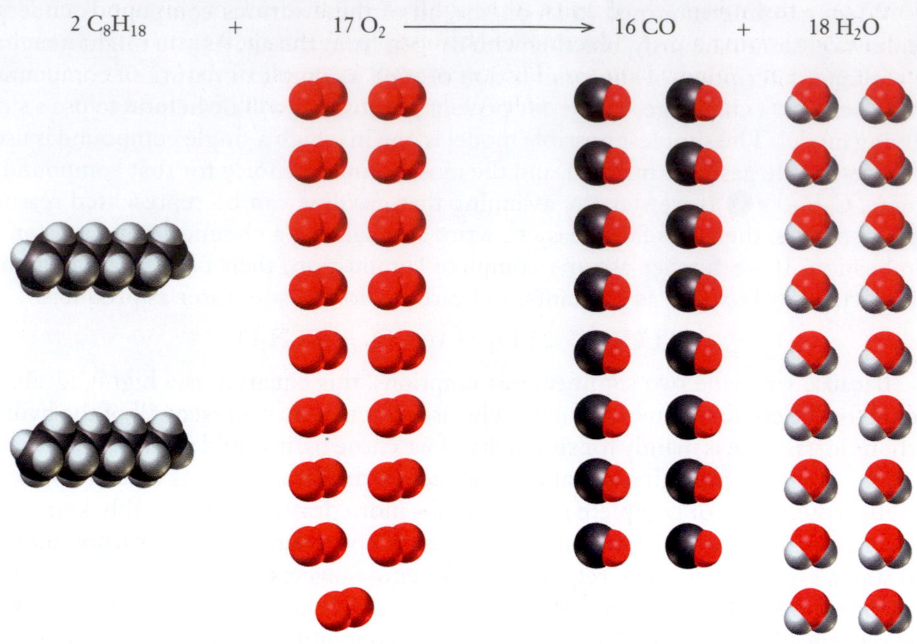

Figure 4.2 ■ Here we show the same sort of molecular picture as in Figure 4.1, but this time for the incomplete combustion of octane to give carbon monoxide and water. Note that both equations balance, but they clearly show different stoichiometry.

undergo? Under most conditions, both reactions will occur to some extent, but the relative importance of each reaction depends on the reaction conditions. The ratio of fuel to oxygen is of particular importance and is closely monitored by the fuel injection system in modern car engines. Other factors, including engine temperature, are also important. A properly tuned engine ensures that the extent of complete combustion is maximized, limits hazardous CO emissions, and also improves gas mileage.

The equations we have written provide us with a good starting point in our consideration of the chemistry of gasoline or other hydrocarbon fuels. But many of the important questions surrounding the chemistry of fuels are quantitative. How much oxygen is needed to allow complete combustion to occur? How does the amount of CO_2 emitted relate to the amount of fuel burned? To begin to answer questions such as these, we will need to consider the stoichiometry of the underlying chemical reactions.

4-2 Fundamentals of Stoichiometry

The word stoichiometry derives from two Greek words: stoicheion (meaning "element") and metron (meaning "measure").

Stoichiometry is a term used to describe quantitative relationships in chemistry. Any chemistry question that asks "How much?" of a particular substance will be consumed or formed in a chemical reaction is a stoichiometry problem. And at the heart of every such stoichiometry problem, you'll always find a balanced chemical equation.

We have seen already that chemical equations are always written in terms of the numbers of particles involved. Whether we interpret them in terms of individual molecules or moles of molecules, the stoichiometric coefficients that balance a chemical equation refer to numbers of particles and not to masses. Usually, we can't measure the number of particles directly in the laboratory; masses and volume of liquids are the quantities that are more likely to be measurable. Thus if we want to make quantitative calculations for a chemical reaction, frequently we need to convert between the measured value of a mass or volume and the desired value of a number of moles. Because such calculations are common and important, chemists have developed a standard approach to overcome this mismatch of units. Although you might think of this approach as an algorithm for solving a particular class of chemistry problems, it is instructive to understand its conceptual underpinnings. The key concept is the use of the balanced chemical equation to establish ratios among moles of the various materials in the reaction.

Obtaining Ratios from a Balanced Chemical Equation

The simplest combustible hydrocarbon—and therefore one of the simplest of all fuels—is methane, CH_4. So the combustion of methane will provide a good place for us to begin our exploration of reaction stoichiometry. We start by writing a balanced chemical equation:

$$CH_4(g) + 2O_2(g) \longrightarrow CO_2(g) + 2H_2O(\ell)$$

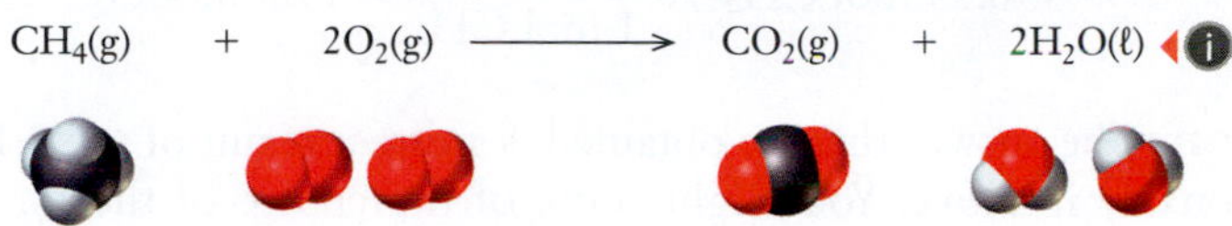

The physical state of the water formed in combustion will depend on the reaction conditions.

In Chapter 3, we have already considered how we might read this chemical equation. We could think of the equation as representing one molecule of methane gas reacting with two molecules of oxygen or a mole of methane molecules reacting with 2 moles of oxygen molecules. In either case, we must have a 1:2 ratio of methane molecules to oxygen molecules, a 1:1 ratio of methane molecules to carbon dioxide molecules, a 1:2 ratio of methane molecules to water molecules, and a 2:2 (or 1:1) ratio of oxygen molecules to water molecules.

As we discussed in Section 3-4, the coefficients in a chemical equation can be interpreted as molar relationships as well as molecular relationships. So we can say that the equation shows that 1 mole of methane reacts with 2 moles of oxygen to produce 1 mole of carbon dioxide and 2 moles of water. From the chemical equation, we can write various **mole ratios,** including the following ones:

$$1 \text{ mol } CH_4 : 2 \text{ mol } O_2$$
$$1 \text{ mol } CH_4 : 1 \text{ mol } CO_2$$
$$1 \text{ mol } CH_4 : 2 \text{ mol } H_2O$$
$$2 \text{ mol } O_2 : 2 \text{ mol } H_2O$$

We could also write the last one as 1:1.

When doing stoichiometric calculations, mole ratios are very often written to look like fractions. They are used analogously to unit conversion factors, relating the amount of one substance to that of another:

$$\frac{1 \text{ mol } CH_4}{2 \text{ mol } O_2} \quad \frac{1 \text{ mol } CH_4}{1 \text{ mol } CO_2} \quad \frac{1 \text{ mol } CH_4}{2 \text{ mol } H_2O} \quad \frac{2 \text{ mol } O_2}{2 \text{ mol } H_2O} \quad \left(\text{or } \frac{1 \text{ mol } O_2}{1 \text{ mol } H_2O}\right)$$

Just like unit conversion factors, these mole ratios can be inverted as needed to carry out a particular calculation. Example Problem 4.1 shows how to use mole ratios in a stoichiometry problem.

EXAMPLE PROBLEM 4.1

In the combustion of methane, how many moles of O_2 are required if 6.75 mol of CH_4 is to be completely consumed?

Strategy We start with the balanced chemical equation and use the stoichiometric coefficients to establish the mole ratio between methane and oxygen. Then, we can use that ratio to relate the amount of methane in the reaction to the amount of oxygen needed.

Solution The balanced equation was given earlier in the chapter:

$$CH_4(g) + 2\ O_2(g) \rightarrow CO_2(g) + 2\ H_2O(\ell)$$

The coefficients from this equation give us the ratio between CH_4 and O_2, which can be expressed in either of the following forms:

$$\frac{1\text{ mol } CH_4}{2\text{ mol } O_2} \text{ or } \frac{2\text{ mol } O_2}{1\text{ mol } CH_4}$$

To calculate the desired amount of O_2 from the known amount of CH_4, we should use the second form. This gives us the result needed:

$$6.75\text{ mol } CH_4 \times \frac{2\text{ mol } O_2}{1\text{ mol } CH_4} = 13.5\text{ mol } O_2$$

Discussion The answer that we obtained is a direct result of the 2:1 mole ratio and so is reasonably intuitive. You might think of the choice of the form of mole ratio as an application of dimensional analysis. Many stoichiometry problems seem more complicated than this because the known quantity is usually not in moles. But the simple use of the mole ratio illustrated here is always the pivotal step in any reaction stoichiometry calculation.

Check Your Understanding How many moles of H_2O are formed in the reaction above?

The balanced chemical equation provides all the mole ratios needed to relate the amounts of the compounds in a reaction. To carry out the reaction in the laboratory, however, we often use a balance to measure out the materials needed, giving us a measurement of mass in grams. The ratios from the chemical equation are in numbers of molecules or moles. So we need to be able to convert between grams and moles. As we have seen, the molar mass provides the way to do this.

The use of ratios to convert from one unit to another is no more complicated in stoichiometry than in simple physical measurements. The important difference, however, is that in many stoichiometry problems, ratios are used three or more times. Each individual use of the ratio follows principles you already know, and if you keep that in mind, you should be able to find your way through even the most complex stoichiometry problems. The main challenge is to determine which ratio to use and when to use it. One way to keep track of the steps needed is to create a flow chart using blocks to indicate when information is derived from a relationship, such as the balanced chemical equation or the molar mass.

In **Figure 4.3**, we can see that if we begin with the mass of one species in the reaction and we want to know the mass of some other species consumed or produced, we have to use three ratios: a ratio that uses molar mass of the substance given, a mole ratio from the balanced chemical equation, and a ratio using the molar mass of the substance whose mass we wish to find. The simplest and most reliable way to identify which ratio to use for each step is to write all quantities with their complete units attached. In setting up the simple calculation in Example Problem 4.1, we wrote the

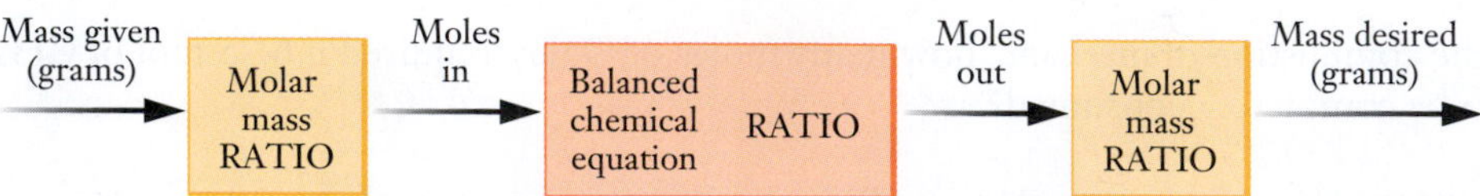

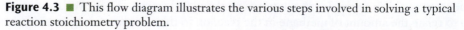

Figure 4.3 ■ This flow diagram illustrates the various steps involved in solving a typical reaction stoichiometry problem.

initial quantity with units of "mol CH_4," and this helped us choose the correct form of the mole ratio for the calculation. Of course, keeping proper units attached to all the quantities in a calculation is *always* a good idea, whether it is a chemical stoichiometry problem or an engineering calculation. Example Problems 4.2 and 4.3 apply this idea to slightly more common stoichiometry calculations, where amounts are measured in mass rather than numbers of moles.

EXAMPLE PROBLEM 4.2

In Section 3-2, we considered the reaction between hydrogen and oxygen to form water. How many grams of water can be produced if sufficient hydrogen reacts with 26.0 g of oxygen?

Strategy First, it will help if we recognize that this is a reaction stoichiometry problem. Two signs of this are that it asks "How much . . ." and that there is obviously a chemical reaction involved. Once we see that, we should immediately realize that we need a balanced chemical equation to give us the mole ratio between water produced and oxygen reacted. We will convert the mass of oxygen given to moles of oxygen, use the mole ratio between oxygen and water to provide moles of water, and then use the molar mass of water to find the mass of water produced.

Solution Recall that both hydrogen and oxygen are diatomic gases. So the balanced equation is

$$2\ H_2(g) + O_2(g) \rightarrow 2\ H_2O(g)$$

The molar mass of oxygen is 32.0 g/mol, and the molar mass of water is 18.0 g/mol. Using the diagram in Figure 4.3, we write sequential ratios, starting with the mass of oxygen given and keeping in mind that the answer should be in grams of water:

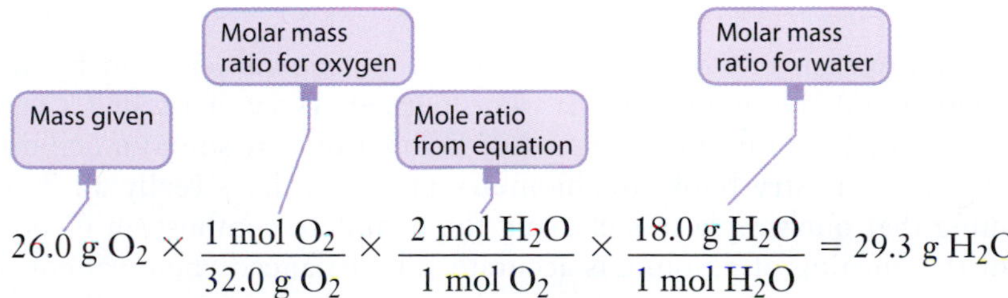

$$26.0\text{ g } O_2 \times \frac{1\text{ mol } O_2}{32.0\text{ g } O_2} \times \frac{2\text{ mol } H_2O}{1\text{ mol } O_2} \times \frac{18.0\text{ g } H_2O}{1\text{ mol } H_2O} = 29.3\text{ g } H_2O$$

Analyze Your Answer These two masses—for the O_2 reactant and the H_2O product—are similar. Does this make sense? Looking at the equation for the reaction, we can see that all of the oxygen atoms end up in water molecules. Because hydrogen is such a light element, the mass of water formed will not be too much greater than the mass of oxygen reacting. Our answer of 29.3 g of H_2O is slightly larger than the 26.0 g of oxygen we started with, so it seems reasonable. (Note that this similarity between the reactant and product masses will not always be the case.)

Discussion Notice that we start with the substance whose quantity was given (26.0 g O_2). The first and last steps are just conversions between mass and moles, with which you should be comfortable. The use of the mole ratio from the balanced equation is the only new concept here.

This calculation can also be carried out in a more stepwise fashion, in which you might first convert the 26.0 g O_2 to 0.813 mol O_2 and so forth. The result will be the same as long as all of the calculations are carried out correctly.

Check Your Understanding Calculate the mass of hydrogen needed to produce 29.2 grams of water.

EXAMPLE PROBLEM 4.3

Tetraphosphorus trisulfide, P_4S_3, is used in the manufacture of "strike anywhere" matches. Elemental phosphorus and sulfur react directly to form P_4S_3:

$$8\ P_4 + 3\ S_8 \rightarrow 8\ P_4S_3$$

If we have 153 g of S_8 and an excess of phosphorus, what mass of P_4S_3 can be produced by this reaction?

Strategy The heart of any stoichiometry problem is the balanced chemical equation that provides the mole ratio we need. We must convert between masses and the number of moles in order to use this ratio, first for the reactant, S_8, and at the end of the problem for the product, P_4S_3. Molar masses provide the needed conversion factors. The phrase "excess of phosphorus" tells us that we have more than enough P_4 to consume 153 g of S_8 completely.

Solution The molar mass of S_8 is 256.6 g/mol and that of P_4S_3 is 220.1 g/mol. Along with the balanced equation, those masses allow us to set up the needed relationships:

$$153\ \text{g S}_8 \times \frac{1\ \text{mol S}_8}{256.5\ \text{g S}_8} \times \frac{8\ \text{mol P}_4\text{S}_3}{3\ \text{mol S}_8} \times \frac{220.1\ \text{g P}_4\text{S}_3}{1\ \text{mol P}_4\text{S}_3} = 3.50 \times 10^2\ \text{g P}_4\text{S}_3$$

Check Your Understanding The smell of a burning match is due to the formation of sulfur oxides. We can represent the relevant reaction as $S + O_2 \rightarrow SO_2$. If 4.8 g of sulfur is burned in this manner, what mass of sulfur dioxide is produced?

Now, we can relate the amount of carbon dioxide emitted by an automobile engine and the amount of gasoline burned. What mass of CO_2 will be produced for every gallon of gasoline used? While this may sound much different from a typical chemistry book stoichiometry problem, that's really all it is. Start by assuming that our simple model using the complete combustion of octane to represent the burning of gasoline is accurate. The balanced chemical equation is one we saw earlier:

$$2\ C_8H_{18} + 25\ O_2 \rightarrow 16\ CO_2 + 18\ H_2O$$

We'll need to know the mass of a gallon of gasoline. As you might guess, this depends on the exact composition of the gasoline in question. But because we are basing our calculation on octane, it might be reasonable to use the density of octane, which is 0.7025 g/mL. Use that and some unit conversion factors to find the mass in grams of a gallon of octane:

$$1\ \text{gallon} \times \frac{3.7854\ \text{L}}{1\ \text{gallon}} \times \frac{1000\ \text{mL}}{1\ \text{L}} \times \frac{0.7025\ \text{g octane}}{\text{mL}} = 2659\ \text{g}$$

Now we've reduced our original question to something that looks a lot more like the previous examples: What mass of CO_2 is produced by the complete combustion of 2659 g of C_8H_{18}? Use molar masses and the mole ratio from the balanced equation to set up the calculation:

$$2659\ \text{g C}_8\text{H}_{18} \times \frac{1\ \text{mol C}_8\text{H}_{18}}{114.23\ \text{g C}_8\text{H}_{18}} \times \frac{16\ \text{mol CO}_2}{2\ \text{mol C}_8\text{H}_{18}} \times \frac{44.010\ \text{g}}{1\ \text{mol CO}_2} = 8196\ \text{g CO}_2$$

To put that in more familiar units, use the fact that a pound is approximately 454 g. Our calculation then says that burning a gallon of gasoline will produce 18 pounds of carbon dioxide. Although we made a few simplifying assumptions, our result should certainly be a reasonable estimate.

4-3 Limiting Reactants

When we carry out chemical reactions, the available supply of one reactant is often exhausted before the other reactants. As soon as we run out of one of the reactants, the reaction stops. The reactant completely consumed has determined how far the reaction can go, limiting the quantity of product produced. We say that the reactant completely consumed in the reaction is the **limiting reactant**.

To illustrate this important concept, let's imagine that somehow we can do an experiment involving just a few molecules. Suppose we start with six molecules each of H_2 and O_2 and allow them to react. How much water can be formed? We know the equation for this reaction from Example Problem 4.2:

$$2\ H_2(g) + O_2(g) \rightarrow 2\ H_2O(g)$$

Figure 4.4 shows what such an experiment might look like. The reaction will proceed as long as both reactants are present. But after six molecules of water have been formed, all of the H_2 will have been consumed. Even though three more molecules of oxygen remain, they cannot react. So the amount of product formed is limited by the amount of hydrogen that was available, and we say that hydrogen was the limiting reactant.

In many cases, we manipulate the amounts of reactants to make certain that one compound is the limiting reactant. If the synthesis of a particular compound or material involves one scarce or expensive reactant, then a clever process engineer would most likely be sure to make that the limiting reactant by using large excesses of other reactants. Combustion reactions like burning gasoline in a car engine provide a ready example. A large excess of oxygen is always available because it can simply be drawn from the air. Your car may run out of gas, but it is not likely to run

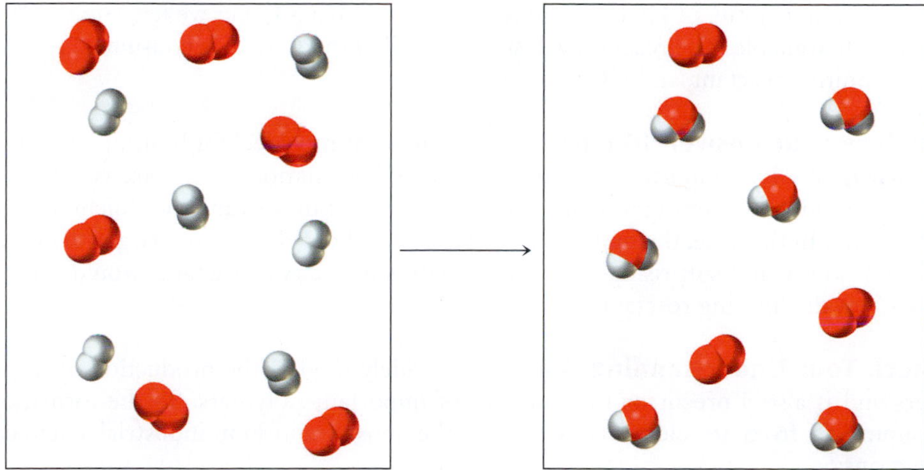

Figure 4.4 ■ This molecular-scale picture illustrates the concept of the limiting reactant. The box on the left represents the mixture of reactants in our hypothetical experiment: six H_2 molecules (shown in white) and six O_2 molecules (in red). The right-hand side shows the molecules present once the reaction has gone to completion: six H_2O molecules and three unreacted O_2 molecules. Here H_2 was the limiting reactant. Because no more H_2 molecules are available, no further reaction is possible.

out of oxygen. So in any calculation related to fuel consumption or exhaust emissions, it would be reasonable to assume that gasoline would always be the limiting reactant.

In other circumstances, the identity of the limiting reactant may not be so obvious. In a laboratory experiment, we might simply mix conveniently available quantities of substances to begin to study their reaction. Determining the limiting reactant requires comparing the amount of each reactant present. As always, we should keep in mind that balanced chemical equations tell us the ratios in which moles of one substance react with moles of another. So we cannot compare masses to determine the limiting reactant; we must do the comparison using the relevant mole ratios. Example Problems 4.4 and 4.5 show how we can determine which substance is in excess and which is the limiting reactant.

EXAMPLE PROBLEM 4.4

A solution of hydrochloric acid contains 5.22 g of HCl. When it is allowed to react with 3.25 g of solid K_2CO_3, the products are KCl, CO_2, and H_2O. Which reactant is in excess?

Strategy As for any reaction stoichiometry problem, we should start with a balanced equation. One way to proceed from there is to calculate the amount of one reactant that would combine with the given amount of the second reactant. Comparing that with the amount actually available will reveal the limiting reactant.

There are a variety of possible strategies for identifying a limiting reactant.

Solution We know the reactants and products, so we can write a skeleton equation easily:

$$HCl + K_2CO_3 \rightarrow KCl + CO_2 + H_2O$$

Balancing this gives us the chemical equation we need for our stoichiometry calculation:

$$2\ HCl + K_2CO_3 \rightarrow 2\ KCl + CO_2 + H_2O$$

Let's use the quantity of HCl given and calculate the amount of K_2CO_3 that would react completely with it:

$$5.22 \text{ g HCl} \times \frac{1 \text{ mol HCl}}{36.46 \text{ g HCl}} \times \frac{1 \text{ mol } K_2CO_3}{2 \text{ mol HCl}} \times \frac{138.2 \text{ g } K_2CO_3}{1 \text{ mol } K_2CO_3} = 9.89 \text{ g } K_2CO_3$$

So the given quantity of HCl (5.22 g) requires 9.89 g K_2CO_3, but we have only 3.25 g of K_2CO_3 available. The reaction will stop once all of the K_2CO_3 is consumed. K_2CO_3 is the limiting reactant, and HCl is in excess.

Analyze Your Answer It's often hard to gain an intuitive feel for limiting reactants, so that is why it is important to learn this type of calculation. There are two factors that determine the limiting reactant—the relative molar masses and the stoichiometric mole ratio. In this case, the molar mass of K_2CO_3 is about three times larger than that of HCl, so it is not surprising that similar masses of the two reactants would lead to K_2CO_3 as the limiting reactant.

Check Your Understanding Ammonia is widely used in the production of fertilizers and is also a precursor to a number of important polymers. So the formation of ammonia from its elements is one of the most important industrial chemical reactions:

$$N_2 + 3\ H_2 \rightarrow 2\ NH_3$$

This process is usually carried out using an excess of hydrogen. If a reaction vessel contains 1.5×10^3 moles of N_2, how much hydrogen would be needed to ensure that it will be in excess?

EXAMPLE PROBLEM 4.5

In Example Problem 4.3, we used the reaction that produces P_4S_3, one of the reactants in the combustion of a match:

$$8\ P_4 + 3\ S_8 \rightarrow 8\ P_4S_3$$

If 28.2 g of P_4 is allowed to react with 18.3 g of S_8, which is the limiting reactant?

Strategy We can choose either reactant and determine how much of the other reactant is required to consume it entirely. Comparing the amount calculated with the amount given in the problem will let us determine which reactant is limiting.

Solution Let's begin with P_4:

$$28.2\ \text{g P}_4 \times \frac{1\ \text{mol P}_4}{123.9\ \text{g P}_4} \times \frac{3\ \text{mol S}_8}{8\ \text{mol P}_4} \times \frac{256.5\ \text{g S}_8}{1\ \text{mol S}_8} = 21.9\ \text{g S}_8$$

So, 28.2 g of P_4 requires 21.9 g of S_8 to react completely. We have only 18.3 g of S_8, so there is not enough S_8 to react with all of the P_4. Therefore, S_8 is the limiting reactant.

Check Your Understanding To convince yourself that you can start with either reactant, redo the example starting with the quantity of S_8 given.

The determination of the limiting reactant is typically just a piece of a larger puzzle. In most limiting reactant stoichiometry problems, the real goal is to determine how much product could be formed from a particular reactant mixture. Perhaps the first critical step in such a problem is to correctly recognize the fact that it is a limiting reactant situation. Whenever a particular amount of more than one reactant is specified in a stoichiometry problem, it should tell us to begin by finding the limiting reactant. Once we know which substance runs out first, we are left with an ordinary stoichiometry calculation to solve the problem. Example Problem 4.6 shows how this type of problem can be approached.

EXAMPLE PROBLEM 4.6

MTBE (methyl *tert*-butyl ether) has been used as an additive in gasoline. The compound is produced by reacting methanol and isobutene, according to the following equation:

We will discuss the use of such additives in the closing section of this chapter.

$$\underset{\text{Methanol}}{CH_3OH} + \underset{\text{Isobutene}}{(CH_3)_2C{=}CH_2} \rightarrow \underset{\text{MTBE}}{(CH_3)_3COCH_3}$$

If 45.0 kg of methanol is allowed to react with 70.0 kg of isobutene, what is the maximum mass of MTBE that can be obtained?

Strategy We are given amounts of two different reactants (methanol and isobutene). So we should realize this is a potential limiting reactant situation. We can identify the limiting reactant as in the previous examples: choose one reactant and find the amount of the other that would be needed to react with it. Comparing that result with the given amount, we can determine which of the reactants is limiting. Once we have done that, it will be straightforward to calculate the expected amount of product.

Solution We'll need the molar masses of all three compounds. Taking care to interpret the parentheses in the molecular formulas correctly, we find the following:

Methanol (CH_3OH): 32.042 g/mol

Isobutene ($(CH_3)_2C{=}CH_2$): 56.107 g/mol

MTBE ($(CH_3)_3COCH_3$): 88.149 g/mol

To identify the limiting reactant, we can calculate the mass of isobutene that would react with 45.0 kg (or 45,000 g) of methanol:

$$45{,}000 \text{ g methanol} \times \frac{1 \text{ mol methanol}}{32.042 \text{ g methanol}} \times \frac{1 \text{ mol isobutene}}{1 \text{ mol methanol}} \times \frac{56.107 \text{ g isobutene}}{1 \text{ mol isobutene}}$$
$$= 7.88 \times 10^4 \text{ g isobutene} = 78.8 \text{ kg isobutene}$$

This tells us that to use 45.0 kg of methanol, we need at least 78.8 kg of isobutene. But we have only 70.0 kg of isobutene. So the amount of isobutene available will determine the amount of MTBE that can be formed. Isobutene is the limiting reactant and methanol is in excess.

Now the problem is reduced to a simpler stoichiometry calculation starting from the mass of isobutene available:

$$70{,}000 \text{ g isobutene} \times \frac{1 \text{ mol isobutene}}{56.107 \text{ g isobutene}} \times \frac{1 \text{ mol MTBE}}{1 \text{ mol isobutene}} \times \frac{88.149 \text{ g MTBE}}{1 \text{ mol MTBE}}$$
$$= 1.10 \times 10^5 \text{ g MTBE} = 1.10 \times 10^2 \text{ kg MTBE}$$

Analyze Your Answer This value is larger than the original amount of either reactant. Does this make sense? If we look closely at the reaction, it is a synthesis where the two molecules become one. If our answer had been greater than the sum of the two reactant masses, we would have cause for concern. But this value is certainly plausible.

Discussion This problem combines two types of calculations that we've seen in earlier examples. The first step is identifying the limiting reactant, and the second is the actual stoichiometry calculation. Note that here isobutene was the limiting reactant even though the available mass of isobutene was greater than that of methanol. This is a reminder that reaction stoichiometry calculations should always be handled in terms of moles, not masses.

Check Your Understanding The compound diborane, B_2H_6, was once proposed for use as a rocket propellant. It can be produced by the following reaction:

$$3\ LiBH_4 + BF_3 \rightarrow 2\ B_2H_6 + 3\ LiF$$

If 24.6 g of $LiBH_4$ is combined with 62.4 g of BF_3, what mass of diborane can be formed?

The concept of limiting reactants has some important implications for engineering applications where chemical reactions are involved. We've already mentioned that a good design for a chemical process would ensure that a scarce or expensive substance would be the limiting reactant, so that none of it would go to waste. A rather different situation exists for the design of rocket engines. Here the total mass of the rocket is a major consideration, so the aim should be to use the minimum mass of fuel that could deliver the needed thrust. Many rocket engines rely on so-called two-component fuel mixtures, where energy is released as two compounds are allowed to react. The optimal design for such a rocket would generally ensure that the two compounds are present in a stoichiometric mixture, so that neither is in excess. Example Problem 4.7 illustrates how such a mixture might be chosen.

EXAMPLE PROBLEM 4.7

The solid fuel booster rockets used on the space shuttle were based on the following reaction between ammonium perchlorate and aluminum:

$$3\ NH_4ClO_4(s) + 3\ Al(s) \rightarrow Al_2O_3(s) + AlCl_3(g) + 3\ NO(g) + 6\ H_2O(g)$$

If either reactant is in excess, unnecessary mass will be added to the shuttle, so a stoichiometric mixture is desired. What mass of each reactant should be used for every kilogram of the fuel mixture?

Strategy We want to ensure that there will be no excess of either reactant. Because we are asked for the composition "for every kilogram" of fuel, we might start by assuming the total mass of fuel will be 1000 g. We can write that as an equation with two unknowns:

$$m_{NH_4ClO_4} + m_{Al} = 1000\ g$$

Because we have two unknowns, we will need a second relationship between the two masses. The balanced chemical equation shows us that the *mole* ratio between the two reactants should be 3:3, or 1:1. We can use molar masses to convert that to a ratio in terms of *masses* and then use that ratio as a second equation to solve the problem.

Solution The molar mass of NH_4ClO_4 is 117.49 g/mol, and that of Al is 26.98 g/mol. To ensure that we have a stoichiometric mixture, we need equal molar amounts of each reactant. So if we were to use 117.49 g (1 mole) of NH_4ClO_4, we would need 26.98 g (also 1 mole) of Al. We can write this as an equation:

$$\frac{m_{NH_4ClO_4}}{m_{Al}} = \frac{117.49\ g}{26.98\ g} = 4.355$$

And although our actual amounts will need to be larger than those, that ratio of masses will still be right. So now we have two equations for the two unknown masses, and we can solve by doing a little bit of algebra. Our two equations are

$$m_{NH_4ClO_4} + m_{Al} = 1000\ g$$

$$m_{NH_4ClO_4} = 4.355\ m_{Al}$$

Substituting the right-hand side of the second equation in the first equation gives

$$4.355\ m_{Al} + m_{Al} = 1000\ g$$

Simplifying that and solving, we get the mass of aluminum needed for each kilogram of the fuel mixture:

$$5.355\ m_{Al} = 1000\ g$$

$$m_{Al} = \frac{1000\ g}{5.355} = 186.8\ g\ Al$$

Now the mass of NH_4ClO_4 is easy to obtain:

$$m_{NH_4ClO_4} = 1000\ g - m_{Al}$$

$$= 1000\ g - 186.8\ g = 813.2\ g$$

So each kilogram of fuel should comprise 186.8 g Al and 813.2 g NH_4ClO_4.

Discussion The mass of fuel required for a space shuttle engine would clearly be much larger than a kilogram, but this result could easily be scaled up to the needed mass. As for most problems, there are a variety of other approaches that could be used to solve this. The particular strategy that we've used is based on the general idea that if we have two unknown quantities (like the two masses here), we need to find two separate relationships between them to solve the problem.

Check Your Understanding A fuel mixture comprising 600.0 g of NH_4ClO_4 and 400.0 g of Al is used in a test of a laboratory-scale mock-up of a shuttle engine. When the engine stops burning, some unused fuel remains. What substance is this unburned fuel, and what mass of it should be found?

4-4 Theoretical and Percentage Yields

For chemical reactions, and especially for reactions that produce commercial products, we would generally prefer to have reactions that are efficient, producing as much of the desired products as possible while minimizing any unwanted byproducts. Why is this important? In many industrial situations, one big reason is economics. If a business can increase the amount of product obtained from a given amount of reactants, it should become more profitable. Many factors such as temperature of the reaction, the possibility of **side reactions**, or further reaction of the product to form something else can decrease the amount of desired product obtained. Minimizing these unwanted reactions could also reduce the amount of waste produced, thereby lowering environmental impacts and associated costs.

We can rate the efficiency of a reaction by calculating how much product would form under perfect or ideal conditions and then comparing the actual measured result with this ideal. The ideal amount of product is called the **theoretical yield**, and it is obtained by working a stoichiometry problem. Measuring the amount of product formed gives us the **actual yield**. From the ratio of the actual yield to the theoretical yield, we can calculate the **percentage yield**:

$$\text{Percentage yield} = \left(\frac{\text{actual yield}}{\text{theoretical yield}}\right) \times 100\% \qquad (4.1)$$

Example Problem 4.8 shows the use of this equation to find the percentage yield for a reaction. Increasing this percentage yield is one of the foremost goals for a chemical engineer working on an industrial process.

EXAMPLE PROBLEM 4.8

The Solvay process is important in the commercial production of sodium carbonate (Na_2CO_3), which is used in the manufacture of most glass. The last step in the Solvay process is the conversion of $NaHCO_3$ (sodium bicarbonate, or baking soda) to Na_2CO_3 by heating:

$$2\ NaHCO_3(s) \xrightarrow{\text{heat}} Na_2CO_3(s) + CO_2(g) + H_2O(g)$$

In a laboratory experiment, a student heats 42.0 g of $NaHCO_3$ and determines that 22.3 g of Na_2CO_3 is formed. What is the percentage yield of this reaction?

Strategy We know the actual yield from the experiment. To calculate the percentage yield, first we need to find the theoretical yield. We can do that by calculating the maximum quantity of product that could form, based on the stoichiometry of the reaction. Once we have both the theoretical yield and the actual yield, finding the percentage yield is straightforward.

Solution Begin by working the stoichiometry problem:

$$42.0\text{ g} \times \frac{1\text{ mol NaHCO}_3}{84.0\text{ g NaHCO}_3} \times \frac{1\text{ mol Na}_2\text{CO}_3}{2\text{ mol NaHCO}_3} \times \frac{106.0\text{ g Na}_2\text{CO}_3}{1\text{ mol Na}_2\text{CO}_3} = 26.5\text{ g Na}_2\text{CO}_3$$

Then compute the percentage yield:

$$\begin{aligned}\text{Percentage yield} &= \frac{\text{actual yield}}{\text{theoretical yield}} \times 100\% \\ &= \frac{22.3\text{ g}}{26.5\text{ g}} \times 100\% \\ &= 84.2\%\end{aligned}$$

Analyze Your Answer A quick estimate of the answer can tell us that we have done the stoichiometry correctly. Because 42 is half of 84, we have 0.5 mol of $NaHCO_3$. From the equation, the mole ratio of $NaHCO_3$ to Na_2CO_3 is 2:1, so we would expect to obtain 0.25 mol of Na_2CO_3. One-fourth of 106 is about 26, so we know that the stoichiometry is correct. Because the actual yield is less than the theoretical yield, an answer less than 100% is plausible. (We may sometimes see yields greater than 100% for reactions like this, but that indicates that the product is not completely dry or contains some impurity.)

Discussion We might ask whether or not this is a "good" percentage yield. It isn't possible to set an absolute numerical target for percentage yields. If a reaction leads to a highly desired (i.e., expensive) product for which there is no alternative source, then a yield of just a few percent might be considered adequate. In the present case, the reaction is rather simple and the product is relatively inexpensive. So a higher percentage yield would probably be needed for an economically viable plant.

Check Your Understanding In the "Check Your Understanding" of Example Problem 4.6, you considered a reaction to produce diborane. An alternative route to this propellant is shown below:

$$3\ NaBH_4 + 4\ BF_3 \rightarrow 2\ B_2H_6 + 3\ NaBF_4$$

If 173.2 g of BF_3 reacts with excess $NaBH_4$ to give 28.6 g of B_2H_6, what is the percentage yield?

How does the idea of percentage yield apply to the chemistry of fuels? For a hydrocarbon such as gasoline, complete combustion liberates more energy than incomplete combustion or other side reactions. Complete combustion also reduces the levels of several potentially harmful compounds, including CO and soot. So we might define the efficiency in terms of the extent of complete combustion. If we wanted to monitor this experimentally, one option would be to monitor the amount of CO_2 in the exhaust stream. Similarly, the presence of CO, CH_2O, or other carbon-containing compounds in the exhaust would allow us to monitor the extent of various side reactions. Such measurements are routinely carried out on test engines to determine the optimum operating conditions.

4-5 Solution Stoichiometry

In the stoichiometry problems we have discussed so far, the numbers of moles involved have been obtained by using the ratio established by the molar mass of the substances. Now we can expand the range of problems that we can consider to include reactions involving solutions. The amount of a solution is typically measured as a volume rather than a mass. And we can then use Equation 3.2 to relate the number of moles to the volume and concentration of a solution. The heart of the stoichiometry problem, however, remains the same. The balanced chemical equation still provides the critical ratios among numbers of moles of various species in the reaction. Again, a flow diagram for solving a stoichiometry problem shows the manipulations we need to consider (**Figure 4.5**).

The only complication added by considering reactions in solution is the need to relate three variables, using Equation 3.2, rather than two variables using the molar mass ratio. Example Problem 4.9 shows how to approach these solution stoichiometry problems.

Figure 4.5 ■ This flow diagram shows the important steps in a typical solution stoichiometry calculation.

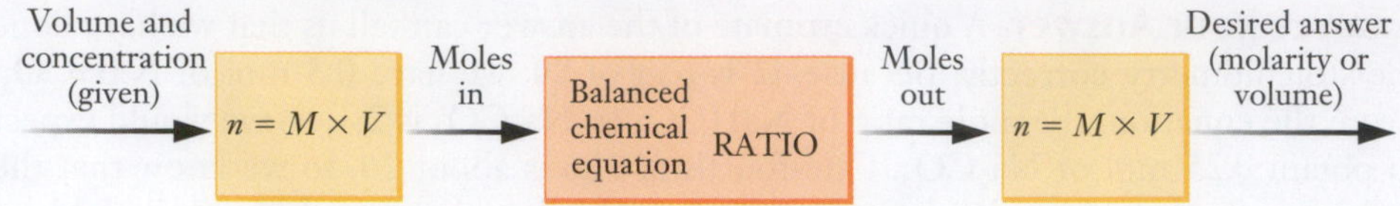

EXAMPLE PROBLEM 4.9

The fuel hydrazine can be produced by the reaction of solutions of sodium hypochlorite and ammonia. The relevant chemical equation is

$$NaClO(aq) + 2\ NH_3(aq) \rightarrow N_2H_4(aq) + NaCl(aq) + H_2O(\ell)$$

If 750.0 mL of 0.806 M NaClO is mixed with excess ammonia, how many moles of hydrazine can be formed? If the final volume of the resulting solution is 1.25 L, what will be the molarity of hydrazine?

Strategy We are asked for the expected amount of a product, so this is a reaction stoichiometry problem. Because NH_3 is said to be in excess, we know that NaClO will be the limiting reactant. So we will use the given volume and concentration to find the number of moles of NaClO reacting. Then use the mole ratio from the balanced equation to find the number of moles of N_2H_4 that can be formed. Finally, we can use that number of moles and the given final volume to obtain the molarity of the N_2H_4 product.

Solution Convert 750.0 mL to 0.7500 L and determine the number of moles of NaClO reacting:

$$n_{\text{NaClO}} = M \times V = 0.806 \text{ mol/L} \times 0.7500 \text{ L} = 0.605 \text{ mol NaClO}$$

Use that along with the 1:1 mole ratio from the balanced equation to find the corresponding number of moles of hydrazine:

$$0.605 \text{ mol NaClO} \times \frac{1 \text{ mol } N_2H_4}{1 \text{ mol NaClO}} = 0.605 \text{ mol } N_2H_4$$

To answer the second part of the problem, we just need to use this number of moles along with the given final volume to find the molarity:

$$M = \frac{n}{V} = \frac{0.605 \text{ mol } N_2H_2}{1.25 \text{ L}} = 0.484 \text{ M } N_2H_4$$

Discussion Notice that the stoichiometric calculation is done using the mole ratio from the balanced equation, just as in all of our earlier examples. Only the conversions to and from numbers of moles have changed because this reaction is in solution.

Check Your Understanding What volume (in milliliters) of 0.150 M HCl is required to react completely with 0.503 g of dry Na_2CO_3?

Many indicators are examples of weak acids or bases, as defined in Chapter 3. ⓘ

A common laboratory technique, called **titration**, requires understanding solution stoichiometry. A solution-phase reaction is carried out under controlled conditions so that the amount of one reactant can be determined with high precision. A carefully measured quantity of one reactant is placed in a beaker or flask. A dye called an **indicator** can be added to the solution. ◀ⓘ The second reactant is added in a controlled fashion, typically by using a burette (**Figure 4.6**). When the reaction is complete, the indicator changes color. When the indicator first changes color, we have a

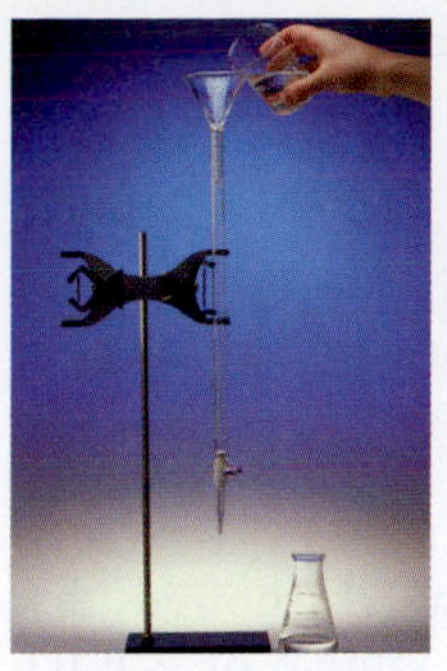
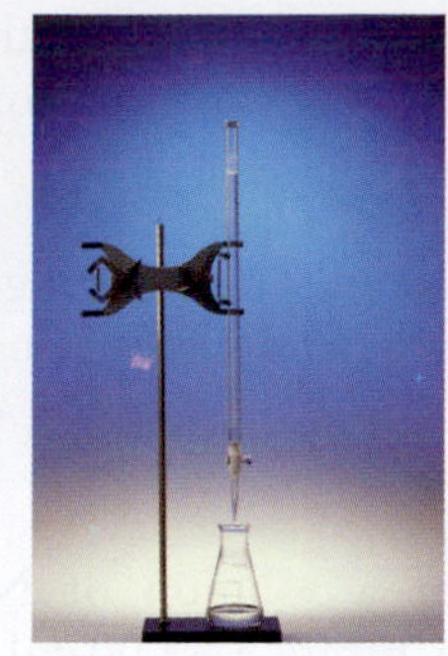

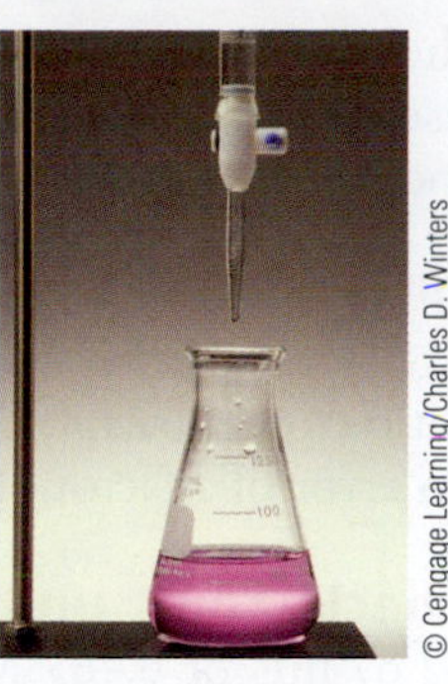

Figure 4.6 ■ The photo sequence shows some of the steps in a typical titration. One reactant solution is poured into a burette in the left-hand panel. The second reactant solution, also containing an indicator, is placed into an Erlenmeyer flask, and the burette is positioned above the flask in the second panel. The valve on the burette allows controlled addition of solution, and the shape of the Erlenmeyer flask permits easy mixing by swirling. In the final panel, the solution in the flask turns pink when the endpoint is reached.

stoichiometric mixture of reactants. We know the number of moles of the first reactant (or the molarity and volume) and the volume of the second reactant used. So as long as we know the balanced equation for the reaction, we can find the unknown concentration of the second reactant, as shown in Example Problem 4.10.

EXAMPLE PROBLEM 4.10

Many common titrations involve the reaction of an acid with a base. If 24.75 mL of 0.503 M NaOH solution is used to titrate a 15.00-mL sample of sulfuric acid, H_2SO_4, what is the concentration of the acid?

Strategy A titration problem is an applied stoichiometry problem, so we will need a balanced chemical equation. We know the molarity and volume for the NaOH solution, so we can find the number of moles reacting. The mole ratio from the balanced equation lets us calculate moles of H_2SO_4 from moles of NaOH. Because we know the volume of the original H_2SO_4 solution, we can find its molarity.

Solution The acid–base reaction will produce sodium sulfate and water. The balanced equation is

$$H_2SO_4(aq) + 2\ NaOH(aq) \rightarrow Na_2SO_4(aq) + 2\ H_2O(\ell)$$

Determine the number of moles of NaOH from the given molarity and volume:

$$0.02475\ \text{L solution} \times \frac{0.503\ \text{mol NaOH}}{1\ \text{L solution}} = 0.0124\ \text{mol NaOH}$$

Use the balanced chemical equation to determine the number of moles of H_2SO_4:

$$0.0124\ \text{mol NaOH} \times \frac{1\ \text{mol}\ H_2SO_4}{2\ \text{mol NaOH}} = 6.22 \times 10^{-3}\ \text{mol}\ H_2SO_4$$

Determine the concentration of H_2SO_4:

$$M = \frac{6.22 \times 10^{-3}\ \text{mol}\ H_2SO_4}{0.01500\ \text{L solution}} = 0.415\ \text{M}\ H_2SO_4$$

Analyze Your Answer The molarity we found for sulfuric acid is smaller than that we had for sodium hydroxide. Does this make sense? The point to remember here is the stoichiometry that provides the 2:1 mole ratio. Even though the volume of NaOH used is greater than the volume of the acid, it is not twice the volume, so our H_2SO_4 molarity should be smaller.

Discussion This is the type of problem that students get wrong if they use the dilution formula (Eq. 3.3) by mistake. The dilution formula does not include the mole ratio from a balanced chemical equation, and in this case the 2:1 ratio would result in an error of a factor of 2 if you tried to use Eq. 3.3.

Check Your Understanding Analysis of a solution of NaOH showed that 42.67 mL of 0.485 M HNO_3 was needed to titrate a 25.00-mL sample of NaOH. What is the concentration of the NaOH solution?

INSIGHT INTO

4-6 Alternative Fuels and Fuel Additives

In the opening insight, we pointed out that gasoline is a rather complicated mixture of chemical compounds. Most of these are hydrocarbons of one sort or another, whose origins can be traced back to the oil from which the gasoline was refined. But the gasoline you buy is also likely to contain a range of **fuel additives**. Specific reasons for using these fuel additives generally fall under three broad categories: improving engine performance, reducing undesirable engine emissions, or reducing dependence on imported petroleum products. Some additives can achieve more than one of these goals simultaneously.

Let's first examine how additives can improve performance. In the gasoline industry, different grades of fuel are usually described in terms of their octane ratings; higher octane numbers are expected to deliver better engine performance. The octane number is really a measure of how highly the gas can be compressed in the engine cylinder before it will spontaneously ignite. This allows an engine to operate at higher compression without unwanted "knocking," which occurs if the fuel mixture in a cylinder ignites prematurely. Some carmakers recommend that only premium gas be used in their cars because the engines are designed to run at higher than usual compression ratios. The term "octane number" has its origins in the fact that octane itself is highly compressible; one way to achieve a higher octane number is to increase the fraction of octane in a particular gasoline. But what is important is the compressibility of the gas, not the amount of octane present. There are a number of ways to improve compressibility that are easier and cheaper than boosting the actual octane content.

One of the first widely used fuel additives was tetraethyl lead, which has the structure shown in **Figure 4.7**. From the 1920s into the 1970s, practically all gasoline sold in the United States contained lead additives, which substantially increase

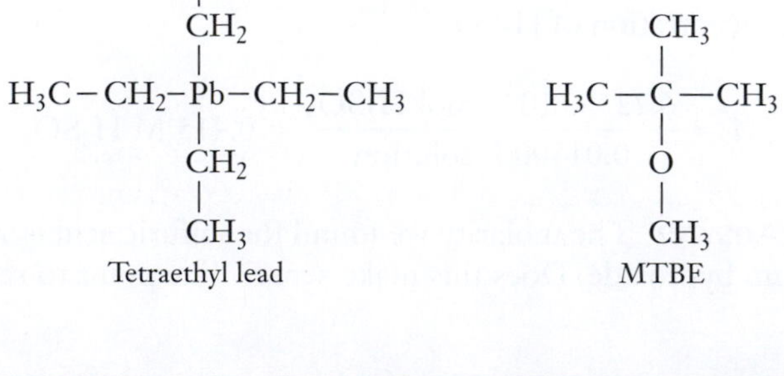

Figure 4.7 ■ Structural formulas of tetraethyl lead and MTBE.

```
              CH3
              |
              CH2                          CH3
              |                            |
H3C − CH2 − Pb − CH2 − CH3           H3C − C − CH3
              |                            |
              CH2                          O
              |                            |
              CH3                          CH3
        Tetraethyl lead                   MTBE
```

compressibility. But the increased performance and reduced fuel costs came at a price. Lead is toxic, and most of the lead introduced into gasoline was ultimately emitted into the atmosphere and settled into soil and water. Although these health concerns were known early, they did not receive widespread attention until heightened environmental interest took hold in the 1970s. The final blow to leaded gas came with the introduction of the catalytic converter in the 1970s. As you may know, the catalytic converter is designed to reduce the emission of carbon monoxide and other pollutants. The important chemistry of the converter takes place on a platinum metal surface. But the microscopic lead particles in the exhaust stream "poison" the platinum surface by binding to it so strongly that the catalytic reaction cannot take place. So "unleaded" gas was introduced and gradually took over the market; since 1996 it has been illegal to sell gasoline containing lead additives for use as a motor fuel in the United States.

We will discuss catalysis and catalytic converters in Chapter 11.

Other additives have stepped in to take on the octane-boosting role formerly played by tetraethyl lead. The most widely used in the United States today is ethanol. Because adding ethanol increases the oxygen content of gasoline, it is often referred to as an **oxygenate**, and gasoline containing ethanol is known as **oxygenated fuel**. The added oxygen helps to ensure more complete combustion and therefore reduces the emission of carbon monoxide, hydrocarbons, and soot. Gasoline containing at least 2% oxygen by weight is known as **reformulated gasoline (RFG)**, and the use of such fuels is mandatory in some areas where pollution problems are severe. Ethanol for use as a fuel additive is produced from biomass, as we learned in Chapter 3. Because the feedstocks are agricultural, ethanol production does not rely on imported petroleum products. Gasoline containing up to 10% ethanol can be burned in any modern automobile, and mixtures containing up to 85% ethanol can be used in specially designed engines. When used at such high levels, ethanol would more properly be thought of as an alternative fuel rather than a gasoline additive. Ethanol use is most common in the midwestern United States.

Gasoline formulations are typically seasonal. This helps explain the commonly observed increase in gasoline prices in late spring when refineries are changing from winter formulations to summer formulations and demand outpaces supply during the transition.

Another additive, MTBE (methyl *tert*-butyl ether, $(CH_3)_3COCH_3$), was widely used in the late 1990s. (See Example Problem 4.6.) Like ethanol, MTBE increases both the octane rating and oxygen content of gasoline, and fuel with as much as 15% MTBE can be used in any modern engine. After the passage of the 1990 Clean Air Act, the use of MTBE increased dramatically. The law mandated the use of oxygenated fuels, and the fuel industry saw that MTBE offered advantages over ethanol in cost of production and transportation. So the information available at the time suggested that introducing MTBE made sense in both environmental and business terms. Later, however, possible health concerns over MTBE called that choice into doubt because MTBE is a possible carcinogen, meaning it may cause some forms of cancer. As the usage of MTBE in gasoline increased, reports of detectable levels of the compound in groundwater also became more widespread. The source of this MTBE was generally believed to be leaks from pipelines or storage tanks. (Unlike lead, MTBE and ethanol undergo combustion along with the other components of gasoline, so no significant levels of these additives are found in vehicle exhaust streams.) Although the extent of MTBE's impact on human health is still somewhat uncertain, some areas (including the state of California) have banned the use of MTBE in gasoline.

A decision on the introduction of a new compound like MTBE into widespread use is always made on the basis of incomplete information. Engineers and scientists must rely on the best data available when the decision is being made.

FOCUS ON PROBLEM SOLVING

In most stoichiometry problems, we begin with a balanced equation showing us the proper reaction ratios. But sometimes the detailed stoichiometry may not be known, and we can use experimental data to determine those reaction ratios. One tactic that can be used is the *method of continuous variation*, and it also illustrates the usefulness of graphical methods in problem solving.

Question Suppose you are working in a lab and you find a bottle labeled simply as "0.1 M copper chloride," but you are not sure if the solution contains copper(I) chloride (CuCl) or copper(II) chloride ($CuCl_2$). How might you be able to determine which compound is actually present?

Strategy One way to approach this problem would be to examine the stoichiometry as the solution undergoes a reaction. Treating the solution with phosphate ions, for example, should lead to the precipitation of the copper as a phosphate salt, and if you could determine the formula of that salt, then you would know which copper ion was present. The method of continuous variation provides a way to determine that formula graphically.

Suppose that you prepare a 0.10 M solution of Na_3PO_4, where the concentration is chosen to match that of the copper chloride solution. You could then carry out a series of experiments in which you use a fixed total volume—say, 60 mL—of the two solutions but vary the volume of each reactant. The first experiment might involve 10 mL of the copper solution and 50 mL of the phosphate solution, for example, followed by 20 mL and 40 mL, and so on. For each reaction mixture, you could determine the amount of precipitate formed.

Solution If you plot that amount of precipitate as a function of the volumes of each solution, you might get a result like that shown in **Figure 4.8**.

On the left-hand side of this graph, there is an excess of phosphate ions present and the copper ions are the limiting reactant. Moving from left to right, the amount of copper increases, so the yield of the precipitate also increases. On the right-hand side of the graph, there is an excess of copper ions present and the phosphate ions are the limiting reactant. At the point where the maximum amount of precipitate is formed, the two ions are present in the correct stoichiometric ratio. Here that occurs when the volume ratio is 36 mL:24 mL, or 3:2. And because the concentrations of the two reacting solutions were equal, this volume ratio must also be the correct mole ratio in the stoichiometric mixture. So the ratio of copper ions to phosphate ions is 3:2, and the precipitate must be $Cu_3(PO_4)_2$. The original solution contains copper(II) chloride, $CuCl_2$. The equation for the precipitation reaction is

$$3\ CuCl_2(aq) + 2\ Na_3PO_4(aq) \rightarrow Cu_3(PO_4)_2(s) + 6\ NaCl(aq)$$

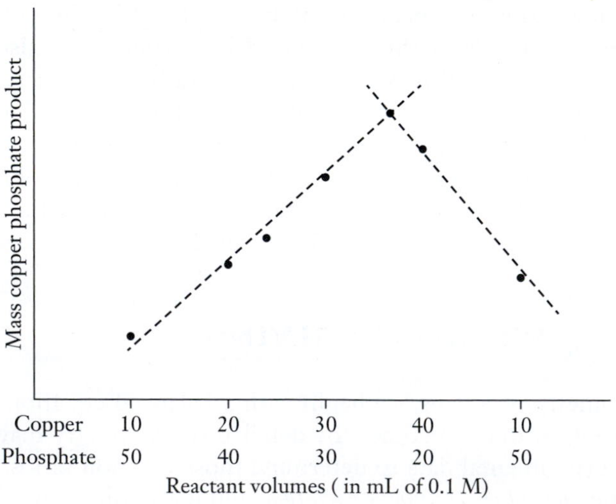

Figure 4.8 ■ Plot of product yield for the reaction of a copper salt with sodium phosphate using the method of continuous variations.

SUMMARY

The quantitative relationships associated with chemical reactions are vitally important for many applications of chemistry—including engineering of internal combustion engines and the fuels that power them. A balanced chemical equation is the key to any quantitative understanding of a reaction. Quantitative calculations for chemical reactions are referred to as stoichiometry problems, and the most important idea in this chapter is that the heart of every such problem is the balanced chemical equation.

The balanced chemical equation provides mole ratios for all reactants and products. To use these ratios, we usually have to convert information from something that can be easily measured—such as the mass or volume of a reactant or product—to moles. This establishes the main pattern for solving stoichiometry problems. First, we must look at the information we know about a reactant or product and convert that to moles of that substance. We use the balanced chemical equation and the mole ratios it contains to convert from the known molar amount of one substance to the number of moles of the substance we wish to know about. Finally, we usually convert from moles to mass (or volume) of the unknown substance.

All stoichiometry problems can be approached with this general pattern. In some cases, however, additional calculations may be needed. For example, if we are given (or able to measure) known amounts of two or more reactants, we must determine which of them will be completely consumed (the limiting reactant). Once again, the mole ratios in the balanced equation hold the key. Another type of calculation that can be considered in a stoichiometry problem is determining the percentage yield of a reaction. In this case, the amount of product determined in the problem represents the theoretical yield of the reaction—a value that is not often attained when a reaction is carried out in the laboratory or in industry. If the actual yield can be measured separately, the ratio of the actual yield to the theoretical yield provides the percentage yield.

KEY TERMS

actual yield (4-4)
alkanes (4-1)
fuel additives (4-6)
hydrocarbons (4-1)
indicator (4-5)
isomers (4-1)
limiting reactant (4-3)
methyl *tert*-butyl ether (MTBE) (4-6)
mole ratio (4-2)
oxygenated fuels (4-6)
percentage yield (4-4)
reformulated gasoline (RFG) (4-6)
side reactions (4-4)
stoichiometry (4-2)
theoretical yield (4-4)
titration (4-5)

PROBLEMS AND EXERCISES

INSIGHT INTO Gasoline and Other Fuels

4.1 List at least two factors that make it difficult to describe the combustion of gasoline accurately. What assumptions can be made to address these complications?

4.2 What is an *alkane*?

4.3 Explain the difference between complete and incomplete combustion.

4.4 Automobile exhaust often contains traces of formaldehyde (CH_2O), which is another possible product of incomplete combustion. Write a balanced equation for the formation of formaldehyde during the combustion of octane. (Water will also be formed as a product.)

4.5 Methane, ethane, and propane are also hydrocarbons, but they are not major components of gasoline. What prevents them from being part of this mixture?

4.6 Use the web to research prices of gasoline at the pump for consumers in your area during the past year. Also find information about the price of crude oil for the same period. Discuss any correlations you observe between the two prices.

Fundamentals of Stoichiometry

4.7 For the following reactions, write the ratios that can be established among molar amounts of the various compounds.

(a) $2\ H_2 + O_2 \rightarrow 2\ H_2O$

(b) $2\ H_2O_2 \rightarrow 2\ H_2O + O_2$

(c) $P_4 + 5\ O_2 \rightarrow P_4O_{10}$

(d) $2\ KClO_3 \rightarrow 2\ KCl + 3\ O_2$

4.8 In an experiment carried out at very low pressure, 1.3×10^{15} molecules of H_2 are reacted with acetylene, C_2H_2, to form ethane, C_2H_6, on the surface of a catalyst. Write a balanced chemical equation for this reaction. How many molecules of acetylene are consumed?

4.9 Sulfur, S_8, combines with oxygen at elevated temperatures to form sulfur dioxide. **(a)** Write a balanced chemical equation for this reaction. **(b)** If 200 oxygen molecules are used up in this reaction, how many sulfur molecules react? **(c)** How many sulfur dioxide molecules are formed in part (b)?

4.10 How many moles of oxygen can be obtained by the decomposition of 7.5 mol of reactant in each of the following reactions?

(a) $2\ KClO_3 \rightarrow 2\ KCl + 3\ O_2$

(b) $2\ H_2O_2 \rightarrow 2\ H_2O + O_2$

(c) $2\ HgO \rightarrow 2\ Hg + O_2$

(d) $2\ NaNO_3 \rightarrow 2\ NaNO_2 + O_2$

(e) $KClO_4 \rightarrow KCl + 2\ O_2$

4.11 MTBE, $C_5H_{12}O$, is one of the additives that replaced tetraethyl-lead in gasoline. (See Example Problem 4.6 and Section 4-6.) How many moles of O_2 are needed for the complete combustion of 1.50 mol of MTBE?

4.12 In petroleum refining, hydrocarbons are often manipulated by reacting them with $H_2(g)$. If hexene, C_6H_{12}, is reacted with hydrogen to form hexane, C_6H_{14}, how many moles of hydrogen are needed to react with 453 moles of hexene?

4.13 For the following reactions, determine the value of x.

(a) $4\ C + S_8 \rightarrow 4\ CS_2$

3.2 mol S_8 yields x mol CS_2

(b) $CS_2 + 3\ O_2 \rightarrow CO_2 + 2\ SO_2$

1.8 mol CS_2 yields x mol SO_2

(c) $N_2H_4 + 3\ O_2 \rightarrow 2\ NO_2 + 2\ H_2O$

7.3 mol O_2 yields x mol NO_2

(d) $SiH_4 + 2\ O_2 \rightarrow SiO_2 + 2\ H_2O$

1.3×10^{-3} mol SiH_4 yields x mol H_2O

4.14 The combustion of liquid chloroethylene, C_2H_3Cl, yields carbon dioxide, steam, and hydrogen chloride gas. **(a)** Write a balanced equation for the reaction. **(b)** How many moles of oxygen are required to react with 35.00 g of chloroethylene? **(c)** If 25.00 g of chloroethylene reacts with an excess of oxygen, how many grams of each product are formed?

4.15 What mass of the unknown compound is formed in the following reactions? (Assume that reactants for which amounts are not given are present in excess.)

(a) $C_2H_4 + H_2 \rightarrow C_2H_6$

3.4 g C_2H_4 reacts to produce x g C_2H_6

(b) $CS_2 + 3\ Cl_2 \rightarrow CCl_4 + S_2Cl_2$

5.78 g Cl_2 reacts to produce x g S_2Cl_2

(c) $PCl_3 + 3\ H_2O \rightarrow H_3PO_3 + 3\ HCl$

3.1 mg PCl_3 reacts to produce x mg HCl

(d) $B_2H_6 + 3\ O_2 \rightarrow B_2O_3 + 3\ H_2O$

4.6 kg B_2H_6 reacts to produce x kg B_2O_3

4.16 Many metals react with halogens to give metal halides. For example, iron reacts with chlorine to give iron(II) chloride, $FeCl_2$:

$$Fe(s) + Cl_2(g) \rightarrow FeCl_2(s)$$

Beginning with 10.0 g iron, what mass of Cl_2, in grams, is required for complete reaction? What quantity of $FeCl_2$, in moles and in grams, is expected?

4.17 Phosgene is a highly toxic gas that has been used as a chemical weapon at times in the past. It is now used in the manufacture of polycarbonates, which are used to make compact discs and plastic eyeglass lenses. Phosgene is produced by the reaction, $CO + Cl_2 \rightarrow COCl_2$. Given an excess of carbon monoxide, what mass of chlorine gas must be reacted to form 4.5 g of phosgene?

4.18 A gas is either CH_4 or C_2H_6. A 1.00-g sample of this gas produces 1.80 g of water when combusted in excess O_2. Assuming complete combustion to CO_2 and H_2O, which gas is it?

4.19 How many metric tons of carbon are required to reduce 7.83 metric tons of Fe_2O_3 according to the following reaction?

$$2\ Fe_2O_3 + 3\ C \rightarrow 3\ CO_2 + 4\ Fe$$

How many metric tons of iron are produced?

4.20 Assuming a charcoal briquette is composed entirely of carbon, what mass of oxygen is needed to completely burn a 37.3-g briquette?

4.21 Ammonium nitrate, NH_4NO_3, will decompose explosively to form N_2, O_2, and H_2O, a fact that has been exploited in terrorist bombings. What mass of nitrogen is formed by the decomposition of 2.6 kg of ammonium nitrate?

Limiting Reactants

4.22 If 3.4 mol Al and 6.2 mol Fe_2O_3 are mixed, what is the limiting reactant?

$$2\ Al + Fe_2O_3 \rightarrow Al_2O_3 + 2\ Fe$$

4.23 If 8.4 moles of disilane, Si_2H_6, is combined with 15.1 moles of O_2, which is the limiting reactant?

$$2\ Si_2H_6 + 7\ O_2 \rightarrow 4\ SiO_2 + 6\ H_2O$$

4.24 Generally, an excess of O_2 is needed for the reaction $Sn + O_2 \rightarrow SnO_2$. What is the minimum number of moles of oxygen required to oxidize 7.3 moles of tin?

4.25 In the reaction of arsenic with bromine, $AsBr_5$ will form only when excess bromine is present. Write a balanced chemical equation for this reaction. Determine the minimum number of moles of bromine that are needed if 9.6 moles of arsenic is present.

4.26 Ammonia gas can be prepared by the reaction

$$CaO(s) + 2\ NH_4Cl(s) \rightarrow 2\ NH_3(g) + H_2O(g) + CaCl_2(s)$$

If 112 g of CaO reacts with 224 g of NH_4Cl, how many moles of reactants and products are there when the reaction is complete?

4.27 When octane is combusted with inadequate oxygen, carbon monoxide may form. If 100 g of octane is burned in 200 g of O_2, are conditions conducive to forming carbon monoxide?

4.28 The equation for one of the reactions in the process of turning iron ore into the metal is

$$Fe_2O_3(s) + 3\ CO(g) \rightarrow 2\ Fe(s) + 3\ CO_2(g)$$

If you start with 2.00 kg of each reactant, what is the maximum mass of iron you can produce?

4.29 Copper reacts with sulfuric acid according to the following equation:

$$2\ H_2SO_4 + Cu \rightarrow CuSO_4 + 2\ H_2O + SO_2$$

How many grams of sulfur dioxide are created by this reaction if 14.2 g of copper reacts with 18.0 g of sulfuric acid?

4.30 One of the steps in the manufacture of nitric acid is the oxidation of ammonia shown in this equation:

$$4\ NH_3(g) + 5\ O_2(g) \rightarrow 4\ NO(g) + 6\ H_2O(g)$$

If 43.0 kg of NH_3 reacts with 35.4 kg of O_2, what mass of NO forms?

4.31 When $Al(OH)_3$ reacts with sulfuric acid, the following reaction occurs:

$$2\ Al(OH)_3 + 3\ H_2SO_4 \rightarrow Al_2(SO_4)_3 + 6\ H_2O$$

If 1.7×10^3 g of $Al(OH)_3$ is combined with 680 g of H_2SO_4, how much aluminum sulfate can form?

4.32 Copper reacts with nitric acid via the following equation:

$$3\ Cu(s) + 8\ HNO_3(aq) \rightarrow 3\ Cu(NO_3)_2(aq) + 2\ NO(g) + 4\ H_2O(\ell)$$

What mass of NO(g) can be formed when 10.0 g of Cu reacts with 115 g of HNO_3?

4.33 How much HNO_3 can be formed in the following reaction if 3.6 kg of NO_2 gas is bubbled through 2.5 kg of water?

$$3\ NO_2(g) + H_2O(\ell) \rightarrow 2\ HNO_3(aq) + NO(g)$$

4.34 Hydrogen and oxygen are reacted and the water formed is collected at 25°C, where it has a density of 0.997 g/mL. If 36.8 g of H_2 and 168 g of O_2 are reacted, how many mL of water will be collected?

4.35 Silicon carbide, an abrasive, is made by the reaction of silicon dioxide with graphite (solid carbon):

$$SiO_2 + C \xrightarrow{\text{heat}} SiC + CO \text{ (balanced?)}$$

We mix 150.0 g of SiO_2 and 101.5 g of C. If the reaction proceeds as far as possible, which reactant is left over? How much of this reactant remains?

Theoretical and Percentage Yields

4.36 Define the term *theoretical yield.*

4.37 What types of events happen in chemical reactions that lead to percentage yields of less than 100%?

4.38 Sometimes students in chemistry labs determine percentage yields greater than 100%. Assuming there is no calculation error, how could this happen?

4.39 The theoretical yield and the actual yield for various reactions are given below. Determine the corresponding percentage yields.

	Theoretical Yield	Actual Yield
Reaction 1	35.0 g	12.8 g
Reaction 2	9.3 g	120 mg
Reaction 3	3.7 metric tons	1250 kg
Reaction 4	40.0 g	41.0 g

4.40 A reaction that produced 4.8 mg of taxol, an anticancer drug, is reported to have a yield of 38%. What was the theoretical yield?

4.41 Methanol, CH_3OH, is used in racing cars because it is a clean-burning fuel. It can be made by this reaction:

$$CO(g) + 2\ H_2(g) \rightarrow CH_3OH(\ell)$$

What is the percentage yield if 5.0×10^3 g H_2 reacts with excess CO to form 3.5×10^4 g CH_3OH?

4.42 When iron and steam react at high temperatures, the following reaction takes place:

$$3\ Fe(s) + 4\ H_2O(g) \rightarrow Fe_3O_4(s) + 4\ H_2(g)$$

How much iron must react with excess steam to form 897 g of Fe_3O_4 if the reaction yield is 69%?

4.43 The percentage yield of the following reaction is consistently 87%.

$$CH_4(g) + 4\ S(g) \rightarrow CS_2(g) + 2\ H_2S(g)$$

How many grams of sulfur would be needed to obtain 80.0 g of CS_2?

4.44 Sulfur hexafluoride is a very stable gas useful in electrical generators and switches. It is formed by direct synthesis from the elements:

$$S(s) + 3\ F_2(g) \rightarrow SF_6(g)$$

If 92 g of SF_6 is produced from the reaction of 115 g of sulfur in excess fluorine, what is the percentage yield?

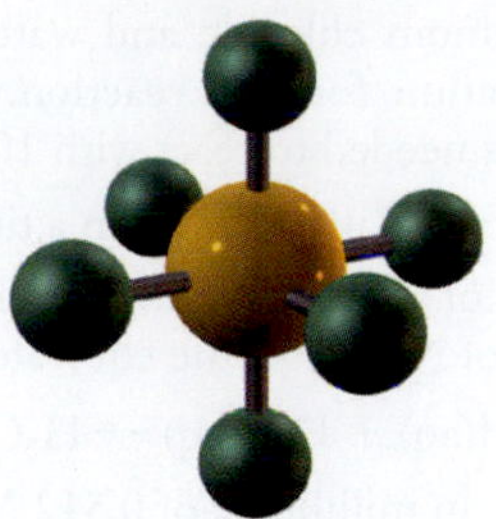

Sulfur hexafluoride, SF_6

4.45 Magnesium nitride forms in a side reaction when magnesium metal burns in air. This reaction may also be carried out in pure nitrogen.

$$3\ Mg(s) + N_2(g) \rightarrow Mg_3N_2(s)$$

If 18.4 g of Mg_3N_2 forms from the reaction of 20.0 g of magnesium with excess nitrogen, what is the percentage yield?

4.46 Industrial production of hydrogen gas uses the reaction shown below. If 1.00 metric ton of propane reacting with excess water yields 270 kg of H_2, what is the percentage yield?

$$C_3H_8(g) + 3\ H_2O(\ell) \rightarrow 3\ CO(g) + 7\ H_2(g)$$

4.47 If 21 g of H_2S is mixed with 38 g of O_2 and 31 g of SO_2 forms, what is the percentage yield?

$$2\ H_2S + 3\ O_2 \rightarrow 2\ SO_2 + 2\ H_2O$$

4.48 A mixture of 10.0 g of NO and 14.0 g of NO_2 results in the production of 8.52 g of N_2O_3. What is the percentage yield?

$$NO(g) + NO_2(g) \rightarrow N_2O_3(\ell)$$

4.49 Silicon carbide is an abrasive used in the manufacture of grinding wheels. A company is investigating whether they can be more efficient in the construction of such wheels by making their own SiC via the reaction

$$SiO_2 + 3\ C \rightarrow SiC + 2\ CO$$

(a) If this reaction consistently has a yield of 85%, what is the minimum amount of both silicon dioxide and carbon needed to produce 3400 kg of SiC for the manufacture of cutting wheels? (b) The silicon dioxide is to be obtained from sand and the carbon is derived from coal. If the available sand is 95% SiO_2 by weight and the coal is 73% C by weight, what mass of coal is needed for each metric ton of sand used?

4.50 Elemental phosphorous is used in the semiconductor industry. It can be obtained from an ore called fluoroapatite via reaction with SiO_2 and C:

$$4\ Ca_5(PO_4)_3F + 18\ SiO_2 + 30\ C \rightarrow 3\ P_4 + 30\ CO + 18\ CaSiO_3 + 2\ CaF_2$$

Suppose a particular semiconductor production plant requires 1500 kg of P_4. If the recovery of P_4 from this reaction is 73% efficient, what mass of fluoroapatite is needed?

Solution Stoichiometry

4.51 Small quantities of hydrogen gas can be prepared by the following reaction:

$$Zn(s) + H_2SO_4(aq) \rightarrow ZnSO_4(aq) + H_2(g)$$

How many grams of H_2 can be prepared from 25.0 mL of 6.00 M H_2SO_4 and excess zinc?

4.52 Aluminum hydroxide can be dissolved in hydrochloric acid to form aluminum chloride and water. Write a balanced chemical equation for this reaction. How many mL of 3.37 M HCl is needed to react with 10.0 g of $Al(OH)_3$?

4.53 What is the role of an indicator in a titration?

4.54 What volume of 0.812 M HCl, in milliliters, is required to titrate 1.45 g of NaOH to the equivalence point?

$$NaOH(aq) + HCl(aq) \rightarrow H_2O(\ell) + NaCl(aq)$$

4.55 What volume, in milliliters, of 0.512 M NaOH is required to react completely with 25.0 mL 0.234 M H_2SO_4?

4.56 What is the molarity of a solution of nitric acid if 0.216 g of barium hydroxide is required to neutralize 20.00 mL of nitric acid?

4.57 Hydrazine, N_2H_4, is a weak base and can react with an acid such as sulfuric acid:

$$2\ N_2H_4(aq) + H_2SO_4(aq) \rightarrow 2\ N_2H_5^+(aq) + SO_4^{-2}(aq)$$

What mass of hydrazine can react with 250. mL 0.225 M H_2SO_4?

4.58 A metallurgical firm needs to dispose of 2.74×10^3 gallons of waste sulfuric acid whose concentration has been determined to be 1.53 M. Environmental regulations require them to neutralize the waste before releasing the wastewater, and they choose to do so with slaked lime, $Ca(OH)_2(s)$. The neutralization reaction is

$$H_2SO_4(aq) + Ca(OH)_2(s) \rightarrow CaSO_4(s) + 2\ H_2O(\ell)$$

If the slaked lime costs $0.52 per pound, how much will it cost to neutralize the waste sulfuric acid?

4.59 Many mining operations produce tailings that can be oxidized to form acids, including sulfuric acid. One chemical that reacts in this way is chalcopyrite, $CuFeS_2$, whose net oxidation can be described by the following equation:

$$4\ CuFeS_2(s) + 17\ O_2(g) + 10\ H_2O(\ell) \rightarrow 4\ Fe(OH)_3(s) + 4\ CuSO_4(aq) + 4\ H_2SO_4(aq)$$

In a laboratory experiment to study weathering, a mining engineer put a 2.3-kg sample of $CuFeS_2(s)$ near an empty 40-L container. Simulated weathering experiments were conducted and the runoff was collected, leading to 35.8 L of solution. A 25.00-mL aliquot of this solution was analyzed with 0.0100 M NaOH(aq). The titration required 38.1 mL to reach the equivalence point. (a) What was the concentration of sulfuric acid? (b) Assuming the only acid in the container comes from the reacted chalcopyrite, what mass of the chalcopyrite has reacted? (c) What percentage of the chalcopyrite reacted in this experiment?

4.60 A student dissolves 17.3 g of a mixture of Na_2O and BaO in water. To try to determine the amount of barium present, the student uses a volumetric pipette to add 10.00 mL of 0.550 M sulfuric acid, which causes a precipitate of barium sulfate to form. The student then dries and weighs the precipitate. If the mixture is 50% BaO by mass, will this experiment yield a good estimate? Explain your answer, and suggest how the procedure might be improved.

4.61 Aluminum dissolves in HCl according to the equation written below, whereas copper does not react with HCl.

$$2\ Al(s) + 6\ HCl(aq) \rightarrow 2\ AlCl_3(aq) + 3\ H_2(g)$$

A 35.0-g sample of a copper–aluminum alloy is dropped into 750 mL of 3.00 M HCl, and the reaction above proceeds as far as possible. If the alloy contains 77.1% Al by mass, what mass of hydrogen gas would be produced?

INSIGHT INTO Alternative Fuels and Fuel Additives

4.62 Why are fuel additives used?

4.63 What is actually measured by the octane ratings of different grades of gasoline?

4.64 Offer an explanation as to how oxygenated fuels might increase the extent of combustion.

4.65 Perform a web search to learn what restrictions (if any) on the use of MTBE in gasoline might exist in your local area.

4.66 Using the web, find information about the amount of lead in the environment during the past 50 years. Correlate what you observe with the presence or absence of tetraethyl lead in gasoline.

4.67 Using the web, find out how lead "poisons" the catalyst in a catalytic converter.

Additional Problems

4.68 Estimate the mass of CO_2 emitted by a typical automobile in a year.

4.69 You have 0.954 g of an unknown acid, H_2A, which reacts with NaOH according to the balanced equation

$$H_2A(aq) + 2\ NaOH(aq) \rightarrow Na_2A(aq) + 2\ H_2O(\ell)$$

If 36.04 mL of 0.509 M NaOH is required to titrate the acid to the equivalence point, what is the molar mass of the acid?

4.70 In this chapter, we used a simplified equation for the combustion of gasoline. But the products generated in the operation of real automobile engines include various oxides of nitrogen, and it is desirable to remove these compounds from the exhaust stream rather than release them into the atmosphere. Currently, this is done by the "catalytic converter," but other technologies are being explored. One possible method involves the use of isocyanic acid, HNCO, which reacts with NO_2 to produce N_2, CO_2, and H_2O, as shown in the equation below:

$$8\ HNCO + 6\ NO_2 \rightarrow 7\ N_2 + 8\ CO_2 + 4\ H_2O$$

What mass of HNCO would be required to react completely with the estimated 1.7×10^{10} kg of NO_2 produced each year by U.S. automobiles? How might the fact that the reaction above produces CO_2 affect its feasibility for this application?

4.71 Phosphoric acid (H_3PO_4) is important in the production of both fertilizers and detergents. It is distributed commercially as a solution with a concentration of about 14.8 M. Approximately 2.1×10^9 gallons of this concentrated phosphoric acid solution is produced annually in the United States alone. Assuming that all of this H_3PO_4 is produced by the reaction below, what mass of the mineral fluoroapatite ($Ca_5(PO_4)_3F$) would be required each year?

$$Ca_5(PO_4)_3F + 5\ H_2SO_4 \rightarrow 3\ H_3PO_4 + 5\ CaSO_4 + HF$$

4.72 The reaction shown below is used to destroy Freon-12 (CF_2Cl_2), preventing its release into the atmosphere. What mass of NaF will be formed if 250.0 kg of CF_2Cl_2 and 400.0 kg of $Na_2C_2O_4$ are heated and allowed to react to completion?

$$CF_2Cl_2 + 2\ Na_2C_2O_4 \rightarrow 2\ NaF + 2\ NaCl + C + 4\ CO_2$$

4.73 Blood alcohol levels are usually reported in mass percentages: a level of 0.10 means that 0.10 g of alcohol (C_2H_5OH) was present in 100 g of blood. This is just one way of expressing the concentration of alcohol dissolved in the blood. Other concentration units could be used. Find the molarity of alcohol in the blood of a person with a blood alcohol level of 0.12. (The density of blood is 1.2 g/mL.)

4.74 One way of determining blood alcohol levels is by performing a titration on a sample of blood. In this process, the alcohol from the blood is oxidized by dichromate ions ($Cr_2O_7^{2-}$) according to the following net ionic equation:

$$C_2H_5OH + 2\ Cr_2O_7^{2-} + 16\ H^+ \rightarrow 2\ CO_2 + 4\ Cr^{3+} + 11\ H_2O$$

A 10.00-g sample of blood was drawn from a patient, and 13.77 mL of 0.02538 M $K_2Cr_2O_7$ was required to titrate the alcohol. What was the patient's blood alcohol level? (See the previous problem for definition of blood alcohol level. $K_2Cr_2O_7$ is a strong electrolyte, so it dissociates completely in solution.)

4.75 Ammonium sulfate ($(NH_4)_2SO_4$) is a common fertilizer that can be produced by the reaction of ammonia (NH_3) with sulfuric acid (H_2SO_4). Each year about 2×10^9 kg of ammonium sulfate is produced worldwide. How many kilograms of ammonia would be needed to generate this quantity of $(NH_4)_2SO_4$? (Assume that the reaction goes to completion and no other products formed.)

4.76 The pictures below show a molecular-scale view of a chemical reaction between the compounds AB_2 and B_2. (Green balls represent B atoms and orange balls are A atoms). The box on the left represents the reactants at the instant of mixing, and the box on the right shows what is left once the reaction has gone to completion.

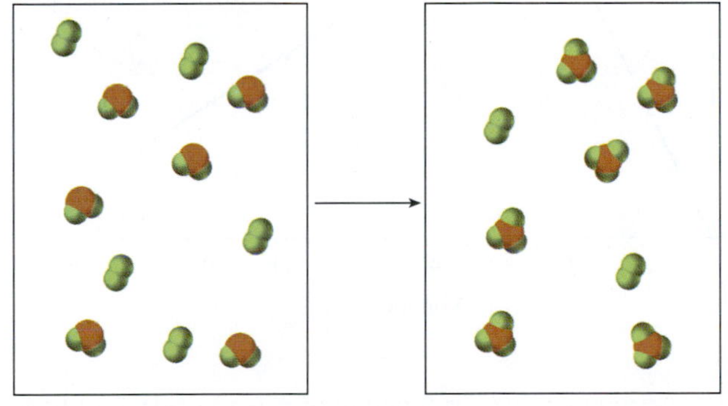

Was there a limiting reactant in this reaction? If so, what was it? Write a balanced chemical equation for this reaction. As usual, your equation should use the smallest possible whole number coefficients for all substances.

4.77 The pictures below show a molecular-scale view of a chemical reaction between H_2 and CO to produce methanol, CH_3OH. The box on the left represents the reactants at the instant of mixing, and the box on the right shows what is left once the reaction has gone to completion.

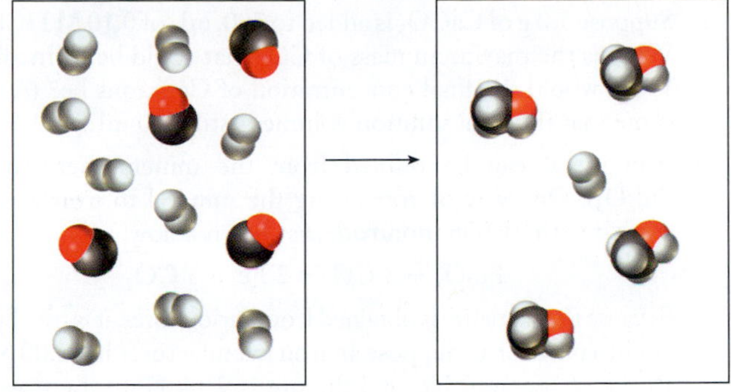

Was there a limiting reactant in this reaction? If so, what was it? Write a balanced chemical equation for this reaction. As usual, your equation should use the smallest possible whole number coefficients for all substances.

4.78 Aluminum chloride ($AlCl_3$) is used as a catalyst in the production of polyisobutylene, which is used in automobile tires. Scrap aluminum metal reacts with chlorine gas (Cl_2) to produce $AlCl_3$. Suppose that 2.70 g of Al and 7.10 g of Cl_2 are mixed. What is the maximum mass of $AlCl_3$ that could be formed?

4.79 In the cold vulcanization of rubber, disulfur dichloride (S_2Cl_2) is used as a source of sulfur atoms, and those sulfur atoms form "bridges," or cross-links, between polymer chains. S_2Cl_2 can be produced by reacting molten sulfur ($S_8(\ell)$) with chlorine ($Cl_2(g)$). What is the maximum mass of S_2Cl_2 that can be produced by reacting 32.0 g of sulfur with 71.0 g of chlorine?

4.80 Styrene, the building block of polystyrene, is a hydrocarbon. If 0.438 g of the compound is burned and produces 1.481 g of CO_2 and 0.303 g of H_2O, what is the empirical formula of the compound?

4.81 Cryolite (Na_3AlF_6) is used in the commercial production of aluminum from ore. Cryolite itself is produced by the following reaction:

$$6\ NaOH + Al_2O_3 + 12\ HF \rightarrow 2\ Na_3AlF_6 + 9\ H_2O$$

A mixture containing 800.0 kg of NaOH, 300.0 kg of Al_2O_3, and 600.0 kg of HF is heated to 950°C and reacts to completion. What is the maximum mass of Na_3AlF_6 formed?

4.82 A quality control technician needs to determine the percentage of arsenic (As) in a particular pesticide. The pesticide is dissolved and all of the arsenic present is converted to arsenate ions (AsO_4^{3-}). Then the amount of AsO_4^{3-} is determined by titrating with a solution containing silver ions (Ag^+). The silver reacts with the arsenate according to the following net ionic equation:

$$3\ Ag^+(aq) + AsO_4^{3-}(aq) \rightarrow Ag_3AsO_4(s)$$

When a 1.22-g sample of pesticide was analyzed this way, it required 25.0 mL of 0.102 M Ag^+ solution to precipitate all of the AsO_4^{3-}. What was the mass percentage of arsenic in the pesticide?

4.83 Calcium carbonate (limestone, $CaCO_3$) dissolves in hydrochloric acid, producing water and carbon dioxide as shown in the following *unbalanced* net ionic equation:

$$CaCO_3(s) + H_3O^+(aq) \rightarrow H_2O(\ell) + CO_2(g) + Ca^{2+}(aq)$$

Suppose 5.0 g of $CaCO_3$ is added to 700. mL of 0.10 M HCl. What is the maximum mass of CO_2 that could be formed? What would the final concentration of Ca^{2+} ions be? (Assume that the final solution volume is still 700 mL.)

4.84 Iron metal can be refined from the mineral hematite (Fe_2O_3). One way of converting the mineral to iron is to react it with carbon monoxide, as shown below:

$$Fe_2O_3 + 3\ CO \rightarrow 2\ Fe + 3\ CO_2$$

Because the hematite is obtained from various ores, it is usually not in a pure form. Suppose an iron manufacturer has 2.00×10^5 kg of ore available, and the ore is 93% Fe_2O_3 by mass. (There is no iron in the remaining 7% of the ore.) How many moles of Fe_2O_3 are present in this ore? How many kg of pure iron could be obtained from this sample of ore? Assume that the process has a 100% yield and that excess CO is available.

4.85 Calcium sulfate is the essential component of plaster and sheet rock. Waste calcium sulfate can be converted into quicklime, CaO, by reaction with carbon at high temperatures. The following two reactions represent a sequence of reactions that might take place:

$$CaSO_4(s) + 4\ C(s) \rightarrow CaS(\ell) + 4\ CO(g)$$

$$CaS(\ell) + 3\ CaSO_4(s) \rightarrow 4\ CaO(s) + 4\ SO_2(g)$$

What mass of sulfur dioxide could be obtained from 1.250 kg of calcium sulfate?

4.86 Iron–aluminum alloys are useful in some applications because they become magnetized in a magnetic field but are easily demagnetized when the field is removed. The composition of an iron–aluminum alloy can be determined chemically by reacting it with hydrochloric acid:

$$2\ Al(s) + 6\ HCl(aq) \rightarrow 2\ AlCl_3(aq) + 3\ H_2(g)$$

$$Fe(s) + 2\ HCl(aq) \rightarrow FeCl_2(aq) + H_2(g)$$

When a 7.264-g sample of a particular iron–aluminum alloy was dissolved in excess hydrochloric acid, 0.3284 g of $H_2(g)$ was produced. What was the mass percentage of aluminum in the alloy?

4.87 A mixture of methane (CH_4) and propane (C_3H_8) has a total mass of 29.84 g. When the mixture is burned completely in excess oxygen, the CO_2 and H_2O products have a combined mass of 142.97 g. Calculate the mass of methane in the original mixture.

4.88 A mixture of hydrochloric and sulfuric acids is prepared so that it contains 0.100 M HCl *and* 0.200 M H_2SO_4. What volume of 0.250 M NaOH would be required to completely neutralize *all* of the acid in 425 mL of this solution?

4.89 A solution contains both Ca^{2+} and Pb^{2+} ions. A 50.0-mL sample of this solution was treated with 0.528 M NaF, and 53.8 mL of the NaF solution was needed to precipitate all of the Ca^{2+} and Pb^{2+} as $CaF_2(s)$ and $PbF_2(s)$. The precipitate was dried and weighed, and had a total mass of 3.141 g. What was the concentration of Ca^{2+} in the original solution?

FOCUS ON PROBLEM SOLVING EXERCISES

4.90 The carbon dioxide exhaled by astronauts must be removed from the air. One way to do this is through the use of CO_2 "scrubbers" based on metal hydroxides. CO_2 reacts with the hydroxide to form the corresponding metal carbonate and water. Because these scrubbers need to be sent into space, weight is a significant concern in their design. If the goal is to minimize the mass of metal hydroxide needed to remove a given amount of CO_2 from the air, which would you recommend using: sodium hydroxide, magnesium hydroxide, or aluminum hydroxide? Explain your answer using arguments based on stoichiometry.

4.91 You are designing a process for a wastewater treatment facility intended to remove aqueous HCl from the input water. Several bases are being considered for use, and their costs are listed in the table below. If cost is the primary concern, which base would you recommend using? Use stoichiometric arguments to justify your answer.

Material	$CaCO_3$	$Ca(OH)_2$	NH_3	NaOH
Cost per metric ton	\$270	\$165	\$255	\$315

4.92 Iron(II) ions react with phenanthroline ($C_{12}H_8N_2$) to form a complex ion with the general formula $Fe_x(C_{12}H_8N_2)_y^{2x+}$. A student uses the method of continuous variation to determine the number of phenanthroline molecules bound to each iron ion (i.e., the values of x and y in the formula). The complex ion is strongly colored, so its concentration can be monitored using a spectrometer. Working with 2.4×10^{-3} M solutions of both Fe^{3+} and $C_{12}H_8N_2$, the data shown in the graph below were obtained. What is the correct formula for the complex ion?

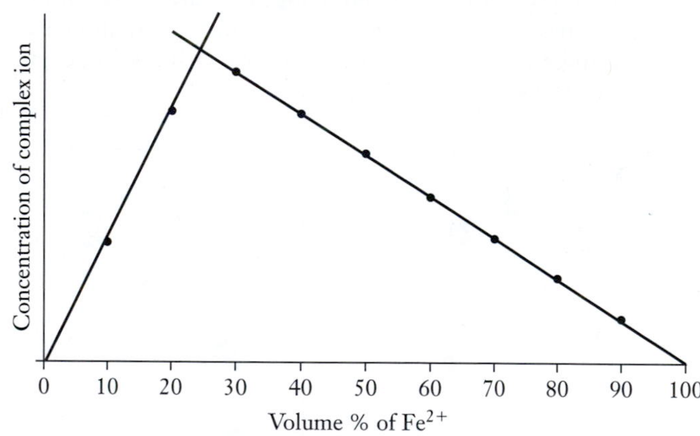

4.93 Chromium metal is found in ores as an oxide, Cr_2O_3. This ore can be converted to metallic Cr by the Goldschmidt process, $Cr_2O_3(s) + 2\ Al(s) \rightarrow 2\ Cr(\ell) + Al_2O_3(s)$. Suppose that you have 4.5 tons of chromium(III) oxide. Describe how you would determine how much aluminum metal you would need to carry out this process. List any information you would have to look up.

4.94 Hydrochloric acid reacts with many metals to form hydrogen gas, H_2. If you needed to produce hydrogen as a source of fuel in some design, how would you determine which metal—Mg, Al, or Zn—would produce the greatest amount of hydrogen per gram of metal? Describe the steps you would need to take in the decision making process.

4.95 When heated in air, some element, X, reacts with oxygen to form an oxide with the chemical formula X_2O_3. If a 0.5386-g sample of this unknown element yields an oxide with a mass of 0.7111 g, describe how you would determine the atomic mass of the element.

4.96 Chlorofluorocarbons were used for many years as refrigerants. One such molecule, $C_2Cl_2F_2$, is synthesized from the reaction of CCl_4 with HF. Suppose that you have 150 tons of carbon tetrachloride as a starting point for this reaction, show how you would determine the minimum amount of HF needed to carry out the reaction.

4.97 Bismuth selenide, Bi_2Se_3, is used in semiconductor research. It can be prepared directly from its elements by the reaction, $2\ Bi + 3\ Se \rightarrow Bi_2Se_3$. Suppose that you need 5 kg of bismuth selenide for a prototype system you are working on. You find a large bottle of bismuth metal but only 560 g of selenium. Describe how you can calculate if you have enough selenium on hand to carry out the reaction or if you need to order more.

4.98 When 2.750 g of the oxide of lead Pb_3O_4 is heated to a high temperature, it decomposes to produce 0.0640 g of oxygen gas and 2.686 g of some new lead oxide compound. How can you use this data to determine the formula of the new compound?

4.99 The best way to generate small amounts of oxygen gas in a laboratory setting is by decomposing potassium permanganate:

$$2\ KMnO_4(s) \rightarrow K_2MnO_4 + MnO_2 + O_2$$

Suppose that you are testing a new fuel for automobiles and you want to generate specific amounts of oxygen with this reaction. For one test, you need exactly 500 g of oxygen. Describe how you would determine the mass of potassium permanganate you need to obtain this amount of oxygen.

4.100 An ore sample with a mass of 670 kg contains 27.7% magnesium carbonate, $MgCO_3$. If all of the magnesium carbonate in this ore sample is decomposed to form carbon dioxide, describe how to determine what mass of CO_2 is evolved during the process.

4.101 Existing stockpiles of the refrigerant Freon-12, CF_2Cl_2, must be destroyed under the terms of the Montreal Protocol because of their potential for harming the ozone layer. One method for doing this involves reaction with sodium oxalate:

$$CF_2Cl_2 + 2\ Na_2C_2O_4 \rightarrow 2\ NaF + 2\ NaCl + C + 4\ CO_2$$

If you had 150 tons of Freon-12, describe how you would know how much sodium oxalate you would need to make that conversion.

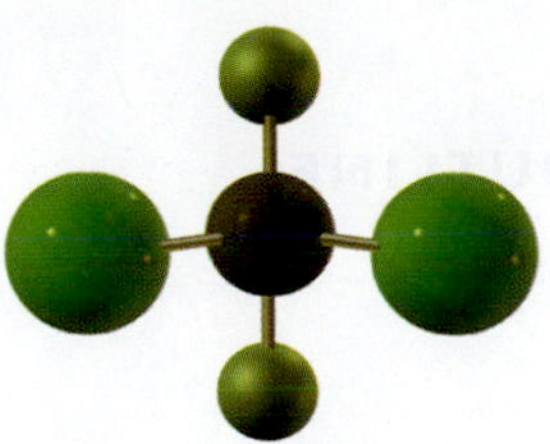

Freon-12, CF_2Cl_2

Cumulative Problems

4.102 Elemental analysis is sometimes carried out by combustion of the sample. For a hydrocarbon, the only products formed are CO_2 and H_2O. If a 1.36-g sample of an unknown hydrocarbon is burned and 2.21 g of H_2O is produced along with 4.07 g of CO_2, what is the empirical formula of the hydrocarbon?

4.103 Benzopyrene is a hydrocarbon that is known to cause cancer. Combustion analysis (see Problem 4.102) of a 2.44-g sample of this compound shows that combustion produces 8.54 g of CO_2 and 1.045 g of H_2O. What is the empirical formula of benzopyrene?

4.104 Write the balanced chemical equation for the combustion of ethane, C_2H_6, and answer these questions. **(a)** How many molecules of oxygen would combine with 14 molecules of ethane in this reaction? **(b)** If 13.0 mol of oxygen is consumed in a reaction, how many moles of water are produced? **(c)** How many grams of ethane are burned if 4.20×10^{22} molecules of CO_2 are produced?

4.105 Aluminum metal reacts with sulfuric acid to form hydrogen gas and aluminum sulfate. **(a)** Write a balanced chemical equation for this reaction. **(b)** Suppose that a 0.792-g sample of aluminum that contains impurities is reacted with excess sulfuric acid and 0.0813 g of H_2 is collected. Assuming that none of the impurities reacts with sulfuric acid to produce hydrogen, what is the percentage of aluminum in the sample?

4.106 A metallurgical firm wishes to dispose of 1300 gallons of waste sulfuric acid whose molarity is 1.37 M. Before disposal, it will be reacted with calcium hydroxide (slaked lime), which costs $0.23 per pound. **(a)** Write the balanced chemical equation for this process. **(b)** Determine the cost that the firm will incur from this use of slaked lime.

5

Gases

CHAPTER OUTLINE

5-1 *INSIGHT INTO* Air Pollution

5-2 Pressure

5-3 History and Application of the Gas Law

5-4 Partial Pressure

5-5 Stoichiometry of Reactions Involving Gases

5-6 Kinetic–Molecular Theory and Ideal Versus Real Gases

5-7 *INSIGHT INTO* Gas Sensors

Mike Carillo, Pacificcoastnews/Newscom

Urban smog results from a complex sequence of chemical reactions in the atmosphere. Knowledge of the properties of gases is essential for a meaningful understanding of smog and other issues involving air pollution.

In many ways, the essence of chemistry is considering things on the atomic and molecular level. This idea applies when we study chemical reactions and also when we examine the properties of matter. So as we begin to look at matter and materials from a chemist's perspective, we will always want to focus on the way the properties of matter are related to the properties and behavior of its constituent molecules. Gases provide an excellent starting point for this because the observable macroscopic properties of gases are very direct results of the behavior of individual molecules within the gas.

Most of the earliest investigations of gas properties were carried out using air. The reasons for this are clear: air is both readily available and obviously important to our lives. Although the sophistication of the measurements we make has increased tremendously, the availability and importance of air have ensured that it remains a very carefully studied gas. Nowadays, much of our interest in air is actually driven by concerns over the levels and effects of various pollutants. The U.S. Environmental Protection Agency (EPA) defines pollutants as "unwanted chemicals or other materials found in the air." Although this definition does include a variety of small solid particles such as dust or soot, most air pollutants are gases. Thus an exploration of some of the issues surrounding the monitoring and control of air pollution will form the backdrop for our study of gases and their properties.

Chapter Objectives

After mastering this chapter, you should be able to

- describe the physical properties of gases.
- identify several gaseous compounds or classes of compounds that are important in urban air pollution.
- use the ideal gas law for calculating changes in the conditions of a gas.
- use the concept of partial pressure to work with mixtures of gases.
- perform stoichiometric calculations for reactions involving gases as reactants or products.
- state the postulates of the kinetic theory of gases.
- describe qualitatively how the postulates of the kinetic theory account for the observed behavior of gases.
- describe the Maxwell-Boltzmann distribution of speeds and the effects of temperature and molar mass on molecular speed.
- identify conditions under which gases might not behave ideally.
- use the van der Waals equation to perform calculations for gases under nonideal conditions.
- describe the principles of operation for some pressure-measuring devices.

INSIGHT INTO

5-1 Air Pollution

Air pollution is one of the unintended but unavoidable side effects of modern society. The issues surrounding the monitoring, prevention, and remediation of air pollution involve an exceedingly complex interplay between science, business, and public policy. Our discussion here will be fairly limited, but we will try to explore some of the ways in which chemical science can help us to understand the complexity of the situation.

We should begin by looking at the chemical composition of clean air. As you probably know, air is a mixture rather than a pure substance. Nitrogen and oxygen account for the majority of the molecules present. Other important components include carbon dioxide and water vapor. The amount of water in the air varies from place to place and day to day, as described by the humidity in a weather report. So it will be convenient to use dry air as a reference point in our discussion. **Table 5.1** lists the

Table ■ 5.1

The composition of a one-cubic-meter sample of dry air at 25°C and normal atmospheric pressure

Gas	# Moles Present
N_2	31.929
O_2	8.567
Ar	0.382
CO_2	0.013
Other trace gases	0.002
Total	40.893

composition of a one-cubic-meter sample of dry air at 25°C and normal atmospheric pressure. The four gases listed account for at least 99.99% of the molecules present; N_2 and O_2 alone make up roughly 99%.

In contrast to the short list in Table 5.1, if we were to collect a sample of air in a modern urban area and analyze it carefully, we might detect dozens of compounds at trace levels. Some of these additional trace components can be linked to natural sources, but most are directly attributable to human activities like transportation and manufacturing. Of the large number of chemical compounds found in urban air, the EPA has identified six principal **criteria pollutants**: CO, NO_2, O_3, SO_2, Pb, and particulate matter (PM). These six substances are commonly found throughout the country, causing a variety of negative effects on health, the environment, and property. The term "criteria pollutants" comes from the fact that the EPA has established a set of science-based standards, or criteria, for the acceptable levels of these pollutants. **Primary standards** are intended to protect our health, and **secondary standards** are intended to protect our environment and property. In many cases, the allowable levels of these pollutants correspond to less than one **part per million**, which means one molecule per every million molecules of air. Levels of these pollutants are monitored periodically throughout the country. If the detected concentrations exceed the primary standard in a given region, that area is designated as a **nonattainment area** for that particular pollutant and additional regulatory action may be taken.

Particulate matter dispersed in the air includes soot, dust, and other small solid particles.

As an example to illustrate the complexity of pollution issues, let's consider one of the criteria pollutants, NO_2. We might begin by asking where this pollutant comes from. Nitrogen dioxide is emitted as part of the exhaust gas from our automobiles. At the high temperatures at which engines operate, the nitrogen and oxygen gases present in air react with each other to form a variety of nitrogen oxides. Such reactions do not occur at lower temperatures but are practically unavoidable in an engine. When engines run hotter, from standing in a traffic jam, for example, they tend to produce even more nitrogen oxides. Once it is present in the air, NO_2 poses a health risk because it attacks the membranes in our lungs. The brown color of smog is also largely due to NO_2. Nitrogen dioxide molecules absorb some portion of the visible light of the sun, leaving the colors that produce brown to pass through.

The various oxides of nitrogen are often referred to collectively as "NO_X."

The detrimental effects of NO_2 in the air extend well beyond this rather unappealing color. Light energy from the sun can also initiate chemical changes through what is known as a **photochemical reaction**. For nitrogen oxides, these photochemical processes lead to products that can react further to trigger the formation of ground-level ozone.

But nitrogen oxides cannot generate ozone on their own. The key additional components that are needed are molecules collectively referred to as **volatile organic chemicals**, or VOCs. As we learned in Section 2-6, organic chemicals contain predominantly carbon and hydrogen. Volatile compounds tend to evaporate easily. VOCs combine these two traits, meaning they are carbon-based molecules that evaporate readily. Many different VOCs are present in urban air. The primary source of these chemicals is fuels. In Section 4-1, we learned that the gasoline we burn in our automobiles and lawn mowers is a complex mixture of organic molecules. All combustion engines release some VOCs when they are operating, and older vehicles tend to release more. Our noses detect this phenomenon every time we are near a vehicle that produces large amounts of VOCs.

Chemical reaction between photochemically activated nitrogen oxides and VOCs can give rise to a mixture of gases that is collectively referred to as smog. Many of the compounds in smog are irritants to human lungs. Ozone (O_3) is the most significant problem. When present at ground level, ozone poses a range of health hazards, even for healthy young adults. People with asthma or other respiratory problems are particularly susceptible to problems associated with ozone. **Figure 5.1** shows a graph of concentrations of various chemicals associated with photochemical smog as a function of time of day in a large city. Notice that the nitrogen oxides and VOCs are both present prior to the formation of ozone. If either of these molecules is missing, ozone will not form. As we will see, this observation provides a way to combat the formation of ozone.

We will look at the desirable effects of ozone in the upper atmosphere in Chapter 11.

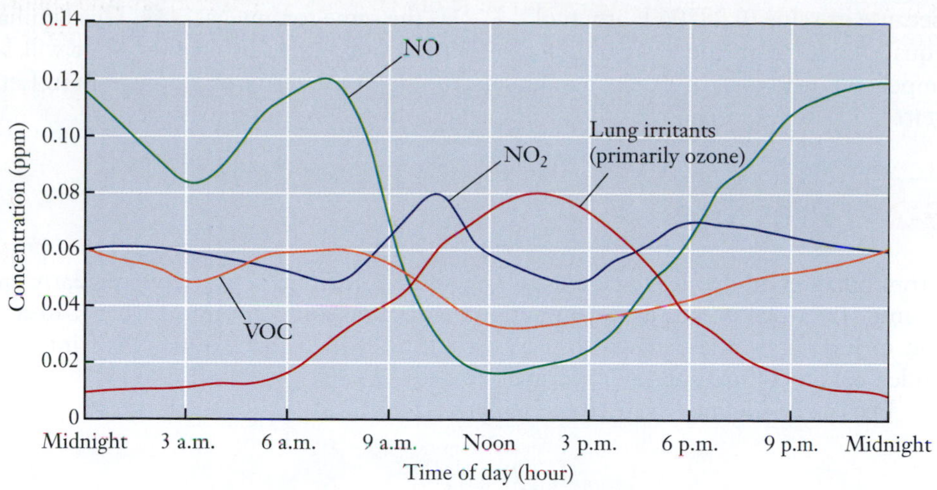

Figure 5.1 ■ The graph shows how the levels of various air pollutants in an urban area vary during a day. The levels of nitrogen oxides and VOCs rise in the morning hours, mainly due to emissions from vehicles. The presence of both NO_x and VOCs leads to increased production of lung irritants such as ozone, whose levels go through a maximum during the afternoon hours.

Many cities have adopted programs that target one or more of the key constituents in photochemical smog, hoping to avoid the formation of its most hazardous components.

Even in this brief overview, we see that a single pollutant—NO_2 in this case—can have a complex impact on the air around us. We also see that the effects of various pollutants can be coupled in unexpected ways; the simultaneous presence of NO_2 and VOCs leads to the generation of O_3, another of the criteria pollutants. Clearly, the chemistry of air pollution is quite complex, so we will not be able to delve into many aspects of it in great detail. But it should be clear that to improve our understanding of these important issues, we must first explore the properties of gases, gas mixtures, and chemical reactions involving gases.

Properties of Gases

Several observable properties are common to all gases and distinguish them from solids and liquids:

- Gases expand to fill the volume of any container.
- Gases have much lower densities than solids or liquids.
- Gases have highly variable densities, depending on the conditions.
- Gases mix with one another readily and thoroughly.
- Gases change volume dramatically with changing temperature.

Some of these properties help explain why air pollution is difficult to combat. The fact that gases expand to fill the available volume, for example, means that exhaust gases released from a car or an industrial plant will readily spread throughout the surrounding area. Similarly, the fact that gases mix readily with one another means that once they are present, traces of gaseous pollutants will usually be difficult to separate from the air.

As we work toward developing an accurate model for gas behavior, we will want to be sure that it can account for all of these observations. We'll want our model to be based on a molecular-level understanding of what a gas is like, and we'll also want it to enable us to make accurate numerical predictions about how a gas will respond to different conditions.

As it turns out, you probably already know an important quantitative part of the model we use most often to describe gases. If there is just one equation that you remember from your high school chemistry class, chances are it is the ideal gas law:

$$PV = nRT \tag{5.1}$$

P is the gas pressure, V is the volume, n is the number of moles of gas present, and T is the temperature. R is a proportionality constant, called the **universal gas constant**

R can also be expressed in other units, as discussed in Section 5-3.

because its value (0.08206 L atm mol^{-1} K^{-1}) is the same for any gas. This familiar equation will form the basis of much of our discussion in this chapter, so it will be important that we have a thorough understanding of the terms that appear in it. Let's start by taking a close look at the meaning and physical origin of gas pressure.

5-2 Pressure

From inflation guidelines for car or bicycle tires to weather reports, we regularly encounter the word pressure, so you undoubtedly have some notion of what it means. But to investigate the behavior of gases scientifically, we need to make our intuitive understanding of pressure more formal.

Pressure is defined as force exerted per unit area:

$$P = \frac{F}{A} \tag{5.2}$$

The obvious question arising from this definition is, "What force causes the pressure of a gas?"

For atmospheric pressure, the force can be attributed to the weight of the molecules of air that are attracted to the surface of the planet by gravity. The existence of atmospheric pressure may not be apparent but can readily be demonstrated by removing the air from a container. Even an apparently sturdy container collapses under the pressure of the atmosphere when the air inside is no longer supporting the container walls. Because the total weight of the molecules increases as the number of molecules increases, the atmospheric pressure changes as a function of altitude. At sea level, 20,320 feet more of atmosphere are above us than at the summit of Mount McKinley, and the atmospheric pressure at sea level is much higher than at the mountaintop. The variation of pressure with altitude is shown in **Figure 5.2**.

But what about gas trapped in a container? We can think about this by considering the air in a balloon. It exerts pressure uniformly on all the walls of the balloon without regard to direction. Clearly, this can't be due to gravity. How does a gas exert force on the walls of its container? To answer this, we need to zoom in to the molecular level, as in **Figure 5.3**, where we will see individual gas molecules colliding with the walls and bouncing off. Each molecular collision imparts a small force to the container. And when these tiny forces are summed across the vast number of gas molecules present, they can produce the macroscopic force that leads to gas pressure. Later we will see how several observable gas properties can readily be explained by considering the behavior of gas molecules on the microscopic scale.

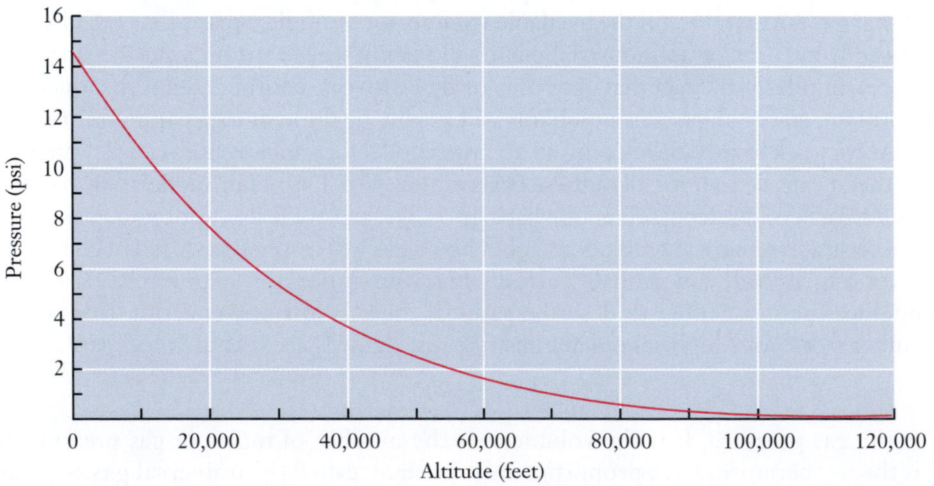

Figure 5.2 ■ Atmospheric pressure varies as a function of altitude. Pressure is given here in units of pounds per square inch. Other units of pressure are discussed later in this section.

Figure 5.3 ■ The inset in this figure shows a microscopic scale "close-up" of the gas molecules in a balloon. As molecules move through the volume of the balloon, occasionally they collide with the rubber walls. In every such collision, the gas molecules exert a small force on the balloon wall, pushing it outward. When summed over the large number of gas molecules present, these small forces give rise to the observed pressure of the gas and are responsible for the shape of the balloon.

Measuring Pressure

Because pressure is one of the important measurable properties of a gas, several different ways have been devised to measure it, beginning with the **barometer**. The essential feature of a standard barometer is a long tube that is closed at one end and is filled with mercury. (Mercury is used because it has a high density and does not react chemically with common gases.) When the mercury-filled tube is inverted in a pool of mercury, some of the liquid, but not all, runs out of the tube. The closed end of the tube is then under vacuum, which means that the pressure in this region is greatly reduced. If this end of the tube is opened, all of the mercury pours out. What prevents the movement of mercury when the tube is closed? We can understand how the pressure of air is responsible by referring to **Figure 5.4**.

Scientists use the word "vacuum" to refer to any region of reduced pressure, not just the idealized case of zero pressure.

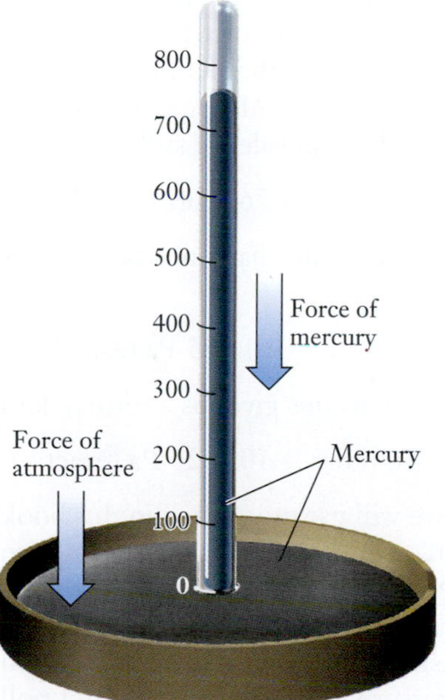

Figure 5.4 ■ A schematic diagram of the forces at work in a mercury barometer. Atmospheric pressure exerts a force pushing down on the surface of the mercury and tends to push the column of mercury in the barometer higher up the tube. The force due to the weight of the mercury in the column tends to push the column of mercury lower in the tube. When the two forces are equal, the column height remains constant.

The pool of mercury is exposed to the atmosphere and thereby experiences the pressure of air weighing down on it. The column of mercury in the tube has a vacuum over it and so has no pressure weighing down on it. Now consider a point at the mouth of the tube in the pool of mercury. This point experiences pressure from two places, the atmospheric pressure on the pool *and* the weight of the mercury from the column. When these two pressures are equal, the height of mercury suspended in the tube remains constant and provides a way to determine the atmospheric pressure. When the atmospheric pressure decreases (from weather variations), the decrease will cause the column to shorten to maintain the equality of pressure at that point. Near sea level, the column of mercury is typically between 730 mm and 760 mm in height.

We'll look at some more modern ways of measuring pressure in the closing section of this chapter.

Units of Pressure

The SI unit for pressure is a derived unit that reflects the definition of pressure. The SI units for force and area are the newton and the square meter, respectively. So pressure is measured in N m^{-2}. This unit has also been named the **pascal (Pa)**. One newton is roughly the force exerted by gravity on a one-quarter-pound object. When spread out over a square meter, this force does not produce very much pressure, so the Pa is a relatively small quantity. Typical atmospheric pressures are on the order of 10^5 Pa. Thus a more practical unit is kPa, which is the unit used in weather forecasts in many countries that employ metric units.

The construction of the barometer has influenced the units commonly used for reporting pressures. Most weather forecasts in the United States report pressure in inches (inches of mercury). Most barometers in school laboratories provide measurements in millimeters of mercury (mm Hg). Both of these units appear to express pressure as a length, which can be confusing because we defined pressure as force per unit of area. The explanation of this contradiction comes directly from the barometer. The height of the mercury column in a barometer is directly proportional to the applied pressure; so the higher the pressure exerted by the atmosphere, the higher the column of mercury. As a result, it became common to report pressure directly in terms of the measured height. In spite of this confusion, the mm Hg unit is still widely used and has been given an additional name, the **torr**. We can consider these two units equivalent:

The torr is named after Evangelista Torricelli (1608–1647), who was the first person to suggest building a mercury barometer.

$$1 \text{ torr} = 1 \text{ mm Hg}$$

Another unit of pressure arises from the importance of the atmosphere. The average pressure at sea level is close to 760 torr. The common observation of an apparent average atmospheric pressure led to the definition of a unit called an **atmosphere (atm)**, which was set equal to 760 torr. Although pressure measurements have become more precise over the years, this equivalence still remains:

$$1 \text{ atm} = 760 \text{ torr (exactly)}$$

But the definition of the unit atm has since been changed to reflect SI units. It is now defined in terms of the pascal:

$$1 \text{ atm} = 101{,}325 \text{ Pa (exactly)}$$

Combining the last two relationships gives us a formal definition of the torr as well:

$$760 \text{ torr} = 101{,}325 \text{ Pa (exactly)}$$

The three units that we will use most often in this book are the atm, torr, and kPa. In some engineering fields, you are also likely to encounter other units, including pounds per square inch. None of these should cause you particular difficulty as long as you are careful that any calculations you do properly account for the units used.

Now, we might ask how gas pressure plays a role in air pollution. Gases naturally flow from higher pressure to lower pressure, and pollutants are often produced in a high-pressure environment, such as in the cylinder of an auto engine or a reaction vessel in an industrial processing plant. When gases are released from these high-pressure environments, they disperse rapidly into the surrounding air. The pressure difference helps speed the spread of the pollutants.

5-3 History and Application of the Gas Law

Unlike solids and liquids, gases change significantly when the conditions in which they are found are altered. Changing the temperature, pressure, or volume of a gas's container results in a dramatic response from the gas. All gases respond to these physical changes in the same way despite their chemical differences, and the gas law equation allows us to calculate the impact of a particular change. Historically, the study of gases preceded most of the important discoveries of modern chemistry. The relationships between physical properties of gases summarized by the ideal gas law equation were initially based on empirical observations and often still carry the name of the investigator who first characterized the relationship.

Jacques Charles first investigated the relationship between temperature and volume in the early 19th century. He noted that when he plotted volume versus temperature for different amounts of gas, the resulting lines converged to the same temperature at zero volume (see **Figure 5.5**). This fact was used to establish the Kelvin temperature scale. Once the temperature is expressed on the Kelvin scale, the volume and temperature are directly proportional, an observation now known as **Charles's law**.

Other early scientists examined the connections between pairs of gas variables and arrived at gas laws to express their observations. Robert Boyle studied the relationship between gas pressure and volume and noted that for any gas sample, when pressure increases, volume decreases, or when volume increases, pressure decreases. Thus **Boyle's law** states that at constant temperature, pressure and volume are inversely proportional. Similarly, **Avogadro's law** states that the volume of a gas is proportional to the number of molecules (or moles) of gas present.

These laws are frequently used to calculate the effect of some particular change in gas conditions. Fortunately, we will not need to memorize specific equations for all of these individual laws, because they are all incorporated into the ideal gas law. As long as we know the ideal gas equation and know how to work with it, we can easily solve problems that might also be approached using any of the individual gas laws. To do this, just start by identifying all of the parameters from the gas law—P, V, T, or n—that do *not* change. (R is a universal constant, so it never changes!) Then rearrange the

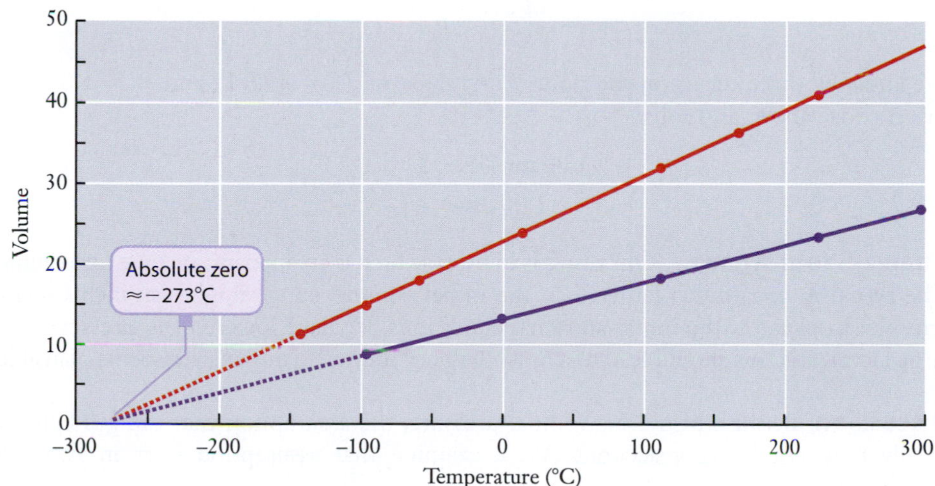

Figure 5.5 ■ The graph shows volume as a function of temperature for two different samples of gas. The sample represented by the upper red line contains more gas molecules than that represented by the lower blue line. Note that for both samples, the relationship between volume and temperature is linear. Also note that when both lines are extrapolated back toward lower temperatures, they have the same x intercept at −273.15°C. This extrapolation defines absolute zero temperature, or 0 K.

gas law equation to collect all of those terms onto the same side, leaving those that *do* change on the other side. If pressure and volume change while the number of moles and the temperature remain constant, for example, we'd get

$$PV = nRT = \text{constant}$$

Because the product nRT remains constant, we know that PV must also be constant. This tells us that the product of P times V must be the same before and after the change, so we can write

$$P_1V_1 = P_2V_2 \tag{5.3}$$

Here the subscripts "1" and "2" indicate the conditions before and after the change takes place. Now if we know any three of those four quantities, we can solve for the missing one, as seen in Example Problem 5.1.

EXAMPLE PROBLEM 5.1

In addition to ongoing sources, such as car engines, pollutants can also be introduced into the air through isolated incidents like the rupture of a gas storage tank. Most gases are stored and transported at high pressures. A common laboratory cylinder of methane, for example, has a volume of 49.0 L and is filled to a pressure of 154 atm. Suppose that all of the CH_4 from this cylinder is released and expands until its pressure falls to 1.00 atm. What volume would the CH_4 occupy?

Strategy The problem involves a change in the conditions, here pressure and volume, of a sample of gas. We will want to use the ideal gas law because it is our model that relates the various properties of a gas to one another. So we begin by assuming that the gas will obey the ideal gas law under both the initial and final conditions. We will further assume that temperature is constant because we are not given any data to suggest otherwise. Because we are working with the same sample of gas throughout, we also know that the number of moles (n) is constant. Collect the constant terms together and work toward a solution.

In Section 5-6, we will look at gases that do not behave ideally.

Solution In this case, n and T are constant, and R is always constant. So,

$$PV = nRT = \text{constant}$$

Therefore,

$$P_1V_1 = P_2V_2$$

Solving this for the final volume (V_2) gives

$$V_2 = \frac{P_1V_1}{P_2}$$

We know all three terms on the right: $P_1 = 154$ atm, $V_1 = 49.0$ L, and $P_2 = 1.00$ atm. Insert those above and solve:

$$V_2 = \frac{(154 \text{ atm})(49.0 \text{ L})}{(1.00 \text{ atm})} = 7550 \text{ L}$$

Analyze Your Answer Our result is in liters, which is an appropriate unit for volume. The actual value is much larger than the initial volume, so we might ask if this seems sensible. Looking at the initial and final conditions, we have lowered the pressure by a large factor, and this should lead to a much larger volume. So our answer seems plausible.

Discussion Because of the high initial pressure, the assumption that the gas behaves ideally may not be very accurate. We'll examine that assumption later in Example Problem 5.8.

Check Your Understanding A small sample of gas is generated in a laboratory. In a 125-mL vessel, it exerts 115 torr of pressure. Then a valve is opened and the gas expands until its volume is 175 mL. If the temperature is constant throughout this process, what is the new pressure of the gas?

In other situations, we might find that V and T vary, whereas n and P are constant. That would lead us to write the following relationship:

$$\frac{V}{T} = \frac{nR}{P} = \text{constant}$$

This, in turn, leads to Equation 5.4:

$$\frac{V_1}{T_1} = \frac{V_2}{T_2} \tag{5.4}$$

Example Problem 5.2 illustrates this relationship.

EXAMPLE PROBLEM 5.2

A balloon is filled with helium, and its volume is 2.2 L at 298 K. The balloon is then dunked into a thermos bottle containing liquid nitrogen. When the helium in the balloon has cooled to the temperature of the liquid nitrogen (77 K), what will the volume of the balloon be?

Strategy We should start by recognizing this as a change in conditions for a gas and realize that we will want to use the ideal gas law. So we assume that the gas will behave ideally under both the initial and final conditions. We will further assume that the pressure is constant since we are not given any data to suggest otherwise. And since the balloon is tied off throughout, we can also assume that n is constant. Collect the constant terms together and work toward a solution.

Solution As in the discussion above, we have

$$\frac{V}{T} = \frac{nR}{P} = \text{constant}$$

So

$$\frac{V_1}{T_1} = \frac{V_2}{T_2}$$

Solve this for the final volume (V_2):

$$V_2 = \frac{V_1 T_2}{T_1}$$

Because the three terms on the right are known, we can just insert values and solve:

$$V_2 = \frac{(2.2 \text{ L})(77 \text{ K})}{(298 \text{ K})} = 0.57 \text{ L}$$

Analyze Your Answer We have decreased the temperature from 298 K to 77 K, which means that it is smaller by a factor of nearly 4. This should lead to a volume that is also smaller by a factor of nearly 4. So our answer seems likely to be correct.

Check Your Understanding The balloon in the example above will burst if its volume exceeds 2.3 L. At what temperature would you expect the balloon to burst?

For *any* change in the conditions of a gas sample, we should always be able to employ this same idea by separating those parameters that are constant from those that are changing. Because you are very unlikely to forget the ideal gas law, this approach is almost certainly more reliable than attempting to remember a list of equations covering all the possible combinations of variables.

Units and the Ideal Gas Law

Suppose that we had expressed the temperature in °C rather than in kelvins in the above example? It should take you only a little work to convince yourself that the equation we used will not hold true. Changing the temperature of a gas from 10 to 20°C, for example, will not double its volume. The ideal gas law itself is only valid when we use an **absolute temperature** scale where $T = 0$ corresponds to absolute zero temperature. To see why this must be so, notice that putting $T = 0$ into the gas law implies that either P or V or both must also go to zero, meaning that there must be no gas present. This simply can't be correct for $T = 0°C$ or $T = 0°F$; that's obvious because the winter temperature routinely reaches these levels in many areas, and yet the air in the atmosphere obviously does not vanish. Absolute zero (0 K), on the other hand, is a very special temperature, and it is reasonable that there might be peculiar consequences if we actually cooled a gas to this level.

Although absolute zero is a fairly abstract concept, temperatures well below 1 K can be achieved routinely in many laboratories.

In contrast to the temperature, which *must* be expressed on an absolute scale, we can use *any* convenient units for pressure and volume, so long as we then reconcile the units on those quantities with those for the gas constant, R. Why this difference? The common temperature scales are fundamentally different from other kinds of units because converting among them involves not only multiplying by a conversion factor but also adding or subtracting. This, in turn, results directly from the fact that different temperature scales (K, °C, or °F) have different zero points. In all the other quantities that we work with, the meaning of "zero" is the same no matter what units we use. If you say the mass of a sample is zero, for example, it doesn't really matter much what units you measured in—there's still nothing there. But for temperature, the differences among 0°C, 0°F, and 0 K are obvious. In some engineering fields, you may encounter the Rankine temperature scale (°R). This is also an absolute scale, meaning that 0°R = 0 K. But one degree on the Rankine scale is the same size as 1°F. Because they are both absolute scales, Rankine and Kelvin temperatures can be interconverted by using a multiplicative unit factor: 1°R = 1.8 K.

No degree sign is used when writing Kelvin temperatures.

If you look at a table like the one in Appendix B of this book, you will find values for R in several combinations of units:

$$\begin{aligned} R &= 0.08206 \text{ L atm mol}^{-1} \text{ K}^{-1} \\ &= 8.314 \text{ J mol}^{-1} \text{ K}^{-1} \\ &= 62.37 \text{ L torr mol}^{-1} \text{ K}^{-1} \end{aligned}$$

Students often ask how to decide which of these values is "the right R" to use in a given situation. The answer is that they are all really the same quantity, just expressed in different units. (If you are curious, you can do unit conversions to verify this for yourself.) If pressure is given in atm and volume in L, then it will be convenient to use R as 0.08206 L atm mol^{-1} K^{-1}, for example. But if pressures are given in torr, it might be easier to insert 62.37 L torr mol^{-1} K^{-1} for R. Or, if you always prefer to use the same value of R, you could simply convert pressures from any other units into atm before doing any calculations. It doesn't matter which set of units we choose, as long as we are careful to reconcile the units in all of our calculations. Example Problem 5.3 looks at the use of the gas law with quantities in some unusual units.

EXAMPLE PROBLEM 5.3

A sample of CO_2 gas has a volume of 575 cm^3 at 752 torr and 72°F. What is the mass of carbon dioxide in this sample?

Strategy We are trying to find the mass of a sample of gas. If we knew how many moles of CO_2 were in the sample, we could easily use the molar mass to find the mass of the sample. And since the problem involves a gas, we can attempt to use the ideal gas law to find the number of moles. We are given values for V, P, and T, so we will be able to solve for n. Some of the units given are unusual, so we will need to be careful to handle them properly.

Solution We can begin by converting some of the given information into units more commonly used for gases:

$$T = (72 - 32) \times 5/9 = 22°\text{C} = 295 \text{ K}$$

$$V = 575 \text{ cm}^3 = 575 \text{ mL} = 0.575 \text{ L}$$

For P, we will use the value in torr and choose the corresponding value for R. We could also convert from torr to atm, of course.

$$n = \frac{PV}{RT} = \frac{(752 \text{ torr})(0.575 \text{ L})}{(62.37 \text{ L torr mol}^{-1} \text{ K}^{-1})(295 \text{ K})} = 0.0235 \text{ mol}$$

The gas in our sample is CO_2, with a molar mass of 44.0 g/mol, so

$$0.0235 \text{ mol} \times \frac{44.0 \text{ g}}{\text{mol}} = 1.03 \text{ g } CO_2$$

Analyze Your Answer The temperature and pressure are near typical room conditions, and the volume is a little more than a half liter. So this should correspond to a small balloon filled with CO_2. One gram is a pretty small mass but seems reasonable for the gas in a small balloon.

Check Your Understanding Find the volume in ft^3 of 1.0 mole of an ideal gas at −25°C and 710 torr.

5-4 Partial Pressure

By using the ideal gas law, we can determine many properties of an individual gas. In many instances, however, we encounter interesting situations with more than one gas present. Our consideration of the levels of pollutants in air is an obvious example. Even if we ignore any pollutants present, clean air is already a mixture of gases. Aside from the possibility of chemical reactions, how do the observed properties of a gas mixture differ from those of pure gases?

The answer to this question is actually suggested by the gas law. In particular, note that none of the terms in the gas law equation depends on the identity of the gas. Consider a sample of air, and for simplicity assume that it consists entirely of N_2 and O_2. Assuming that the air behaves ideally, its pressure can be expressed using the gas law:

$$P = \frac{nRT}{V}$$

Here n is the total number of moles of air present. But this total number of moles will simply be the sum of the moles of each component:

$$n_{\text{total}} = n_{N_2} + n_{O_2}$$

We can insert this into the expression above, giving

$$P = \frac{nRT}{V} = \frac{(n_{N_2} + n_{O_2})RT}{V} = \frac{n_{N_2}RT}{V} + \frac{n_{O_2}RT}{V}$$

Each of those last two nRT/V terms has a special significance; they are the pressures that we would see if only one of the components of the gas mixture were present. If the nitrogen were the only gas present, n_{O_2} would be zero, and we would be left with

$$P = P_{N_2} = \frac{n_{N_2}RT}{V}$$

If only oxygen were present, we could write a similar result for P_{O_2}. For a mixture, the observed pressure is the sum of the hypothetical **partial pressures** of the individual gases.

This concept, first established by John Dalton, is sometimes referred to as **Dalton's law of partial pressures**: *The pressure of a mixture of gases is the sum of the partial pressures of the component gases.* In simple terms, the partial pressure is what the pressure would be if only that component were present. As long as the component gases behave ideally and do not actually react with one another, this relationship will hold for any gas mixture.

John Dalton is also known for his model proposing the existence and nature of atoms.

For an arbitrary gas mixture, we can write the partial pressure of each component as

$$P_i = \frac{n_iRT}{V} \tag{5.5}$$

where the index, i, indicates which gas we are talking about (for a mixture of four gases, i would vary from 1 to 4 yielding, P_1, P_2, P_3, and P_4) and n_i is the number of moles of component i. The total pressure of the mixture, P, is the sum of these partial pressures:

$$P = \sum_i P_i = \sum_i n_i \frac{RT}{V} \tag{5.6}$$

The uppercase Greek letter sigma (Σ) is used to indicate the sum of a number of similar terms.

The total number of moles is the sum of the numbers of moles of the individual components:

$$P = n_{\text{total}}\frac{RT}{V} \tag{5.7}$$

These expressions suggest another way to understand partial pressures. We might choose to express the concentration of each component gas in terms of its **mole fraction**, defined as

$$X_i = \frac{n_i}{n_{\text{total}}} \tag{5.8}$$

For a pure gas, $n_i = n_{\text{tot}}$ and the mole fraction is one. Any mixture will have $n_i < n_{\text{tot}}$, so the value of the mole fraction has a range from zero to one. If we take the ratio of a partial pressure to total pressure, using Equations 5.5, 5.7, and 5.8,

$$\frac{P_i}{P} = \frac{n_i(RT/V)}{n_{\text{total}}(RT/V)} = \frac{n_i}{n_{\text{total}}} = X_i \tag{5.9}$$

So the partial pressure is given by the mole fraction times the total pressure:

$$P_i = X_iP \tag{5.10}$$

This set of equations provides the numerical means for us to use partial pressures, as shown in Example Problems 5.4 and 5.5.

EXAMPLE PROBLEM 5.4

Not all pollution is due to human activity. Natural sources, including volcanoes, also contribute to air pollution. A scientist tries to generate a mixture of gases similar to those found in a volcano by introducing 15.0 g of water vapor, 3.5 g of SO_2, and 1.0 g of CO_2 into a 40.0-L vessel held at 120.0°C. Calculate the partial pressure of each gas and the total pressure.

Strategy As usual, we will begin by assuming that all the gases behave ideally. Then we can treat each gas separately to determine its partial pressure. Finally, sum these partial pressures to calculate the total pressure.

Solution For all of the gases, T = 120°C = 393 K, and V = 40.0 L.

$$H_2O\text{: } 15.0 \text{ g } H_2O \times \frac{1 \text{ mol } H_2O}{18.0 \text{ g } H_2O} = 0.833 \text{ mol}$$

$$P_{H_2O} = \frac{(0.833 \text{ mol})(0.08206 \text{ L atm mol}^{-1} \text{ K}^{-1})(393 \text{ K})}{40.0 \text{ L}} = 0.672 \text{ atm}$$

$$SO_2\text{: } 3.5 \text{ g } SO_2 \times \frac{1 \text{ mol } SO_2}{64.1 \text{ g } SO_2} = 0.055 \text{ mol}$$

$$P_{SO_2} = \frac{(0.055 \text{ mol})(0.08206 \text{ L atm mol}^{-1} \text{ K}^{-1})(393 \text{ K})}{40.0 \text{ L}} = 0.044 \text{ atm}$$

$$CO_2\text{: } 1.0 \text{ g } CO_2 \times \frac{1 \text{ mol } CO_2}{44.0 \text{ g } CO_2} = 0.023 \text{ mol}$$

$$P_{CO_2} = \frac{(0.023 \text{ mol})(0.08206 \text{ L atm mol}^{-1} \text{ K}^{-1})(393 \text{ K})}{40.0 \text{ L}} = 0.018 \text{ atm}$$

Total pressure is the sum of these:

$$P_{tot} = P_{H_2O} + P_{SO_2} + P_{CO_2} = 0.672 + 0.044 + 0.018 = 0.734 \text{ atm}$$

Note that you could also have used the total number of moles to find P_{tot} and then used mole fractions to find the partial pressures.

Analyze Your Answer It is possible to gain a feeling for quantities such as pressure and number of moles if you work a lot of these problems. In this case, we have almost a mole of gas in a volume of 40 L, and the temperature is somewhat high. Our result of 0.7 atm is not unreasonable.

Check Your Understanding A mixture containing 13.5 g of oxygen and 60.4 g of N_2 exerts a pressure of 2.13 atm at 25°C. What is the volume of the container, and what are the partial pressures of each gas?

EXAMPLE PROBLEM 5.5

An experiment is designed to determine the effect of sulfur dioxide, one of the EPA criteria pollutants, on plants. Among the variations used is a mixture that has the mole fractions given in the following table.

Gas	N_2	O_2	H_2O	SO_2
Mole fraction	0.751	0.149	0.080	0.020

If the desired total pressure is 750 torr, what should the partial pressures be? If the gas is to be in a 15.0-L vessel held at 30°C, how many moles of each substance are needed?

Strategy We know the mole fractions and the desired total pressure. So we can calculate partial pressures using the relationship defined in Equation 5.10. From the total pressure and the volume, we can calculate the total number of moles and thereby the number of moles of each gas.

Solution

$$P_i = X_i P_{\text{total}}$$

$$P_{N_2} = (0.751)(750.\text{ torr}) = 563 \text{ torr}$$

$$P_{O_2} = (0.149)(750.\text{ torr}) = 112 \text{ torr}$$

$$P_{H_2O} = (0.080)(750.\text{ torr}) = 60 \text{ torr}$$

$$P_{SO_2} = (0.020)(750.\text{ torr}) = 15 \text{ torr}$$

The desired total pressure of 750 torr can also be expressed as 0.987 atm. So

$$n_{\text{total}} = \frac{PV}{RT} = \frac{(0.987 \text{ atm})(15.0 \text{ L})}{(0.08206 \text{ L atm mol}^{-1}\text{ K}^{-1})(303 \text{ K})} = 0.595 \text{ mol}$$

$$n_i = X_i n_{\text{total}}$$

$$n_{N_2} = (0.751)(0.595) = 0.447 \text{ mol}$$

$$n_{O_2} = (0.149)(0.595) = 8.87 \times 10^{-2} \text{ mol}$$

$$n_{H_2O} = (0.080)(0.595) = 4.8 \times 10^{-2} \text{ mol}$$

$$n_{SO_2} = (0.020)(0.595) = 1.2 \times 10^{-2} \text{ mol}$$

Check Your Understanding A mixture of $SO_2(g)$ and $SO_3(g)$ is to be prepared with a total pressure of 1.4 atm. If the mole fractions of the gases are 0.70 and 0.30, respectively, what are the partial pressures? If the mixture is to occupy 2.50 L at 27°C, what mass of each gas is needed?

5-5 Stoichiometry of Reactions Involving Gases

When we discussed quantitative aspects of chemical reactions in Chapter 4, we emphasized the importance of ratios of moles. The ideal gas law provides a relationship between the number of moles of a gas and some easily measurable properties: pressure, volume, and temperature. So when gases are involved in a chemical reaction, the ideal gas law often provides the best way to determine the number of moles. Using the ideal gas law in a stoichiometry problem really doesn't involve any new ideas. It just combines two kinds of calculations that you've already been doing. We'll still do the stoichiometric calculation in terms of mole ratios, as always, and we'll use the gas law to connect the number of moles of a gas with its temperature, pressure, and volume. Example problem 5.6 demonstrates this approach.

EXAMPLE PROBLEM 5.6

When an experiment required a source of carbon dioxide, a student combined 1.4 g of sodium bicarbonate ($NaHCO_3$) with excess hydrochloric acid. If the CO_2 produced is collected at 722 torr and 17°C, what volume will the gas occupy?

Strategy We are asked to find the volume of a gas, and we are given its pressure and temperature. We'll assume that the gas behaves ideally. So if we knew the number of moles, we could easily use the gas law to get the volume we need. Looking a little closer, we should recognize this as a reaction stoichiometry problem because it asks us how much CO_2 will be produced. The new wrinkle here is that it asks us to express the answer as a volume rather than as a mass or a number of moles. So we will first do a stoichiometric calculation to find the number of moles of CO_2 produced and then use the gas law to find the volume of that amount of gas at the indicated temperature and pressure. As in any stoichiometry problem, we'll need to start with a balanced equation for the reaction to be sure we use the correct mole ratio.

Solution

$$NaHCO_3 + HCl \rightarrow NaCl + H_2O + CO_2$$

$$1.4 \text{ g } NaHCO_3 \times \frac{1 \text{ mol } NaHCO_3}{84.0 \text{ g } NaHCO_3} \times \frac{1 \text{ mol } CO_2}{1 \text{ mol } NaHCO_3} = 0.017 \text{ mol } CO_2$$

Now use the ideal gas law to solve for the volume. As always, we'll need to remember to convert the temperature from °C to K and to be careful with pressure and volume units:

$$V = \frac{nRT}{P} = \frac{(0.017 \text{ mol})(0.08206 \text{ L atm mol}^{-1} \text{ K}^{-1})(290 \text{ K})}{(0.95 \text{ atm})} = 0.42 \text{ L} = 420 \text{ mL}$$

Analyze Your Answer Our result of 420 mL is a relatively small volume—does this make sense? In this experiment, the reaction is similar to the one that occurs between baking soda and vinegar. Because 1.4 g of baking soda is not a large amount, it's not surprising that the amount of gas generated is also small.

Check Your Understanding Suppose that 270 mL of CO_2 at 1.03 atm and 25°C is produced by this reaction. Assuming that an excess of hydrochloric acid was present, how many grams of sodium bicarbonate were used?

STP Conditions

From our discussion so far, it should be obvious that to describe an amount of gas properly, we need to specify its temperature, volume, and pressure. If we want to compare various quantities of gas, it is convenient to use some standard set of conditions. To set a standard for an ideal gas, we might specify a temperature and a pressure; in that way, the volume would be directly proportional to the number of moles of gas. The conventional choices for the **standard temperature and pressure** of a gas, or **STP**, are a temperature of 0°C or 273.15 K and a pressure of 1 atm. Inserting these values into the ideal gas equation, we find the molar volume (volume of one mole) of an ideal gas at STP:

This standard molar volume provides a handy way to estimate gas properties or to help judge the reasonableness of a calculated value.

$$V = \frac{nRT}{P} = \frac{(1 \text{ mol})(0.08206 \text{ L atm mol}^{-1} \text{ K}^{-1})(273.15 \text{ K})}{(1 \text{ atm})} = 22.41 \text{ L}$$

This number provides a conversion factor for stoichiometric problems that include gases, *provided that STP conditions are maintained.* Example Problem 5.7 shows this type of problem.

EXAMPLE PROBLEM 5.7

One way to reduce air pollution is to remove potential pollutant gases from an exhaust stream before they are released into the air. Carbon dioxide can be removed from a stream of gas by reacting it with calcium oxide to form calcium carbonate. If we react 5.50 L of CO_2 at STP with excess CaO, what mass of calcium carbonate will form?

Strategy You should recognize this as a reaction stoichiometry problem because it is asking us how much $CaCO_3$ will be produced. As for any stoichiometry problem, we should write a balanced chemical equation for the reaction. Then we convert the volume of gas into moles and proceed as usual. Because the gas volume given is at STP, we can use the molar volume we calculated above as a conversion factor.

Solution

$$CO_2(g) + CaO(s) \rightarrow CaCO_3(s)$$

$$5.50\text{ L } CO_2 \times \frac{1\text{ mol } CO_2}{22.41\text{ L } CO_2} \times \frac{1\text{ mol } CaCO_3}{1\text{ mol } CO_2} \times \frac{100.1\text{ g } CaCO_3}{1\text{ mol } CaCO_3} = 24.6\text{ g } CaCO_3$$

Analyze Your Answer From the molar volume at STP, we can see that 5.5 L would be about one-quarter of a mole of CO_2. So we should be making about one-quarter mole of $CaCO_3$, and our answer agrees with that.

Discussion Note that the conversion factor of 22.4 L = 1 mole is valid *only* when the gas in question is at STP. You should be very careful not to use that molar volume for any other conditions! We could also have worked this problem by using the ideal gas law to find the volume of the CO_2, using the appropriate temperature and pressure for STP conditions.

In this example, a gas is converted to a solid product rather than being emitted into the air. What implications might this have for engineering design? Can you think of situations in which this type of pollution control scheme might cause problems?

Check Your Understanding SO_2 can be generated in the laboratory by heating ZnS with O_2. (The other product is zinc oxide.) If 14.5 L of oxygen at STP is reacted with excess zinc sulfide, what volume of SO_2 at STP is generated?

At this point, we have seen how to carry out a range of calculations involving gases, all based on the ideal gas law. But though we have repeatedly said that we will assume the gas behaves ideally, we have not yet looked at what that statement implies in terms of molecular behavior. We have also not yet considered the many practical circumstances in which we might make large errors by assuming ideal gas behavior. We take up these challenges in the next section.

5-6 Kinetic–Molecular Theory and Ideal Versus Real Gases

The ability to model the behavior of gases numerically was an important achievement that has broad applicability in many areas of science and engineering. The similarity of the behavior of different gases commonly encountered in the laboratory allowed the construction of models to describe how molecules must behave. But in many important practical settings, gases do not always behave ideally. So the simple ideal gas law calculations that we have been doing may not always be an accurate model of reality. Nonideal behavior is frequently seen in gases at very high pressures, such as might be found in the cylinder of an engine or in an industrial processing plant. Can an atomic perspective developed for ideal gases help explain these real world cases? Perhaps more importantly, can an improved model define quantitative methods for dealing with real gases such as those encountered in a running engine? In this section, we will develop a very useful model called the **kinetic–molecular theory**, which provides connections between the observed macroscopic properties of gases, the gas law equation, and the behavior of gas molecules on a microscopic scale. Once we understand this model, we can use it to see how gases at high pressure might deviate from ideal behavior.

The kinetic–molecular theory is also called the kinetic theory of gases or simply the kinetic theory.

Postulates of the Model

Because observation of individual gas molecules is not generally feasible, we must begin with postulates that are accepted as reasonable when we construct a model.

For the kinetic–molecular theory of gases, the following postulates form the bedrock of the model:

1. A gas is made up of a vast number of particles, and these particles are in constant random motion.
2. Particles in a gas are infinitely small; they occupy no volume.
3. Particles in a gas move in straight lines except when they collide with other molecules or with the walls of the container. Collisions with each other and with the walls of the container are elastic, so that the total kinetic energy of the particles is conserved.
4. Particles in a gas interact with each other only when collisions occur.
5. The average kinetic energy of the particles in a gas is proportional to the absolute temperature of the gas and does not depend on the identity of the gas.

As we consider the implications of these postulates, we will also comment on their validity or the conditions under which they are not particularly accurate. Starting from these postulates, it is actually possible to derive the ideal gas law. Although that result is a great example of the synergy between scientific theory and experimental observations, it is a bit beyond our reach here. So instead we will consider the connections between the postulates of the kinetic theory and our practical experience with gases. How do these postulates help to explain the observed behavior of various gases?

The statement that a gas consists of a large number of particles is consistent with what we know about the size of individual molecules. And the postulate that the molecules are in constant random motion is consistent with the observation that gases readily expand to fill their containers. The postulates are also consistent with what we said at the outset of this chapter about the origin of gas pressure. The molecules in a gas exert force when they collide with the wall of the container, and this accounts for the relationship between pressure and volume. As the volume of the container increases, the walls move farther apart, and so each molecule will strike the wall less often. Thus each molecule will exert less force on the wall in a given time interval, reducing the pressure. So the kinetic theory is consistent with the inverse relationship between P and V that we've seen in the gas law equation.

How can the kinetic theory account for the effect of changes in temperature? The situation here is a little more complicated. We need to consider postulate #5 carefully: *The kinetic energy of the particles in a gas is proportional to the absolute temperature of the gas and does not depend on the identity of the gas.* We know from physics that kinetic energy is related to speed and to mass: $\text{KE} = 1/2\ mv^2$. So if the kinetic energy depends on the temperature, as stated in the postulate, then the speed of the molecules must be a function of temperature. More specifically, increasing the temperature must increase the speed of the molecules. When the molecules move faster, they will collide with the walls more often, and they will also impart a greater force on the wall during each collision. Both of these effects will lead to an increase in gas pressure with increasing temperature, as predicted by the gas law.

If you were paying close attention to the statement of postulate #5, though, you may be wondering about the inclusion of the word "average" in "average kinetic energy." That word is included because some molecules in a collection of randomly moving gas molecules move very fast, whereas others are virtually standing still at any given moment. A moment later the identity of fast-moving and slow-moving particles may have changed due to collisions, but there will still be both kinds. So it isn't really correct to think about the molecules as having a particular speed, and we should actually say that the average molecular speed increases with increasing temperature. This situation is best approached using a **distribution function** to describe the range of speeds.

Distribution functions are common mathematical devices used in science and also in economics, social sciences, and education. Your percentile scores on the SAT and other standardized tests are predicated on a distribution function for the students who take the exam. If you scored in the 85th percentile, for example, you would be expected to score higher than 85% of those taking the exam. This ranking is expected to hold independent of the number of students who have actually taken the test, so that your percentile is not revised every time another round of tests is administered to more students.

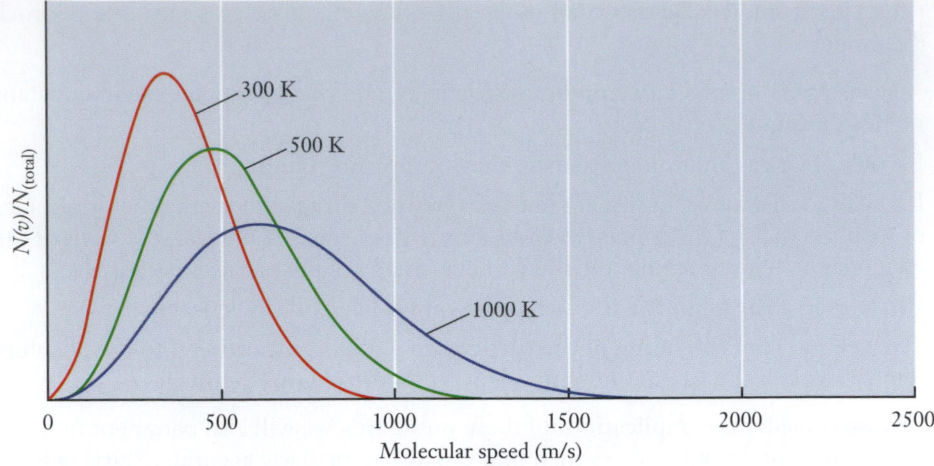

Figure 5.6 ■ The distributions of molecular speeds for CO_2 molecules at three different temperatures are shown. The quantity on the y axis is the fraction of gas molecules moving at a particular speed. Notice that as the temperature increases, the fraction of molecules moving at higher speed increases.

Because we do not care in which direction the particles are going, our discussion here is based on speed rather than velocity. Speed does not depend on direction and is always positive.

The distribution function that describes the speeds of a collection of gas particles is known as the **Maxwell-Boltzmann distribution** of speeds. This function, which was originally derived from a detailed consideration of the postulates of the kinetic theory, predicts the fraction of the molecules in a gas that travel at a particular speed. It has since been verified by a variety of clever experiments that allow measuring the speed distribution in the laboratory.

Let's examine a plot of the Maxwell-Boltzmann distribution for a sample of CO_2 gas at various temperatures, as shown in **Figure 5.6**. As with any unfamiliar graph, you should start by looking at what is plotted on each axis. Here the x axis shows a fairly familiar quantity: speed in meters per second. The y axis is a little less obvious. It shows the fraction of all the molecules in the sample that would be traveling at a particular speed under the given conditions. (The actual mathematical form of the distribution function is discussed in the MathConnections section on the next page.) We've already established that the distribution of speeds must be a function of temperature, and the figure bears this out. At high temperatures, we see that the peak in the distribution shifts toward higher speeds. This is entirely consistent with postulate #5, which states that the kinetic energy of the molecules must increase with temperature.

Furthermore, postulate #5 also says that the kinetic energy does not depend on the identity of the gas. But the mass of the gas molecules clearly depends on their identity. For the kinetic energy to remain fixed as the mass changes, the speed distribution must also be a function of mass. **Figure 5.7** illustrates this by showing the speed distribution for various gases at the same temperature. You can readily see that on average lighter molecules travel faster than heavier molecules.

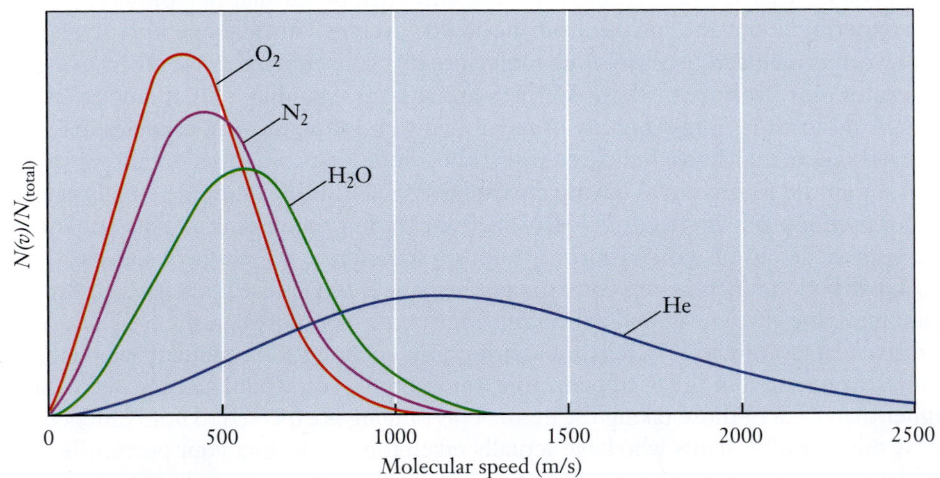

Figure 5.7 ■ Distributions of molecular speeds for four different gases at the same temperature (300 K). As in Figure 5.6, the quantity on the y axis is the fraction of gas molecules moving at a particular speed. Notice that as the molecular mass decreases, the fraction of the molecules moving at higher speed increases.

MathConnections

The actual equation for the Maxwell-Boltzmann distribution of speeds is somewhat complicated, but it consists entirely of terms in forms that you should recognize from your math classes:

$$\frac{N(v)}{N_{\text{total}}} = 4\pi(M/2\pi RT)^{3/2}v^2e^{-Mv^2/2RT} \tag{5.11}$$

$N(v)$ is the number of molecules moving with speeds between v and $v + \Delta v$, where Δv is some small increment of speed. N_{total} is the total number of molecules. So the left-hand side is the fraction of the molecules having speeds between v and $v + \Delta v$, which is also the quantity plotted on the y axis in Figures 5.6 and 5.7. M is the molar mass of the gas, R is the gas constant, and T is the temperature. (As in the gas law, this must be on an absolute scale.) Units for all of these quantities must be chosen consistently, with particular care taken to ensure that the exponent ($-Mv^2/2RT$) has no units. Now look at Figure 5.6 or 5.7 again and notice that the independent variable on the x axis is the speed. So we want to think of the right-hand side of the distribution function as an elaborate function of the speed, v. We see two terms containing v. The first is just v^2, which will obviously increase as v increases. The exponential term, though, varies as $\exp(-v^2)$, which will decrease as v increases. So the overall shape of the distribution function reflects a competition between these two terms. At small values of v, the v^2 term dominates, and the distribution function increases. At larger values of v, the exponential term takes over, and the overall function decreases. So, between those two limits, the distribution function goes through a maximum.

The speed at which the peak in the plot occurs is known as the **most probable speed**, for fairly obvious reasons: more molecules travel at this speed than at any other. Although we will not do the derivation here, the most probable speed is given by Equation 5.12:

$$v_{\text{mp}} = \sqrt{\frac{2RT}{M}} \tag{5.12}$$

If you know some calculus, you can do the derivation. Remember that the most probable speed is the maximum in the distribution function. To find this maximum, we would take the derivative of Equation 5.11 and set it equal to zero.

We might also want to describe our distribution by talking about the **average speed**. Because the distribution function is not symmetric, the most probable speed and the average speed will not be the same. In particular, the existence of the "tail" on the distribution curve at high speeds will pull the average to a speed higher than the most probable value. The average speed actually turns out to be 1.128 times the most probable speed.

Finally, sometimes we characterize the distribution by talking about something called the **root-mean-square speed**. This value is actually the square root of the average value of v^2. At first glance, it might sound like that should be the same as the average speed, but for the Boltzmann distribution, the root-mean-square speed is actually 1.085 times the average speed. (If this sounds wrong to you, try a simple test. Pick a few numbers at random and calculate their average. Then square all of your numbers and take the average of the squares. Finally, take the square root of that result and you'll see that it is larger than the average of the original numbers.) The root-mean-square speed is useful if we are interested in kinetic energy because the average kinetic energy is given by

$$\text{KE}_{\text{avg}} = \frac{1}{2}mv_{\text{rms}}^2 \tag{5.13}$$

Real Gases and Limitations of the Kinetic Theory

In all of our calculations so far, we have begun by assuming that any gas or gas mixture behaves ideally. But the postulates of the kinetic theory—and the gas law itself—are strictly true only for purely hypothetical ideal gases. How good is the assumption that gases behave ideally, and under what kinds of conditions might we want to question this assumption? The answers to these questions can help us to improve our understanding of gases at the molecular level. By learning how real gases deviate from ideal behavior, we will understand the nature of the kinetic–molecular theory better. We'll start by taking a more critical look at some of the postulates of the theory.

The assertion that gas molecules occupy no volume may seem at odds with reality because we know that all matter occupies space. What is actually implied by this postulate? The essential idea is that, compared to the volume of empty space between particles, the volume of the particles themselves is not significant. One way to assess the validity of this assertion is to define the mean free path of the particles. The **mean free path** is the average distance a particle travels between collisions with other particles. In air at room temperature and atmospheric pressure, the mean free path is about 70 nm, a value 200 times larger than the typical radius of a small molecule like N_2 or O_2. (For comparison, the molecules in a liquid typically have a mean free path roughly the same as a molecular radius.) When we consider the cubic relationship of volume to distance, the difference in volume is on the order of 200^3, or 8×10^6. So the volume of empty space in a gas at room temperature and pressure is on the order of 1 million times greater than the volume of the individual molecules; the assumption in the kinetic theory that the volume of the molecules is negligible seems reasonable under these conditions.

But under what circumstances would this assumption break down? We cannot generally increase the size of molecules substantially by changing conditions, so the only way the volumes become more comparable is to squeeze the molecules closer together by decreasing the volume. Thus at high pressure, when gases are highly compressed, the volume of individual molecules may become significant and the assumption that the gas will behave ideally can break down (see **Figure 5.8a**). This makes sense if we realize that one way to condense a gas into a liquid is to compress it, and of course we don't expect the ideal gas law to hold for a liquid.

The postulates also assert that gas molecules always move in straight lines and interact with each other only through perfectly elastic collisions. Another way to put

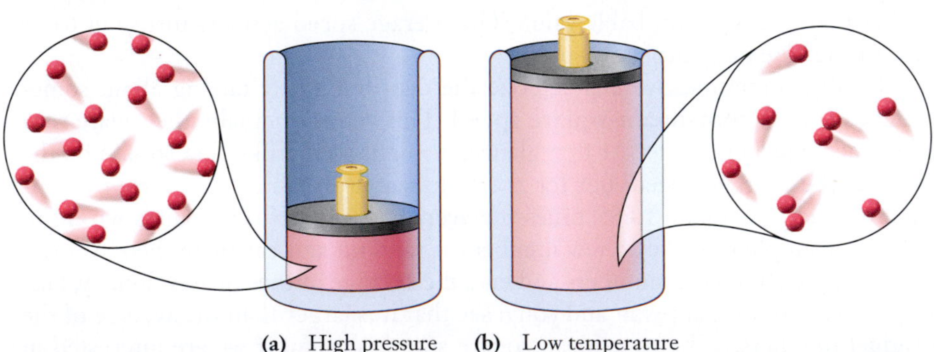

Figure 5.8 ■ The ideal gas model breaks down at high pressures and low temperatures. **(a)** At high pressures, the average distance between molecules decreases, and the assumption that the volume of the molecules themselves is negligible becomes less valid. **(b)** At low temperatures, molecules move more slowly and the attractive forces between molecules can cause "sticky" collisions, as shown here by the paired molecules.

this is to say that the molecules neither attract nor repel one another. If this were strictly true, then molecules could never stick together, and it would be impossible to condense a gas into a liquid or a solid. Clearly, there must be some attractive forces between molecules, a topic we will explore further in Chapter 8. The ideal gas model works because under many conditions these attractive forces are negligibly small compared to the kinetic energy of the molecules themselves. The fact that these forces lead to the formation of liquids and solids points us to the conditions under which the forces become important. We know that gases will condense if we lower the temperature sufficiently. The reason for this is that at lower temperatures the kinetic energy of the molecules decreases, and eventually the strength of the attractive forces between molecules will become comparable to the kinetic energy. At that point, molecules will begin to experience sticky collisions, as shown in **Figure 5.8b**. This means that pairs or small clusters of molecules might stay in contact for a short time, rather than undergoing simple billiard ball collisions. This tendency for molecules to stick together, however fleetingly, will reduce the number of times that the molecules collide with the walls of the container, so the pressure of the gas will be lower than that predicted by the ideal gas equation.

Correcting the Ideal Gas Equation

When the assumptions reflected in the postulates of the kinetic theory are not valid, the observed gas will not obey the ideal gas equation. In many cases, including a variety of important engineering applications, gases need to be treated as nonideal, and empirical mathematical descriptions must be devised. There are many equations that may be used to describe the behavior of a real gas; the most commonly used is probably the **van der Waals equation**:

An empirical equation contains one or more adjustable parameters that are found from a best fit to observed data rather than from a theoretical model.

$$\left(P + \frac{an^2}{V^2}\right)(V - nb) = nRT \tag{5.14}$$

Here, a and b are called van der Waals constants. Unlike the universal gas constant (R), the values of these van der Waals constants must be set for each specific gas. If a gas behaves ideally, both a and b are zero, and the van der Waals equation reverts to the ideal gas equation. Investigating the connection of this equation to the ideal gas equation further, we see that the term involving a accounts for the attraction between particles in the gas. The term involving b adjusts for the volume occupied by the gas particles. **Table 5.2** provides the values of van der Waals constants for some common gases. Notice that larger molecules have larger values for b, consistent with its role in accounting for the volume of the molecules themselves. The largest value of a listed is that for HF. This means that compared to other gases, HF molecules interact with each other more and require larger corrections from ideal behavior. When we consider chemical bonding in the next few chapters, we will begin to understand why this is true. Example Problem 5.8 demonstrates the use of the van der Waals equation.

EXAMPLE PROBLEM 5.8

In Example Problem 5.1, we considered a CH_4 storage tank with a volume of 49.0 L. When empty, the tank has a mass of 55.85 kg, and when filled, its mass is 62.07 kg. Calculate the pressure of CH_4 in the tank at an ambient temperature of 21°C using both the ideal gas equation and the van der Waals equation. What is the percentage correction achieved by using the more realistic van der Waals equation?

Strategy We will use two different models to describe the same gas, and each model will be represented by its own equation. So we will do two independent calculations,

Table ■ 5.2

Van der Waals constants for several common gases

Gas	a (atm L^2 mol^{-2})	b (L mol^{-1})
Ammonia, NH_3	4.170	0.03707
Argon, Ar	1.345	0.03219
Carbon dioxide, CO_2	3.592	0.04267
Helium, He	0.034	0.0237
Hydrogen, H_2	0.2444	0.02661
Hydrogen fluoride, HF	9.433	0.0739
Methane, CH_4	2.253	0.04278
Nitrogen, N_2	1.390	0.03913
Oxygen, O_2	1.360	0.03183
Sulfur dioxide, SO_2	6.714	0.05636
Water, H_2O	5.464	0.03049

solving for P in each case. We can find the mass of CH_4 from the given data and then convert that mass into the number of moles by using the molar mass. And the van der Waals constants for CH_4 can be found in Table 5.2. To calculate the percentage difference, we will divide the difference of the two numbers by the value for the ideal gas case.

Solution

$$\text{Mass of CH}_4 = \text{mass of full cylinder} - \text{mass of empty cylinder}$$
$$= 62.07\text{ kg} - 55.85\text{ kg} = 6.22\text{ kg} = 6220\text{ g}$$

$$\text{Moles of gas} = 6220\text{ g CH}_4 \times \frac{1\text{ mol CH}_4}{16.04\text{ g CH}_4} = 388\text{ mol CH}_4$$

Ideal Gas:

$$P = \frac{nRT}{V} = \frac{(388\text{ mol})(0.08206\text{ L atm mol}^{-1}\text{ K}^{-1})(294\text{ K})}{49.0\text{ L}} = 191\text{ atm}$$

van der Waals:

$$a = 2.253\text{ L}^2\text{ atm mol}^{-2}, b = 0.04278\text{ L mol}^{-1}$$

$$P = \frac{nRT}{V - nb} - \frac{n^2a}{V^2}$$
$$= \frac{(388\text{ mol})(0.08206\text{ L atm mol}^{-1}\text{ K}^{-1})(294\text{ K})}{49.0\text{ L} - (388\text{ mol})(0.0428\text{ L mol}^{-1})}$$
$$- \frac{(388\text{ mol})^2(2.25\text{ L}^2\text{ atm mol}^{-2})}{(49.0\text{ L})^2}$$
$$= 148\text{ atm}$$

Percentage correction:

$$\frac{191 - 148}{191} \times 100\% = 23\%$$

Discussion The correction here is quite significant. If you look at the calculated pressures, you should realize that they are quite high. Recall that the ideal gas law works best at low pressures and high temperatures. In this example, the pressure is sufficiently high that the ideal gas model is a rather poor description of the actual behavior of the gas. Still, the simplicity of the ideal gas law makes it a good starting point for many calculations. In a practical application, it would be crucial for an engineer to recognize the conditions under which the assumption of ideal gas behavior might not be reasonable.

Check Your Understanding Under less extreme conditions, the corrections for nonideal behavior are smaller. Calculate the pressure of 0.500 mol of methane occupying 15.0 L at 25.7°C using both the ideal gas law and the van der Waals equation. Compare the percentage correction with that calculated in the example above.

INSIGHT INTO

5-7 Gas Sensors

The U.S. Environmental Protection Agency lists more than 180 substances as toxic air pollutants and specifies acceptable levels that can be less than 1 ppm in some cases. To track the effects of pollutants at such trace levels, first we must be able to measure their concentrations reliably. One way to describe the concentration of a gas, of course, is to report its pressure or partial pressure. Although a mercury barometer quite similar to one that might have been used hundreds of years ago will allow us to monitor the weather, it will not be of much help here. Instead, we will need to rely on a number of more modern instruments that offer fundamental advantages over a barometer. A variety of devices now can allow us to measure exceedingly small pressures or changes in pressure. Some can also allow us to measure selectively the partial pressure of a particular component in a gas mixture. In this section, we will look at a couple of representative examples and try to show how their clever designs exploit the basic properties of gases to make incredibly sensitive measurements.

Capacitance Manometer

One sensor frequently used to measure pressure in the laboratory is the capacitance manometer. This type of pressure gauge, shown schematically in **Figure 5.9**, directly exploits the fact that pressure is force per unit area.

The term manometer *refers to any pressure-measuring device.*

A capacitor is an electronic component, as you may know from a physics class. One way to create a capacitor is to position two parallel metal plates with a space between them. In such an arrangement, the distance between the plates determines the capacitance. A capacitance manometer is based on a capacitor in which one of these plates is actually a thin metal diaphragm and the other is fixed and rigid. The space between the plates is kept under vacuum. We connect the sensor so that the other side of the diaphragm is exposed to the gas whose pressure we want to measure. The pressure of the gas produces a force on the diaphragm, causing it to bend toward the second (fixed) plate of the capacitor. This, in turn, changes the capacitance of the device. If we construct an electric circuit in which the capacitor is one

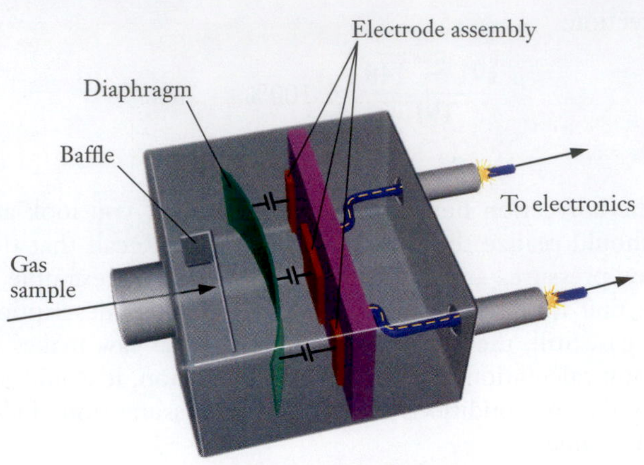

Figure 5.9 ■ Schematic diagram of a capacitance manometer. The tube on the left is connected to the gas sample whose pressure is to be measured. As the gas molecules strike the diaphragm, the force of the collisions determines the space between the diaphragm and the fixed electrode assemblies.

component, we can measure a voltage in that circuit that will be related to the pressure applied to the diaphragm. To make the sensor easy to use, that circuit can be designed so that we can measure a single voltage whose value is directly proportional to the pressure. Monitoring the voltage allows us to see any changes in pressure.

Capacitance manometers are most often used to measure pressures in the range of 0.001–1000 torr. This range includes moderate levels of vacuum frequently encountered in the laboratory. One advantage of this type of gauge is that it has virtually no moving parts; as long as the diaphragm is not damaged, the gauge should work reliably for very long periods of time.

Thermocouple Gauge

Both the barometer and the capacitance manometer respond fairly directly to the force exerted by gas molecules. But not all pressure sensors work that way. A useful pressure-measuring device can be based on any gas property that varies predictably with pressure.

Pressures between 0.01 and about 1.0 torr, for example, are frequently measured using a thermocouple vacuum gauge (**Figure 5.10**). A thin filament is heated by passing a fixed electric current through it, and the temperature of the filament is then measured using a device called a thermocouple. So how does this measurement tell us

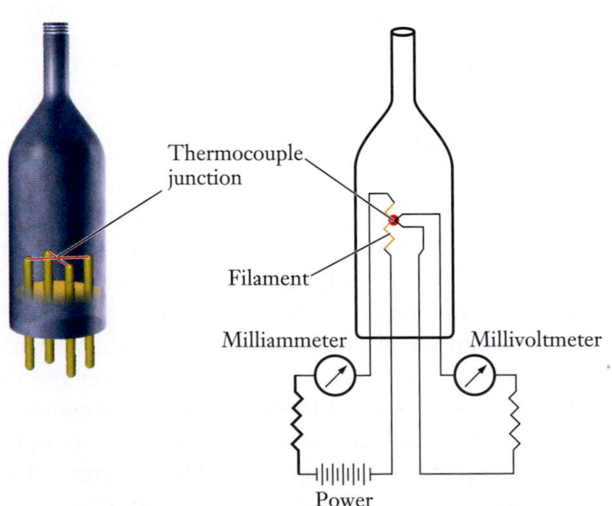

Figure 5.10 ■ Schematic diagram of a thermocouple gauge. An electric current heats the filament, and a thermocouple monitors the filament's temperature. Gas molecules collide with the filament and cool it. So the higher the pressure, the lower the filament temperature will be. A readout circuit usually converts the measured filament temperature into pressure units for display.

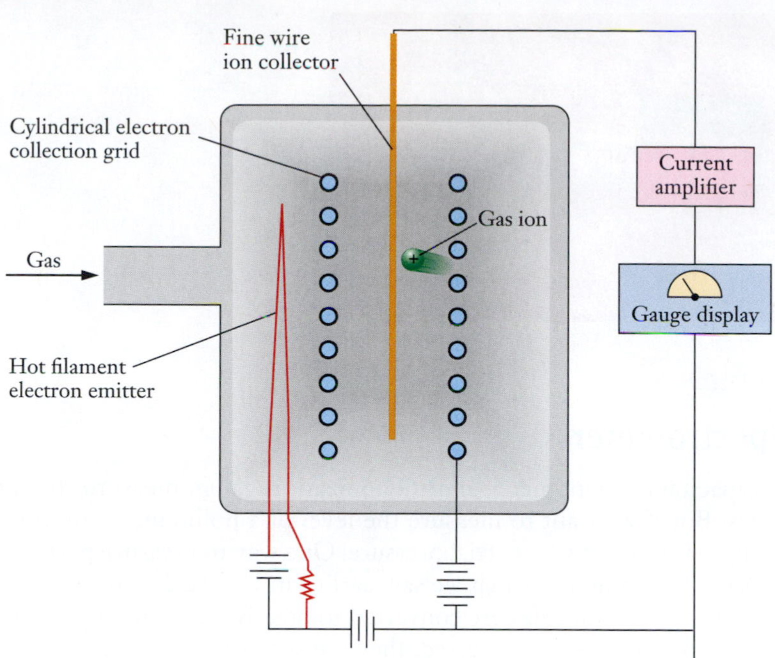

Figure 5.11 ■ Schematic diagram of an ionization gauge. Electrons emitted from the hot filament collide with gas molecules and knock an electron free. The resulting cations are collected at the center of the gauge, and the current they produce is a measure of the gas pressure. (The drawing is not to scale and greatly exaggerates the size of the gas ion.)

anything about pressure? The filament is in contact with the gas whose pressure we want to determine. As the molecules in that gas collide with the hot filament, they will tend to carry away some energy, reducing the filament temperature. The higher the pressure, the more molecules will collide with the filament, and the lower its temperature will be. The thermocouple produces a voltage that is related to the filament temperature and therefore to the gas pressure. A readout device connected to the gauge translates the voltage output into pressure units.

Ionization Gauge

Capacitance manometers can be effective to pressures as low as about 0.001 torr. That is a very low pressure, just over a millionth of an atmosphere. But in certain types of experiments, such as studies of extremely clean metal surfaces, laboratory vacuums as low as 10^{-11} torr are necessary. ◀ⓘ At such low pressures, it is no longer feasible to measure the force produced by molecular collisions, so other types of gauges are needed. The most common of these is the *ionization gauge*, shown in **Figure 5.11**.

Pressures on the order of 10^{-14} torr are found in outer space. ⓘ

A fixed electric current flows through the filament of the ionization gauge, heating it until it glows red. At these high temperatures, the metal atoms in the filament emit electrons. Some of these electrons then collide with gas molecules, and the resulting impact can knock an electron off the molecule, producing a positive ion. Suppose some or all of the gas present is N_2. Electrons from the filament would then produce N_2^+ ions, through a process that we could write as

$$e^- + N_2 \rightarrow N_2^+ + 2\ e^-$$

The resulting positive ions are then collected by a wire. Measuring the current flowing into this wire will measure how many ions were collected. The number of ions formed will be proportional to the number of gas molecules in the region of the gauge: more gas will allow more ions to be formed. Finally, we simply have to realize that the number of gas molecules present is proportional to the pressure. So the current, which is fairly easy to measure, will be proportional to pressure. In practice, the same electronics that read the current will also be calibrated so that they can display the reading directly as pressure.

The photo shows (from left to right) a thermocouple gauge, a capacitance manometer, and an ionization gauge. The electronics needed to operate the gauges are not shown.

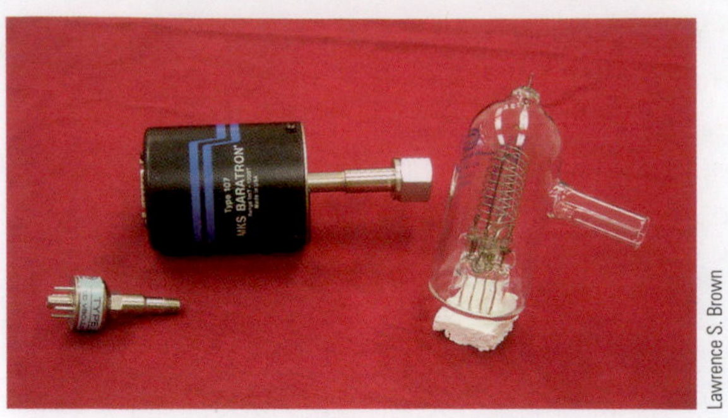

Mass Spectrometer

Both the capacitance manometer and the ionization gauge measure the total pressure of a gas. But if we want to measure the level of a pollutant in air, for example, we will really want to know its partial pressure. One way to measure partial pressures is with a mass spectrometer, which we saw earlier in Figure 2.4 (page 33). As in the ionization gauge, gas molecules are converted to positive ions by electron bombardment. But before the ions are collected, they are sent through a magnetic field that serves as a mass filter and sorts the stream of ions by mass. In this arrangement, the current that is measured consists only of ions with a chosen mass. This lets us monitor the partial pressure of a gas by looking at the ions of the corresponding mass. Control electronics allow these instruments to scan continuously over a range of masses and thus can tell us the partial pressure of a number of components in the same gas mixture.

FOCUS ON PROBLEM SOLVING

Question A balloon is to be used to float instruments for observation into the upper atmosphere. At sea level, this balloon occupies a volume of 95 L at a temperature of 20°C. If helium gas is used to fill the balloon, how would you determine the maximum height that this balloon and its instruments will attain?

Strategy This is the type of question that the engineers who design apparatus such as this must answer. The law of nature here is that objects that are less dense float on those that are denser. The balloon will float upward until it reaches a height where its density is equal to that of the atmosphere. With this understanding, we can begin to define the variables we need to understand in this problem and note the relationships between those variables that would allow us to estimate an answer to such a problem.

Solution First, identify what is known about the variables in the problem. In this case, we have a gas that will be at low pressures, so we can assume it will behave as an ideal gas. We note that we will release this balloon from sea level where the pressure can be estimated as 1 atm. Therefore, we can establish the initial pressure, volume, and temperature. We also know the molar mass of helium.

The key relationship is provided by the ideal gas law, which allows us to calculate density. To see this, think about n/V as the number of moles per unit volume. This is equal to the mass density (ρ) divided by the molar mass. Replacing n/V with that relationship gives us a gas law in terms of density:

$$P = \frac{\rho RT}{MM}, \text{ or } \rho = \frac{MM \cdot P}{RT}$$

Looking at this set of equations, the importance of temperature and pressure is apparent.

Because the key condition is the point at which the density of the helium in the balloon is equal to that of the atmosphere, we must look up or determine the density of the atmosphere as a function of altitude. The source of this information might also provide the temperature and pressure of the atmosphere as a function of altitude.

Knowing T and P provides the density of helium at various altitudes via the equation above. Having looked up the density of the atmosphere as a function of altitude, we must simply find the altitude at which this calculation equals the looked up value to estimate the maximum height that can be achieved by the balloon. We must recognize, however, that the mass of the instruments must be considered if we wish an accurate answer to this type of problem.

SUMMARY

The study of gases was extremely important in the development of a molecular description of matter, and it still offers us the most direct glimpse into the connections between molecular behavior and macroscopically observable properties. In our everyday experience, no gas is more important than air. For those who live in urban areas, the reality of pollution plays a critical role in that experience.

To understand gases, we need to think in terms of four key variables: pressure, temperature, volume, and number of moles. Pressure can be measured over wide ranges by various devices, ranging from traditional barometers to modern mass spectrometers. To understand the origin of gas pressure, we must realize that the molecules in the gas exert a force when they collide with objects. Although any individual molecular collision imparts only a tiny force, the vast number of such collisions in a macroscopic sample of gas is enough to sum to a measurable value.

For most applications, the ideal gas law, $PV = nRT$, provides the machinery necessary for quantitative understanding of gases. This equation includes the universal gas constant, R, in addition to the four variables we just noted. There are a number of ways in which this equation can be used, but it is always important to watch the units used carefully. The equation itself is only valid if temperature is expressed on an absolute scale, usually the Kelvin scale. Although pressure and volume can be expressed in any convenient units, we must express R in units that are compatible with the other variables. The ideal gas law can be used to determine changes in a variable for a single gas—essentially recreating historic gas laws such as Boyle's law or Charles's law. It also allows us to extend our methodology for stoichiometry problems to include reactions involving gases. By applying the ideal gas law to each component of mixtures of gases, we can understand the concept of partial pressure and calculate the contribution of each gas in the mixture.

The postulates of the kinetic–molecular theory provide a molecular-scale explanation of what it means for a gas to be ideal. A careful examination of those postulates also points us to the conditions where gases are most likely to deviate from ideal behavior: high pressures and low temperatures. Under these conditions, empirical equations, such as the van der Waals equation, can account for observed deviations from ideal behavior.

KEY TERMS

absolute temperature (5-6)
atmosphere (atm) (5-2)
average speed (5-6)
Avogadro's law (5-3)
barometer (5-2)
Boyle's law (5-3)
Charles's law (5-3)
criteria pollutant (5-1)
Dalton's law of partial pressures (5-4)
distribution function (5-6)
ionization gauge (5-7)
kinetic–molecular theory (5-6)
manometer (5-7)
Maxwell-Boltzmann distribution (5-6)
mean free path (5-6)
mole fraction (5-4)
most probable speed (5-6)
nonattainment area (5-1)
part per million (5-1)
partial pressure (5-4)
pascal (Pa) (5-2)
photochemical reaction (5-1)
pressure (5-2)
primary standard (5-1)
root-mean-square speed (5-6)
secondary standard (5-1)
standard temperature and pressure (STP) (5-5)
thermocouple gauge (5-7)
torr (5-2)
universal gas constant (R) (5-1)
van der Waals equation (5-6)
volatile organic chemical (VOC) (5-1)

PROBLEMS AND EXERCISES

INSIGHT INTO Air Pollution

5.1 List two types of chemical compounds that must be present in the air for photochemical smog to form. What are the most common sources of these compounds?

5.2 When ozone levels in urban areas reach unhealthy levels, residents are typically urged to avoid refueling their cars during daylight hours. Explain how this might help to reduce smog formation.

5.3 In the production of urban air pollution shown in Figure 5.1, why does the concentration of NO decrease during daylight hours?

5.4 VOCs are not criteria pollutants. Why are they monitored to understand urban air pollution?

5.5 Asphalt is composed of a mixture of organic chemicals. Does an asphalt parking lot contribute directly to the formation of photochemical smog? Explain your answer.

5.6 Do a web search to identify which (if any) of the EPA's criteria pollutants are prevalent in your city or state.

5.7 One observable property of gases is the variability of density based on conditions. Use this observation to explain why hot air balloons rise.

Pressure

5.8 If you are pounding a nail and miss it, the hammer may dent the wood, but it does not penetrate it. Use the concept of pressure to explain this observation.

5.9 How do gases exert atmospheric pressure?

5.10 Why do mountain climbers need to wear breathing apparatuses at the tops of high mountains such as Denali in Alaska?

5.11 If you had a liquid whose density was half that of mercury, how tall would you need to build a barometer to measure atmospheric pressure in a location where the record pressure recorded was 750 mm Hg?

5.12 Water has a density that is 13.6 times less than that of mercury. How high would a column of water need to be to measure a pressure of 1 atm?

5.13 Water has a density that is 13.6 times less than that of mercury. If an undersea vessel descends to 1.5 km, how much pressure does the water exert in atm?

5.14 Does the vacuum above the mercury in the column of a barometer affect the reading it gives? Why or why not?

5.15 Gas pressures can be expressed in units of mm Hg, atm, torr, and kPa. Convert these pressure values. **(a)** 722 mm Hg to atm, **(b)** 1.25 atm to mm Hg, **(c)** 542 mm Hg to torr, **(d)** 745 mm Hg to kPa, **(e)** 708 kPa to atm

5.16 If the atmospheric pressure is 97.4 kPa, how much is it in mm Hg? In atm?

5.17 Why do your ears "pop" on occasion when you swim deep underwater?

The Gas Law

5.18 When helium escapes from a balloon, the balloon's volume decreases. Based on your intuition about stretching rubber, explain how this observation is consistent with the gas law.

5.19 A sample of CO_2 gas has a pressure of 56.5 mm Hg in a 125-mL flask. The sample is transferred to a new flask, where it has a pressure of 62.3 mm Hg at the same temperature. What is the volume of the new flask?

5.20 A gas has an initial volume of 39 mL at an unknown pressure. If the same sample occupies 514 mL at 720 torr, what was the initial pressure?

5.21 When you buy a Mylar™ balloon in the winter months in colder places, the shopkeeper will often tell you not to worry about it losing its shape when you take it home (outside) because it will return to shape once inside. What behavior of gases is responsible for this advice?

5.22 Why should temperature always be converted to kelvins when working with gases in problem solving?

5.23 What evidence gave rise to the establishment of the absolute temperature scale?

5.24 Calculate the missing variable in each of these sets:

$V_1 = 2.0$ L, $T_1 = 15°C$, $V_2 = ?$, $T_2 = 34°C$

$V_1 = ?$, $T_1 = 149°C$, $V_2 = 310$ mL, $T_2 = 54°C$

$V_1 = 150$ L, $T_1 = 180$ K, $V_2 = 57$ L, $T_2 = ?$

5.25 Why is it dangerous to store compressed gas cylinders in places that could become very hot?

5.26 A balloon filled to its maximum capacity on a chilly night at a carnival pops without being touched when it is brought inside. Explain this event.

5.27 A gas bubble forms inside a vat containing a hot liquid. If the bubble is originally at 68°C and a pressure of 1.6 atm with a volume of 5.8 mL, what will its volume be if the pressure drops to 1.2 atm and the temperature drops to 31°C?

5.28 A bicycle tire is inflated to a pressure of 3.74 atm at 15°C. If the tire is heated to 35°C, what is the pressure in the tire? Assume the tire volume doesn't change.

5.29 A balloon filled with helium has a volume of 1.28×10^3 L at sea level where the pressure is 0.998 atm and the temperature is 31°C. The balloon is taken to the top of a mountain where the pressure is 0.753 atm and the temperature is −25°C. What is the volume of the balloon at the top of the mountain?

5.30 How many moles of an ideal gas are there if the volume of the gas is 158 L at 14°C and a pressure of 89 kPa?

5.31 A newly discovered gas has a density of 2.39 g/L at 23.0°C and 715 mm Hg. What is the molar mass of the gas?

5.32 Calculate the mass of each of the following gases at STP: **(a)** 1.4 L of SO_2, **(b)** 3.5×10^5 L of CO_2

5.33 What are the densities of the following gases at STP? **(a)** CF_2Cl_2, **(b)** CO_2, **(c)** HCl

5.34 A cylinder containing 15.0 L of helium gas at a pressure of 165 atm is to be used to fill party balloons. Each balloon must be filled to a volume of 2.0 L at a pressure of 1.1 atm. What is the maximum number of balloons that can be inflated? Assume that the gas in the cylinder is at the same temperature as the inflated balloons. (HINT: The "empty" cylinder will still contain helium at 1.1 atm.)

5.35 A cylinder is filled with toxic COS gas to a pressure of 800.0 torr at 24°C. According to the manufacturer's specifications, the cylinder may rupture if the pressure exceeds 35 psi (pounds per square inch; 1 atm = 14.7 psi). What is the maximum temperature to which the cylinder could be heated without exceeding this pressure rating?

5.36 Cylinders of compressed gases are often labeled to show how many "SCF" or "standard cubic feet" of gas they contain. 1 SCF of gas occupies a volume of 1 ft^3 at a standard temperature and pressure of 0°C and 1 atm. A particular cylinder weighs 122 lb. when empty and 155 lb. when filled with krypton gas at 26°C. How many SCF of Kr does this cylinder contain?

Partial Pressure

5.37 Define the term *partial pressure*.

5.38 Define the term *mole fraction*.

5.39 How does the mole fraction relate to the partial pressure?

5.40 What is the total pressure exerted by a mixture of 1.50 g of H_2 and 5.00 g of N_2 in a 5.00-L vessel at 25°C?

5.41 What is the total pressure (in atm) of a 15.0-L container at 28.0°C that contains 3.5 g of N_2, 4.5 g of O_2, and 13.0 g of Cl_2?

5.42 For a gas sample whose total pressure is 740 torr, what are the partial pressures if the gas present consists of 1.3 mol of N_2, 0.33 mol of O_2, and 0.061 mol of Ar?

5.43 A sample containing only NO_2 and SO_2 has a total pressure of 120. torr. Measurements show that the partial pressure of NO_2 is 43 torr. If the vessel has a volume of 800.0 mL and the temperature is 22.0°C, how many moles of each gas are present?

5.44 A sample of gas is made entirely of carbon dioxide and water, and there are 259 moles of CO_2 and 513 moles of water. If the total pressure of the sample is 21 atm, what is the partial pressure of each gas?

5.45 A sample of a smokestack emission was collected into a 1.25-L tank at 752 mm Hg and analyzed. The analysis showed 92% CO_2, 3.6% NO, 1.2% SO_2, and 4.1% H_2O by mass. What is the partial pressure exerted by each gas?

5.46 Air is often modeled as dry air, ignoring the water present. If the atmospheric pressure outside is 0.944 atm and the pressure associated with water is 0.039 atm, what are the partial pressures of nitrogen and oxygen—assuming the composition given in Table 5.1.

5.47 In an experiment, a mixture of gases occupies a volume of 3.00 L at a temperature of 22.5°C. The mixture contains 14.0 g of water, 11.5 g of oxygen, and 37.3 g of nitrogen. Calculate the total pressure and the partial pressure of each gas.

5.48 An experiment is being carried out to test the removal of sulfur dioxide from gases being released by a power plant. The initial sample, which contains only carbon dioxide and sulfur dioxide, occupies a volume of 35 L at a temperature of 41°C and a pressure of 715 torr. After all of the SO_2 has been removed, the sample occupies a volume of 23.5 L at a temperature of 29°C and a pressure of 715 torr. Determine the partial pressures of both CO_2 and SO_2 in the initial sample.

5.49 Use the web to determine the range of partial pressures that oxygen sensors must measure in the exhaust manifold of automobile engines. Why is it important for an engineer to know the amount of oxygen in an exhaust stream?

5.50 Use the web to research the types of fuel-to-air mixtures present in regular consumer driven automobiles. Calculate the partial pressures of the gases present based on the information you find.

Stoichiometry with Gases

5.51 HCl(g) reacts with ammonia gas, $NH_3(g)$, to form solid ammonium chloride. If a sample of ammonia occupying 250 mL at 21°C and a pressure of 140 torr is allowed to react with excess HCl, what mass of NH_4Cl will form?

5.52 Hydrogen gas is generated when acids come into contact with certain metals. When excess hydrochloric acid reacts with 2.5 g of Zn (the metal product is Zn^{2+}), what volume of hydrogen gas is collected at a pressure of 0.93 atm and a temperature of 22°C?

5.53 If you need 400.0 mL of hydrogen gas to conduct an experiment at 20.5°C and 748 torr, how many grams of Zn should be reacted with excess HCl to obtain this much gas?

5.54 The first step in processing zinc metal from its ore, ZnS, is to react it with O_2 according to the reaction

$$2\ ZnS(s) + 3\ O_2(g) \rightarrow 2\ ZnO(s) + 2\ SO_2(g)$$

If 620 kg of ZnS is to be reacted, what volume of oxygen at 0.977 atm and 34.0°C is needed (at a minimum) to carry out this reaction?

5.55 What volume of oxygen at 24°C and 0.88 atm is needed to completely react via combustion with 45 g of methane gas?

5.56 If boron hydride, B_4H_{10}, is treated with pure oxygen, it burns to give B_2O_3 and H_2O:

$$2\ B_4H_{10}(s) + 11\ O_2(g) \rightarrow 4\ B_2O_3(s) + 10\ H_2O(g)$$

If a 0.050-g sample of the boron hydride burns completely in O_2, what will be the pressure of the gaseous water in a 4.25-L flask at 30.0°C?

5.57 N_2O_5 is an unstable gas that decomposes according to the following reaction:

$$2\ N_2O_5(g) \rightarrow 4\ NO_2(g) + O_2(g)$$

What would be the total pressure of gases present if a 10.0-L container at 22.0°C begins with 0.400 atm of N_2O_5 and the gas completely decomposes?

5.58 One way to generate oxygen is to heat potassium chlorate, $KClO_3$ (the other product is potassium chloride). If 386 mL of oxygen at 41°C sand 97.8 kPa is generated by this reaction, what is the minimum mass of $KClO_3$ used?

5.59 Ammonia is not the only possible fertilizer. Others include urea, which can be produced by the reaction, $CO_2(g) + 2\ NH_3(g) \rightarrow CO(NH_2)_2(s) + H_2O(g)$.

A scientist has 75 g of dry ice to provide the carbon dioxide. If 4.50 L of ammonia at 15°C and a pressure of 1.4 atm is added, which reactant is limiting? What mass of urea will form?

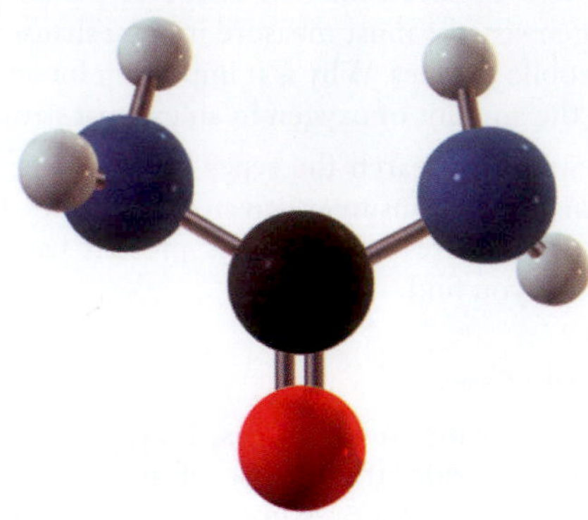

Urea, $CO(NH_2)_2$

5.60 Consider the following reaction:

$$6\ NiO(s) + 4\ ClF_3(g) \rightarrow 6\ NiF_2(s) + 2\ Cl_2(g) + 3\ O_2(g)$$

What mass of NiO will react with a sample of ClF_3 gas that has a pressure of 250 torr in a 2.5-L flask at 20°C?

5.61 What volume of hydrogen gas, in liters, is produced by the reaction of 3.43 g of iron metal with 40.0 mL of 2.43 M HCl? The gas is collected at 2.25 atm of pressure and 23°C. The other product is $FeCl_2$.

5.62 Magnesium will burn in air to form both Mg_3N_2 and MgO. What mass of each product would be found if burning a 3.11-g sample of magnesium to completion produces a combined total of 5.09 g of the two products?

5.63 During a collision, automobile air bags are inflated by the N_2 gas formed by the explosive decomposition of sodium azide, NaN_3:

$$2\ NaN_3 \rightarrow 2\ Na + 3\ N_2$$

What mass of sodium azide would be needed to inflate a 30.0-L bag to a pressure of 1.40 atm at 25°C?

5.64 Automakers are always investigating alternative reactions for the generation of gas to inflate air bags, in part because the sodium produced in the decomposition of NaN_3 (see Problem 5.63) presents safety concerns. One system that has been considered is the oxidation of graphite by strontium nitrate, $Sr(NO_3)_2$, as shown in the equation below:

$$5\ C(s) + 2\ Sr(NO_3)_2(s) \rightarrow 2\ SrO(s) + 2\ N_2(g) + 5\ CO_2(g)$$

Suppose that a system is being designed using this reaction and the goal is to generate enough *gas* to inflate a bag to 61 L and 1.3 atm at 23°C. What is the minimum mass of graphite that should be used in the design of this system?

5.65 As one step in its purification, nickel metal reacts with carbon monoxide to form a compound called nickel tetracarbonyl, $Ni(CO)_4$, which is a gas at temperatures above about 316 K. A 2.00-L flask is filled with CO gas to a pressure of 748 torr at 350.0 K, and then 5.00 g of Ni is added. If the reaction described occurs and goes to completion at constant temperature, what will the final pressure in the flask be?

5.66 Ammonium dinitramide (ADN), $NH_4N(NO_2)_2$, was considered as a possible replacement for aluminum chloride as the oxidizer in the solid fuel booster rockets used to launch the space shuttle. When detonated by a spark, ADN rapidly decomposes to produce a gaseous mixture of N_2, O_2, and H_2O. (This is *not* a combustion reaction. The ADN is the only reactant.) The reaction releases a lot of heat, so the gases are initially formed at high temperature and pressure. The thrust of the rocket results mainly from the expansion of this gas mixture. Suppose a 2.3-kg sample of ADN is detonated and decomposes completely to give N_2, O_2, and H_2O. If the resulting gas mixture expands until it reaches a temperature of 100.0°C and a pressure of 1.00 atm, what volume will it occupy? Is your answer consistent with the proposed use of ADN as a rocket fuel?

5.67 Clouds of hydrogen molecules have been detected deep in interstellar space. It is estimated that these clouds contain about 1×10^{10} hydrogen molecules per m^3 and have a temperature of just 25 K. Using these data, find the approximate pressure in such a cloud.

NASA

Interstellar hydrogen clouds, as seen by the Hubble Space Telescope

Kinetic Theory and Real Gases

5.68 State the postulates of the kinetic theory of gases.

5.69 Under what conditions do the postulates of the kinetic theory break down?

5.70 Which postulates of the kinetic theory are the most likely to break down under the conditions of high pressure present inside volcanoes?

5.71 Place these gases in order of increasing average molecular speed at 25°C: Kr, CH_4, N_2, CH_2Cl_2.

5.72 Use kinetic theory to explain what happens to the pressure exerted by a gas as its temperature is increased.

5.73 The figure below shows the distribution of speeds for two samples of N_2 gas. One sample is at 300 K, and the other is at 1000 K. Which is which?

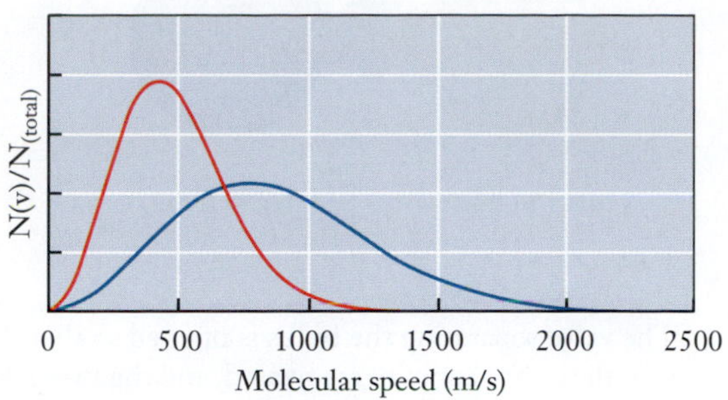

5.74 Why do heavier gases move more slowly than light gases at the same temperature?

5.75 Define the term *mean free path*.

5.76 There are two parameters in the van der Waals equation for real gases. Which parameter corrects for the volume of gas molecules? Use the equation to explain your answer.

5.77 Calculate the pressure of 15.0 g of methane gas in a 1.50-L vessel at 45.0°C using **(a)** the ideal gas law and **(b)** the van der Waals equation using constants in Table 5.2.

5.78 You want to store 165 g of CO_2 gas in a 12.5-L tank at room temperature (25°C). Calculate the pressure the gas would have using **(a)** the ideal gas law and **(b)** the van der Waals equation. (For CO_2, a = 3.59 atm L^2/mol^2 and b = 0.0427 L/mol.)

5.79 Consider a sample of N_2 gas under conditions in which it obeys the ideal gas law exactly. Which of these statements are true?

(a) A sample of Ne(g) under the same conditions must obey the ideal gas law exactly.

(b) The speed at which one particular N_2 molecule is moving changes from time to time.

(c) Some N_2 molecules are moving more slowly than some of the molecules in a sample of O_2(g) under the same conditions.

(d) Some N_2 molecules are moving more slowly than some of the molecules in a sample of Ne(g) under the same conditions.

(e) When two N_2 molecules collide, it is possible that both may be moving faster after the collision than they were before.

Pressure Sensors

5.80 Scientists routinely use several different types of pressure gauges, and sometimes a single vacuum chamber may be equipped with more than one type of gauge. Why do you suppose that this is so?

5.81 What is actually measured in an ionization gauge pressure sensor? How is this actual measurement related to pressure?

5.82 Use the web to find out what type of sensor is used to determine the oxygen pressure in the exhaust manifold of automotive fuel injection system engines.

5.83 Do a web search to learn about at least one other type of pressure sensor not described in the chapter. Describe how the gauge operates in a way that your roommate could understand.

Additional Problems

5.84 You have four gas samples:

1. 1.0 L of H_2 at STP
2. 1.0 L of Ar at STP
3. 1.0 L of H_2 at 27°C and 760 mm Hg
4. 1.0 L of He at 0°C and 900 mm Hg

(a) Which sample has the largest number of gas particles (atoms or molecules)?

(b) Which sample contains the smallest number of particles?

(c) Which sample represents the largest mass?

5.85 A gas mixture contains 10.0% CH_4 and 90.0% Ar by moles. Find the density of this gas at 250.0°C and 2.50 atm.

5.86 Liquid oxygen for use as a rocket fuel can be produced by cooling dry air to −183°C, where the O_2 condenses. How many liters of dry air at 25°C and 750 torr would need to be processed to produce 150 L of *liquid* O_2 at −183°C? (The mole fraction of oxygen in dry air is 0.21, and the density of liquid oxygen is 1.14 g/mL.)

5.87 A number of compounds containing the heavier noble gases, and especially xenon, have been prepared. One of these is xenon hexafluoride (XeF_6), which can be prepared by heating a mixture of xenon and fluorine gases. XeF_6 is a white crystalline solid at room temperature and melts at about 325 K. A mixture of 0.0600 g of Xe and 0.0304 g of F_2 is sealed into a 100.0-mL bulb. (The bulb contains no air or other gases.) The bulb is heated, and the reaction goes to completion. Then the sealed bulb is cooled back to 20.0°C. What will be the final pressure in the bulb, expressed in torr?

5.88 You work in a semiconductor production plant that relies on several chlorofluorocarbons in its manufacturing process. One day, you find an unlabeled gas cylinder, and you are assigned to figure out what is in the tank. First, you fill a 1.000-L flask with the gas. At a pressure of 250.0 torr and a temperature of 25.00°C, you determine that the mass of the gas in the flask is 2.2980 g. Then, you send the flask to an outside lab for elemental analysis, and they report that the gas contains 14.05% C, 44.46% F, and 41.48% Cl by mass. Find the molecular formula of this gas.

5.89 A 0.2500-g sample of an Al–Zn alloy reacts with HCl to form hydrogen gas:

$$2\,Al(s) + 6\,H^+(aq) \rightarrow Al^{3+}(aq) + 3\,H_2(g)$$

$$Zn(s) + 2\,H^+(aq) \rightarrow Zn^{2+}(aq) + H_2(g)$$

The hydrogen produced has a volume of 0.147 L at 25°C and 755 mm Hg. What is the percentage of zinc in the alloy?

5.90 A mixture of Ar (0.40 mol), O_2 (0.50 mol), and CH_4 (0.30 mol) exerts a pressure of 740 mm Hg. If the methane and oxygen are ignited and complete combustion occurs, what is the final pressure of Ar, CO_2, H_2O, and the remainder of the excess reactant? What is the total pressure of the system?

5.91 The complete combustion of octane can be used as a model for the burning of gasoline:

$$2\ C_8H_{18} + 25\ O_2 \rightarrow 16\ CO_2 + 18\ H_2O$$

Assuming that this equation provides a reasonable model of the actual combustion process, what volume of air at 1.0 atm and 25°C must be taken into an engine to burn 1 gallon of gasoline? (The partial pressure of oxygen in air is 0.21 atm, and the density of liquid octane is 0.70 g/mL.)

5.92 Mining engineers often have to deal with gases when planning for the excavation of coal. Some of these gases, including methane, can be captured and used as fuel to support the mining operation. For a particular mine, 2.4 g of CH_4 is present for every 100.0 g of coal that is extracted. If 45.6% of the methane can be captured and the daily production of the mine is 580 metric tons of coal, how many moles of methane could be obtained per day?

5.93 Some engineering designs call for the use of compressed air for underground work. If water containing iron(II) ions is present, oxygen in the compressed air may react according to the following unbalanced net ionic equation:

$$Fe^{2+} + H^+ + O_2 \rightarrow Fe^{3+} + H_2O$$

(a) Write the balanced net ionic equation. Remember that the amounts of each substance and the charges must balance. **(b)** Assume all of the oxygen from 650 L of compressed air at 15°C and 6.5 atm is lost by this reaction. What mass of water would be produced? (The mole fraction of oxygen in air is about 0.21.) **(c)** What will be the final pressure after the loss of the oxygen?

5.94 You are designing a warning system for a semiconductor fabrication facility that must have fewer than 1000 oxygen molecules per cm^3 to operate successfully. While researching possible design options, you learn about a new sensor that can detect partial pressures of oxygen in the range of 1–2 μbar. Could this sensor be used as the basis for your warning system? Explain your answer.

5.95 Homes in rural areas where natural gas service is not available often rely on propane to fuel kitchen ranges. The propane is stored as a liquid, and the gas to be burned is produced as the liquid evaporates. Suppose an architect has hired you to consult on the choice of a propane tank for such a new home. The propane *gas* consumed in 1.0 hour by a typical range burner at high power would occupy roughly 165 L at 25°C and 1.0 atm, and the range chosen by the client will have six burners. If the tank under consideration holds 500.0 gallons of *liquid* propane, what is the minimum number of hours it would take for the range to consume an entire tankful of propane? The density of *liquid* propane is 0.5077 kg/L.

5.96 Consider the apparatus shown below, in which two 1.00-L bulbs are initially separated from one another by a closed valve. One bulb is filled with 500 torr of sulfur dioxide and the other with 400 torr of oxygen, and the entire apparatus is at an initial temperature of 27°C.

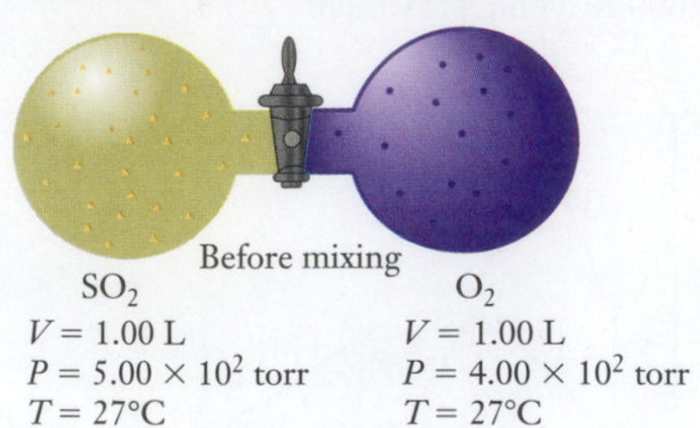

The valve separating the bulbs is opened so that the gases mix, the entire container is heated, and the two gases react to produce the maximum possible amount of SO_3. If the container is at a final temperature of 600 K when the reaction reaches completion, what is the total pressure?

5.97 Pure gaseous nitrogen dioxide (NO_2) cannot be obtained, because NO_2 *dimerizes*, or combines with itself, to produce a mixture of NO_2 and N_2O_4. A particular mixture of NO_2 and N_2O_4 has a density of 2.39 g/L at 50°C and 745 torr. What is the partial pressure of NO_2 in this mixture?

5.98 One of the noble gases (Group 18 in the periodic table) has a density of 1.783 g/L at 0°C and 1 atm. Which noble gas is this?

5.99 A mixture of CH_4(g) and C_3H_8(g) with a total pressure of 425 torr is prepared in a reaction vessel. Just enough O_2(g) is added to allow complete combustion of the two hydrocarbons. The gas mixture is ignited, and the resulting mixture of CO_2(g) and H_2O(g) is cooled back to the original temperature. If the total pressure of the products is 2.38 atm, what was the mole fraction of C_3H_8 in the original mixture?

5.100 A mixture of helium and neon gases has a density of 0.285 g/L at 31.0°C and 374 torr. Find the mole fraction of neon in this mixture.

5.101 Aerospace engineers sometimes write the gas law in terms of the mass of the gas rather than the number of moles:

$$PV = mR_{\text{specific}}T$$

In such a formulation, the molar mass of the gas must be incorporated into the value of the gas constant, which means that the gas constant would differ from one gas to another. (We have written the gas constant here as R_{specific} to emphasize this point. Many engineering texts use different notations to distinguish between the universal and specific gas constants.) **(a)** Suggest a reason why this approach might be particularly attractive in aerospace engineering. **(b)** Assume that the mole fractions of O_2 and N_2 in air are 0.21 and 0.79, respectively. Calculate the average molar mass of air (i.e., the mass of one mole of air). **(c)** Use your result from (b) to determine the value of R_{specific} for air, and express it in SI units of $m^2\ s^{-2}\ K^{-1}$.

5.102 Consider a sample of an ideal gas with n and T held constant. Which of the graphs below represents the proper relationship between P and V? How would the graph differ for a sample with a larger number of moles?

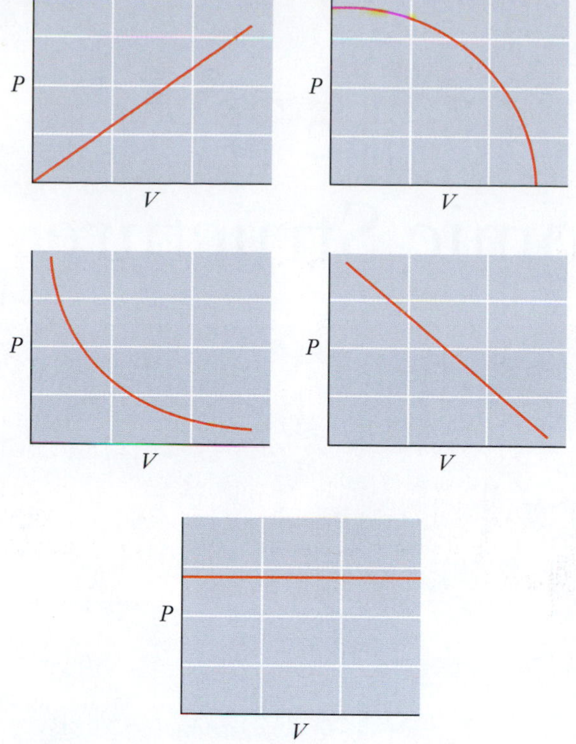

5.103 The decomposition of mercury (II) thiocyanate produces an odd brown snake-like mass that is so unusual the process was once used in fireworks displays. There are actually several reactions that take place when the solid $Hg(SCN)_2$ is ignited:

$$2Hg(SCN)_2(s) \rightarrow 2HgS(s) + CS_2(s) + C_3N_4(s)$$

$$CS_2(s) + 3O_2(g) \rightarrow CO_2(g) + 2SO_2(g)$$

$$2C_3N_4(s) \rightarrow 3(CN)_2(g) + N_2(g)$$

$$HgS(s) + O_2(g) \rightarrow Hg(l) + SO_2(g)$$

A 42.4-g sample of $Hg(SCN)_2$ is placed into a 2.4-L vessel at 21°C. The vessel also contains air at a pressure of 758 torr. The container is sealed and the mixture is ignited, causing the reaction sequence above to occur. Once the reaction is complete, the container is cooled back to the original temperature of 21°C. **(a)** Without doing numerical calculations, predict whether the final pressure in the vessel will be greater than, less than, or equal to the initial pressure. Explain your answer. **(b)** Calculate the final pressure and compare your result with your prediction. (Assume that the mole fraction of O_2 in air is 0.21.)

FOCUS ON PROBLEM SOLVING EXERCISES

5.104 Before the introduction of unleaded gasoline, the amount of lead in severe pollution episodes in air could be measured at levels of 2.85 μg/m^3. Describe how you could determine the number of lead atoms in the active space of your lungs (the air that is actively exchanged during a breath) if the volume of that space is 120 cm^3. What, if anything, would need to be looked up?

5.105 A soft drink can label indicates that the volume of the soda it contains is 12 oz or 355 mL. There is probably some empty space at the top of the can. Describe what you can measure and how that measurement allows you to determine the actual density of the soda.

5.106 The best way to generate small amounts of oxygen gas in a laboratory setting is by the decomposition of potassium permanganate, $2\ KMnO_4(s) \rightarrow K_2MnO_4(s) + MnO_2(s) + O_2(g)$. Suppose that you are carrying out an experiment for which you need exactly 2.00 L of oxygen. Describe how you would determine the mass of potassium permanganate needed to obtain this amount of oxygen. What would you have to measure in your experiment in addition to this mass?

5.107 An ore sample with a mass of 670 kg contains 27.7% magnesium carbonate, $MgCO_3$. If all of the magnesium carbonate in this ore sample is decomposed to form carbon dioxide, describe how to determine what volume of CO_2 is evolved during the process. What would have to be measured to predict the needed volume in advance?

5.108 Suppose that you are going to collect H_2 gas produced from the reaction of an acid with a metal such as zinc. This reaction gives off heat, so the temperature of the solution will increase as the reaction proceeds. You need to know the partial pressure of the H_2 you generate as precisely as possible. What will you need to account for in your calculation? What things would you need to measure to make accurate calculations? What information would you need to look up?

Cumulative Problems

5.109 Table 5.1 provides the mole percentage composition of dry air. What is the mass percentage composition of dry air?

5.110 Use the composition of dry air from Table 5.1 to calculate the average density of dry air at 25°C and 1 atm pressure.

5.111 Consider a room that is 14 ft × 20 ft with an 8-ft ceiling. **(a)** How many molecules of air are present in this room at 20°C and 750 torr? **(b)** If a pollutant is present at 2.3 ppm, how many pollutant molecules are in this room?

5.112 Nitric acid can be prepared by bubbling dinitrogen pentoxide into water.

$$N_2O_5(g) + H_2O(\ell) \rightarrow 2\ H^+(aq) + 2\ NO_3^-(aq)$$

(a) How many moles of H^+ are obtained when 1.50 L of N_2O_5 at 25°C and 1.00 atm pressure is bubbled into water?

(b) The solution obtained in (a) after reaction is complete has a volume of 437 mL. What is the molarity of the nitric acid obtained?

5.113 A 0.0125-g sample of a gas with an empirical formula of CHF_2 is placed in a 165-mL flask. It has a pressure of 13.7 mm Hg at 22.5°C. What is the molecular formula of the compound?

5.114 Analysis of a gaseous chlorofluorocarbon, CCl_xF_y, shows that it contains 11.79% C and 69.57% Cl. In another experiment, you find that 0.107 g of the compound fills a 458-mL flask at 25°C with a pressure of 21.3 mm Hg. What is the molecular formula of the compound?

6 The Periodic Table and Atomic Structure

CHAPTER OUTLINE

NASA/JPL-Caltech

NASA's Mars Science Laboratory *Curiosity* rover, shown here in an artist's rendition, landed on the surface of Mars in August 2012. The rover contains an impressive array of scientific instruments, including X-ray fluorescence instruments to perform trace analysis and determine the chemical composition of Martian soil and rocks.

From the fuels that power our planes and cars to the generation and control of air pollution, chemical reactions are central to our existence. What drives these reactions that are so much a part of who we are and what we do? How can we predict which reactions are likely to happen? To answer questions such as these, we will need an understanding of the chemical bonds that link atoms into molecules. To understand the bonds between atoms, first we must refine our descriptions of the atoms themselves. Fortunately, there are tools that allow us to observe atoms and other tools to help us organize what we have observed and learned. No tool is more important to building our understanding of chemistry than the periodic table. Its structure is closely related to the nature of atoms. Our understanding of atomic structure, in turn, has influenced the development of countless modern technologies, including sensitive methods of chemical analysis.

Chapter Objectives

After mastering this chapter, you should be able to

- describe trace analysis and explain its role in materials testing.
- describe waves in terms of frequency, wavelength, and amplitude.

- interconvert between the frequency and wavelength of light.
- relate the frequency, wavelength, and amplitude of light to characteristics such as color and brightness.
- describe the photoelectric effect by stating what sort of experiment is involved and what results are seen.
- explain how the results of the photoelectric effect experiment are consistent with a photon model of light.
- use Planck's equation to calculate the energy of a photon from its wavelength or frequency.
- use ideas about conservation of energy to explain how the observation of atomic spectra implies that electrons in atoms have quantized energies.
- use an energy-level diagram to predict the wavelengths or frequencies of light that an atom will absorb or emit, or use the observed wavelengths or frequencies to determine the allowed energy levels.
- describe similarities and differences between the Bohr model and the quantum mechanical model of atomic structure.
- recognize how quantum numbers arise as a consequence of the wave model.
- define the term *orbital*.
- identify an orbital (as 1*s*, 3*p*, etc.) from its quantum numbers, or vice versa.
- list the number of orbitals of each type (1*s*, 3*p*, etc.) in an atom.
- sketch the shapes of *s* and *p* orbitals and recognize orbitals by their shapes.
- rank various orbitals in terms of size and energy.
- use the Pauli exclusion principle and Hund's rule to write electron configurations for atoms and ions of main group elements.
- explain the connection between valence electron configurations and the periodic table.
- define the following properties of atoms: atomic radius, ionization energy, and electron affinity.
- state how the above properties vary with position in the periodic table.

INSIGHT INTO

6-1 Trace Analysis

The role of chemistry in the production of new and improved materials is one visible connection between chemistry and engineering. It is not, however, the only place where chemistry and materials science overlap. Perhaps just as critical as the development of new alloys, for example, is the chemical analysis of alloys that are routinely used in the production of products. In many cases, such analysis is used for quality assurance (QA) purposes, confirming that parts are meeting their specifications. In other cases, analysis is undertaken to determine the cause of failures, particularly in manufacturing environments. Quite often, in order to understand why a particular material, such as an alloy, failed during the manufacture or use of a product requires **trace analysis** to detect chemicals that are present at extremely low levels in the alloy.

There are many components to failure analysis, but we will focus on those that are related to chemical analysis.

A number of techniques can be used to perform trace analysis and determine the composition of an alloy. Many of these analytical methods require that the sample to be analyzed first be decomposed chemically. Other methods, referred to collectively as **nondestructive testing**, can be carried out without the need to decompose or

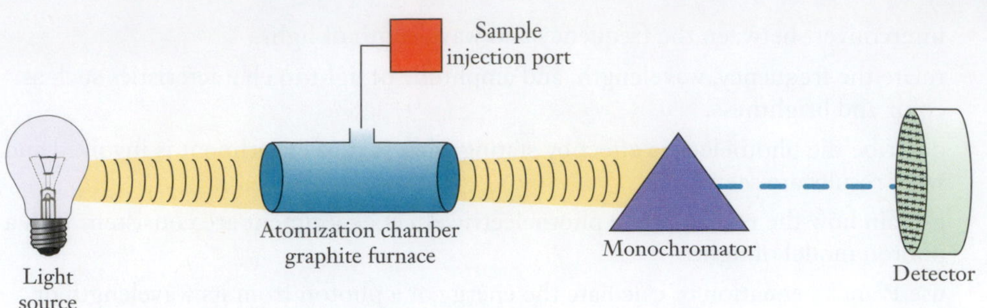

Figure 6.1 ■ The important components of an atomic absorption spectroscopy apparatus are shown. The sample to be analyzed is placed in a graphite furnace, where it is heated to very high temperatures. The sample decomposes into individual atoms in the gas phase, and light from a special lamp is passed through that atomic gas. The light then passes through a monochromator, which separates it into individual wavelengths or frequencies, and a detector measures the intensity of the light at each wavelength to determine how much light has been absorbed.

damage the sample. Although nondestructive testing sometimes is more difficult to carry out, there are obvious advantages in being able to analyze a sample without having to destroy it in the process. Many trace analysis techniques depend on either microscopy or spectroscopy, and take advantage of the fact that atoms of different elements have very specific and characteristic properties. A successful analysis scheme can tell us not only what is present but also how much of each element or compound is present. What features of an atom can be exploited to achieve these goals?

There are many possible ways that trace analysis can be done in the laboratory, but we will focus on only two here. The first method, which requires sample digestion, is called **atomic absorption spectroscopy (AAS)** and is shown schematically in **Figure 6.1**. For solid materials like alloys, a sample is put into a graphite furnace where it is atomized, or decomposed into atoms, and then ultraviolet or visible light is focused through the sample and collected. Atoms from the sample will absorb light, but only at specific frequencies. The particular frequencies absorbed depend on the chemical identity of the atoms and reveal the elemental composition of the alloy. The amount of light absorbed at each frequency also provides information on the amount of each element that is present. In order to understand AAS, we must first address some key questions in this chapter. Why do atoms absorb light only at selected frequencies? And how are the frequencies absorbed related to the structure of the atom?

For many engineering applications, the atomization required for AAS is not feasible. If information about the elemental composition of a part that is being used in a construction project is required, the part cannot be destroyed to determine its composition. An important tool for nondestructive trace analysis in this circumstance is **X-ray fluorescence (XRF)**, a technique that has been developed to the point that it can be carried out using handheld units like the one shown in **Figure 6.2**.

In XRF, the sample to be examined is first illuminated with X-rays. Atoms in the sample absorb those X-rays, and in the process are excited to high-energy states. Atoms in such excited states must lose their excess energy, and they can do that by emitting light. This re-emitted light is called fluorescence, and, once again, different atoms will fluoresce at different frequencies, allowing for the identification of the atoms present. ◂ ⓘ The intensity of fluorescence can also be measured, and this can tell us how much of a given element is present. Just as in AAS, the usefulness of XRF stems from the fact that atoms of a particular element interact with light in ways that are a characteristic of that element. As we develop our understanding

Because these frequencies are unique, they are sometimes referred to as an atomic fingerprint, an analogy to the idea that fingerprints in humans are also unique. ⓘ

Figure 6.2 ■ Instrumentation for X-ray fluorescence analysis has evolved to the point that handheld instruments like the one shown here are used routinely in many applications.

of atomic structure, we will begin to see why different elements absorb or emit different frequencies of light. But first we must look more closely at the properties of light itself.

6-2 The Electromagnetic Spectrum

Visible light is a more accurate term for what we usually refer to simply as light. The light that our eyes can detect comprises only a small portion of the **electromagnetic spectrum** and accounts for only a small part of the emission of most light bulbs. Other familiar forms of electromagnetic radiation include radio waves, microwaves, and X-rays. The origin of the word *electromagnetic* lies in the nature of light. Historically, light has been described as a wave traveling through space. One component is an electric field, and another is a magnetic field (**Figure 6.3**). To understand the wave nature of light, then, we should probably be familiar with the general features of waves.

The Wave Nature of Light

Many of the features of light that we encounter can be explained as properties of waves. The central characteristics of a wave can be defined by four variables: wavelength, frequency, velocity, and amplitude. **Figure 6.4** illustrates the definitions of these terms. The **wavelength** of any wave is the distance between corresponding points on adjacent waves. In the illustration, for example, the wavelength (designated as λ and measured in units of length) is defined as the distance between peaks. We could also define the wavelength as the distance between valleys, and the value would be the same. The **amplitude** is the size or height of the wave. The lower wave in Figure 6.4 has a larger amplitude than the upper wave. The **frequency** is the number of complete cycles of the wave passing a given point per second. Frequency is usually

λ is the Greek letter lambda.

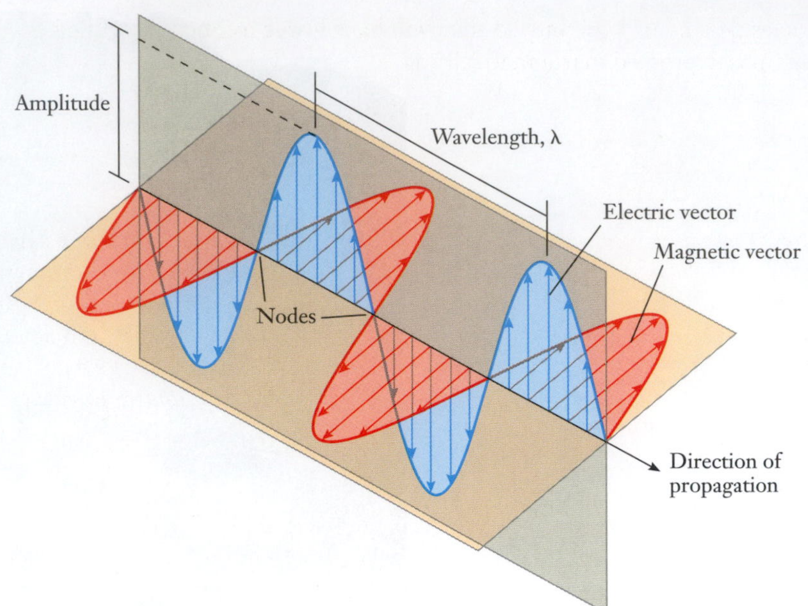

Figure 6.3 ■ Light is one form of electromagnetic radiation, which consists of oscillating electric and magnetic fields.

ν is the Greek letter nu.

designated as ν and measured in units of 1/s, or hertz (Hz). Frequency and wavelength are not independent of one another. The speed of light in a vacuum (designated by c and expressed in units of distance/time) is a constant of nature that has been measured with impressive precision and is now defined to have an exact value:

$$c = 2.99792458 \times 10^8 \text{ m s}^{-1}$$

Because the speed of light is a constant, the number of waves that will pass a certain point in time will be inversely related to the wavelength. Long wavelengths will have

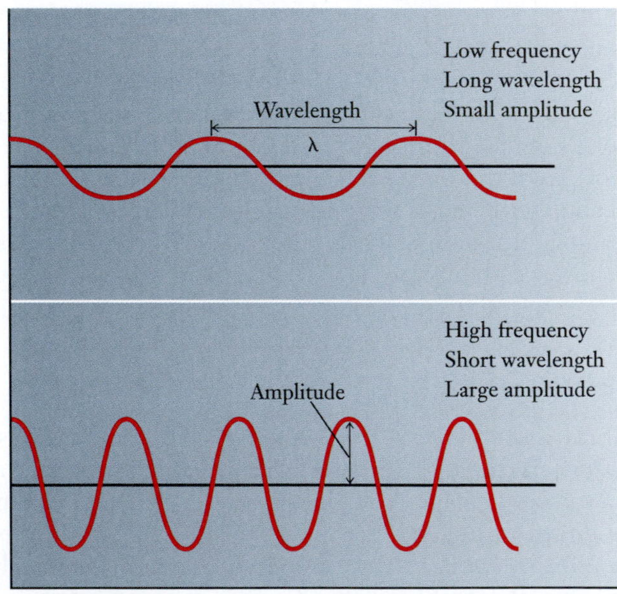

Figure 6.4 ■ Two sine waves are shown, and the meanings of some important terms are illustrated. Because light has a fixed speed, its frequency and wavelength are always inversely related. Amplitude is independent of frequency; the choice of which of these waves would have larger amplitude was made arbitrarily.

fewer cycles pass by in 1 second, so they will have lower frequencies (Figure 6.4). This relationship is expressed mathematically as

$$c = \lambda \times \nu \tag{6.1}$$

MathConnections

If we want to describe the detailed shape of a wave, we need to write a mathematical equation for it. Let's consider a very simple example in which the wave happens to follow the shape of a sine function. (This turns out to be true for many naturally occurring waves.) Imagine that we take a snapshot of the wave at a particular instant and picture it as an x–y graph. The wave originates at $x = 0$ and travels along the x axis. The y values represent the height of the wave at each point. We could write a very general equation for this wave:

$$y = A \sin (bx)$$

A is clearly the amplitude of the wave, but what about b? Let's assume that we want to evaluate the sine function in terms of radians. That means that the quantity bx must have no units. Because x is the distance from the origin, it would have units of length. So the b in our equation must have units of 1/length. If we know the wavelength of the wave—which we could read off the graph—then we can also fix the value of b. The wavelength of the sin (x) function is 2π. So, suppose that the wavelength of our wave is 450 nm, in the range of blue light. For our equation to match the wave, we require that

$$b \times 450 \text{ nm} = 2\pi$$

or

$$b = 2\pi/450 \text{ nm}$$

More generally, we can see that the b in our equation must always be $2\pi/\lambda$.

$$y = A \sin\left(\frac{2\pi x}{\lambda}\right)$$

If we wanted to describe a more complicated wave, we might have to use a combination of several sine (or cosine) functions, each of the same general form as we see here but with different wavelengths. By combining enough such functions, we could describe a waveform of arbitrary complexity. If we wanted to describe a wave as a function of time rather than position, we could proceed in a similar fashion.

Speed, wavelength, frequency, and amplitude are characteristics that can describe any wave. But because we are interested in light, it should help to compare these quantities with some familiar properties of light. If someone asks you to compare two different lights, one of the things you would probably consider is how bright each one appears. How does this relate to the wave parameters we've defined? In the wave model for light, the amplitude determines the brightness of the light: the larger the amplitude of the wave, the brighter the light will appear. ◀ⓘ Another property that you would surely consider in comparing two lights is color. In the wave model, both wavelength and frequency correspond to the color of the light. How can these two seemingly different properties of the wave both correspond to color? The answer can be seen in Equation 6.1. Because the speed of light must remain constant, specifying *either* the frequency *or* the wavelength automatically fixes the value of *both* of those properties of a light wave. So, we could say that the frequency and the wavelength are simply two different ways of expressing the same information.

The intensity of a light wave is actually proportional to the square of its amplitude. ⓘ

The wave nature of light also helps to explain many of the phenomena we experience with light. **Refraction** is the bending of a wave when it passes from one medium

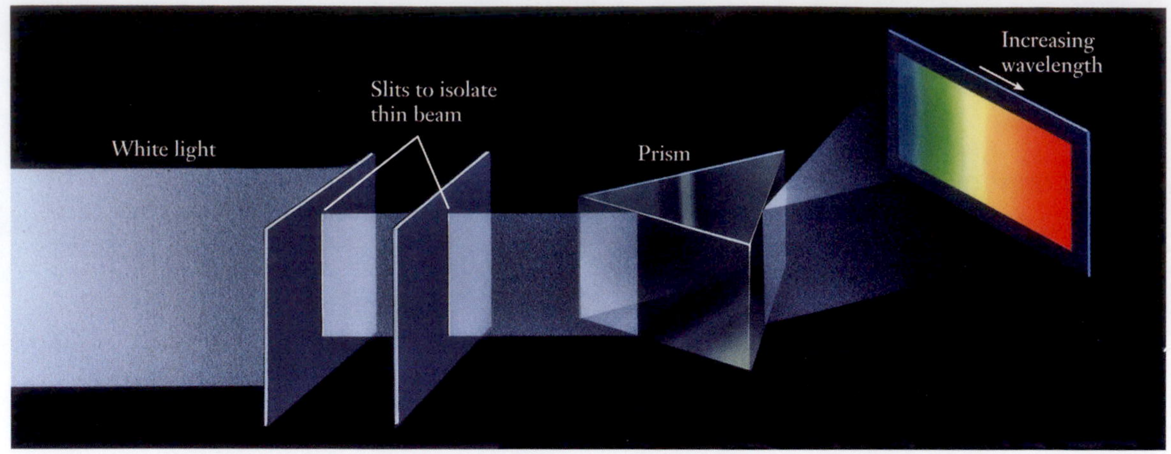

Figure 6.5 ■ When white light passes through a prism, it is separated into its constituent colors by refraction.

to another of different refractive index. One common manifestation of refraction is the rainbow, in which white light is separated into colors. In passing from air to water or glass, the speed of light changes, and the light bends at an angle that depends on its wavelength. Thus the visible spectrum is separated into its colors, and the colors are dispersed in order of their corresponding wavelengths. **Figure 6.5** illustrates this phenomenon.

The X-rays used in XRF analysis have relatively short wavelengths.

Because of wave-like phenomena such as refraction, it is common to categorize electromagnetic radiation in terms of its wavelength or frequency. **Figure 6.6** shows the entire spectrum, including information about both frequency and wavelength. In addition to visible light, the spectrum includes X-ray, ultraviolet (UV), infrared (IR), microwave, and radio wave radiation; the visible falls between UV and IR. This order lists the regions of the spectrum in increasing wavelength. We will often find that we can measure or are provided with the wavelength of light but need to express it as a frequency. Example 6.1 shows this type of conversion.

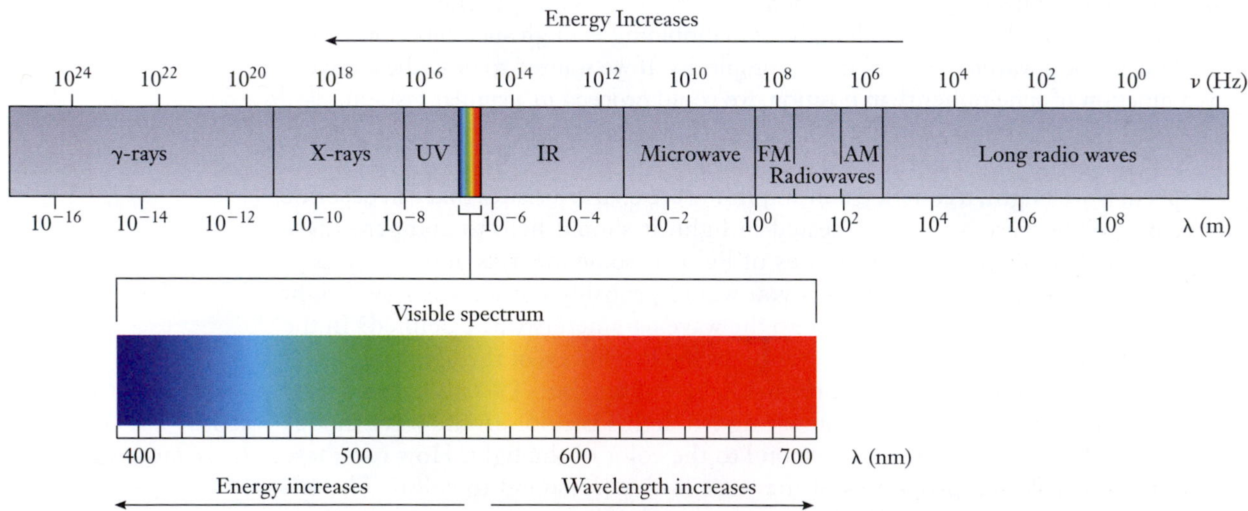

Figure 6.6 ■ Visible light is just the small portion of the entire electromagnetic spectrum that our eyes can detect. Other regions of the spectrum are used in many technological applications, including remote controls, cell phones, and wireless networks.

EXAMPLE PROBLEM 6.1

Carbon can be detected in an X-ray fluorescence experiment by monitoring the emission at a wavelength of 4.47 nm. What is the frequency of this light?

Strategy We know that wavelength and frequency are related through Equation 6.1. In using this equation, we will need to be careful of units. Because the wavelength is given in nanometers, we will convert it to meters so that it is consistent with the meters per second in the speed of light.

Solution

$$4.47 \text{ nm} \times \frac{1 \text{ m}}{1 \times 10^{9} \text{ nm}} = 4.47 \times 10^{-9} \text{ m}$$

We know that $c = \lambda\nu$, so, rearranging, $\nu = \frac{c}{\lambda}$. Then we simply substitute the values for the speed of light and the wavelength in the equation:

$$\nu = \frac{2.998 \times 10^{8} \text{ m s}^{-1}}{4.47 \times 10^{-9} \text{ m}} = 6.71 \times 10^{16} \text{ s}^{-1}$$

Analyze Your Answer It is difficult to have a physical intuition about something like the frequency of light. But with some practice we can develop an idea of the expected order of magnitude. First, notice that to find the frequency, we took a large number (the speed of light) and divided it by a small number (the wavelength). So we should expect a very large numerical answer. It is also handy to become familiar with the typical magnitudes of some quantities. A glance at Figure 6.6, for example, will show us that this frequency is in the range between UV and X-ray radiation, so our answer seems plausible. These kinds of checks can be very handy because the most common errors in problems like this involve mishandling of units, leading to results that can be off by several orders of magnitude.

Check Your Understanding Trace amounts of silicon are often found in steel alloys. X-ray fluorescence detection of silicon focuses on a wavelength of 712.6 pm. What is the frequency of this radiation?

The word "light" in general conversation usually implies visible light, but we've seen that the electromagnetic spectrum includes radiation from many regions. Chemical analysis can take advantage of the specific types of radiation found in these different regions. The trace analysis techniques we have focused on thus far, XRF and AAS, rely on X-rays and visible light, respectively. But there are other techniques that use other regions of the spectrum as well. Infrared (IR) radiation is used widely in chemical analysis. Unlike XRF or AAS, though, IR spectroscopy does not directly determine what elements are present in a sample. Instead, an instrument called a Fourier Transform Infrared (FTIR) spectrometer detects the presence of particular functional groups in molecular systems. Because it has molecular rather than elemental sensitivity, FTIR is often used for trace analysis in things like polymers, paints, and adhesives whose properties depend on what compounds are present and not just what elements. In addition to chemical analysis, many consumer products also use different parts of the spectrum. Microwave ovens, radios, televisions, cell phones, and wireless computer networks all function by exploiting radiation from specific regions of the electromagnetic spectrum.

You may have heard of a Fourier transform in a math class. In spectroscopy, it allows for measuring a signal as a function of time and transforming that signal so it can be expressed in terms of frequency.

The Particulate Nature of Light

The wave model of electromagnetic radiation can explain many observed properties of light. In the early 1900s, however, scientists developed more sophisticated apparatus and carried out experiments that challenged the wave model of light. One pivotal group of experiments involved the observation of what came to be known as the **photoelectric effect**, in which light strikes a piece of metal and causes electrons to be ejected. **Figure 6.7** is a schematic diagram of a typical photoelectric effect experiment. At the heart of the experiment is light shining on a piece of metal. If the metal is enclosed in a container under vacuum so that there are very few surrounding gas molecules to interfere, then we can detect electrons emitted from the metal under certain conditions. (The metal itself is grounded so that it does not develop a large positive charge during this process.) As we learned in Section 2-2 (page 31), electrons carry a negative charge. This makes it fairly easy to measure how many electrons are ejected and even how fast they are going. So in studying the photoelectric effect, the experimenter could vary the color or intensity of the light and see how the number of electrons emitted or their kinetic energy might change. The experiment could also be repeated using different metals.

How can this photoelectric effect be understood? The simplest explanation is that energy from the light is transferred to the electrons in the metal. If an electron picks up enough energy, it can break free from the surface of the metal. The more energy given to the electron, the faster it will travel after it leaves the metal.

Figure 6.8 summarizes the sort of results obtained in these experiments. A close look at these graphs reveals that the detailed observations are not what we would expect based on the wave model for light. In a wave description of a light source, the amplitude of the wave or the intensity of the light determines the energy. So we would expect that as the light gets brighter, the kinetic energy of the outgoing electrons should increase. But Figure 6.8(d) shows that the kinetic energy of the electrons is actually independent of the intensity of the incoming light. Only the number of electrons detected increases when the intensity goes up (graph b). But the kinetic energy does depend on the

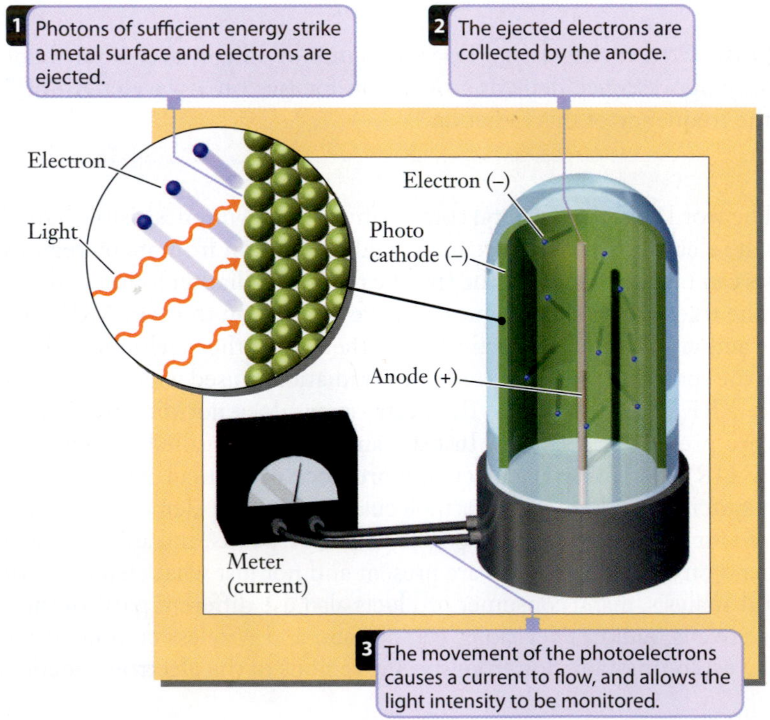
1 Photons of sufficient energy strike a metal surface and electrons are ejected.
2 The ejected electrons are collected by the anode.
Electron
Light
Electron (–)
Photo cathode (–)
Anode (+)
Meter (current)
3 The movement of the photoelectrons causes a current to flow, and allows the light intensity to be monitored.

Figure 6.7 ■ In the photoelectric effect, light strikes the surface of a metal and electrons are ejected, as shown in the inset. This is exploited in photoelectric cells: light strikes a surface called the photocathode, and a wire (the anode) collects the emitted electrons. (A positive voltage applied to the anode ensures that the electrons travel in the desired direction.) The resulting current can be measured easily and is related to the intensity of the incident light.

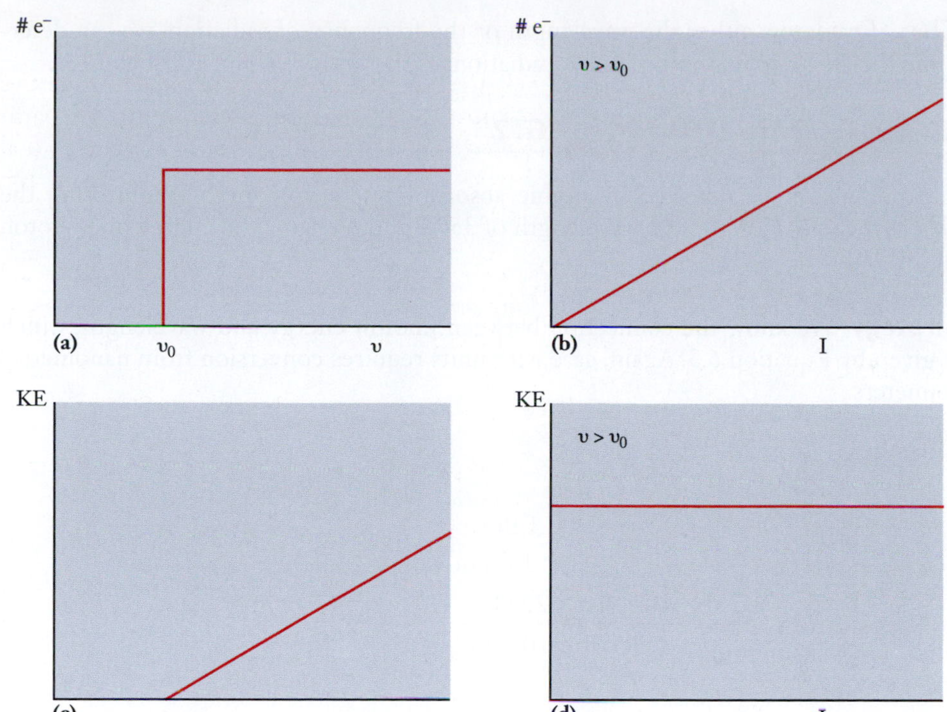

Figure 6.8 ■ The graphs sketched here show the most important observations from experiments on the photoelectric effect. Below some threshold frequency (ν_0), no photoelectrons are emitted. For frequencies greater than ν_0, the number of electrons emitted is independent of frequency (graph a) but increases with increasing light intensity (graph b). The kinetic energy of the emitted electrons, on the other hand, increases linearly with frequency (graph c) but is independent of intensity (graph d).

frequency or color of the light used (graph c). This is also at odds with the wave model. So these experiments (and others) caused scientists to reexamine the nature of light.

Despite elaborate efforts to make these observations of the photoelectric effect conform to existing wave concepts, consistent explanations of light based solely on waves ultimately failed. The only way to explain all of the experimental results was to invoke the notion of **wave–particle duality**, which says that, in some situations, light is best described as a wave, whereas in other cases, a particle description works better. It's important to realize that this does not mean that there are two different kinds of light! It simply means that neither the wave model nor the particle model provides an accurate description of all of the properties of light. So in a given situation, we use whichever model works best to describe the properties we are interested in. Someday, advances in our understanding of light may produce a new model that reconciles the wave–particle conflict. But until that happens, we take a pragmatic approach and use the model that's best suited to the problem at hand.

The context in which we normally think of light as a particle is when it is imparting energy to another object. Albert Einstein proposed that light could be described as a collection of packets of energy, called **photons**. Bright light has many photons, whereas dim light has few. The energy of a photon of light has been shown to be proportional to its frequency, leading to a simple equation:

$$E = h\nu \tag{6.2}$$

Recall that in Chapter 3 (p. 63) we used hν to indicate that light is required in a photochemical reaction.

Here E represents the photon energy, and ν is the frequency of the light. The h term is a constant, called Planck's constant, after German scientist Max Planck. Planck's constant can be determined experimentally, and its value is

$$h = 6.62606957 \times 10^{-34}\ \text{J s}.$$

Because frequency is related to wavelength, we can substitute for ν in this equation, using Equation 6.1:

$$E = \frac{hc}{\lambda} \tag{6.3}$$

Thus, if we know either the wavelength or the frequency of radiation, we can determine the energy of a photon of that radiation, as shown in Example Problem 6.2.

EXAMPLE PROBLEM 6.2

Chromium can be detected in atomic absorption spectroscopy by monitoring the absorbance of UV light at a wavelength of 357.8 nm. What is the energy of a photon of this light?

Strategy We know the connection between photon energy and wavelength, which is given by Equation 6.3. Again, care with units requires conversion from nanometers to meters.

Solution

$$357.8 \text{ nm} \times \frac{1 \text{ m}}{1 \times 10^9 \text{ nm}} = 3.578 \times 10^{-7} \text{ m}$$

$$E = \frac{6.626 \times 10^{-34} \text{ J s} \times 2.998 \times 10^8 \text{ m s}^{-1}}{3.578 \times 10^{-7} \text{ m}} = 5.552 \times 10^{-19} \text{ J}$$

Analyze Your Answer The result is a very small number. But we should realize that Planck's constant is incredibly small—many orders of magnitude smaller than the other quantities in the problem. So a very small energy is likely. Beyond that, we can again rely on a general sense of the magnitude of the quantity we are calculating. For visible light, photon energies are typically on the order of 10^{-19} J, which makes our answer seem plausible because the light we are dealing with is close to the visible region.

Check Your Understanding The specifications for a handheld XRF analyzer indicate that the X-ray source emits photons with an energy of 8048.0 eV (electron-volt; 1 eV = 1.6022×10^{-19} J). What is the wavelength of these X-rays?

The idea that light is made up of a collection of particles, or "chunks of energy," does not seem consistent with our everyday experience. To understand why, look at the magnitude of the photon energy calculated in Example Problem 6.2: 10^{-19} J. Compare that with the amount of energy emitted by an ordinary lamp. Even a modest 60-watt bulb emits 60 J of energy each second. So the lights that you are familiar with all consist of vast numbers of photons. As the light gets brighter, the number of photons emitted per second increases. But because each photon is such a small additional amount of energy, it appears to us as if the energy coming from the lamp varies continuously rather than in discrete increments or steps.

Be careful not to confuse the photon energy with the intensity or brightness of the light. If you remember that photon energy is another way of specifying the color of the light, you will see that it is independent of intensity.

Though not readily discernible to us, the energy of a single photon can be very significant from an atomic or molecular vantage point. Atoms and molecules are very small, after all, and their typical energies are small as well. Consider the results presented earlier for the photoelectric effect experiment. How can the concept of photons help us to explain those observations? Imagine that we could do the experiment by shooting a single photon at the metal. That means we would be supplying an amount of energy (equal to $h\nu$) to the metal. If this energy is enough to overcome the force that the metal atoms exert to hold onto their electrons, then when the metal absorbs the photon, it will be able to eject one electron. If the energy of the photon is smaller than the **binding energy** holding the electrons to the metal, then no electron can be emitted. The fact that there is a threshold frequency (ν_0 in Figure 6.8), below which no electrons are detected, is evidence that the photoelectric process is carried out by individual photons.

Below the threshold frequency, a photon does not have enough energy to overcome the binding energy and eject an electron. Why does the electron kinetic energy depend on the frequency of the light? This is really a question of conservation of energy. For a particular metal, the electron binding energy is fixed. For any frequency above ν_0, a portion of the energy of the absorbed photon will be used to overcome the binding energy. Then, whatever energy remains will be transferred to the electron as kinetic energy. So, as the frequency increases, the photon energy increases. This leads to an increase in the observed kinetic energy of the ejected electrons. Example Problem 6.3 shows how we can use the photoelectric effect to measure the electron binding energy in a metal.

EXAMPLE PROBLEM 6.3

In a photoelectric effect experiment, ultraviolet light with a wavelength of 337 nm was directed at the surface of a piece of potassium metal. The kinetic energy of the ejected electrons was measured as 2.30×10^{-19} J. What is the electron binding energy for potassium?

Strategy We know that energy must be conserved, and we can use this idea to set up an equation relating the photon energy to the kinetic energy and the binding energy. Because the photon is absorbed in this process, we must be able to account for its energy. Some of that energy does work to overcome the electron binding energy, and the rest appears as the kinetic energy of the ejected electron. We can write this as a simple equation:

$$E_{\text{photon}} = \text{Binding } E + \text{Kinetic } E$$

We know how to find the photon energy from the wavelength, and we know the kinetic energy, so we can solve for the binding energy. As always, we must be careful with the units on all quantities involved.

Solution First, find the photon energy:

$$337 \text{ nm} \times \frac{1 \text{ m}}{10^9 \text{ nm}} = 3.37 \times 10^{-7} \text{ m}$$

$$E_{\text{photon}} = \frac{hc}{\lambda} = \frac{(6.626 \times 10^{-34} \text{ J s})(2.998 \times 10^8 \text{ m s}^{-1})}{3.37 \times 10^{-7} \text{ m}} = 5.89 \times 10^{-19} \text{ J}$$

Then, use the conservation of energy relationship to find the desired binding energy:

$$E_{\text{photon}} = \text{Binding } E + \text{Kinetic } E$$

So,

$$\begin{aligned}\text{Binding } E &= E_{\text{photon}} - \text{Kinetic } E \\ &= 5.89 \times 10^{-19} \text{ J} - 2.30 \times 10^{-19} \text{ J} = 3.59 \times 10^{-19} \text{ J}\end{aligned}$$

Analyze Your Answer The wavelength used here is slightly shorter than those for visible light, so a photon energy in the mid-to-upper 10^{-19} J range seems reasonable. The binding energy we found is smaller than the photon energy, as it must be. So it appears likely that we have done the calculation correctly.

Check Your Understanding The electron binding energy for chromium metal is 7.21×10^{-19} J. Find the maximum kinetic energy at which electrons can be ejected from chromium in a photoelectric effect experiment using light at 266 nm.

Our discussion of the photoelectric effect can help us begin to see how trace analysis works. When exposed to the same light source, different metals will emit different photoelectrons with different kinetic energies because the metals have different binding energies. We will soon see that this type of behavior is seen not only for large collections of atoms like the surface of a metal. Even single atoms have characteristic binding energies for the attraction between their nuclei and electrons. Methods like XRF and AAS exploit this fact to obtain the unique fingerprints for the atoms of a particular element. To explore this further, we will now consider what happens when an individual atom absorbs or emits light. With just one proton and one electron, hydrogen is the simplest possible atomic system, so it is a good place to begin. Once we have an understanding of the connection between the structure of an atom and the way that atom interacts with light, we will be able to look more closely at how trace analysis techniques like XRF and AAS reveal the composition of materials.

6-3 Atomic Spectra

In describing trace analysis techniques in Section 6-1, we noted that both AAS and XRF are based on the premise that each element has its own characteristic way of absorbing or emitting light. Now that we know a little more about light, we can look more closely at this idea and begin to see how we might explain it. Because we are interested in the interaction of light with individual atoms, the photon model will play a key role in our explanation.

The plural of spectrum is spectra.

The particular pattern of wavelengths absorbed and emitted by any element is called its **atomic spectrum.** The existence of such spectra has been known for more than a century. What is most peculiar is that excited atoms of any particular element emit only specific frequencies. Because these frequencies tend to be well separated, the spectra are said to be *discrete*. This behavior is quite distinct from the light given off by the sun or by an incandescent light bulb, which contains a continuous range of wavelengths.

Some surprising observations emerged from investigations of the spectra emitted by various atoms. All elements display discrete spectra, with relatively few wavelengths appearing. The wavelengths that are emitted vary from one element to the next. Discrete spectra were first observed in the late 19th century and posed an enormous challenge for scientists to interpret. Although data were gathered from many elements, including mercury, the spectrum of hydrogen played the most important role. Hydrogen gas can be excited in a glass tube by passing an electric arc through the tube (**Figure 6.9**). The energy from the arc breaks the H_2 molecules into individual hydrogen atoms and excites some of those atoms. The excited hydrogen atoms then emit light as they relax

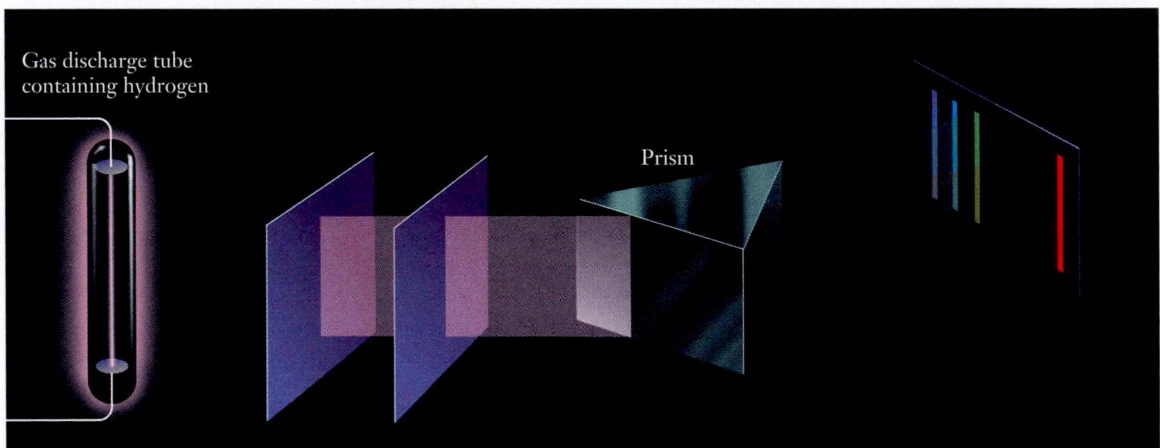

Figure 6.9 ■ When an electric discharge is applied to a sample of hydrogen gas, H_2 molecules dissociate into atoms, and those atoms then emit light. That light passes through a set of slits to create a narrow beam of light, which then passes through a prism. The prism separates the light into its constituent colors and shows that only a few discrete wavelengths are emitted.

toward lower energy states. The light given off appears blue, and when passed through a prism, it reveals four lines visible to the eye.

What conclusions can be drawn from the fact that hydrogen (or any other atom) emits only certain wavelengths of light? When an atom emits light, it is releasing energy to the surrounding world. So we should think about this situation in terms of the conservation of energy. When light is emitted, the atom goes from a higher energy state to a lower energy state, and the emitted photon carries away the energy lost by the atom. The fact that only a few wavelengths of light are emitted from a particular atom is direct evidence that the atom can exist in only a few states with very specific energies. To see that this must be true, imagine what would happen if the atom could take on any arbitrary energy. In that case, we would expect to see a continuous spectrum of wavelengths emitted because each atom could lose an arbitrary amount of energy. The notion that an atom could take on only certain allowed energies is extremely counterintuitive because no ordinary macroscopic objects seem to behave similarly. To explain the observed discrete emission spectra, scientists had to develop a model for the behavior of atoms that is fundamentally different from the ways that we describe macroscopic objects.

Example Problem 6.4 shows how the different wavelengths of light emitted by an atom are related to the separation between energy levels.

EXAMPLE PROBLEM 6.4

The first four energy levels for the hydrogen atom are shown in the diagram below. The lowest energy state is labeled E_1, and the higher states are labeled E_2, E_3, and E_4, respectively. When a hydrogen atom undergoes a transition from E_3 to E_1, it emits a photon with $\lambda = 102.6$ nm. Similarly, if the atom undergoes a transition from E_3 to E_2, it emits a photon with $\lambda = 656.3$ nm. Find the wavelength of light emitted by an atom making a transition from E_2 to E_1.

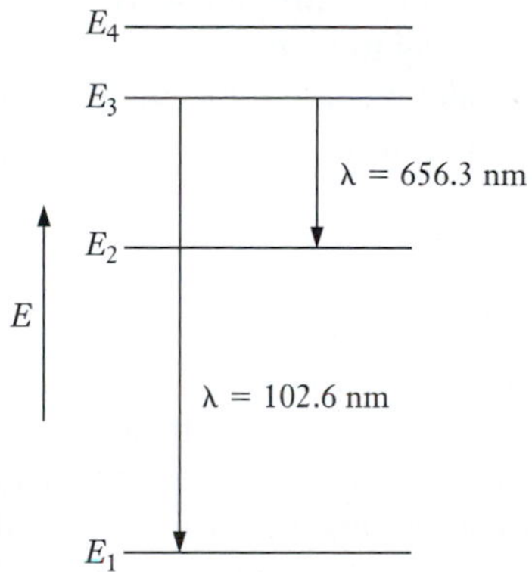

Strategy The energy-level diagram shows us the relationship between the various energy levels involved. The vertical direction in the diagram is energy. So, if we represent each transition by an arrow, as shown, then the length of that arrow will be proportional to the energy lost by the atom in that transition. From the diagram, we can see that the lengths of the arrows for the E_3 to E_2 transition and for the E_2 to E_1 transition must add up to the length of the arrow for the E_3 to E_1 transition. (Another way of thinking about this is that an atom going from the E_3 state to the E_1 state must lose the same amount of energy whether or not it stops at E_2 along the way.) Because we know the wavelengths for two of the transitions, we can find the corresponding photon energies. That will let us find the energy for the third transition, and then finally we can convert that back to wavelength.

Solution We first convert the given wavelengths to energies. For the $E_3 \rightarrow E_1$ transition:

$$E_{3\rightarrow 1} = \frac{hc}{\lambda} = \frac{(6.626 \times 10^{-34}\ \text{J s})(2.998 \times 10^{8}\ \text{m s}^{-1})}{102.6\ \text{nm}} \times \frac{10^{9}\ \text{nm}}{1\ \text{m}} = 1.936 \times 10^{-18}\ \text{J}$$

For the $E_3 \rightarrow E_2$ transition:

$$E_{3\rightarrow 2} = \frac{hc}{\lambda} = \frac{(6.626 \times 10^{-34}\text{ J s})(2.998 \times 10^{8}\text{ m s}^{-1})}{656.3\text{ nm}} \times \frac{10^{9}\text{ nm}}{1\text{ m}} = 3.027 \times 10^{-19}\text{ J}$$

From the diagram, we can see that

$$E_{3\rightarrow 1} = E_{3\rightarrow 2} + E_{2\rightarrow 1}$$

So,

$$E_{2\rightarrow 1} = E_{3\rightarrow 1} - E_{3\rightarrow 2} = 1.936 \times 10^{-18}\text{J} - 3.027 \times 10^{-19}\text{J} = 1.633 \times 10^{-18}\text{J}$$

Now we just need to convert from energy to wavelength to reach the desired answer:

$$\lambda_{2\rightarrow 1} = \frac{hc}{E_{2\rightarrow 1}} = \frac{(6.626 \times 10^{-34}\text{ J s})(2.998 \times 10^{8}\text{ ms}^{-1})}{1.633\text{ nm} \times 10^{-18}\text{ J}} = 1.216 \times 10^{-7}\text{m}$$

Because we know that wavelengths are most often given in nanometers, we might choose to convert that to 121.6 nm.

Analyze Your Answer Our result lies between the two wavelengths we were originally given, which is consistent with the spacing of the energy levels in the diagram.

Check Your Understanding A hydrogen atom undergoing a transition from the E_4 level to the E_3 level emits a photon with a wavelength of 1.873 μm. Find the wavelength of light emitted by an atom making the transition from E_4 to E_1.

But why do atoms have different energy levels? Scientists began asking that question when atomic spectra, like that of hydrogen, were first observed. Finding an answer took some time, and along the way models were proposed that turned out to be only partially correct. One such model proposed by Niels Bohr had the greatest influence on our subsequent understanding of the atom.

Bohr won the Nobel prize for physics in 1922.

The Bohr Atom

As the structure of atoms was being deduced, experiments suggested that the nucleus of the atom was positively charged, whereas electrons surrounding that nucleus have a negative charge. You probably learned this model of the atom early in your time as a student, perhaps even in grade school, and so you are likely to accept it without much thought. Yet when first proposed, it was perplexing. In particular, the attraction between oppositely charged particles would seem to indicate that the electrons should collapse into the positively charged nucleus. Bohr addressed this concern as well as other more subtle ones in devising his model for the atom. His most creative contribution was to suggest that the electrons occupy stable orbits from which they cannot deviate without first absorbing or emitting energy in the form of light. One effect of this assumption is that the energy of an electron in an atom is **quantized**, or restricted to certain allowed values. If only certain distances between the electron and the nucleus were allowed, then Coulomb's law would suggest that only certain energies would be allowed. The supposition of such orbits had never been made before, but through this model, Bohr was able to explain observations no previous hypothesis could. The influence of the Bohr model is evident in modern society, as the atom is still commonly depicted with electrons orbiting around the nucleus.

For molecules, the concepts of quantized energies and excited states can be extended to vibrational and rotational motions. But we will consider only electronic energy levels.

Now we know that the Bohr model for the atom is not completely correct, but it does provide useful intuition as well as a convenient way to visualize the process of exciting an atom. In any atom, the oppositely charged electrons and protons attract each other. Therefore, the closer to the nucleus an electron is, the stronger the attraction and the lower the energy. For an electron to move from an inner orbit to an outer orbit, as depicted in **Figure 6.10**, energy must be absorbed. The new distribution or configuration of electrons consists of a grouping that is not at the lowest possible energy and is therefore referred to as an **excited state**. One of the most powerful driving forces in nature is the tendency to move toward states of lower energy. So an excited state of an atom cannot exist indefinitely. It will return to the lowest energy state, or **ground state**, by emitting radiation. Both the excitation and emission processes are understandable in terms of the **Bohr model** of the atom. So this model has some usefulness even though it is not really an accurate representation of atomic structure.

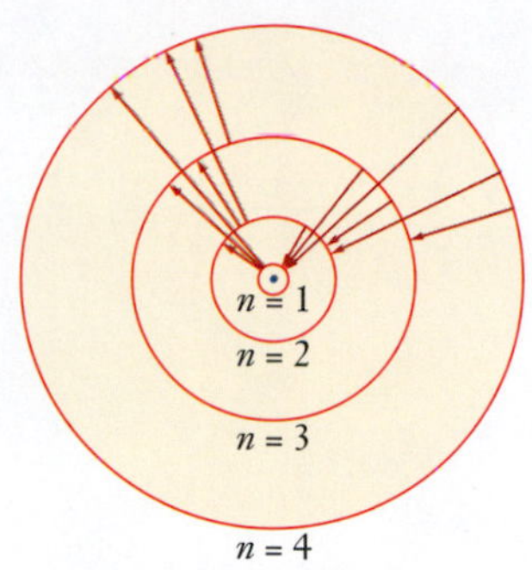

Figure 6.10 ■ In the Bohr model, electrons move between allowed orbits when an atom absorbs or emits light. Important elements of this model, including the idea of fixed orbital paths, are not correct. But it played an essential role in the development of our understanding of atomic structure. Note the similarity between this figure and the energy level diagram we used in Example Problem 6.4; both reflect the concept that electrons in atoms are restricted to certain allowed energies.

6-4 The Quantum Mechanical Model of the Atom

The observation of atomic spectra and the development of the Bohr model were important advances in the understanding of atoms. But further experiments in the early part of the 20th century revealed still more unexpected properties of atoms, and important limitations of the Bohr model became apparent. So efforts to develop a more robust model of the atom continued. Although these efforts were initially driven by fundamental scientific curiosity, they eventually facilitated many important practical applications. The development of modern light sources such as light emitting diodes (LEDs) and lasers would not have been possible without improved models of atoms. The picture that emerged and ultimately enabled these technological advances is the quantum mechanical model of the atom. As we'll see in Chapter 7, this new model also allows us to describe chemical bonding more accurately, which in turn allows for the design of improved materials.

Because the quantum mechanical picture of the atom is both complex and non-intuitive, we begin by pointing out some key comparisons between it and the more familiar Bohr model. In the Bohr model, electrons are viewed as particles traveling along circular orbits of fixed radius. In the quantum model, electrons are viewed as waves rather than particles, and these waves are considered to be spread out, or delocalized, through a region of space called an **orbital**. The most important similarity between the two models is that in either case the energy of an electron is quantized, meaning it is restricted to certain allowed values. A detailed treatment of the quantum mechanical model is well beyond our grasp here, but in the rest of this section we will present some of the conceptual and mathematical ideas from which the model arises. As we do this, it will help if you keep in mind the simple comparison with the Bohr atom we've made here.

The essential and peculiar feature of the quantum mechanical model of the atom lies in its description of electrons as waves rather than particles. It is far more intuitive to think of electrons as particles, perhaps resembling tiny marbles, than to envision them as waves. But just as we've seen for light, experimental observations led to the idea that electrons can exhibit wave-like behavior. The first evidence of the wave nature of electrons came through diffraction experiments in 1927. **Diffraction** was already a well-understood phenomenon of waves, so the observation of electron diffraction strongly suggested the need for a wave-based treatment of the electron.

Electron diffraction techniques are now commonly used to probe the structures of solid surfaces.

A mathematical description of a wave generally begins with a function that describes the periodic behavior. Such a function is referred to as a **wave function**, for fairly obvious reasons. The simplest such functions are sin x and cos x, which can be used to describe simple oscillatory motion. Although the functions that describe electron waves tend to be more complicated, the idea is the same. The notion of using the mathematics of waves to treat electrons was first put forward by Erwin Schrödinger. The so-called **Schrödinger equation** is summarized as

$$H\psi = E\psi \tag{6.4}$$

but this form is deceptively simple. That's because the H term is an **operator**, which designates a complicated series of mathematical operations to be carried out. E is the energy and ψ (the Greek letter psi) is the wave function for the electron. If we write out the entire operator H, we find that the Schrödinger equation is actually a second-order differential equation. (See MathConnections section.) It turns out that in most cases the equation cannot be solved exactly, so the study of quantum mechanics invariably involves approximation techniques. Because of the complexity of the mathematics involved, we will not make any attempt to solve this equation. But we will describe the results of its solution and look at the way these results form the foundation of our modern understanding of atoms. The concept of the wave function, in particular, is central to the quantum mechanical model of the atom. The wave function contains all the information we could ever know about the electron, if we can determine the means by which to extract the information.

MathConnections

Written as $H\psi = E\psi$, the Schrödinger equation appears deceptively simple. If we write out the operator H for a hydrogen atom, we get the more intimidating form below:

$$-\frac{h^2}{8\pi^2\mu}\left\{\frac{\partial^2\psi}{\partial x^2}+\frac{\partial^2\psi}{\partial y^2}+\frac{\partial^2\psi}{\partial z^2}\right\}+V(x, y, z)\psi = E\psi$$

Although this is a complicated equation, we can relate it to things with which you are probably familiar. First, let's realize that the purpose of this equation is to describe the total energy of an electron as it moves through the three-dimensional space surrounding the atom's nucleus. Because the nucleus is the center of the atom, it makes sense to define this as the origin of our x, y, z coordinate system. We might expect that there should be terms for the kinetic energy and for the potential energy, and in fact there are. The $V(x, y, z)$ term accounts for the potential energy, which is a function of position. The other term on the left-hand side accounts for the kinetic energy. So the Schrödinger equation expresses the total energy as the sum of the kinetic energy and the potential energy.

Next, let's look more closely at the terms that appear. If you are familiar with calculus, you might recognize the "$\delta^2/\delta x^2$" notation as a derivative. Here it is written with deltas (δ) rather than with d's because x (or y or z) is only one of several variables on which our wave function ψ depends. The superscript 2's tell us this is a second derivative. So for $\delta^2\psi/\delta x^2$, we would take the derivative of ψ with respect to x and then take the derivative of the result with respect to x as well. As long as we know how to take derivatives, we could evaluate the kinetic energy term. The potential energy term is also understandable. For an electron in an atom, the potential energy results primarily from the attraction between the positive charge on the nucleus and the negative charge on the electron. So the function V in the equation would represent the way that potential varies as the position of the electron changes. From physics, you may know this as the Coulomb potential, which depends on the distance between charges.

You will not be expected to work directly with the Schrödinger equation in a class at this level. But if we think about it step by step like this, you should begin to see the connections between this complicated equation and ideas you work with in your math and physics classes.

Potential Energy and Orbitals

The Schrödinger wave equation describes the energy of electrons. As for any object or particle, that total energy includes both kinetic and potential energy contributions. But the potential energy is the most important in describing the structure of atoms. The potential energy of electrons in atoms is associated with the coulombic attraction between the positively charged nucleus and negatively charged electrons as well as the repulsion between like-charged electrons. The periodic nature of waves allows many solutions to the wave equation for any given potential interaction. The Schrödinger equation is often rewritten as

We introduced the Coulomb potential in Section 2-3, page 37.

$$\mathrm{H}\psi_n = \mathrm{E}_n\psi_n \tag{6.5}$$

to emphasize the fact that multiple solutions can exist. Here n is an index that labels the different solutions. The wave functions, ψ_n, that provide solutions to this equation for an atom can be written in terms of two components: a radial component, depending only on the distance of the electron from the nucleus, and an angular component that may depend on the direction or orientation of the electron with respect to the nucleus. Just like the simple sin x function, these wave functions may have positive and negative signs in different regions. The most physically meaningful way to interpret a wave function is to look at its square, ψ^2. This quantity is always positive and tells us the probability of finding an electron at any particular point. This probability interpretation will seem odd to you, but that's because you are more accustomed to dealing with discrete objects (or particles) rather than with waves. If you think about the standing wave that exists in a vibrating guitar string, you will realize that the wave is not at any one point along the string; the wave is "found" all along the string simultaneously. That's the same idea we need to apply to electrons in the quantum mechanical model.

The indexing system that is represented by the "n" subscript in Equation 6.5 provides important language for discussing atomic structure. The index is typically written to contain both a number and a letter, and these labels identify solutions to the wave equation. Because of their historic correspondence to the position of electrons in the orbits proposed by Bohr, each solution is referred to as an orbital. An orbital is the quantum mechanical equivalent of the "location" of an electron. But because we are treating the electron as a wave, this location is actually a region of space rather than a particular point. It's quite possible you have heard the names of these orbitals before: 1*s*, 2*s*, 2*p*, 3*s*, 3*p*, 3*d*, 4*s*, 4*p*, 4*d*, 4*f*, etc. What is the origin of these "names" of orbitals, and what information do those names provide?

Quantum Numbers

Although we don't need detailed treatments of quantum mechanics to understand most introductory chemical concepts, we will find that the vocabulary established by the theory is important. For example, the names attached to atomic orbitals come from the functions that solve the wave equations. Collectively, they are referred to as **quantum numbers**.

Before examining the specific quantum numbers we need to describe an atomic orbital, let's look at a simpler problem that can help show where these numbers come from. Suppose we take a guitar string, pin down both ends of it so they can't move, and then think about what sorts of waves might be produced if we pluck the string so it vibrates (see **Figure 6.11**). In the simplest case, the length of the string would just be one-half the wavelength of a sine function. In mathematical terms, we could write the following equation for this wave (A is the amplitude of the wave, x is the position along the string, and L is the length of the string):

$$\psi(x) = A \sin \frac{\pi x}{L} \tag{6.6}$$

Other waves are also possible without moving the ends of the string. If a guitarist wants to play what is called the harmonic, a finger is placed very lightly at the midpoint of the string. The string can still vibrate on both sides of the midpoint, so now we can form a sine wave where the length of the string is one complete wavelength, as represented by the following equation:

$$\psi(x) = A \sin \frac{2\pi x}{L} \tag{6.7}$$

This wave has a higher frequency, and we would hear it as a higher pitched note. Our guitarist could also produce additional waves of different frequencies by pinning down appropriate positions along the string. The important restriction is that the ends of the string must remain fixed; this means that the length of the string must contain an integral multiple of half-wavelengths.

Figure 6.11 ■ A string fixed at both ends—as in a guitar—is an example of how multiple waves can satisfy a particular set of conditions. In the upper figure, the length of the string is equal to one-half of the wavelength of the wave. Additional cases where the length of the string is equal to one wavelength or to one and one-half wavelengths are also shown. A standing wave can be formed as long as the length of the string is equal to an integral multiple of half wavelengths. But waves that do not satisfy this condition are not possible because they would violate the condition that the ends of the string stay fixed.

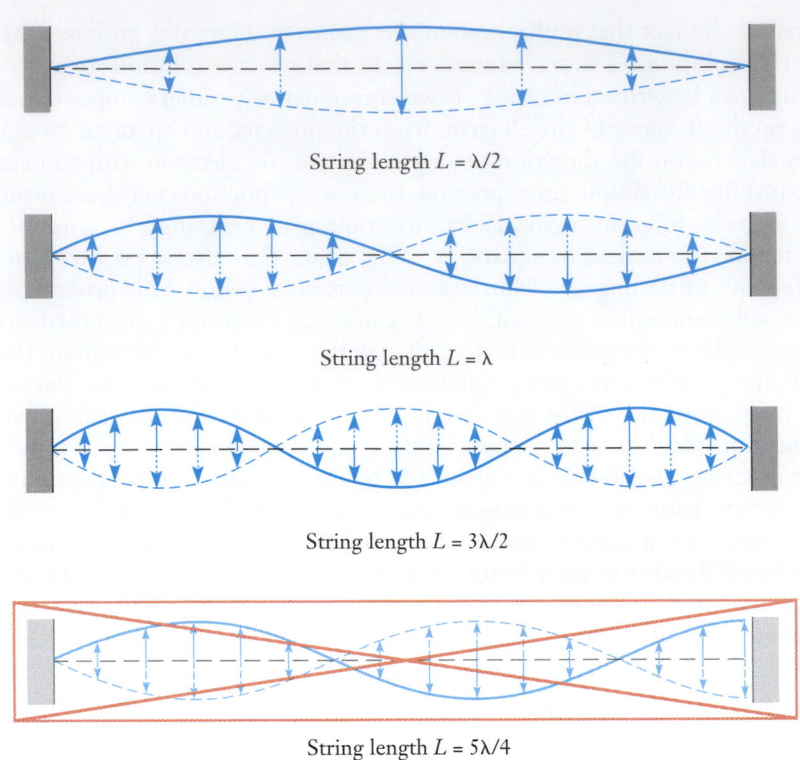

Suppose that we wanted to write a general equation for the waves that can be formed in our string. Working from the cases above, you should be able to see that the necessary form will be

$$\psi_n(x) = A \sin \frac{n\pi x}{L} \tag{6.8}$$

where n must be a positive integer. This n is a quantum number. If we substitute an allowed (positive integer) value for n in the general equation, we will get an equation for a particular wave that can exist in the string.

When we solve the Schrödinger equation for an atom, the resulting wave functions are much more complicated than these sine waves. To write the mathematical equations for the wave functions of the atomic orbitals, we need to use three different quantum numbers, which will be described individually below: the **principal quantum number** (designated n), the **secondary quantum number** (designated ℓ), and the **magnetic quantum number** (designated m_ℓ). But just as in our vibrating string problem, these quantum numbers are a natural consequence of the mathematics of the wave equation. Furthermore, when we solve the wave equation for the atomic orbitals, we find that the values which these three quantum numbers are allowed to take on are all interrelated. As we'll see, this has significant consequences for the kinds of orbitals that exist.

The principal quantum number defines the **shell** in which a particular orbital is found and must be a positive integer ($n = 1, 2, 3, 4, 5, \ldots$). When $n = 1$, we are describing the first shell; when $n = 2$, the second shell; and so on. Because there is only one electron in hydrogen, all orbitals in the same shell have the same energy, as predicted by the Bohr model. However, when atoms have more than one electron, the negative charges on the electrons repel one another. This repulsion between electrons causes energy differences between orbitals in a shell. This difficulty was one reason the Bohr model had to be replaced.

The secondary quantum number provides a way to index energy differences between orbitals in the same shell of an atom. Like the principal quantum number, it assumes integral values, but it may be zero and is constrained to a maximum value

Table ■ 6.1

Letter designations for the secondary quantum number

ℓ-value	0	1	2	3	4
Letter Designation	*s*	*p*	*d*	*f*	*g*

of $n-1$: $\ell = 0, 1, 2, \ldots, n-1$. The secondary quantum number specifies **subshells**, smaller groups of orbitals within a shell. When $n = 1$, the only possible value of ℓ is zero, so there is only one subshell. When $n = 2$, $\ell = 0$ and $\ell = 1$ are both possible, so there are two subshells. Each shell contains as many subshells as its value of n.

The secondary quantum number also provides the key to the letter designations of orbitals. **Table 6.1** provides the letter designation of the first five values of the ℓ quantum number.

The origin of these letters is from the language used to describe the lines seen in early studies of atomic spectra: *s* was "sharp," *p* was "principal," *d* was "diffuse," and *f* was "fundamental." Beyond *f*, the letter designations proceed in alphabetical order. There is no known atom that has electrons in any subshell higher than *f* when it is in its ground state, but higher orbitals may be occupied in excited states, and additional "superheavy" elements in which these orbitals are populated may yet be discovered.

For atoms under normal conditions, the energies of the orbitals are specified completely using only the n and ℓ quantum numbers. In the laboratory, it is possible to place atoms in magnetic fields and observe their spectra. When this is done, some of the emission lines split into three, five, or seven components. This observation points to the need for a third quantum number. This index, called the magnetic quantum number, is labeled m_ℓ and is also an integer. It may be either positive or negative, and its absolute value must be less than or equal to the value of ℓ for that orbital. Thus when $\ell = 0$, m_ℓ is also zero. When $\ell = 1$, m_ℓ may be -1, 0, or $+1$, and when $\ell = 2$, m_ℓ may be -2, -1, 0, $+1$, or $+2$. The relationships among the three quantum numbers are summarized in **Table 6.2** and illustrated further in Example Problem 6.5.

Table ■ 6.2

Relationships among values of the different quantum numbers are illustrated. This table allows us to make another observation about quantum numbers. If we count the total number of orbitals in each shell, it is equal to the square of the principal quantum number, n^2.

Value of n	Values for ℓ (letter designation)	Values for m_ℓ	Number of Orbitals
1	0 (*s*)	0	1
2	0 (*s*)	0	1
	1 (*p*)	−1, 0, 1	3
3	0 (*s*)	0	1
	1 (*p*)	−1, 0, 1	3
	2 (*d*)	−2, −1, 0, 1, 2	5
4	0 (*s*)	0	1
	1 (*p*)	−1, 0, 1	3
	2 (*d*)	−2, −1, 0, 1, 2	5
	3 (*f*)	−3, −2, −1, 0, 1, 2, 3	7

EXAMPLE PROBLEM 6.5

Write all of the allowed sets of quantum numbers (n, ℓ, and m_ℓ) for a $3p$ orbital.

Strategy The $3p$ designation tells us the values of the n and ℓ quantum numbers, and we can use the relationship between ℓ and m_ℓ to determine the possible values of m_ℓ.

Solution The "3" in "$3p$" tells us that the orbital must have $n = 3$, and the "p" tells us that $\ell = 1$. Possible values for m_ℓ range from $-\ell$ to $+\ell$; in this case, that gives us $\ell = -1, 0$, or $+1$. So we have three possible sets of quantum numbers for a $3p$ orbital: $n = 3, \ell = 1, m_\ell = -1$; $n = 3, \ell = 1, m_\ell = 0$; and $n = 3, \ell = 1, m_\ell = 1$.

Check Your Understanding An orbital has quantum numbers of $n = 4$, $\ell = 2$, $m_\ell = -1$. Which type of orbital ($1s$, etc.) is this?

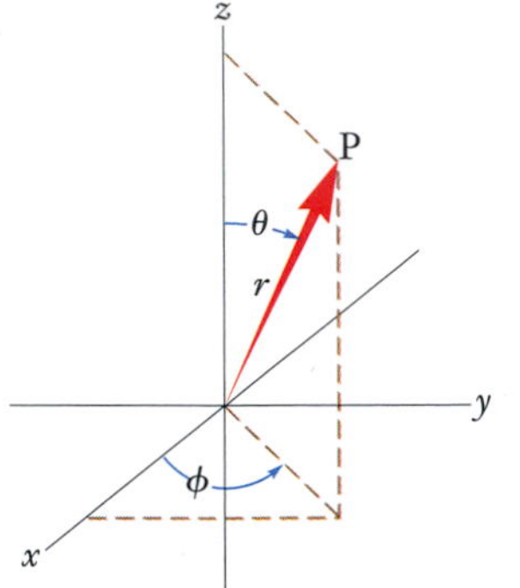

Figure 6.12 ■ A point P can be specified using three coordinates. Because atoms are spherically symmetric, the spherical polar coordinates r, θ, and φ, are often used rather than the more familiar x, y, z coordinate system.

Visualizing Orbitals

Table 6.3 lists the actual wave functions for the first few orbitals of a hydrogen atom. (Don't worry—you are *not* expected to memorize these functions! They are given only to show that they are made up of several standard mathematical functions with which you are already familiar.) Because we expect the atom to be spherical, it is customary to write these functions in spherical polar coordinates, with the nucleus of the atom located at the origin. You are probably familiar with two-dimensional polar coordinates, where a point in a plane is specified by a radius (r) and an angle (θ). Spherical polar coordinates are the three-dimensional equivalent. To specify a point in three-dimensional space, we need a radius (r) and two angles (θ and φ), as shown in **Figure 6.12**. So, the terms that contain r determine the size of the orbital, and those that contain θ or φ determine the shape.

Most chemists usually think of orbitals in terms of pictures rather than mathematical functions and quantum numbers. The pictorial depiction of orbitals requires an arbitrary decision. The wave function provides information about the probability of finding an orbital in a given region of space. If we wish to have a 100% probability of finding an electron, the amount of space needed is quite large and accordingly fails to provide insight about the tendency of that electron to be involved in chemical bonding. Conventionally, the space within which there is a 90% probability of finding an

The a_0 appearing in these equations is a constant called the Bohr radius, with a value of 53 pm.

Table ■ 6.3

Wave functions for the $n = 1$ and $n = 2$ shells of a hydrogen atom are listed. There is no need to memorize these functions—they are shown merely to point out that they are made up of ordinary mathematical terms.

Orbital	Wave Function
$1s$	$\psi_{1s} = \pi^{-1/2} a_0^{-3/2} e^{-(r/a_0)}$
$2s$	$\psi_{2s} = (4\pi)^{-1/2}(2a_0)^{-3/2}\left(2 - \frac{r}{a_0}\right) e^{-(r/2a_0)}$
$2p_x$	$\psi_{2p_x} = (4\pi)^{-1/2}(2a_0)^{-3/2} \frac{r}{a_0} e^{-(r/2a_0)} \sin\theta \cos\phi$
$2p_y$	$\psi_{2p_y} = (4\pi)^{-1/2}(2a_0)^{-3/2} \frac{r}{a_0} e^{-(r/2a_0)} \sin\theta \sin\phi$
$2p_z$	$\psi_{2p_z} = (4\pi)^{-1/2}(2a_0)^{-3/2} \frac{r}{a_0} e^{-(r/2a_0)} \cos\theta$

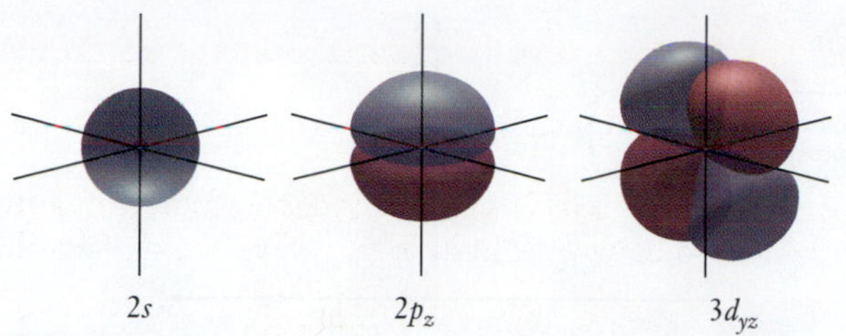

Figure 6.13 ■ Shapes of (from left to right) 2*s*, 2*p*, and 3*d* orbitals are shown. Although we sometimes think of figures like these as "pictures," they are obtained by plotting the corresponding wave functions, as described in the text. The different colored shadings indicate regions in which the wave function has opposite signs.

electron is chosen as a boundary for depicting orbitals. Within this convention, a plot of the angular part of the wave function will show us the shape of the corresponding orbital. Pictures of prototype *s*, *p*, and *d* orbitals are shown in **Figure 6.13**.

The *s* orbital is spherical, whereas the *p* orbital has two lobes and the *d* orbital has four lobes. The two lobes of the *p* orbital are separated by a plane in which there is no probability of finding an electron. This plane is called a **node** or a nodal plane. The *d* orbital has two nodes. These nodes give rise to one of the more intriguing philosophical questions of science. If the electron can be on either side of the node, but never in the nodal plane, how does it get from one side to the other? We can suggest an answer to this question in two ways. One is to remember that electrons behave as waves and even classical wave motion such as in the string of a guitar can have nodes. The second and more technical version of the answer lies in a peculiar concept of quantum mechanics called the **uncertainty principle**. This concept, stated by Werner Heisenberg, expresses the difficulty of observing an electron. Simply put, the uncertainty principle says it is impossible to determine both the position and momentum of an electron simultaneously and with complete accuracy. The uncertainty principle imposes a fundamental limit on observing the microscopic world and forces us to accept the idea that orbitals provide nothing more than the probability of finding an electron.

The actual degree of uncertainty demanded by the uncertainty principle is exceedingly small, so its effects are not detectable for macroscopic objects.

The shapes in Figure 6.13 do not tell us anything about the way the probability of finding the electron varies with distance from the nucleus. For that, we need to consider the radial part of the wave function. Depicting this while also considering the angular portion can be somewhat challenging. **Figure 6.14** shows cross-sectional "slices" through three different *s* orbitals corresponding to the principal quantum numbers $n = 1$, 2, and 3. Note that there are nodes in the radial portion of the wave function just as there are in many of the angular functions (*p*, *d*, and *f* orbitals). There is one less node than the value of the quantum number. Thus there is one node in the 2*s* orbital,

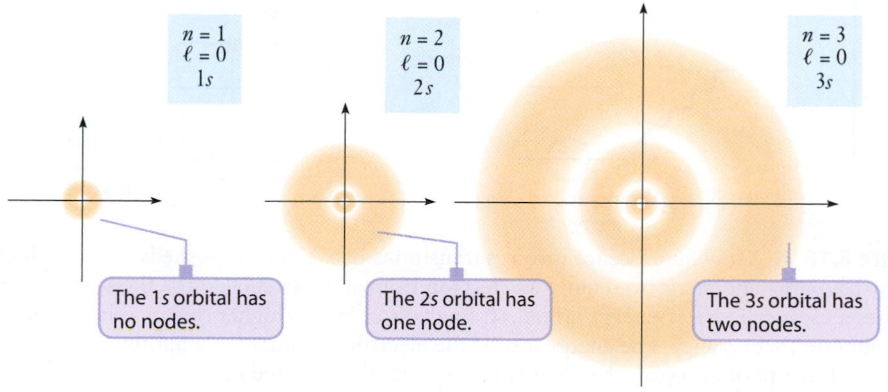

Figure 6.14 ■ Cross-sectional views of the 1*s*, 2*s*, and 3*s* orbitals are shown. In this type of figure, the density of dots indicates the probability of finding an electron. The greatest electron density will be found in those areas that appear darkest; the white regions within the 2*s* and 3*s* orbitals represent nodes.

Figure 6.15 ■ Electron density plots for $3s$, $3p$, and $3d$ orbitals show the nodes in each orbital. Although this figure is a sketch, accurate plots can be calculated from wave functions.

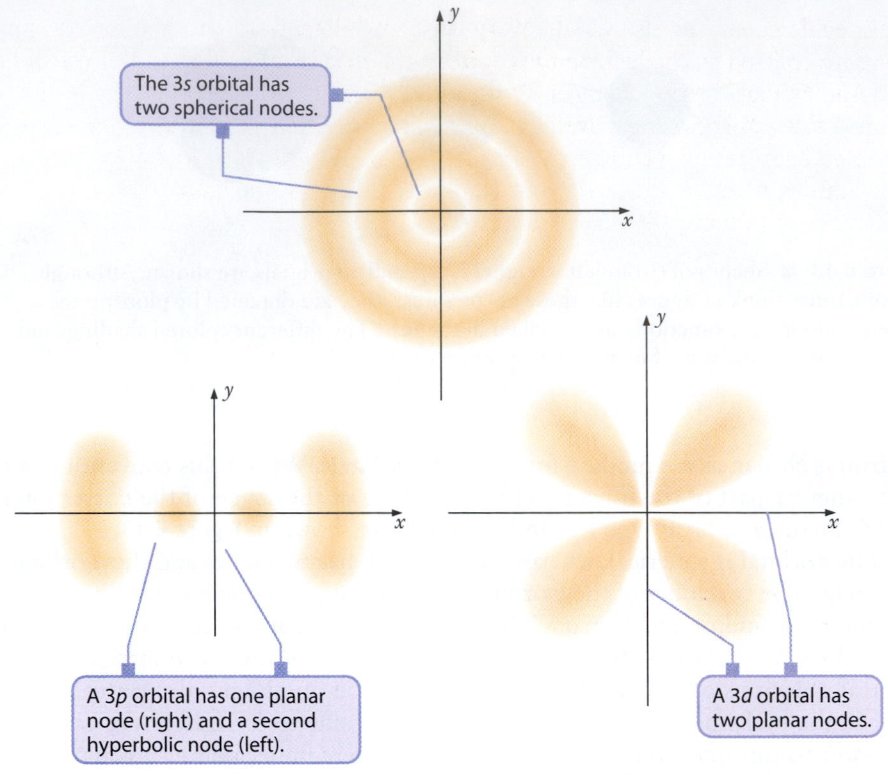

two in the $3s$ orbital, and so on. Some p and d orbitals, such as $3p$, $4p$, and $4d$, also have radial nodes. **Figure 6.15** shows pictures of orbitals in the n = 3 shell.

In spite of its inherent complexity, the quantum model of the atom became accepted largely because it provided a good way to understand the origin of energy levels in atoms. And the quantum model and its orbitals provide the framework we need in order to understand how X-ray fluorescence occurs and how it can identify what elements are present. In Section 6-1, we said that the first step in XRF is that high-energy X-ray radiation is directed at the sample. This radiation has enough energy to excite an electron, not only from a higher energy orbital, but also from lower energy levels in an inner shell as well. When an electron is missing from an inner shell, electrons from higher energy levels will tend do drop down to fill that vacancy. As we saw for atomic spectra, when an electron moves to a lower energy level the atom must release the excess energy as radiation. This two-step process is depicted in **Figure 6.16**.

X-ray radiation is sometimes labeled according to the shell that the relaxing electron falls into. K-radiation involves filling a vacancy in the first shell, and L-radiation involves filling a vacancy in the second shell.

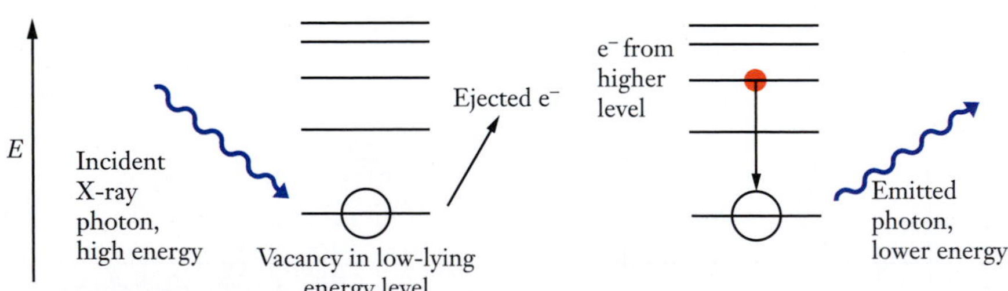

Figure 6.16 ■ X-ray fluorescence involves transitions between various shells. The incident X-ray has sufficient energy to remove an electron from an inner shell, often the $1s$ shell. This creates a vacancy in a low-energy orbital, so an electron from a higher shell will lose energy and fall into that vacancy. The energy lost by the electron is emitted as a photon, and the energy of that photon reveals the identity of the atom that emitted it.

The specific energy levels available vary from one element to another, so the amount of energy emitted as an electron moves from a higher to a lower energy level will also vary. And as that energy changes, so does the wavelength of the emitted X-ray. The data in **Table 6.4** show the wavelengths of X-rays emitted by some commonly found trace elements in steel. While these wavelengths might not seem all that different, they can be easily distinguished by the detectors used in XRF spectrometers, so it is possible to identify what elements are present in a sample.

Table ■ 6.4

X-ray wavelengths used to detect selected elements in XRF analysis

Element	XRF Wavelength (nm)
Al	0.834
Co	0.1789
Cu	0.1541
Mo	0.07094
Mn	0.2102
Nb	0.07462
Si	0.7126
Ti	0.2749
V	0.2504
W	0.1476

6-5 The Pauli Exclusion Principle and Electron Configurations

According to the quantum mechanical model of the structure of atoms, each electron in an atom can be described as occupying a particular orbital. To try to understand the chemical behavior of atoms of different elements, then, we might try to understand how electrons are distributed in orbitals. Let's begin by asking a very basic question: How many electrons can occupy an orbital? The answer requires the introduction of one more quantum number, the spin quantum number (designated m_s).

When placed in a strong magnetic field, like those used in magnetic resonance imaging (MRI), electrons behave like tiny magnets. When electric charges move in loops they create magnetic fields. So scientists postulated that electrons (which are negatively charged) must be spinning. This postulate gave rise to the idea and the name of the spin quantum number. Because our modern model describes the electron as a delocalized wave, the spinning particle picture is no longer considered literally correct. But the magnetic properties of electrons are consistent with the notion that they are spinning in one of two directions, and we often refer to electrons as being either "spin up" or "spin down." We use the quantum number m_s to distinguish between these two possible spins, and so it can assume only two possible values, $+1/2$ or $-1/2$. Using this quantum number, now, we can specify any electron in an atom by using a set of four quantum numbers.

In MRI, the "magnets" are protons in hydrogen atoms rather than electrons.

The specific values of $+1/2$ or $-1/2$ are related to the magnitude of the magnetic moment of the electron.

We are constrained, however, in the way we can specify the quantum numbers. We've already seen that the values of n, ℓ, and m_ℓ must conform to certain rules that determine which orbitals exist. With the introduction of the m_s quantum number, we also need an additional restriction. The **Pauli exclusion principle** states that *no two electrons in an atom may have the same set of four quantum numbers.* (Loosely speaking, this is the wave equivalent of saying that two particles cannot occupy the same space.) If two electrons have the same values for n, ℓ, and m_ℓ, then they must have different spin quantum numbers (m_s). The major consequence of this principle is that no more than two electrons can occupy any orbital. If two electrons occupy the same orbital, they must be **spin paired**, one with spin up and the other with spin down. Now that we know that orbitals can hold two electrons, how do we know which orbitals in a given atom are occupied?

Orbital Energies and Electron Configurations

We have already noted that in the ground state of an atom, electrons will occupy the lowest energy orbitals available. But how do we know which orbitals are lower in energy, and what causes them to be lower? This question is best considered in two parts. First, realize that a negatively charged electron is attracted to a positively charged nucleus. Electrons in smaller orbitals are held more tightly to the nucleus, so they have lower energies. We've already seen from Figure 6.14 that the size of orbitals tends to increase as the value of n increases. Putting these two ideas together, we conclude that the energy of atomic orbitals will increase as the value of the n quantum number increases.

For the hydrogen atom, this is the entire story. But for multielectron atoms, we must think about this interaction further. The first difference is the size of the nuclear charge; it will be larger for other elements than for hydrogen. Larger nuclear charges exert stronger attractive forces on electrons, so the size of the orbitals will tend to decrease for higher atomic numbers. More importantly, the electrons in the orbitals interact with other electrons and not only with the positively charged nucleus. For electrons in larger orbitals, therefore, the charge they "feel" is some combination of the actual nuclear charge and the offsetting charge of electrons located in smaller orbitals closer to the nucleus. This masking of the nuclear charge by other electrons is referred to as **shielding**. Instead of interacting with the full nuclear charge, electrons in multielectron atoms interact with an **effective nuclear charge** that is smaller than the full charge because of shielding effects.

Referring to **Figure 6.17**, we can see how the concept of effective nuclear charge allows us to understand the energy ordering of orbitals. The graphs show the probability of finding an electron as a function of the distance from the nucleus for each subshell up to $n = 3$. It is obvious that there are substantial size differences between the shells, such that 1*s* is notably smaller than 2*s* and 2*p*, which in turn are smaller than 3*s*, 3*p*, and 3*d*. This is totally consistent with the argument we just made that orbital energy should increase as n increases. When we consider subshells, however, the distinctions are subtler. Compare the 2*s* and 2*p* wave functions. Although the overall sizes of these orbitals are similar, the 2*s* wave function shows a small local maximum close to the nucleus. No such feature appears in the 2*p* wave function. This means that a 2*s* electron has a greater probability of being found very close to the nucleus, and an electron closer to the nucleus will experience a larger effective nuclear charge. As a result, the 2*s* orbital is lower in energy than the 2*p* orbitals. Similarly, the 3*s* orbital, on average, interacts with a larger charge because of the probability it will be close to the nucleus. Even though the first two lobes

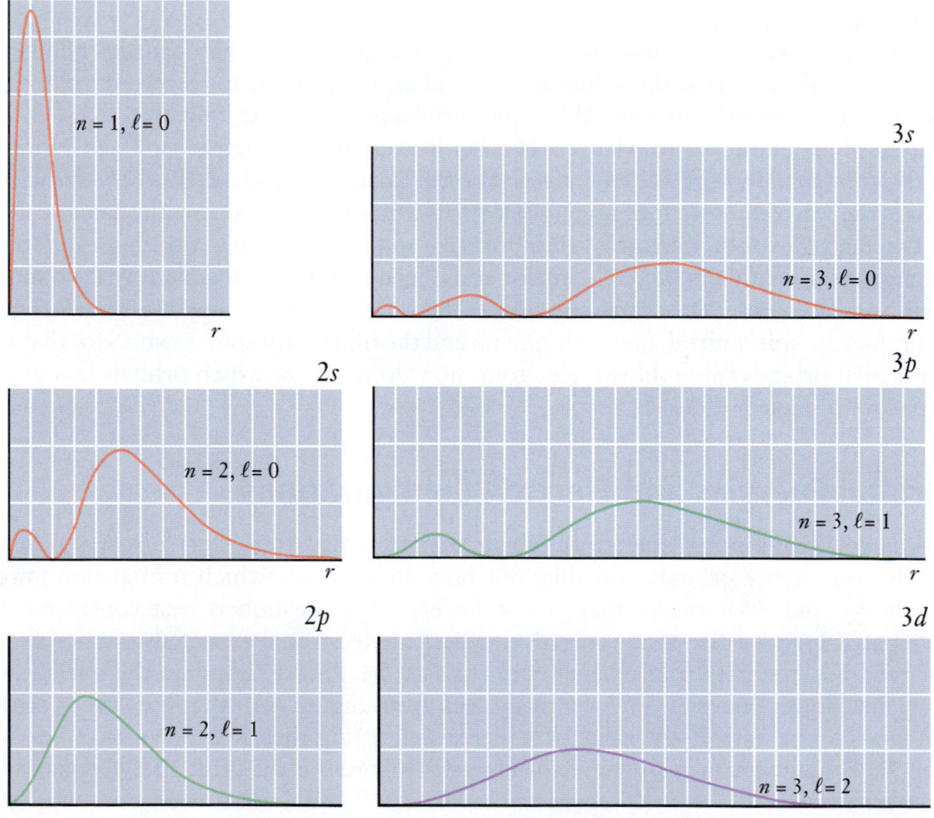

Figure 6.17 ■ Graphs of the radial probability function are shown for hydrogen orbitals with $n = 1, 2$, and 3. The increase in orbital size with increasing n is obvious. More subtle variations for orbitals with the same n value are also important in understanding the sequence of orbital energies.

of the radial wave function are small, they provide the opportunity for electrons in the 3*s* orbital to "penetrate" through the shielding of other electrons. This makes the energy of the 3*s* orbital lower than that of the 3*p* or 3*d* orbitals. Similar penetration in the 3*p* orbitals lowers their energy in comparison to the nonpenetrating 3*d* orbitals.

As we reach the fourth shell of orbitals, the comparison of orbital size to penetrating probability becomes even more difficult. We will not attempt to carry the conceptual reasoning any further, but detailed calculations can be carried out to determine ordering. The accepted order of orbital energies that arises from these calculations is 1*s*, 2*s*, 2*p*, 3*s*, 3*p*, 4*s*, 3*d*, 4*p*, 5*s*, 4*d*, 5*p*, 6*s*, 4*f*, 5*d*, 6*p*, 7*s*, 5*f*, 6*d*, and 7*p*. There are no elements currently known whose ground state has electrons that occupy orbitals higher in energy than 7*p*. You may have learned some device to remember this sequence of orbitals. We will reinforce this ordering in the next section by relating it to the structure of the periodic table. Now that we've established the energy sequence of the various subshells, we are ready to decide which orbitals should be occupied in each element. Now we will consider rules for determining these **electron configurations**.

Hund's Rule and the Aufbau Principle

The process of deciding which orbitals hold electrons in an atom requires that we "fill" orbitals starting with the lowest energy and proceeding to higher energy. In some ways, this concept resembles constructing a building from the ground up, and it is referred to as the **aufbau principle.** This process is uncomplicated for the first few atoms.

Aufbau comes from the German word for "build up."

The hydrogen atom has one electron, and it occupies the 1*s* orbital. Helium has two electrons. Both can occupy the 1*s* orbital, as long as they are spin paired. Lithium has three electrons. Only two can occupy the 1*s* orbital, and so one electron must go into the next lowest energy orbital, the 2*s*. A shorthand way to designate this orbital occupation is to write the orbital name followed by a superscript indicating the number of electrons in the orbital. Thus lithium has an electron configuration of $1s^2 2s^1$. Beryllium is the next element and its atoms contain four electrons. The 2*s* orbital can also hold two electrons, so the fourth electron is placed there to give a configuration of $1s^2 2s^2$. With five electrons, boron will also fully populate both the 1*s* and 2*s* orbitals, but we must account for an additional electron. The fifth electron goes into the next available orbital, 2*p*, to give $1s^2 2s^2 2p^1$.

Upon reaching carbon we are faced with a choice. There are three different 2*p* orbitals, all with equal energy. So two electrons could pair and occupy a single 2*p* orbital, or they could occupy two different 2*p* orbitals with one electron in each. The correct choice is motivated by the fact that all electrons are negatively charged particles. Two negatively charged electrons should repel one another, and so the most energetically favorable configuration would have them occupy separate orbitals where they can be farther apart from each other. Measurements of the magnetic moments of atoms also show that in such cases the unpaired electrons will have the same spin as one another. The concept is summarized in **Hund's rule**, which states that within a subshell, electrons occupy orbitals individually and with parallel spins whenever possible. To depict this type of distinction in electron configuration requires a device more detailed than the representation we used for the first five elements. One possible choice is to use diagrams with boxes representing orbitals, as shown in **Figure 6.18**. Another option is to identify individual orbitals with subscripts, such as p_x, p_y, and p_z. Once we are familiar with electron configurations, though, the simple notation of $1s^2 2s^2 2p^2$ for carbon will suffice. Knowing Hund's rule will allow whoever reads this configuration to envision the correct orbital occupation explicitly shown in more detailed representations. Practice with this concept is provided in Example Problem 6.6.

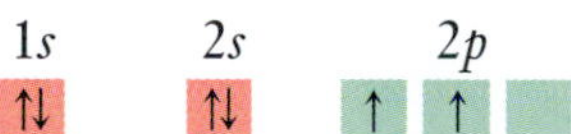

Figure 6.18 ■ Electron configurations are sometimes depicted using diagrams of the type shown here for carbon. Such diagrams emphasize the presence of unpaired electrons and may be helpful as you become familiar with electron configurations.

EXAMPLE PROBLEM 6.6

The presence of sulfur in steel alloys can have a negative effect on ductility and other properties. So trace analysis is often used to monitor the level of sulfur present. What is the electron configuration of a sulfur atom?

Strategy We use the energy ordering and number of orbitals for each subshell to arrive at the overall electron configuration. Then, using Hund's rule, we can predict the detailed electron configuration for an atom of sulfur.

Solution Sulfur has 16 electrons, so we can place two in 1*s*, two in 2*s*, six in 2*p*, two in 3*s*, and four in 3*p* to get 16. The electron configuration is $1s^22s^22p^63s^23p^4$.

For the detailed configuration, we must consider how to distribute the four electrons in the 3*p* subshell. We imagine placing the first three electrons each into separate 3*p* orbitals and then filling one of those orbitals by adding the fourth electron. The resulting configuration is shown below:

1*s*	2*s*	2*p*	3*s*	3*p*
↑↓	↑↓	↑↓ ↑↓ ↑↓	↑↓	↑↓ ↑ ↑

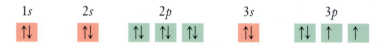

Check Your Understanding Silicon is found in some steels, and its presence can make it harder to machine the steel. What is the electron configuration of a silicon atom?

Elements in the same column of the periodic table will have very similar electron configurations. To illustrate this, let's consider argon and krypton, both of which are found among the noble gases in Group 18. What are the electron configurations of these two elements? Argon is $1s^22s^22p^63s^23p^6$ and krypton is $1s^22s^22p^63s^23p^64s^23d^{10}4p^6$. These two elements are chemically similar (both are essentially inert) precisely because they have similar electron configurations. In each case, the last electron fills the *p* subshell. This correspondence between chemical inertness and a filled *p* subshell provides the motivation for a shorthand notation for electron configurations.

This shorthand notation accomplishes two things. First, it makes writing electron configurations less unwieldy. More importantly, it introduces a very important concept that relates electronic structure to chemical bonding. The electrons in the outermost occupied subshells give rise to the chemical reactivity of an element. Electrons in filled lower shells will have little impact on an element's chemical properties. The inner electrons, which lie closer to the nucleus, are referred to as **core electrons**. They can be represented by the atomic symbol of the noble gas element that has the corresponding electron configuration. Thus, although the complete electronic configuration for potassium is $1s^22s^22p^63s^23p^64s^1$, the outermost $4s^1$ electron determines the chemical properties of potassium. We can emphasize this fact by writing the configuration as $[Ar]4s^1$. Similarly, tin (Sn) can be written as $[Kr]5s^24d^{10}5p^2$. The electrons that are explicitly noted in this notation occupy orbitals that are further from the nucleus. Although it is appropriately descriptive to designate them as outer electrons, they are usually referred to as **valence electrons**. When we take up the study of chemical bonding in Chapter 7, we will see how important it is to know the valence electron configuration of elements. Meanwhile, Example Problem 6.7 provides some practice in using the shorthand notation we just defined.

We will discuss the connections between electron configuration and chemical bonding in detail in the next chapter.

EXAMPLE PROBLEM 6.7

Rewrite the electron configuration for sulfur (Example Problem 6.6) using the shorthand notation introduced above.

Strategy We found the electron configuration in Example Problem 6.6. Now we simply need to identify the core electrons and replace them with the symbol for the corresponding noble gas.

Solution The full electron configuration of sulfur is $1s^22s^22p^63s^23p^4$. The valence electrons occupy orbitals with $n = 3$, and the core electrons fill the $n = 1$ and $n = 2$ shells. Neon is the noble gas element whose electron configuration is $1s^22s^22p^6$. So we can abbreviate the configuration for sulfur as $[Ne]3s^23p^4$.

Check Your Understanding Trace amounts of arsenic are sometimes found in steel. Use the shorthand notation to write the electron configuration for arsenic.

6-6 The Periodic Table and Electron Configurations

We noted that as members of the same group in the periodic table, argon and krypton will have similar electron configurations. And similar observations could be made for the halogens or any of the other groups. When we first encountered these groups, however, it was in conjunction with the periodic table (in Section 2-5). And the periodic table was first established by considering chemical properties, long before the ideas of quantum mechanics or electron configurations were proposed. The relationship between the electron configuration predicted by the quantum mechanical model of the atom and the periodic table was vital to the acceptance of quantum mechanics as a theory. We will take advantage of this important correlation between electron configurations and the periodic table to make it relatively easy to remember configurations for the ground state of any element.

The outline of the periodic table in **Figure 6.19** includes the elements that are usually shown separately at the bottom. The coloring of regions in this figure divides the table into four blocks of elements. The elements in each block are similar in that the electron in the highest energy orbital comes from the same subshell. Thus all the elements in the section at the left edge of the table, shaded in red, have their highest energy electron in an *s* orbital. These are sometimes referred to as the *s* block. The

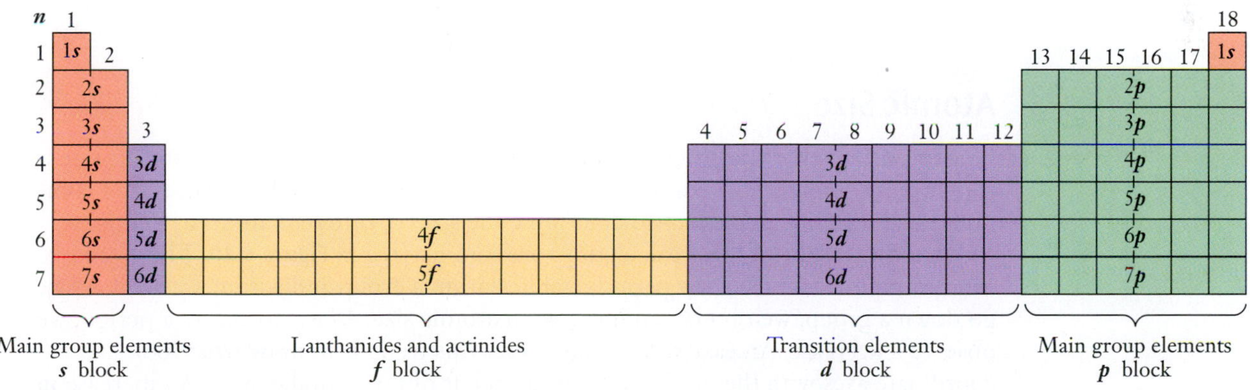

Figure 6.19 ■ The shape of the periodic table can be broken down into blocks according to the type of orbital occupied by the highest energy electron in the ground state of each element. The *f*-block elements are usually broken out of the table and displayed separately at the bottom. Here we have placed them in the body of the table to emphasize the sequence in which the various orbitals are filled.

green section at the far right of the table is the *p* block because according to the aufbau principle, the last orbital occupied is a *p* orbital. The transition metals are *d*-block elements (shown in purple), and the lanthanides and actinides make up the *f* block (shown in yellow).

Because of this structure, we can use the periodic table to determine electron configurations for most elements. (A few exceptions arise, mainly among the transition metals.) We find the element of interest in the periodic table and write its core electrons using the shorthand notation with the previous rare gas element. Then we determine the valence electrons by noting where the element sits within its own period in the table. Example Problem 6.8 demonstrates this procedure.

EXAMPLE PROBLEM 6.8

Tungsten alloys are often used for parts that must withstand high temperatures. Use the periodic table to determine the electron configuration of tungsten (W).

Strategy Begin by finding tungsten in the periodic table. Use the rare gas element from the previous period to abbreviate the core electrons. Then determine the correct valence electron configuration by counting across the period in which tungsten is located.

Solution The rare gas element preceding tungsten is xenon. When counting the valence electrons, don't forget the *f*-block elements. The correct configuration is [Xe] $6s^2 4f^{14} 5d^4$.

Check Your Understanding A number of corrosion resistant "superalloys" can be formed by combining nickel with varying amounts of other metals, including tungsten. Use the periodic table to determine the electron configuration of nickel (Ni).

6-7 Periodic Trends in Atomic Properties

When we first introduced the concept of the periodic table, we noted that various observations about the behavior of atoms led to its development. The reactivity of halogens and the inertness of rare gases are examples of the periodic law. Using the understanding of orbitals and atomic structure we have just gained, it is possible to explain some of these periodic variations in greater detail. These atomic properties will then be used to help understand reactivity in later chapters.

Atomic Size

One important characteristic of atoms is their size. The measurement of atomic size is sometimes complicated, because atoms do not have well-defined boundaries. But computational techniques allow us to determine the sizes of different atoms in a consistent way, and the results of these calculations are summarized in **Figure 6.20**. There are two ways to explore trends in the periodic table, within a group and within a period. As we go down a group, we observe an increase in atomic size. As we go across a period, we observe a decrease. An easy way to remember this trend is to draw what look like x, y coordinate axes with the origin at the bottom left of the periodic table. An increase in either the x or y direction on these axes corresponds to a decrease in atomic size. How do we understand this trend?

The size of the atom is determined largely by its valence electrons because they occupy the outermost orbitals. Two factors are important: (1) the shell in which the valence electrons are found and (2) the strength of the interaction between the nucleus

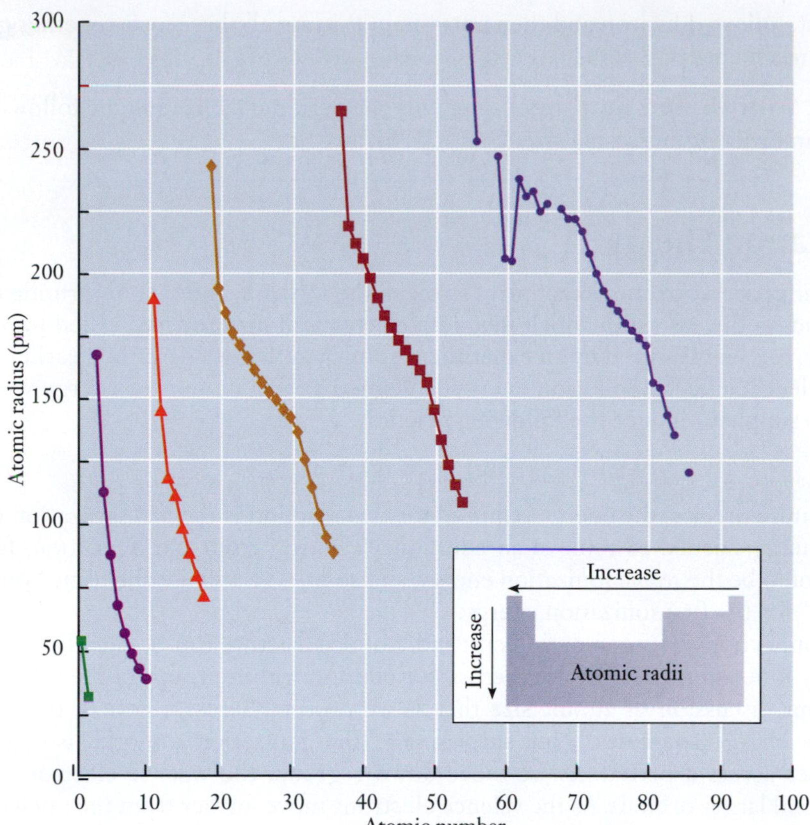

Figure 6.20 ■ The graph shows atomic radius as a function of atomic number. The different colors and data point symbols represent the different rows in the periodic table. The decrease in size from left to right across each period is clearly seen. (Calculated radii taken from E. Clementi, D.L. Raimondi, and W.P. Reinhardt, J. Chem. Phys. 1963, 38, 2686.)

and the valence electrons. As we go down a group in the table, the first of these factors is at work. The size of the valence orbitals increases with *n*, so the size of the atom increases as we go down the period. The trend in moving across a row is less intuitive. We have already discussed the concept of effective nuclear charge in Section 6-5 (page 164). The distance between a valence electron and the nucleus will tend to be greater if there is a small effective nuclear charge. As we go across a period, we add electrons to the same subshell. These electrons do not shield each other very well, and the effective nuclear charge increases because each increment in atomic number adds a positively charged proton to the nucleus. The higher effective nuclear charge produces a stronger attraction between the electrons and the nucleus. So the valence electrons are drawn closer to the nucleus and the size of the atom decreases. Example Problem 6.9 shows how to apply these ideas to rank various atoms by size.

In Chapter 7 we will use similar arguments to consider the sizes of ions. In general, cations are smaller than their corresponding neutral atoms, whereas anions are larger than their corresponding neutral atoms.

EXAMPLE PROBLEM 6.9

Using only the periodic table, rank the following elements in order of increasing size: Fe, K, Rb, S, and Se.

Strategy We want to arrange the elements from smallest to largest. Begin by finding each element in the periodic table. Then use the general trends for atomic size to place them in the requested order.

Solution Atomic radius increases as we go from top to bottom or right to left in the periodic table. Sulfur and selenium are both in Group 16; because sulfur is above selenium, it will be smaller. Similarly, the positions of rubidium and potassium in Group 1 tell us that potassium is the smaller of the two. Finally, potassium, iron, and selenium are in the fourth period. Selenium is the rightmost of the three, so it will be the smallest, followed by iron and then potassium. Putting all these facts together gives us the requested order: S < Se < Fe < K < Rb.

Check Your Understanding Using only the periodic table, rank the following elements in order of increasing size: Cr, Cs, F, Si, and Sr.

Ionization Energy

Another property of atoms that plays a role in the way they interact in forming chemical bonds is the ease with which they lose electrons. This property, called **ionization energy**, can be measured by an experiment similar to that we described earlier for the photoelectric effect. Ionization energy is defined as the amount of energy we would have to supply to induce the following reaction:

$$X(g) \rightarrow X^+(g) + e^-$$

To be more precise, the energy required for this reaction is the *first* ionization energy. The energy needed to remove an additional electron from the X^+ cation, forming X^{2+}, would be the *second* ionization energy, and so on. We will use the term "ionization energy" for the first ionization energy.

Connections between periodicity and atomic structure can be seen by looking at the way ionization energy varies across periods and within groups of the table. Recall from our discussion of atomic size that as we proceed across a period, the effective nuclear charge increases. Not surprisingly, the ionization energy also generally increases across a period. As we move down a group, the valence electrons occupy larger and larger orbitals. As the valence electrons move farther from the nucleus, they become easier to remove. So ionization energy decreases as we move down a column in the periodic table. As we proceed up and toward the right in the periodic table, the ionization energy increases, as shown in **Figure 6.21**.

Additional clues about atomic structure are revealed by the exceptions to the trends in ionization energy. Looking closely at Figure 6.21, we see that from nitrogen to oxygen there is a slight *decrease* in ionization energy rather than the increase expected based on the periodic trend. This observation can be rationalized by considering electron pairing. Nitrogen has a half-filled *p* subshell, with one electron in each 2*p* orbital. To accommodate its p^4 configuration, oxygen must pair two electrons in one of the 2*p* orbitals. Removing one electron, therefore, leaves only unpaired electrons in all three 2*p* orbitals. Such a change leads to notably less electron–electron

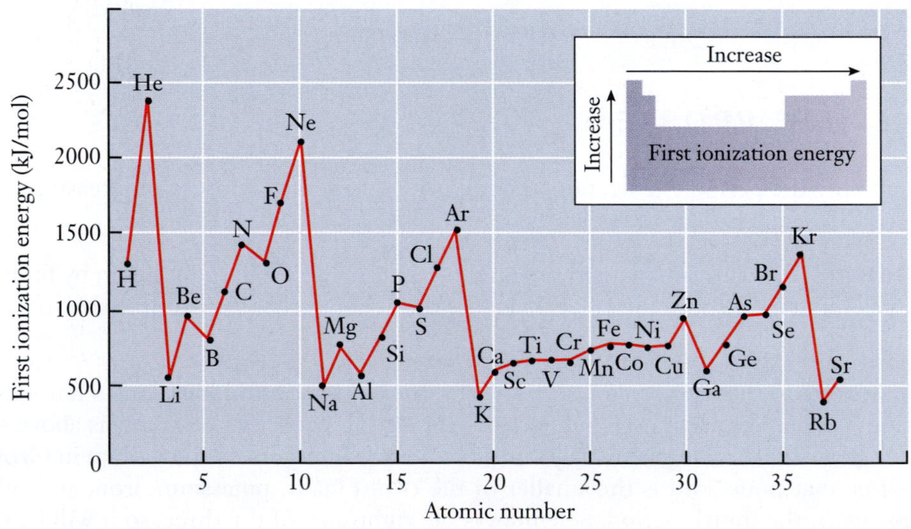

Figure 6.21 ■ The graph shows the first ionization energy (in kJ/mol) vs. atomic number for the first 38 elements in the periodic table. The inset in the upper right emphasizes the general periodic trend: Ionization energy increases from left to right and bottom to top in the periodic table.

repulsion, enough so that the ionization energy of oxygen is lower than that of nitrogen. Similar exceptions to trends are seen for sulfur in the third period and selenium in the fourth. Example Problem 6.10 shows how to use the periodic table to rank atoms by ionization energy.

EXAMPLE PROBLEM 6.10

Using only the periodic table, rank the following elements in order of increasing ionization energy: Br, F, Ga, K, and Se.

Strategy We want to arrange the elements from the smallest ionization energy to the largest. Begin by finding each element in the periodic table. Then use the general trends for ionization energy to place them in the requested order.

Solution Ionization energy increases as we go up or to the right in the periodic table. From our list of elements, all but fluorine are in the fourth period. So we expect their ionization energies to increase from left to right: K < Ga < Se < Br. Finally, fluorine lies above bromine in Group 17, so it will have the largest ionization energy. This gives us the requested order: K < Ga < Se < Br < F.

Check Your Understanding Using only the periodic table, rank the following elements in order of increasing ionization energy: He, Mg, N, Rb, and Si.

We can also gain insight into atomic structure by looking at successive ionization energies for the same element, as summarized in **Table 6.5**. For sodium, the first ionization energy is among the lowest, but the second is much larger than that for any neutral atom. For magnesium, the situation is quite different. The second ionization is still larger than the first because the effective nuclear charge experienced by a valence electron is larger in the ion than in the neutral atom. But the more dramatic change in ionization energy lies between the second and third ionizations. The point at which this large jump occurs for each element is highlighted throughout the table. In each case, the largest increase in ionization energy occurs when an electron is removed from a completely filled *p* subshell. From this observation, we can see the unusual stability of that filled *p* subshell, and we can begin to appreciate why noble gas elements are so unreactive. Filled subshells of electrons are difficult to break up, and as we'll see, this fact will help us to understand chemical bonding.

Electron Affinity

Ionization energy is defined as the energy required to remove an electron from an atom, forming a positively charged cation. But we also know that some atoms will routinely pick up an electron to form a negatively charged anion:

$$X(g) + e^- \rightarrow X^-(g)$$

We can define a property called **electron affinity** as the amount of energy we would have to supply to induce this reaction. Removing an electron from a neutral atom always requires an input of energy because the atom will not simply eject electrons on its own. So ionization energies are always positive. But electron affinity values can be either positive or negative. If the resulting anion is stable, then energy will actually be released as the electron is added. But our definition of electron affinity says it is the amount of energy that must be *supplied* to form the anion. If energy is being released, we might say that we must "supply a negative amount of energy," so the electron affinity value will be negative. If the electron affinity is positive, that means that the anion formed is not stable; in this case, we would need to supply

Table ■ 6.5

The first four ionization energies (all in kJ/mol) for the elements of the first three periods. Those ionizations with values shown in shaded cells with **bold** print involve removing the last electron from a particular shell. Further ionization requires removing an electron from a more stable filled shell, and this leads to a very large increase in ionization energy.

Z	Element	IE_1	IE_2	IE_3	IE_4
1	H	**1312**	—	—	—
2	He	2372	**5250**	—	—
3	Li	**520.2**	7298	11,815	—
4	Be	899.4	**1757**	14,848	21,007
5	B	800.6	2427	**3660**	25,026
6	C	1086	2353	4620	**6223**
7	N	1402	2856	4578	7475
8	O	1314	3388	5300	7469
9	F	1681	3374	6050	8408
10	Ne	2081	3952	6122	9370
11	Na	**495.6**	4562	6912	9544
12	Mg	737.7	**1451**	7733	10,540
13	Al	577.6	1817	**2745**	11,578
14	Si	786.4	1577	3232	**4356**
15	P	1012	1908	2912	4957
16	S	999.6	2251	3357	4564
17	Cl	1251	2297	3822	5158
18	Ar	1520	2666	3931	5771

energy to force the atom to accept the electron. Most electron affinity values are negative.

This definition of electron affinity may seem a little twisted because it is at odds with the normal meaning of the word "affinity." Under our definition, if the atom "wants" the electron, its electron affinity value is negative. But we define the signs in this way because it is consistent with the conventions usually followed when discussing energy in chemistry, physics, and engineering. Energy added to an atom is positive, and energy released by an atom is negative.

The study of energy changes in chemical reactions is called thermodynamics and is presented in Chapters 9 and 10.

Electron affinity values display periodic trends based on electron configuration, much like the other properties we've been considering. The most immediate conclusion from **Figure 6.22** might be that electron affinity values vary more erratically than ionization energies. Still, it should be evident from the figure that as we move from bottom to top or from left to right in the periodic table, the electron affinity tends to become more negative. Thus in terms of the *x*, *y* scheme we have been using for remembering general trends, electron affinity becomes more negative with increasing *x* or *y*, much like ionization energy. It is customary to think of the trend in electron affinity in this way, even though the value is actually becoming algebraically smaller as

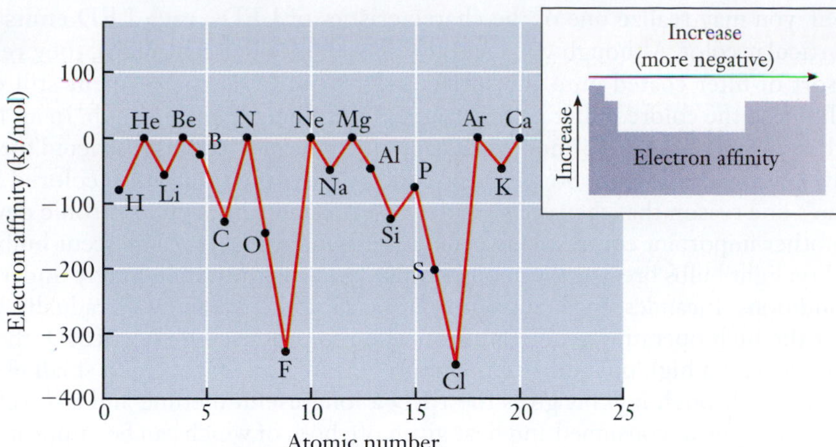

Figure 6.22 ■ The graph shows the electron affinity (in kJ/mol) vs. atomic number for the first 20 elements of the periodic table. The inset at the upper right emphasizes the general periodic trend: electron affinity "increases"—meaning that formation of the anion becomes more favorable—from left to right and bottom to top in the periodic table.

it gets more negative. That's because the more negative the electron affinity becomes, the more stable the anion will be.

Closer examination of the electron affinity values reveals that there are exceptions in the trends associated with electron pairing, just as there were for ionization energy. Thus nitrogen does not have a negative electron affinity, because adding an electron would force the pairing of electrons in a *p* orbital, a process that increases the electron–electron repulsion substantially.

There are other periodic trends in the properties of elements, including the variation in density that we noted back in Section 2-5. The X-ray emission lines used to identify elements in XRF vary in a predictable way. Consider the first five transition metals. Looking at the fluorescence that arises from electrons falling back into the 1*s* shell (K-shell in X-ray notation), the respective wavelengths are 30.32, 27.50, 25.05, 22.91, and 21.03 nm for scandium (Sc), titanium (Ti), manganese (Mn), vanadium (V), and chromium (Cr). This trend of decreasing wavelengths continues across the *d*-block elements. To explain this trend we need to realize that the attraction between the 1*s* electrons and the nucleus becomes stronger as the number of protons in the nucleus increases. So the 1*s* energy level gets lower as we move from left to right in the periodic table. The energies of outer shells are affected less because of shielding, so the energy difference between the 1*s* and outer shells will increase as we move across a row.

INSIGHT INTO

6-8 Modern Light Sources: LEDs and Lasers

In many cases, trace analysis is used to detect small levels of unwanted contaminants that might degrade the performance of a material. Modern engineering is just as likely, however, to intentionally introduce small levels of an element in order to adjust the properties of a material. One example is the development of materials used to produce **light-emitting diodes (LEDs)** and **lasers**. Both of these modern light sources owe their existence to the improved understanding of atomic structure that emerged over the course of the 20th century.

How does the light produced by an LED or a laser differ from that given off by an incandescent or fluorescent lamp? And what particular attributes of these devices and the light they emit make them so well suited for particular tasks?

You can almost certainly find light-emitting diodes in many of the appliances you use every day. The power indicators on everything from cell phones to TVs and the displays on many digital alarm clocks and microwave ovens are just a few examples of common uses for LEDs. As you think about some of the applications in which they are used, you may realize one of the characteristics of LEDs: each LED emits light of a particular color. Although you can buy colored incandescent bulbs, they rely on some sort of filter coated onto the surface of the bulb. But the filament still emits white light, so the colored filter greatly reduces the efficiency of the bulb. In an LED, the light emitted is usually **monochromatic**, which means that it is restricted to a single wavelength or color. So LEDs provide an easy way to produce colored light, and that's one reason that they are so widely used as indicators on electronic devices.

Some LEDs emit two or more wavelengths simultaneously, but usually they are so closely spaced that the human eye cannot separate them.

Another important contrast between an LED and a small incandescent bulb is in durability: light bulbs break or burn out, but LEDs last almost indefinitely under normal conditions. Incandescent bulbs burn out because their filaments gradually break down at the high operating temperatures needed to produce light. An LED, though, does not require a high operating temperature. It functions as long as a small electric current flows through it. This gives the LED a long useful lifetime and also reduces the amount of power consumed and heat given off, both of which can be important design considerations in modern electronics. Because they lose very little energy as heat, LEDs can operate at very high efficiency, converting nearly all of the input energy into light. This is especially important for small portable devices, such as cell phones or MP3 players, where battery lifetime is a major consideration. The same combination of long life and high efficiency are now being exploited in LED bulbs for use in household lighting, and you are likely to see such bulbs become much more popular in the next few years.

LEDs are an example of what are called solid-state devices, in which the primary component is simply a piece of some solid material (**Figure 6.23**). The important functional properties of such devices are determined primarily by the composition of this solid material—and that's where chemistry comes in. The color of the light emitted by an LED depends on the details of the chemical bonding between the atoms in the solid. As you might guess, that bonding depends on the identity of the atoms involved. So adjusting the composition of the solid can change the color of light emitted. We'll look at some of the details behind this in Section 8-3. But the practical implication is that we can produce an LED that emits almost any color of light.

For many years blue LEDs were not available because their higher photon energy made them more challenging to produce.

Lasers share some traits with LEDs: they produce monochromatic light, and the wavelength of that light depends on the specific properties of the laser. But lasers also possess other important properties. Perhaps most obviously, the light emitted by a laser is highly directional and emerges as a well-defined beam. You've almost certainly seen laser pointers or other lasers, so you may know that the beam is generally not even visible from the side because all of the light is traveling in the same direction. Laser light is also **coherent**, which means that all of its waves are perfectly in phase and go through maxima and minima together. Many varieties of lasers are known, ranging from small devices used in DVD players, laser printers, and supermarket price scanners to very high-powered systems used in industrial machining.

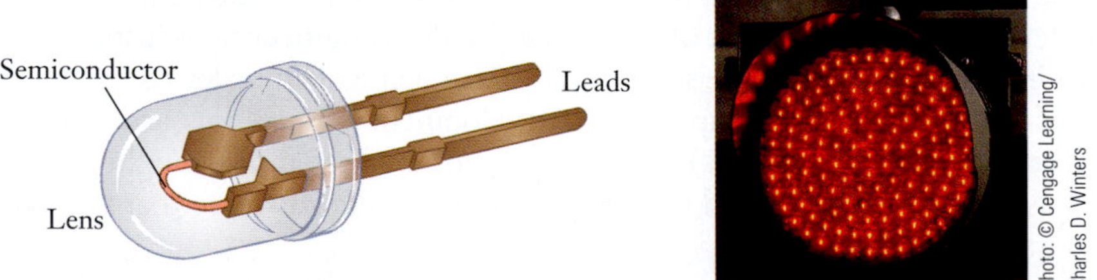

Figure 6.23 ■ The standard design of an LED is extremely simple. Metallic leads allow a current to pass through a small piece of a semiconductor material, which then emits light. A plastic enclosure serves as both a lens and a case for the device. The color of light emitted is determined by the chemical composition of the semiconductor. Because LEDs are superior to incandescent lights in efficiency and durability, they are now routinely used in traffic lights, like the one shown here.

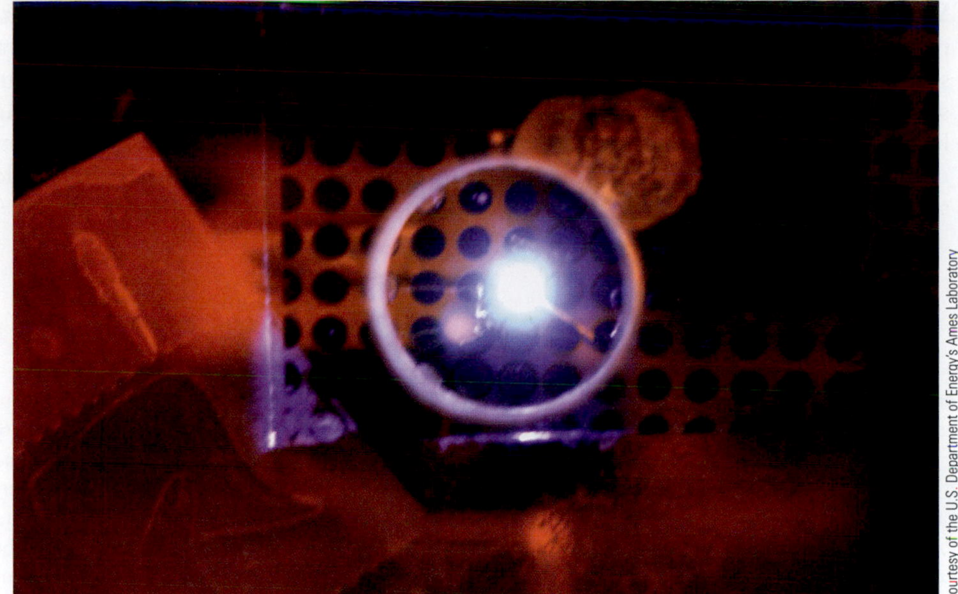

The bright blue light in this photo is given off by an organic light emitting diode (OLED). These emerging devices hold great promise for use in small, flexible displays, and versions emitting white light may also find use as light sources.

The invention of the laser is a direct result of the emergence of the quantum mechanical model of the atom. Laser light is emitted as atoms or molecules go from a higher energy level down to a lower energy level, just as we saw in atomic spectra. So every laser's design is ultimately based on a detailed understanding of the energy levels available to the atoms or molecules in what is called the laser medium. The operation of a laser generally relies on a phenomenon called stimulated emission, which becomes important when the population of one energy level becomes greater than the population of a lower energy level. Many of the most common lasers, like those in your DVD players, are solid-state devices in which this population inversion is established for the electrons in a solid. In other lasers, atoms or molecule in a gas or in solution emit the light. The primary engineering challenges in designing a laser are to identify a medium that can emit the desired wavelength of light and then to devise a way to establish and maintain the population inversion. So an understanding of the energy levels involved is clearly essential.

As new lighting technologies emerge in the 21st century, the connections between chemistry and engineering continue to play a major role. Recent improvements in liquid crystal displays used in flat panel computer monitors, for example, are due in part to the development of new chemical compounds whose light absorbing properties are tailored for their tasks. Research into organic light emitting diodes (OLEDs) has been moving forward rapidly, and OLED screens are being touted as the next step in television technology. Advances in all of these areas require the close cooperation of research chemists and design engineers.

FOCUS ON PROBLEM SOLVING

Question **(a)** Based on your understanding of atoms and their spectra, speculate on why streetlights that use sodium lamps seem yellow/orange in color. **(b)** What device or instrument could be used to make a measurement that would test your hypothesis from (a)?

Strategy This question involves conceptual problem solving. We can approach questions like this by asking what atomic-or molecular-level information must be inferred to answer it. Then we need to think about what sort of measurement or experiment could provide the atomic or molecular information needed to confirm or refute our idea.

Solution **(a)** The key words in this question are "sodium lamp." This description implies that there is something important about the lamp that is related to the element sodium. We saw in this chapter that atoms have distinctive spectra, so it is reasonable to surmise that in the case of sodium atoms enough colors are missing that the remaining light appears yellow/orange. **(b)** We could test this hypothesis by using a spectrometer or prism to view the light from the lamp. If our assertion is correct, the spectrum should show lines in the yellow-orange region but not in other parts of the visible spectrum.

SUMMARY

To develop an understanding of matter at the microscopic level, first we must have a sound model for the structure of atoms. Much of our current understanding of atoms grew out of studies of the interaction of atoms with electromagnetic radiation, or light. Many properties of light can be explained by a wave model in which the frequency or wavelength defines the color of the light and the amplitude defines its intensity. But we can only understand some aspects of the interaction of light with atoms by using a particulate model in which we think of light as packets of energy, called photons. We thus encounter the concept of wave–particle duality for the first time when we consider the nature of light.

The most direct connection between atoms and light is the observation of discrete atomic spectra. In this context, discrete means that when atoms absorb or emit light they do so at only a few frequencies rather than across the entire spectrum. This observation leads to the hypothesis that the electrons in an atom can occupy only a limited number of energy levels: the energies of electrons in atoms are said to be quantized. Over time, various models have been introduced and improved to account for these quantized energies. Our best current model is the quantum mechanical theory of the atom.

In quantum theory, electrons are treated as waves that can be described by mathematical wave functions. (This is another instance of wave–particle duality.) The detailed mathematics of the quantum mechanical model is responsible for much of the vocabulary we use to describe atomic structure. Thus, electrons occupy orbitals that are characterized by quantum numbers. For the purpose of understanding chemistry, there are two important things to remember about this model. First, orbitals have sizes and shapes that can be correlated with the labeling scheme derived from quantum numbers. Second, for atoms under most ordinary conditions, we can assign the electrons to the lowest energy orbitals available.

These two ideas lead to the concept of electron configurations for atoms—a concept that ultimately will help us gain an understanding of chemical bonding. At this point, we have already seen that the quantum model and electron configuration can explain many facets of the structure of the periodic table. The familiar shape of the periodic table arises from the fact that the *s*-block elements in Groups 1 and 2 have electron configurations where the highest energy electron is in an *s* orbital. Similarly, Groups 13–18 comprise the *p*-block elements because the highest energy electron is in a *p* orbital. In addition to this overall structure of the periodic table, trends in atomic properties such as size or ionization energy can also be explained by considering the quantum mechanical model of the atom and the electron configurations that arise from it.

KEY TERMS

amplitude (6-2)
atomic absorption spectroscopy (AAS) (6-1)
atomic spectrum (6-3)
aufbau principle (6-5)
binding energy (6-2)
Bohr model (6-3)
coherent light (6-8)
core electrons (6-5)
diffraction (6-4)
effective nuclear charge (6-5)
electromagnetic spectrum (6-2)
electron affinity (6-7)
electron configuration (6-5)
excited state (6-3)
frequency (6-2)
ground state (6-3)
Hund's rule (6-5)
ionization energy (6-7)
laser (6-8)
light-emitting diode (LED) (6-8)
magnetic quantum number (m_ℓ) (6-4)
monochromatic (6-8)
node (6-4)
nondestructive testing (6-1)
operator (6-4)
orbital (6-4)
Pauli exclusion principle (6-5)
photoelectric effect (6-2)
photon (6-2)
principal quantum number (n) (6-4)
quantized (6-3)
quantum number (6-4)
refraction (6-2)
Schrödinger equation (6-4)
secondary quantum number (ℓ) (6-4)
shell (6-4)
shielding (6-5)
spin paired (6-5)
spin quantum number (m_s) (6-5)
subshell (6-4)
trace analysis (6-1)
uncertainty principle (6-4)
valence electrons (6-5)
visible light (6-2)
wave function (6-4)
wavelength (6-2)
wave–particle duality (6-2)
X-ray fluorescence (XRF) (6-1)

PROBLEMS AND EXERCISES

INSIGHT INTO Trace Analysis

6.1 Trace analysis may be carried out via nondestructive testing. How does this differ from other types of trace analysis? Give an example of a situation in which non-destructive analysis might be important.

6.2 Unlike XRF, AAS cannot be used for nondestructive testing. Explain why not.

6.3 In analysis by atomic absorption spectroscopy, the wavelengths of light being absorbed reveal the identity of the elements present in the sample. What information is used to determine the relative amounts of the different elements?

6.4 An X-ray fluorescence instrument must include a source of high-energy X-rays. Explain the role of these X-rays in the fluorescence process.

6.5 The fluorescence emitted in XRF is also in the X-ray range of the spectrum, but is always at lower energies than the X-rays used to initiate the process. Explain why this must be true.

6.6 As part of its analytical capabilities, the *Curiosity* Mars rover carries a novel instrument called ChemCam. Do a web search to learn what ChemCam does and how it works, and write a short paragraph comparing ChemCam to an AAS instrument like the one shown in Figure 6.1.

The Electromagnetic Spectrum

6.7 Explain why light is referred to as electromagnetic radiation.

6.8 Which of the waves depicted here has the highest frequency? Explain your answer.

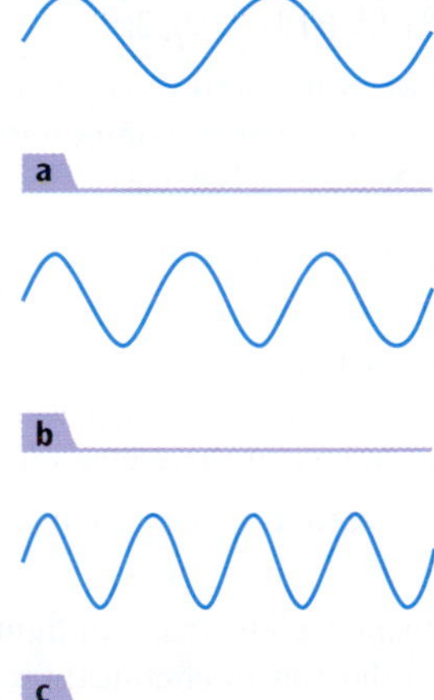

6.9 Arrange the following regions of the electromagnetic spectrum in order of increasing frequency: IR, UV, radio wave, visible.

6.10 Calculate the wavelengths, in meters, of radiation of the following frequencies. **(a)** 5.00×10^{15} s^{-1}, **(b)** 2.11×10^{14} s^{-1}, **(c)** 5.44×10^{12} s^{-1}

6.11 Decorative lights, such as those found on Christmas trees, often achieve their colors by painting the glass of the bulb. If a string of lights includes bulbs with wavelengths of 480, 530, 580, and 700 nm, what are the frequencies of the lights? Use Figure 6.6 to determine which colors are in the set.

6.12 Define the term *refraction*.

6.13 Define the term *photon*.

6.14 What is the relationship between the number of photons and the intensity of light?

6.15 Find the energy of a photon with each of the following frequencies. **(a)** 15.3 THz, **(b)** 1.7 EHz (see Table 1.2 if needed), **(c)** 6.22×10^{10} Hz

6.16 Place these types of radiation in order of increasing energy per photon. **(a)** green light from a mercury lamp, **(b)** X-rays from a dental X-ray, **(c)** microwaves in a microwave oven, **(d)** an FM music station broadcasting at 89.1 MHz

6.17 For photons with the following energies, calculate the wavelength and identify the region of the spectrum they are from. **(a)** 3.5×10^{-20} J, **(b)** 8.7×10^{-26} J, **(c)** 7.1×10^{-17} J, **(d)** 5.5×10^{-27} J

6.18 For photons with the following energies, calculate the wavelength and identify the region of the spectrum they are from. **(a)** 6.0×10^{-19} J, **(b)** 8.7×10^{-22} J, **(c)** 3.2×10^{-24} J, **(d)** 1.9×10^{-28} J

6.19 Various optical disk drives rely on lasers operating at different wavelengths, with shorter wavelengths allowing a higher density of data storage. For each of the following drive types, find the energy of a single photon at the specified wavelength. **(a)** CD, λ = 780 nm, **(b)** DVD, λ = 650 nm, **(c)** Blu-ray® disc, λ = 405 nm

6.20 The laser in most supermarket barcode scanners operates at a wavelength of 632.8 nm. What is the energy of a single photon emitted by such a laser? What is the energy of one mole of these photons?

6.21 Assume that a microwave oven operates at a frequency of 1.00×10^{11} s^{-1}. **(a)** What is the wavelength of this radiation in meters? **(b)** What is the energy in joules per photon? **(c)** What is the energy per mole of photons?

6.22 Fill in the blanks below to complete a description of the photoelectric effect experiment. (You should be able to do this with just one or two words in each blank.)

A beam of __________ strikes __________,

causing __________ to be emitted.

6.23 When light with a wavelength of 58.5 nm strikes the surface of tin metal, electrons are ejected with a maximum kinetic energy of 2.69×10^{-18} J. What is the binding energy of these electrons to the metal?

6.24 The electron binding energy for copper metal is 7.18×10^{-19} J. Find the longest wavelength of light that could eject electrons from copper in a photoelectric effect experiment.

Atomic Spectra

6.25 What is the difference between continuous and discrete spectra?

6.26 What was novel about Bohr's model of the atom?

6.27 Describe how the Bohr model of the atom accounts for the spectrum of the hydrogen atom.

6.28 According to the Bohr model of the atom, what happens when an atom absorbs energy?

6.29 Define the term *ground state*.

6.30 The figure below depicts the first four energy levels in a hydrogen atom. The three transitions shown as arrows emit ultraviolet light and occur at wavelengths of 121.566 nm, 102.583 nm, and 97.524 nm, respectively. Find the frequency of light that would be emitted in a transition from the state labeled as $n = 4$ to the state labeled as $n = 3$.

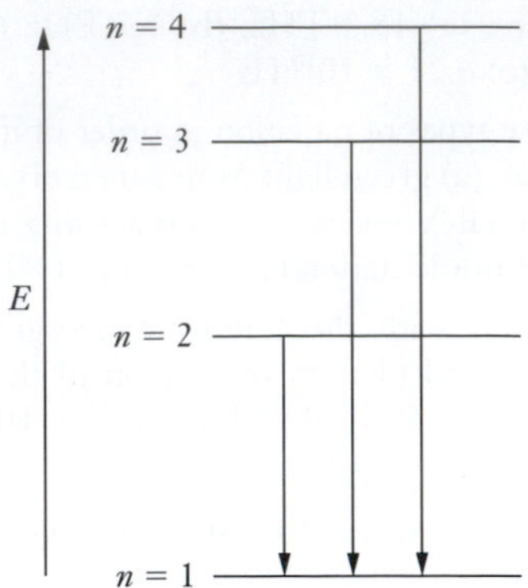

6.31 Refer to the energy-level diagram shown above in Problem 6.30, and find the wavelength of light that would be emitted when a hydrogen atom undergoes the transition from the state labeled as $n = 4$ to the state labeled as $n = 2$. Express your answer in nm.

6.32 A neon atom emits light at many wavelengths, two of which are at 616.4 and 638.3 nm. Both of these transitions are to the same final state. **(a)** What is the energy difference between the two states for each transition? **(b)** If a transition between the two higher energy states could be observed, what would be the frequency of the light?

6.33 A mercury atom emits light at many wavelengths, two of which are at 435.8 and 546.1 nm. Both of these transitions are to the same final state. **(a)** What is the energy difference between the two states for each transition? **(b)** If a transition between the two higher energy states could be observed, what would be the frequency of the light?

The Quantum Mechanical Model of the Atom

6.34 How did the observation of electron diffraction affect the development of the quantum mechanical model of the atom?

6.35 Why do we use a wave function to describe electrons?

6.36 What are the mathematical origins of quantum numbers?

6.37 What are the allowed values for the principal quantum number? For the secondary quantum number?

6.38 Which of the following represent valid sets of quantum numbers? For a set that is invalid, explain briefly why it is not correct. **(a)** $n = 3, \ell = 3, m_\ell = 0$, **(b)** $n = 2, \ell = 1, m_\ell = 0$, **(c)** $n = 6, \ell = 5, m_\ell = -1$, **(d)** $n = 4, \ell = 3, m_\ell = -4$

6.39 A particular orbital has $n = 4$ and $\ell = 2$. What must this orbital be? **(a)** $3p$, **(b)** $4p$, **(c)** $5d$, or **(d)** $4d$

6.40 Why are there no $2d$ orbitals?

6.41 What is the maximum number of electrons in an atom that can have the following quantum numbers? **(a)** $n = 2$; **(b)** $n = 3$ and $\ell = 1$; **(c)** $n = 3, \ell = 1$, and $m_\ell = 0$; **(d)** $n = 3$, $\ell = 1, m_\ell = -1$, and $m_s = -1/2$.

6.42 How many orbitals correspond to each of the following designations? **(a)** $3p$, **(b)** $4p$, **(c)** $4p_x$, **(d)** $6d$, **(e)** $5d$, **(f)** $5f$, **(g)** $n = 5$, **(h)** $7s$

6.43 The p and d orbitals are sometimes referred to as directional. Contrast them with s orbitals to offer an explanation for this terminology.

6.44 Define the term *nodal plane* (or *node*).

6.45 Referring to Figure 6.15, draw a $4p$ orbital, showing all of its nodes.

The Pauli Exclusion Principle and Electron Configurations

6.46 Does the fact that the fourth quantum number is called the spin quantum number prove that electrons are spinning? What observations gave rise to this name?

6.47 Define the term *spin paired.*

6.48 On what does the Pauli exclusion principle place a limit?

6.49 Define the term *shielding*.

6.50 Why does the size of an orbital have an effect on its energy?

6.51 What is effective nuclear charge and how does it influence the size of an orbital?

6.52 Depict two ways to place electrons in the $2p$ orbitals for a nitrogen atom. Which depiction is correct according to Hund's rule?

6.53 How does the charge of electrons provide some rationale for Hund's rule?

6.54 Write the ground state electron configuration for **(a)** B, **(b)** Ba, **(c)** Be, **(d)** Bi, **(e)** Br.

6.55 Which of these electron configurations are for atoms in the ground state? In excited states? Which are impossible? **(a)** $1s^2 2s^1$, **(b)** $1s^2 2s^2 2p^3$, **(c)** [Ne] $3s^2 3p^3 4s^1$, **(d)** [Ne] $3s^2 3p^6 4s^3 3d^2$, **(e)** [Ne] $3s^2 3p^6 4f^4$, **(f)** $1s^2 2s^2 2p^4 3s^2$

6.56 From the list of atoms and ions given, identify any pairs that have the same electron configurations and write that configuration: Na^+, S^{2-}, Ne, Ca^{2+}, Fe^{2+}, Kr, I^-.

The Periodic Table and Electron Configurations

6.57 Halogen lamps rely on the predictable chemistry of the halogen elements. Compare the electron configurations of three halogens and use this comparison to justify the similarity of the chemistry of these elements.

6.58 Distinguish between the terms *core electrons* and *valence electrons.*

6.59 Describe how valence electron configurations account for some of the similarities in chemical properties among elements in a group.

6.60 Why is there no element to the immediate right of magnesium in the periodic table?

6.61 If another 50 elements were discovered, how would you expect the appearance of the periodic table to change?

6.62 Which blocks of the periodic table comprise the main group elements?

6.63 Explain why the s block of the periodic table contains two columns, the p block contains six columns, and the d block contains ten columns.

6.64 Look at the table of electron configurations in Appendix C. Which elements have configurations that are exceptions to the aufbau principle? Propose a reason why these elements have these exceptions.

Periodic Trends of Atomic Properties

6.65 How does the effective nuclear charge help explain the trends in atomic size across a period of the periodic table?

6.66 Use the electronic configuration of the alkali metals and your knowledge of orbitals and quantum numbers to explain the trend in atomic size of alkali metal atoms.

6.67 Using only a periodic table as a guide, arrange each of the following series of atoms in order of increasing size. **(a)** Na, Be, Li, **(b)** P, N, F, **(c)** I, O, Sn

6.68 Define the term *ionization energy*. What is the difference between the first ionization energy and the second ionization energy?

6.69 At which ionization for chlorine would you expect the first large jump in ionization energy? Would this be the only large jump in energy if you continued to ionize the chlorine?

6.70 Arrange the following atoms in order of increasing ionization energy: Li, K, C, and N.

6.71 How do we explain the fact that the ionization energy of oxygen is less than that of nitrogen?

6.72 Which element would you expect to have the largest second ionization energy, Na, C, or F? Why?

6.73 Answer each of the following questions. **(a)** Of the elements S, Se, and Cl, which has the largest atomic radius? **(b)** Which has the larger radius, Br or Br^-? **(c)** Which should have the largest difference between the first and second ionization energy: Si, Na, P, or Mg? **(d)** Which has the largest ionization energy: N, P, or As? **(e)** Which of the following has the largest radius: O^{2-}, N^{3-}, or F^-?

6.74 Indicate which species in each pair has the more favorable (more negative) electron affinity. Explain your answers. **(a)** Cl or S, **(b)** S or P, **(c)** Br or As

6.75 Compare the elements Na, B, Al, and C with regard to the following properties. **(a)** Which has the largest atomic radius? **(b)** Which has the most negative electron affinity? **(c)** Place the elements in order of increasing ionization energy.

6.76 Rank the following in order of decreasing ionization energy: Cl, F, Ne^+, S, S^-.

INSIGHT INTO Modern Light Sources: LEDs and Lasers

6.77 Several excited states of the neon atom are important in the operation of a helium-neon laser. In these excited states, one electron of the neon atom is promoted from the $2p$ level to a higher energy orbital.

An excited neon atom with a $1s^22s^22p^55s^1$ electron configuration can emit a photon with a wavelength of 3391 nm as it makes a transition to a lower energy state with a $1s^22s^22p^54p^1$ electron configuration. Other transitions are also possible. If an excited neon atom with a $1s^22s^22p^55s^1$electron configuration makes a transition to a lower energy state with a $1s^22s^22p^53p^1$ electron configuration, it emits a photon with a wavelength of 632.8 nm. Find the wavelength of the photon that would be emitted in a transition from the $1s^22s^22p^54p^1$ electron configuration to the $1s^22s^22p^53p^1$ electron configuration. (It should help if you start by drawing an energy-level diagram.)

6.78 LED bulbs offer a fairly new lighting alternative, and are likely to have a large impact on home lighting. These bulbs combine the energy efficiency of traditional fluorescent lights with the smaller size, "warmer" light, and dimming ability of incandescent lights. Although LED bulbs are expensive, manufacturers claim that subsequent savings will more than offset the initial costs. Analyze the relative cost of incandescent versus LED lighting, assuming that the information in the table below is accurate. Are industry claims of cost savings justified?

	LED	Incandescent
Initial cost	\$15.00	\$0.75
Electricity usage	8 watts	40 watts
Electricity cost	\$0.04/kwh	\$0.04/kwh
Lifetime	25,000 hours	2,500 hours

6.79 How much energy could be saved each year by replacing incandescent bulbs with LED bulbs? Assume that six 40-watt lamps are lit an average of three hours a night and they are each replaced with 8-watt LED bulbs. (See Problem 6.78.)

Additional Problems

6.80 The photoelectric effect can be used to measure the value of Planck's constant. Suppose that a photoelectric effect experiment was carried out using light with $\nu = 7.50 \times 10^{14}\ s^{-1}$ and ejected electrons were detected with a kinetic energy of 2.50×10^{-11} J. The experiment was then repeated using light with $\nu = 1.00 \times 10^{15}\ s^{-1}$ and the *same metal target*, and electrons were ejected with kinetic energy of 5.00×10^{-11} J. Use these data to find a value for Planck's constant. HINTS: These data are fictional and will give a result that is quite different from the real value of Planck's constant. Be sure that you do *not* use the real value of Planck's constant in any calculations here. It may help to start by thinking about how you would calculate the metal's binding energy if you already knew Planck's constant.

6.81 A mercury atom is initially in its lowest possible (or ground state) energy level. The atom absorbs a photon with a wavelength of 185 nm and then emits a photon with a frequency of 4.924×10^{14} Hz. At the end of this series of transitions, the atom will still be in an energy level above the ground state. Draw an energy-level diagram for this process and find the energy of this resulting excited state, assuming that we assign a value of $E = 0$ to the ground state. (This choice of $E = 0$ is *not* the usual convention, but it will simplify the calculations you need to do here.)

6.82 When a photoelectric effect experiment was carried out using a metal "M" and light at wavelength λ_1, electrons with a kinetic energy of 1.6×10^{-19} J were emitted. The wavelength was reduced to one-half its original value and the experiment was repeated (still using the same metal target). This time electrons with a kinetic energy of 6.4×10^{-19} J were emitted. Find the electron binding energy for metal M. (The actual wavelengths used were not recorded, but it is still possible to find the binding energy.)

6.83 A metallic sample is known to be barium, cesium, lithium, or silver. The electron binding energies for these metals are listed in the following table.

Metal	Electron Binding Energy (J)
Barium	4.30×10^{-19}
Cesium	3.11×10^{-19}
Lithium	3.94×10^{-19}
Silver	7.59×10^{-19}

One way to identify the element might be through a photoelectric effect experiment. The experiment was performed three times, each time using a different laser as the light source. The results are summarized below. (The kinetic energy of the ejected photoelectrons was not measured.)

Laser Wavelength	Photoelectrons Seen?
532 nm	No
488 nm	Yes
308 nm	Yes

Based on this information, what conclusions can be drawn as to the identity of the metal?

6.84 When a helium atom absorbs light at 58.44 nm, an electron is promoted from the 1*s* orbital to a 2*p* orbital. Given that the ionization energy of (ground state) helium is 2372 kJ/mol, find the longest wavelength of light that could eject an electron from the excited state helium atom.

6.85 Arrange the members of each of the following sets of cations in order of increasing ionic radii. **(a)** K^+, Ca^{2+}, Ga^{3+}, **(b)** Ca^{2+}, Be^{2+}, Ba^{2+}, Mg^{2+}, **(c)** Al^{3+}, Sr^{2+}, Rb^+, K^+, **(d)** K^+, Ca^{2+}, Rb^+

6.86 Arrange the following sets of anions in order of increasing ionic radii. **(a)** Cl^-, S^{2-}, P^{3-}, **(b)** O^{2-}, S^{2-}, Se^{2-}, **(c)** N^{3-}, S^{2-}, Br^-, P^{3-}, **(d)** Cl^-, Br^-, I^-

FOCUS ON PROBLEM SOLVING EXERCISES

6.87 The photoelectric effect can be used in engineering designs for practical applications. For example, infrared goggles used in night-vision applications have materials that give an electrical signal with exposure to the relatively long wavelength IR light. If the energy needed for signal generation is 3.5×10^{-20} J, what is the minimum wavelength and frequency of light that can be detected?

6.88 Some spacecraft use ion propulsion engines. These engines create thrust by ionizing atoms and then accelerating and expelling them. According to Newton's laws, this leads to thrust in the opposite direction. Minimizing mass is always important for space applications. Discuss what periodic trends you would need to consider for choosing a material to ionize for engineering designs with this application.

6.89 Laser welding is a technique in which a tightly focused laser beam is used to deposit enough energy to weld metal parts together. Because the entire process can be automated, it is commonly used in many large-scale industries, including the manufacture of automobiles. In order to achieve the desired weld quality, the steel parts being joined must absorb energy at a rate of about 10^4 W/mm^2. (Recall that 1 W = 1 J/s.) A particular laser welding system employs a Nd:YAG laser operating at a wavelength of 1.06 μm; at this wavelength steel will absorb about 80% of the incident photons. If the laser beam is focused to illuminate a circular spot with a diameter of 0.02 inch, what is the minimum power (in watts) that the laser must emit per second to reach the 10^4 W/mm^2 threshold? How many photons per second does this correspond to? (For simplicity, assume that the energy from the laser does not penetrate into the metal to any significant depth.)

6.90 Ionization gauges for pressure measurement (see Section 5-7) can be manipulated by changing the voltage of the electron collection grid (see Figure 5.11). Referring to Table 6.5, if the voltage is set to yield electron energies of 1200 kJ/mol, determine what elements the gauge would *not* be able to detect. Given the ability to vary this energy, describe how you would design an apparatus that could distinguish between sodium and magnesium atoms.

6.91 Atomic absorption spectroscopy is based on the atomic spectra of the elements being studied. It can be used to determine the impurities in a metal sample. If an element is present, light at the appropriate wavelength is absorbed. You are working with a metal stamping company and the rolled steel you use to form panels for automobile doors is failing at an alarming rate. There is some chance that the problem is unacceptably high levels of manganese in the steel. Given that the atomic spectrum of manganese has three lines near 403 nm, how could you use a spectrometer to determine the amount of manganese in the steel?

6.92 The red color in fireworks is the result of having strontium-containing salts in the fireworks bomb. Similarly, the green/blue colors sometimes seen in fireworks arise from copper salts. Based on your understanding of atomic spectra and the colors in fireworks, describe which atom, copper or strontium, has more widely separated energy levels.

Cumulative Problems

6.93 When we say that the existence of atomic spectra tells us that atoms have specific energy levels, are we using deductive or inductive reasoning?

6.94 When Bohr devised his model for the atom, was he using deductive or inductive reasoning? Explain your answer.

6.95 The photochemical reaction that initiates the production of smog involves the decomposition of NO molecules, and the energy needed to break the N—O bond is 1.04×10^{-18} J. **(a)** Which wavelength of light is needed? **(b)** How many photons are needed to decompose 0.32 mg of NO?

Chemical Bonding and Molecular Structure

7

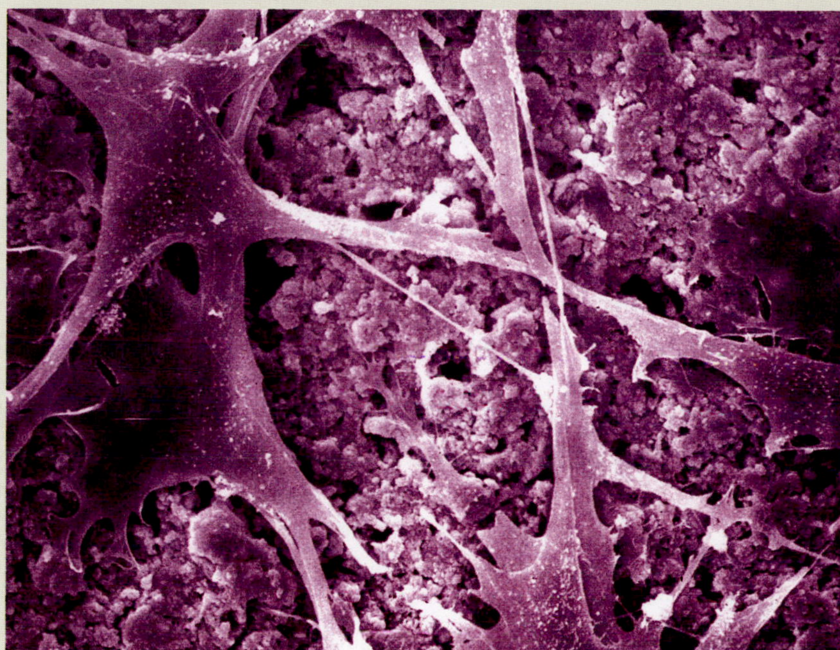

Biomaterials are designed for use in living systems, so the interactions between such materials and living cells are vital. In this microscope image, the biomaterial is a form of cement used to help repair bones. The smooth structures in the image consist of bone cells that have adhered to the more jagged-looking cement. Here this adhesion is the desired effect. But in other cases, strong interactions between cells and biomaterials can have serious negative consequences.
H. Xu/American Dental Association Foundation/National Institute of Standards and Technology

CHAPTER OUTLINE

The study of atomic structure in Chapter 6 provided an interesting detour into modern physics, and there's no doubt that the paradoxes of quantum mechanics are intellectually intriguing. But quite apart from pure scientific curiosity, the concepts we developed for atomic structure offer important insights into some of the most fundamental questions in chemistry. How and why do atoms combine to form molecules? What do those molecules actually look like, and which factors determine their shapes? In our technological society, the answers to these seemingly abstract questions can have tremendous practical and economic importance. Knowing how to control or selectively exploit the formation of chemical bonds has played a crucial role in the development of many new materials and devices.

Many facets of new materials could provide examples of chemical bonding behavior, but here we will highlight the development of materials for use in biomedical applications. Because these materials are designed for use within the human body, they must conform to important constraints in order to work properly. Virtually all of these constraints are associated with the chemical nature of the materials. In this chapter, we will look at the reasons atoms form bonds with one another, the different types of chemical bonds, and the connections between bonding and molecular shape. An examination of these characteristics can help us understand the properties of many substances and is crucial to understanding how molecules interact and react with one another. We'll also improve our connection between the microscopic and symbolic views of chemistry as we become able to visualize a molecule's three-dimensional shape based on its formula.

Chapter Objectives

After mastering this chapter, you should be able to

- list some factors influencing the biocompatibility of materials and explain how those factors are related to chemical bonding.
- use electron configurations to explain why metals tend to form cations whereas nonmetals tend to form anions.
- describe the energy changes in the formation of an ionic bond.
- define electronegativity and state how electronegativity varies with position in the periodic table.
- identify or predict polar, nonpolar, and ionic bonds by comparing electronegativities.
- write Lewis electron structures for molecules or ions.
- describe chemical bonding using a model based on the overlap of atomic orbitals and recognize some of the limitations of this simple model.
- explain how hybridization reconciles observed molecular shapes with the orbital overlap model.
- predict the geometry of a molecule from its Lewis structure.
- use models (real or software) to help visualize common molecular shapes.
- explain the formation of multiple bonds in terms of the overlap of a combination of hybridized and unhybridized atomic orbitals.
- identify sigma and pi bonds in a molecule and explain the difference between them.

 ***INSIGHT INTO***

7-1 Materials for Biomedical Engineering

From replacement hip joints to heart valves to new lenses for eyes, the collaboration of chemistry and engineering annually provides medical miracles for millions of patients who need to replace worn-out or damaged tissues. Materials used in the human body must be capable of many things, including having physical properties that are at least similar to the biological materials, such as bone, that they are replacing. Just as important, however, is **biocompatibility**, which is the ability of materials to interact with the natural biological materials without triggering a response from the immune system.

Understanding biocompatibility begins with understanding how molecules and materials are put together. For biomedical applications, systems must be engineered to provide the required physical attributes, such as strength and durability, while also allowing the material to be enmeshed with bone or muscle tissues. There are several strategies for achieving biocompatibility. In some cases, the key need is strong cement.

The first bone cements used were based on poly(methyl methacrylate) (PMMA). The polymerization reaction, which goes to completion after the cement has been applied, causes the cement to harden. More recently, cements based on calcium phosphate or a composite of calcium phosphate and calcium silicate have been developed. In what ways are these materials similar, and how do they differ in terms of their chemistry? A better understanding of chemical bonding can help us to address these questions.

Methyl methacrylate was one of the examples we used to introduce line structures in Chapter 2.

In some cases, materials that could provide a desired physical property may not offer an acceptable level of biocompatibility. The modern engineer, however, need not abandon such materials, because it is possible to establish biocompatible coatings. A number of possible methods start with smaller molecules referred to as precursors and react them with each other or with the surface of the object to be coated. Choosing appropriate precursors for coating reactions requires knowledge of chemical bonding not only in these starting materials but also in the resultant coating.

The importance of surfaces points to the fact that biomedical engineers must also understand how biomaterials interact with cells in the body. Cell adhesion, like that shown in the chapter opening photo, depends critically on the nature of electrical charges on the surfaces, a property called **polarity**. Polarity is important not only on the surface of extended systems but also in small molecules in the body, particularly water. We can understand the fundamentals of polarity by looking at the nature of chemical bonding in these systems.

The design and implementation of biomaterials clearly involves many complicated factors. Yet the basics of chemical bonding explored in this chapter will allow us to explore many of the questions and challenges posed in this emerging field of engineering.

7-2 The Ionic Bond

All chemical bonds arise from the interaction of oppositely charged particles. The simplest example of such an interaction is found in an ionic bond, formed through the electrostatic attraction of two oppositely charged ions. Many bioengineering applications include ionic substances such as the calcium phosphate or calcium silicate cements that we've already mentioned. Ionic compounds can also act as precursors for the growth of ceramic coatings. To understand these systems, we need to examine the forces that dictate which ions will form and how they interact with one another.

Binary ionic compounds always form between a metal and a nonmetal, and the more different two elements are in their metallic properties, the more likely it is that a compound formed from these elements will be ionic. So ionic compounds usually consist of elements that are widely separated in the periodic table: a metal from the left-hand side and a nonmetal from the right. As we attempt to gain a better understanding of ionic compounds, we'll want to consider a number of fundamental questions.

- In Section 2-5 (page 42), we noted that metallic elements form cations, whereas nonmetals form anions. Why should this be?
- How does the energy of attraction between ions compensate for the energy costs of ion formation?
- How does the ionic radius affect the strength of the bond between an anion and a cation?

Formation of Cations

We've already learned that there are energy changes in the formation of ions, as represented by ionization energies and electron affinities. We also know how those energy changes vary as we move through the periodic table. Those trends correlate very well with the location of metals and nonmetals in the table, and this correlation explains the observations that metals form cations and nonmetals form anions. In each case, the ions we find in compounds are those whose formation is not too costly in energy.

Recall that ionization energy is the amount of energy required to remove an electron and form a cation. Electron affinity is the amount of energy required to add an electron and form an anion. See Section 6-7, pages 168–173.

By looking beyond the first ionization energy, we can also see why some metals form 1+ cations, whereas others have higher charges. As pointed out in our discussion of Table 6.5 (page 172), there is always a large jump in ionization energy whenever removing an additional electron disrupts an electron configuration ending in np^6. This large jump in ionization energy explains whether the cation formed will have a charge of 1+, 2+, etc. For instance, sodium occurs naturally only as singly charged Na^+. When we look at the progression in ionization energies for sodium, this is not surprising. The first, second, and third ionization energies for sodium are 496, 4562, and 6912 kJ/mol, respectively. When we see the huge amount of energy required to take a second or third electron from a sodium atom, we can see why nature might choose to stop at only one! Example Problem 7.1 explores the relationship between metal cations and ionization energies.

EXAMPLE PROBLEM 7.1

Using data from Table 6.5 (page 172), predict the ions that magnesium and aluminum are most likely to form.

Strategy Scan the successive ionization energies for each element to find a point where removing one additional electron causes a dramatic increase in the value. It is not likely that sufficient energy would be available to compensate for such a large increase in ionization energy, so the ion formed will be dictated by the number of electrons lost before that jump in ionization energy.

Solution For magnesium, the first significant jump occurs between the second ionization energy (1451 kJ/mol) and the third (7733 kJ/mol). We would expect two electrons to be removed, and the most commonly found ion should be Mg^{2+}. For aluminum, the third ionization energy is 2745 kJ/mol, whereas the fourth is 11,578 kJ/mol. We would expect three ionizations to occur, so aluminum should form Al^{3+}.

Discussion This type of question is designed to help you see the reasoning behind some of the facts you may already know. In many cases, the way we learn chemistry is by learning information as facts first and then finding out the reason those facts are evident. In the examples from this problem, the answers are consistent with the idea that the ion formed is dictated by the number of electrons lost before the jump in ionization energy. They are also consistent with facts you may already have known, for example, that magnesium will form a 2+ ion because it is an alkaline earth element in Group 2 of the periodic table.

Check Your Understanding The calcium in calcium phosphate exists as Ca^{2+} ions. Based on this fact, what would be your qualitative prediction for the values of the successive ionization energies for calcium?

The idea of s, p, d, and f blocks in the periodic table was introduced in Figure 6.19, page 167.

If a shell is half-filled, no electrons are paired in any orbitals. Pairing of electrons is allowed (with opposite spins), but it still increases electron–electron repulsions.

As our examples show, the most common cation for metals in the *s* or *p* blocks of the periodic table is generally one with an np^6 configuration. But this idea does not carry over to transition metals. As we noted in Chapter 2, many transition metals form more than one stable cation. Transition metals have partially filled *d* subshells, but ionization usually begins with loss of electrons from the *s* subshell. When iron forms Fe^{2+}, for example, the electron configuration changes from $1s^22s^22p^63s^23p^64s^23d^6$ in the neutral atom to $1s^22s^22p^63s^23p^63d^6$ in the ion. The loss of one additional electron from a *d* orbital leaves Fe^{3+} with a half-filled *d* subshell. This half-filled subshell is a fairly stable arrangement, so the Fe^{2+} and Fe^{3+} ions are both stable. With many transition metals, the situation can become quite complicated, and we will not explore this in greater detail.

Formation of Anions

If we look at the periodic trends for electron affinity (Figure 6.22, page 173) we see that they are similar to those for ionization energies. Recall that electron affinity is the energy change for the formation of an anion from a neutral atom in the gas phase:

$$X(g) + e^- \rightarrow X^-(g)$$

This process is often energetically favored, which means that energy is released when the anion is formed. The amount of energy released when atoms acquire additional electrons tends to increase as we go from left to right across the periodic table. Looking at this trend, those atoms that gain electrons most readily also would need to lose large numbers of electrons to produce cations with stable np^6 electron configurations. Thus nonmetals tend to form anions rather than cations.

For the formation of anions from nonmetals, we do not find any of the complexities we encountered with the transition metals, because all of the nonmetals belong to the main groups and do not have partially filled *d* subshells. They behave like the main group metals, in that they tend to acquire electron configurations with filled *p* subshells. The primary difference is that they do so by gaining, rather than losing, electrons. Thus, all the halogens form singly charged (1−) anions, resulting in np^6 electron configurations. Similarly, oxygen and sulfur form anions with a 2− charge, resulting in the same np^6 configuration.

In addition to charge, the size of an ion ultimately has an effect on the strength of an ionic bond, largely due to geometric considerations. In Section 6-7, we noted that the atomic radius increases going down a group and decreases going across a period. What trend should we expect for the radii of cations and anions? The loss of an electron to form a cation decreases electron–electron repulsion, leaving the remaining electrons more tightly bound to the nucleus. So we would expect positive ions to be smaller than their respective parent atoms. On the other hand, adding an electron to form an anion will lead to increased repulsion among the valence electrons. So anions should be larger than their corresponding neutral atoms. Ionic radii can be determined very accurately by measuring the spacing in crystal lattices, and the data are consistent with these expectations. With an ionic radius of 102 pm, Na^+ is significantly smaller than the sodium atom, which has an atomic radius of 190 pm. ◀ⓘ And the 133-pm ionic radius of F^- is significantly larger than the 42-pm atomic radius of neutral fluorine.

1 picometer (pm) = 10^{-12} m ⓘ

Figure 7.1 shows atomic and ionic radii, and the observed trends are clear. When we compare similarly charged ions going down a group, ionic size increases in the

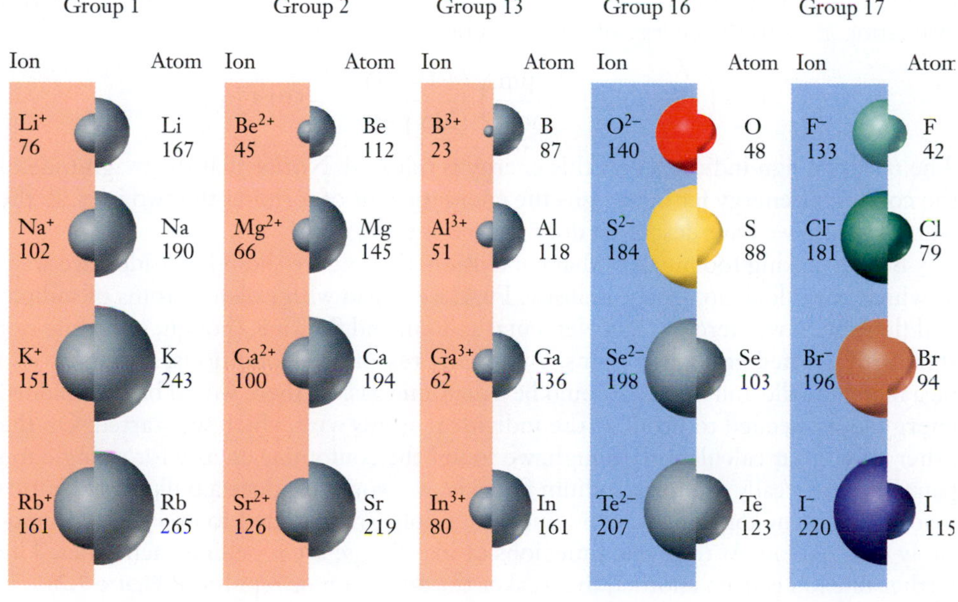

Figure 7.1 ■ The nested spheres in the figure illustrate how the sizes of the indicated ions compare to the corresponding neutral atoms. Metals and their cations are shown on the left and nonmetals and their anions on the right. All radii are given in picometers. (Radii from E. Clementi, D.L. Raimondi, and W.P. Reinhardt, J. Chem. Phys. 1963, 38, 2686, and R.D. Shannon, Acta Cryst., 1976, A32, 751.)

same way that we saw for atomic size. For cations, the trend in ionic size going across a period also mimics the trend we saw for neutral atoms—decreasing as we go from left to right. This is easily understood. Considering Na^+ and Mg^{2+}, for example, we see that both ions have the same electron configuration. But Mg^{2+} has one additional proton in its nucleus, and so it will hold its electrons more tightly, leading to a smaller ion. We can apply the same sort of reasoning to the common anions in the *p* block. The oxide ion, O^{2-}, and the fluoride ion, F^-, have the same electron configuration. But fluoride has one additional proton in its nucleus, and so it is the smaller ion. Similarly, Cl^- is slightly smaller than S^{2-}.

The formation of isolated ions, of course, does not constitute an ionic bond. Indeed, starting with a neutral metal atom and a neutral nonmetal atom, it will usually *cost* energy to form an ion pair (i.e., a cation and an anion). Ionization energies are always positive, and the energy input to produce a cation from a metal is not offset by the energy released when an anion is formed from a nonmetal. But once formed, the oppositely charged ions will attract one another, leading to a significant lowering of their overall energy. As we saw in Section 2-3, Coulomb's law says that the *force* between two charges is proportional to the product of the magnitude of their charges divided by the square of the distance between them:

$$F \propto \frac{q_1 q_2}{r^2} \tag{7.1}$$

Here F is the force, q_1 and q_2 are the charges, and r is the distance between them (in this case, the distance between the nuclei of the two ions). It can also be shown that the *potential energy*, V, of the system is given by

$$V = k\frac{q_1 q_2}{r} \tag{7.2}$$

Here k is a proportionality constant, whose value is 1.389×10^5 kJ pm/mol if the charges are expressed in units of electron charge.

Let's look at a specific case by evaluating the energy change in the formation of a bound NaF ion pair from isolated atoms of sodium and fluorine. The ionization energy of sodium is 496 kJ/mol, and the electron affinity of fluorine is −328 kJ/mol. So we need to supply 496 kJ/mol to ionize the sodium, but 328 kJ/mol will be released in the formation of the fluoride ion. Combining these two numbers, we see that forming the individual ions will cost 168 kJ/mol (496 − 328 = 168). Next, we need to use Equation 7.2 to determine the energy released by the coulombic attraction between the anion and cation. The radii of Na^+ and F^- are 102 pm and 133 pm, respectively, so the internuclear distance should be 235 pm. Inserting this into Equation 7.2 as r and evaluating gives us the energy of attraction between the ions:

$$V = \left(1.389 \times 10^5\ \frac{\text{kJ pm}}{\text{mol}}\right)\frac{(+1)(-1)}{(235\ \text{pm})} = -591\ \text{kJ/mol}$$

The negative sign indicates that this energy is released. Notice that the magnitude of the coulombic energy far outweighs the energetic cost of forming the two ions, so the whole process we have described does release energy.

Before making too much of that conclusion, though, we should consider two ways in which our calculation is not realistic. First, we began with isolated atoms of sodium and fluorine. If we were to react elemental sodium and fluorine, though, we would not start with isolated gas phase atoms. Under any reasonable conditions, sodium would be a metallic solid and fluorine would be a diatomic gas. So there would be some added energy input needed to produce the individual atoms with which we started. On the other end of our calculation, though, we found the coulombic energy for a single ion pair. But if we really did form sodium fluoride, we would produce a bulk crystal lattice made up of many ions. In such an ionic lattice, ions are attracted to every other oppositely charged ion. At the same time, ions of like charge are repelling each other. The farther one ion is from another, the weaker the attraction or repulsion (**Figure 7.2**).

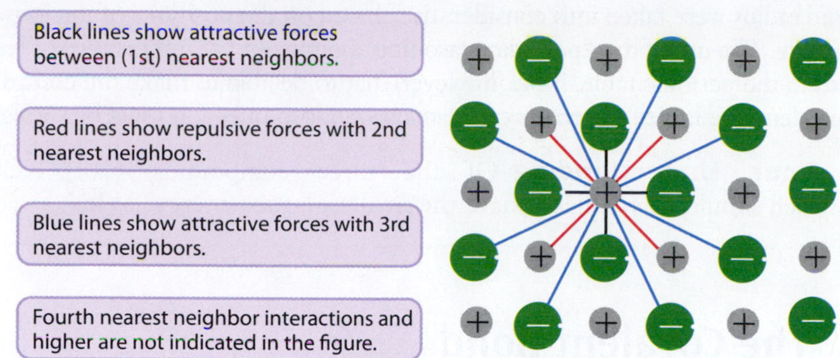

Black lines show attractive forces between (1st) nearest neighbors.
Red lines show repulsive forces with 2nd nearest neighbors.
Blue lines show attractive forces with 3rd nearest neighbors.
Fourth nearest neighbor interactions and higher are not indicated in the figure.

Figure 7.2 ■ In a solid lattice, any given ion has a large number of interactions with other ions, some attractive and some repulsive. This two-dimensional representation of a lattice shows the interactions of the nearest neighbors, second nearest neighbors, and third nearest neighbors.

Ultimately, the **lattice energy** of a crystal contains contributions from all of these attractions and repulsions. In the NaF structure, each sodium ion is surrounded by six fluoride ions, and each fluoride has six neighboring sodium atoms. To get an accurate indication of the coulombic energy of the lattice, we would need to account for the interaction of each ion with its many neighbors. So, although starting with isolated atoms leads to an overestimation of the energy released in forming the ionic bond, using a single ion pair instead of an extended lattice greatly underestimates the energy. The actual energy released in the reaction between sodium metal and diatomic fluorine gas to produce solid sodium fluoride turns out to be 577 kJ/mol.

The overall result of all attractive and repulsive forces in an ionic crystal is estimated by calculating a Madelung constant, a mathematical series that accounts for all of the interactions in a particular crystal geometry.

From the form of Equation 7.2, we can see that small, highly charged ions will tend to form ionic compounds with large lattice energies, and large ions with small charges will tend to form compounds with small lattice energies. Example Problem 7.2 explores this connection.

EXAMPLE PROBLEM 7.2

In each of the following pairs of ionic solids, the ions are arranged into similar lattices. Which one has the largest lattice energy? **(a)** CaO or KF **(b)** NaF or CsI

Strategy Questions like this require us to find some basis for comparison. In this particular case, the factors associated with lattice energy have to do with the ions that make up the compound. We have two factors to consider. Smaller ions tend to have stronger coulombic interactions because the distance between the nuclei is smaller, and ions with larger charges have stronger interactions. An examination of these two factors may allow us to predict which substance has the larger lattice energy.

Solution

(a) In CaO, calcium exists as Ca^{2+} and oxygen as the oxide ion, O^{2-}. In KF, the constituent ions are K^+ and F^-. Because both are from elements in the same row of the periodic table, the sizes of the two cations should not be too different. The same is true for the anions, so to a first approximation, we might discount the size factor. CaO should have the larger lattice energy because the charges on its ions are larger than the charges on the ions in KF. (Note that if we wanted to look more closely at the sizes, we would expect that K^+ would be slightly larger than Ca^{2+} and O^{2-} would be slightly larger than F^-. Because the larger cation is in KF and the larger anion is in CaO, our assumption that size would not be very important seems good.)

(b) All of the ions in this problem are singly charged, so the size difference must be the key issue. Na^+ and F^- are smaller than Cs^+ and I^-, so the lattice spacing in NaF must be smaller. Therefore, NaF should have the larger lattice energy.

Discussion To get started here, we needed to determine the identities of the ions involved. One way to do this is to consider the jump in ionization energy we explored in Example Problem 7.1. However, with just a little practice, it is easy to determine the likely charges for most main group elements based on their location in the periodic table. Ionic charge and radius were taken into consideration, based on the positions of the ions in the periodic table. The use of this "periodic reasoning" points out the importance of learning the trends in the periodic table. Note, however, that to be able to make the comparisons in this problem, the lattice structures of the compounds in question must be similar.

Check Your Understanding Of the three compounds—CaF_2, KCl, and RbBr—which should we predict to have the smallest lattice energy? Why?

7-3 The Covalent Bond

Consideration of ionization energy, electron affinity, and coulombic potential is fine for ionic compounds. But what about covalent bonds like those found in polymers? We noted back in Section 2-4 that **covalent bonds** are based on the sharing of pairs of electrons between two atoms. How does covalent bonding occur, and how does it influence the behavior of compounds like the poly(methyl methacrylate) used in bone cements?

Chemical Bonds and Energy

When chemical bonds are formed between nonmetal elements, valence electrons are shared rather than transferred. But just as for ions, we must expect that the driving force behind bond formation should be a lowering of the overall energy. The energetic advantage of forming a covalent bond can be illustrated by considering how the potential energy would vary in a thought experiment in which two atoms approach each other. If the atoms are sufficiently far apart, they won't exert any discernible force on one another. (You might say that neither one knows the other is there.) But as the atoms begin to come closer together, eventually the electrons from one atom can become attracted to the positively charged nucleus of the other atom. This added attraction lowers the potential energy of the system. Of course, the two nuclei will repel each other, as do the bonding electrons, so the nuclei cannot approach too close to one another. At some point, a balance between these attractions and repulsions will occur, and the energy of the system is minimized (**Figure 7.3**). This point of minimum energy

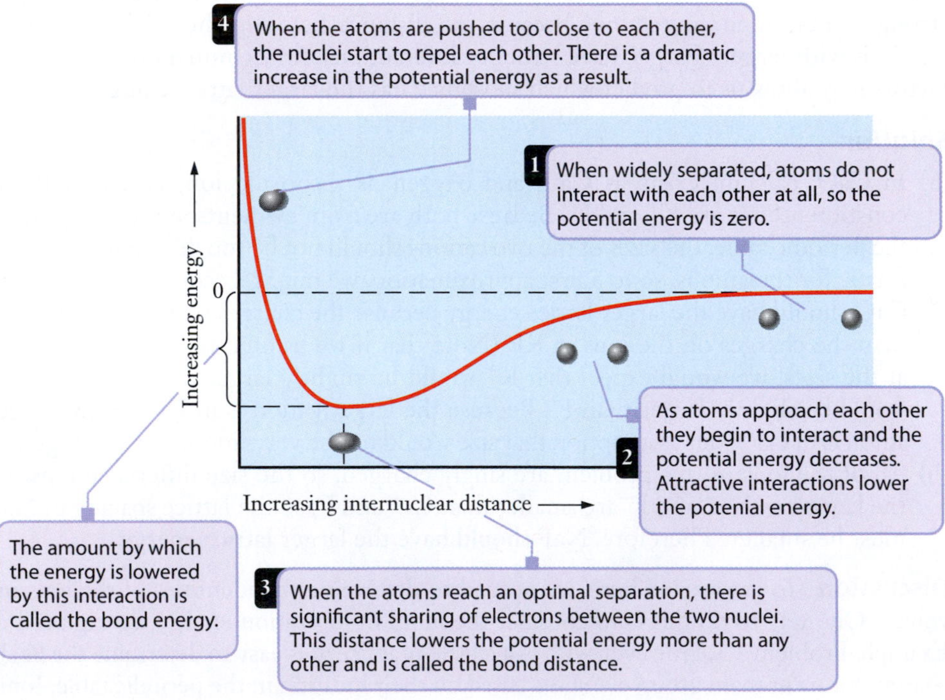

4 When the atoms are pushed too close to each other, the nuclei start to repel each other. There is a dramatic increase in the potential energy as a result.
1 When widely separated, atoms do not interact with each other at all, so the potential energy is zero.
Increasing energy
0
As atoms approach each other they begin to interact and the potential energy decreases. Attractive interactions lower the potenial energy.
2
Increasing internuclear distance
The amount by which the energy is lowered by this interaction is called the bond energy.
3 When the atoms reach an optimal separation, there is significant sharing of electrons between the two nuclei. This distance lowers the potential energy more than any other and is called the bond distance.

Figure 7.3 ■ A schematic diagram shows how the potential energy of a pair of atoms might vary as the atoms draw closer together and form a covalent bond. The minimum in this "potential well" corresponds to the normal bond length and bond energy for the diatomic molecule.

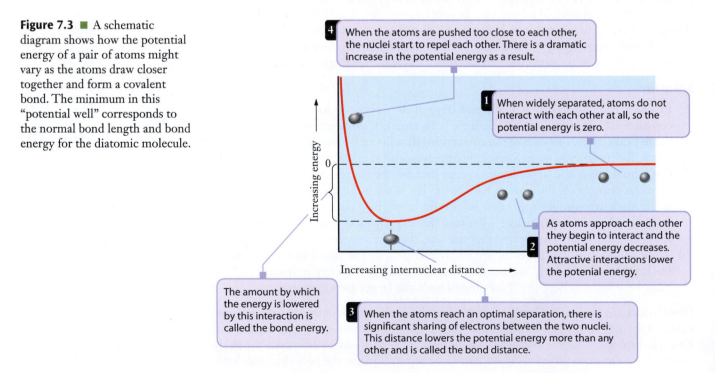

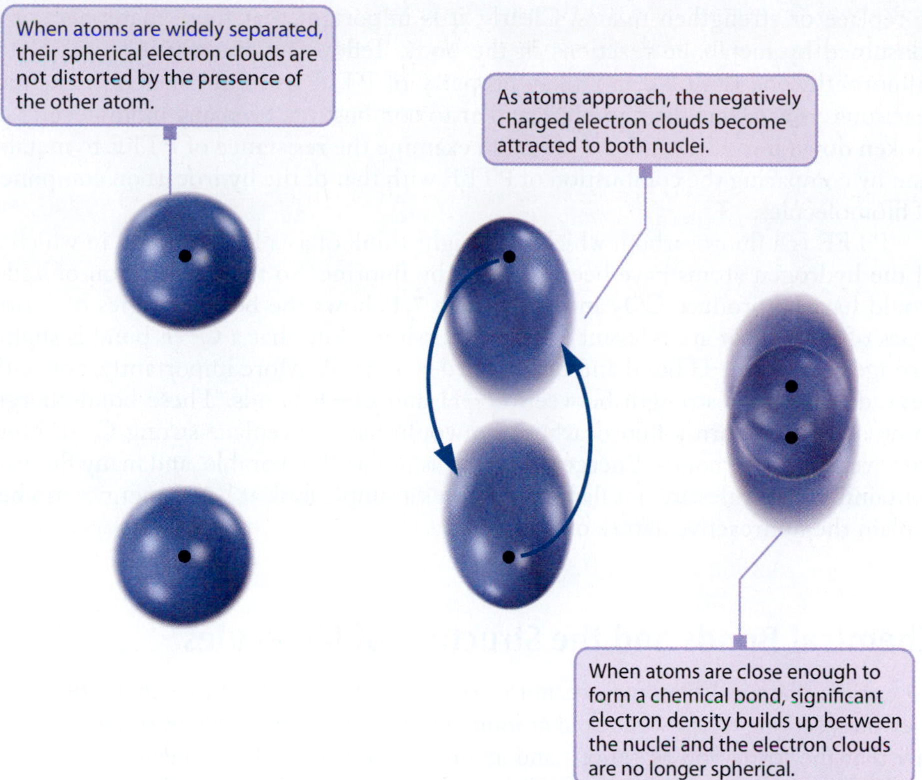

When atoms are widely separated, their spherical electron clouds are not distorted by the presence of the other atom.
As atoms approach, the negatively charged electron clouds become attracted to both nuclei.
When atoms are close enough to form a chemical bond, significant electron density builds up between the nuclei and the electron clouds are no longer spherical.

Figure 7.4 ■ When two atoms approach one another, the negatively charged electron clouds are each attracted to the other atom's positively charged nucleus. This attractive interaction potential is largely responsible for the reduction of energy in the formation of a chemical bond.

corresponds to the formation of a covalent bond. The energy released when isolated atoms form a covalent bond is called the **bond energy**, and the distance between the nuclei of the bonded atoms is called the **bond length**.

It is important to remember that bond formation ALWAYS releases energy.

When electrons are shared between two atoms, their distribution is different from that in the individual atoms. In **Figure 7.4**, we see how electron density in a molecule differs from the spherically symmetrical distribution seen in isolated atoms. In a covalently bonded molecule, there is a buildup of electron density, or a concentration of negative charge, in the region between the two bonded atoms. The increase in electron density need not be particularly large; even a slight increase of negative charge in this region lowers the potential energy of the system and can stabilize the molecule.

The formation of a chemical bond *always* releases energy. Once a bond is formed, that same amount of energy—the bond energy—must be supplied to break the bond apart. The magnitude of this bond energy depends on the atoms involved. For example, when 2 moles of fluorine atoms combine to form 1 mole of molecular fluorine, 156 kJ of energy is released. So we say that the F—F bond energy is 156 kJ per mole. Compared to many other molecules, F_2 is held together by a relatively weak covalent bond. Hydrogen fluoride, for example, has a bond energy of 565 kJ/mol. This means that its bond is more than three times stronger. Because more energy is required to break its atoms apart, we say that the bond in HF is more stable than the bond in F_2. When we want to assess the stability of a compound, some general ideas about the different types of bonds present are useful.

Chemical Bonds and Reactions

Chemical reactions involve rearranging bonds, so that reactants are transformed into products. A reaction will be energetically favored if the energy required to break bonds is less than that released in making new bonds. Some biomaterials such as a Teflon® mesh used to reinforce an abdominal wall after surgery are intended to replace or strengthen tissues. Clearly, it is important that these materials not be consumed by metabolic reactions in the body. Teflon is a trade name for polytetrafluoroethylene (PTFE), and a key property of PTFE is its resistance to chemical reactions. Metabolism is fairly similar to combustion. So many biomolecules are broken down into H_2O and CO_2. We can examine the resistance of PTFE to metabolism by comparing the combustion of PTFE with that of the hydrocarbon component of biomolecules.

The familiar use of Teflon in cooking products shows that PTFE tends not to react even at relatively high temperatures.

PTFE is a fluorocarbon, which you might think of as a hydrocarbon in which all of the hydrogen atoms have been replaced by fluorine. So the combustion of Teflon would have to produce CO_2 and OF_2. **Table 7.1** shows the bond energies of various types of bonds that are relevant to this discussion. Note that a C—F bond is slightly stronger than a C—H bond and thus is harder to break. More importantly, notice the large difference in strength between O—H and O—F bonds. These bond energies show us that to burn a fluorocarbon, we would have to replace strong C—F bonds with weaker O—F bonds. Energetically, this is not at all favorable, and many fluorine-containing molecules are not flammable. Thus a simple look at bond energies can help explain the nonreactive nature of Teflon.

Table 7.1

Bond energies of some types of bonds important in the combustion of hydrocarbons and fluorocarbons

Type of Bond	Bond Energy (kJ/mol)
C—F	485
C—H	415
O—F	190
O—H	460

Chemical Bonds and the Structure of Molecules

To explain chemical bonding, we must go beyond the discussion of energetics in the previous section. When we looked at ionic bonding between metals and nonmetals, we saw that the formation of cations and anions is consistent with achieving np^6 electron configurations that resemble those of the noble gases. For nonmetal atoms, a noble gas configuration can be achieved by gaining the electrons from metal atoms, leading to ionic bonds. When nonmetals interact with other nonmetals, however, they share electrons, effectively giving each atom access to all the shared (bonding) electrons. Thus, by sharing a suitable number of electrons, an atom can achieve an electron configuration that resembles that of a noble gas. For many molecules composed of elements from the main groups, this feature leads to eight valence electrons around each atom. This observation is the basis for what is known as the **octet rule**: *An atom will form covalent bonds to achieve a complement of eight valence electrons.* Recall that a noble gas atom has a valence shell electron configuration of ns^2np^6, for a total of eight valence electrons. Because it has no *p* orbitals in its $n = 1$ valence shell, hydrogen is an exception to the octet rule and can share only two electrons. There are other exceptions as well, which we will note as we encounter them.

The basic concept of the octet rule can be conveniently depicted using a **Lewis dot symbol**, as developed by the American chemist G. N. Lewis in the early 20th century. These Lewis symbols help keep track of valence electrons, making it easier to predict bonding in molecules. They are particularly useful for compounds of main group elements. To draw a Lewis symbol for an element, we use its atomic symbol and place dots around it on four sides to depict the number of valence electrons. As we go across a period in the main groups, the first four dots are placed singly around the symbol. Starting with the fifth dot, they are paired. Thus the elements in the second period would have the following Lewis symbols:

Lewis dot symbols of atoms help explain the valence of elements. The original definition of valence was derived from the number of chemical bonds that an element would form. Counting unpaired electrons in the Lewis symbol provides the valence of elements in the s and p blocks of the periodic table.

·Li ·Be· ·Ḃ· ·Ċ· ·N· ·O· :F· :Ne:

Because elements in the same group have the same number of valence electrons, the Lewis symbols for subsequent periods are the same, except for the identity of the elements. For example, the Lewis dot symbols for carbon, silicon, and germanium are as follows:

·Ċ· ·Si· ·Ge·

Group	1	2	13	14	15	16	17	18
Number of electrons in valence shell	**1**	**2**	**3**	**4**	**5**	**6**	**7**	**8 (except He)**
Period 1	H·							Ḧe
Period 2	Li·	·Be·	·Ḃ·	·Ċ·	·N̈·	·Ö·	:F̈·	:N̈e:
Period 3	Na·	·Mg·	·Ȧl·	·Si·	·P̈·	·S̈·	:C̈l·	:Är:
Period 4	K·	·Ca·	·Ġa·	·Ge·	·Äs·	·S̈e·	:B̈r·	:K̈r:
Period 5	Rb·	·Sr·	·İn·	·Sn·	·S̈b·	·T̈e·	:G̈·	:Ẍe:
Period 6	Cs·	·Ba·	·Ṫl·	·Pb·	·B̈i·	·P̈o·	:Ät·	:R̈n:
Period 7	Fr·	·Ra·						

Figure 7.5 ■ The Lewis symbols for the main group elements of the periodic table are shown. Notice the similarities among elements in the same group. Lewis structures help keep track of electrons involved in chemical bonds.

Lewis symbols for the main group elements are shown in **Figure 7.5**.

These symbols for individual atoms are of fairly little use, but similar representations for molecules are much more powerful. When we draw Lewis dot representations of molecules, we add one more level of symbolism to denote bonds between atoms. A line between two atomic symbols represents a shared pair of electrons forming a bond; thus when counting electrons, one line is equivalent to two dots. **Figure 7.6** shows the Lewis symbol for the hydrogen atom and the Lewis structure of the simplest molecule, H_2.

H· + ·H ⟶ H:H or H—H

Figure 7.6 ■ By sharing the electron from each atom, two hydrogen atoms can form a chemical bond. In doing so, each atom has enough valence electrons around it to achieve a filled $n = 1$ shell.

A **Lewis structure** shows how electrons are shared in a molecule. The Lewis structure for F_2 is shown in **Figure 7.7**. The sharing of electrons to form a bond effectively gives each fluorine atom a full octet of electrons. When fluorine is part of a covalently bonded system, it virtually always forms a single bond by sharing an electron pair, thereby satisfying the octet rule. A pair of electrons shared by two atoms is referred to as a **bonding pair**. The other paired electrons that are associated with a single atom are referred to as nonbonding electrons, or as **lone pairs**. A fluorine molecule has one bonding pair and six lone pairs.

:F̈· + ·F̈: ⟶ :F̈:F̈: or :F̈—F̈:

Figure 7.7 ■ When two fluorine atoms combine, they form a stable chemical bond. The Lewis structures for the F atoms show why just one bond forms. By sharing one pair of electrons, each atom is surrounded by eight valence electrons.

Of course, most molecules are more complicated than H_2 or F_2 and will contain more than one pair of shared electrons. If two oxygen atoms share a single pair of electrons, for example, then neither atom has an octet of valence electrons. However, if *two pairs* of electrons are shared, then both oxygen atoms conform to the octet rule. For diatomic nitrogen, *three pairs* of electrons must be shared to satisfy the octet rule. Multiple bonding results from the sharing of more than one electron pair. When two pairs of electrons are shared, the result is a **double bond**, and three shared pairs result in a **triple bond**. Double bonds are stronger than single bonds, and triple bonds are stronger than double bonds. However, a double bond, although stronger than a single bond, is generally not twice as strong. The bond energies for carbon–carbon bonds provide a good example:

Type of Bond	Bond Energy (kJ/mol)
C—C	346
C═C	602
C≡C	835

The relative strengths of multiple bonds have important consequences. The polymerization of methyl methacrylate to form PMMA provides a good example. The methyl methacrylate monomer contains a carbon–carbon double bond. In the polymerization reaction, two of the electrons forming this double bond are used to create a bond to a second monomer unit. The end result is that the double bond in the reactant is transformed into two single bonds in the product. Because the energy released in forming two single bonds is greater than the energy needed to break a double bond, the reaction to form PMMA is an energetically favorable process. This is true for many similar reactions in which a reactant with a double bond is converted

This polymerization reaction is analogous to the formation of polyethylene, which was discussed in Section 2-8.

into a product with two single bonds. It explains why the common reaction of the C=C functional group is addition. The same is true for triple bonds, although those are much less common.

7-4 Electronegativity and Bond Polarity

Having considered both ionic and covalent bonding, we can begin to understand the chemical properties of both polymers and ceramics. One key biomaterial is hydroxyapatite, $Ca_{10}(PO_4)_6(OH)_2$, which involves both ionic and covalent bonding. How can we predict which bonding interactions will be ionic and which will be covalent?

The sharing of electrons to form covalent bonds and the outright transfer of electrons that occurs in ionic bonding represent two ends of a bonding continuum. To describe compounds whose bonding falls somewhere in the middle of this continuum, we will introduce two additional concepts: electronegativity and bond polarity.

Electronegativity

We've seen that ionic bonds form between a metal and a nonmetal, with one or more electrons being transferred from the metal to the nonmetal. Compared to metals, nonmetals are relatively "greedy" for electrons. Even in covalent bonds, the small size and high electron affinity of many nonmetals mean that electrons tend to be strongly attracted to those atoms. The attraction of an atom for the shared electrons in a covalent bond is called **electronegativity**.

Although this concept may seem simple enough, quantifying it is very difficult. Unlike ionization energy or electron affinity, electronegativity is not an energy change that can be determined experimentally. Instead, electronegativity values are established by considering a number of factors, such as atomic size, electron affinity, and ionization energy, and a scale of arbitrary units has been chosen. **Figure 7.8** gives the resulting values for the electronegativities of most elements. The higher the electronegativity value, the more likely an element is to attract extra electron density when it forms chemical compounds. Electronegativity tends to increase as we go up a group (decreasing atomic number) or across a period from left to right. Thus the trends in electronegativity parallel those in ionization energy and electron affinity.

Fluorine, the most electronegative element, is assigned a value of 4.0. Oxygen and nitrogen also have relatively high electronegativities. When these elements take part in covalent bonding, they tend to take "more than their fair share" of the bonding electron pair. What are the implications for a molecule when electrons are shared unequally?

Bond Polarity

The high electronegativity of fluorine plays an important role in a number of processes. The fluoridation of teeth, for example, is one of the earliest manipulations of biomaterials and involves the replacement of an —OH group in hydroxyapatite with a fluorine atom to form fluorapatite. To understand how fluorine participates in chemical bonds from covalent to ionic, a convenient starting point is to compare the electron distribution around the atoms in a molecule with the electron distribution around the isolated atoms.

If we observe a fluorine atom from a distance, it appears to be electrically neutral because the nuclear and electronic charges offset each other. Because fluorine is the most electronegative element, when a fluorine atom participates in a covalent bond with another element, such as hydrogen, more than half of the electron distribution

Metals
Nonmetals
Metalloids

	1	2	3	4	5	6	7	8	9	10	11	12	13	14	15	16	17	18
1	1 H 2.1																	2 He
2	3 Li 1.0	4 Be 1.5											5 B 2.0	6 C 2.5	7 N 3.0	8 O 3.5	9 F 4.0	10 Ne
3	11 Na 1.0	12 Mg 1.2											13 Al 1.5	14 Si 1.8	15 P 2.1	16 S 2.5	17 Cl 3.0	18 Ar
4	19 K 0.9	20 Ca 1.0	21 Sc 1.3	22 Ti 1.4	23 V 1.5	24 Cr 1.6	25 Mn 1.6	26 Fe 1.7	27 Co 1.7	28 Ni 1.8	29 Cu 1.8	30 Zn 1.6	31 Ga 1.7	32 Ge 1.9	33 As 2.1	34 Se 2.4	35 Br 2.8	36 Kr
5	37 Rb 0.9	38 Sr 1.0	39 Y 1.2	40 Zr 1.3	41 Nb 1.5	42 Mo 1.6	43 Tc 1.7	44 Ru 1.8	45 Rh 1.8	46 Pd 1.8	47 Ag 1.6	48 Cd 1.6	49 In 1.6	50 Sn 1.8	51 Sb 1.9	52 Te 2.1	53 I 2.5	54 Xe
6	55 Cs 0.8	56 Ba 1.0	57 La 1.1	72 Hf 1.3	73 Ta 1.4	74 W 1.5	75 Re 1.7	76 Os 1.9	77 Ir 1.9	78 Pt 1.8	79 Au 1.9	80 Hg 1.7	81 Tl 1.6	82 Pb 1.7	83 Bi 1.8	84 Po 1.9	85 At 2.1	86 Rn
7	87 Fr 0.8	88 Ra 1.0	89 Ac 1.1	104 Rf	105 Db	106 Sg	107 Bh	108 Hs	109 Mt	110 Ds	111 Rg	112 Cn	113 Uut	114 Fl	115 Uup	116 Lv	117 Uus	118 Uuo

58 Ce 1.1	59 Pr 1.1	60 Nd 1.1	61 Pm 1.1	62 Sm 1.1	63 Eu 1.1	64 Gd 1.1	65 Tb 1.1	66 Dy 1.1	67 Ho 1.1	68 Er 1.1	69 Tm 1.1	70 Yb 1.0	71 Lu 1.2
90 Th 1.2	91 Pa 1.3	92 U 1.5	93 Np 1.3	94 Pu 1.3	95 Am 1.3	96 Cm 1.3	97 Bk 1.3	98 Cf 1.3	99 Es 1.3	100 Fm 1.3	101 Md 1.3	102 No 1.3	103 Lr 1.5

Figure 7.8 ■ Electronegativity values for the elements are shown. Notice how the values increase toward the right and top of the periodic table.

of the bonding electron pair is associated with the fluorine atom. The result is that the fluorine atom becomes a center of partial negative charge because it has more electron density than is necessary to balance its own nuclear charge. At the same time, the hydrogen atom at the opposite end of the bond takes on a partial positive charge because it has insufficient electron density to balance its nuclear charge **(Figure 7.9)**. Thus covalent bonding between atoms of different electronegativity results in a bond in which there is a separation of positive and negative charge centers. Because the molecule has one region of positive charge and another of negative charge, the molecule has an electric field associated with it. These two points of positive and negative charge constitute a **dipole**, and such a bond is referred to as a **polar bond**. Because the electrons are still shared rather than transferred, these bonds are often called polar covalent bonds.

Polar covalent bonding completes the continuum of chemical bonding. The polarity of covalent bonds varies from completely nonpolar (as in F_2 and H_2), through varying intermediate degrees of polarity (as in HF), to completely ionic (as in NaF). The type of bond formed depends on the electronegativity difference between the bonding

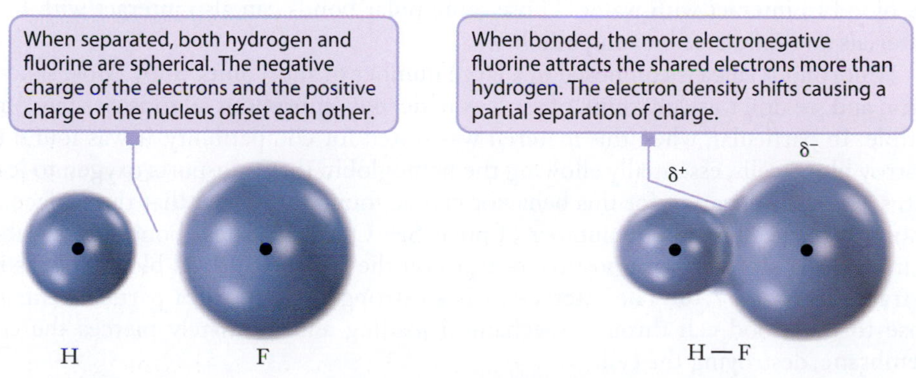

When separated, both hydrogen and fluorine are spherical. The negative charge of the electrons and the positive charge of the nucleus offset each other.
When bonded, the more electronegative fluorine attracts the shared electrons more than hydrogen. The electron density shifts causing a partial separation of charge.
δ+
δ−
H
F
H—F

Figure 7.9 ■ The highly electronegative fluorine atom attracts electrons more strongly than hydrogen does. So the two form a polar covalent bond.

The relationship between the difference in electronegativity and the ionic character of a bond is actually continuous, but using a cutoff to label a bond as ionic or polar simply makes the language more tractable.

atoms. *The greater the difference in electronegativity, the more polar the bond.* At some point, typically when the difference in electronegativity exceeds about 2.0, the bond is classified as ionic. Using the electronegativity values listed in Figure 7.8, we see that NaF, with an electronegativity difference of 3.0, is clearly ionic. The C—F bond in a fluorocarbon such as Teflon is polar covalent (electronegativity difference = 1.5), and the F—F bond is completely covalent (electronegativity difference = 0). Example Problem 7.3 explores this concept further.

EXAMPLE PROBLEM 7.3

Which bond is the most polar: C—H, O—H, or H—Cl? For each of these bonds, determine which atom has a partial positive charge.

Strategy Bond polarity depends on the difference in electronegativity between the two bonded atoms. So we can look up the electronegativity values and subtract to find the difference in each case. Negatively charged electrons are drawn more strongly toward an atom with higher electronegativity, so the less electronegative atom has a partial positive charge.

Solution From Figure 7.8, we find the following electronegativities: H = 2.1, C = 2.5, O = 3.5, and Cl = 3.0. The electronegativity differences are, therefore, C—H = 0.4, O—H = 1.4, and H—Cl = 0.9. So the O—H bond is the most polar of the three. In each of the three bonds considered here, H is the less electronegative atom, so it carries the positive charge.

Discussion This question may appear simple because we are using electronegativity values and differences in a straightforward way. Despite this simplicity, the results can provide important insight into material properties. Our focus at this point is on single molecules, but invariably in laboratory observations in the real world, we encounter huge collections of molecules. The way these samples of molecules behave is often influenced strongly by the polarity of the bonding within the individual molecules. Therefore, understanding a bond dipole and predicting its presence using electronegativity will be useful in a number of ways as we move forward in our study of chemistry.

Check Your Understanding Which bond should be nonpolar: C—N, N—Cl, or H—Br?

Bond polarity plays an important role in several aspects of biomaterials, particularly with regard to biocompatibility. Materials that will be deployed in living system always interact with those systems. A key example is the interaction of biomaterials with blood. Polar bonds in molecules along the outer wall of blood cells help to allow the blood to interact with water. Those same polar bonds can also interact with biomaterials and influence biocompatibility.

Amorphous silica is composed of a large number of silica units, SiO_4. Most sand is silica, and we don't usually think of sand as toxic, but amorphous silica is not biocompatible. In particular, when this material was tested for compatibility, it was found to destroy blood cells, essentially allowing the hemoglobin that transports oxygen to leak out. An explanation for this behavior can be found in the fact that the surface of amorphous silica has a large number of polar Si—O bonds. These bonds are capable of interacting strongly with positive charges on the surface of a red blood cell, as illustrated in **Figure 7.10.** The interaction is so strong that the silica particle remains close to the blood cell through mechanical jostling and ultimately pierces the cell membrane, destroying the cell.

The term used to describe the breaking open of red blood cells is hemolysis.

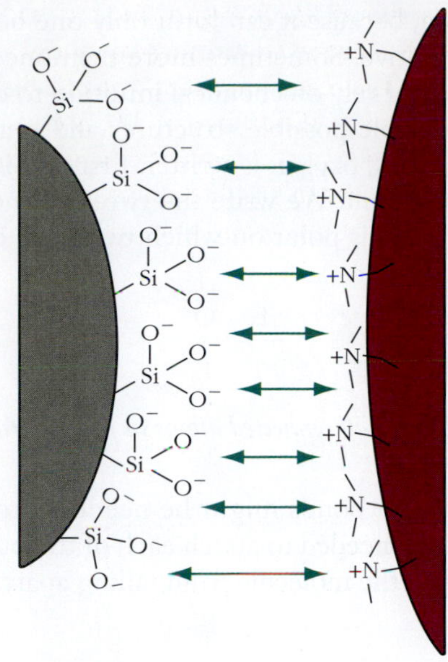

Figure 7.10 ■ Cells often have polar bonds on their exterior because they interact with polar water molecules. A substance like amorphous silica (left) that also has polar surface species can interact strongly with the polar bonds on the surface of a cell such as the blood cell depicted on the right. These strong interactions can lead to cell damage, causing biocompatibility concerns.

7-5 Keeping Track of Bonding: Lewis Structures

We've now surveyed some important aspects of chemical bonding, and we've gained an understanding of a number of key concepts. But if we want to think about the connections between bonding and the properties of compounds, we often need to know not just *that* bonds have formed, but *how many* bonds have formed and between which elements. The Lewis structures we introduced in Section 7-3 provide a powerful tool for doing this. In drawing Lewis structures for very simple molecules like H_2 and F_2, we drew each atom separately and then brought them together to form a bond. This approach obviously assumes that we know which atoms are to be attached to one another. The choice is obvious when we are dealing with a diatomic molecule but can become problematic for larger molecules. If we want to be able to draw Lewis structures for more complicated molecules or ions, it is useful to use an algorithmic approach. By applying such an algorithm carefully, we can predict which bonds are likely to be formed by examining and comparing the possible structures for a specific molecule.

As we introduce our algorithm for drawing Lewis structures, it will help if we use a simple molecule to illustrate each step. We could choose water, but the fact that hydrogen does not follow the octet rule means that is not the best example. Instead, we will use a very similar molecule, oxygen difluoride (OF_2), as our first example.

Step 1: *Count the total number of valence electrons in the molecule or ion.*

Although it may seem routine after a little practice, it is always vitally important to carry out Step 1 of this process. Using the wrong number of valence electrons is among the most common errors that students make in drawing Lewis structures.

We can do this by summing the numbers of valence electrons for each atom in a molecule. (If we were trying to draw the structure for a polyatomic ion, we would also need to account for the ion's charge, as we'll see in Example Problem 7.5.) For OF_2, each fluorine atom has seven valence electrons, and the oxygen atom has six. So we will have a total of 20 valence electrons in the molecule:

F	$2 \times 7 = 14$
O	$1 \times 6 = 6$
	Total = 20

Step 2: *Draw the skeletal structure of the molecule.*

Here we can rely on some general rules. The element written first in the formula is usually the central atom in the molecule, except when hydrogen is listed first. (Hydrogen can never be a central atom, because it can form only one bond!) Usually, the central atom is the least electronegative. Sometimes more than one possible skeleton can be drawn. In such cases, we might rely on chemical intuition to choose likely alternatives. Or we could also draw multiple possible structures and evaluate those to see which seemed more plausible. In OF_2, oxygen is written first and is the less electronegative element, so it is the central atom. We write the two fluorine atoms surrounding the oxygen. (It isn't important at this point on which two sides of the oxygen symbol we choose to place the fluorine atoms.)

```
F  O

   F
```

Step 3: *Place single bonds between all connected atoms in the structure by drawing lines between them.*

We don't yet know how many bonds might be needed in our molecule, but we do know that at least one bond is needed to attach each of the outer fluorine atoms to the central oxygen atom, to keep the molecule from falling apart. So we can go ahead and draw two single bonds:

```
F—O
  |
  F
```

Step 4: *Place the remaining valence electrons not accounted for in Step 3 on individual atoms until the octet rule is satisfied. Place electrons as lone pairs wherever possible.*

It is usually best to begin placing electrons on the outer atoms first, leaving the central atom(s) for last. Remember that each bond contains two electrons, so in drawing two single bonds, we have accounted for four electrons. Our molecule has 20 valence electrons, so 16 remain to be added to our structure. Placing 6 of these electrons around each fluorine atom brings us up to 16, and placing the remaining 4 on the central oxygen atom accounts for all 20 available electrons. All 16 of these electrons should be drawn as lone pairs:

```
 ..  ..
:F—O:
 ..  |
    :F:
     ..
```

At this point, we know that all of the available valence electrons appear somewhere in our structure. To decide whether this is the optimal arrangement of the valence electrons, we check to see if each atom satisfies the octet rule. For OF_2, our structure now has a full octet of electrons on each atom, so this is the final structure. In other cases, we might find one or more atoms that have less than a full octet; then, we would take an additional step.

Step 5: *Create multiple bonds by shifting lone pairs into bonding positions as needed for any atoms that do not have a full octet of valence electrons.*

Our structure for OF_2 already satisfies the octet rule, so multiple bonds are not required in this molecule. We will work with this added step in future structures.

Example problems 7.4, 7.5, and 7.6 will further illustrate our algorithm. Of course, the best way to become familiar with these steps is simply to practice drawing Lewis structures yourself.

EXAMPLE PROBLEM 7.4

Chlorofluorocarbons had been used as refrigerants since the 1930s, but their use has largely been phased out worldwide because of concerns about atmospheric ozone depletion. Draw the Lewis structure of dichlorodifluoromethane, CF_2Cl_2, also known as DuPont's Freon-12.

Strategy Use the algorithm for drawing Lewis structures.

Solution

Step 1: *Count the total number of valence electrons.* Carbon has four, fluorine has seven, and chlorine has seven:

C	$1 \times 4 = 4$
F	$2 \times 7 = 14$
Cl	$2 \times 7 = 14$
	Total = 32

Step 2: *Draw the skeletal structure of the molecule.* Carbon is listed first in the formula, and it is the least electronegative, so use it as the central atom:

```
   Cl
F  C  F
   Cl
```

Step 3: *Place single bonds between all connected atoms:*

```
    Cl
    |
F — C — F
    |
    Cl
```

Step 4: *Place the remaining valence electrons not accounted for in Step 3 on individual atoms until the octet rule is satisfied.* Four single bonds have been drawn, and they require eight valence electrons. We have a total of 32 available electrons, so we are left with 32 − 8 = 24. This gives each of the four halogen atoms three lone pairs (6 electrons):

```
      ..
     :Cl:
  ..  |   ..
 :F — C — F:
  ..  |   ..
     :Cl:
      ..
```

Step 5: *Create multiple bonds for any atoms that do not have a full octet of valence electrons.* Check to make sure each atom has an octet of eight electrons. Carbon has four bonding pairs, or eight electrons. Each halogen has one bonding pair and three lone pairs, so they also have octets. Multiple bonds are not needed, and the final structure is the one arrived at after Step 4.

Discussion Because we are using an algorithm to draw these structures, it is important that each step be carried out correctly because any errors may be propagated through the rest of the process. The first step is the most critical. If we lose track of the number of valence electrons we need to include, we will have little chance of arriving at the correct structure. Many students have missed exam questions by adding a lone pair of electrons they didn't have rather than making multiple bonds.

Check Your Understanding Draw the Lewis structure for C_2H_6.

EXAMPLE PROBLEM 7.5

Calcium phosphate is an important precursor for the formation of bioceramic coatings. Draw the Lewis structure of the phosphate ion, PO_4^{3-}.

Strategy Follow the steps of the Lewis structure algorithm. Because phosphate is an anion, we'll need to account for its charge in determining the number of valence electrons.

Solution

Step 1: *Count the total number of valence electrons.* The phosphorus atom has five valence electrons, and each oxygen atom also has six. The ion as a whole carries a charge of 3−, which means it must contain three additional electrons.

$$\begin{array}{lr} \text{P} & 1 \times 5 = 5 \\ \text{O} & 4 \times 6 = 24 \\ +3 \text{ extra } e^- & = 3 \\ \hline & \text{Total} = 32 \end{array}$$

Step 2: *Draw the skeletal structure of the ion.* Phosphorus is the least electronegative atom and appears first in the formula, so it should be the central atom. We'll draw square brackets around the entire ion and indicate its charge:

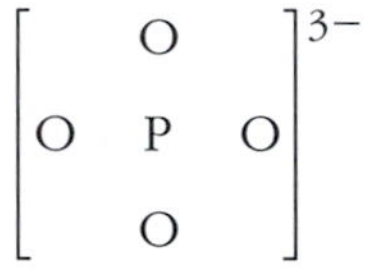

$$\left[\begin{array}{ccc} & \text{O} & \\ \text{O} & \text{P} & \text{O} \\ & \text{O} & \end{array}\right]^{3-}$$

Step 3: *Place single bonds between all connected atoms:*

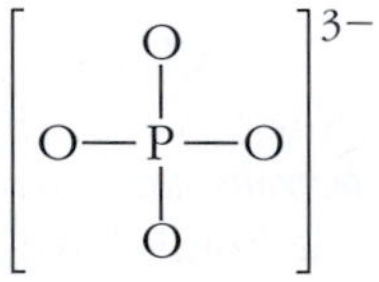

$$\left[\begin{array}{ccccc} & & \text{O} & & \\ & & | & & \\ \text{O} & — & \text{P} & — & \text{O} \\ & & | & & \\ & & \text{O} & & \end{array}\right]^{3-}$$

Step 4: *Place the remaining valence electrons not accounted for in Step 3 on individual atoms until the octet rule is satisfied.* By drawing bonds to the oxygen atoms, we have accounted for 8 of the 32 electrons, and 24 remain to be assigned. Placing three lone pairs around each oxygen atom will account for all of the available electrons:

$$\left[\begin{array}{ccccc} & & :\ddot{\text{O}}: & & \\ & & | & & \\ :\ddot{\text{O}} & — & \text{P} & — & \ddot{\text{O}}: \\ & & | & & \\ & & :\ddot{\text{O}}: & & \end{array}\right]^{3-}$$

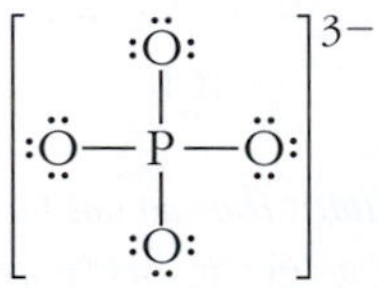

Step 5: *Create multiple bonds for any atoms that do not have a full octet of valence electrons.* We have fulfilled the octet rule and have accounted for all of the available electrons. Multiple bonds are not required in this ion.

Discussion This is similar to the earlier examples, but the fact that we are working with an ion requires extra care in determining how many valence electrons to include.

Check Your Understanding Draw Lewis structures for the NH_4^+ and ClO_2^- ions.

EXAMPLE PROBLEM 7.6

Poly(vinyl alcohol) is used in several biomaterials applications, including surgical sutures. Draw the Lewis structure of vinyl alcohol, CH_2CHOH, the monomer from which poly(vinyl alcohol) is made.

Strategy Follow the steps of the Lewis structure algorithm. Because we have a larger molecule, we will have multiple central atoms in this case.

Solution

Step 1: *Count the total number of valence electrons.* Each carbon atom has four, each hydrogen atom has one, and the oxygen atom has six. So there are 18 valence electrons:

$$\text{Number of valence electrons} = 2(4) + 4(1) + 1(6) = 18$$

Step 2: *Draw the skeletal structure of the molecule.* This is an organic compound, and our knowledge of organic chemistry suggests that the carbon atoms will be in the center of the molecule and will be linked by a chemical bond. So we'll place both carbon atoms in the middle, surrounded by the remaining atoms. We also saw in Chapter 2 that alcohols contain the —OH functional group, so we should expect vinyl alcohol to show that feature. And we know that hydrogen is always an outer atom, forming just one chemical bond. Combining those ideas, we should have three hydrogen atoms and one —OH group bound to the carbon atoms. The way the formula is written above tells us that we want two hydrogens bound to one carbon atom and then a hydrogen and the hydroxyl group bound to the other.

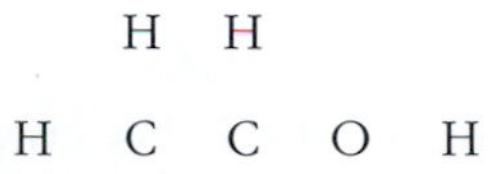

```
   H  H

H  C  C  O  H
```

Step 3: *Place single bonds between all connected atoms.* To hold the molecule together, we need at least one bond between each pair of adjacent atoms.

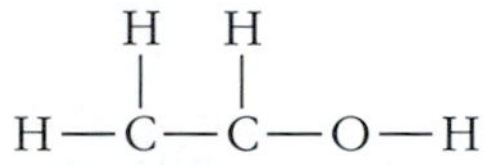

```
   H  H
   |  |
H—C—C—O—H
```

Step 4: *Place the remaining valence electrons not accounted for in Step 3 on individual atoms until the octet rule is satisfied.* Making 6 bonds requires 12 electrons, so we still have 18 − 12 = 6 electrons available. We cannot assign lone pairs to the hydrogen atoms. Oxygen is more electronegative than carbon, so we start by assigning two lone pairs to the oxygen atom so that it has an octet. There are two remaining valence electrons, which are placed for the moment on one of the carbon atoms. (At this point the choice of which carbon gets those electrons is arbitrary.)

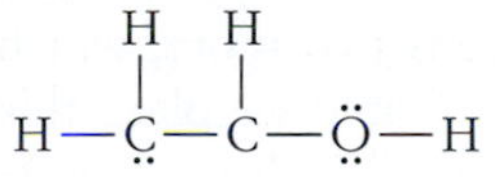

```
   H  H
   |  |
H—C̣—C—Ö—H
        ··
```

Step 5: *Create multiple bonds for any atoms that do not have a full octet of valence electrons.* The carbon atom that did not receive a lone pair in the last step does not have an octet, so we'll need to convert a lone pair from an adjacent atom into a bonding pair, creating a double bond. But should this double bond be formed between the two carbon atoms or between the carbon atom and the oxygen atom? At least two factors point toward the carbon–carbon choice. First, we have already seen that carbon often forms multiple bonds. Second, placing the double bond between the two carbon atoms gives the molecule a more symmetrical structure, and such arrangements are usually preferable:

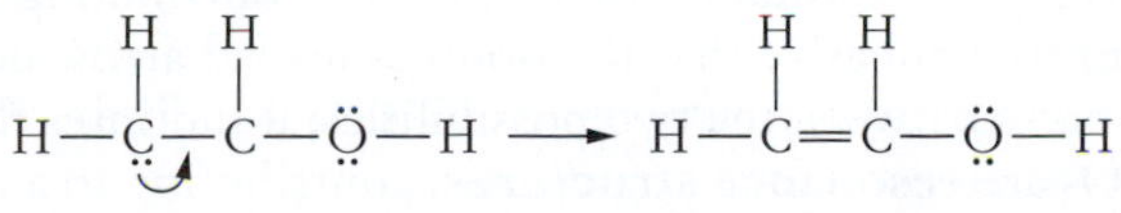

```
   H  H                H  H
   |  |                |  |
H—C̣—C—Ö—H  ⟶  H—C=C—Ö—H
        ··                     ··
```

The C=C double bond also leads to a structure in which each carbon atom forms four bonds, consistent with the rules we learned in Chapter 2 for writing line structures.

Discussion This example shows that using our algorithm for obtaining a Lewis structure sometimes leaves us needing to make choices. As you grow more experienced in chemistry, decisions such as where to place a double bond become easier. Often you can look for clues, as we did in this case. Carbon routinely forms double (or even triple) bonds. Further, we know that the presence of the carbon—carbon double bond allows the addition reaction that would be needed to produce the polymer from the monomer. But what happens when the choice is less obvious? Such a situation is discussed in the next subsection.

Check Your Understanding Draw the Lewis structure for carbon dioxide.

Chemists are quite fond of Lewis structures, and the reason for this is clear. The structures are generally simple to draw; yet they can often provide powerful insight into the bonding within a molecule. On occasion, however, when the final step of our algorithm is carried out, the choice of where to put a double bond appears entirely arbitrary. The concept of resonance has been developed to help describe the bonding in such molecules within the framework of Lewis structures.

Resonance

Let's consider sulfur dioxide, one of the EPA's criteria pollutants we discussed in Chapter 5. What should its Lewis structure look like? Both sulfur and oxygen have 6 valence electrons, so we have a total of 18 valence electrons. Sulfur occurs first in the formula and is less electronegative, so we place it at the center of the skeletal structure:

$$\mathrm{O \quad S \quad O}$$

We next place bonding pairs between the sulfur atom and each oxygen atom:

$$\mathrm{O{-}S{-}O}$$

We still have 14 valence electrons to add, starting with the oxygen atoms. When we place all 14 electrons in the structure as lone pairs, we have the following structure:

$$:\ddot{\underset{\cdot\cdot}{\mathrm{O}}}{-}\ddot{\mathrm{S}}{-}\ddot{\underset{\cdot\cdot}{\mathrm{O}}}:$$

Checking for octets, we find that both oxygen atoms have filled octets, but the sulfur has only six electrons. It needs two more, which can be obtained by forming a double bond. But on which side should we place this double bond? Unlike Example Problem 7.6, here we are confronted with two equivalent choices for the position of the double bond. We could draw either of the two structures:

$$:\ddot{\mathrm{O}}{=}\ddot{\mathrm{S}}{-}\ddot{\underset{\cdot\cdot}{\mathrm{O}}}: \longleftrightarrow :\ddot{\underset{\cdot\cdot}{\mathrm{O}}}{-}\ddot{\mathrm{S}}{=}\ddot{\mathrm{O}}:$$

Each of these structures has the same fundamental weakness: it implies that the central sulfur atom somehow interacts more strongly with one oxygen atom than with the other. The explanation for this contradiction is that the structure of the molecule cannot be accurately represented by a single Lewis structure. Instead, we say that the real structure corresponds to an average of all the valid Lewis structures, and we call this average a **resonance hybrid** of the contributing Lewis structures. It is very important to emphasize that the double-headed arrow does *not* mean that the structure alternates between the two possibilities; it indicates that the two structures shown for SO_2 are **resonance structures,** contributing to a resonance hybrid.

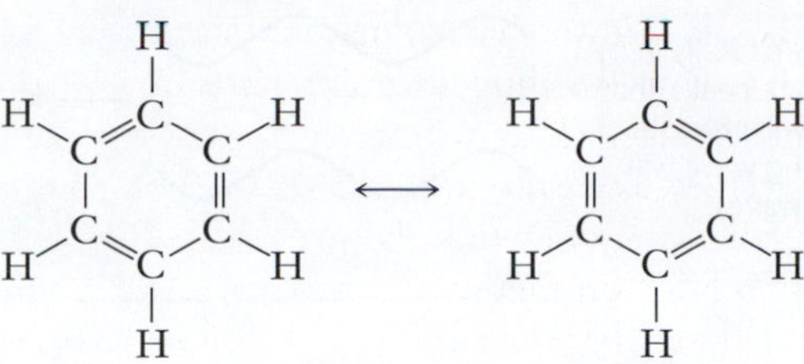

Figure 7.11 ■ The two resonance structures of benzene show there are two ways to distribute three double bonds among the six carbon atoms. Benzene does not behave chemically like a compound with three double bonds, however, because the true structure is an average between these two resonance structures.

In resonance structures, the positions of all atoms are identical; only the positions of the electrons are different.

The key point to emphasize here is that a resonance hybrid has a structure that is an average of the contributing resonance structures. To explain this statement better, let's take another look at the two possible structures we drew for SO_2. Either resonance structure taken alone would suggest that an SO_2 molecule contains one single bond and one double bond. If this were true, then one bond (the single bond) should be longer than the other (the double bond). But bond lengths are experimentally measurable, and such measurements clearly show that the two bonds are equal in length. Moreover, the bond lengths measured for SO_2 are intermediate between those typically seen for an S—O single bond and an S═O double bond. These experimental results confirm the idea that neither resonance structure alone gives a good picture of the SO_2 molecule and reinforce the view that the real structure is best thought of as an average of the possible resonance forms. *When we can draw resonance structures, the actual structure is a hybrid, an average of the contributing structures, and NOT a mixture of them.* The bonds can be thought of as "spread out" or delocalized over the molecule. Thus for SO_2, we might think of having two "one and a half bonds" rather than a single bond and a double bond. Given its simplicity, it is not really surprising that the Lewis dot model can't adequately describe every molecule we come across. The concept of resonance can be thought of as a patch that attempts to rationalize the experimental facts while preserving most of the simplicity of the Lewis model.

We have considered the structural implications of resonance, but there are also energetic consequences of resonance. This means that molecules with resonance structures behave differently from what one might expect on the basis of the individual resonance structures, which show a combination of single and double bonds. Benzene is the classic example of a substance that displays resonance, and in fact the concept of resonance originated from a need to explain the chemistry of benzene. Benzene, C_6H_6, is a volatile liquid obtained from coal tar. It was accepted in the latter half of the 19th century that benzene is a cyclic compound, whose carbon atoms form a six-membered ring. The only structures that could be drawn for it, however, had alternating single and double bonds (**Figure 7.11**). If these structures really represented the structure of benzene, the substance should have shown the addition reactions typical of compounds with C═C bonds. But benzene is actually much less reactive than alkenes. This means that the molecule is more stable than any one Lewis structure would predict. Over time, it became apparent from examining many compounds that this result is more general: when a molecule can be represented by multiple resonance structures, it is more stable (i.e., has lower energy and is less reactive) than would be predicted on the basis of any one of the contributing resonance structures. There are ways to understand this energy reduction based on quantum mechanics, but we will not present them here.

The two identical sine waves on the left are totally in phase with one another, meaning that their peaks and valleys line up. Adding these two waves will produce constructive interference, and the resulting wave will have twice the amplitude of the original waves.

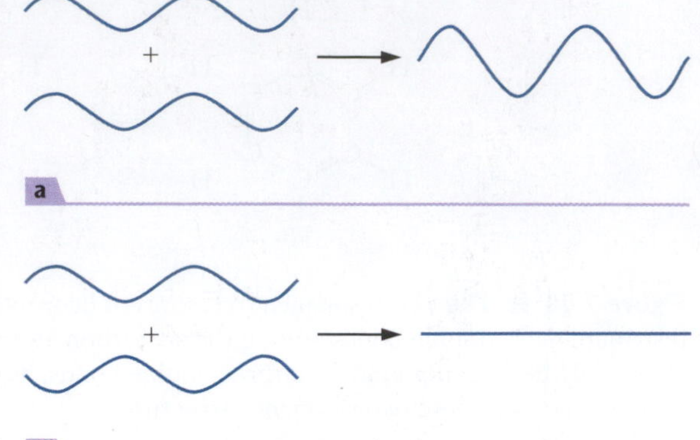

The two waves are out of phase, so that the peaks of one correspond to the valleys of the other. Combining these waves will result in destructive interference, where the two waves effectively cancel one another and the result is a flat line.

b

Figure 7.12 ■ Simple sine waves illustrate the concept of interference between waves.

7-6 Orbital Overlap and Chemical Bonding

Lewis structures provide us with a convenient way to gain some insight into a molecule's bonding. But they don't really tell us much about *how* chemical bonds might be formed, other than through the general notion of electron sharing. Scientific curiosity leads us to want to know more about covalent bonds, as does a sense that understanding bonding will pay off in practical applications. Because we recognize shared electrons as the "glue" that holds covalent bonds together, we should expect any deeper explanation to focus on electrons. In Chapter 6, we learned that the best current model of the atom views electrons as delocalized waves, described by orbitals. If electrons behave as waves, then to explain how electrons can form bonds between atoms, we must consider how waves interact with one another.

To explain how two waves—or two orbitals—might interact with one another, we will need to invoke the concept of interference, which provides a clear contrast between waves and particles. When two particles occupy the same space, they collide, but when two waves occupy the same space, they interfere. As you may have learned in a high school physics or physical science class, this interference may be either constructive or destructive. In constructive interference, two interacting waves reinforce one another to form a larger wave. In destructive interference, the two interacting waves offset or "cancel" one another to produce a smaller wave or no wave at all. Both constructive and destructive interference are illustrated in **Figure 7.12.**

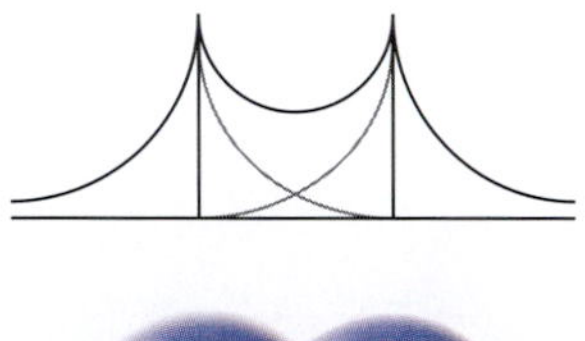

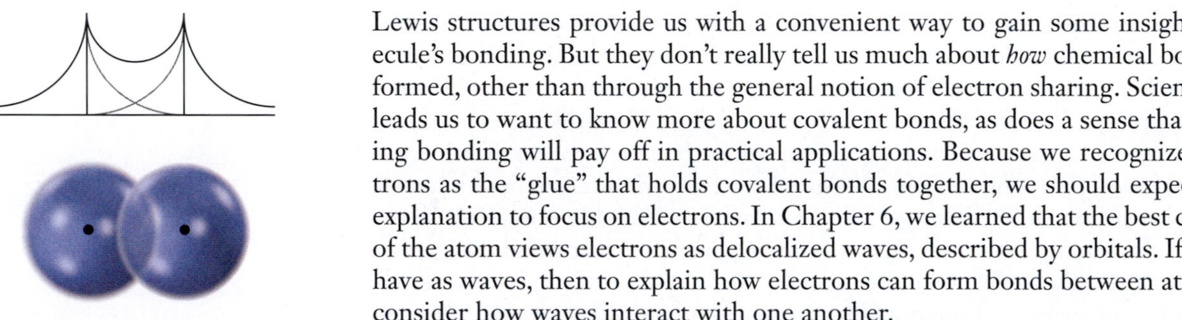

Figure 7.13 ■ The covalent bond in molecular hydrogen can be described in terms of the overlap between the 1*s* orbitals from the two hydrogen atoms. The upper diagram in the figure shows this overlap by plotting the wave functions for the 1*s* orbitals. Constructive interference between these wave functions in the region between the nuclei results in bond formation. The lower diagram uses shading to represent the same buildup of electron density.

In the quantum mechanical model, electrons are simply one type of wave, and we can picture the formation of chemical bonds as an example of constructive interference between electron waves. But for electron waves from different atoms to interact, they must first occupy the same region of space. Another way of saying this is that the atoms must be positioned so that the valence orbitals from one atom must overlap those of the other atom if a bond is to form. This idea forms the basis for the **valence bond model** of chemical bonding, in which all bonds are seen as the result of overlap between atomic orbitals. Let's examine this idea of **orbital overlap** by looking at the simplest possible molecule, H_2.

The name valence bond model arises from the fact that only valence electrons are considered.

Hydrogen atoms have a single valence electron in a 1*s* orbital. To form a diatomic hydrogen molecule, the 1*s* orbitals from two different hydrogen atoms must overlap. **Figure 7.13** shows what this overlap might look like and how interference between the wave functions of the two electrons produces a bond between the atoms. Note that because 1*s* orbitals are spherical, there is no preferred direction of approach. But when *p* orbitals are involved, their more complex shape will lead to different geometries for orbital interaction. These different geometries can be used to distinguish between types of chemical bonds.

Figure 7.14 ■ When two *p* orbitals approach one another and overlap end to end, as shown here, they form a sigma bond. (This is often denoted by the lowercase Greek letter sigma: σ.)

Using Lewis dot structures, we can show easily that N_2, for example, must have a triple bond between the two nitrogen atoms:

$$:N{\equiv}N:$$

Although this large buildup of negative charge density between the nuclei allows very strong attraction between them, it also raises a question. How do six electrons occupy the same region of space without strongly repelling each other? We can address this question by considering the shape of the orbitals involved and their arrangement as overlap occurs. The valence electron configuration for a nitrogen atom is $2s^22p^3$. Because the $2p$ orbitals are all singly occupied, overlap of a $2p$ orbital from one atom with a corresponding $2p$ orbital from the other atom should enable bond formation. But because of their shapes, *p* orbitals can overlap in different orientations.

Recall that the three $2p$ orbitals from any atom can be described as lying along the *x*, *y*, and *z* axes. (See Figure 6.13, p. 161.) Let's assume that two nitrogen atoms approach one another along the *z* axis. As seen in **Figure 7.14**, the $2p_z$ orbitals can overlap end to end. This leads to a buildup of electron density along a line drawn between the two nuclei, just as we saw for H_2. This type of constructive interference produces what are called **sigma bonds**. The accumulation of electron density in a sigma bond sits along the line connecting the two nuclei. This type of overlap can also occur in bonds in which an *s* orbital overlaps a *p* orbital or when two *s* orbitals overlap.

Now let's consider the $2p_x$ and $2p_y$ orbitals of the nitrogen atoms. As the nuclei draw closer together, these orbitals can overlap in a side-to-side orientation, as shown in **Figure 7.15.** Once again the overlap leads to constructive interference. This allows the buildup of electron density between nuclei that is crucial to chemical bonding. When this sideways overlap occurs, however, the buildup of electron density will not be localized along the line between the nuclei. As shown in Figure 7.15, the electron density is localized above and below or in front and in back of this line. Such bonds are called **pi bonds**. In the N_2 molecule, then, we can envision one pi bond formed by the overlap of the $2p_x$ orbitals from each atom and another pi bond formed by the overlap of the $2p_y$ orbitals. The triple bond thus comprises one sigma bond and two pi bonds.

The idea that bonding involves the overlap and interaction of orbitals from different atoms is a fundamental building block for a more detailed understanding of the chemical bond. For this reason, the valence bond model continues to play an

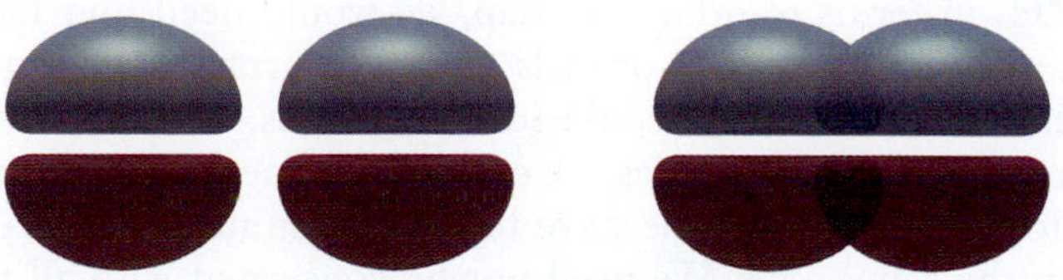

Figure 7.15 ■ When two *p* orbitals approach one another and overlap side to side, as shown here, they form a pi bond. (This is often denoted by the lowercase Greek letter pi: π.) In the example shown, orbital overlap will produce a buildup of electron density above and below the line defined by the two nuclei.

important role in the way scientists envision bonding. We do, however, encounter serious difficulties in describing bonding through the overlap of atomic orbitals. Because *p* orbitals form 90° angles with one another, we might predict that many molecules, including H_2O and OF_2, would show 90° bond angles. But such 90° angles are quite rare. Furthermore, looking at the electron configuration of carbon, we might expect compounds such as CH_2 or CF_2 to be stable and commonplace. But instead we find CH_4 and CF_4, and the simple overlap of atomic orbitals cannot help us to rationalize this. So we must extend our valence bond model to allow more extensive interactions between orbitals.

7-7 Hybrid Orbitals

Our notion of orbitals arose from the quantum mechanical treatment of individual atoms. Despite the obvious conceptual challenges that quantum mechanics poses, we accept the existence of *s*, *p*, *d*, and *f* orbitals because, among other things, doing so provides such a reasonable way to understand the shape of the periodic table. When atoms combine to form molecules, however, the observations noted above suggest that often those same orbitals cannot explain the observed facts. We just noted that electrons in *p* orbitals would be expected to lead to an H—O—H bond angle of 90° in water, for example, and this does not agree with the observed geometry of the molecule. And yet the relative simplicity of the idea that bonds result from orbital overlap remains alluring. This leads to the introduction of the concept of orbital **hybridization**, which is intended to reconcile the notion of orbital overlap with a wide range of observations on molecular shapes and structures. At its heart, the idea of hybridization acknowledges the fact that when two or more atoms approach one another closely enough to form chemical bonds, the interactions can be strong enough to reshape the orbitals on those atoms. The simplest stable binary compound formed by carbon and hydrogen is methane, CH_4, for which the Lewis structure is shown below along with a molecular model:

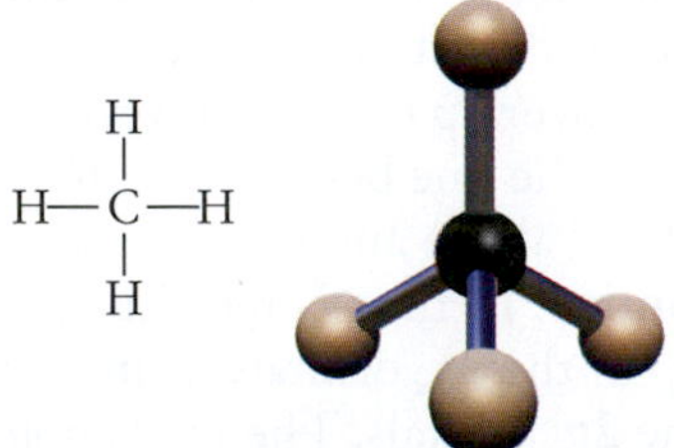

Bond lengths and strengths can be measured experimentally, and such measurements clearly show that all four C—H bonds in CH_4 are identical. Furthermore, molecular shapes can also be measured by observing the rotation of molecules, and such measurements show that the H—C—H bond angles in CH_4 are all 109.5°, as shown in the molecular model above. These data suggest that if we wanted to describe the bonds in CH_4 in terms of orbital overlap, we would need four identical orbitals on the carbon atom and that these orbitals should be arranged at angles of 109.5° from one another. When we introduced the idea of orbitals, we noted that they were mathematical in origin. In mathematics, we can manipulate functions and transform them according to accepted rules. The wave functions for atomic orbitals can also be manipulated in the same ways. We need not be concerned with all the details of such manipulations, but one rule will be important; the number of orbitals we have must not change due to the manipulation. If we have two orbitals to begin with, we must still have two orbitals when we finish. Among other things, this ensures that we can still accommodate all the electrons needed. If we take linear combinations of atomic

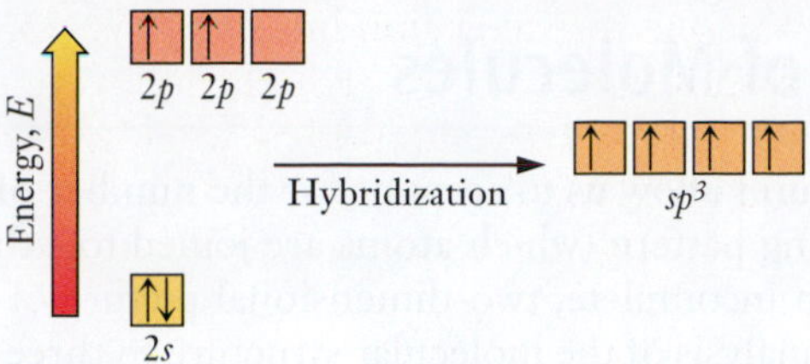

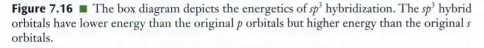

Figure 7.16 ■ The box diagram depicts the energetics of sp^3 hybridization. The sp^3 hybrid orbitals have lower energy than the original p orbitals but higher energy than the original s orbitals.

orbitals, we can devise new orbitals that can be used as the basis for a model of bonding. The resulting orbitals are called **hybrid orbitals** because they are mixtures of two or more atomic orbitals.

Orbital theory began as a model. As long as adjustments, such as the invention of hybrid orbitals, do not change the premise of the model, such changes can be made—and are worth making if they provide new insight.

One way to envision this process is to use boxes to represent orbitals, as we did in Chapter 6 when we introduced electron configurations. **Figure 7.16** shows this for the hybridization of a carbon atom. The *s* orbital and the three *p* orbitals are mixed to form four new hybrid orbitals, all having the same energy. The name given to these orbitals is sp^3, where the superscript indicates that three *p* orbitals are included. Placing four electrons in these four equivalent sp^3 orbitals means that each will be singly occupied. Thus these orbitals readily explain the ability of carbon to form four identical bonds. If we were to examine the details of the mathematics, we would also discover that each of these sp^3 orbitals is strongly oriented in a particular direction at angles of 109.5° between orbitals. These hybrid orbitals thus make it easy to explain the observations for CH_4. The overlap of an sp^3 orbital from carbon and an *s* orbital from hydrogen forms each of the four C—H bonds.

Other hybridization schemes are also possible, as shown in **Table 7.2.** When only one *p* orbital is included in the scheme, we form *sp* hybridized atoms, and when two *p* orbitals are included, the hybrid orbitals are sp^2. All of these schemes can be used to help explain bonding in covalent molecules.

Table ■ 7.2

The three common orbital hybridization schemes are shown. The names for these hybrid orbitals are derived from the type and number of orbitals that combine to form them. Thus sp^2 hybrids result from combining an *s* orbital with a pair of *p* orbitals. The orbital geometries shown in the right-hand column give rise to the common molecular shapes described in the next section.

Orbitals Combined	Hybridization	Orbital Geometry
s, *p*	*sp*	A
s, *p*, *p*	sp^2	A
s, *p*, *p*, *p*	sp^3	A

In the figures shown here, each lobe represents a different hybrid orbital. This is an important contrast to the unhybridized p orbitals, where each orbital consists of two lobes of electron density.

7-8 Shapes of Molecules

Although Lewis structures allow us to account for the number of bonds in a molecule and show us the bonding pattern (which atoms are joined together and by how many bonds), they give us an incomplete, two-dimensional picture of the molecular structure. If we extend our analysis of the molecular structure to three dimensions, we need to examine the order in which the atoms are joined to one another and also how they are arranged in space—in other words, the **molecular shape**. In addition to being of theoretical interest, the shapes of molecules are important because they affect molecular properties, including reactivity. Being able to predict the shapes of molecules is crucial to understanding the properties and reactivity of a polymer like PMMA, the polymeric proteins and nucleic acids that are the operating systems in living organisms, the catalysts used in industrial processes, and the molecules synthesized for medicines and consumer goods. One thing that sets experienced chemists apart from beginning students is the ability to read a molecular formula and visualize what the molecule would look like on the microscopic scale.

To arrive at an understanding of the shapes of molecules, we need to analyze Lewis structures in more detail. A very useful technique for doing this relies on the **valence shell electron pair repulsion (VSEPR) theory**, which states that *molecules assume a shape that allows them to minimize the repulsions between electron pairs in the valence shell of the central atom.* Lewis structures emphasize the importance of pairs of electrons around atoms, either as bonding pairs or as lone pairs. If we consider that these pairs occupy localized regions of space, we would expect that the various regions would be as far away from each other as possible. After all, these regions are centers of negative charge, and like charges repel one another. Couple this *electron pair repulsion* with the fact that Lewis structures include only electrons in the *valence shell*, and the origin of the name VSEPR is apparent. The theory provides a very simplistic view of the electrons in molecules, but it leads to surprisingly accurate predictions of molecular geometries.

VSEPR is an application of concepts from geometry. In particular, we want to answer this question: If the bond lengths are fixed, what geometric shapes allow the regions of negative charge represented by various electron pairs to be as far away from one another as possible? The answer depends on the number of electron pairs that surround the atom. **Table 7.3** shows the basic geometry for distributing two through six electron pairs to minimize repulsions. Examining this table, you should realize that most of the shapes involved are fairly simple. It should be intuitively clear that when only two regions of electron density are present, for example, a linear arrangement would maximize their separation. Of all the shapes shown, the tetrahedral and trigonal bipyramidal arrangements are probably the least familiar. You should be sure that you have a real understanding of what these different geometries look like. (Models—either physical or software-based—can be very helpful.) ◀ⓘ Most of the bonding arrangements we will encounter can be analyzed in terms of the five basic shapes shown here.

If you don't have access to molecular models, you can build your own by using small balloons and holding them together with a clip of some sort. The balloons must stay out of each other's way physically, analogous to the electronic repulsions of electrons. ⓘ

We can predict molecular geometries systematically. Because our approach centers on the arrangement of electron pairs around the central atom, we will always start by drawing a Lewis structure. Once we've done that, it is simple to count the number of bonding and nonbonding electron pairs around the central atom. We'll start with the most straightforward examples, in which all the electrons around the central atom are in single bonds. Then we'll find out how to determine the molecular geometry when there are also nonbonding pairs or multiple bonds involved.

If the central atom is surrounded only by single bonds, the process of determining the molecular geometry is fairly simple. We count the number of single bonds emanating from the central atom and use the information in Table 7.3 to correlate that number of electron pairs with the geometry predicted by VSEPR, and this describes the shape of the molecule. Example Problem 7.7 demonstrates the application of these ideas to determine the shape of molecular or ionic species.

Table ■ 7.3

Each of the geometrical arrangements shown in the table minimizes the electron pair repulsions for the indicated number of electron pairs. To visualize the shapes of molecules, it is essential that you have a sound mental picture of each of these geometries.

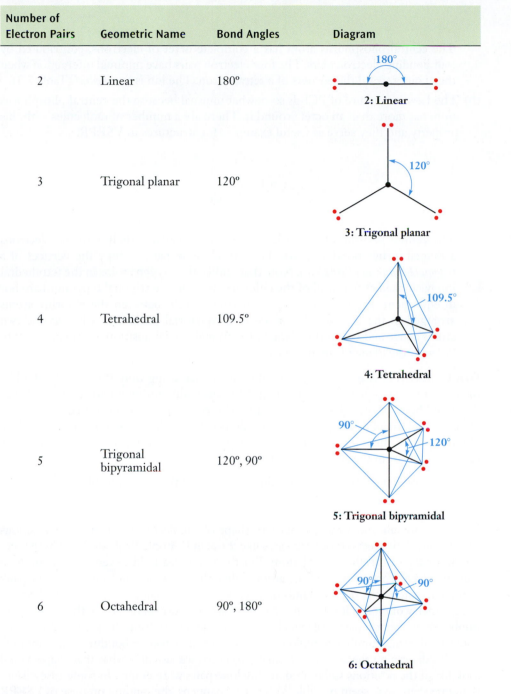

Number of Electron Pairs	Geometric Name	Bond Angles	Diagram
2	Linear	180°	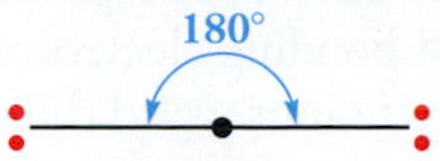 2: Linear
3	Trigonal planar	120°	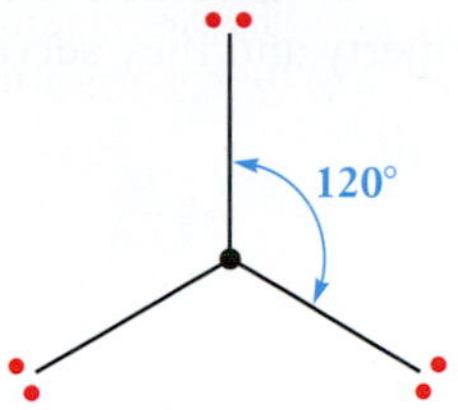 3: Trigonal planar
4	Tetrahedral	109.5°	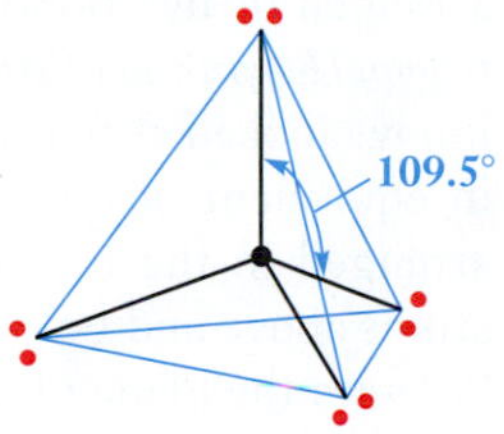 4: Tetrahedral
5	Trigonal bipyramidal	120°, 90°	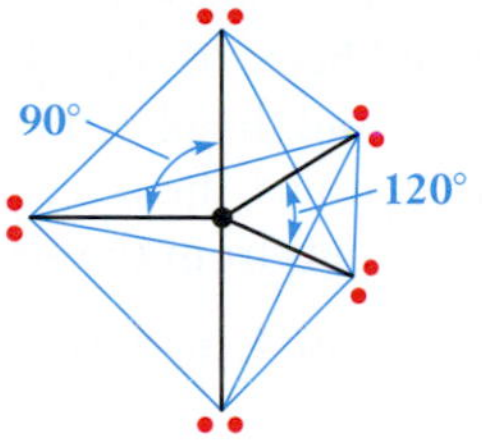 5: Trigonal bipyramidal
6	Octahedral	90°, 180°	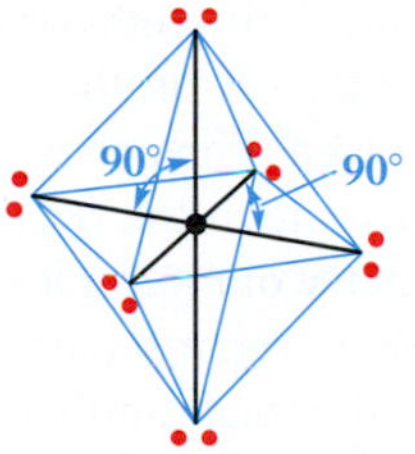 6: Octahedral

EXAMPLE PROBLEM 7.7

Determine the shape of each of the following species. (a) PO_4^{3-}, (b) PCl_5

Strategy Draw the Lewis structures. Each of these molecules has only single bonds around its central atom. So we can count the number of bonding pairs around the central atom and assign the geometry by consulting Table 7.3 as needed.

Solution

(a) The Lewis structure of PO_4^{3-}, which we drew in Example Problem 7.5, is

$$\left[\begin{array}{c} :\ddot{O}: \\ | \\ :\ddot{O}-P-\ddot{O}: \\ | \\ :\ddot{O}: \end{array}\right]^{3-}$$

The central phosphorus atom has a complete octet of electrons, contributed by four bonding electron pairs. The four electron pairs have minimal interaction when they point toward the corners of a tetrahedron. The ion is *tetrahedral* (Table 7.3).

(b) The Lewis structure of PCl_5 is somewhat unusual because the central phosphorus atom has more than an octet around it. There are a number of molecules with this property and they serve as useful examples for structures in VSEPR.

$$\begin{array}{c} :\ddot{Cl}: \\ | \quad \ddot{Cl}: \\ :\ddot{Cl}-P \\ | \quad \ddot{Cl}: \\ :\ddot{Cl}: \end{array}$$

The central phosphorus atom has an expanded valence shell with ten electrons arranged in five bonding pairs. The five electron pairs occupy the vertices of a *trigonal bipyramid* (Table 7.3). Note that, unlike the oxygen atoms in the tetrahedral ion we looked at in (a), all of the chlorine atoms in the trigonal bipyramid are not in equivalent positions. There is a spacing of 120° between the chlorine atoms arranged in the horizontal triangle (the equatorial positions), whereas the two atoms above and below the plane of the triangle (axial positions) form an angle of 90° with the plane of the triangle.

Discussion These examples can be difficult to see using only the text. Model kits or software models can help you visualize the three-dimensional nature of the shapes described here. In this example, we had no choice but to draw more than eight electrons around the phosphorus atom. If we do not have at least one bond to each of the chlorine atoms, the molecule would not stay together. Such structures never occur for central atoms from the second row of the periodic table.

Check Your Understanding Predict the shape of the SF_6 molecule.

Whenever we talk about the geometry or shape of a molecule, we refer to the positions of its atoms. This seems obvious in cases like those in Example Problem 7.7, where there are no lone pairs on the central atom. But the same idea holds, even if there are lone pairs involved. The positions of the atoms define the shape of a molecule. But lone pairs also contribute to the overall repulsion between electron pairs, and so they do *influence* the geometry of molecules. When the central atom is surrounded by both bonding and nonbonding electron pairs, we can choose a base geometry from the list in Table 7.3 by counting the total number of regions of high electron density—bonding pairs plus lone pairs. To determine the actual molecular geometry, we visualize what that shape would look like if the positions to be occupied by lone pairs were empty. In some cases, more than one shape may seem possible. When that happens, the guiding premise of VSEPR is that the correct shape will be the one that minimizes the overall repulsion between electron pairs. In deciding which structure that is, the order of repulsive interactions is lone pair–lone pair > lone pair–bond pair > bond pair–bond pair. This order can be explained in terms of the fact that lone pairs are attracted only to one nucleus. This means that a lone pair is less tightly localized, so it occupies a larger space. Examples of the various possible geometries are shown in **Table 7.4.** Determining the geometry of a molecule with both bonding and nonbonding electrons is illustrated in Example Problem 7.8.

Table ■ 7.4

The molecular shapes resulting from various combinations of the total number of electron pairs around the central atom and the number of lone pairs

Note that as long as you have a good grasp of the basic geometries (Table 7.3), it is not necessary to memorize this extensive listing. Instead you should be able to start from the basic geometry for the number of electron pairs present and visualize the shape that would result from "emptying" a number of positions corresponding to the lone pairs.

Number of Electron Pairs	Number of Lone Pairs	Shape	Ball and Stick Model
3	0	Trigonal planar	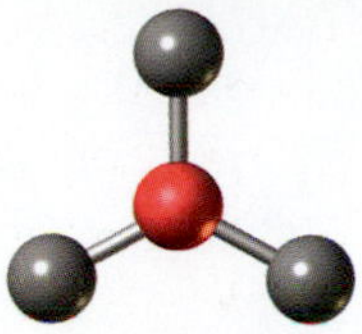
3	1	Bent (120°)	
4	0	Tetrahedral	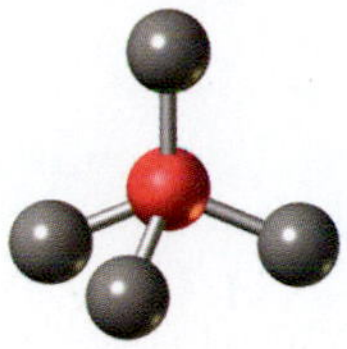
4	1	Trigonal pyramidal	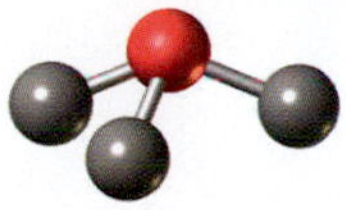
4	2	Bent (109.5°)	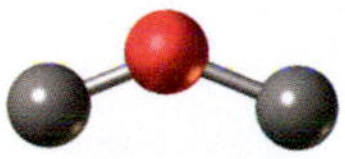
5	0	Trigonal bipyramidal	
5	1	Seesaw	
5	2	T-shape	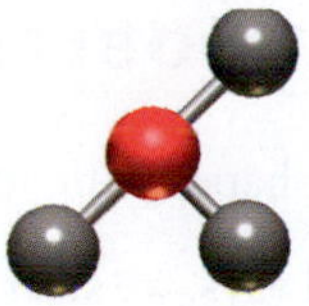

Table ■ 7.4 *(continued)*

The molecular shapes resulting from various combinations of the total number of electron pairs around the central atom and the number of lone pairs

Number of Electron Pairs	Number of Lone Pairs	Shape	Ball and Stick Model
5	3	Linear	
6	0	Octahedral	
6	1	Square pyramidal	
6	2	Square planar	
6	3	T-shape	
6	4	Linear	

EXAMPLE PROBLEM 7.8

Determine the shape of the following molecules using VSEPR theory. (a) SF_4, (b) BrF_5

Strategy As always, we start by drawing the Lewis structures. Then count the number of electron pairs around the central atom and determine the spatial arrangement of electrons pairs, consulting Table 7.3 as necessary. Place the lone pairs in positions where the electron repulsions are minimized and describe the resulting geometric arrangement of the atoms.

Solution

(a) The Lewis structure of SF_4 is as shown:

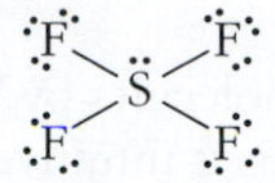

There are five pairs of electrons around sulfur, so the shape with minimum interaction is a trigonal bipyramid. Unlike the previous example, though, here the central sulfur atom has one lone pair. The five positions of a trigonal bipyramid are not equivalent, so we need to determine the best location for the lone pair. Electron pairs in the planar triangle (equatorial positions) have two near neighbors at an angle of 120° and two at 90°, whereas the axial positions have three near neighbors, each at an angle of 90°. Thus there is "more space" for electron pairs in the equatorial positions. Because lone pair–bond pair repulsion is larger than bond pair–bond pair repulsion, this means that the lone pair should occupy an equatorial position. This reduces the overall repulsive interactions. (Only the two closest bonding pairs are 90° away, whereas if the lone pair was in an axial position, there would be three bonding pairs 90° away.)

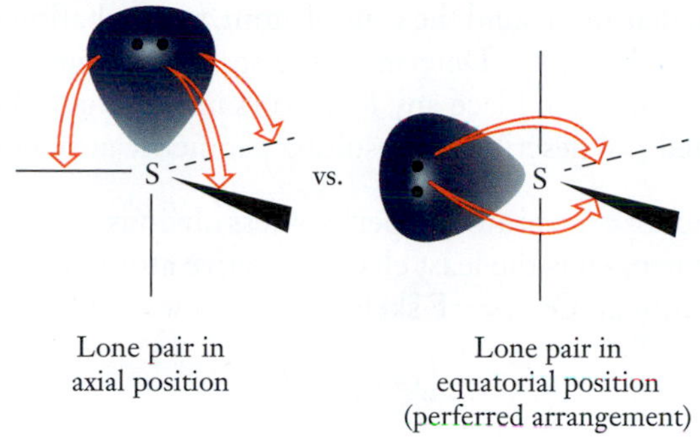

Lone pair in axial position

Lone pair in equatorial position (perferred arrangement)

The resulting shape is usually described as a seesaw. If you don't see why, try looking at a model.

(b) The Lewis structure of BrF_5 is as shown:

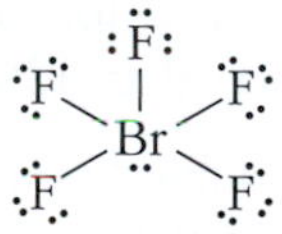

There are six electron pairs around the central bromine atom, so the distribution of electrons is based on an octahedron. A lone pair occupies one of the six positions, so we must consider where to place it. But because an octahedron is completely symmetrical and all of its vertices are equivalent, the lone pair can occupy any corner. The resulting shape of the molecule is a square pyramid.

Discussion For many students, the confusing part of this type of problem lies in the appearance that a lone pair of electrons is considered for part of the problem but not for all of it. There are two things to remember: (1) The shape of a molecule is determined by the positions of its atoms. This is a result of the ways we measure shapes. Because nuclei are so much more massive than electrons, our experiments detect them more readily, and thus we consider that the shape is defined by the positions of the nuclei. (2) Whether or not it is involved in bonding, every electron pair in the Lewis structure constitutes a region in space where there is a relatively high concentration of negative charge. The key concept of VSEPR theory is that these negatively charged regions will repel each other and therefore must be arranged as far apart as possible. So the electrons dictate the shape of a molecule, but they are not directly measured. That's why we have a process where we appear to consider them at some stages but not at others.

Check Your Understanding What are the shapes of the following molecules? (a) ClF_3, (b) XeF_2

In determining the geometry of molecules by VSEPR, multiple bonds are treated as a single region of electron density. This is intuitively reasonable because the position of greatest electron density for a multiple bond lies in the region between the two bonding atoms, just as for a single bond. To put this in another way, it wouldn't make sense to say that the electron pairs making up a double or triple bond should point in totally different directions from one another. Thus for purposes of determining molecular shapes, any double or triple bonds behave as if they were a single electron pair. Example problem 7.9 shows how we determine the shape of a molecule containing a double or triple bond.

EXAMPLE PROBLEM 7.9

Use VSEPR theory to determine the shape of the NOF molecule.

Strategy Once again, we start by drawing the Lewis structures. Then count the number of regions of electron density around the central atom, remembering to count any double or triple bonds as a single region. Determine the spatial arrangement of electron pairs, consulting Table 7.3 as needed. Place any lone pairs in positions where the electron repulsions are minimized and describe the resulting geometric arrangement of the atoms.

Solution The choice of central atom is perhaps less obvious here than in some cases we have looked at. But nitrogen is the least electronegative atom in this molecule, so it will be central. Starting from an O—N—F skeleton, the Lewis structure is as shown below:

$$\ddot{\underset{\cdot\cdot}{O}}{=}\ddot{N}{-}\ddot{\underset{\cdot\cdot}{F}}{:}$$

There are three regions of electron density around the nitrogen atom, so the shape with minimum interaction is trigonal planar. The central nitrogen atom has one lone pair. In the trigonal planar geometry, all three positions are equivalent, so it does not matter which one the lone pair occupies. The molecule will be bent, with an O—N—F bond angle of approximately 120°. Because lone pairs provide stronger repulsion, the actual bond angle should be slightly less than 120°.

Discussion Much like lone pairs, double bonds can cause confusion for many students. When counting the number of valence electrons to draw the Lewis structure, we must be sure to count the double bond as four electrons. (And a triple bond, if one were present, would count as six electrons.) But in determining the shape, we treat the double (or triple) bond as a single region of electron density. It may help if you realize that both of the bonds in the double bond must "point" in the same direction because they connect the same pair of atoms.

Check Your Understanding What are the shapes of the following molecules? (a) COF_2, (b) COS

Applying the same principles as for smaller molecules, VSEPR theory can also be used to predict the geometry of larger molecules. In predicting the structure of a molecule with more than one "central" atom, we look at each atom within the backbone of the molecule and predict the bonding about that atom, using the same procedures that we have just outlined. For example, let's see what happens when we apply VSEPR theory to the structure of vinyl alcohol, $H_2C{=}CHOH$. The Lewis structure we drew in Example Problem 7.6 is a good place to start:

The order in which we look at different central atoms is not important.

$$\begin{array}{c} \quad\ \ H\quad H \\ \quad\ \ |\quad\ \ | \\ H{-}C{=}C{-}\ddot{\underset{\cdot\cdot}{O}}{-}H \end{array}$$

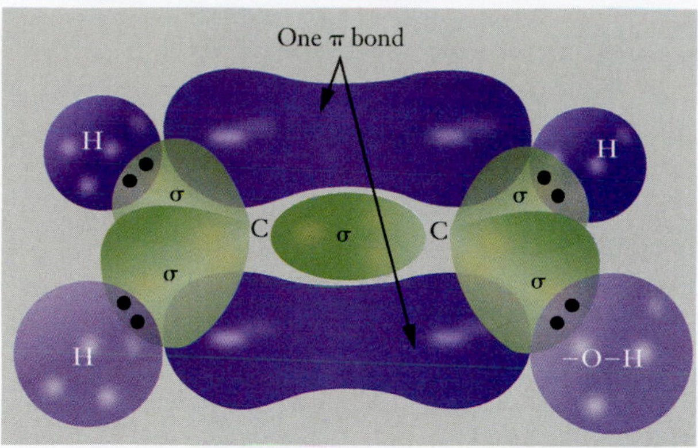

Figure 7.17 ■ The orbitals involved in the bonding in $H_2C{=}CHOH$ are shown. The σ-bonding framework results from the overlap of sp^2 hybrid orbitals on the carbon atoms with **(a)** each other, **(b)** the s orbitals on hydrogen atoms, and **(c)** an sp^3 orbital on the oxygen. The π bond is due to overlap of the remaining unhybridized p orbitals of the carbon atoms. Thus the electron density of the π bond would be above and below the plane of the molecule. The shape at each carbon atom is trigonal planar, and the carbon atoms and the three hydrogen atoms bound to them always lie in the same plane. Although not shown in the figure, the shape around the oxygen atom is bent, and the hydrogen atom in the OH group is free to rotate out of the plane of the molecule.

Each of the two carbon atoms is bound to three other atoms and has no lone pairs, implying trigonal planar geometry. Next, we need to ask whether or not those two trigonal planar ends of the molecule are coplanar. Here it helps to invoke hybridization and think about the orbitals involved in forming the double bond. Each carbon atom can be described as using sp^2 hybrid orbitals. In each case, two of those hybrid orbitals will overlap with orbitals from the hydrogen or oxygen atoms, and the third overlaps an sp^2 orbital from the other carbon. This gives us a single σ bond between the carbon atoms so far. But each carbon also still has one remaining unhybridized p orbital, and that p orbital is perpendicular to the plane of the sp^2 hybrids, as shown in **Figure 7.17.** If these p orbitals are aligned parallel to one another, they will be able to overlap side to side, forming a π bond. This explains the existence of the double bond and also tells us that the carbon atoms and the atoms bound to them must all lie in the same plane. If one end of the molecule were to rotate with respect to the other, the p orbitals would no longer be in the right orientation to form the π bond.

INSIGHT INTO

7-9 Molecular-Scale Engineering for Drug Delivery

Throughout this chapter, we have seen a number of ways in which aspects of chemical bonding are important factors in the design of materials for biomedical engineering applications. It is increasingly possible to consider doing engineering design at the molecular level, and one active area for this work lies in more targeted delivery of drugs for therapeutic benefits.

Many pharmaceuticals, particularly those used in the treatment of cancer through chemotherapy, have significant side effects. One reason why side effects occur is because the drug molecules affect healthy cells as well as diseased ones. Thus a strategy that uses new biomaterials to deliver drugs exclusively to the desired locations (in tumor cells, for example) would represent an important advance in biomedical engineering. This must be accomplished within the constraints of biocompatibility that have already been emphasized in this chapter.

A promising method for this type of drug delivery system uses a material called mesoporous silica nanoparticles (MSN). ◂ As we saw earlier, silica is composed

Nanotechnology will be discussed further in Chapter 8.

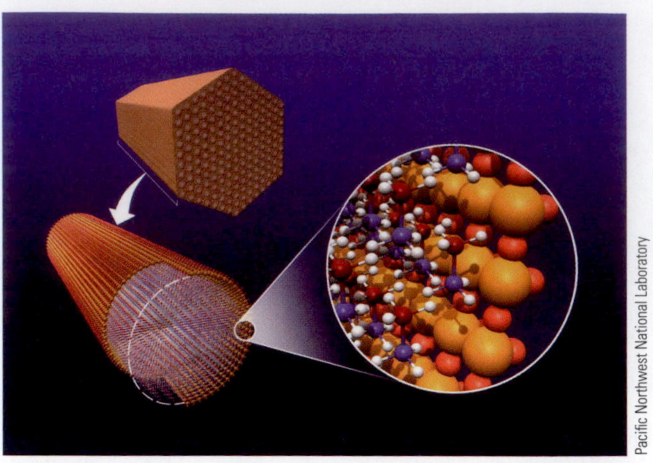

Figure 7.18 ■ This drawing shows the honeycomb structure found in mesoporous silica. The overall honeycomb structure is formed from an array of individual tubular pores like the one shown at the lower left of the figure. The inset at the right of the figure shows how an individual pore might be loaded with small molecules in a drug delivery application.

of networks of SiO_4 units, and those SiO_4 units can form a honeycomb structure as shown in **Figure 7.18.** An MSN is simply a very small particle with this honeycomb arrangement. Because of this structure, these particles have enormous surface to volume ratios: one gram of the material has roughly the same surface area as a football field. Moreover, the pores in an MSN can be used to store drug molecules. Once loaded with the desired therapeutic agents, the pore can be capped with another molecule, and the whole nanoparticle is delivered to the target.

Earlier we noted that amorphous silica particles can destroy red blood cells and so are not biocompatible. In light of this knowledge, it may seem surprising that these MSNs could be biocompatible. The difference arises from the honeycomb structure of the MSNs. Although the overall surface area of the MSN is very large, the majority of that surface area represents the interior walls of the honeycomb pores. Relatively little of the surface area is actually on the outside of the particle itself. And since the pores are smaller than red blood cells, those SiO_4 groups lining the inner walls of the pores cannot damage blood cells.

The development of nanoparticles like these MSNs for drug delivery represents an engineering design project at the molecular scale. Perhaps the most critical element in this design is how the cap molecules at the end of the pores are to be held in place. The materials chosen for this task must be capable of staying in place until an agent in a target cell somehow breaks the bond. To undertake this type of work, biologists, engineers, and chemists work together to tune the type and strength of bonding used, showing once again how the collaboration between experts in different fields moves modern technology forward.

FOCUS ON PROBLEM SOLVING

Question An unidentified solid is dissolved in water to produce a clear, colorless solution, and this solution conducts electricity. A second solution is added, and a precipitate forms. Once the precipitate has settled, the liquid above it does not conduct electricity perceptibly. What does this experiment tell you about the bonding in the initial solid and in the precipitate, and how does it tell you that?

Strategy This question is another example of conceptual problem solving. We have seen three types of chemical bonding in this chapter, and from the observable behavior of this experiment, we need to infer which one is important in the unknown solid. The other piece of information we need to include involves what allows a solution to conduct electricity, as discussed in Chapter 3.

Answer For any substance to conduct electricity, it must allow a charge to move. The charged particles in solution are usually ions, so this flow of charge arises from the movement of ions. This strongly suggests that the original solid was ionically bonded. How does this explain the observations? When the ionic solid is dissolved, its constituent ions dissociate from one another, allowing the solution to conduct electricity. The addition of the second solution forms a precipitate. This tells us a couple of things. First, because it was formed from two solutions, the precipitate must be an insoluble ionic compound. Moreover, one of the constituent ions in the precipitate must have come from each solution. If the anion came from the first solution, the cation must have come from the second. We can also note that because the final solution no longer conducts electricity, it must not contain significant concentrations of any ions. This points toward a few possibilities. Each of the individual solutions must have contained both anions and cations to be electrically neutral. When the two solutions were mixed, these ions must have recombined to form two products: the cation from the first solution combining with the anion from the second and vice versa. So we have two products to account for. One possibility is that two different precipitates may have formed, removing all of the ions. Alternatively, the second product could be water, formed from the neutralization reaction between an acid and a base.

SUMMARY

Building on our understanding of atomic structure, we have now established a better picture of how atoms "stick together" in chemical bonds. The forces involved are those between charged particles, as is most easily seen in ionic bonding. When a metallic atom forms a cation and a nonmetal forms an anion, it is not at all surprising that these oppositely charged particles attract each other and produce what is referred to as an ionic bond. We can go further in our understanding of ionic bonds by viewing bond formation as a series of steps. Forming a cation requires an energy input equal to the ionization energy, and the formation of an anion generally releases energy, as indicated by the negative electron affinity. Finally, the arrangement of the oppositely charged particles into a lattice results in significantly stronger attractive forces than repulsive ones, so metals and nonmetals routinely form ionic compounds.

When bonds form between nonmetal atoms, electrons are shared in pairs rather than wholly transferred to form ions. This sharing of electron pairs is called covalent bonding. Different atoms possess varying capacities to attract electrons within a covalent bond, as measured by the electronegativity. So the distribution of shared electrons is not always symmetrical and leads to the idea of polar bonds. There is a continuum of behavior from even sharing (as between two identical atoms) in covalent bonds, to uneven sharing to form polar covalent bonds, and to complete transfer of electrons from one atom to another to form ionic bonds.

The propensity of atoms to share electrons can be summarized visually in a Lewis structure. By following a fairly simple set of rules, we can draw Lewis structures for many ions or molecules. We can then use those Lewis structures to predict the nature of the bonding involved. Most importantly, Lewis structures help us understand whether one or more pairs of electrons need be shared—leading to predictions of single versus double or triple bonds for molecules.

Beyond predicting what types of bonds are present in a molecule, however, the Lewis structure tells us fairly little about the details of a chemical bond. To understand just how electrons are shared, we must realize that electrons behave as waves and that when waves overlap, they interfere with each other. When the waves "buildup," they are said to interfere constructively and the chemical implication is the formation of a chemical bond. Overlap can be achieved in more than one way, so that we can distinguish types of bonds as either sigma or pi bonds.

Finally, in addition to simply representing a pair of shared electrons, a chemical bond has structural implications as well. Because electrons are negatively charged, when there are several distinct bonds, they will tend to be physically separated from each other. This idea is the basis for a method to predict the geometry of molecules called the Valence Shell Electron Pair Repulsion (VSEPR) theory. Using this theory, the general shape of molecules and ions can be predicted.

KEY TERMS

biocompatibility (7-1)
bond energy (7-3)
bond length (7-3)
bonding pair (7-3)
covalent bond (7-3)
dipole (7-4)
double bond (7-3)
electronegativity (7-4)
hybrid orbitals (7-7)
hybridization (7-7)
ionic bond (7-2)
lattice energy (7-2)
Lewis dot symbol (7-3)
Lewis structure (7-3)
lone pair (7-3)
molecular shape (7-8)
octet rule (7-3)
orbital overlap (7-6)
pi bond (7-6)
polar bond (7-4)
resonance (7-5)
sigma bond (7-6)
triple bond (7-3)
valence bond model (7-6)
VSEPR theory (7- 8)

PROBLEMS AND EXERCISES

INSIGHT INTO Materials for Biomedical Engineering

7.1 Define the term *biocompatibility*.

7.2 List some properties associated with biomaterials used for joint replacements.

7.3 Describe how PMMA functions as bone cement. How does chemistry play a role in this process?

7.4 The formulas of sodium phosphate and sodium orthosilicate are Na_3PO_4 and Na_4SiO_4 respectively. What are the chemical formulas of calcium phosphate and calcium silicate?

7.5 Why do biomedical engineers sometimes need to use coatings on materials they use? When coatings are generated, what are the materials used as reactants called?

7.6 Use the concept of polarity of water and the basic composition of the body to explain why the polarity of biomaterials is important.

The Ionic Bond

7.7 Why is the Na^{2+} ion not found in nature?

7.8 Why do nonmetals tend to form anions rather than cations?

7.9 Select the smaller member of each of the following pairs. **(a)** N and N^{3-}, **(b)** Ba and Ba^{2+}, **(c)** Se and Se^{2-}, **(d)** Co^{2+} and Co^{3+}

7.10 Arrange the members of each of the following sets of cations in order of increasing ionic radii. **(a)** K^+, Ca^{2+}, Ga^{3+}, **(b)** Ca^{2+}, Be^{2+}, Ba^{2+}, Mg^{2+}, **(c)** Al^{3+}, Sr^{2+}, Rb^+, K^+, **(d)** K^+, Ca^{2+}, Rb^+

7.11 Arrange the following sets of anions in order of increasing ionic radii. **(a)** Cl^-, P^{3-}, S^{2-}, **(b)** S^{2-}, O^{2-}, Se^{2-}, **(c)** Br^-, N^{3-}, S^{2-}, **(d)** Br^-, Cl^-, I^-

7.12 Which pair will form a compound with the larger lattice energy: Na and F or Mg and F? Why?

7.13 In a lattice, a positive ion is often surrounded by eight negative ions. We might reason, therefore, that the lattice energy should be related to eight times the potential of interaction between these oppositely charged particles. Why is this reasoning too simple?

7.14 Use the concept of lattice energy to rationalize why sodium fluoride dissolves in water, whereas calcium fluoride does not. Extending this reasoning, would you expect magnesium fluoride to be soluble?

7.15 Figure 7.2 depicts the interactions of an ion with its first nearest neighbors, second nearest neighbors, and third nearest neighbors in a lattice. **(a)** Would the interactions with the fourth nearest neighbors be attractive or repulsive? **(b)** Based on Coulomb's law, how would the relative sizes of the terms compare if the potential energy were expressed as $V = V_{1st} + V_{2nd} + V_{3rd} + V_{4th}$?

7.16 What type of bond is likely to form between one element with low ionization energy and another element with high electron affinity? Explain your answer.

The Covalent Bond

7.17 Describe the difference between a covalent bond and an ionic bond.

7.18 Considering the potential energy of interacting particles, how does covalent bonding lower the energy of a system?

7.19 Sketch a graph of the potential energy of two atoms as a function of the distance between them. On your graph, indicate how bond energy and bond distance are defined.

7.20 When a covalent bond forms, is energy absorbed or released? Explain how your answer is related to the graph you sketched in the previous problem.

7.21 Coulombic forces are often used to explain ionic bonding. Are coulombic forces involved in covalent bonding as well? Explain.

7.22 In terms of the strengths of the covalent bonds involved, why are combustion reactions exothermic?

7.23 If the formation of chemical bonds always releases energy, why don't all elements form dozens of bonds to each atom?

7.24 Draw the Lewis dot symbol for each of the following atoms. **(a)** boron, **(b)** fluorine, **(c)** selenium, **(d)** indium

7.25 Theoretical models for the structure of atomic nuclei predict the existence of superheavy elements that have not yet been discovered and suggest that such elements might be fairly stable if they could be produced. So researchers are currently trying to synthesize these superheavy elements to test these theories. Suppose that element 117 has been synthesized and given the atomic symbol Lt. What would the Lewis dot symbol be for this new element?

7.26 Use Lewis dot symbols to explain why chlorine bonds with only one hydrogen atom.

7.27 Define the term *lone pair*.

7.28 How many electrons are shared between two atoms in **(a)** a single covalent bond, **(b)** a double covalent bond, and **(c)** a triple covalent bond?

7.29 How does the bond energy of a double bond compare to that of two single bonds between the same elements? How does this relationship explain the types of reactions that compounds with double bonds undergo?

Electronegativity and Bond Polarity

7.30 How is electronegativity defined?

7.31 Distinguish between electron affinity and electronegativity.

7.32 Certain elements in the periodic table shown in Figure 7.8 had no electronegativity value defined. Based on the definition of electronegativity and the identity of these elements, hypothesize as to why they have no electronegativity value.

7.33 When two atoms with different electronegativities form a covalent bond, what does the electron distribution in the bond look like?

7.34 The bond in HF is said to be polar, with the hydrogen carrying a partial positive charge. For this to be true, the hydrogen atom must have less than one electron around it. Yet the Lewis dot structure of HF attributes two electrons to hydrogen. Draw a picture of the electron density distribution for HF and use it to describe how the hydrogen atom can carry a partial positive charge. How can these two models of the HF bond (the electron density and the Lewis structure) seem so different and yet describe the same thing?

7.35 Why is a bond between two atoms with different electronegativities called a polar bond?

7.36 Based on the positions in the periodic table of the following pairs of elements, predict whether bonding between the two would be primarily ionic or covalent. Justify your answers. **(a)** Ca and Cl, **(b)** P and O, **(c)** Br and I, **(d)** Na and I, **(e)** Si and Br, **(f)** Ba and F.

7.37 In each group of three bonds, which bond is likely to be the most polar? Which will be the least polar? **(a)** C—H, O—H, S—H, **(b)** C—Cl, Cl—Cl, H—Cl, **(c)** F—F, O—F, C—F, **(d)** N—H, N—O, N—Cl

7.38 Considering the trends in electronegativity shown in Figure 7.8, explain why alloys that form between two or more transition metals are not ionic substances.

7.39 Can fluorine atoms ever carry a partial positive charge in a molecule? Why or why not?

7.40 Which one of the following contains *both* ionic and covalent bonds? **(a)** $BaCO_3$, **(b)** $MgCl_2$, **(c)** BaO, **(d)** H_2S, **(e)** SO_4^{2-}

7.41 Using only halogen atoms, what would be the most polar diatomic molecule that could be formed? Explain your reasoning.

7.42 Suppose that you wanted to make a diatomic molecule using two different halogen atoms. What combination of halogens would be the least polar?

Keeping Track of Bonding: Lewis Structures

7.43 Draw the Lewis structure for each of the following molecules. **(a)** CO, **(b)** H_2S, **(c)** SF_6, **(d)** NCl_3

7.44 Draw a Lewis structure for each of the following molecules or ions. **(a)** CS_2, **(b)** BF_4^-, **(c)** HNO_2 (where the bonding is in the order HONO), **(d)** $OSCl_2$ (where S is the central atom)

7.45 Write Lewis structures for these molecules. **(a)** Tetrafluoroethylene, C_2F_4, the molecule from which Teflon is made, **(b)** Acrylonitrile, CH_2CHCN, the molecule from which Orlon® is made

7.46 Why is it impossible for hydrogen to be the central atom in the Lewis structure of a polyatomic molecule?

7.47 In the context of Lewis structures, what is resonance?

7.48 Draw resonance structures for **(a)** NO_2^-, **(b)** NNO, and **(c)** HCO_2^-.

7.49 How does the structure of a molecule that exhibits resonance justify the use of the term *resonance hybrid*?

7.50 Draw the Lewis structure of two resonance forms for ozone, O_3. What prediction can you make about the relative lengths of the O—O bonds?

7.51 How does the existence of resonance structures for a molecule such as benzene explain its unusually unreactive chemical behavior?

7.52 Consider the nitrogen–oxygen bond lengths in NO_2^+, NO_2^-, and NO_3^-. In which ion is the bond predicted to be longest? In which is it predicted to be the shortest? Explain briefly.

7.53 Which of the species listed has a Lewis structure with only one lone pair of electrons? F_2, CO_3^{2-}, CH_4, PH_3

7.54 Identify what is incorrect in the Lewis structures shown for BBr_3 and SO_2.

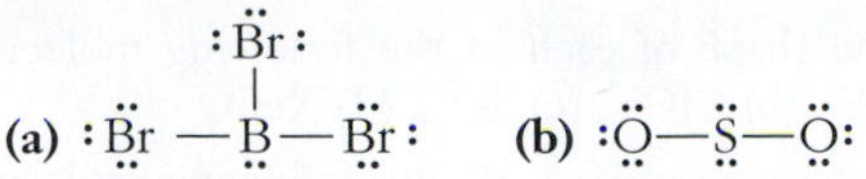

(a) :B̤r̈ —B(—:B̤r̈:)—B̤r̈: (b) :Ö—S̤̈—Ö:

7.55 Identify what is incorrect in the Lewis structures shown for O_3 and XeF_4.

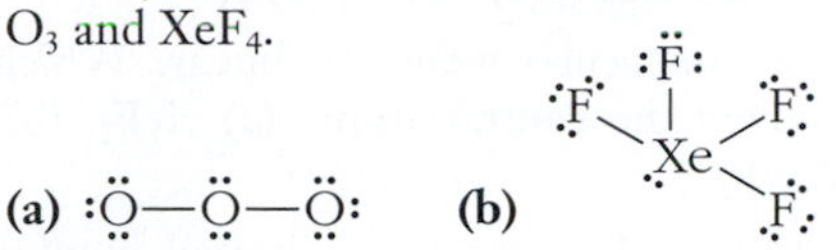

(a) :Ö—Ö—Ö: (b) Xe bonded to four F atoms (:F:, :F:, :F:, :F:)

7.56 Chemical species are said to be *isoelectronic* if they have the same Lewis structure (regardless of charge). Consider these ions and write a Lewis structure for a neutral molecule that is isoelectronic with them. **(a)** CN^-, **(b)** NH_4^+, **(c)** CO_3^{2-}

Orbital Overlap and Chemical Bonding

7.57 Explain the concept of wave interference in your own words.

7.58 Distinguish between constructive and destructive interference.

7.59 How is the concept of orbital overlap related to the wave nature of electrons?

7.60 How does overlap explain the buildup of electron density between nuclei in a chemical bond?

7.61 How do sigma and pi bonds differ? How are they similar?

7.62 CO, CO_2, CH_3OH, and CO_3^{2-}, all contain carbon–oxygen bonds. Draw Lewis structures for these molecules and ions. Given that double bonds are stronger than single bonds and triple bonds are stronger than double bonds, rank the four species in order of increasing C—O bond strength.

7.63 Draw the Lewis dot structure of the following species and identify the number of pi bonds in each. **(a)** CS_2, **(b)** CH_3Cl, **(c)** NO_2^-, **(d)** SO_2

7.64 Draw the Lewis dot structures of the following compounds and identify the number of pi bonds in each. **(a)** Cl_2O, **(b)** H_2CCH_2, **(c)** HCCCN, **(d)** SiO_2

Hybrid Orbitals

7.65 What observation about molecules compels us to consider the hybridization of atomic orbitals?

7.66 The number of hybrid orbitals formed must always be equal to the number of atomic orbitals combined. Suggest a reason why this must be true.

7.67 What type of hybrid orbital is generated by combining the valence *s* orbital and all three valence *p* orbitals of an atom? How many hybrid orbitals result?

7.68 Considering only *s* and *p* atomic orbitals, list all the possible types of hybrid orbitals that can be used in the formation of **(a)** single bonds, **(b)** double bonds, and **(c)** triple bonds.

7.69 What hybrid orbitals would be expected for the central atom in each of the following molecules? **(a)** $GeCl_4$, **(b)** PBr_3, **(c)** BeF_2, **(d)** SO_2

7.70 What type of hybridization would you expect for the carbon atom in each of the following species? **(a)** CO, **(b)** CO_2, **(c)** H_2CO, **(d)** CH_2F_2

Shapes of Molecules

7.71 What physical concept forms the premise of VSEPR theory?

7.72 Predict the geometry of the following species. **(a)** SO_2, **(b)** $BeCl_2$, **(c)** $SeCl_4$, **(d)** PCl_5

7.73 Predict the shape of each of the following molecules or ions. **(a)** IF_3, **(b)** ClO_3^-, **(c)** TeF_4, **(d)** XeO_4

7.74 Predict the shape of each of the following molecules or ions. **(a)** PH_4^+, **(b)** OSF_4, **(c)** ClO_2^-, **(d)** I_3^-

7.75 Which of these molecules would be linear? Which have lone pairs around the central atom? **(a)** XeF_2, **(b)** CO_2, **(c)** BeF_2, **(d)** OF_2

7.76 Give approximate values for the indicated bond angles. **(a)** Cl—S—Cl in SCl_2, **(b)** N—N—O in N_2O, **(c)** bond angles marked as 1, 2, and 3 in the following structure for vinyl alcohol:

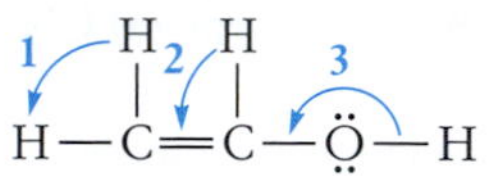

```
 1   H 2 H    3
     |   |   ..
 H — C ═ C — O — H
             ..
```

7.77 Propene has the chemical formula $H_2C{=}CH{-}CH_3$. Describe the overall shape of the molecule by considering the geometry around each carbon atom.

7.78 Describe why a central atom with four bonding pairs and one lone pair will have the lone pair in an equatorial rather than an axial location.

7.79 Describe what happens to the shape about the carbon atoms when a C═C double bond undergoes an addition reaction in which it is converted into a C—C single bond.

INSIGHT INTO Molecular Scale Engineering for Drug Delivery

7.80 Describe why treatments like chemotherapy for cancer patient often have serious side effects. What is a strategy that could lessen such side effects?

7.81 Use the fundamental structure of mesoporous silica nanoparticles (MSN) to explain why they have such a large surface area.

7.82 How does an MSN differ from amorphous silica so that it has improved biocompatibility?

7.83 Draw a schematic picture of an MSN being used as a molecular engineering design selective drug delivery vehicle. Label all key components of the design.

Additional Problems

7.84 In 1999, chemists working for the Air Force Research Laboratory reported that they had produced a compound containing an unusual ion with the formula N_5^+. This ion had not been expected to exist, and its instability makes it both difficult and dangerous to prepare.

Consider the possible Lewis structure below. Indicate the hybridization expected for each nitrogen atom and the expected bond angles.

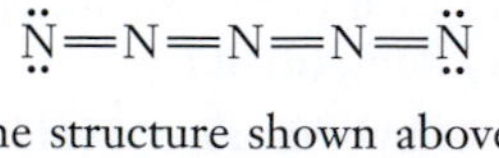

```
 ..               ..
 N ═ N ═ N ═ N ═ N
 ..               ..
```

Assuming that the structure shown above is correct, how many of the five nitrogen atoms would *always* lie in the same plane?

7.85 The following molecules have similar formulas, but each has a different shape. Draw Lewis structures for these compounds and then determine the molecular shape. **(a)** SiF_4, **(b)** KrF_4, **(c)** SeF_4

7.86 Consider the Lewis structure below. Is this an ion? If so, what is its charge?

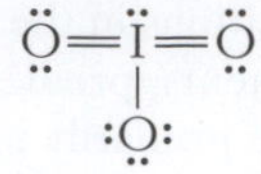

```
 ..   ..   ..
 O ═  I  ═ O
 ..   |    ..
     :O:
      ..
```

7.87 A Lewis structure for the oxalate ion is shown below. (One or more other resonance forms are also possible.)

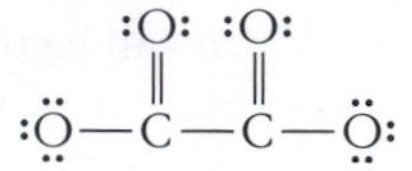

```
      :O:  :O:
       ‖    ‖
 ..    |    |    ..
:O  —  C  — C  — O:
 ..              ..
```

What is the correct charge on the oxalate ion? What type of orbital hybridization is expected for each of the carbon atoms in this structure? How many sigma bonds and how many pi bonds does the structure contain?

7.88 Methyl cyanoacrylate, $C_5H_5NO_2$, is the compound commonly sold as super glue. The glue works through a polymerization reaction, in which molecules of methyl cyanoacrylate form strong chemical bonds to one another and to the surfaces being glued. This process is initiated by traces of water on the surfaces.

The skeletal arrangement of the atoms in methyl cyanoacrylate is shown below. Complete the Lewis structure of the compound by adding lone pairs of electrons or multiple bonds where appropriate.

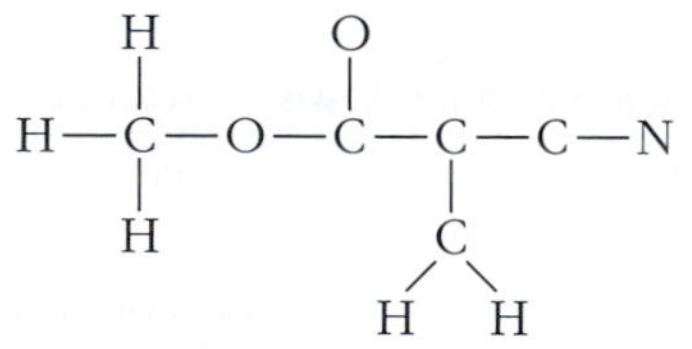

```
     H         O
     |         |
 H — C — O  —  C — C — C — N
     |             |
     H             C
                  / \
                 H   H
```

Using your Lewis structure, predict the hybridization for the six central atoms (the five carbon atoms and the leftmost oxygen atom). How many sigma bonds and how many pi bonds does your structure contain?

7.89 An unknown metal M forms a chloride with the formula MCl_3. This chloride compound was examined and found to have a trigonal pyramidal shape. Draw a Lewis structure for MCl_3 that is consistent with this molecular geometry. Use your structure to propose an identity for the metal M and explain how you have made your choice.

7.90 Nearly all of the other elements form binary compounds with hydrogen. Based on the electronegativity values shown in Figure 7.8, with what category of elements will hydrogen form bonds in which the hydrogen atom carries a partial negative charge? With what category of elements will hydrogen form bonds in which the hydrogen atom carries a partial positive charge? How does hydrogen's electronegativity value help explain its ability to form both of these types of bonds?

7.91 Although we often classify bonds between a metal and a nonmetal as ionic, not all such bonds have the same degree of ionic character. Consider the series of tin(IV) halides: SnF_4, $SnCl_4$, $SnBr_4$, and SnI_4. Rank these compounds in terms of expected ionic character, from most ionic to least ionic. What property do you need to consider in making this ranking?

7.92 Consider the hydrocarbons whose structures are shown below. Which of these molecules would be planar, meaning that all of the atoms must always lie in the same plane? Explain your answer in terms of orbital hybridizations.

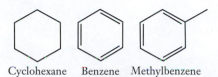

Cyclohexane Benzene Methylbenzene

7.93 Consider the structure shown below for N_2O_3 as well as any other important resonance structures.

$$:\ddot{O}=\ddot{N}-\ddot{N}(-\ddot{O}:)=\ddot{O}:$$

(a) What is the expected O—N—O bond angle in this structure? **(b)** The N_2O_3 molecule contains N—O bonds of two different lengths. How many *shorter* N—O bonds would be present?

FOCUS ON PROBLEM SOLVING EXERCISES

7.94 Electrical engineers often use lithium niobate, $LiNbO_3$, in designs for surface acoustic wave filters in devices such as cellular phones. What type(s) of bonding would you expect to be present in this compound?

7.95 One of the "grand challenges" for engineering identified by the U.S. National Academy of Engineering is the capture of carbon dioxide produced by the burning of fossil fuels. Although carbon and oxygen have significantly different electronegativities, CO_2 cannot easily be separated from air by applying a voltage in the smokestack of power plant where it is generated. How does the geometry of carbon dioxide account for this observation?

7.96 Another "grand challenge" for engineering is managing the nitrogen cycle. Use the nature of the chemical bond in diatomic nitrogen to explain why any management is needed for this biogeochemical cycle. With so much nitrogen available in the atmosphere, why are engineering solutions needed to manage nitrogen for agricultural uses?

7.97 Lead selenide nanocrystals may provide a breakthrough in the engineering of solar panels to be efficient enough to be an economical source of electricity. Selenium is generally considered a nonmetal while lead is considered a metal. Is this distinction enough to suggest that this compound should be ionic? Explain your answer.

7.98 Both ozone, O_3, and oxygen, O_2, absorb UV light in the upper atmosphere. The ozone, however, absorbs slightly longer wavelengths of UV light—wavelengths that would otherwise reach the surface of the planet. **(a)** How can Lewis structures help explain the bonding difference between ozone and oxygen that gives rise to the difference in the absorption wavelength? **(b)** What relationship or relationships between light and energy do you need to know to understand this photochemical application of Lewis structures?

7.99 Nitrogen is the primary component of our atmosphere. It is also used as an inert reagent to fill containers of chemicals that might react with the oxygen in air. Draw a Lewis structure of nitrogen and use this drawing to help explain why nitrogen does not react readily with other molecules.

7.100 Hydrogen azide, HN_3, is a liquid that explodes violently when subjected to shock. In the HN_3 molecule, one nitrogen–nitrogen bond length has been measured at 112 pm and the other at 124 pm. Draw a Lewis structure that accounts for these observations. Explain how your structure reflects these experimental facts.

7.101 If leads are attached to the opposite sides of a crystal of sodium chloride, it does not conduct electricity. Distilled water does not conduct electricity either. Yet, if the sodium chloride crystal is dissolved in distilled water, the resulting solution does conduct electricity. Use pictures at the molecular level of detail to describe why this occurs.

Cumulative Problems

7.102 How does electronegativity relate to the periodic trends of atomic properties? How does effective nuclear charge, the concept invoked to understand those trends, also help explain the trends in electronegativity?

7.103 How do the Lewis symbols for C, Si, and Ge reflect the similarity in their electron configurations?

7.104 Use Lewis symbols to describe how free radical polymerization occurs.

7.105 When free radical polymerization occurs, how does the hybridization change the carbon atoms involved?

8 Molecules and Materials

CHAPTER OUTLINE

Original illustration courtesy of Raymond Schaak

The drawing shows the crystal structure for bimetallic FePt. Nanoscale particles of this material are magnetic and hold promise for extremely high density information storage devices.

The chemical bond is clearly one of the unifying concepts in chemistry, and our understanding of chemical bonds continues to improve. For as long as we have known that atoms join together to form molecules, scientists have continually refined the models we use to describe bonding. Active fields in modern chemical research include both computational modeling of bonds and experimental efforts to manipulate individual bonds highly selectively. Chemical bonding is the "glue" that holds molecules together and also the concept that holds together the study of chemistry.

The concepts of chemical bonding that we have introduced so far may seem rather detached from the properties of the materials used in engineering designs. To bridge the gap between isolated molecules and those materials, our vantage point must be extended. In this chapter, we will look at materials such as polymers and metals and see how ideas related to chemical bonding can be applied to understand their important engineering properties. We begin by considering the various forms of carbon, an element that plays a role in many of these materials.

Chapter Objectives

After mastering this chapter, you should be able to

- describe the structures of graphite and diamond and explain how the properties of each substance arise from its structure.

- describe the arrangement of atoms in the common cubic crystal lattices and calculate the packing efficiency for a lattice.
- use band theory to describe bonding in solids.
- draw band diagrams for metals, insulators, and semiconductors (including n- and p-type materials).
- identify a material as a metal, insulator, or semiconductor from its band diagram.
- explain how the electrical properties of metals, insulators, and semiconductors are related to their chemical bonding.
- identify the types of intermolecular forces likely to be most important for a particular substance.
- explain the connection between intermolecular forces and properties such as boiling point and vapor pressure.
- describe the growth of polymers through addition and condensation reactions and predict which of these processes is likely to be important for a given monomer.
- describe the connection between polymer properties and molecular structure.

8-1 Carbon

When you think about the elements in the periodic table, you probably assume that most things are known about them. For decades, chemistry textbooks said that there were two forms of the element carbon: graphite and diamond. In 1985, that picture changed overnight. A team of chemists at Rice University discovered a new form of carbon, whose 60 atoms formed a framework that looks like a tiny soccer ball. Because the structure resembled the geodesic domes popularized by the architect Buckminster Fuller, it was given the whimsical name *buckminsterfullerene*. The discovery of C_{60} and other related molecules, now collectively known as the fullerenes, helped usher in a new form of carbon and also a new branch of science—nanotechnology. New carbon structures are certainly not the only area where nanotechnology research is flourishing, but the story of C_{60} is an excellent example of the way the study of chemistry affects the development of new materials. More traditional forms of carbon have long been used as materials in important applications as well.

The 1996 Nobel prize in chemistry was awarded to Professors Curl, Kroto, and Smalley for the discovery of C_{60}.

The physical properties of diamond, for example, have led to a wide range of uses. Among those properties, the impressive appearance of cut diamonds is responsible for their most familiar use in jewelry. From an engineering perspective, however, the hardness of the diamond probably represents a more significant property. Diamond drill bits find uses in many areas, including the drilling of oil wells in some locations. (The choice of drill bits depends on the nature of the rock that must be traversed.) The hardness of diamonds is sufficiently useful that industrial processes to convert graphite into small diamonds were sought for many years. Although a number of processes for making synthetic diamonds for certain applications are now viable, their use remains rather limited. Considering the cost of naturally occurring diamonds, this immediately suggests that making diamonds must be both difficult and expensive.

Figure 8.1 indicates why it is so hard to make a diamond. The figure shows what is known as a **phase diagram** for carbon. It is a map showing which state of the element will be most stable at a given combination of temperature and pressure. If you pay close attention to the units on the vertical scale, you will see that diamond becomes the preferred form of the element only at extraordinarily high pressures—higher than about 20,000 atm! Fortunately, the process by which diamond would be converted to graphite is so slow as to be essentially nonexistent, so at least the diagram does not imply that our existing diamonds will soon be turned to graphite. But it does

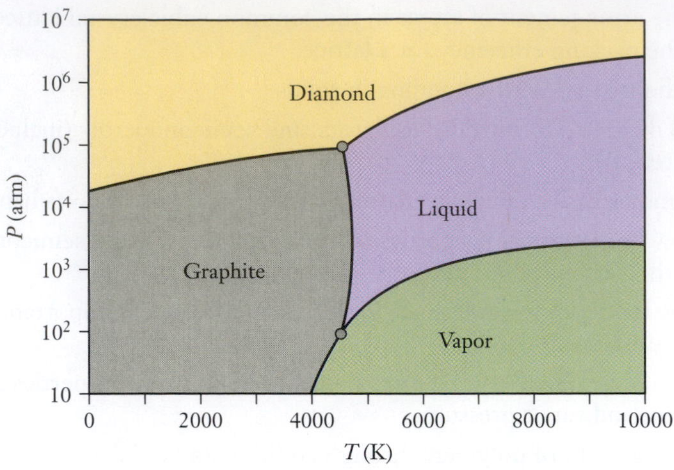

Figure 8.1 ■ This phase diagram for carbon shows which form of the element is most stable as a function of temperature and pressure. Note that pressure is shown on a logarithmic scale. It is clear that graphite is the favored form at all ordinary combinations of temperature and pressure and that the formation of diamonds requires extremely high pressure.

Diamond films and wafers can also be grown at very low pressure by a process called chemical vapor deposition (CVD).

tell us that to convert bulk graphite into diamond, we will need to achieve very high pressures. (As it turns out, high temperatures are also needed to speed up the conversion process to a useful rate.) Generating and maintaining these high temperatures and pressures is difficult, expensive, and dangerous, and that's why we don't see synthetic diamonds in every jewelry store. As we explore bonding in solids, we will want to look into some fundamental questions about why these extreme conditions might be needed to form diamonds. How is the form of an element related to variables such as pressure and temperature? How do graphite and diamond differ in the first place so that an observable transformation from one to the other can be affected?

Of course, graphite is not just a raw material for the production of diamonds; it has important applications of its own. If you happen to be writing notes in the margin of this book using a pencil, you are using graphite. What are the properties of graphite that allow it to be useful in a lead pencil, and how does the bonding in graphite give rise to those properties? Graphite also is used as a lubricant. How do lubricants work, and what is their impact on engineering designs?

Other modern materials also rely on graphite as a key, if not exclusive, component. Various carbon fiber materials offer an appealing combination of high strength and low weight and are increasingly popular in sports equipment. Graphite shafts on golf clubs or carbon fibers in snow skis confirm the durability and flexibility of this material. It's difficult to imagine making a golf club out of a bundle of pencil leads, so what is required to convert the graphite into a material that can be used in this application? For a composite material such as carbon fiber, what factors will control the properties?

Moving from the practical to the possible, what do the new discoveries about C_{60} hold for materials that engineers might use in the 21st century? Although the small sphere of C_{60} may not have any immediate uses, manipulation of experimental conditions allows the growth of tubes of carbon, called **nanotubes**, as shown in **Figure 8.2**. These materials are not yet viable for large-scale design projects, but they have remarkable properties. Their tensile strength is significantly higher than that of steel, and futurists have already begun to contemplate their potential for things as speculative as space elevators. Carbon nanotubes grown with a metallic element enclosed within them might represent a way to build wires that are one molecule wide. The implications for miniaturization of electronic devices are among the most immediate and exciting opportunities promised by new nanotechnologies. So scientists and engineers are actively exploring fundamental questions related to nanotechnology. How does the inclusion of metal atoms within a nanotube provide properties that might lead to a molecular wire? This question points toward the need to understand the nature of bonding in bulk samples of metals as well.

The connections between molecular-level characteristics and the macroscopic properties of materials are not always easy to discern, but current research in materials

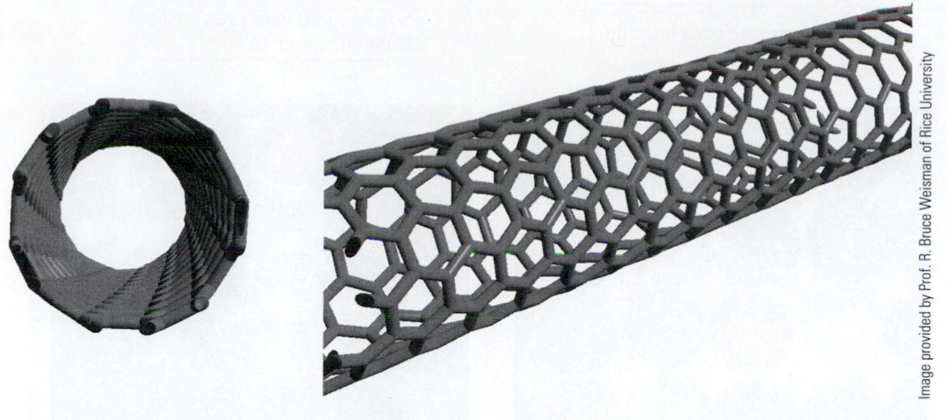

Image provided by Prof. R. Bruce Weisman of Rice University

Figure 8.2 ■ Nanotubes like the one pictured here are the most recently discovered form of the element carbon. The structure of the tubes has similarities to both graphite and buckminsterfullerene. Wide ranging applications of these tubes are already beginning to emerge.

science and computer modeling are advancing our ability to make them. In this chapter, we will introduce ideas that can be used to infer the molecular-scale explanations of why materials behave the way they do. Along the way, we will be able to answer at least some of the questions we have raised about carbon-based materials and the emerging field of nanotechnology.

8-2 Condensed Phases—Solids

The various forms of carbon that we have just described share one important physical property—they are all solids at normal laboratory temperature and pressure. The vast majority of elements are solids under such conditions. What factors contribute to the stability of condensed phases? Forces between atoms and molecules certainly play a role in the answer to this question, but the basic structure of condensed phases also contributes to an understanding of their stability. We will begin our treatment from this structural perspective.

The term condensed phases *refers to both solids and liquids.*

Back in Chapter 1 in our first description of the microscopic view of matter, we noted that the atoms or molecules in a solid or liquid are packed together much more closely than they would be in a gas. When we consider how those atoms or molecules are arranged into a solid, there are two broad categories. Many substances assume regular, repeating geometric arrangements and are referred to as **crystalline solids**. Other substances solidify into random arrangements, and are known as **amorphous solids**. Though both categories include important examples, much more is known about crystalline materials simply because their regular periodic arrays can be studied much more easily with available experimental tools. We will, therefore, focus on crystalline solids in our description here.

Because atoms can be thought of as spheres, arranging them into a crystalline solid is loosely analogous to stacking a collection of balls into a box. It is intuitively clear that no matter how we arrange the balls in the box, there will always be some gaps between them; the spheres can never completely fill the volume of the box. But it should also be clear that the amount of empty space among the balls could be minimized by carefully arranging them. We can illustrate the differences in the way the balls—or atoms—may be packed by using a two-dimensional model, as shown in **Figure 8.3** using marbles. It is easy to see that the gaps can be reduced if the rows of marbles are offset from one another. This idea can be quantified by thinking about the **packing efficiency** of a structure, which represents the percentage of space that is occupied in a given arrangement. We'll save the calculation of this result for a problem at the end of the chapter, but the offset structure has a packing efficiency of 90.7%, whereas the structure in which the rows all line up has a packing efficiency of 78.5%.

Nonspherical molecules can also be packed into crystals, but we will limit our discussion to spherical atoms for the sake of simplicity.

Figure 8.3 ■ The marbles in the picture on the left are arranged in a very orderly pattern, but significant gaps remain. If our aim were to place as many marbles as possible into a given area, we would do better by staggering the rows, as shown in the right-hand panel. Both pictures are the same size, but the arrangement on the right allows more marbles to fit in that area.

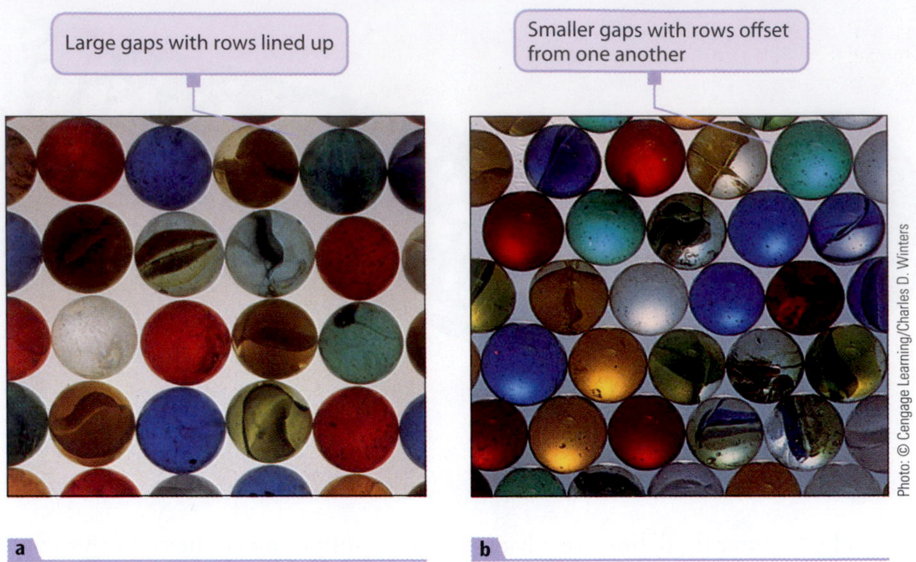

Now let's shift our thinking from marbles to atoms. The packing efficiency of the atoms will clearly be related to the density of a material because an increase in the packing efficiency will put more atoms into the same volume. Experiments on crystal structures show that many elements condense into solids that fill space as completely as possible. Such arrangements are called close-packed structures and are the three-dimensional equivalent of the staggered rows of marbles in Figure 8.3(b).

In moving from circles in two dimensions to spheres in three dimensions, though, we'll have some additional choices to make. Let's continue with the analogy of stacking marbles or balls into a box and suppose that we have completed a single layer of balls in the close-packed arrangement of Figure 8.3(b). When a second layer of balls is added (**Figure 8.4**), we could place them directly above the balls in the first layer, or we could again offset them so that each ball rests in a gap between balls in the lower layer. It is probably clear that this offset choice again will let us achieve higher packing efficiency and fit more balls into the box. When the second layer of balls is complete and we contemplate where to place a third layer, we face an additional choice. Again, we'll want to offset the third layer from the second, but now there are two possible ways of doing this. The third layer of balls could be placed directly above those in the first layer, or they could be placed above hollows between the first layer balls. If the third layer is directly over the first layer, the structure is said to have *hexagonal close-packing* (hcp), whereas the offset third layer results in *cubic close-packing* (ccp). Both of these structures occupy roughly 74% of the available space.

The cubic close-packed structure is often referred to as **face-centered cubic (fcc)**, for reasons that are apparent when we look at the structure on the right in **Figure 8.5**. One atom is located at each corner of the cube, and additional atoms are placed at the center of each of the six faces of the cube. (Admittedly, it is difficult to see that this is the same arrangement as we saw in Figure 8.4. The best way to see it is to build models with a set of marbles or small balls.) The atoms at the four corners of a face all "touch" the atom at the center of that face. To fit an atom into the center of the face, the edge length of the cube must be larger than the diameter of the atoms, so the atoms at the corners do not touch one another. The single cube we have shown in Figure 8.5 is known as the **unit cell** for the face-centered cubic structure. It is the smallest collection of atoms that displays all of the features of this structure. To move from this unit cell to the structure of an actual sample of an fcc metal, we would need to extend the crystal lattice by repeating the unit cell in all directions.

As you are likely to discover in a subsequent materials science course, not all unit cells are cubic, but we will restrict our discussion to those that are.

Some substances do not occupy space with the level of efficiency afforded by close-packed structures. For example, some metals assume a **body-centered cubic (bcc)** structure, as shown in Figure 8.5. Here again, we have atoms at the corners of a cube, but now an atom is placed at the center of the cube instead of at the center of each face. The corner atoms touch the center atom but not each other, similarly to the face-centered structure.

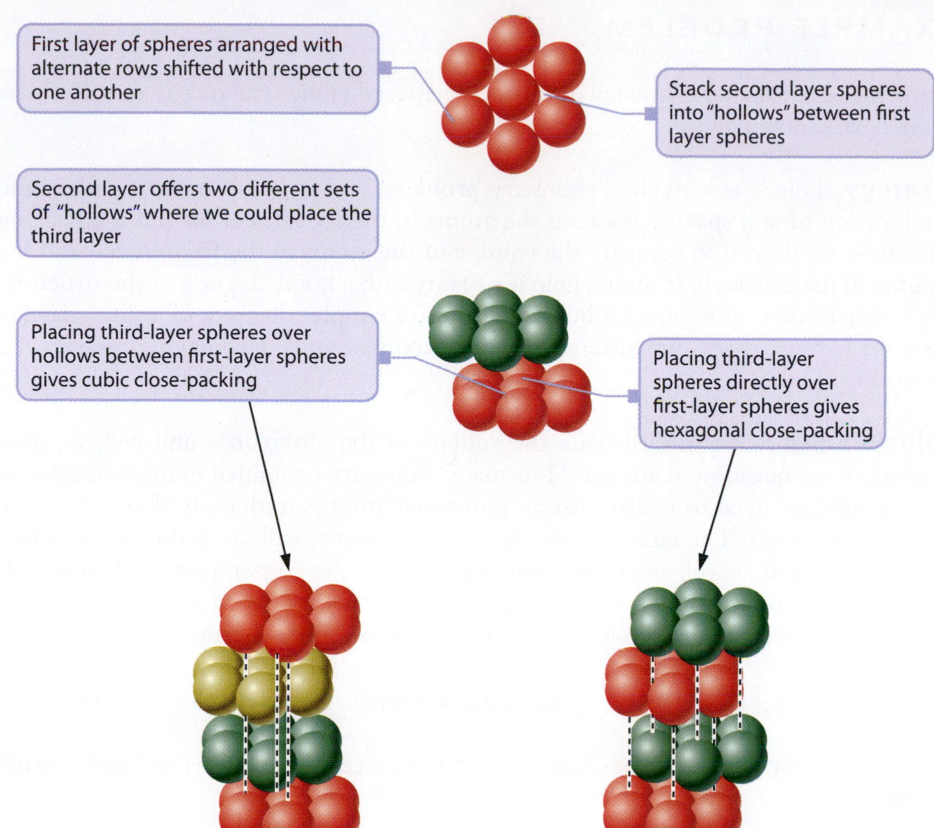
First layer of spheres arranged with alternate rows shifted with respect to one another
Stack second layer spheres into "hollows" between first layer spheres
Second layer offers two different sets of "hollows" where we could place the third layer
Placing third-layer spheres over hollows between first-layer spheres gives cubic close-packing
Placing third-layer spheres directly over first-layer spheres gives hexagonal close-packing

Figure 8.4 ■ Stacking of spheres offers two ways to achieve maximum packing density. The structure on the left is called cubic close-packing, and that on the right is called hexagonal close-packing. The two arrangements differ in the way that the layers line up with one another. The cubic structure is said to feature an "*a*,*b*,*c*,*a*,*b*,*c* . . ." stacking pattern, whereas the hexagonal structure has an "*a*,*b*,*a*,*b*, . . ." pattern.

The packing efficiency in this structure is lower; the atoms occupy only about 68% of the space. In rare cases, the central atom in this bcc structure is missing. This means that the edge length of the cube becomes equal to the diameter of one atom, allowing the corner atoms to touch one another. The resulting structure is called **simple cubic (sc)**. Although it is the simplest looking cubic arrangement, it is also the least efficient way to fill space. The packing efficiency is only 52%. Example Problem 8.1 illustrates how these packing efficiencies are calculated.

The simple cubic lattice is also sometimes called "primitive cubic."

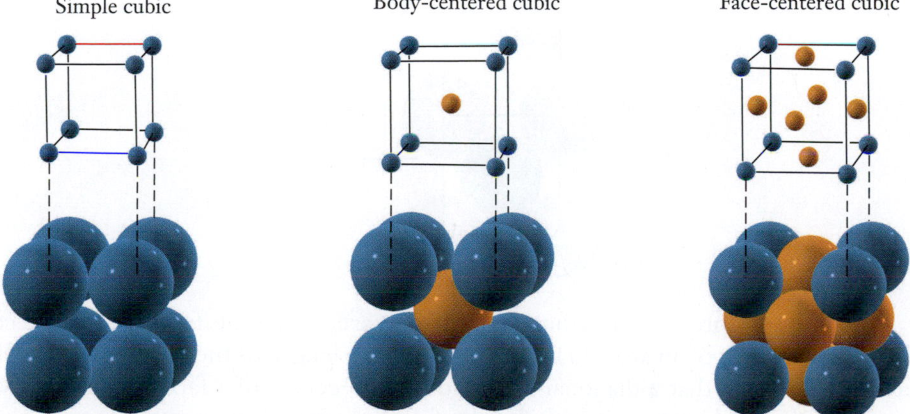
Simple cubic
Body-centered cubic
Face-centered cubic

Figure 8.5 ■ The three cubic crystal lattices are shown. In a simple cubic crystal, atoms are located at each of the corners of a cube. In a body-centered cubic crystal, an additional atom sits at the center of the cube, and in a face-centered cubic crystal, atoms are found at the center of each face of the cube. Each of these arrangements repeats throughout the crystal. All of the atoms in each of these structures are identical; different colors are used only to help you to see the different positions in the lattice.

EXAMPLE PROBLEM 8.1

Show that the packing efficiency of the face-centered cubic structure is actually 74%, as we have claimed above.

Strategy This is essentially a geometry problem, and it requires us to understand the location of and spacing between the atoms in the fcc lattice. To find the packing efficiency, we'll need to compare the volume of the atoms in the fcc unit cell and the volume of the cell itself. It should help if we start with a good diagram of the structure. We'll rely in part on Figure 8.5 but will also use a simpler diagram of a single face of the cube because three-dimensional crystal structures can be challenging to draw and interpret.

Solution Before we can calculate the volume of the atoms in a unit cell, we have one very basic question to answer: How many atoms are contained in the unit cell? To answer that, we need to realize that an individual atom is frequently shared between multiple unit cells. The atoms on the faces, for example, will be in the unit cell that we are looking at as well as the adjacent one. Each of the corner atoms will appear in a total of eight unit cells because the corners of eight cubes would meet if we were to stack the unit cells up like blocks. That gives us the following result:

$$\text{\# atoms per unit cell} = \frac{1}{2}(\text{\# face-center atoms}) + \frac{1}{8}(\text{\# of corner atoms})$$

A cube has eight corners and six faces, giving us six face-center atoms and eight corner atoms.

$$\text{\# atoms per unit cell} = \frac{1}{2}(6) + \frac{1}{8}(8) = 4$$

So each unit cell will contain four complete atoms. If we let r be the radius of an atom, then the volume of each atom is $\frac{4}{3}\pi r^3$. So the total volume occupied by the atoms themselves is the number of atoms multiplied by the volume of each atom:

$$4 \text{ atoms} \times \left(\frac{4}{3}\pi r^3\right) = \frac{16}{3}\pi r^3$$

Now we just need to find the volume of the unit cell in terms of the radius of the atoms. Here, a drawing of a single face of the cube will come in handy:

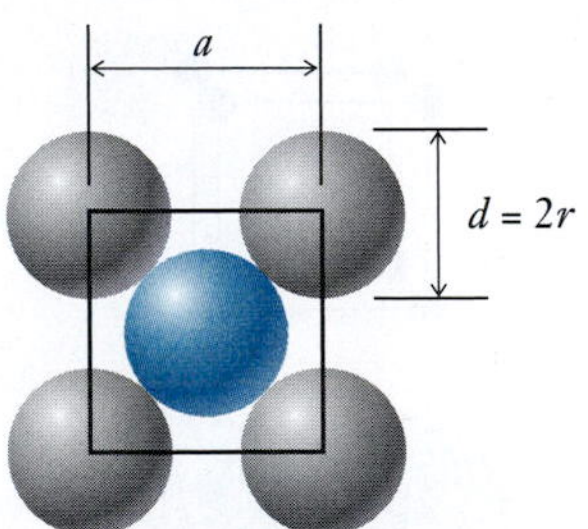

The circles represent the atoms, and the square shows the face of the unit cell cube. The diameter of an atom (d) is $2r$, and the edge length of the cube is a. From the picture, we can see that a diagonal line through the center of a face of the cube will have a length of $2d$ because it includes the full diameter of the center atom and the radius of each of two corner atoms. Because the face of the cube is a square, we also know that the diagonal line makes a 45° angle with the sides. So, recalling a little bit of trigonometry, we can write the following relationship:

$$\frac{a}{2d} = \sin 45° = \frac{1}{\sqrt{2}} = 0.707$$

So,

$$a = 1.414\,d = 2.828\,r$$

The volume of the cube is a^3, and now we can express this in terms of r:

$$V_{\text{unit cell}} = a^3 = (2.828\,r)^3 = 22.63\,r^3$$

Combine this with the volume filled by atoms from the equation above:

$$\text{Packing efficiency} = \frac{\text{volume of 4 atoms}}{\text{volume of unit cell}} \times 100\%$$

$$= \frac{\frac{16}{3}\pi r^3}{22.63\,r^3} \times 100\% = \frac{16.76}{22.63} \times 100\% = 74.05\%$$

This verifies the number cited in the text.

Analyze Your Answer Because we have confirmed the result we gave earlier, it is easy to conclude that we have done the problem correctly. If we did not know the answer in advance, how might we assess our result? First, we are calculating packing efficiency, so the result must be less than 100% to be physically meaningful. The 74% efficiency of the close-packed structures is actually an upper bound. So if we did the same calculation for a different lattice, we should expect a result less than 74%.

Discussion This simple calculation helps to point out the importance of visualizing crystal structures when working with solids. Once we constructed an accurate diagram, the calculation here required only high school geometry.

Check Your Understanding Calculate the packing efficiencies for simple cubic and body-centered cubic structures. Verify your answers by comparing them with the numbers given on page 217.

Packing efficiency is not the only important factor associated with solid structures. From the perspective of bonding, an equally important concept is the **coordination number**—the number of atoms immediately adjacent to any given atom. Looking at the structures we have introduced thus far, we see that a simple cubic structure has a coordination number of 6, a body-centered cubic structure has a coordination number of 8, and close-packed structures (both ccp and hcp) have coordination numbers of 12. This suggests a reason that close-packed structures are much more common: by increasing the coordination numbers, they allow each atom in the lattice to interact with more nearest neighbor atoms. The attractive forces of those interactions are ultimately responsible for holding the crystal together.

The periodic table in **Figure 8.6** shows the preferred crystal lattice for the solid phase of each element. Scanning this table, we see that metals typically assume one of three structures: bcc, fcc, or hcp. There is no discernible pattern as to which structure is most likely for any particular metal. Nonetheless, knowledge of the structure that a metal assumes can be very important in developing new materials. Steel, for example, is predominantly iron, but by adding relatively small amounts of other elements, its properties can be fine-tuned. Deciding what atoms might usefully combine within the solid framework of iron requires that the materials engineer know what that structure is and how it might be altered by exchanging iron atoms for those of some other element.

Some elements, including carbon, form more than one solid phase. (In such cases, the structure shown in Figure 8.6 represents the most stable form.) The familiar diamond and graphite forms of carbon have dramatically different physical properties, and those properties are very directly related to differences in crystal structure. The crystal structures of diamond and graphite are both rather unusual, in large part because they are

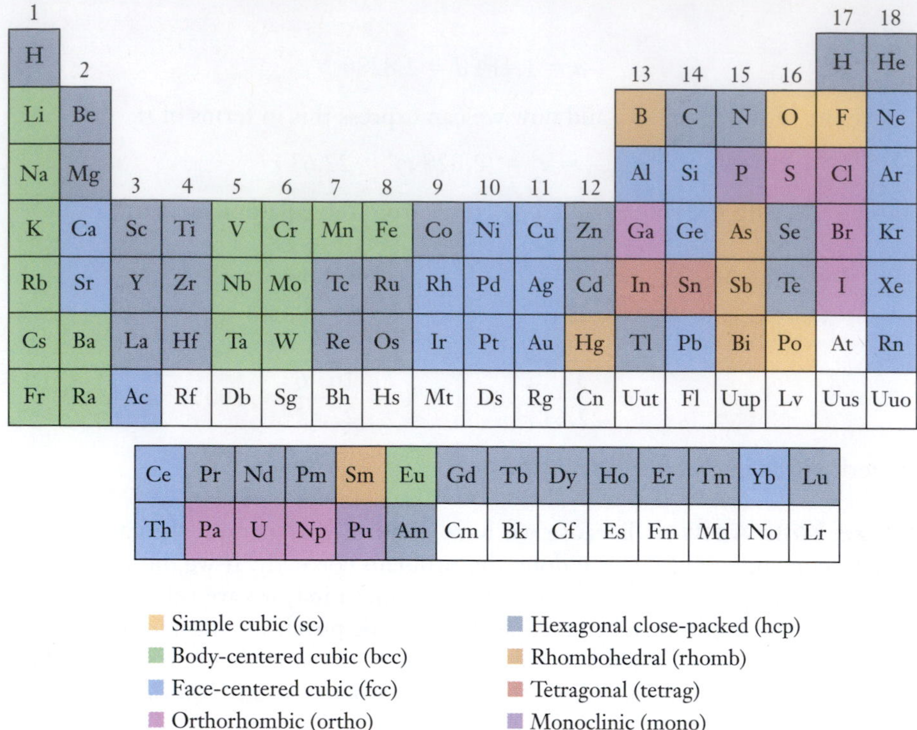

1 2 3 4 5 6 7 8 9 10 11 12 13 14 15 16 17 18
H He
Li Be B C N O F Ne
Na Mg Al Si P S Cl Ar
K Ca Sc Ti V Cr Mn Fe Co Ni Cu Zn Ga Ge As Se Br Kr
Rb Sr Y Zr Nb Mo Tc Ru Rh Pd Ag Cd In Sn Sb Te I Xe
Cs Ba La Hf Ta W Re Os Ir Pt Au Hg Tl Pb Bi Po At Rn
Fr Ra Ac Rf Db Sg Bh Hs Mt Ds Rg Cn Uut Fl Uup Lv Uus Uuo
Ce Pr Nd Pm Sm Eu Gd Tb Dy Ho Er Tm Yb Lu
Th Pa U Np Pu Am Cm Bk Cf Es Fm Md No Lr
Simple cubic (sc)
Body-centered cubic (bcc)
Face-centered cubic (fcc)
Orthorhombic (ortho)
Hexagonal close-packed (hcp)
Rhombohedral (rhomb)
Tetragonal (tetrag)
Monoclinic (mono)

Figure 8.6 ■ The shading in this periodic table indicates the most stable crystal structure for the solid state of the elements. Most of the elements display one of the structures we have discussed, but a few less common crystal arrangements are also known.

based to one degree or another on covalent bonding. The diamond structure, illustrated in **Figure 8.7**, is consistent with what we already know about bonding in carbon compounds. As we saw in Section 7-7, a carbon atom with four single bonds will have a tetrahedral structure. If this geometry is extended in all directions, the result is the diamond structure in which each carbon atom is bound to four neighboring atoms. Figure 8.7 emphasizes the local tetrahedral geometry for one carbon atom. The fact that each carbon atom is held in place by four strong C—C bonds explains why diamond is such a hard material. To break a diamond, we must break many covalent bonds, which would require a prohibitively large amount of energy. Silicon also displays the diamond structure.

As we've already noted in the opening "Insight" section, though, diamond is not the only form of solid carbon, or even the most common. Graphite obviously doesn't share the hardness of diamond, so its structure must be very different. The layered structure

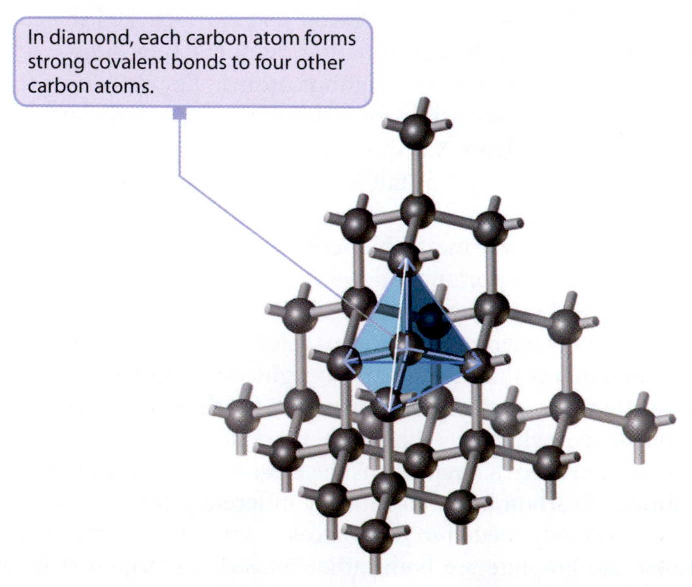

In diamond, each carbon atom forms strong covalent bonds to four other carbon atoms.

Figure 8.7 ■ The carbon atoms in diamond are joined by covalent bonds, and each atom displays the same tetrahedral geometry we saw in Chapter 7 for molecules like methane. Fracturing a diamond crystal requires breaking many covalent bonds, and this explains the inordinately high strength of diamond.

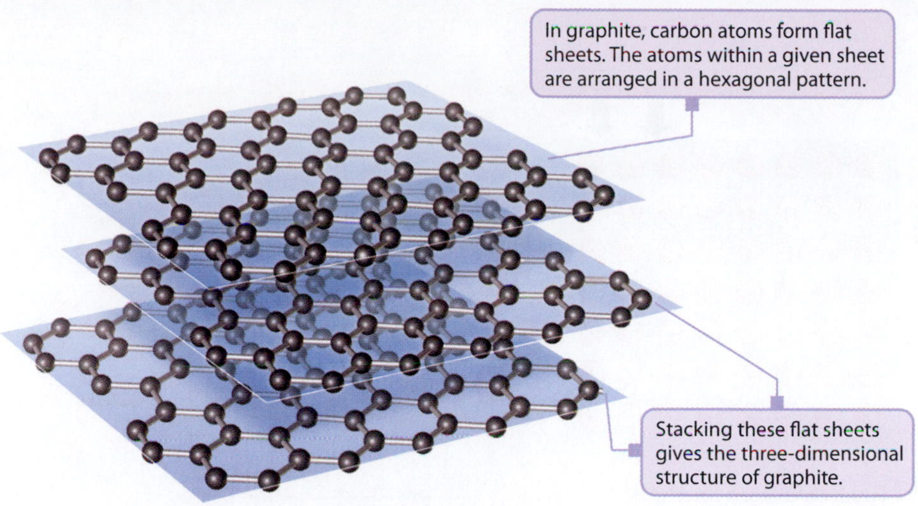

In graphite, carbon atoms form flat sheets. The atoms within a given sheet are arranged in a hexagonal pattern.
Stacking these flat sheets gives the three-dimensional structure of graphite.

Figure 8.8 ■ Graphite has a layered structure. The carbon atoms within each planar sheet are held together by covalent bonds, but much weaker forces hold the different sheets together. As we'll see in Section 8-4, many of the useful properties of graphite arise from the fact that the sheets can slide across one another.

of graphite (**Figure 8.8**) is somewhat more complicated than that of diamond. In graphite, the distance between carbon atoms within a single plane is relatively short—on the same scale as the bond length of carbon–carbon covalent bonds. Between the planes (or layers), however, the distances are much larger—more than double that within the layer. In Section 8-4, we will see that this distinction arises from fundamental differences in the forces operating within a layer versus those between layers, and it helps explain the unusual and useful physical properties of graphite.

The geometric structures of solids are a very rich topic of study, and we are only able to scratch the surface of this field here. But as an engineering student, you may take an entire course later exploring the structures and properties of materials. For our present purposes, we want to introduce the structural concepts associated with solids, so that we can extend our knowledge of chemical bonding and use it to understand solids better.

8-3 Bonding in Solids: Metals, Insulators, and Semiconductors

The physical properties of metals that lead to their widespread use in countless designs can be listed readily. You are probably familiar with the basic properties of metals from science classes dating back at least to middle school. Metals are **malleable**, meaning they can be formed into useful shapes or foils. They are **ductile**, meaning they can be pulled into wires. Metals are good conductors of electricity and heat. How does the bonding in metals help explain these properties? Can the inclusion of metal atoms inside a nanotube provide the same properties? By looking at a model of metallic bonding, we can gain significant insight into these questions.

Models of Metallic Bonding

In considering what we know about metallic elements, we might choose to start from the periodic trends noted in Chapter 6. Looking at metals in general and transition metals in particular, there is little difference in electronegativity from element to element. So metals and alloys are not likely to undergo ionic bonding. They are also a long way from having filled subshells, so they would require a tremendous number of covalent bonds to achieve complete octets. The experience we've gained so far suggests that individual metal atoms are not likely to form such large numbers of covalent bonds. If neither ionic nor covalent bonding can be expected for a metal, which models can account for the fact that metal atoms do bind to one another?

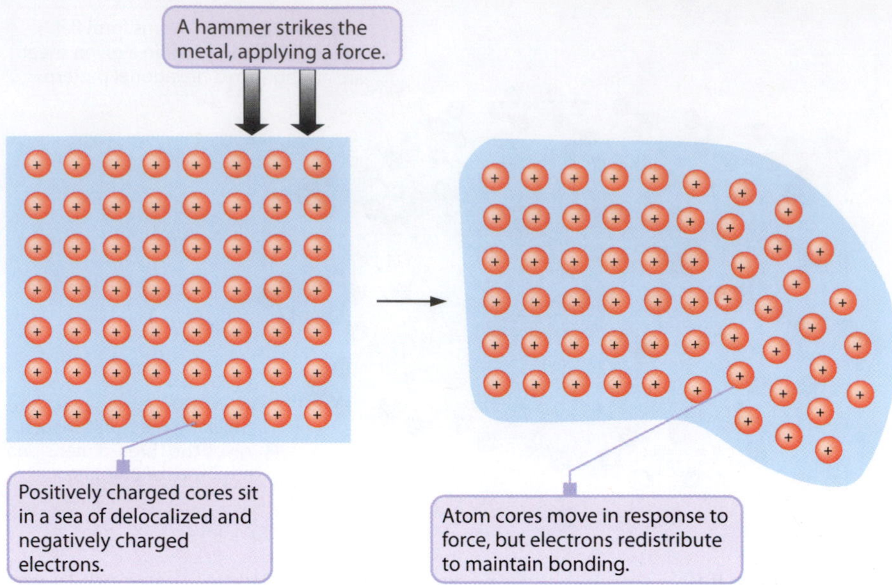
A hammer strikes the metal, applying a force.
Positively charged cores sit in a sea of delocalized and negatively charged electrons.
Atom cores move in response to force, but electrons redistribute to maintain bonding.

Figure 8.9 ■ The sea of electrons model of metallic bonding is somewhat crude but does account for the malleability and ductility of metals, as well as their electrical conductivity. The blue area represents the delocalized (and negatively charged) valence electrons spread throughout a piece of metal. The circles depict the positively charged cores of the metal atoms. In the left-hand picture, the atoms are arranged in a regular array. A force is applied as indicated by the large arrows—perhaps by striking the metal with a hammer. The cores of the metal atoms move in response to this force, and the sample is deformed as shown in the right-hand picture. But the free flow of electrons maintains the bonding throughout this process. A similar argument could be made for the process of drawing a metal through a die to make a wire.

We first introduced this model in Section 2-4, page 42.

The first and simplest explanation of metallic bonding is often referred to as the **sea of electrons model**. The essential feature of this model is that the valence electrons of metal atoms are delocalized and move freely throughout the solid rather than being tied to any specific atom, as depicted in **Figure 8.9**. This model accounts for several observed properties of metals. To understand malleability, consider what would happen at the atomic level if a metal were struck with a hammer. At least some of the atoms would respond to this force by moving relative to others. If the electrons are moving freely, however, as postulated in the sea of electrons model, they can adjust to the new position of the atoms and bonding is relatively unaffected. Similarly, pulling a metal into a wire changes the positions of the atoms, but delocalized electrons can adjust and maintain metallic bonding. Finally, these delocalized electrons can move in response to any applied electrical field, so the conductivity of metals can also be understood from the model.

The sea of electrons model provides qualitative understanding, but quantitative models for metallic bonding also exist. The most important such model is **band theory**, and we can use our example of putting metal atoms into a carbon nanotube to explain the origins of this theory.

If you could put lithium atoms inside a suitably sized nanotube, the metal atoms could be made to line up into a "wire" just one atom wide. (We use lithium for the simplicity of the *s* orbital overlap, not because this is the most practical choice.) Because the valence electron configuration of lithium is $2s^1$, any bonding interactions between the metal atoms will have to be based on the 2*s* orbitals. Those orbitals, of course, are waves, and so we can use the concept of wave interference to think about how they will interact with one another. We'll start with the simplest case: two lithium atoms placed close enough together for their 2*s* orbitals to overlap one another and interact. Now recall that waves can interfere either constructively or destructively. If the pair of 2*s*

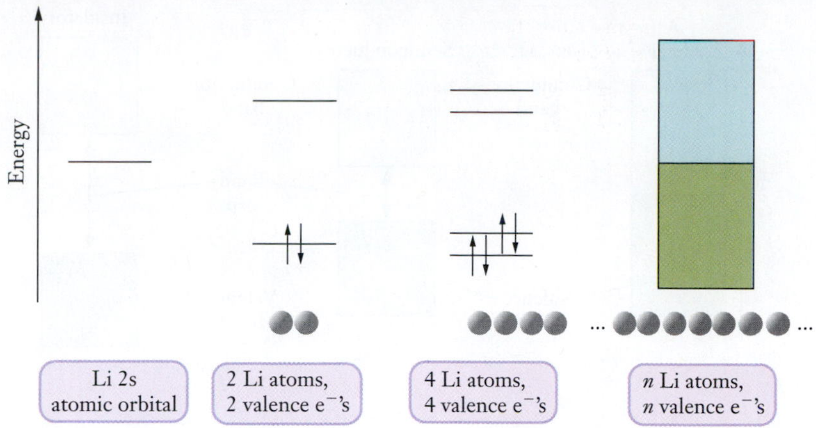

Figure 8.10 ■ Here we examine the bonding in a one-dimensional row of lithium atoms, as might be assembled in a nanotube. In the second panel, we have just two lithium atoms, each of which contributes one valence electron. With four atoms, as in the third panel, we have four electrons. As the number of atoms becomes very large, the available orbitals stack up very close together in energy. With one electron from each atom, the lower half of the available orbitals will be filled, creating the valence band. The higher energy states in the conduction band remain empty.

orbitals interacts in phase with one another, the interference will be constructive. The resulting wave, which will have increased amplitude (or electron density) between the nuclei, is referred to as a **bonding molecular orbital**. On the other hand, if the pair of 2*s* orbitals interacts out of phase with one another, the resulting wave will have a node between the nuclei. This is known as an **antibonding molecular orbital**. Rather than being associated with either atom individually, these new molecular orbitals can be thought of as belonging to the pair of lithium atoms together. **Figure 8.10** shows how the energies of these molecular orbitals are related to that of the original 2*s* orbitals. Looking at Figure 8.10, we can think through what would happen as more and more lithium atoms are loaded into the nanotube. With two atoms, there would be one bonding and one antibonding orbital. With four lithium atoms, there would be two bonding and two antibonding orbitals, as shown in the third panel of Figure 8.10. (With an odd number of atoms, one nonbonding orbital would be formed, keeping the number of bonding and antibonding orbitals equal to each other.) As we proceed further to very large numbers of atoms, the number of molecular orbitals formed becomes so large that there is virtually no energy difference from one orbital to the next. At this point, the orbitals have merged into a band of allowed energy levels, and this is the origin of the term *band theory*. Although our presentation here is entirely qualitative, it is important to note that a more thorough development of these ideas leads to a quantitative model.

The term antibonding *is used because electrons in this orbital will actually weaken the bond between the atoms.*

In this development, we used lithium atoms within a nanotube because doing so provided a plausible one-dimensional system. A three-dimensional bulk metal is slightly more complicated, and the way that the orbitals will combine has some dependence on the structure of the solid. For metals with valence electrons beyond the *s* subshell, the orbitals that ultimately form bands will include *p* or *d* orbitals. These will form additional bands, and the energy of the resulting *s* bands, *p* bands, or *d* bands may overlap one another. To understand properties such as conductivity, the band structure of the material provides a very powerful model.

Band Theory and Conductivity

The band structure of a material plays much the same role as atomic orbital energy levels did for atoms in Chapter 6. Just as the aufbau principle dictates that electrons occupy the lowest energy orbitals, electrons in bulk materials fill the bands starting

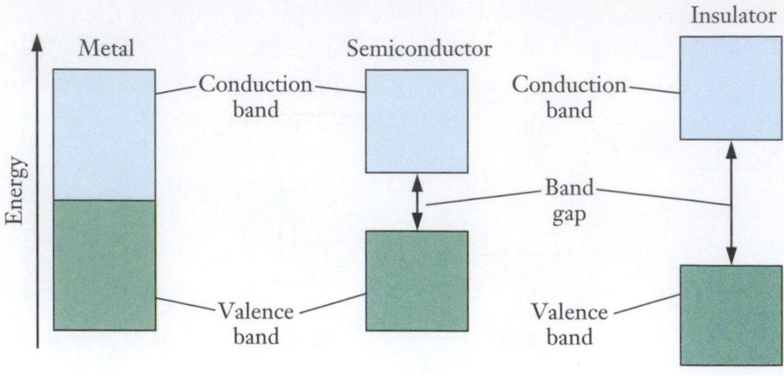

Figure 8.11 ■ These band energy diagrams point out the key differences among metals, insulators, and semiconductors. The green-shaded portions represent energy levels filled with electrons, and the blue portions depict unoccupied energy levels. Differences in conductivity can be explained in terms of the band gap and the relative difficulty of promoting electrons from filled to unfilled levels.

at the lowest energy. The highest energy with electron occupation and the energy gap between this energy level and the next available band provide a means for understanding the electrical conductivity of a material. Electrons carry a current by moving through a material, and this motion can be thought of in terms of electrons moving from one orbital to another. But electrons within a filled band cannot move readily to conduct electricity. To have mobile electrons, we will need to have an unfilled band or portion of a band that is close in energy to a filled band or portion of a band.

Let's consider these ideas for the band diagram for lithium shown on the right-hand side of Figure 8.10. The band populated by valence electrons is referred to as the **valence band**, and the unoccupied band above it is known as the **conduction band**. The conduction band lies directly above the valence band in energy, and the highest occupied energy level sits at the boundary between these two bands. So an electron sitting at the top of the valence band needs to gain only a very small amount of energy to move into the conduction band. This means that lithium should be a good electrical conductor, which it is.

The energy below which an energy level is likely to be filled is called the Fermi level. For a metal, the Fermi level is at the top of the valence band.

Figure 8.11 shows some general features of the bands for three different classes of materials: metals, semiconductors, and **insulators**. For metals, we see the situation we just described in which addition of even an infinitesimal amount of energy can promote an electron into the conduction band. The nonmetal, on the other hand, has a large energy gap between the filled valence band and the empty conduction band. So it takes a significant amount of energy to reach that empty band, where electron movement would be allowed. Under ordinary circumstances, electrons will not have enough energy to overcome this large **band gap**, so nonmetals do not conduct and are referred to as insulators. The diamond form of carbon is an excellent example of a nonconductive material.

The intermediate case is a semiconductor, shown in the center panel of the figure. The highest filled energy level is at the top of the valence band, but the next band is only modestly higher in energy. In semiconductors, some electrons have enough thermal energy to reach the upper band even at room temperature, and therefore they have at least limited electrical conductivity. If the temperature is increased, more electrons should occupy the upper band, and the conductivity of semiconductors increases at higher temperatures. Silicon, from the same group as carbon, is the prototypical example of a pure element that behaves as a semiconductor.

Semiconductors

As an engineering student, you are undoubtedly aware of the pervasive importance of semiconductors in our 21st-century lives. Semiconductor technology is at the heart of the electronics industry. Now that microprocessors have found their way into our cars and appliances, the scope of that industry has never been broader. The rise of the semiconductor is an excellent example of the interplay between the atomic-scale view of chemistry and the macroscopic-scale thinking of engineering. It is fair to say that the semiconductor revolution could not have taken place without the ingenious development of a host of new materials not found in nature.

Pure silicon provides an interesting textbook example of a semiconductor, but it is not useful in most applications. To generate significant conductivity in pure silicon, somehow we must promote electrons from the valence band into the conduction band. (This is the origin of the term *conduction band*.) This means that we must add energy, and the two most likely ways to do that are by exposing the silicon to significant heat or by allowing it to absorb light with sufficient photon energy to excite the electrons across the band gap. Neither of those would be easy to achieve or sustain in most electronic devices. Looking elsewhere in the periodic table doesn't really offer better options, though. There are relatively few elements that can be classified as semiconductors, and none of them is suitable for widespread use in its pure form. Instead, the semiconductor industry relies on the idea that by cleverly combining elements, we can produce materials with the specific semiconducting properties desirable for a particular application. One way to do this is through a process called **doping** in which carefully controlled trace amounts of another element are incorporated into a semiconductor like silicon.

Doping a material to derive useful properties is not limited to semiconductors. Many alloys, including specialty steels, are created by intentional doping.

To understand how doping works, let's start with the band diagram of silicon shown on the left-hand side of **Figure 8.12**. Suppose that we wanted to increase the conductivity of pure silicon slightly. From what we have learned about the conductivity of metals, we might conclude that to do this we might put some electrons into the conduction band. So if we had some way to "inject" an extra electron, it might go into the upper band. Then it could move around in that band. If we could add enough electrons, then we might get a good conductor. How might we go about adding electrons to silicon? Each silicon atom has four valence electrons. To increase the number of valence electrons, we might try adding a small amount of an element with more than four valence electrons. Suppose we took a sample of pure solid silicon and somehow replaced one silicon atom with a phosphorus atom. Because we changed only one atom, we could assume that the overall structure of the crystal and its band diagram shouldn't be disrupted. But now we do have one extra electron. Because the valence band is full, that additional electron must go into a higher energy orbital. Because this orbital is associated with the phosphorus atom, rather than with silicon, it will have slightly lower energy than that of the silicon conduction band. We say that the presence of the phosphorus atom introduces a **donor level** that lies close to the energy of the conduction band. Because the gap between this donor level and the conduction band is very small, the electron from the donor level can be promoted into the conduction band fairly easily. So in this thought experiment, adding one phosphorus atom into our silicon crystal will boost its conductivity ever so slightly.

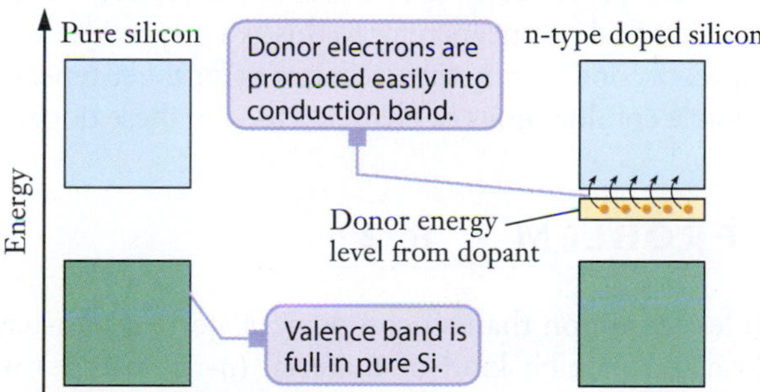

Figure 8.12 ■ Imagine that we add electrons to pure silicon by incorporating a trace of phosphorus. The added electrons will occupy a donor level just below the conduction band. This leads to an n-type semiconductor. The donor electrons do not need to cross the band gap to be promoted into unfilled levels of the conduction band, so they boost conductivity. We can control the number of these donor electrons by adjusting the extent of doping.

Figure 8.13 ■ In a p-type semiconductor, the addition of electron-deficient dopant atoms provides an acceptor level just above the top of the valence band. Electrons can be promoted into this level, leaving behind vacancies or "holes" in the valence band.

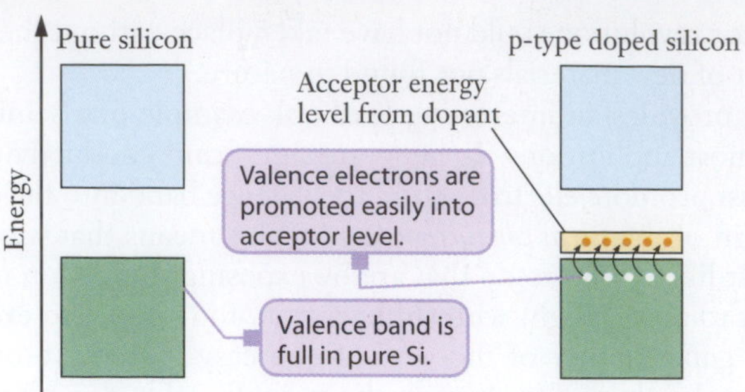

It is not possible to carry out such a single atom replacement, of course. Even if it were, the resulting change in conductivity would be immeasurably small. But the same idea will hold true if we add more phosphorus atoms. The fifth valence electron from each phosphorus atom will go into a new band similar to the donor level we introduced above. Then the donor level electrons can be promoted into the conduction band, leading to a band diagram like that shown on the right-hand side of Figure 8.12. The donor level of the phosphorus atoms provides a source of electron density near the conduction band, increasing conductivity. Moreover, the extent of that conductivity can be adjusted by controlling the amount of phosphorus present. This makes what is known as an **n-type semiconductor**. The name comes from the fact that we have added additional electrons, each of which has a negative charge. (Note that the overall lattice remains neutral because each phosphorus atom also has an extra proton compared to the silicon atom it replaces.) Those negatively charged electrons allow the semiconductor to carry a current.

A typical n-type semiconductor might contain on the order of 0.00001% phosphorus added to a silicon crystal. (Note that to be able to do this, the silicon itself must be incredibly pure.) We say that the silicon is doped with phosphorus, and this doping usually takes place as the material is grown. As long as the dopant atoms are randomly dispersed throughout the solid, the overall effect on the crystal structure will be minimal. Other elements with more than four valence electrons could also be used. But phosphorus is a convenient choice because it has an atomic radius close to that of silicon, so it won't disrupt the crystal lattice appreciably.

For a macroscopic analogy of the importance of size, think of a pile of oranges at a grocery store. If you were to replace an orange near the bottom and still maintain the pile, which would be easier to use—a tangerine (about the same size) or a large grapefruit?

An alternative doping scheme leads to a **p-type semiconductor**. The idea here is to provide a way to decrease the number of electrons in the valence band, so that conductivity can be achieved without needing to promote electrons across the band gap. The strategy is similar to what we described for the n-type, except that now we want to use a dopant with *fewer* than four valence electrons. This introduces an **acceptor level** that is slightly higher in energy than the top of the valence band. The most common choice for the dopant in this case is aluminum, and **Figure 8.13** shows how the band diagram will change as the material is doped. Electrons from the valence band can easily be promoted into the acceptor level, leaving behind vacancies or "holes" in the valence band. These holes behave as if they had a positive charge, so electrical engineers refer to this as a p-type material. The extent of doping controls the level of conductivity achieved, just as for the n-type. Example Problem 8.2 looks at some possible combinations of elements used in these doped materials.

EXAMPLE PROBLEM 8.2

Although much less common than silicon devices, germanium-based semiconductors can also be fabricated. Which kind of material (n- or p-type) would result if pure germanium were doped with (a) gallium, (b) arsenic, or (c) phosphorus?

Strategy Like silicon, germanium has four valence electrons per atom. So a dopant with fewer than four valence electrons will produce a p-type material, whereas a dopant with more than four valence electrons will give an n-type material. We can use the periodic table to find the number of valence electrons in each element.

Solution

(a) Gallium is in Group 13, so it has three valence electrons. Germanium doped with gallium will be a p-type semiconductor.
(b) Arsenic is in Group 15, so it has five valence electrons. Germanium doped with arsenic will be an n-type semiconductor.
(c) Phosphorus is also in Group 15, so germanium doped with phosphorus will be an n-type semiconductor, just as in (b).

Check Your Understanding A materials engineer needs to have an n-type semiconductor made from silicon. Which atoms might be used as dopants? Sketch a band diagram for the resulting material.

From what we've said so far, it probably isn't clear why we need both types of materials or how we might make any useful devices out of them. In practice, all important circuit functions in a chip take place at the junctions between different types of materials. The point at which a piece of p-type material meets a piece of n-type material is called a **p-n junction**. The flow of electrons across this junction can be regulated easily by applying voltage from a battery or other power supply. Clusters of these junctions can be made to function as switches and gates of varying degrees of complexity. We won't attempt to explain much of this here, but we can describe qualitatively how a simple p-n junction works.

Figure 8.14 shows a diagram of a p-n junction. The p-type material on the left is shown with some positive holes drawn. Remember these are just missing electrons. The n-type material on the right has its extra electrons shown. Now imagine that we connect a battery across the ends of the joined materials, as shown in part (b) of the figure. The electrons from the p-type material will drift toward the positive side of the battery, whereas the holes will move toward the negative side. (Holes act as if they have positive charges.) This means that current flows across the junction when the battery is connected in this way. But suppose that we connect the voltage in the other direction, as in part (c) of the figure. Now the electrons and holes are both pulled away

For a common type of device—the metal oxide semiconductor field effect transistor (MOSFET), the voltage needed is in the range of 1–3 V.

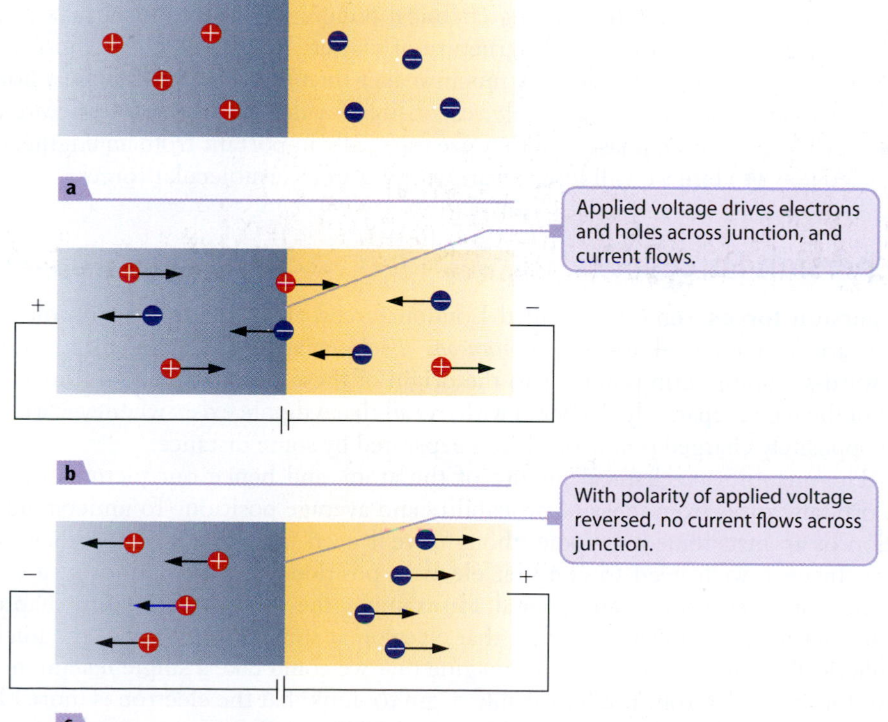
a
Applied voltage drives electrons and holes across junction, and current flows.
+
−
b
With polarity of applied voltage reversed, no current flows across junction.
−
+
c

Figure 8.14 ■ A p-n junction can serve as a simple switch, allowing current to pass when a voltage with the correct polarity is applied. This simple functionality can be exploited to build a variety of complex circuit functions.

from the junction, so no current flows across the boundary. Our junction can serve as a voltage-controlled switch: we can turn the current across the junction on or off by applying the voltage in the right direction. This simple idea can be applied to develop circuits of great complexity. That takes a lot of junctions of various sorts and a lot of clever design, but it can be done. You may go on to take a materials properties course in engineering, where you will probably see these ideas from a different perspective. But for our purposes here, the main aim is to show how important bulk properties can be manipulated by adjusting the atomic composition of a solid.

The features of metallic conductors, semiconductors, and insulators that we've outlined here apply to more than just elemental forms of substances. Most alloys are combinations of metallic elements (with an occasional minor component that is a nonmetal), and their bonding resembles what we have just described for metals. Semiconductors are often made from combinations of elements, too. Gallium arsenide is a particularly important example. Insulators, too, are often made from combinations of elements, and many ceramic materials are insulators.

Not all materials have chemical bonds that are responsible for holding a solid together. Now we will look at other forces that play roles in condensed matter.

8-4 Intermolecular Forces

In Chapter 7, we saw that the formation of chemical bonds is always governed by minimization of the energy of the atoms involved. As we turn our attention to the structures of solids, we'll see that the same principle still holds. The structure of any solid, whether crystalline or amorphous, is determined by the balance of attractive and repulsive forces between the atoms or molecules involved. In this section, we will consider the nature of **intermolecular forces**—the forces between molecules.

Forces Between Molecules

In Chapter 7, we learned how interactions between atoms lead to bonds that hold molecules together. There's no doubt that those interactions are an extremely important facet of chemistry. But understanding chemical bonding still leaves us a bit short of understanding many aspects of the structure of matter. The forces *between* molecules are also important. Though typically much weaker than the forces in chemical bonds, these intermolecular forces are largely responsible for determining the structure and properties of condensed phases. So they are especially important from an engineering perspective. ◂ⓘ Here we will look at various types of intermolecular forces.

Intermolecular forces are also critically important in biochemical systems, including several of the biomaterials we considered in Chapter 7. ⓘ

Dispersion Forces

Dispersion forces, sometimes called London forces, are common to all molecules. They are also referred to as *instantaneous dipole–induced dipole forces*. This rather awkward-sounding term points us to the origin of these forces, so let's consider each part of the name separately. To begin with, recall that a dipole exists whenever we have two oppositely charged points or objects separated by some distance.

The quantum mechanical picture of the atom, and hence our pictures of various orbitals, arises from ideas of probability and average position. To understand the notion of an instantaneous dipole, though, we'll need to step away from that viewpoint. Instead, we'll need to consider electron positions or distributions at a single instant. For an electron in an *s* orbital, for example, the ideas presented in Chapter 6 tell us that the probability of finding that electron at some distance from the nucleus is spherically symmetrical. But now imagine that we could take a single instantaneous snapshot of the electron. You'll probably begin to think that the electron is more likely to be found on one side of the nucleus. (A similar snapshot a moment later might show the electron on the other side, so the average position will still be consistent with the spherical shape we ascribe to the orbital.) This dynamic fluctuation in the distribution of electrons holds in molecules as well. Thus, at any given instant, we may find more electrons on one end of a molecule than on the other, even if, on average, they are evenly dispersed throughout. Whenever such a fluctuation occurs, there will be a momentary asymmetry of charge; one end of the molecule will have a small, transient negative charge, and the other will have a small, transient positive charge. This immediately implies a short-lived charge separation, which gives rise to a dipole moment. This dipole differs from those described earlier in our consideration of polar bonds because it lasts only for that instant that the electrons happen to be unevenly distributed. Accordingly, it is referred to as an *instantaneous dipole*. The lifetime of this instantaneous dipole is fleeting, and it soon disappears only to form again as the momentary distribution of electrons becomes asymmetrical again.

This dynamic distribution of electrons is not the only way to achieve an uneven distribution of electron density. Because electrons have very little mass and are charged, they can also be influenced by electric fields external to the molecule, such as those from a dipole in a neighboring molecule. When a molecule encounters an external electric field, the negatively charged electrons will tend to migrate away from the negative pole of the applied field. This response establishes a new dipole within the molecule itself. Thus the external field has forced a dipole to exist where it had not existed previously; it has *induced* the dipole. This process is shown schematically in **Figure 8.15**.

Now we combine the ideas of the instantaneous and induced dipoles. Consider two molecules that are separated by a small distance. One molecule experiences a fluctuation in its electron density, creating an instantaneous dipole. The second molecule will then experience that dipole as an external electric field, creating an induced dipole in the second molecule. This combination is the instantaneous dipole–induced dipole interaction, or dispersion. Because the dipoles paired in this way are aligned so that the positive end of one is near the negative end of the other, dispersion forces are attractive. You may be thinking that such fleeting forces between pairs of molecules must be incredibly weak—and if so, you're right. But because macroscopic samples contain on the order of Avogadro's number of molecules, the sum of all these weak interactions can be a significant amount of energy. This energy holds many liquids and some solids together.

Many polymers, including polyethylene, form solids in which dispersion forces account for the most important intermolecular attractions.

How does one estimate the relative size of dispersion forces for different materials? The key concept needed to answer this question is **polarizability**. This term indicates how susceptible a molecule is to having its charge distribution perturbed by an external field. This feature of molecules can be quantified experimentally by measuring the strengths of interactions between molecules, and there are some general conclusions that we can draw. The most important of them is that large molecules tend to be more polarizable than small ones. Because the interaction between positively charged nuclei and negatively charged electrons decreases as the distance increases, it makes sense that large molecules are more polarizable. The electrons are less tightly held and thus

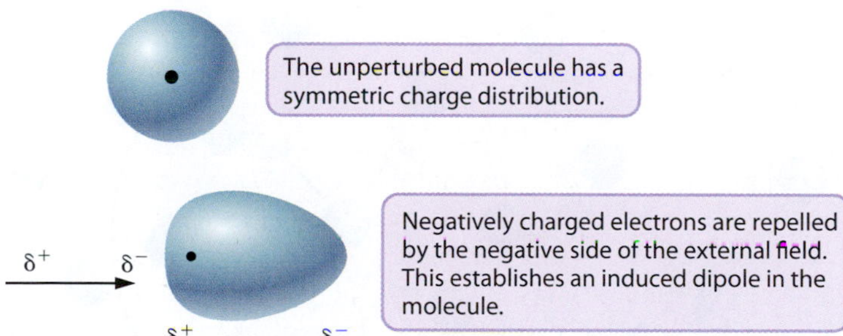
The unperturbed molecule has a symmetric charge distribution.
δ+
δ−
Negatively charged electrons are repelled by the negative side of the external field. This establishes an induced dipole in the molecule.
δ+
δ−

Figure 8.15 ■ An external electric field distorts the shape of the electron cloud around a molecule or atom, creating a transient or induced dipole. The distorted molecule can then act as an external field on a nearby molecule, propagating the effect. Dispersion forces result from the mutual attractions among such induced dipoles in a collection of molecules.

more easily influenced by the external field. The fact that larger molecules are more polarizable means that they tend to experience stronger dispersion forces.

Dipole–Dipole Forces

When molecules that have permanent dipoles interact with each other, there is a tendency for those dipoles to align. This alignment, particularly in a liquid, is not rigid. But the positive region of one molecule prefers to be near the negative end of other molecules and vice versa. The charge at these poles, however, is seldom more than a fraction of the charge of an electron. Thus, although the extent of the attraction of opposite charges or repulsion of the same charges is more than that in dispersion forces, it is quite modest compared to the strength of chemical bonds. With only small attractions and repulsions operating, the alignment of the dipoles will not be absolute. At any given time, there will be pairs or groups of molecules in a material that experience repulsive interactions, but there will be more attractive ones. **Figure 8.16** shows this situation for a snapshot of a 50-molecule sample. Each molecule will interact with its several nearest neighbors. If the oppositely charged ends of two dipoles approach one another, the interaction will be attractive. If similarly charged ends of the dipoles approach one another, the interaction will be repulsive. The attractive interactions will draw the molecules together, whereas repulsive interactions will tend to push the molecules apart. As a result, the molecules will tend to favor arrangements that maximize the attractions and minimize repulsions. Although the positions of individual molecules are constantly changing, the attractive interactions will consistently outweigh the repulsive ones. Because they are based on permanent rather than transient dipoles, these dipole–dipole interactions are typically stronger than dispersion forces.

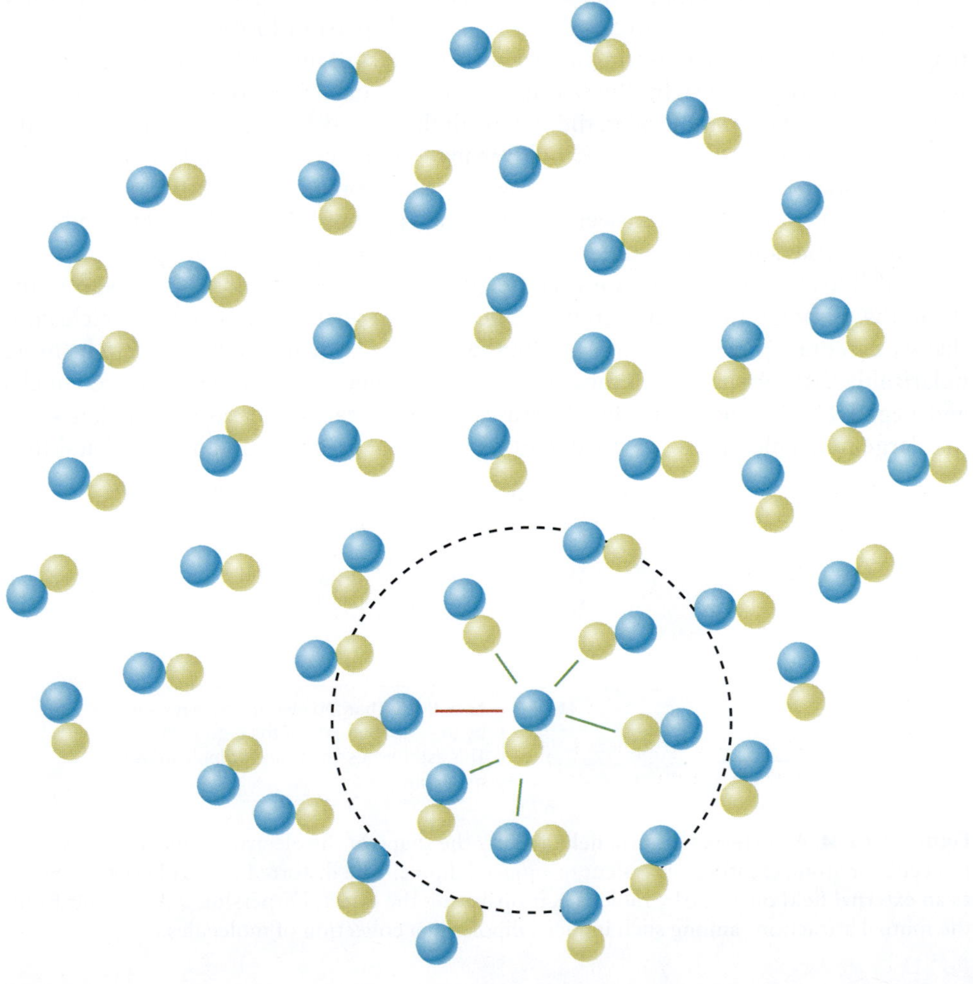

Figure 8.16 ■ This drawing shows an arrangement of 50 polar molecules, as might be found in an instantaneous snapshot of a liquid. The circle highlights the interactions between one molecule and its closest neighbors. Some of these interactions will be attractive, as shown by the green lines. Others will be repulsive, as shown by the red line. The number and strength of the attractions outweigh those of the repulsions, so there is a net force holding the molecules together.

Hydrogen Bonding

The remaining intermolecular interaction is among the most poorly labeled terms in all of science. Although they are frequently called "hydrogen bonds," hydrogen bonding interactions are not really bonds at all. They do not involve a chemical bond in the same formal sense in which we used that term in Chapter 7. They are, rather, an especially strong form of dipole–dipole interactions that are observed when certain combinations of elements are present in compounds. Hydrogen bonding is typically the strongest of all intermolecular interactions. Because hydrogen bonds play a prominent role in biological chemistry, these interactions have a profound effect on life as we know it. There are some specific criteria that must be met for hydrogen bonding interactions to form.

Among the many roles for hydrogen bonding, perhaps the most important is the linking of two strands of DNA into a double helix.

The "hydrogen" in hydrogen bonding is always a hydrogen atom bound to a highly electronegative atom, typically N, O, or F. The highly electronegative atom will tend to attract the electron pair it shares with hydrogen, and so the hydrogen atom will take on a partial positive charge. Because hydrogen has no other valence electrons, this positive charge will be especially "exposed" to other molecules. If an atom with a partial negative charge and an available lone pair of electrons approaches this exposed positive charge, a strong dipole-based attraction will result. The list of atoms best able to provide this source of electron density is the same as mentioned before: N, O, and F. In virtually all of their compounds, these elements tend to have one or more lone pairs of electrons available, and these electrons can be at least partially shared with a nearby hydrogen atom from another molecule.

The simplest cases of hydrogen bonding involve pure compounds whose molecules contain both the hydrogen atom and the highly electronegative atom with which it will interact. Thus in hydrogen fluoride, the H atom of one molecule can be attracted toward the F atom of a second molecule. A similar situation exists for water, where the oxygen atoms provide the lone pairs to which the hydrogen atoms will be attracted. It is crucial to remember that these hydrogen bonding interactions are much weaker than the chemical bonds within a single molecule. The hydrogen bonds in water, for example, are responsible for holding the molecules to one another in the solid or liquid phases. When liquid water is heated, the hydrogen bonds break down, but the chemical bonds do not. Thus as the water boils, the vapor driven off is still made up of H_2O molecules. (Recall Example Problem 1.1, page 8.) In more complicated cases, including many that arise in biological molecules, hydrogen bonding can occur between different molecules or even between different parts of large molecules like proteins. The common thread in all of these instances is the presence of hydrogen atoms bound to highly electronegative atoms. Example Problem 8.3 looks at the types of intermolecular forces present in some simple compounds.

EXAMPLE PROBLEM 8.3

Which type(s) of intermolecular forces need to be overcome to convert each of the following from liquids to gases? (a) CH_4, (b) CH_3F, (c) CH_3OH

Strategy Dispersion forces will exist for any substance but will be the most important forces only in the absence of stronger dipole–dipole or hydrogen bonding interactions. To see if those stronger forces are possible, we need to look at the structure of each compound. So we might start by using Lewis structures to establish the molecular geometries. In this case, we are dealing with methane, whose structure we know well, and two closely related compounds.

Solution

(a) We are quite familiar with the structure of CH_4 from Chapter 7: it is a tetrahedral molecule with four identical C—H bonds:

```
      H
      |
  H — C — H
      |
      H
```

Because the structure is totally symmetrical, no dipole–dipole forces are possible. Because the H atoms in CH_4 are not bound to N, O, or F, no hydrogen bonding is possible. Only dispersion forces need to be overcome to vaporize liquid methane.

(b) Methyl fluoride also is tetrahedral, but one of the hydrogen atoms in methane has been replaced by fluorine:

$$\begin{array}{c} :\ddot{F}: \\ | \\ H-C-H \\ | \\ H \end{array}$$

With only one fluorine atom present, the molecule is no longer symmetrical. The C—F bond is highly polar and will give the molecule a dipole moment. So to vaporize CH_3F, dipole–dipole forces must be overcome.

(c) The structure of methanol is also closely related to that of methane. Here we replace one of the hydrogen atoms in CH_4 with an —OH group:

$$\begin{array}{c} H \\ | \\ H-C-\ddot{\underset{\cdot\cdot}{O}}-H \\ | \\ H \end{array}$$

As in CH_3F, the molecule will have a dipole moment, but the dipole arises from the —OH group. That same functional group means that hydrogen bonding will be possible here. The hydrogen atom in the —OH group of one molecule can be strongly attracted to a lone pair on the oxygen atom in another molecule. So to vaporize methanol, we need to overcome hydrogen bonding interactions.

Discussion Identifying the forces involved lets us predict the order of boiling points for these liquids. With only weak dispersion forces involved, CH_4 should have a very low boiling point. Dipole–dipole interactions should make it more difficult to vaporize CH_3F, giving it a higher boiling point. (Dispersion forces would also be slightly stronger than for CH_4 because of the increased mass.) Finally, hydrogen bonding in CH_3OH should give it the highest boiling point among this group. These predictions are correct, as the boiling points are −164°C (CH_4), −78°C (CH_3F), and 65°C (CH_3OH).

Check Your Understanding Rank the following in order of increasing boiling points by considering the important intermolecular forces in each: Ne, CO, CH_4, NH_3.

The forces we've considered here—dispersion, dipole–dipole interactions, and hydrogen bonding—give rise to many of the interactions that can occur between molecules and allow us to understand both the existence and properties of condensed phases. None of these forces is even as strong as fairly weak chemical bonds. Hydrogen bonding interactions are typically the strongest of the intermolecular interactions, but even they are an order of magnitude weaker than the average covalent bond. Despite this apparent weakness, when considering a macroscopic sample, the sheer number of such interactions can make them a key factor in determining bulk properties.

The comparison between intermolecular and intramolecular forces can be illustrated very clearly by looking more closely at graphite. (Recall the general features of the structure from Figure 8.8.) Within a plane, the carbon atoms are bound by strong covalent bonds, as shown in the left-hand panel of **Figure 8.17**. A large number of resonance structures could be drawn, but it is reasonably clear that, on average, the strength of each carbon–carbon bond will be between those of a single bond and a double bond. Between the sheets, however, the interactions are due to dispersion forces, making them dramatically weaker. So the carbon atoms within a given sheet

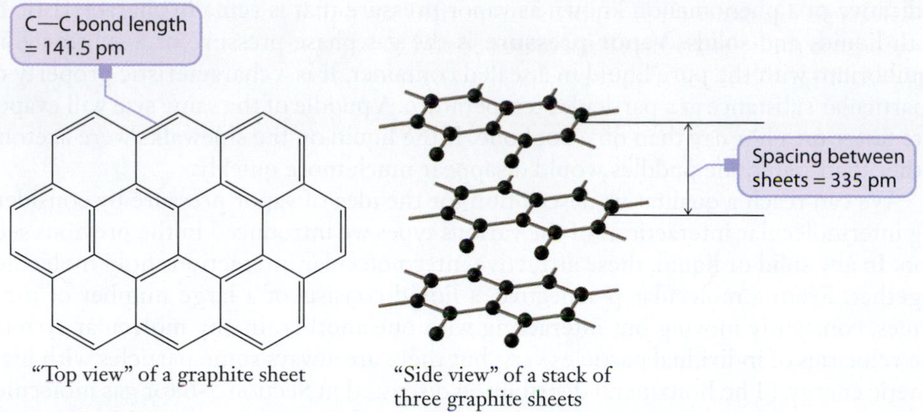

Figure 8.17 ■ Carbon atoms within each sheet in the graphite structure are joined by strong covalent bonds, with a fairly short C—C distance. Because each carbon atom forms four bonds within its own sheet, intermolecular forces must be responsible for holding the sheets together. This is apparent from the much larger spacing between layers. Because the forces holding the layers to one another are weak, sliding of one layer relative to the next is fairly easy, and this explains the softness and lubricating properties of graphite.

are held together by relatively strong chemical bonds, whereas much weaker dispersion forces hold pairs of adjacent sheets to one another. The result is that sheets of graphite can slide past each other relatively easily and this helps explain why graphite is useful as a lubricant. Lubricants must withstand large forces, and the strong bonds within a sheet provide that strength. They must also be capable of viscous flow, and the movement of one sheet relative to the next makes that possible.

A single sheet with this structure is known as graphene, and many applications of this two-dimensional material are being explored.

Though fairly simple, the structure of graphite points to the importance of both chemical bonding and intermolecular forces in determining the properties of a solid. Next, we turn our attention to the influence of intermolecular forces on the physical properties of liquids.

8-5 Condensed Phases—Liquids

Solids are not the only condensed phase of matter. Intermolecular forces of the type we just described often result in the formation of liquids as well as solids. From a structural perspective, liquids are not as orderly as solids, but we can look into some of their important characteristics.

The key difference between liquids and solids lies in the much greater mobility of the atoms or molecules in a liquid. For a solid, the average position of each particle remains essentially unchanged; even though there may be significant vibration about that average position, the atoms remain in a fixed arrangement over time. By contrast, particles in a liquid are free to move with respect to one another, and they do so constantly. This constant motion plays an important role in the fact that liquids often serve as solvents that can dissolve other chemicals. The dissolved particles, like those of the liquid itself, also become capable of movement. This allows dissolved particles to encounter other dissolved materials (or solids in contact with the liquid), and such encounters may allow chemical reactions to take place.

Confining our discussion for the moment to pure liquids, though, what can we learn about their properties from looking at intermolecular forces? Let's consider some physical properties that might be important in understanding the behavior of liquids.

Vapor Pressure

After a springtime rain shower, the sidewalks are covered with puddles. Those puddles disappear but clearly the water never boils. The evaporation of these puddles is indicative of a phenomenon known as vapor pressure that is actually characteristic of both liquids and solids. **Vapor pressure** is the gas phase pressure of a substance in equilibrium with the pure liquid in a sealed container. It is a characteristic property of a particular substance at a particular temperature. A puddle of the same size will evaporate faster on a hot day than on a cool one. If the liquid on the sidewalks were acetone rather than water, the puddles would disappear much more quickly.

We can reach a qualitative description for the idea of vapor pressure by considering intermolecular interactions of the various types we introduced in the previous section. In any solid or liquid, these attractive intermolecular interactions hold molecules together. From a molecular perspective, a liquid consists of a large number of molecules, constantly moving but interacting with one another. In any molecular system, the velocities of individual particles vary, but there are always some particles with high kinetic energy. (The Boltzmann distribution discussed in Section 5-6 for gas molecules is an example of this. Similar ideas apply to molecules in liquids, too, although their exact form is more complicated.) If an especially energetic molecule happens to find its way to the surface of the liquid, it may have enough energy to overcome the attractive tugs of its neighboring molecules and escape into the vapor phase. That is the molecular origin of evaporation.

This molecular viewpoint lets us predict the behavior of vapor pressure with temperature changes. As temperature increases, the number of molecules with high kinetic energy also increases. So, more molecules will be capable of escaping the attractive forces of intermolecular interaction. Thus vapor pressure should increase as temperature increases, and this is seen experimentally. Our molecular description also allows us to understand the relative vapor pressures of different molecules. To escape from a liquid, molecules must have kinetic energy sufficiently high to overcome the intermolecular interactions in the liquid. So the strength of those intermolecular forces sets the bar that evaporating molecules must overcome to escape. If a system has strong intermolecular interactions, molecules will need particularly large amounts of kinetic energy. There are simply fewer molecules capable of escaping when intermolecular interactions are strong, and so the vapor pressure will be low. Liquids with high vapor pressures evaporate rapidly and are said to be **volatile**. Many engineering designs, including those of gasoline-powered engines, must take into account the volatility of some component, such as the fuel. Volatility is sometimes an asset–for example, when fuel is being injected into a cylinder for combustion. It can also be a liability–for example, when trying to minimize the release of unburned fuel from a storage tank into the atmosphere.

This gives us a reasonably useful qualitative view of vapor pressure. But how can it be quantified? Water has a lower vapor pressure than acetone, but how much lower? To measure the vapor pressure of a liquid or a solid requires that the system reach equilibrium. In other words, the molecules that are escaping from the liquid must be replaced by other molecules that are being captured, as shown in **Figure 8.18**. Once

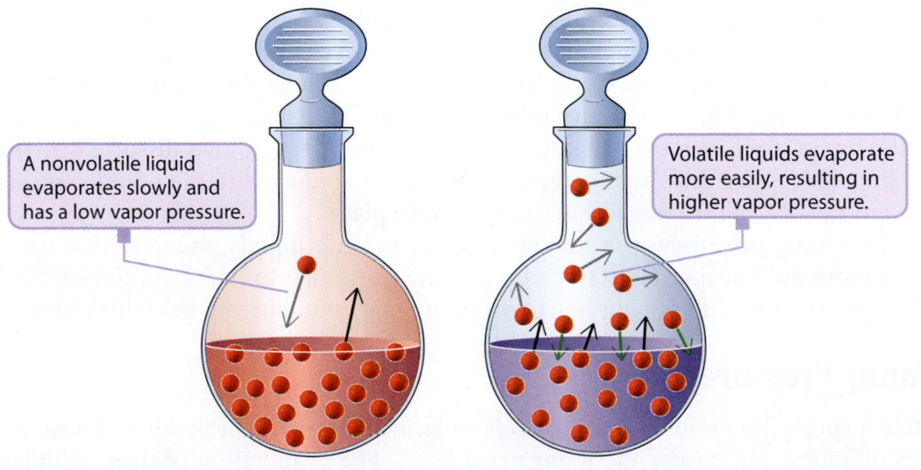

A nonvolatile liquid evaporates slowly and has a low vapor pressure.
Volatile liquids evaporate more easily, resulting in higher vapor pressure.

Figure 8.18 ■ When a liquid is held in a closed container, it will establish equilibrium with its vapor phase. The vapor pressure provides a measure of the ease with which a particular liquid evaporates.

Table ■ 8.1

Vapor pressures at 295 K and normal boiling points of a variety of substances

Substance	Vapor Pressure (torr)	Normal Boiling Point (°C)
Acetone	202	56.2
Br_2	185	58.8
$CClF_3$	24,940	–81.1
CCl_2F_2	4448	–29.8
CCl_3F	717	23.8
CCl_4	99.0	76.54
HCN	657	26
Formaldehyde	3525	–21
Methanol	108	64.96
n-Pentane	455	36.07
Neopentane	1163	9.5
Isobutane	2393	–11
n-Butane	1658	–0.5
Propane	6586	–42.07
Ethane	29,380	–88.63
Water	19.8	100

Recall that 760 torr = 1 atm.

this equilibrium has been established, the pressure of the molecules in the gas phase will remain constant and can be measured fairly easily. This process, in which the net state of the system is unchanged but there are still molecules escaping and being captured by the liquid, is called **dynamic equilibrium.** We'll study this idea in detail in Chapter 12 because it is important in many more circumstances than just vapor pressure. The values of the vapor pressure for a number of liquids near room temperature are provided in **Table 8.1.**

Vapor pressure measurmenets are made in evacuated vessels, so the only source of pressure is from the vapor.

Boiling Point

The boiling point of a liquid also depends on the strengths of the intermolecular interactions and is closely related to vapor pressure. As just noted, the vapor pressure of a material increases as temperature increases. This phenomenon is true over all temperature ranges, so that eventually as temperature continues to increase, the vapor pressure of the liquid equals the external pressure. When that happens, small bubbles of vapor form within the liquid and the system boils. The **normal boiling point** of a liquid is defined as the temperature at which its vapor pressure is equal to one atmosphere. The well-known normal boiling point of water is 100°C. If the external pressure is below 1 atmosphere, the temperature at which boiling will occur is lower. This idea gives rise to the special cooking instructions for people who live in the mountains, where atmospheric pressure is reduced. If one wishes to boil an egg at high altitude, it will take longer than at sea level because the water is boiling at a lower temperature. The vapor pressure of water as a function of temperature is shown in **Figure 8.19.** The difference in the boiling temperature between

Formation of gas bubbles in a liquid is not automatic, so it is possible to "superheat" a liquid above its boiling point under certain circumstances.

Figure 8.19 ■ The vapor pressure of water varies dramatically as a function of temperature.

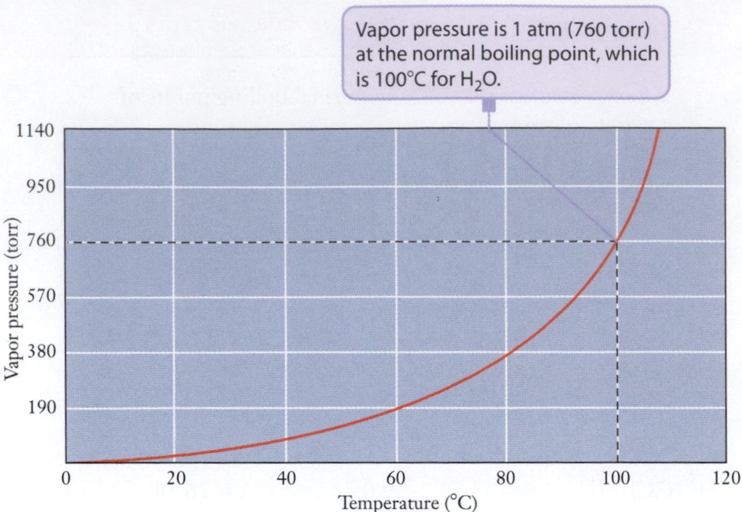

sea level and the mountains is a direct result of the temperature dependence of vapor pressure. At higher altitude, atmospheric pressure is lower, and boiling occurs whenever the vapor pressure is equal to the external pressure. Note also the shape of the curve. The pressure increases rapidly as temperatures exceed 100°C, where water typically boils.

The fact that vapor pressure continues to increase past the boiling point is important for things like popcorn or pressure cookers.

The normal boiling points of several liquids are listed in Table 8.1. Note that those liquids with low vapor pressures have high boiling points. It is possible therefore to rate the strength of intermolecular interactions by observing the normal boiling point of liquids: the higher the boiling point, the stronger the intermolecular forces. It is also possible to separate molecules of different intermolecular interaction strengths by distillation, which is an important step in oil refining.

Surface Tension

The attractive nature of intermolecular interactions means that a liquid will, in general, behave to maximize the number of interactions. One effect of this is that the number of molecules at the liquid's surface is minimized. Such surface molecules will find themselves in a different environment than those in the bulk of the liquid. If the surface of the liquid is in contact with a gas, there are fewer molecules available to interact with these surface molecules because gases are less dense than solids (**Figure 8.20**). Because the typical intermolecular interactions are attractive, the fact that these surface molecules have fewer neighbors means they find themselves in a state of higher energy than bulk molecules. The system will naturally minimize the number of molecules in this higher energy environment, giving rise to a phenomenon called **surface tension**. Something under tension is in a higher energy state. Here, the origin of the high energy state is the existence of the surface and the fact that molecules at the surface have fewer nearest neighbors than molecules in the bulk of the liquid.

Surface tension explains a variety of observable phenomena, including the fact that water tends to bead up into spherical (or nearly spherical) drops. The energy of any given sample of liquid will be reduced if fewer molecules sit at the surface. The shape that encloses the greatest volume with the least amount of surface is a sphere. The fact that water experiences strong intermolecular interactions due to hydrogen bonding means that its surface tension is relatively high and provides a strong force to form spherical drops. This effect is best seen on a waxed surface because the polar water molecules do not interact strongly with the nonpolar wax. On other surfaces, including a dirty car hood, strong interactions between the water molecules and substances on the surface can overcome the surface tension, producing a more uniform film of water.

In engineering applications where surface tension can create problems, molecules called surfactants are often added to decrease the surface tension. Surfactants also rely on an understanding of intermolecular forces in their design.

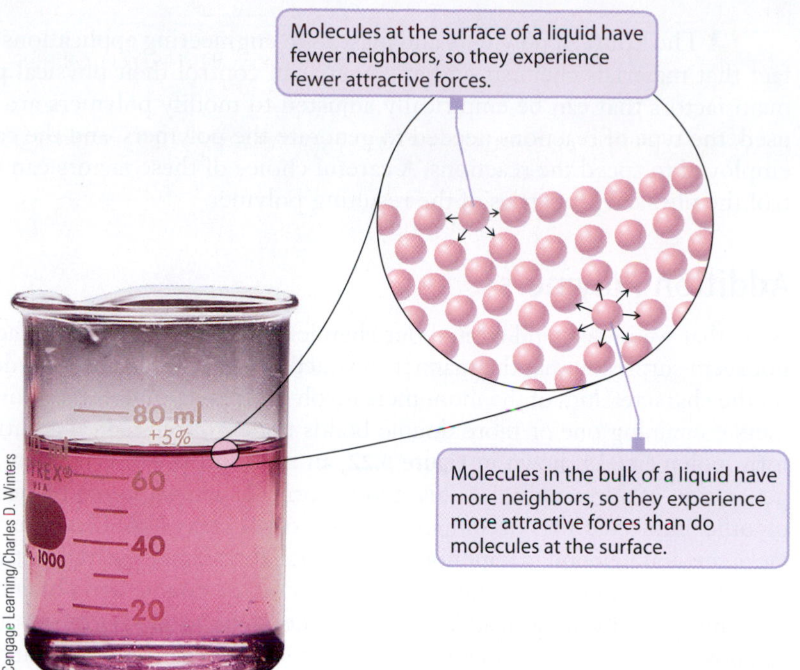
Molecules at the surface of a liquid have fewer neighbors, so they experience fewer attractive forces.
Molecules in the bulk of a liquid have more neighbors, so they experience more attractive forces than do molecules at the surface.
80 ml
+5%
60
40
20
©Cengage Learning/Charles D. Winters

Figure 8.20 ■ Molecules at the surface of a liquid have fewer nearby neighbors than molecules in the bulk of the liquid. Surface tension arises because of this difference in the molecular scale environment.

When the surface of a liquid interacts with a solid, we must consider competing interactions. The liquid molecules interact with each other, but they also interact with the molecules in the solid. The liquid–liquid interactions are referred to as **cohesion**, and the liquid–solid interactions are called **adhesion**. The relative strengths of these forces will dictate the shape of the curved surface, or meniscus, that forms in the liquid. **Figure 8.21** shows a comparison of two different liquids in a glass tube. Water (on the top) has strong interactions with the glass, which has numerous polar bonds on its surface. So the adhesion forces are larger than cohesion and the meniscus is concave up. The liquid in the bottom of the tube is mercury, which does not interact strongly with the polar glass surface. So adhesion is low relative to cohesion and the resulting meniscus is concave down. A liquid in which the adhesive and cohesive forces are equally strong would be expected to show a horizontal surface, with virtually no meniscus.

The carbon structures we noted in the opening "Insight" section in this chapter do not readily exist in liquid phases. To melt diamond or graphite requires temperatures that are high enough that there are no practical consequences of the liquid phase. There are, however, carbon-based materials that have important properties in both the solid and liquid states, and they comprise a sizable portion of the materials used in modern engineering designs. Before we can move on past our study of materials, we need to take another look at polymers—particularly carbon-based polymers.

© Cengage Learning/Charles D. Winters

Figure 8.21 ■ The graduated cylinder shown in this photograph contains a layer of water on top of a layer of mercury. The meniscus at the top of the mercury column is concave down, meaning that the mercury is higher in the center of the cylinder than at the walls. The meniscus at the surface of the water is concave up, with the water higher along the walls than in the center of the cylinder. These shapes result from the differences in strength of the adhesive and cohesive interactions between the two liquids and the glass of the graduated cylinder.

8-6 Polymers

The importance of carbon in chemistry and materials science is due to the myriad chemical compounds it can form, not to the elemental forms we have noted thus far in this chapter. Though there are more than enough organic compounds to provide a complete course devoted to their properties and reactions, we will limit our consideration here to carbon-based polymers. As we saw back in Chapter 2, polymers are giants among molecules, constructed by the sequential stringing together of smaller molecules called monomers. They constitute the largest component of the materials we call plastics—though plastics are defined by their ability to be molded and usually include additives.

The utility of polymers and plastics in engineering applications arises from the fact that materials chemists and engineers can control their physical properties. The main factors that can be empirically adjusted to modify polymers are the monomers used, the type of reactions needed to generate the polymers, and the catalysts that are employed to speed the reactions. A careful choice of these factors can ultimately control the physical properties of the resulting polymer.

Addition Polymers

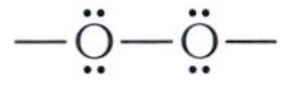

Now that we understand a bit about chemical bonds and chemical reactions, it should not seem surprising that the manner in which polymers are generated depends strongly on the characteristics of the monomers involved. In Section 2-8, we showed how monomers containing one or more double bonds undergo free radical addition reactions to form polymers. As shown in **Figure 8.22**, an **addition polymerization** begins with the generation of a free radical. (This step is normally accomplished by heating a peroxide or other molecule that decomposes into two radicals.) This free radical then attacks the double bond in a monomer molecule, forming a new radical that now includes one monomer unit. The resulting radical attacks the double bond of another monomer, breaking it and forming an additional single bond. Importantly, the product is still a radical, now containing two monomer units. This radical can attack yet another monomer,

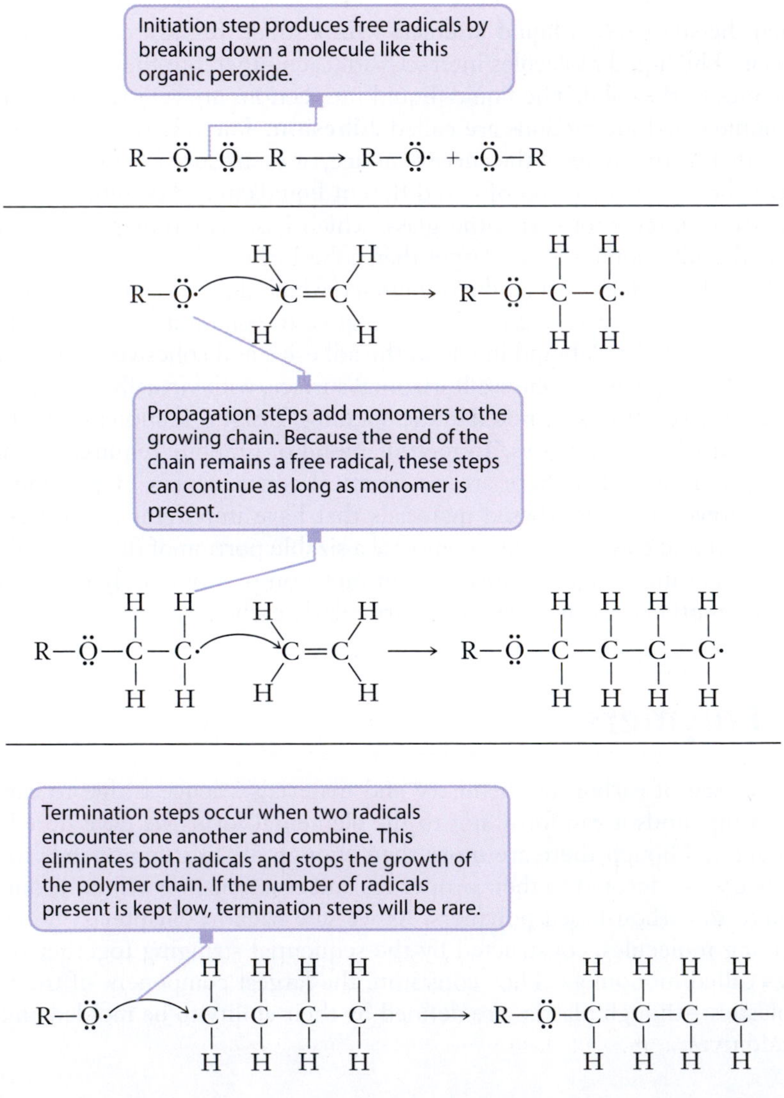

Peroxides contain a pair of oxygen atoms connected by a single bond.

$$\text{—}\ddot{\underset{\cdot\cdot}{O}}\text{—}\ddot{\underset{\cdot\cdot}{O}}\text{—}$$

Initiation step produces free radicals by breaking down a molecule like this organic peroxide.

$$R{-}\ddot{O}{-}\ddot{O}{-}R \longrightarrow R{-}\ddot{O}\cdot + \cdot\ddot{O}{-}R$$

$$R{-}\ddot{O}\cdot + H_2C{=}CH_2 \longrightarrow R{-}\ddot{O}{-}CH_2{-}CH_2\cdot$$

Propagation steps add monomers to the growing chain. Because the end of the chain remains a free radical, these steps can continue as long as monomer is present.

$$R{-}\ddot{O}{-}CH_2{-}CH_2\cdot + H_2C{=}CH_2 \longrightarrow R{-}\ddot{O}{-}CH_2{-}CH_2{-}CH_2{-}CH_2\cdot$$

Termination steps occur when two radicals encounter one another and combine. This eliminates both radicals and stops the growth of the polymer chain. If the number of radicals present is kept low, termination steps will be rare.

$$R{-}\ddot{O}\cdot + \cdot CH_2{-}CH_2{-}CH_2{-}CH_2{-} \longrightarrow R{-}\ddot{O}{-}CH_2{-}CH_2{-}CH_2{-}CH_2{-}$$

Figure 8.22 ■ Addition polymerization often proceeds by a free radical mechanism, featuring initiation, propagation, and termination steps. Conditions are chosen so that the initiation and termination steps occur infrequently. Each propagation step adds one monomer unit to the growing polymer chain.

and with enough of these steps, the polymer can be formed. Just how many times this reaction repeats itself controls the length of the polymer chain. Eventually, a radical will be eliminated, either by encountering another radical or by some other reaction. This stops the polymer molecule from growing longer and is referred to as *chain termination*. Because the occurrence of these termination steps is somewhat random, not all of the polymer chains will grow to the same length. Instead, there is normally a range of lengths and hence a distribution of molecular masses. This distribution can be roughly described in terms of the **degree of polymerization**, which is the average number of repeating units in the polymer. This is usually calculated in terms of mass:

$$\text{Degree of polymerization} = \frac{\text{molar mass of polymer}}{\text{molar mass of monomer}}$$

One reason that polymers are such common engineering materials is that their properties can be tailored in a number of ways. The most obvious is the choice of the monomer unit because we know that different polymers can offer vastly different characteristics. But it is also possible to have distinctly different properties in polymers derived from the same monomer. One variable is the degree of polymerization, which can be controlled by varying parameters used in the synthesis reaction. Another approach is to control the way in which the monomers are linked together. Perhaps the most important example of this is the formation of both low-density polyethylene (LDPE) and high-density polyethylene (HDPE), which we explored in Section 2-8.

When the monomer is less symmetrical than ethylene, there is additional variability that might result from the polymerization process. For example, propylene has a structure similar to that of ethylene but with a methyl ($-CH_3$) group replacing one hydrogen atom. This change means that the resulting polymer may have methyl groups attached to the carbon backbone in three different ways, resulting in the various geometric arrangements shown in **Figure 8.23**. It is important to realize that the geometry at each carbon atom in the chain is tetrahedral, so the methyl groups are either in front

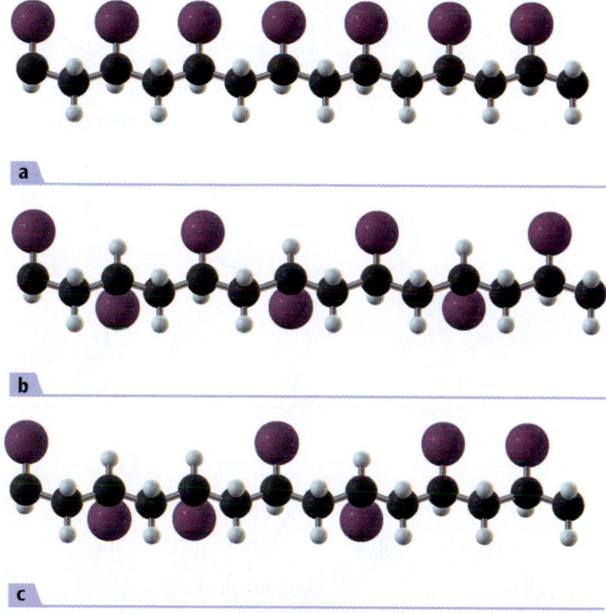

Figure 8.23 ■ For monomers with more complicated structures, different ways of linking them together become possible. Here we see polypropylene, which differs from polyethylene in that every other carbon atom has a methyl group in place of one hydrogen atom. The large purple balls here represent those methyl groups. In (a), all of the methyl groups are arranged on the same side of the polymer chain, giving isotactic polypropylene. In (b), the methyl groups are on alternate sides of the chain, making syndiotactic polypropylene. Finally, the random arrangement of the methyl groups in (c) is atactic polypropylene.

of or behind the plane of the page in this illustration. If the methyl groups all point forward (as in Figure 8.23(a)) or backward, then the polymer is said to be **isotactic**. If the position of the methyl groups systematically alternates between forward and backward, the configuration is called **syndiotactic**. In the third case, the arrangement of the methyl groups is random, sticking "into" and "out of" the page with no regular pattern, and the polymer is said to be **atactic**.

Polystyrene is an example of a commercial polymer whose properties depend on whether it is syndiotactic or atactic.

The physical properties of a polymer can be strongly influenced by this type of variability. Because the structure of atactic polymers is not very well defined, their physical properties can vary considerably from one sample to another. The regular arrangement of the monomers in isotactic and syndiotactic polymers generally leads to more predictable and controllable properties. So our ability to produce polymers selectively in one of these arrangements offers a powerful tool to the materials scientist or engineer. This ability to control the specific configuration of a polymer was first achieved by Karl Ziegler and Giulio Natta in the 1950s. These scientists discovered new catalysts for the addition polymerization reaction that increased the rate of reaction and also controlled the structure. The catalysts came to be known as Ziegler-Natta catalysts. Their discovery invigorated the study of polymers, and they were awarded the Nobel prize in chemistry in 1963 for their efforts.

Condensation Polymers

The second commonly observed type of reaction for the formation of polymers is called *condensation*. In this reaction, there are two functional groups on the monomer, which can react together to form small molecules. Those small molecules then split off, and the remaining parts of the two monomers are joined together. (The term **condensation polymer** originates from the fact that the small molecule eliminated is often water.) **Figure 8.24** shows the condensation reactions that produce two polymers

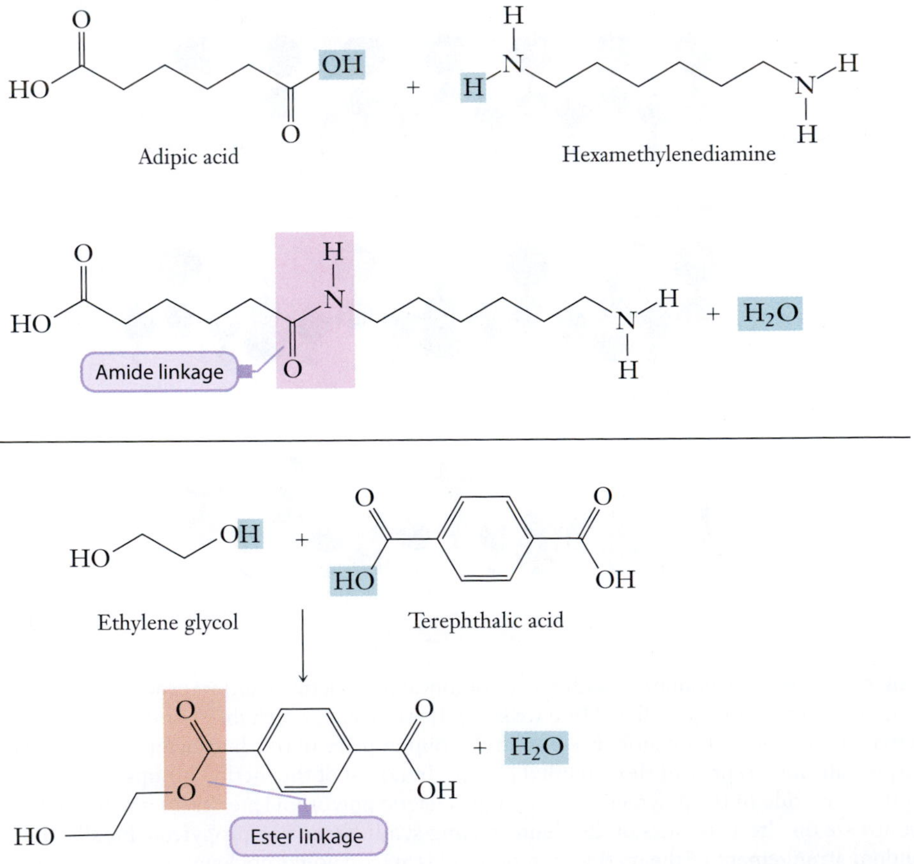

Figure 8.24 ■ In condensation polymerization, a molecule of water is eliminated as each monomer is added to the chain. Here we show the first steps in the formation of Nylon (upper panel) and Dacron (lower panel). Both are copolymers, with two different monomer molecules combining to form the polymer. Because each monomer contributes its own functional group to the condensation step, regular alternation of monomers is ensured. Other condensation polymerizations can involve a single monomer with different functional groups on each end.

that are important in fiber manufacturing, Nylon® and Dacron®. Nylon is referred to as a polyamide, and Dacron is a polyester. These names refer to the characteristic amide and ester linkages found in the molecules, as highlighted in Figure 8.24. Looking at these two examples, we can see the functional groups that must be present in monomers for condensation polymerization. Example Problem 8.4 looks at some other condensation polymers.

EXAMPLE PROBLEM 8.4

Each of the following monomers or pairs of monomers can undergo condensation polymerization reactions. Draw the structures showing the repeat units and linkages in each of the resulting polymers.

(a) glycine:

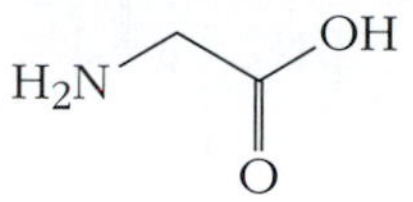

H_2N OH O

(b) 6-aminocaproic acid:

O H_2N OH

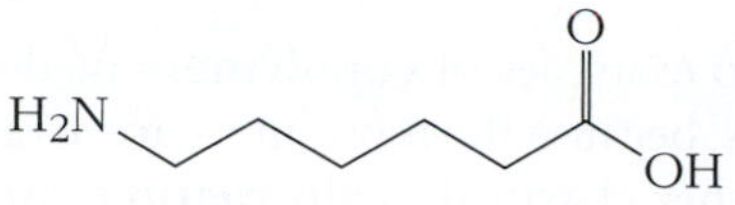

(c) *p*-phenylenediamine and terephthalic acid:

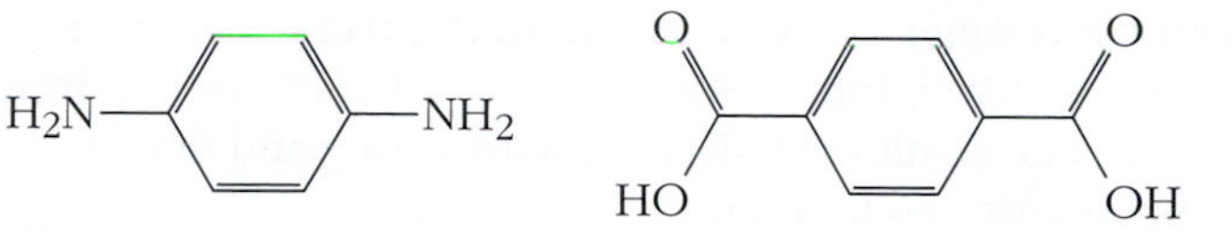

H_2N NH_2 O O HO OH

Strategy Condensation polymerization involves the elimination of a small molecule—usually water—as each monomer is added to the growing chain. We can look for functional groups on the ends of the monomers that can undergo such a reaction.

Solution

(a) We have an amine group (—NH_2) on one end of the monomer and a carboxyl group (—COOH) on the other. Elimination of H_2O from these groups will produce an amide linkage. The polymer will have the following structure, with the monomers connected through amide linkages:

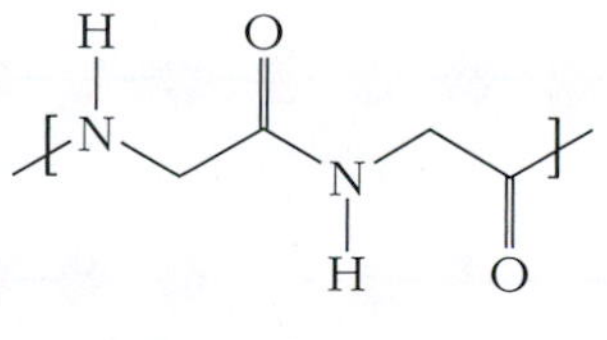

H O N N H O

(b) This is similar to (a):

H O N N H O

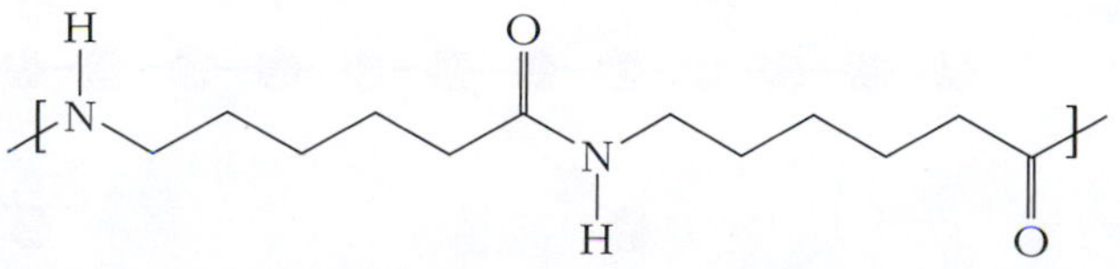

(c) Here, we have amide groups on one monomer and carboxyl groups on the second. We will still form amide linkages, here with the different monomers alternating. (The polymer formed here is Kevlar®.)

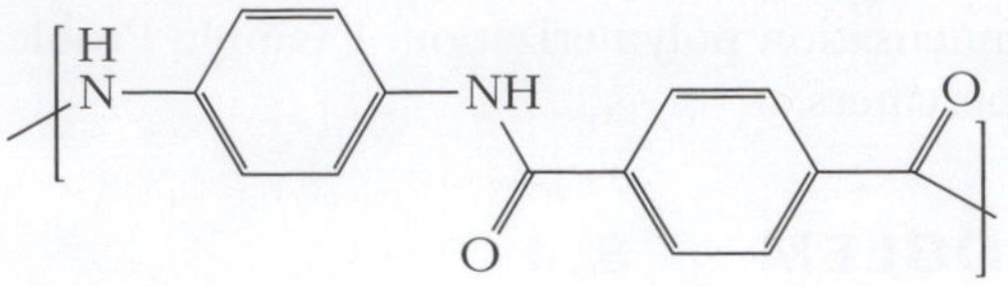

Check Your Understanding The polymer whose structure is shown below is produced by a condensation reaction in which water is eliminated. Draw the structure of the corresponding monomer.

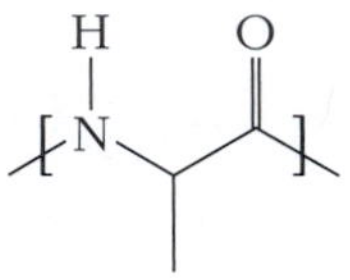

Copolymers

Nylon and Dacron are two examples of **copolymers** made from more than one type of monomer. In these cases, because the monomers are arranged in a regular, alternating series, they can be further classified as **alternating copolymers**. There are other important types of copolymers as well (**Figure 8.25**). One of the more creative ways to engineer materials is to design block copolymers. A **block copolymer** has regions in the material where a single monomer unit is repeated, interspersed with other regions where a different monomer is the repeating unit. Spandex is an example of a block copolymer. In this material, some regions of the polymer are relatively stiff, whereas others are flexible. The resulting material is both strong and flexible, leading to widespread use in sports gear and fashion.

Except for the alternating copolymer, these diagrams are schematic. The actual segments of each type of monomer would often be much longer than what is shown here.

Another engineering material that is a copolymer is commonly called ABS because the monomer units involved are acrylonitrile, butadiene, and styrene. This material is a **graft copolymer** in which both butadiene and acrylonitrile are attached to a polystyrene backbone. The resulting polymer offers a combination of the properties of its three constituents. Polystyrene provides ease of processing, glossiness, and rigidity. Acrylonitrile adds chemical resistance and hardness, and butadiene provides impact resistance. The specific composition can be tailored to a particular application. ABS is the standard

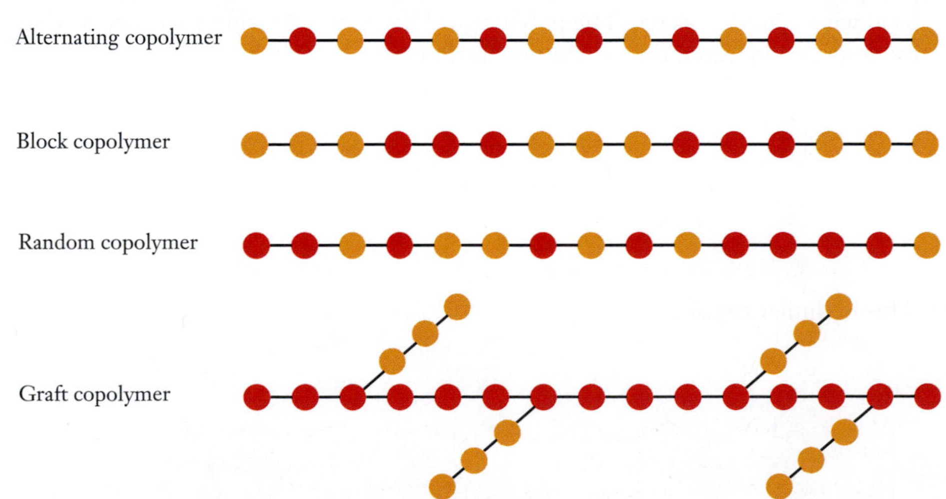

Figure 8.25 ■ Copolymers can be grown with different arrangements of monomers. Here the different colored circles represent two different monomers. For the most part, the names of these types of copolymers are self-explanatory. The different monomers alternate in an alternating copolymer or are randomly combined in a random copolymer. In a block copolymer, each type of monomer tends to be clumped together. Graft copolymers involve side chains of one polymer attached to a backbone of a different polymer.

material of choice for the strong plastic cases of computers, televisions, and other household electronic goods and sometimes is used as an alternative to PVC for pipes.

Physical Properties

There are two key reasons for the importance of polymers as engineering materials. First and most obviously, polymers offer physical properties such as strength and elasticity that can be desirable in a wide range of uses. Second, those properties can be controlled or tailored to a greater degree than is usually possible in metals or other classes of materials. Through the choice of a particular polymer and the details of its synthesis and processing, materials engineers can choose the properties they need with a striking degree of specificity. In this section, we will look at a few examples of the types of choices that can be made.

There are levels of polymer structure, such as the formation of ordered regions with the polymer folded back onto itself, which we are not considering here.

We begin by considering the thermal properties of polymers. In choosing a polymer for a specific design, you would obviously need to make sure that your finished object will be suitable for the temperature range in which it will be used. But an experienced engineer will also recognize the need to think about the way the object will be manufactured. Many plastic components are formed at high temperatures, and different polymers respond in dramatically different ways to heat. Polymers are often divided into two broad categories: thermoplastic and thermosetting. **Thermoplastic polymers** melt or deform on heating. This may seem like a weakness because it means that they are not suitable for high-temperature applications. But a great many plastic objects, including children's toys and bottles of many sorts, are generally used at ambient temperatures. So, the fact that they would melt if heated appreciably is not a major drawback. When we think about how to make these objects, which can come in rather complicated shapes, the ability to melt the polymer at reasonable temperatures becomes a major advantage. Depending on their complexity, products made from thermoplastic polymers are typically extruded or formed in molds or presses. The fact that the material softens or melts when heated allows shaping it into the desired form. Once cooled, the polymer solidifies and regains its structural properties.

If we are trying to design an object that might need to be used at higher temperatures, though, the fact that thermoplastics soften and melt will make such materials a poor choice. Instead, we might turn to **thermosetting polymers**, which can maintain their shape and strength when heated. The name "thermosetting" comes from the fact that these polymers must be heated to set or "lock in" their structures. But once this has been done, the materials offer increased strength and do not lose their shape upon further heating. Rather than being extruded, most thermosetting polymers are molded. The molecular origin of the difference between thermoplastic and thermosetting polymers is shown in **Figure 8.26**. The initial heating and setting of a thermosetting polymer produces a number of links between sites on the carbon backbone of different molecular chains. These links are referred to as cross-links because they *cross* between and *link* individual molecular strands of the polymer. Chemically, these cross-links are additional covalent bonds that join the polymer chains to one another. Like most covalent bonds, they are strong enough that they do not readily fail upon heating. So the cross-linked polymer keeps its shape.

In a thermoplastic polymer, chains interact only through intermolecular forces.
In thermosetting polymers, chains are tied together by actual chemical bonds.

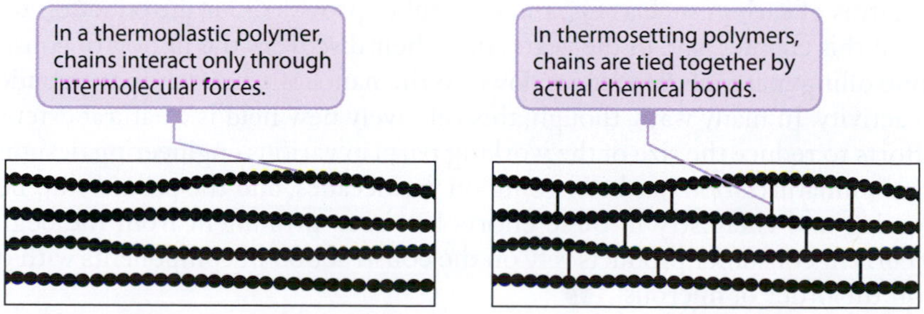

Figure 8.26 ■ The different properties of thermoplastic and thermosetting polymers result from the ways in which the polymer chains interact with one another.

An important example of the engineering importance of cross-linking in American industrial history is the discovery of vulcanization. In vulcanization, natural rubber is heated in the presence of sulfur. This produces cross-linking and leads to a harder material that is markedly more resistant to heat. Until vulcanization was discovered, natural rubber was difficult to use in applications such as automobile tires because it would become sticky when heated. Natural rubber is no longer widely used, having been replaced by synthetic forms. But the large-scale development of vulcanized tires and the design freedoms they afforded automotive engineers was an important component of the growth of the U.S. automobile and tire industries.

Another critical physical property of polymers is elasticity. The ability of many polymeric materials to be stressed and deformed but return to their original shape is a characteristic that is often valuable to engineers. Fibers, in particular, must be elastic. Polymers that are particularly flexible and elastic are sometimes referred to as elastomers. Looking at the molecular structure of elastomers, we find that they tend to share some common traits. In particular, the geometry of the carbon backbone is often such that the elastomer forms an amorphous solid rather than a crystalline one. The forces between polymer molecules in an amorphous solid are not as strong as those in a more crystalline system, so they can be deformed and restored with less force.

Polymers and Additives

Despite their wide range of useful properties, polymers alone often do not have the characteristics needed for a particular design. Fortunately, additives can be included in the material to control polymer properties further. Some additives have only modest mechanical effects. Pigments, for example, can be added solely to change the color of the material and don't influence the underlying molecular structure of a polymer. Other additives play critical roles in the performance of the materials. PVC alone, for example, is too brittle for many applications. To rectify this, relatively small molecules called plasticizers are added to improve its flexibility. A plasticizer has to have several critical features. It must be capable of being incorporated into the solid polymer, so its structure should resemble the polymer being used. It must be nonvolatile, or it will escape from the solid too quickly and no longer impart the desired flexibility. Plasticizers do escape slowly, however, and often the smell associated with plastic materials arises from the release of trace amounts of these additives. Other additives that find regular use include antistatic agents, fillers, fire retardants, and light and heat stabilizers.

There are many more features of polymers that you may learn about as you proceed in your engineering curriculum. From the perspective of chemistry, they serve as a tremendous example of the impressive structural variability possible with carbon-based molecules. Although other elements can form polymeric materials, it is the chemical characteristics of carbon that make the modern plastics industry possible.

INSIGHT INTO

8-7 Micro-Electrical-Mechanical Systems (MEMS)

Exotic forms of carbon such as C_{60} and nanotubes provided our introduction to nanoscience in this chapter, and in the years since their discovery the notion of considering and controlling materials properties down to the nanoscale has driven tremendous research activity. In many ways, though, this relatively new field is a natural extension of past efforts to reduce the size of the working parts in various engineering designs. This drive for miniaturization has been going on for decades, and the partnership between engineering and chemistry in these efforts has been prominent from the beginning. Many current consumer products rely on the construction of components with dimensions on the order of microns.

A micron is 1 μm, or 10^{-6} m.

How are these microscale components most commonly made? And how might the techniques used to generate them be improved to give even smaller nanoscale structures?

One class of these tiny components is called **micro-electro-mechanical systems (MEMS)**, which can be defined as miniaturized mechanical or electro-mechanical devices made by microfabrication techniques. Currently, the size range for MEMS components goes roughly from a few microns to a few millimeters. So although they are quite small, MEMS devices are much larger than the hypothetical nanowire we considered in Section 8-3. Current research, however, is rapidly pushing the application of techniques used to produce MEMS toward the nanoscale. What are these techniques, and how can an understanding of chemistry help improve their usefulness in the future?

Many current MEMS devices begin with silicon as their primary component, and the necessary structures are made through a process called etching. In **etching,** layers of atoms are selectively removed using chemical reactions (wet etching) or bombardment with energetic ions produced in a glow discharge (plasma etching). Etching is analogous to carving a block of wood, in that the initial block is larger than the finished piece because the process removes materials. It's easy to see how an artist controls the carving of wood, but how do chemists and engineers control the removal of material to be etched?

The fundamental concept is to use masking. As depicted in **Figure 8.27**, the original material, called the *substrate*, is covered with a nonreactive material, called the *mask*, which has a desired pattern. When the etching occurs, the part of the substrate under the mask is not exposed to the reactive species and is therefore not removed. After the etching is completed, the mask is removed and the desired architecture for that part of the MEMS device is achieved. For most MEMS devices, the etching process is coupled with other chemical events, such as depositing metals onto selected areas of the substrate, to build up more complicated structures.

The chemistry of microfabrication generally relies on generating chemicals that are highly reactive with the substance being etched. In one type of wet etching, the chemistry uses two acids, nitric acid and hydrofluoric acid. There are, accordingly, two reactions that take place to remove the silicon:

$$Si + HNO_3 + H_2O \rightarrow SiO_2 + HNO_2 + H_2$$

$$SiO_2 + 6\ HF \rightarrow H_2SiF_6 + 2\ H_2O$$

Although there is some variability based on details of the silicon surface exposed, these reactions can remove roughly 1 μm of silicon per minute. The silicon nitride mask, in contrast, is etched at a rate of less than 5 nm/hour. There are many different reactions that are utilized in either wet etching or plasma etching, but the key chemistry lies in taking advantage of a large difference in the rate at which reactions occur. ◀ⓘ

The rates of chemical reactions will be considered in detail in Chapter 11. ⓘ

The variety and elegance of structures that can be created via microfabrication methods are quite impressive. The cover of this book depicts one such MEMS structure, in which very small fluid samples would flow through channels in a silicon substrate as part of a chemical analysis scheme. Among other structures, perhaps the most important devices made include microsensors and microactuators. For example, pressure gauges like those we discussed in Chapter 5 can be made with microsensors and incorporated into designs that could not accommodate pressure sensors made

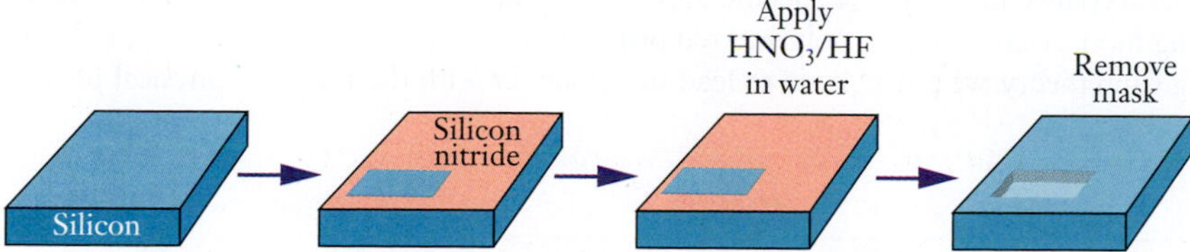

Figure 8.27 ■ Steps in the wet etching of silicon by a mixture of HNO_3 and HF are shown. A silicon nitride mask is applied to the silicon substrate. The mixed acids are applied and react to remove silicon from any areas not covered by the mask. The mask is then removed, leaving the patterend silicon. A sequence of such steps can lead to highly complex structures.

with macroscopic components. Other applications of MEMS technology include the accelerometers found in modern smart phones and "lab on a chip" components for DNA analysis.

FOCUS ON PROBLEM SOLVING

Question The metals in Groups 11 and 12 all have face-centered cubic structures. They are also among the least reactive metals. Aside from looking up the density, what could you look up to determine which metal was the densest, and how would that information help you to determine the density?

Strategy This question has both a conceptual component and a computational component. First, we need to understand how a lattice is related to density, and then we need to identify the computational pathway that relates structures at the atomic level to densities at the macroscopic level.

Solution Knowing that the structure is face-centered cubic tells us about the geometry of the atoms—how they are connected. If we can look up the length of the cube edge, we can calculate the number of atoms in a given volume (related to the volume of the unit cell). Then we would need to look up the atomic mass to convert between number of atoms in the given volume and the mass of those atoms. Density is mass per unit of volume, so these steps would provide a density estimate that could be used to estimate which metal would have the greatest density.

SUMMARY

Understanding chemical bonding is vitally important in being able to devise models for materials, including those important in engineering applications. In most materials, however, the forces that must be considered extend beyond a localized bond between two atoms. Thus an understanding of the interactions that give rise to the bulk properties of materials requires us to consider additional ideas.

For solid materials, the packing of the constituent particles into a regular array, or lattice, often provides the first simplifying component for building a model. There are important disordered, or amorphous, solids, but the order in crystalline solids makes them much easier to understand. A crystalline solid can be described at the atomic or molecular level in terms of its packing efficiency or coordination number.

To understand many properties of solids, including electrical conductivity, we must consider the packing of the atoms and also the way in which the electrons are shared or distributed. Thus we are able to extend our notion of chemical bonding and consider metals and other extended systems, using models with varying levels of sophistication. The simple sea of electrons model accounts for some metallic behaviors, but a more complete bonding model called band theory allows an additional level of detail. Using band theory, we can also understand the differences between metals, semiconductors, and insulators as well. The ability to manipulate materials—particularly semiconductors—by doping them with small levels of impurities to obtain desired characteristics is generally predicated on band theory.

Crystalline solids are not the only form of condensed phase, however, and to consider other solids or liquids, we must define the nature of the interaction between molecules. Several types of intermolecular forces have been discerned, including dispersion, dipole–dipole forces, and hydrogen bonding. The strength of these intermolecular forces helps explain observable macroscopic properties such as vapor pressure, boiling point, and surface tension. From an engineering perspective, one class of materials for which the nature of intermolecular interactions plays an enormous role is polymers. Here, we must consider the chemical bonding within a single polymer molecule and also the interaction between polymer chains via intermolecular forces. Careful consideration of these factors can enable the production of materials with specific physical properties. Thus when a materials engineer needs a plastic for a specific application, the choice of monomers and how they are polymerized provide the two critical variables that can be manipulated to lead to a polymer with the required physical properties.

KEY TERMS

acceptor level (8-3)
addition polymerization (8-6)
adhesion (8-5)
alternating copolymer (8-6)
amorphous solid (8-2)
antibonding molecular orbital (8-3)
atactic polymer (8-6)
band gap (8-3)
band theory (8-3)

block copolymer (8-6)
body-centered cubic (bcc) (8-2)
bonding molecular orbital (8-3)
cohesion (8-5)
condensation polymer (8-6)
conduction band (8-3)
coordination number (8-2)
copolymer (8-6)
crystalline solid (8-2)
degree of polymerization (8-6)
dipole–dipole forces (8-4)
dispersion forces (8-4)
donor level (8-3)
doping (8-3)
ductile (8-3)
dynamic equilibrium (8-5)
etching (8-7)
face-centered cubic (fcc) (8-2)
graft copolymer (8-6)
hydrogen bonding (8-4)
insulator (8-3)
intermolecular forces (8-4)
isotactic polymer (8-6)
malleable (8-3)
micro-electrical-mechanical systems (MEMS) (8-7)
n-type semiconductor (8-3)
nanotube (8-1)
normal boiling point (8-5)
p-n junction (8-3)
p-type semiconductor (8-3)
packing efficiency (8-2)
phase diagram (8-1)
polarizability (8-4)
sea of electrons model (8-3)
simple cubic (sc) (8-2)
surface tension (8-5)
syndiotactic polymer (8-6)
thermoplastic polymer (8-6)
thermosetting polymer (8-6)
unit cell (8-2)
valence band (8-3)
vapor pressure (8-5)
volatile (8-5)

PROBLEMS AND EXERCISES

INSIGHT INTO Carbon

8.1 How many solid forms of elemental carbon are known?

8.2 Why is the C_{60} form of carbon called *buckminsterfullerene*?

8.3 What property of diamond leads to the most engineering applications? Which types of applications would benefit from this property?

8.4 Which variables must be controlled and in what manner to produce diamonds from graphite?

8.5 What is the relationship between the structures of buckminsterfullerene and carbon nanotubes?

8.6 Use the web to look up information on nanotubes. Distinguish between single-walled and double-walled nanotubes.

8.7 Use the web to look up the experimental conditions required to synthesize buckminsterfullerene in an electric arc. Would running an arc in air work?

Condensed Phases—Solids

8.8 What is the difference between an amorphous solid and a crystalline solid?

8.9 Define *packing efficiency*.

8.10 Using circles, draw regular two-dimensional arrangements that demonstrate low packing efficiency and high packing efficiency.

8.11 Using pentagons, draw arrangements that demonstrate low packing efficiency and higher packing efficiency.

8.12 Using circles, show that a cubic structure (squares in 2-D) has a lower coordination number than a hexagonal structure.

8.13 Calculate the packing efficiencies for both two-dimensional arrangements shown in Figure 8.3.

8.14 Explain why hcp and ccp structures have the same coordination number.

8.15 What is the coordination number of atoms in the diamond structure?

8.16 Explain why graphite is described as having a layered structure.

8.17 Polonium is the only metal that forms a simple cubic crystal structure. Use the fact that the density of polonium is 9.32 g/cm^3 to calculate its atomic radius.

8.18 Iridium forms a face-centered cubic lattice, and an iridium atom is 271.4 pm in diameter. Calculate the density of iridium.

8.19 Europium forms a body-centered cubic unit cell and has a density of 4.68 g/cm^3. From this information, determine the length of the edge of the cubic cell.

8.20 Manganese has a body-centered cubic unit cell and has a density of 7.88 g/cm^3. From this information, determine the length of the edge of the cubic cell.

8.21 The spacing in a crystal lattice can be measured very accurately by X-ray diffraction, and this provides one way to determine Avogadro's number. One form of iron has a body-centered cubic lattice, and each side of the unit cell is 286.65 pm long. The density of this crystal at 25°C is 7.874 g/cm^3. Use these data to determine Avogadro's number.

Metals and Metallic Bonding

8.22 How many electrons per atom are delocalized in the sea of electrons model for the following metals? **(a)** iron, **(b)** vanadium, **(c)** silver

8.23 One important physical property of metals is their malleability. How does the sea of electrons model account for this property?

8.24 The sea of electrons model is not generally used for quantitative predictions of properties. What factors are left out of this model that might prevent quantitative precision?

8.25 How does the sea of electrons model of metallic bonding explain why metals are good conductors of electricity?

8.26 What is the key difference between metallic bonding (in the sea of electrons model) and ionic bonding (as described in Chapter 7) that explains why metals conduct electricity and ionic solids do not?

8.27 What is the difference between a bonding orbital and an antibonding orbital?

8.28 Describe how the combination of atomic orbitals gives rise to bands in the limit of large numbers of atoms.

8.29 Draw a depiction of the band structure of a metal. Label the valence band and conduction band.

8.30 In terms of band theory, what is the difference between a conductor and an insulator? Between a conductor and a semiconductor?

8.31 The conductivity of semiconductors increases as temperature is increased. Using band theory, explain this observation.

8.32 Use the web to find out what a "III–V" semiconductor is. How is this type of semiconductor related to silicon?

8.33 Explain how doping affects the electronic properties of a semiconductor.

8.34 What type of atom is needed as a dopant in an n-type semiconductor? Why is it called n-type?

8.35 What type of atom is needed as a dopant in a p-type semiconductor? Why is it called p-type?

8.36 Is an n-type semiconductor actually negatively charged?

8.37 How does a p-n junction serve as a voltage-gated switch?

8.38 Suppose that a device is using a 15.0-mg sample of silicon that is doped with $1 \times 10^{-5}\%$ (by mass) phosphorus. How many phosphorus atoms are in the sample?

Intermolecular Forces

8.39 What is an *instantaneous dipole*?

8.40 Why are dispersion forces attractive?

8.41 If a molecule is not very polarizable, how will it respond to an external electric field?

8.42 What is the relationship between polarizability and dispersion forces?

8.43 Most gaseous compounds consist of small molecules, while polymers are never gaseous at room temperature. Explain this observation based on intermolecular forces.

8.44 Why are dipole–dipole forces typically stronger than dispersion forces?

8.45 Under what circumstances are ion–dipole forces important?

8.46 Which of the following compounds would be expected to form intermolecular hydrogen bonds in the liquid state? **(a)** CH_3OCH_3 (dimethyl ether), **(b)** CH_4, **(c)** HF, **(d)** CH_3CO_2H (acetic acid), **(e)** Br_2, **(f)** CH_3OH (methanol)

8.47 What is the specific feature of N, O, and F that causes them to play a role in hydrogen bonding?

8.48 What type of intermolecular forces must be overcome in converting each of the following from a liquid to a gas? **(a)** CO_2, **(b)** NH_3, **(c)** $CHCl_3$, **(d)** CCl_4

8.49 Identify the kinds of intermolecular forces (London dispersion, dipole–dipole, or hydrogen bonding) that are the most important in each of the following substances. **(a)** methane (CH_4), **(b)** methanol (CH_3OH), **(c)** chloroform ($CHCl_3$), **(d)** benzene (C_6H_6), **(e)** ammonia (NH_3), **(f)** sulfur dioxide (SO_2)

8.50 Rank the following in order of increasing strength of intermolecular forces in the pure substances. Which exists as a gas at 25°C and 1 atm? **(a)** $CH_3CH_2CH_2CH_3$ (butane), **(b)** CH_3OH (methanol), **(c)** He

8.51 Carbon tetrachloride (CCl_4) is a liquid at room temperature and pressure, whereas ammonia (NH_3) is a gas. How can these observations be rationalized in terms of intermolecular forces?

8.52 Explain from a molecular perspective why graphite has properties that are useful for lubrication.

Condensed Phases—Liquids

8.53 Describe how interactions between molecules affect the vapor pressure of a liquid.

8.54 What makes a chemical compound volatile?

8.55 Answer each of the following questions with *increases*, *decreases*, or *does not change*.

(a) If the intermolecular forces in a liquid increase, the normal boiling point of the liquid ________.

(b) If the intermolecular forces in a liquid decrease, the vapor pressure of the liquid ________.

(c) If the surface area of a liquid decreases, the vapor pressure ________.

(d) If the temperature of a liquid increases, the equilibrium vapor pressure ________.

8.56 Why must the vapor pressure of a substance be measured only after dynamic equilibrium is established?

8.57 Which member of each of the following pairs of compounds has the higher boiling point? **(a)** O_2 or N_2, **(b)** SO_2 or CO_2, **(c)** HF or HI, **(d)** SiH_4 or GeH_4

8.58 Predict the order of increasing vapor pressure at a given temperature for the following compounds. **(a)** FCH_2CH_2F, **(b)** $HOCH_2CH_2OH$, **(c)** FCH_2CH_2OH

8.59 A substance is observed to have a high surface tension. What predictions can you make about its vapor pressure and boiling point?

8.60 Suppose that three unknown pure substances are liquids at room temperature. You make vapor pressure measurements and find that substance Q has a pressure of 110 torr, substance R has a pressure of 42 torr, and substance S has a pressure of 330 torr. If you slowly increase the temperature, which substance will boil first and which will boil last?

8.61 Suppose that three unknown pure substances are liquids at room temperature. You determine that the boiling point of substance A is 53°C, that of substance B is 117°C, and that of substance C is 77°C. Based on this information, rank the three substances in order of their vapor pressures at room temperature.

8.62 Rank the following hydrocarbons in order of increasing vapor pressure: C_2H_6, $C_{10}H_{22}$, CH_4, C_7H_{16}, $C_{22}H_{46}$.

8.63 Draw the meniscus of a fluid in a container where the interactions among the liquid molecules are weaker than those between the liquid and the container molecules.

8.64 When water is in a glass tube such as a buret, in which direction does the meniscus curve? What does this observation say about the relative magnitude of adhesion and cohesion in that system?

Polymers

8.65 If shown structures of monomers that might polymerize, what would you look for to determine if the type of polymerization would be addition?

8.66 Why is temperature increased to start most addition polymerization reactions?

8.67 What is meant by the term *chain termination* in a polymerization reaction?

8.68 The monomer of polyvinylchloride is $H_2C{=}CHCl$. Draw an example of an isotactic PVC polymer.

8.69 Why is there no isotactic or syndiotactic form of polyethylene?

8.70 Why are isotactic or syndiotactic polymers often more attractive for materials development?

8.71 What are the products of a condensation polymerization reaction?

8.72 Nylon-6 is made from a single monomer:

O
H_2N OH

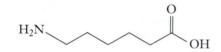

Sketch a section of the polymer chain in Nylon-6.

8.73 Distinguish between a block copolymer and a graft copolymer.

8.74 Use the web to look up the polymers used for six synthetic fibers. Classify the polymers as **(a)** addition or condensation polymers and **(b)** as alternating, block or graft copolymers if they are copolymers.

8.75 What is the key physical property that characterizes a thermoplastic polymer?

8.76 What conditions would prohibit the use of a thermoplastic polymer?

8.77 What happens molecularly in a thermosetting polymer when it is heated?

8.78 What structural characteristics are needed for additives such as plasticizers?

8.79 Use the web to find out what pencil erasers are made of. Would it be possible to form a new shape of an eraser by melting it and pouring it into a mold?

8.80 Use the web to find an application where the presence of isotactic, syndiotactic, or atactic polymers affects the physical properties important for an application of a polymer.

INSIGHT INTO Micro-Electrical-Mechanical Systems (MEMS)

8.81 Define the terms *substrate* and *mask* in the context of microfabrication of MEMS devices.

8.82 Silicon nitride is a commonly used masking material in microfabrication. Based on the positions of silicon and nitrogen in the periodic table, what is the most likely chemical formula for silicon nitride? Explain your answer.

8.83 Silicon dioxide is also used as a masking material in microfabrication. If a silicon substrate with an SiO_2 mask is wet etched by HNO_3/HF, one of the reactions that takes place will be between SiO_2 and HF. Provide a reasonable explanation as to why SiO_2 can be used as a mask even though it will react with the etchant.

8.84 In what way(s) are wet etching and plasma etching similar?

8.85 Estimate the number of atoms that would have to be removed from a silicon substrate to produce a rectangular channel 5 μm wide, 10 μm deep, and 20 μm long. (The density of silicon is 2.330 g cm^{-3}.)

8.86 An important category of MEMS devices is microactuators. These devices produce mechanical motion in response to some electrical signal. Search the web for information and write a short paragraph explaining one way that such a device might work.

Additional Problems

8.87 Use the vapor pressure curves illustrated here to answer the questions that follow.

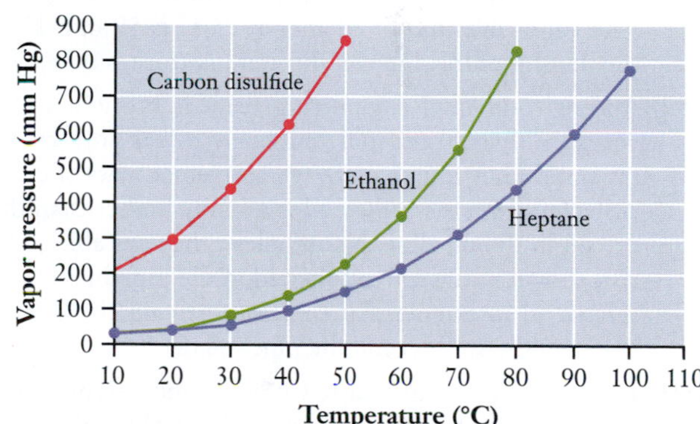

(a) What is the vapor pressure of ethanol (C_2H_5OH) at 60°C?

(b) Considering only carbon disulfide (CS_2) and ethanol, which has the stronger intermolecular forces in the liquid state?

(c) At what temperature does heptane (C_7H_{16}) have a vapor pressure of 500 mm Hg?

(d) What are the approximate normal boiling points of each of the three substances?

(e) At a pressure of 400 mm Hg and a temperature of 70°C, is each substance a liquid, a gas, or a mixture of liquid and gas?

8.88 Cylinders of compressed gas are equipped with pressure gauges that allow the user to monitor the amount of gas remaining. But such gauges are not useful for substances like propane or carbon dioxide, which are stored as liquids. So tanks of propane or carbon dioxide are sometimes mounted on scales to allow the user to see the rate of consumption. Use ideas from this chapter to explain why this is so.

8.89 The following data show the vapor pressure of liquid propane as a function of temperature. **(a)** Plot a vapor pressure curve for propane and use it to estimate the normal boiling point. **(b)** Use your curve to estimate the pressure (in atm) in the propane tank supplying fuel for a gas barbecue grill on a hot summer day when the temperature is 95°F. **(c)** What implications might your answer to **(b)** have for an engineer designing propane storage tanks?

T (°C)	–100	–80	–60	–40	–20	0	20
P (torr)	22.4	100.6	328.8	856.9	1888.6	3665.6	6445.9

8.90 Geckos are known for their ability to climb smooth vertical walls or windows. Recent research has shown that this unusual trait results from van der Waals forces between the surface and the tips of tiny hairlike setae in the Gecko's feet. Many materials science and engineering groups are working to develop tapes and other adhesives designed to mimic this behavior. Do a web search to find information about such efforts and write a short summary explaining how one resulting adhesive is related to ideas we explored in this chapter.

8.91 If you place 1.0 L of ethanol (C_2H_5OH) in a room that is 3.0 m long, 2.5 m wide, and 2.5 m high, will all the alcohol evaporate? If some liquid remains, how much will there be? The vapor pressure of ethanol at 25°C is 59 mm Hg, and the density of the liquid at this temperature is 0.785 g/cm^3.

8.92 An environmental engineering team is brought in to investigate a small oil spill caused by a leak in a crude oil pipeline. They find that the average molar mass of a sample collected from the spilled oil is notably higher than that of the oil in the pipeline. Nonetheless, they are able to verify that the oil in the spill originated in the pipeline. Use ideas related to intermolecular forces to explain the difference between the spill oil and the pipeline oil.

8.93 Do a web search to find out how p- and n-type semiconductors can be used in designing solar panels. Write a short paragraph explaining what you learn. What new types of materials are being considered for use in solar panel systems?

8.94 Hydraulic fluids are liquids moving in a confined space under pressure, and they can serve many purposes in engineering designs. One use of hydraulic fluids is lubrication. Suppose you need to choose a hydrocarbon for use as a lubricant. What size molecules would provide high viscosity for a lubricant in a design?

8.95 It is possible to make syndiotactic polystyrene, although most polystyrene is atactic. Syndiotactic polystyrene melts at 270°C while atactic polystyrene melts closer to 240°C. **(a)** Why might this difference in structure result in different boiling points? **(b)** If your supplier charges 20% more for syndiotactic polystyrene, what factors of a design might call for using this material despite the higher cost?

8.96 A business manager wants to provide a wider range of p- and n-type semiconductors as a strategy to enhance sales. You are the lead materials engineer assigned to communicate with this manager. How would you explain why there are more ways to build a p-type semiconductor from silicon than there are ways to build an n-type semiconductor from silicon?

8.97 The doping of semiconductors can be done with enough precision to tune the size of the band gap in the material. Generally, in order to have a larger band gap, the dopant should be smaller than the main material. If you are a materials engineer and need a semiconductor that has lower conductivity than pure silicon, what element or elements could you use as your dopant? (You do not want either an n- or a p-type material.) Explain your reasoning.

FOCUS ON PROBLEM SOLVING EXERCISES

8.98 If you know the density of a material and the length of the edge of its cubic lattice, how would you determine if it is face-centered cubic, body-centered cubic, or simple cubic? Would you have to look up any information?

8.99 If a pin is run across a magnet a number of times, it will become magnetized. How could this phenomenon be used as an analogy to describe polarization in atoms and molecules?

8.100 You go into the laboratory to look for a squirt bottle containing acetone. You find two unlabeled bottles with different colored tops suggesting they are different liquids. Unfortunately, you have a terrible cold and cannot tell by smell which one might be acetone. What simple test could you use to determine which liquid is acetone and which is water? How does this test tell you this information?

8.101 A materials engineer, with an eye toward cost, wants to obtain a material whose degree of polymerization is high. What types of measurements must be made in the laboratory to determine whether the degree of polymerization is acceptable?

Cumulative Problems

8.102 How is polarizability related to the periodic trends of elements in the periodic table?

8.103 In previous chapters, we have noted that cryolite is used in refining aluminum. Use the web to look up what this addition does for the process, and relate it to the concepts of intermolecular forces from this chapter.

8.104 Use the web to look up the percentage of dopant for a commercially available p-type semiconductor. Imagine that you were setting up a process for doping 1 metric ton of silicon with this dopant. **(a)** What mass would be required? **(b)** What would be the mole fraction of the dopant?

8.105 Use the web to look up the percentage of dopant for a commercially available n-type semiconductor. Imagine that you were setting up a process for doping 1 metric ton of silicon with this dopant. **(a)** What mass would be required? **(b)** What would be the mole fraction of the dopant?

9

Energy and Chemistry

Wind farms like the one shown here are one of several alternative energy technologies that have grown in importance in recent years. Despite these advances, fossil fuels will continue to provide the bulk of our energy for the foreseeable future. © *Kevan O'Meara/Shutterstock.com*

CHAPTER OUTLINE

If you list applications of chemistry that are important in your life, many of them would be likely to involve energy and energy transformation. In Chapter 4, we looked at the burning of fuels as an application of chemical combustion. All the batteries you depend on to power portable electrical devices are based on chemical reactions as well. In this chapter, we will take a closer look at energy and its role in chemistry. We begin by focusing on the production and use of energy in North America, a component of life in the developed world that we'll refer to as the **energy economy**.

Chapter Objectives

After mastering this chapter, you should be able to

- explain the economic importance of conversions between different forms of energy and the inevitability of losses in this process.
- define work and heat using the standard sign conventions.
- define state functions and explain their importance.
- state the first law of thermodynamics in words and as an equation.
- use calorimetric data to obtain values for ΔE and ΔH for chemical reactions.
- define ΔH_f° and write formation reactions for compounds.

- explain Hess's law in your own words.
- calculate $\Delta H°$ for chemical reactions from tabulated data.

INSIGHT INTO

9-1 Energy Use and the World Economy

Few factors reveal the wealth of a country or region more vividly than its energy consumption. The central role of the United States in the world economy is reflected in the fact that the nation accounts for 24% of world energy use. Canada, Japan, and Western Europe are also large users of energy. Energy use is closely connected to a broad array of societal, economic, and political factors. For example, in the early 1990s, Pacific Rim countries such as Thailand and South Korea had growing economies and substantial increases in their energy use. Yet during this time, energy use worldwide was stagnant, as shrinking economies in Eastern Europe offset the increases elsewhere in the world.

GDP is usually defined as the total value of all goods and services produced annually in a region.

Figure 9.1 illustrates the link between economic growth (as measured by gross domestic product) and energy use. Data are plotted for several countries as energy consumption per dollar of GDP from 1980 to 2009. It is clear from the figure that the lines shown are all nearly horizontal. This confirms the correlation between economic performance and energy consumption because a horizontal line in this plot implies that energy use is proportional to GDP. But the figure also shows that some countries are more energy-dependent than others because the various lines are not identical. This is partly due to differences in the efficiency with which different countries use energy and partly due to climate and other factors.

We can learn more about the pivotal role of energy in an economic system by examining the sources and uses of energy. Although there are differences between countries, we'll look at the United States as an example of energy production and use in an industrialized nation.

Figure 9.2 is a visual representation of the pattern of energy production and use in the United States in 2011. At first glance, this figure is somewhat complicated, but it can be analyzed to reveal a great deal of information. Start by looking

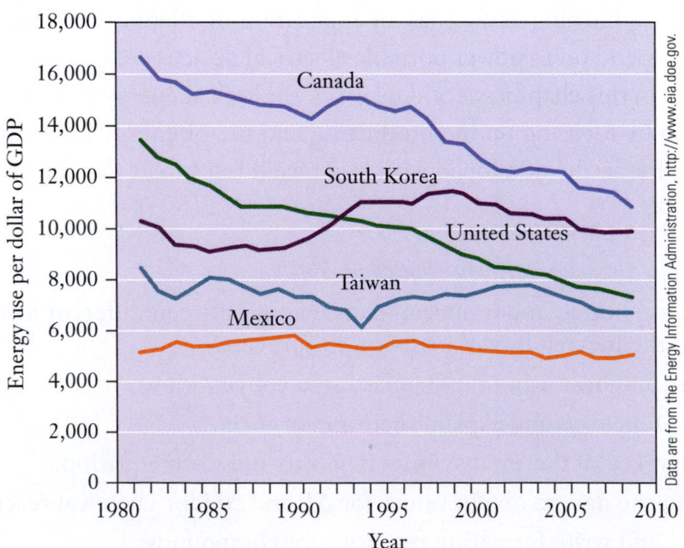

Figure 9.1 ■ The graph demonstrates the link between energy consumption and gross domestic product (GDP), which is a common indicator of economic strength. The data are plotted in units of Btu per U.S. dollar, with all data converted to constant (2005) dollar values. The fact that these lines are all essentially horizontal confirms the connection between energy use and economic strength.

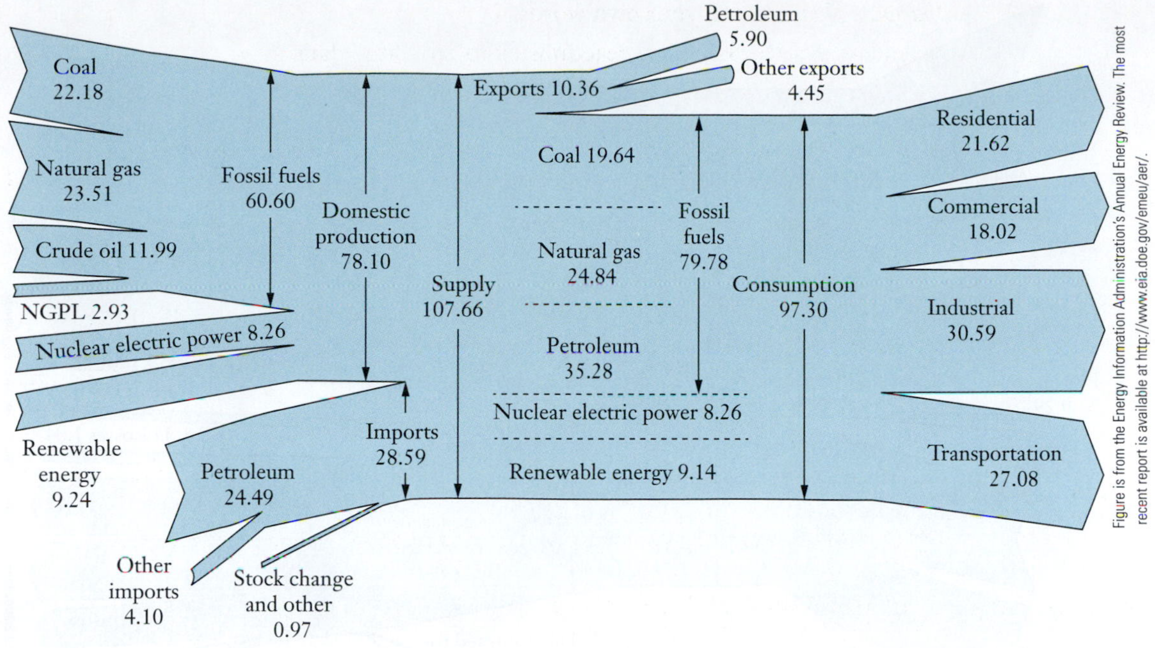

Figure 9.2 ■ Energy production and consumption (in quadrillion Btu) in the United States during the year 2011. The discussion in the text explains how to read this complex figure, which contains an enormous amount of information about the energy economy. Data come from the Department of Energy and do not always add up exactly as expected due to rounding and other issues.

at each end: the left side represents sources of energy, and the right side shows consumption. Thus the whole figure represents the flow of resources through the U.S. energy economy. All numbers on the figure are given in units of quadrillion Btu. (A quadrillion is 10^{15}, and the **Btu** is a unit of energy. The letters stand for **British thermal unit**, and 1 Btu = 1054.35 J.) Starting in the upper left, we can examine the sources of energy production. Coal accounted for 22.18 quadrillion Btu, or 28.4% of the total domestic production of 78.10 quadrillion Btu. Natural gas and crude oil provided 30.1% and 15.4% of the total, respectively. Smaller contributions came from nuclear energy and renewable energy sources, which include hydroelectric power, wood, solar energy, and wind. Energy imports appear at the bottom left of the figure, where we see that the main imports are crude oil and related petroleum products. Combining the domestic production and the imports gives the total supply of energy, which was 107.66 quadrillion Btu for 2011. Moving to the right, the figure divides this supply to show how energy was consumed. The four main components of use are classified as (1) residential, (2) commercial, (3) industrial, and (4) transportation, accounting for roughly 22, 19, 31, and 28% of total energy use, respectively.

Some assumptions are built into the division of uses into these four categories, so absolute numbers are not as important as comparative ones.

Nearly half of all domestic energy use goes into producing electricity. The generation and consumption of electricity is detailed in **Figure 9.3**. (The form of this diagram is similar to that of Figure 9.2, so you should examine it in the same way.) This figure shows that conversion losses account for nearly two-thirds of the energy consumed to generate electricity. These conversion losses in electricity generation mean that more than 25% of the total domestic energy consumption is lost from the economy. This raises some important questions: How is energy from other sources converted to electricity, and why is this conversion necessary? Why is so much energy lost or wasted in this process, and how can we minimize or eliminate this loss?

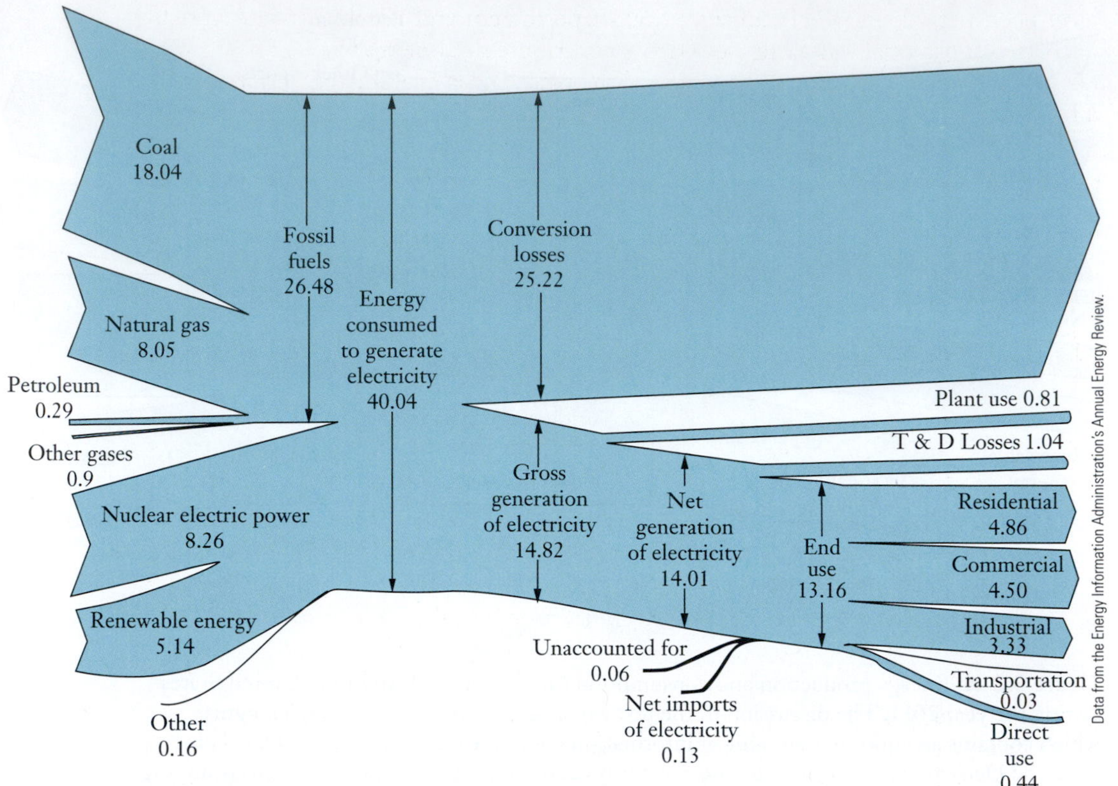

Figure 9.3 ■ Visual summary of the generation and consumption of electricity in the United States during the year 2011, presented in a manner similar to Figure 9.2.

Other questions arise when considering the relative roles of various fossil fuels in the energy economy. **Figure 9.4** shows the recent history of U.S. domestic consumption of energy from various sources. Our consumption of fossil fuels continues to grow, as it has throughout our country's entire history. But this growth can be sporadic, and there

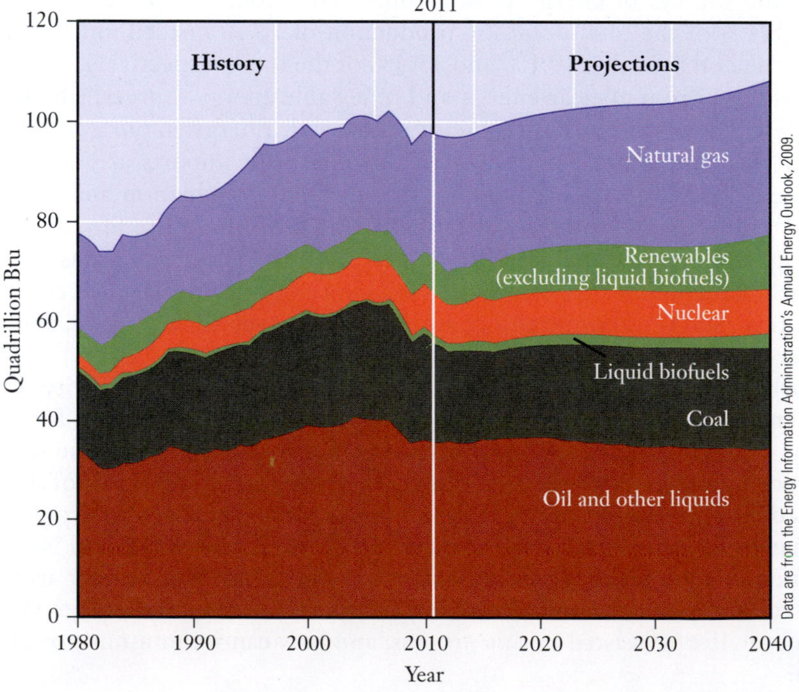

Figure 9.4 ■ U.S. consumption of energy from various sources changes slowly over time. The graph shows actual consumption for 1980–2011 and projected use for 2012–2040.

have been periods in which consumption of petroleum and natural gas has actually decreased substantially. A variety of factors contribute to these trends in consumption, notably the availability of raw material and the price and availability of imported fuels. The decrease in petroleum consumption in the early 1980s, for example, coincides with the recession of that era, a period in which decreased economic activity led to lower demand for oil. More recently, higher oil prices have driven interest in renewable energy sources. Because our interest here is chemistry, though, we will want to ask whether there are more fundamental differences among these fuels in terms of energy production and transformation. What is the role of chemistry in deciding which of these fossil fuels is most attractive in a given set of circumstances? To address all of the questions raised here we must consider some aspects of **thermochemistry**—the study of the energetic consequences of chemistry.

9-2 Defining Energy

The questions raised in Section 9-1 lead us into a study of energy. Most people have an intuitive feel for the concept of energy, and we see the word every day. Breakfast cereals advertise their ability to provide energy in the morning. In sports, substitute players are described as boosting their team's energy. The energy crisis brought on by oil embargoes in the 1970s is a critical component of the history of that decade, and energy remains central to important political issues in the 21st century. This widespread use of the concept of energy underscores its importance. It also suggests that we'll need to be very careful in defining energy and related terms because it is apparent that in casual use the word "energy" conveys many related but subtly different meanings.

Forms of Energy

Most of the energy we encounter can be placed into two broad categories: potential energy and kinetic energy. **Potential energy** is associated with the relative position of an object. For example, a roller coaster gains potential energy as it is pulled up the initial slope because it acquires a higher position relative to the ground and gravity is attracting it downward. But gravity is not the only force responsible for objects having energy associated with their positions. The attraction and repulsion between electrical charges also lead to potential energy, as we've seen in our consideration of the structure of atoms and molecules. **Kinetic energy** is associated with motion. When the roller coaster heads down the first hill, it transforms potential energy into kinetic energy. You probably are familiar with Equation 9.1, which provides a mathematical definition for the kinetic energy of an object in terms of its mass (m) and velocity (v):

$$\text{Kinetic energy} = \frac{1}{2}mv^2 \tag{9.1}$$

The roller coaster example shows kinetic and potential energy on a macroscopic scale. But on a microscopic level, all substances and objects, from fuels to the paper or display on which these words appear, have these same forms of energy. Any substance or object is composed of atoms and molecules. These atoms and molecules have kinetic energy associated with their constant motion, and they have potential energy due to the various forces they exert on one another. The combined kinetic and potential energies of the atoms and molecules that make up an object constitute its **internal energy**. Thus the roller coaster cars have three basic forms of energy: kinetic (from their motion), potential (from their position relative to the ground), and internal (from the molecules that compose the materials from which they are made).

Much of the internal energy of an object, be it a roller coaster car or a piece of coal, is associated with the potential energy arising from the relative positions of the atoms that make up the object. As we saw in Chapter 7, the making or breaking of chemical bonds changes this potential energy. In most chemical reactions, bonds are broken in the reactants and new bonds are formed in the products. If the amount of energy liberated in bond making is greater than that consumed in bond breaking, the overall process releases energy. This energy release is often referred to as **chemical energy**, and the harnessing of chemical energy is an important aspect of the overlap between chemistry and engineering.

Chemical energy is a form of potential energy.

Occasionally, we find it useful to have more specific terms for various forms of energy as well. *Radiant energy* is associated with light or electromagnetic radiation. The radiant energy of the sun is ultimately responsible for the majority of energy resources on this planet, as we saw when considering biofuels in Chapter 3. *Mechanical energy* is associated with the movement of macroscopic objects. *Thermal energy* arises from the temperature of an object. (We can associate thermal energy with the molecular-level motion of atoms and molecules.) *Electrical energy* results from moving charge—usually electrons in a metal. *Nuclear energy*, which can be released in nuclear fusion and fission processes, is a form of potential energy associated with the arrangement of protons and neutrons in atomic nuclei.

Heat and Work

Although we can use a wide variety of classifications for types of energy, all energy flow is either heat or work. We'll need to understand both forms of energy transfer to assess the energy economy of the world and the role of chemistry in that economy. **Heat** is the flow of energy between two objects, from the warmer one to the cooler one, because of a difference in their temperatures. Thus if we are speaking carefully, heat is a *process* and not a *quantity*. Although we routinely hear statements such as "turn up the heat," heat is not an entity we can pump into a room or a cup of coffee. An object does not possess heat. In a strictly scientific sense, a furnace does not produce heat but rather a body of warm air or hot water that has a higher temperature than the cool air in a room. What emerges from the vent on the floor is not "heat," but warm air. Although these distinctions are essentially semantic, they can be very important in many cases.

Work is the second form of energy transfer. **Work** is the transfer of energy accomplished by a force moving a mass some distance against resistance. Lifting a set of roller coaster cars up a hill against the pull of gravity is an example of work. When we consider macroscopic examples, we are typically viewing work in terms of mechanical energy. Work, however, encompasses a wider range of phenomena than just mechanical movement of macroscopic objects. The most common type of work we will encounter in chemical processes is pressure-volume work **(PV-work)**. When a gas expands, it can do work. If an inflated balloon is released before it is tied off, it flies around as the gas inside the balloon expands into the large volume of the room. Because the flying balloon has mass, it is easy to see that the expanding gas is doing work on the balloon: this is pressure-volume work.

For a more productive example of work being done by a chemical reaction, we might look at the burning of gasoline in a car engine. We learned in Section 4-1 that gasoline is actually a complex mixture of hydrocarbons. The energy needed to propel a car is released by the combustion of those hydrocarbons in the engine cylinders:

$$\text{Hydrocarbon} + O_2(g) \rightarrow CO_2(g) + H_2O(g)$$

This combustion produces carbon dioxide and water vapor, and those gases do PV-work as they expand against the piston in the cylinder. This PV-work is then transmitted through the drive train to move the car.

Energy Units

We use a wide variety of units when measuring and discussing energy, and various disciplines of science and engineering tend to favor particular choices of units. The SI unit of energy is the **joule (J)**, and 1 joule is equal to 1 kg m^2/s^2. Realizing that work should have units of energy helps us to make sense of this unit. Work is force times distance, and force is mass times acceleration. Thus work is

$$\text{Mass} \times \text{acceleration} \times \text{distance}$$

or

$$\text{kg} \times \frac{\text{m}}{\text{s}^2} \times \text{m} = \frac{\text{kg m}^2}{\text{s}^2}$$

It's often handy to develop an intuition for the magnitude of a particular unit. A joule is roughly the amount of energy it takes to raise a 1-kg (about 2 lb.) book 10 cm (4 in.) above a tabletop. If we think about chemistry on a molecular level, a joule is a very large amount of energy; breaking a single chemical bond typically requires only about 10^{-18} J. But we seldom break only one bond at a time. If we look at a reaction from the macroscopic viewpoint, we might consider breaking a mole of bonds. In that case, the joule seems like too small a measure of energy, and we might use kJ (or kJ/mol) instead. Breaking 1 mole of C—H bonds, for example, requires about 410 kJ.

Many older units for energy relied on some readily observable property for their original definition. Thus the Btu, which is still widely used in several engineering disciplines, was first defined as the amount of energy needed to raise the temperature of 1 lb. of water by 1°F. Another traditional unit for energy is the **calorie**, which was originally defined as the amount of energy required to heat 1 g of water from 14.5 to 15.5°C.

The joule has become the most widely accepted unit for energy, and so these other units are now defined in terms of it. A calorie is defined as 4.184 J, and a Btu is defined as 1055 J. A source of frequent confusion in comparing energy units is that the Calorie reported for foods is actually a kilocalorie. (The food Calorie should be written with an uppercase "C.") Thus 1 Calorie is actually equivalent to 4184 J, or 4.184 kJ.

The energy typically consumed by large economies is enormous, as we saw when analyzing Figures 9.2 and 9.3.

9-3 Energy Transformation and Conservation of Energy

We have introduced several ways to categorize energy. But these multiple forms of energy are not all equally useful, so in many cases it is desirable to transform energy from one form into another. For example, the lighting in your room is provided by electricity, but that electricity was probably generated by the release of chemical energy through the combustion of coal. Unless you want to try to light your room by burning a chunk of coal, you need a way to harness the chemical energy released as the coal burns and then convert it to electrical energy. That electrical energy must then be conveyed to your room, where your light bulbs convert it into radiant energy. We have already noted the reality of energy waste in these processes in Section 9-1.

Now we begin to consider the laws of nature that apply when one form of energy is converted into another. The first and foremost constraint on energy transformation is that total energy must be conserved. If we account properly for all energy conversion and energy transfer processes, the total amount of energy present must remain constant. To account properly for all types of energy, we will need to define a number of terms quite carefully. First, we must specify precisely what is being studied. The **system** is defined as the part of the universe that is being considered. The remainder of the universe is referred to as the **surroundings**, even though it is not generally necessary to consider everything else in the actual universe. These definitions ensure that the system plus the surroundings must equal the **universe**. The system and the surroundings are separated by a **boundary**. In some cases, this boundary may be a physical container, and in others, it might be a more abstract separation.

Careful definition of the system and its surroundings is also important in many engineering problems.

These ideas may seem obvious, but the choice of system and surroundings is not always clear. For example, if the system to be studied is "the atmosphere of the earth," the definition of the boundary is somewhat challenging. The atmosphere simply gets gradually less dense, and we must make some arbitrary decision as to where it ends. In most cases, the main requirement is that we be consistent. The same choice of system and surroundings *must* be used throughout a particular problem, even if the choice is somewhat arbitrary.

Once an appropriate choice of a system has been made, the concept of conservation of energy immediately becomes useful. Because we said that heat and work are the only possible forms of energy transfer, we can attribute the overall change in energy, E, of a system to these two components. Heat is commonly designated as q and work as w, so we can write

$$\Delta E = q + w \tag{9.2}$$

The symbol Δ (delta) is introduced here as a notation meaning "the change in." This symbol, which will be used frequently in our study of thermodynamics, is always defined as the difference between the final state and the initial state:

$$\Delta E = E_{\text{final}} - E_{\text{initial}} \tag{9.3}$$

Equation 9.2 is deceptively simple because it includes arbitrary choices for the meaning of the signs for the quantities of heat and work. Again, the key is to choose consistent definitions for those signs. Convention dictates that energy transferred *into* a system is given a positive sign and energy flowing *out* of a system carries a negative sign. Thus when heat flows into a system from the surroundings, the value of q is positive, and when work is done on a system, the value of w is positive. Conversely, when heat flows out of a system or work is done by the system on the surroundings, q and w will be negative. Example Problem 9.1 demonstrates the use of this equation.

EXAMPLE PROBLEM 9.1

If 515 J of heat is added to a gas that does 218 J of work as a result, what is the change in the energy of the system?

Strategy Energy flow occurs as either heat or work, and Equation 9.2 relates those to the total energy. We are given the magnitudes for q and w, but we must determine the direction for each process to assign the correct signs.

Solution

Heat added TO the system means that $q > 0$, so $q = +515$ J.
Work done BY the system means that $w < 0$, so $w = -218$ J.

$$\Delta E = q + w = 515\text{ J} + (-218\text{ J}) = +297\text{ J}$$

Note that in most cases when a value is positive, the "+" sign is not placed in front of the number.

Discussion Though fairly simple numerically, this problem points to the need to consider the signs of q and w carefully.

Check Your Understanding If 408 J of work is done on a system that releases 185 J of heat, what is the energy change in the system?

In the example above, we see that ΔE is not zero. How can this be consistent with the idea that energy must be conserved? By following our sign conventions for q and w, we considered the processes of heat and work from the perspective of the system (the unspecified gas in this case). So the value of ΔE we calculated is also the change in internal energy for the system. This change in the internal energy of the system is exactly offset by a change in the surroundings: $\Delta E_{\text{surroundings}} = -\Delta E_{\text{system}}$. (Usually, we just write ΔE rather than ΔE_{system} because, by definition, the system is what we are most interested in.) Thus the energy of the universe remains constant:

$$\Delta E_{\text{universe}} = \Delta E_{\text{surroundings}} + \Delta E_{\text{system}} = 0$$

Energy can be transformed from one form to another but cannot be created or destroyed. This is known as the **first law of thermodynamics**.

Waste Energy

We have already alluded to the conversion of energy from one form to another a number of times in this chapter. Anyone who sits in front of a campfire to ward off the evening chill can argue the virtues of heat, but in most cases, civilization prefers its energy in the form of work. The combustion of gasoline is not inherently useful, but when the heat released is harnessed in the engine of an automobile, the resulting work gets us where we need to go. All available observations, however, point to the idea that it is impossible to convert heat completely to work. (This leads to the second law of thermodynamics, which we'll examine in detail in Chapter 10.) The car's engine gets hot when it runs. The heat that warms the engine does not propel the car toward its destination. So a portion of the energy released by the combustion of gasoline does not contribute to the desired work of moving the car. In terms of the energy economy, this energy can be considered wasted.

Of course, on a cold winter day, some of the waste heat from the engine is used to warm the passenger cabin. So clever engineering can put "wasted" heat to use.

One common way to obtain work from a system is to heat it: heat flows into the system and the system does work. But in practice, the amount of heat flow will always exceed the amount of useful work achieved. The excess heat may contribute to thermal pollution. (*Thermal pollution* is the raising or lowering of water temperature in streams, lakes, or oceans above or below normal seasonal ranges from the discharge of hot or cold waste streams into the water.) The efficiency of conversion from heat to work can be expressed as a percentage.

Typical efficiencies for some common conversion processes are shown in **Table 9.1.** The prospect of improving the efficiencies of appliances is often discussed as an important means of conserving energy in the future. The possible impact of this is shown in **Figure 9.5**, which compares two possible scenarios for the efficiency of technologies in the year 2030. In one case, no new technologies are developed after 2009, but the appliances in use gradually become more efficient as people replace older models. In the second case, it is assumed that consumers always choose the best available technology, independent of cost and other factors. But even with savings in energy associated with efficient appliances, the generation of electricity will still always result in energy being lost as heat. Energy output in the form of electrical energy is always less than the chemical or nuclear energy input at the generating plant.

Table ■ 9.1

Typical efficiencies of some common energy conversion devices

Device	Energy Conversion	Typical Efficiency (%)
Electric heater	Electrical → thermal	~100
Hair dryer	Electrical → thermal	~100
Electric generator	Mechanical → electrical	95
Electric motor (large)	Electrical → mechanical	90
Battery	Chemical → electrical	90
Steam boiler (power plant)	Chemical → thermal	85
Home gas furnace	Chemical → thermal	85
Home oil furnace	Chemical → thermal	65
Electric motor (small)	Electrical → mechanical	65
Home coal furnace	Chemical → thermal	55
Steam turbine	Thermal → mechanical	45
Gas turbine (aircraft)	Chemical → mechanical	35
Gas turbine (industrial)	Chemical → mechanical	30
Automobile engine	Chemical → mechanical	25
Fluorescent lamp	Electrical → light	20
Silicon solar cell	Solar → electrical	15
Steam locomotive	Chemical → mechanical	10
Incandescent lamp	Electrical → light	5

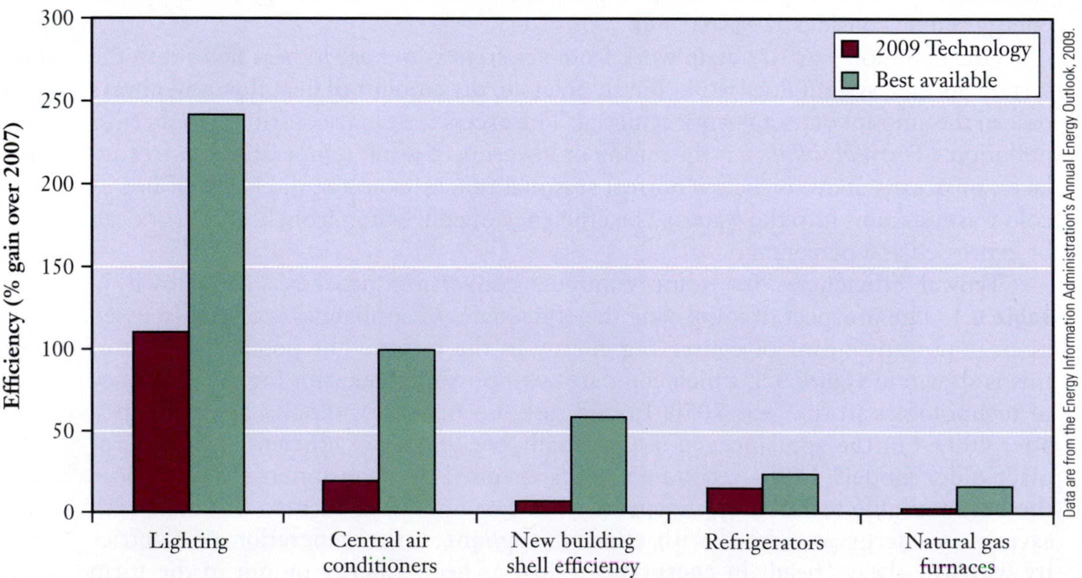

Figure 9.5 ■ The graph shows the predicted efficiency gains by the year 2030 for various technologies. (Data are plotted as percentage gain in efficiency using 2007 as a baseline.)

9-4 Heat Capacity and Calorimetry

General statements about nationwide or worldwide energy resources and their consequences may seem abstract in many respects, in part because the numbers are so large. Making the connection between this global scale and systems we can observe in a laboratory requires a systematic way to measure energy flow. We can do this by observing heat flow into or out of a system through a set of techniques collectively called **calorimetry**.

Heat Capacity and Specific Heat

Suppose that we wish to raise the temperature of two different systems or objects. In general, the different systems will absorb different amounts of energy based on three main factors: the amount of material, the type of material, and the temperature change. One way to note the importance of the amount of material is to compare the behavior of a glass of water and an ocean. On a hot summer day at the beach, a glass of cold water will quickly become warm, whereas the ocean water temperature does not change noticeably over the same time. The small amount of water in the glass behaves differently from the large amount in the ocean. The type of material is also important, and the summer beach provides some insight here, too. The sand at most beaches is predominantly silicon dioxide, and it heats up much more quickly than even the shallowest water. Both materials are exposed to similar amounts of energy from the sunlight, but they behave differently. Finally, the amount of energy supplied and the temperature change are also related. On a cloudy day at the beach, the sand does not get as hot: less energy is supplied to the sand because the clouds absorb some energy from the sun. Summarizing these observations, if we want to calculate the heat associated with a given temperature change, we'll need to account for the amount and identity of the material being heated as well as the extent of the temperature change.

This idea can easily be expressed as an equation, and we will use two slightly different versions of this relationship. We have two options for expressing the amount of material: by mass or by moles. Either choice leads to a usable equation. You already may be familiar with the use of mass because that is commonly done in physics or physical science classes. In that case, the identity of the substance is included through a term called the specific heat capacity (c, usually simply called the specific heat), leading to Equation 9.4:

$$q = mc\Delta T \tag{9.4}$$

The **specific heat** is a physical property of a material that measures how much heat is required to raise the temperature of 1 gram of that material by 1°C. Similarly, the **molar heat capacity** is a physical property that describes how much heat is required to raise the temperature of 1 mole of a substance by 1°C. So if we choose to express the amount of material in terms of moles rather than mass, our equation changes only slightly:

$$q = nC_p\Delta T \tag{9.5}$$

(The subscript "p" on C_p indicates that this is the heat capacity at constant pressure. Under other conditions, such as constant volume, the value of the heat capacity may differ slightly.)

Either equation provides the same information about the heat needed to produce a given change in temperature. As long as we know the molar mass of the substance, it should be simple to convert between the specific heat and the molar heat capacity. **Table 9.2** provides a list of specific heats and molar heat capacities for a few materials. A more extensive table can be found in Appendix D.

Example Problems 9.2 and 9.3 show applications of these equations. As you work through these examples, notice that the heat capacity will have the same numerical

Note that water has a much higher heat capacity than the metals shown in this table. This is true for most metals.

Table ■ 9.2

Specific heat and molar heat capacities for some common substances

Substance	Specific Heat, c ($J\ g^{-1}\ K^{-1}$)	Molar Heat Capacity, C_p ($J\ mol^{-1}\ K^{-1}$)
Al(s)	0.900	24.3
Cu(s)	0.385	24.5
$H_2O(s)$	2.09	37.7
$H_2O(\ell)$	4.18	75.3
$H_2O(g)$	2.03	36.4

value whether the temperature is expressed in K or in °C. Although this may seem odd at first, it makes sense when you realize that the heat depends only on the temperature *difference*, not on the temperature itself. Verify for yourself that if the temperature of a piece of metal rises from 20 to 30°C, ΔT can correctly be written as either 10°C or 10 K.

EXAMPLE PROBLEM 9.2

Heating a 24.0-g aluminum can raises its temperature by 15.0°C. Find the value of q for the can.

Strategy Heat flow depends on the heat capacity of the material being heated, the size of the sample, and the change in temperature. Because the information given is in mass, it will be easiest to use the specific heat rather than the molar heat capacity. The value of the specific heat of aluminum can be found in Table 9.2. Because the temperature change is given in °C, we will write the units as $J\ g^{-1}\ °C^{-1}$.

Solution

$$\begin{aligned} q &= mc\Delta T \\ &= 24.0\ g \times \frac{0.900\ J}{g\ °C} \times 15.0°C \\ &= 324\ J \end{aligned}$$

Analyze Your Answer It may take some practice to gain a feel for the correct magnitude for heat. We do know that q should have energy units, and the units in our calculation are consistent with that. We can also look at the order of magnitude for the quantities involved to make sure our arithmetic makes sense. The mass and the temperature change are both on the order of 10^1, whereas the specific heat is on the order of 10^0, or 1. So the product of the three should be on the order of 10^2, and an answer in hundreds of joules makes sense.

Check Your Understanding A block of iron weighing 207 g absorbs 1.50 kJ of heat. What is the change in the temperature of the iron?

EXAMPLE PROBLEM 9.3

The molar heat capacity of liquid water is 75.3 J mol^{-1} K^{-1}. If 37.5 g of water is cooled from 42.0 to 7.0°C, what is q for the water?

Strategy The heat flow is proportional to the amount of water and the temperature change. As before, we can do the calculation with the amount of water expressed in either grams or moles. Molar heat capacity is given but the amount of water is provided in grams. So we will use the molar mass to convert the amount from grams to moles. (We could also leave the amount in grams and convert the molar heat capacity into the specific heat.) Care must also be taken in defining the change in temperature. Such changes are always the final state minus the initial state. Because the water is cooling here, ΔT will be negative.

Solution

$$q = nC_p\Delta T$$

$$= 37.5 \text{ g} \times \frac{1 \text{ mol}}{18.0 \text{ g}} \times \frac{75.3 \text{ J}}{\text{mol °C}} \times -35.0\text{°C}$$

$$= -5.49 \times 10^3 \text{ J} = -5.49 \text{ kJ}$$

The negative value indicates that the system (in this case, the water) has lost energy to the surroundings. Notice that as long as we correctly express ΔT as $T_{\text{final}} - T_{\text{initial}}$, the correct sign for q will result automatically.

Check Your Understanding If 226 kJ of heat increases the temperature of 47.0 kg of copper by 12.5°C, what is the molar heat capacity of copper?

In these examples, we have simply stated that a given amount of heat was added to or removed from our system, without considering the source of that heat. Although this lets us familiarize ourselves with the equations, it isn't very realistic. In a more common application, we can use the same type of equations to determine heat flow between two objects, as shown in Example Problem 9.4.

EXAMPLE PROBLEM 9.4

A glass contains 250.0 g of warm water at 78.0°C. A piece of gold at 2.30°C is placed in the water. The final temperature reached by this system is 76.9°C. What was the mass of gold? The specific heat of water is 4.184 J g^{-1} $°C^{-1}$, and that of gold is 0.129 J g^{-1} $°C^{-1}$.

Strategy We will need to assume that heat flows only between the gold and the water, with no heat lost to or gained from the glass or the surroundings. Then the heats must balance; the heat gained by the gold was lost by the water. Each of those can be calculated using Equation 9.3, and the two must be equal and opposite in sign.

Solution

$$q_{\text{gold}} = -q_{\text{water}}$$

$$m_{\text{gold}} \times c_{\text{gold}} \times \Delta T_{\text{gold}} = -\text{m}_{\text{water}} \times c_{\text{water}} \times \Delta T_{\text{water}}$$

$$m_{\text{gold}} \times \frac{0.129 \text{ J}}{\text{g °C}} \times 74.6\text{°C} = -250.0 \text{ g} \times \frac{4.184 \text{ J}}{\text{g °C}} \times -1.1\text{°C}$$

Rearranging gives us

$$m_{gold} = \frac{-250.0 \text{ g} \times \frac{4.184 \text{ J}}{\text{g °C}} \times -1.1\text{°C}}{\frac{0.129 \text{ J}}{\text{g °C}} \times 74.6\text{°C}} = 120 \text{ g}$$

Analyze Your Answer From the original problem statement, we can see that the temperature of the hot water doesn't change very much. This small ΔT for the hot water implies that the gold sample must be fairly small. Next, consider that the heat capacity for water is roughly 40 times larger than that for gold, and the temperature change for gold is almost 80 times greater than for water. These two observations suggest that the gold sample should be close to half the size of the water sample, and our calculated result confirms this.

Check Your Understanding A 125-g sample of cold water and a 283-g sample of hot water are mixed in an insulated thermos bottle and allowed to equilibrate. If the initial temperature of the cold water is 3.0°C, and the initial temperature of the hot water is 91.0°C, what will be the final temperature?

These examples provide a means to calculate heat flow or related information. But we have yet to see how we can actually make laboratory measurements of heat flow.

Calorimetry

Calorimetry is the term used to describe the measurement of heat flow. Experiments are carried out in devices called *calorimeters* that are based on the concept established in Example Problem 9.4. The heat evolved or absorbed by the system of interest is determined by measuring the temperature change in its surroundings. Every effort is made to isolate the calorimeter thermally, preventing heat flow between the immediate surroundings and the rest of the universe. **Figure 9.6** shows a typical calorimeter. If the instrument is thermally isolated from the rest of the universe, the only heat flow

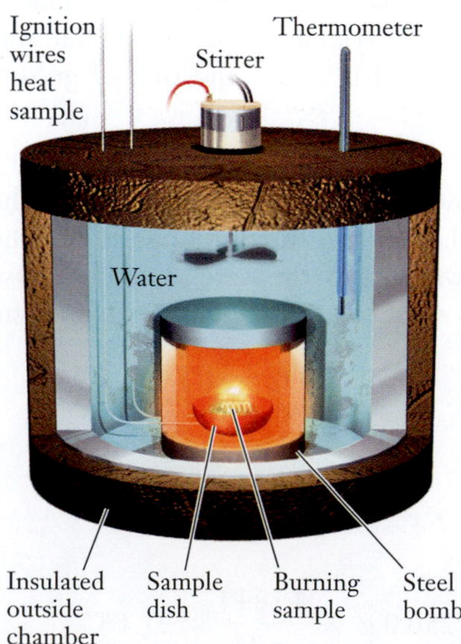

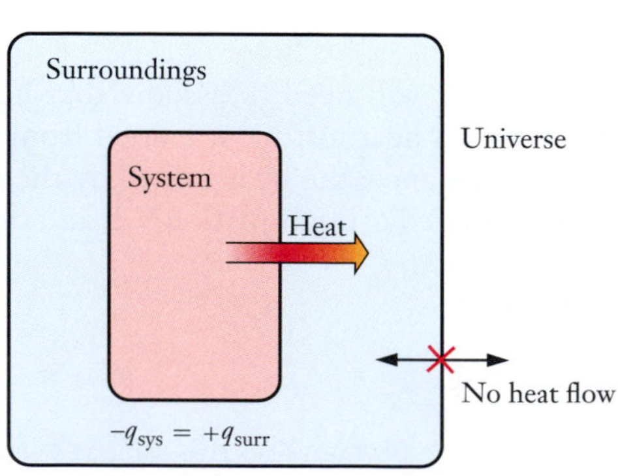

Figure 9.6 ■ A bomb calorimeter is a fairly complicated piece of equipment, as shown in the diagram on the left. But the general premise of the device is simply to carry out a reaction at constant volume and with no heat flow between the calorimeter and the outside world. The diagram on the right shows the standard choice of system and surroundings for a bomb calorimetry experiment. The system consists of the contents of the bomb itself. The surroundings include the bomb and the water bath surrounding it. We assume that no heat is exchanged with the rest of the universe outside the insulated walls of the apparatus.

that must be considered is that between the system being studied and the immediate surroundings, whose temperature can be measured.

A two-step process is used to make a calorimetric measurement. The first step is calibration, in which a known amount of heat is generated in the apparatus. The second step is the actual measurement, in which we determine the amount of heat absorbed or released in the reaction of a known amount of material. The calibration can be done either by burning a known amount of a well-characterized material or by resistive heating, in which a known amount of current is passed through a wire that heats due to its electrical resistance. The heat capacity of the entire calorimeter may be obtained by measuring the change in temperature of the surroundings resulting from a known heat input:

There are many specialized types of calorimetry, but all involve this sort of calibration.

$$\text{Known amount of heat} = \text{calorimeter constant} \times \Delta T$$

or

$$q = C_{\text{calorimeter}} \times \Delta T \tag{9.6}$$

Note that in contrast to our earlier equations relating q and ΔT, there is no mass or number of moles term here for the quantity of material. The calorimeter constant is the heat capacity of a particular *object* (or set of objects) rather than that of a *material*. It may help to think of it as the heat capacity "per calorimeter" and then realize that we have just one calorimeter. For someone who routinely uses the same calorimeter, this approach is much simpler than the alternative, which would be to keep track of the masses of steel, water, and other materials in the calorimeter. In the case of a bomb calorimeter such as that shown in Figure 9.6, the calorimeter constant is largely attributable to the water that surrounds the bomb, but it also includes the heat capacities of the thermometer, the stirring system, and the bomb itself.

Once the calorimeter constant is known, we are ready to use the calorimeter for our actual measurement. We place known amounts of reactant(s) into the calorimeter, initiate the reaction, and then measure the resulting temperature change of the calorimeter. The calorimeter constant allows us to determine the amount of heat released or absorbed in the reaction. Example Problem 9.5 shows how this process works.

EXAMPLE PROBLEM 9.5

A calorimeter is to be used to compare the energy content of some fuels. In the calibration of the calorimeter, an electrical resistance heater supplies 100.0 J of heat and a temperature increase of 0.850°C is observed. Then 0.245 g of a particular fuel is burned in this same calorimeter, and the temperature increases by 5.23°C. Calculate the energy density of this fuel, which is the amount of energy liberated per gram of fuel burned.

We will explore the idea of energy density further in Section 9-7.

Strategy The calibration step allows us to determine the calorimeter constant. Once this is known, the amount of heat evolved from the fuel can be determined by using Equation 9.6. Finally, we divide this heat by the mass of fuel that generated it to arrive at the requested energy density.

Solution

***Step 1:** Calibration*

$$q = C_{\text{calorimeter}} \times \Delta T$$

so

$$C_{\text{calorimeter}} = q/\Delta T$$
$$= 100.0\ \text{J}/0.850^\circ\text{C}$$
$$C_{\text{calorimeter}} = 118\ \text{J}/^\circ\text{C}$$

Step 2: Determination of heat evolved by fuel

$$q_{calorimeter} = C_{calorimeter} \times \Delta T$$
$$= 118\ \text{J/°C} \times 5.23\text{°C}$$
$$= 615\ \text{J}$$

And

$$q_{fuel} = -q_{calorimeter} = -615\ \text{J}$$

Step 3: Calculation of the energy density

$$\text{Energy density} = -q_{fuel}/m$$
$$= -(-615\ \text{J})/0.245\ \text{g}$$
$$= 2510\ \text{J/g} = 2.51\ \text{kJ/g}$$

Discussion This problem illustrates the need to be careful with signs in thermodynamic calculations. Because the burning of fuel releases heat, q for the fuel should be negative. The energy density, though, would be reported as a positive number, resulting in the additional negative sign in the final step.

Check Your Understanding The combustion of naphthalene ($C_{10}H_8$), which releases 5150.1 kJ/mol, is often used to calibrate calorimeters. A 1.05-g sample of naphthalene is burned in a calorimeter, producing a temperature rise of 3.86°C. Burning a 1.83-g sample of coal in the same calorimeter causes a temperature change of 4.90°C. What is the energy density of the coal?

9-5 Enthalpy

The concept of heat flow is reasonably intuitive. Cool objects get warmer at the expense of hotter objects. Scientifically, however, this natural process requires more careful examination. The conditions under which heat flow occurs will have an impact on the measurements that are made. For example, when 1 mole of octane, C_8H_{18}, burns to form gaseous carbon dioxide and liquid water under constant volume conditions, 5.45×10^3 kJ of heat is released. The same reaction under constant pressure conditions releases 5.48×10^3 kJ. The percentage difference between the two conditions is small, but consider the engineering consequences in designing safety systems for large octane storage containers. Such vessels may hold in excess of 1000 moles of octane. Each mole releases 30 kJ more energy under combustion conditions of constant pressure versus constant volume. When thousands of moles burn, the difference in total energy release may be important for safety. Why do the two cases release different amounts of heat, and how can we account for these differences properly?

Defining Enthalpy

Because of the observation that heats of reaction depend on reaction conditions, it has proven advantageous to define two mathematical functions for energy. Internal energy has already been introduced. By using the definition of internal energy, we can show that under constant volume conditions, the change in internal energy equals the heat flow. Start with

$$\Delta E = q + w$$

In chemical reactions, we usually need to consider only PV-work. When a gas expands, it does an amount of work equal to $P\Delta V$ on its surroundings. But if the expanding gas is our system, we want w to be the work done *on* the gas, and that will be $-P\Delta V$. So we can replace w in the equation above with $-P\Delta V$:

$$\Delta E = q - P\Delta V \tag{9.7}$$

This equation now allows us to see how internal energy is related to heat. If the volume is held constant, ΔV is zero, so the second term is zero. All that remains is

$$\Delta E = q_v \tag{9.8}$$

where the subscript "v" is added to denote that the equation is correct *only* under constant volume conditions. Equation 9.8 tells us that if we do a calorimetry experiment at constant volume, we will get a direct determination of ΔE.

But what if we do the experiment at constant pressure instead? It would be useful to have a function that would be equal to the heat flow under constant pressure conditions. This function, known as **enthalpy**, is defined as

$$H = E + PV \tag{9.9}$$

Working from this definition, we can show that the change in enthalpy (ΔH) will be equal to the heat flow under constant pressure conditions. From the above definition, the change in enthalpy (ΔH) must be

$$\Delta H = \Delta E + \Delta(PV)$$

We can expand this by substituting for ΔE using Equation 9.7:

$$\Delta H = (q - P\Delta V) + \Delta(PV)$$

If the pressure is held constant, then the $\Delta(PV)$ term will simply become $P\Delta V$, giving

$$\Delta H = q - P\Delta V + P\Delta V$$

The second and third terms clearly cancel and leave the desired result:

$$\Delta H = q_p \tag{9.10}$$

The enthalpy change therefore equals the heat flow under constant pressure. (This is denoted with a subscript "p," just as we did in Equation 9.8 for constant volume.)

Now we have two ways to define heat flow into a system, under two different sets of conditions. For a process at constant volume, the measurable heat flow is equal to ΔE, the change in internal energy. For a process at constant pressure, the measurable heat flow is equal to the change in enthalpy, ΔH. In many ways, enthalpy is the more useful term because constant pressure conditions are more common. A reaction carried out in a beaker in the chemistry laboratory, for instance, occurs under constant pressure conditions (or very nearly so). Thus, when we refer to the heat of a process, we are typically referring to a change in enthalpy, ΔH. As in previous definitions, ΔH refers to $H_{final} - H_{initial}$.

When heat evolves from a system, the process is said to be **exothermic** and the value of ΔH is less than zero. An exothermic process feels hot: if you pick up the beaker in which an exothermic reaction is taking place, heat will flow from the reacting system into your hand. Conversely, when heat is absorbed by the system, the process is said to be **endothermic**, and the value of ΔH is greater than zero. Endothermic processes feel cold because they draw heat from their surroundings.

ΔH of Phase Changes

Heat flow into a substance does not always raise its temperature. If heat flows into an ice cube at 0°C, for example, the ice will melt to form liquid water at 0°C. (If heat continues to flow into the resulting water, its temperature will begin to rise, of course.) How can the temperature remain constant despite the influx of heat? To understand this, we need to remember that intermolecular forces are more extensive in a solid

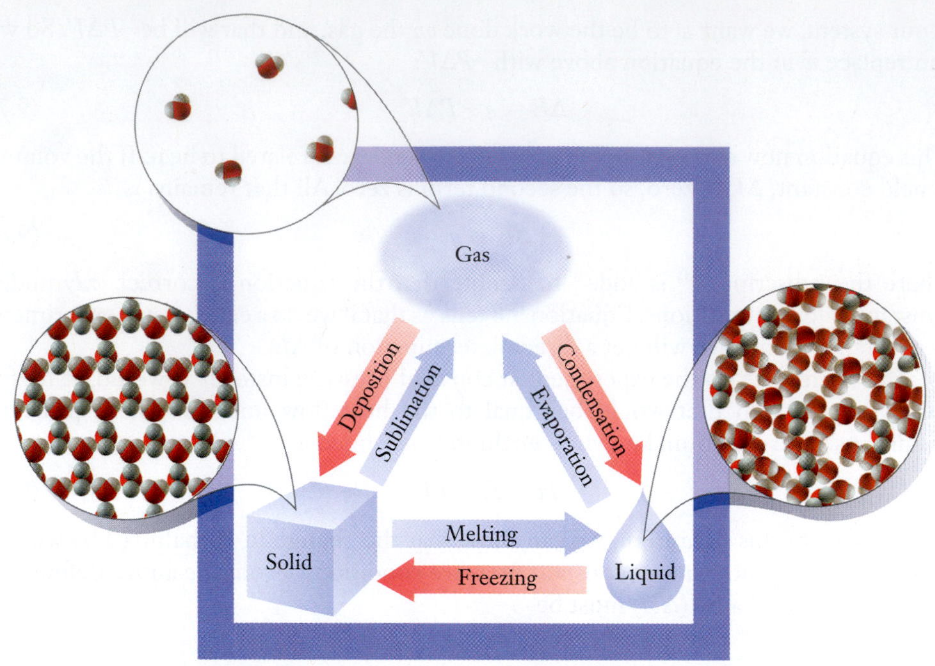

Figure 9.7 ■ Microscopic-level views of solid, liquid, and gas phases are shown, along with the terms commonly used for the possible transitions among these phases.

than in a liquid. So as the ice cube melts, energy must be expended to overcome some of the intermolecular attractions. The internal energy of liquid water is higher than that of ice, even though both are at the same temperature.

For similar reasons, there will be heat flow in any phase change. The names of phase changes among solids, liquids, and gases are summarized in **Figure 9.7**. Because these phase changes generally take place at constant pressure, the corresponding heat flows should be viewed as changes in enthalpy. Some phase changes are so common that their enthalpy changes have specific names and symbols assigned to them. The heat required to melt a substance is the heat of fusion, ΔH_{fus}. The enthalpy change for converting a liquid to a gas is known as the heat of vaporization, ΔH_{vap}. We know that when a liquid is converted to a gas, all the molecules in a sample must overcome whatever intermolecular forces are present. That means that energy must flow into the substance to vaporize it, so the heat of vaporization will always be positive. It follows that the reverse process, condensation, will always release heat. The values of enthalpy changes in opposite directions have the same numeric value and differ only in their signs. Because the strength of the intermolecular forces varies from one substance to another, the magnitude of the enthalpy change for any phase transition will also depend on the substance involved. Values can be found in many standard tables, including one in Appendix D. Data for water are provided in **Table 9.3** as an example.

The term fusion *here means melting.*

For any substance, the heat of fusion is typically much smaller than the heat of vaporization.

Table ■ 9.3

Standard molar enthalpies and temperatures for phase changes of water

Phase Change	Fusion	Freezing	Vaporization	Condensation
Transition Temperature	0°C	0°C	100°C	100°C
ΔH (J/mol)	ΔH_{fus} = 6009.5	ΔH_{freeze} = −6009.5	ΔH_{vap} = 4.07×10^4	ΔH_{cond} = -4.07×10^4

These data would allow us to calculate the heat required for a phase transition involving any particular sample of water. Just as we saw when we considered temperature changes, the heat will depend on the amount of substance undergoing the transition: it will take more heat to melt a large block of ice than a small cube. The values shown in Table 9.3 are given in units of J/mol, so we might want to specify the amount of water in terms of moles, leading to the following equation:

$$\Delta H = n \times \Delta H_{\text{phase change}} \tag{9.11}$$

Here n is the number of moles, as usual. Notice that this relationship does not include ΔT, in contrast to Equations 9.4 and 9.5. This should make sense if we keep in mind that phase changes occur at a constant temperature, so there is no ΔT in the transition. Example Problem 9.6 shows the calculation of the heat associated with a phase change.

EXAMPLE PROBLEM 9.6

Calculate the enthalpy change when 245 g of ice melts.

Strategy The ΔH_{fus} value in Table 9.3 is in J/mol, so the amount of ice must be converted into moles. Multiplying the number of moles by ΔH_{fus} will provide the desired quantity.

Solution

$$245 \text{ g } H_2O \times \frac{1 \text{ mol}}{18.015 \text{ g}} = 13.6 \text{ mol } H_2O$$

$$\begin{aligned} \Delta H &= n \times \Delta H_{\text{fus}} \\ &= 13.6 \text{ mol} \times 6009.5 \text{ J/mol} \\ &= 8.17 \times 10^4 \text{ J} \end{aligned}$$

Analyze Your Answer The enthalpy of fusion is constant at 6 kJ/mol, so the enthalpy change depends on the size of the sample. Because the molar mass of water is 18 g/mol, a sample that is 240 grams is a bit more than 10 moles. That means we should expect an answer that is a bit larger than 60 kJ, which is consistent with our result.

Check Your Understanding Calculate the enthalpy change when 14.5 g of water vapor condenses to liquid water.

Just as for a temperature change, there is no fundamental reason why we could not express the quantity of material in mass rather than moles. But it is essential that we be consistent in the choice of units: if the quantity is in mass, then we will need an enthalpy change value that is expressed in mass as well. Because you know the molar mass of water, it should be easy to convert the values in Table 9.3 from J/mol to J/g. If you repeat the calculations from the last example that way, you should get the same result. In most engineering fields, it is more common to use mass rather than moles to specify the amount of a substance. But the underlying concept is the same no matter which units we use.

At this point, we can combine what we've learned about heat flow for both phase changes and temperature changes to obtain the enthalpy change for converting water from ice to liquid and then to steam. A plot of temperature versus heat flow for such a process is shown in **Figure 9.8**. The heat transfers in sections (1), (3), and (5) in this figure are for changes of temperature, and sections (2) and (4) represent phase changes.

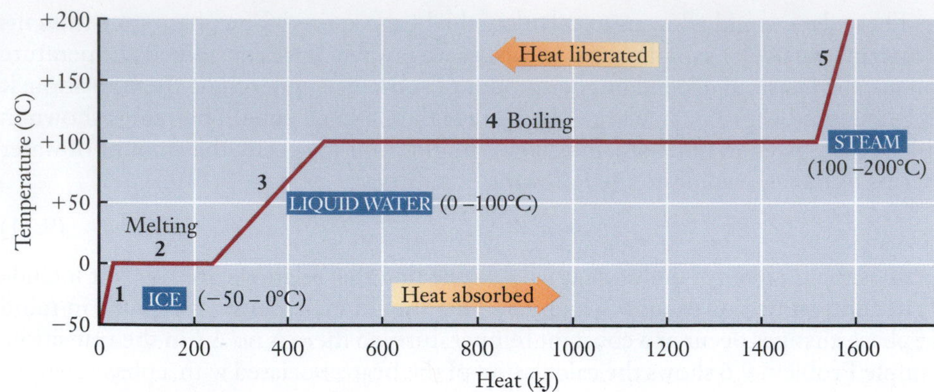

Figure 9.8 ■ The figure shows how the temperature of a 500-g sample of H_2O varies as it absorbs heat. Here we begin with ice at –50°C. In the segment of the graph marked (1), the temperature of the ice rises until it reaches the melting point at 0°C. In region (2), ice melts at a constant temperature of 0°C until it is converted into liquid. In region (3), only liquid water is present, and its temperature rises until it reaches the boiling point at 100°C. In region (4), the water boils at a constant temperature of 100°C until it is converted into steam. Finally, in region (5), the resulting steam continues to absorb heat, and its temperature rises.

Vaporization and Electricity Production

The large amount of energy needed for converting water from a liquid to a gas, as by the series of processes shown in Figure 9.8, is exploited in converting chemical energy into electricity. The basic features of a fossil fuel–powered electricity plant are shown schematically in **Figure 9.9**.

When the fuel—typically coal or natural gas—burns, chemical energy is released as heat. The goal of the power plant is to convert as much of this energy as possible into electricity. The critical step in this process is to trap the heat given off in the combustion reaction. Water is the material of choice for this process because it has a large heat of vaporization. You'll recall from Section 8-4 that intermolecular forces are unusually strong in water because of extensive hydrogen bonding. These hydrogen bonds

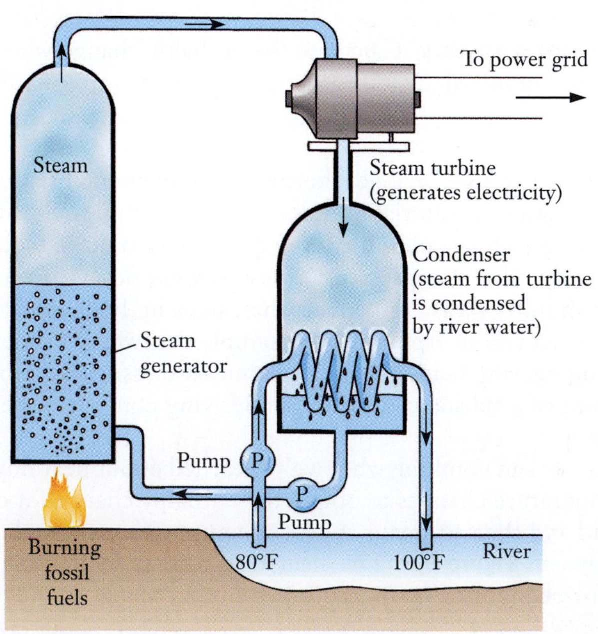

Figure 9.9 ■ The important elements of a standard electric power plant are shown in this schematic diagram. The process exploits the large heat of vaporization of water.

between water molecules in the liquid lead to the large value of ΔH_{vap}. If a material with a small heat of vaporization were substituted for water, much more of it would be required to absorb the same amount of heat. The large heat of vaporization coupled with the relative abundance of water has led to its widespread use in the energy sector of the economy.

Heat of Reaction

So far, we have considered enthalpy changes for simple physical processes such as temperature changes and phase transitions. But the importance of chemistry to the energy economy arises from the fact that there are enthalpy changes in chemical reactions as well. This enthalpy change is commonly referred to as the **heat of reaction**. Because many reactions are carried out under constant pressure conditions, this term is sensible, even if slightly imprecise.

Bonds and Energy

Chemical reactions involve energy changes because chemical bonds are broken and formed when reactants are transformed into products. Consider a fairly simple reaction—the combustion of methane:

$$CH_4(g) + 2\ O_2(g) \rightarrow CO_2(g) + 2\ H_2O(\ell)$$

We are familiar with all of these molecules, so we can easily write Lewis structures for them:

$$\mathrm{H{-}\overset{\displaystyle H}{\underset{\displaystyle H}{\overset{|}{\underset{|}{C}}}}{-}H} + 2\ \ddot{\underset{\cdot\cdot}{O}}{=}\ddot{\underset{\cdot\cdot}{O}} \longrightarrow \ddot{\underset{\cdot\cdot}{O}}{=}C{=}\ddot{\underset{\cdot\cdot}{O}} + 2\ \mathrm{H}{\diagup}\ddot{\underset{\cdot\cdot}{O}}{\diagdown}\mathrm{H}$$

On the reactant side of the equation, we have four C—H single bonds and two O═O double bonds. On the product side, we have two C═O double bonds and four O—H single bonds. Over the course of the reaction, all bonds in the reactants must be broken, and that will require an input of energy. On the other hand, all bonds in the products must be formed, and that will release energy. (Recall that bond formation is *always* exothermic and bond breaking is *always* endothermic.) If the energy released in forming new bonds is greater than the energy required to break the original bonds, then the overall reaction should be exothermic. Conversely, if the energy needed for bond breaking is greater than that released in bond making, then the reaction will be endothermic. Because methane is the principal component of natural gas, we know that burning it must release energy. So we should anticipate that this particular reaction must be exothermic, and calorimetric measurements bear this out; ΔH for the combustion of 1 mole of methane is −890.4 kJ.

We might attempt to use estimates for the individual bond energies to calculate an approximate value of ΔH for a reaction. This is sometimes done to obtain an estimate for reactions involving compounds where no thermochemical data are available. The accuracy of this approach is not very good, however, because tabulated values for bond energies represent averages. As we will see in the following section, a simple and accurate alternative exists in most cases.

The overall energetics of a chemical reaction can be summarized in a **thermochemical equation**. For the combustion of methane, the thermochemical equation is

$$CH_4(g) + 2\ O_2(g) \rightarrow CO_2(g) + 2\ H_2O(\ell) \qquad \Delta H = -890.4\ \text{kJ}$$

This equation tells us two important things. First, we can immediately see whether the reaction is exothermic or endothermic because ΔH carries a sign. We know that the combustion of methane must be exothermic, and the negative sign on ΔH confirms this. Second, the thermochemical equation includes the numerical value of ΔH, so we can tell exactly how much heat will be released. It is important to realize that the heat of reaction shown is for the equation exactly as written; if 1 mole of methane reacts with 2 moles of oxygen, 890.4 kJ will be released. But if more fuel is burned, more heat will be released. So if the stoichiometric coefficients are multiplied by some factor, the heat of reaction must also be multiplied by that factor. Thus another thermochemical equation for the combustion of methane is

$$2\ CH_4(g) + 4\ O_2(g) \rightarrow 2\ CO_2(g) + 4\ H_2O(\ell) \qquad \Delta H = -1780.8\ kJ$$

Heats of Reaction for Some Specific Reactions

Some classes of chemical reactions are sufficiently common or useful that they have been assigned their own label for heats of reaction. The example reaction for methane we have already used falls into one such category, combustion. Because combustion is a common part of the energy economy, enthalpy changes for combustion reactions are used to compare various fuels. Sometimes these heats of combustion are designated as ΔH_{comb}. Similarly, the neutralization reactions between acids and bases have heats of reaction that are called heats of neutralization, ΔH_{neut}.

Another class of reactions, formation reactions, has specially designated heats of reaction known as **heats of formation**, $\Delta H_f°$, even though many of them are never carried out in practice. Their importance is due to the computational utility derived from the following definition. A **formation reaction** is the chemical reaction by which 1 mole of a compound is formed from its elements in their standard states. The *standard state* is the most stable form of the element at room temperature (25°C) and pressure (1 atm). The formation reaction for carbon monoxide is

The "degree sign" appearing on this symbol for the heat of formation indicates that the value is for a process occurring under standard conditions of 25°C and 1 atm.

$$C(s) + \frac{1}{2} O_2(g) \rightarrow CO(g) \qquad \Delta H° = \Delta H_f°[CO(g)]$$

Note that the definition of the formation reaction immediately implies that the heat of formation for any element in its standard state must always be zero. To see this, imagine writing the formation reaction for an element in its standard state, such as $O_2(g)$. We would need 1 mole of $O_2(g)$ as the product and oxygen in its standard state as the reactant. But because the standard state of oxygen *is* $O_2(g)$, our formation reaction is really no reaction at all:

$$O_2(g) \rightarrow O_2(g) \qquad \Delta H° = \Delta H_f°[O_2(g)] = 0$$

Because both sides of this equation are the same, there can be no change in enthalpy. So $\Delta H°$ must be zero. This will be true for any element in its standard state.

Because we require just 1 mole of CO on the right-hand side, we must use a fractional coefficient on the O_2 to balance the equation, even though this may look odd. Two very common errors when writing formation reactions are including elements that are not in their standard states or having more than 1 mole of product compound. For carbon monoxide, these mistakes might produce either of the following *incorrect* "formation" reactions:

$$C(s) + O(g) \rightarrow CO(g) \qquad \Delta H° \neq \Delta H_f°[CO(g)]$$

$$2\ C(s) + O_2(g) \rightarrow 2\ CO(g) \qquad \Delta H° \neq \Delta H_f°[CO(g)]$$

Although both are valid chemical equations, they do *not* describe the formation reaction for CO. The first reaction is not correct because oxygen is not shown in its standard state, as a diatomic molecule. The second reaction is not correct, because 2 moles of carbon monoxide are formed. Formation reactions are quite useful for determining heats of reaction, as we will see in the following section.

9-6 Hess's Law and Heats of Reaction

Enthalpy changes in chemical reactions play an important role in a wide range of situations. Comparisons of possible new fuels or explosives are obvious examples. In many cases, it might be desirable to determine the enthalpy change for a reaction without necessarily carrying out the reaction itself. The synthesis of a new explosive might be very difficult and dangerous, for example, so it would help to know in advance if its potential usefulness might justify the effort. In other cases, it may simply be very difficult to perform a direct calorimetric determination of a heat of reaction. For a variety of reasons, we often need to obtain heats of reaction indirectly.

Hess's Law

To obtain information about heats of reaction indirectly, we use what is known as **Hess's law**: the enthalpy change for any process is independent of the particular way the process is carried out. The underlying concept upon which this idea is built is that enthalpy is a **state function**. A state function is a variable whose value depends only on the state of the system and not on its history. When you drive your car, your location is a state function, but the distance that you have traveled to get there is not. For a chemical reaction, the concept of state functions can be very important. We rarely if ever know the microscopic details of how reactant molecules are actually converted into product molecules. But it is relatively easy to determine what the reactants and the products are. Because enthalpy is a state function, if we can find a way to determine the change in enthalpy for *any* particular path that leads from the reactants to the desired products, we will know that the result will apply for *whatever* actual path the reaction may take. The block diagram in **Figure 9.10** illustrates this concept.

Pressure, volume, and temperature are all state functions.

For the situation shown in Figure 9.10, the desired enthalpy change could be obtained via either of two alternative pathways:

$$\Delta H_{\text{desired}} = \Delta H_{\text{A}_\text{i}} + \Delta H_{\text{A}_\text{f}}$$

$$\Delta H_{\text{desired}} = \Delta H_{\text{B}_\text{i}} + \Delta H_{\text{B}_\text{f}}$$

We can develop this idea further by specifying a system and including an enthalpy axis in the picture. **Figure 9.11** shows such an **enthalpy diagram** for the combustion of methane. As we saw in Section 4-1, this combustion may either be complete, producing carbon dioxide directly, or incomplete, producing carbon monoxide. (The production of carbon monoxide from burning of methane presents a danger for natural gas–burning furnaces that are not vented properly.)

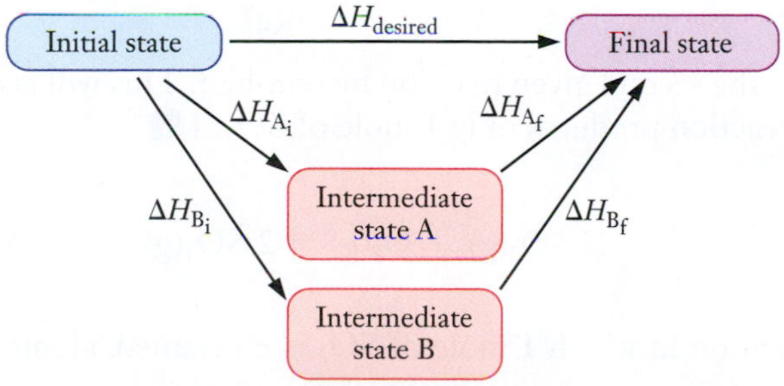

Figure 9.10 ■ Conceptual diagram representing Hess's law. Because enthalpy is a state function, we can choose any convenient path from the initial state to the final state and use that path to calculate the enthalpy change.

Figure 9.11 ■ Enthalpy diagram for the combustion of methane. Here we imagine that 1 mole of CH_4 is first converted to CO (step 1) and that the CO then reacts further to form CO_2 (step 2). If we know the values of ΔH_1 and ΔH_2, we can use them to calculate ΔH_{comb}.

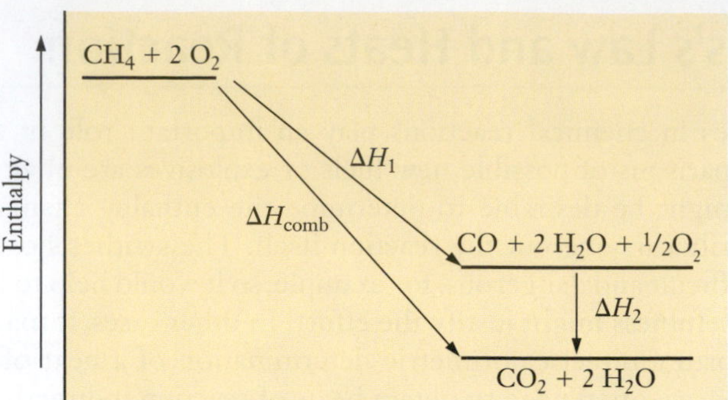

Hess's law has important practical implications. Many times we can break a chemical reaction down into a series of steps whose net result is the same as that of the original reaction. According to Hess's law, in such a case, we can evaluate the enthalpy of the original reaction by using the sequential ones, as shown in Example Problem 9.7.

EXAMPLE PROBLEM 9.7

Sulfur trioxide reacts with water to form sulfuric acid, a major contributor to acid rain. One origin of SO_3 is the combustion of sulfur, which is present in small quantities in coal, according to the following equation:

$$S(s) + \frac{3}{2} O_2(g) \rightarrow SO_3(g)$$

Given the thermochemical information below, determine the heat of reaction for this process:

$$S(s) + O_2(g) \rightarrow SO_2(g) \qquad \Delta H^\circ = -296.8 \text{ kJ}$$

$$2\ SO_2(g) + O_2(g) \rightarrow 2\ SO_3(g) \qquad \Delta H^\circ = -197.0 \text{ kJ}$$

Strategy We need to use the reactions whose enthalpy changes are known to construct a pathway that results in the desired reaction. In this case, we can see that SO_2 is formed in the first reaction we are given and then is consumed in the second. It is critical to pay attention to the precise stoichiometry of the given and desired reactions. The reaction of interest produces just 1 mole of SO_3, whereas the second reaction we are given produces 2 moles. So we will need to correct for that.

Solution We start with the first of the two given reactions:

$$S(s) + O_2(g) \rightarrow SO_2(g) \qquad \Delta H^\circ = -296.8 \text{ kJ}$$

Next, we multiply the second given reaction by one-half. This will account for the fact that the desired reaction produces only 1 mole of SO_3:

Note that the $\frac{1}{2}$ multiplies the stoichiometric coefficients and the enthalpy change.

$$\frac{1}{2} \times [2\ SO_2(g) + O_2(g) \rightarrow 2\ SO_3(g) \qquad \Delta H^\circ = -197.0 \text{ kJ}]$$

This gives an equation in which 1 mole of SO_2 is consumed, along with its enthalpy change:

$$SO_2(g) + \frac{1}{2} O_2(g) \rightarrow SO_3(g) \qquad \Delta H^\circ = -98.5 \text{ kJ}$$

Adding this to the first reaction above gives the desired amount of SO_3 and the heat of reaction:

$$S(s) + O_2(g) \rightarrow SO_2(g) \qquad \Delta H° = -296.8 \text{ kJ}$$

$$SO_2(g) + \frac{1}{2}O_2(g) \rightarrow SO_3(g) \qquad \Delta H° = -98.5 \text{ kJ}$$

$$S(s) + \frac{3}{2}O_2(g) \rightarrow SO_3(g) \qquad \Delta H° = -395.3 \text{ kJ}$$

Analyze Your Answer We may not have any intuition regarding the particular chemical reactions involved, so we'll have to look at the structure of the problem to see if our answer makes sense. Both of the reactions we added are exothermic, so it makes sense that the combination would be more exothermic than either of the individual reactions.

Check Your Understanding Use the following thermochemical equations as needed to find the heat of formation of diamond:

$$C(\text{diamond}) + O_2(g) \rightarrow CO_2(g) \qquad \Delta H° = -395.4 \text{ kJ}$$

$$2\ CO_2(g) \rightarrow 2\ CO(g) + O_2(g) \qquad \Delta H° = 566.0 \text{ kJ}$$

$$C(\text{graphite}) + O_2(g) \rightarrow CO_2(g) \qquad \Delta H° = -393.5 \text{ kJ}$$

$$2\ CO(g) \rightarrow C(\text{graphite}) + CO_2(g) \qquad \Delta H° = -172.5 \text{ kJ}$$

Formation Reactions and Hess's Law

The type of calculation shown in the previous example has occasional application in chemistry. But Hess's law is also useful in another more common approach. Heats of formation for many substances are widely tabulated (see Appendix E). Hess's law allows us to use these tabulated values to calculate the enthalpy change for virtually any chemical reaction. The block diagram in **Figure 9.12** shows how this useful scheme arises.

Heats of formation and other thermodynamic data can be found in reference books such as the Handbook of Chemistry and Physics, *or in online sources such as the NIST WebBook.*

Step 1 is the decomposition of reactants into elements in their standard states. But this is just the opposite of the formation reaction of the reactants, so the enthalpy change of the process is $-\Delta H_f°$(reactants). Similarly, Step 2, the formation of the products from elements in their standard states, has an enthalpy change of $\Delta H_f°$(products). Remember, however, that the formation reaction is defined for the generation of *1 mole* of the compound. Consequently, to use tabulated heats of formation we must multiply by the stoichiometric coefficients from the balanced equation to account for the number of

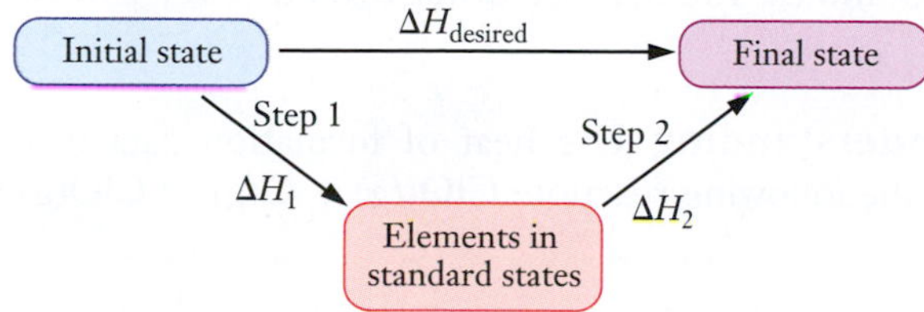

Figure 9.12 ■ This conceptual diagram shows how to use tabulated enthalpies of formation to calculate the enthalpy change for a chemical reaction. We imagine that first the reactants are converted to elements in their standard states, and then those elements recombine to form the products. Because enthalpy is a state function, we do not need to know anything about the actual pathway that the reaction follows.

moles of reactants consumed or products generated. Taking these factors into account leads to one of the more useful equations in thermochemistry:

$$\Delta H^\circ = \sum_i v_i \Delta H_f^\circ\,(\text{product})_i - \sum_j v_j \Delta H_f^\circ\,(\text{reactant})_j \qquad (9.12)$$

In this equation, we have designated the stoichiometric coefficients by the Greek letter ν. The first summation is over all of the products, and the second is over all of the reactants. Example Problems 9.8 and 9.9 will show how heats of formation are used in understanding the thermochemistry of reactions useful in producing energy.

EXAMPLE PROBLEM 9.8

Use tabulated data to find the heat of combustion of 1 mole of propane, C_3H_8, to form gaseous carbon dioxide and liquid water.

Strategy We will need values for the heats of formation of the reactants and products to determine the desired heat of combustion. First, we must write a balanced chemical equation for the process. Then, we can use Equation 9.12 to calculate the heat of the reaction (in this case the heat of combustion) by looking up heats of formation in the table in Appendix E. The stoichiometric coefficients needed will be obtained from the balanced equation. Remember that the heat of formation of an element in its standard state—like the O_2 in this equation—will always be zero.

Solution

$$C_3H_8(g) + 5\,O_2(g) \rightarrow 3\,CO_2(g) + 4\,H_2O(\ell)$$

$$\begin{aligned}\Delta H^\circ &= 3\text{ mol }\Delta H_f^\circ[CO_2(g)] + 4\text{ mol }\Delta H_f^\circ[H_2O(\ell)] - 1\text{ mol }\Delta H_f^\circ[C_3H_8(g)] \\ &\qquad - 5\text{ mol }\Delta H_f^\circ[O_2(g)] \\ &= 3\text{ mol }(-393.5\text{ kJ/mol}) + 4\text{ mol }(-285.8\text{ kJ/mol}) - 1\text{ mol }(-103.8\text{ kJ/mol}) \\ &\qquad -5\text{ mol }(0\text{ kJ/mol}) \\ &= -2219.9\text{ kJ}\end{aligned}$$

Discussion We use units of kJ/mol for the heat of formation of a substance. But in writing the enthalpy change of a chemical reaction, we will use kJ as our preferred unit, not kJ/mol. The reaction in this example illustrates why we do this. The value we calculated, $\Delta H^\circ = -2219.9$ kJ, is for a reaction in which 1 mole of propane reacts with 5 moles of oxygen to form 3 moles of carbon dioxide and 4 moles of water. So if we were to say "−2219.9 kJ/mol," we would need to specify carefully which substance that "mol" refers to. We choose to write the ΔH value in kJ, with the understanding that it refers to the reaction as written. This is also dimensionally consistent with Equation 9.12, provided that we treat the stoichiometric coefficients as carrying units of moles. You may see other texts that refer to values as "per mole of reaction."

Check Your Understanding Use heat of formation data from Appendix E to calculate ΔH° for the following reaction: $ClO_2(g) + O(g) \rightarrow ClO(g) + O_2(g)$

EXAMPLE PROBLEM 9.9

Ethanol, C_2H_5OH, is used to introduce oxygen into some blends of gasoline. It has a heat of combustion of 1366.8 kJ/mol. What is the heat of formation of ethanol?

Strategy We know the relationship between the heat of combustion of a reaction and the heats of formation of the substances involved. We will need a balanced chemical equation for the combustion, and we must recognize that the combustion of ethanol yields the same products as hydrocarbon combustion. We can write the balanced equation and then use Equation 9.12 to determine the desired quantity. In this case, we know the heat of reaction and will be solving for one of the heats of formation.

Solution

$$C_2H_5OH(\ell) + 3\ O_2(g) \rightarrow 2\ CO_2(g) + 3\ H_2O(\ell) \qquad \Delta H° = -1366.8\ kJ$$

Using Equation 9.12:

$$\Delta H° = 2\ mol\ \Delta H_f°[CO_2(g)] + 3\ mol\ \Delta H_f°[H_2O(\ell)] - 1\ mol\ \Delta H_f°[C_2H_5OH(\ell)] - 3\ mol\ \Delta H_f°[O_2(g)]$$

$$-1366.8\ kJ = 2\ mol\ (-393.5\ kJ/mol) + 3\ mol\ (-285.8\ kJ/mol) - \Delta H_f°[C_2H_5OH(\ell)] - 3\ mol\ (0\ kJ/mol)$$

Rearranging and solving gives $\Delta H_f°[C_2H_5OH(\ell)] = -277.6$ kJ/mol

Analyze Your Answer By now, we have seen enough heats of formation to know that values in the hundreds of kJ/mol seem fairly typical. Notice that it is very important to handle signs carefully when solving this type of problem: the fact that we are adding and subtracting quantities that can be positive or negative allows many opportunities for errors.

Check Your Understanding Incomplete combustion of hydrocarbons leads to the generation of carbon monoxide rather than carbon dioxide. As a result, improperly vented furnaces can poison people who live in an affected building because of the toxicity of CO. Calculate the heat of reaction for the incomplete combustion of methane, $CH_4(g)$, to yield liquid water and CO(g).

9-7 Energy and Stoichiometry

The ability to predict the energetic consequences of a chemical reaction is an important skill in chemistry that has many practical applications. Writing a thermochemical equation allows for a treatment of energy that is similar in many ways to the stoichiometry problems we learned to solve in Chapter 4. For an exothermic reaction, we can treat energy as a product, and in an endothermic reaction, energy can be thought of as a reactant. We must keep in mind that the stated value of ΔH corresponds to the reaction taking place exactly as written, with the indicated numbers of moles of each substance reacting.

The emphasis on the importance of the mole as the heart of the stoichiometry problem remains when we solve problems involving the energy of reactions. So if we wanted to calculate the amount of energy released by burning a given mass or volume of methane, we would start by converting that quantity into moles. Then we could use the balanced thermochemical equation to relate the amount of energy to the number of moles of methane actually burned. In solving stoichiometry problems, we have used the balanced chemical equation to convert from the number of moles of one compound to the number of moles of another. Now we can use the thermochemical equation to convert between the number of moles of a reactant or product and the amount of energy released or absorbed. **Figure 9.13** shows this approach schematically. The conversion factors needed to obtain the number of moles are the same as earlier: molar mass, density, gas pressure or volume, etc.

Remember that the thermochemical equation tells us the heat for the reaction of the specific molar amounts written.

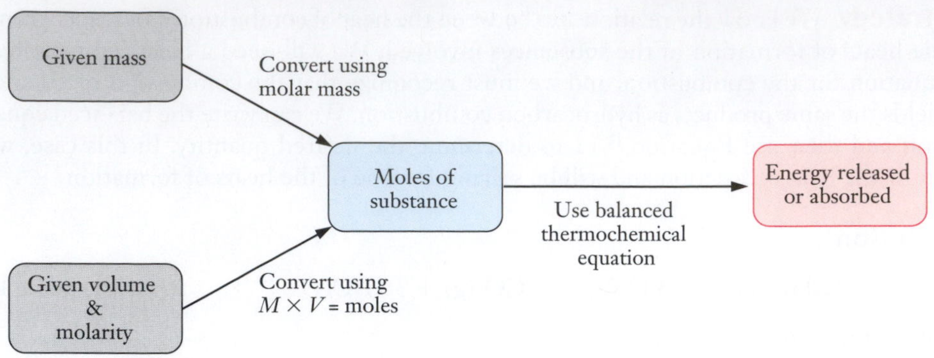

Figure 9.13 ■ This flow chart shows the sequence of steps needed to calculate the amount of energy released or absorbed when a chemical reaction is carried out using a given amount of material.

Let's consider, as an example, the reaction between nitrogen gas and oxygen to form nitric oxide. This occurs regularly as a side reaction when hydrocarbons are burned as fuel. The exothermic combustion reaction that powers an automobile uses air rather than pure oxygen, so there is always a large amount of nitrogen present. At the high temperatures produced in a running engine, some of the nitrogen reacts with oxygen to form NO:

$$N_2(g) + O_2(g) \rightarrow 2\ NO(g) \qquad \Delta H° = 180.5\ kJ$$

The nitric oxide gas formed in this way is an important species in several pollution pathways and is itself an irritant even when present at fairly low levels. For example, NO(g) reacts further with oxygen to form nitrogen dioxide, whose brown color is largely responsible for the dark haze typical of urban smog. Example Problem 9.10 shows how this type of equation can be used to determine heats of reaction of specific quantities of substances.

EXAMPLE PROBLEM 9.10

An engine generates 15.7 g of nitric oxide gas during a laboratory test. How much heat was absorbed in producing this NO?

Strategy The thermochemical equation for this reaction is shown above. This equation provides a link to convert the amount of NO formed to the energy absorbed. As in any other stoichiometry problem, we will work with moles of substance, which we can obtain from the given mass. Note that the stated $\Delta H°$ value is for the production of 2 moles of NO because 2 moles appear in the equation.

Solution

$$15.7\text{g NO} \times \frac{1\text{ mol NO}}{30.0\text{ g NO}} = 0.523\text{ mol NO}$$

$$0.523\text{ mol NO} \times \frac{180.5\text{ kJ}}{2\text{ mol NO}} = 47.2\text{ kJ}$$

Analyze Your Answer The thermochemical equation gives us ΔH for the reaction as written. That means that 180 kJ are absorbed for every 2 moles of NO formed. The molar mass of NO is very close to 30 g/mol, so the amount formed in the problem is a little more than half a mole. That's about a quarter of the amount formed in the thermochemical equation. Our answer is roughly a quarter of the ΔH value from the thermochemical equation as well, so it seems likely to be correct.

Check Your Understanding If 124 kJ of heat is absorbed in a reaction that forms nitric oxide from nitrogen and oxygen, what mass of NO must have been produced? What mass of N_2 was consumed?

This type of manipulation can provide insight into the relative merits of various fuels as well.

Energy Density and Fuels

When considering the economic merits of a particular fuel, several factors should be considered. Among the typical characteristics of a useful fuel are the availability of technology for extracting it, the amount of pollution released by its combustion, and its relative safety. (Because fuels are burned, there is always some danger of unintended or uncontrolled combustion.) From an economic viewpoint, the ease of transporting the fuel is a key factor in this consideration. Fuels that are expensive or dangerous to get to the consumer are less attractive than those that can be delivered at more modest cost. Transportation costs for commodities are dictated largely by the mass that is transported. Because of this important cost, one key feature of a fuel is its **energy density**, the amount of energy that can be released per gram of fuel burned.

We introduced the idea of energy density in Example Problem 9.5.

The energy densities of some fossil fuels are shown in **Table 9.4**, along with those of some alternative fuels. It's easy to see why petroleum is such a prominent source of energy. Consider the positive factors for petroleum use: (1) It is a liquid, which makes it easy to transport and deliver to the customer. (2) It is relatively safe. Explosions can occur, but they are preventable using designs that have a modest cost. (3) The products of its combustion are gases. Where a liquid fuel is easy to transport, gaseous combustion products are easy to dispose of. (4) It has a high energy density.

Table 9.4

Energy densities for a few possible fuels

Fuel	Energy Density (MJ/kg)
Hydrogen	142.0
Methane	55.5
Octane	47.9
Propane	50.3
Aviation gasoline	43.1
Coal, anthracite	31.4
Diesel fuel	45.3
Oil, crude (petroleum)	41.9
Oil, heating	42.5
Gasoline, automotive	45.8
Kerosene	46.3
Wood, oven dry	20.0

In addition to the issues of transportation cost already discussed, this means that automobile engines can be designed to produce relatively large amounts of energy without having the fuel increase the overall weight of the vehicle by very much. Imagine the complications if a tank of gas for your car weighed as much as the car itself. You would use a significant portion of your gas just to carry its own weight, and vehicle safety would be significantly compromised if the mass of the car were increased by a large amount just because the gas tank was full.

9-8 Power Distribution and the Electrical Grid

Throughout this chapter, we have examined ways in which the flow of energy associated with chemical or physical processes can be quantified and potentially put to use. As we saw in Section 9-1, the scale of the global energy economy is vast. In the United States alone there are more than 17,000 electrical generators. And, not surprisingly, there are tremendous technical challenges associated with producing and distributing energy on such a large scale. A particularly important example of this can be found in the generation and delivery of electricity for consumer needs. We expect to have electricity available on demand to power the many appliances in our homes and businesses. But how exactly does the energy released in a power plant find its way to your wall socket?

The complicated, multipart system designed to balance the demand for electricity with its production is referred to as the **electrical grid**. The demand for and supply of electricity both vary in complex ways. Perhaps the most obvious variations occur as we move through the seasons of the year. But demand for electricity also depends on shorter-term factors, including the time of day, weather, and operating conditions in power plants. Not surprisingly, the grid is increasingly dependent on various computer control systems to manage the delicate balance among all of these factors. So although it may seem like the chemistry associated with producing electricity and delivering it to your television is far removed from computer engineering, the connection is really quite important. What are the important components in the grid and how do computers keep them all working together?

Nuclear power plants will be discussed in Chapter 14.

Figure 9.14 shows the key elements in the electrical grid. The system begins with the production of electricity in a *generating station*. As we noted in Figure 9.9, for any plant that uses fossil fuels, the generation of electricity is a transformation of chemical energy released by exothermic reactions into electrical energy. The output from a typical generating station is at fairly high voltages, on the order of thousands of volts. The first component outside of the generation plant is a *transmission substation*. The transmission substation steps up or increases the voltage to levels on the order of 10^5 V, because doing so reduces the loss of energy during transmission. The transmission substation then passes electrical energy along to a *distribution substation*. As their name implies, distribution substations are where electricity can arrive from various points (generation stations) and then be sent off to consumers in industrial, commercial, or residential locations. The distribution substation steps down the voltage before transmitting power over relatively shorter distances to residential or commercial users. All of this movement of electrons takes place through power lines of varying capacities. As we saw in Figure 9.3, the amount of energy "lost" along the way is quite large, so improvements in efficiency in the grid could yield large energy savings.

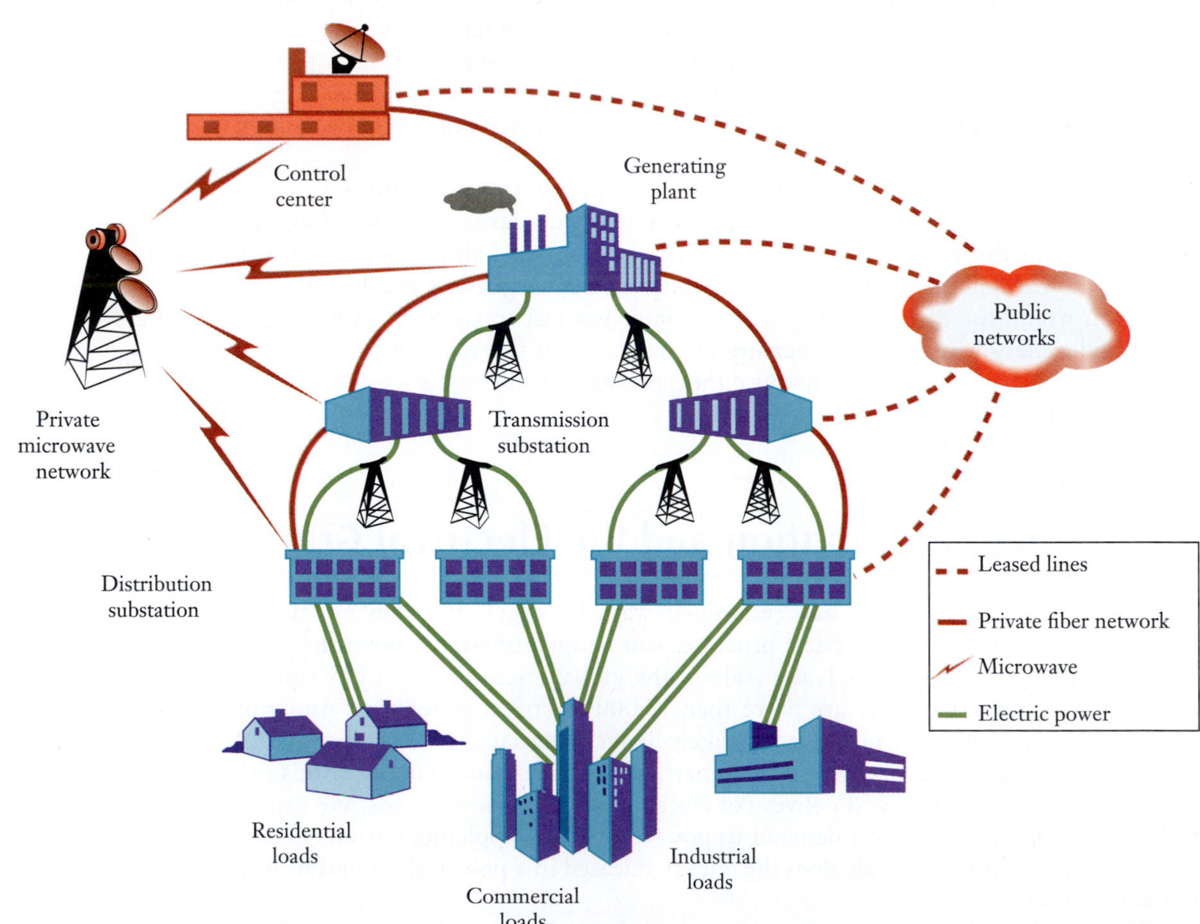

Figure 9.14 ■ The main components of the electrical grid system are shown schematically. As power flows from generators through the various substations to consumers, data flows through a network of control systems that manage the overall operation of the system.

In addition to the general infrastructure for distributing electricity, an overlaid control structure plays a central role in the grid. Control substations act as a sort of nerve center for the overall movement of electricity, implementing decisions about how to match supply and demand. Local utilities must make choices on a daily basis as to how much electrical output they'll need to meet their customers' needs and from where to buy that electricity. As you might imagine with a system like the electrical grid that has been built up gradually over decades, the exact manner in which the control stations communicate with and control the various components can vary greatly from one station to another. For example, some older substations still utilize dial-up modems for access by technical personnel to carry out maintenance. Newer components of the grid are more likely to have wireless access.

Many local utilities in the United States have moved to so-called smart meters that report usage data over the network in real time.

All of this complexity places significant demands on computer hardware and software. The assets of the electrical grid are widely distributed geographically, so network-based communication systems are essential. But these same communication systems also represent a significant vulnerability for cyberattacks. Such attacks can physically destroy generators or other components of the grid. Because electricity in the United States is based on alternating current, all components must operate in phase. In a 2007 experiment by scientists at the Department of Energy's Idaho National Laboratory, a simulated cyberattack was able to cause a generator to go out of phase with the grid to which it was connected, and the generator failed cataclysmically, filling the room with smoke. Modern economies are so strongly linked to the need for electricity that the potential economic impact of a widespread failure of the grid would be enormous.

The need to improve and maintain the security of the grid points to the importance of continued research in computer science on cybersecurity. Finding ways to make sure needed communication between parts of the grid is allowed while any malicious attempts to alter the grid are thwarted represents a high priority for governments. The U.S. Department of Energy has issued the *Roadmap to Achieve Energy Delivery Systems Cybersecurity – 2011* that has established a goal for robust energy delivery to be realized by 2020. This report is available online and shows once again how the technological advances of society depend on the collaborative efforts spanning many different science and engineering fields.

FOCUS ON PROBLEM SOLVING

Question During the nuclear reactor accident at Three Mile Island in 1979, an unknown mass of fuel pellets melted, allowing some of the fuel material to fall into water at the bottom of the reactor. Assume that the melting fuel pellets were pure UO_2 and had resolidified and cooled to 900°C before reaching the water. Assume further that the water was initially at 8°C and that sensors indicated the final temperature of the water was 85°C.

(a) What information would you have to look up to determine the mass of fuel pellets that fell into the water?

(b) What would you have to know to use this information to determine the percentage of the fuel that melted in the accident? How would you calculate the percentage?

Strategy

(a) The first question that must be addressed is, "What sources of heat raised the temperature of the water?" Were the fuel pellets the only thing that melted? Because we have no other information, we will assume that the only important heat flow was between the pellets and the water. If that were not true, then we would need to account for other materials in our handling of this problem. If the assumption turns out to be incorrect, we will overestimate the amount of fuel that melted in the accident.

Once we have made the assumption that the melted fuel is the only source of heat for the water, we would set up a heat balance between the water and the fuel pellets. Thus

$$q_{\text{pellets}} = -q_{\text{water}}$$

and

$$(mc\Delta T)_{\text{pellets}} = -(mc\Delta T)_{\text{water}}$$

We can see there are six possible variables in this equation. Which do we know and which can be looked up? Both ΔT values are known. The initial temperatures of both the water and the fuel pellets were given in the problem and both share the final temperature of 85°C. The specific heat (*c*) is not given for either the water or the fuel pellets. This is information that must be looked up, but it should be available from standard reference sources.

That leaves the two masses, and clearly the goal of the exercise is to find the mass of the melted fuel pellets. But we don't know the mass of the water involved, either. So either we must find some way to look up this value or we must make a reasonable estimate.

(b) This part of the question assumes that we can find the mass from part (a). If that's true, then we would need to know the total mass of fuel pellets initially present. This number must be looked up or estimated based on some reliable information.

Solution

(a) We would have to look up the heat capacity of water and fuel pellets and the mass of water in the reactor (initially at 8°C) to be able to answer this question.

(b) The percentage melted can be calculated using the equation

$$\text{Percent melted} = \frac{m_{\text{melted}}}{m_{\text{total}}} \times 100\%$$

Here we are assuming that we have obtained values for both the mass melted and the total mass of fuel pellets. We should also keep in mind that our result will only be as good as the assumption that the melted fuel was the sole source of heat for the water.

SUMMARY

Energy plays a central role in science and also in economic development and societal advancement. Many energy-related technologies rely on the fact that chemical reactions can release the energy stored in chemical bonds. The study of chemical thermodynamics examines these energy changes in chemical reactions.

In developing a scientific understanding of energy, we must be especially careful to define our terms precisely. For words such as *work* and *heat*, the scientific meanings are much more specific than everyday meanings. In scientific usage, work is the movement of a mass against an opposing force, and heat is the transfer of energy between objects at different temperatures. As for any other calculations, we also must take care to use consistent units when working with energy values.

One familiar concept is that energy is conserved, which is a concise statement of the first law of thermodynamics. To use energy, however, it must often be converted from one form to another. This conversion invariably involves some waste of energy, though, and has a significant impact on the overall energy economy. Experimental measurements of the energy changes in chemical reactions are made by calorimetry, the study of heat flow. Heat flow is important in engineering design, and the ability to relate heat flow to temperature change, for example, is a key skill in both thermochemistry and engineering.

The thermodynamics of chemical reactions is often described in terms of enthalpy. Enthalpy is a state function, meaning that its value depends only on the current state of a system. The enthalpy change is equal to the heat flow for a process at constant pressure. Enthalpy changes for some specific processes, including phase changes, are tabulated. But it would not be practical to attempt to catalog the enthalpy change for every possible chemical reaction. Instead, values for the heat of formation of compounds are tabulated. As long as these values are available for all of the substances involved, they can be used to calculate the enthalpy change for any reaction. The enthalpy change can also be related to the stoichiometry of a reaction, and such calculations can be used to find the amount of fuel needed to produce a given amount of energy.

KEY TERMS

boundary (9-3)
British thermal unit (Btu) (9-1)
calorie (cal) (9-2)
calorimetry (9-4)
chemical energy (9-2)
electrical grid (9-8)
endothermic (9-5)
energy density (9-7)
energy economy (9-1)
enthalpy (9-5)
enthalpy diagram (9-6)
exothermic (9-5)
first law of thermodynamics (9-3)
formation reaction (9-5)
heat (9-2)
heat capacity (9-4)
heat of formation (9-5)
heat of reaction (9-5)
Hess's law (9-6)
internal energy (9-2)
joule (J) (9-2)
kinetic energy (9-2)
molar heat capacity (9-4)
potential energy (9-2)
PV-work (9-2)
specific heat (9-4)
state function (9-6)
surroundings (9-3)
system (9-3)
thermochemical equation (9-5)
thermochemistry (9-1)
universe (9-3)
work (9-2)

PROBLEMS AND EXERCISES

INSIGHT INTO Energy Use and the World Economy

9.1 List reasons why there might be a connection between the amount of energy used by a country and its economic development.

9.2 What are the advantages of electricity as a type of energy that make it worth generating despite the sizable losses that occur during this process?

9.3 What differences (if any) would you expect to find between the energy use patterns of the United States shown in Figure 9.2 and Figure 9.3 and use in **(a)** European countries and **(b)** developing countries?

9.4 The total energy supply for the United States in the year 2011, as shown in Figure 9.2, was 107.66×10^{15} Btu. **(a)** What percentage of that energy was obtained from imports? **(b)** What percentage was used to generate electricity?

9.5 Using the value you calculate from **(a)** in the previous problem and the fact that an average barrel of petroleum has an energy equivalent of 5.85 million Btu, how many barrels of oil, on average, must be imported each day?

Defining Energy

9.6 Distinguish between kinetic and potential energy.

9.7 Define the term *internal energy*.

9.8 How fast (in meters per second) must an iron ball with a mass of 56.6 g be traveling in order to have a kinetic energy of 15.75 J?

9.9 What is the kinetic energy of a single molecule of oxygen if it is traveling at 1.5×10^3 m/s?

9.10 The kinetic energy of molecules is often used to induce chemical reactions. The bond energy in an O_2 molecule is 8.22×10^{-19} J. Can an O_2 molecule traveling at 780 m/s provide enough energy to break the O=O bond? What is the minimum velocity of an O_2 molecule that would give a kinetic energy capable of breaking the bond if it is converted with 100% efficiency?

9.11 Analyze the units of the quantity (pressure × volume) and show that they are energy units, consistent with the idea of PV-work.

9.12 How many kilojoules are equal to 3.27 L atm of work?

9.13 Define the term *hydrocarbon*.

9.14 What are the products of the complete combustion of a hydrocarbon?

9.15 Carry out the following conversions of energy units: **(a)** 14.3 Btu into calories, **(b)** 1.4×10^5 cal into joules, **(c)** 31.6 mJ into Btu

9.16 According to Figure 9.2, the total energy supply in the United States in 2011 was 107.66×10^{15} Btu. Express this value in joules and in calories.

Energy Transformation and Conservation of Energy

9.17 If a machine does 4.8×10^3 kJ of work after an input of 7.31×10^4 kJ of heat, what is the change in internal energy for the machine?

9.18 Calculate **(a)** q when a system does 54 J of work and its energy decreases by 72 J and **(b)** ΔE for a gas that releases 38 J of heat and has 102 J of work done on it.

9.19 If the algebraic sign of ΔE is negative, in which direction has energy flowed?

9.20 State the first law of thermodynamics briefly in your own words.

9.21 Which type of energy, heat or work, is valued more by society? What evidence supports your judgment?

9.22 PV-work occurs when volume changes and pressure remains constant. If volume is held constant, can PV-work be done? What happens to Equation 9.2 when volume is held constant?

9.23 Which system does more work: **(a)** $\Delta E = -436$ J, $q = 400$ J; or **(b)** $\Delta E = 317$ J, $q = 347$ J?

9.24 In which case is heat added to the system: **(a)** $\Delta E = -43$ J, $w = 40$ J; or **(b)** $\Delta E = 31$ J, $w = 34$ J?

9.25 Figure 9.5 shows projections for improved efficiency in many important technologies over the next two decades. Will these efficiency gains necessarily lead to reduced household energy demand? What factors might apply pressure for continued increases in household energy use?

9.26 Gas furnaces have achieved impressive efficiency levels largely through the addition of a second heat exchanger that condenses water vapor that would otherwise escape out the exhaust system attached to the furnace. How does this process improve efficiency?

9.27 When energy conservation programs are promoted, they sometimes include strategies such as turning out lights. Use the web to determine the average percentage of U.S. household energy usage that is attributable to lighting and use the information to comment on how effective the "turn out the lights" strategy might be.

9.28 When an electrical appliance whose power usage is X watts is run for Y seconds, it uses $X \times Y$ joules of energy. The energy unit used by electrical utilities in their monthly bills is the *kilowatt-hour* (kWh, that is, 1 kilowatt used for 1 hour). How many joules are there in a kilowatt-hour? If electricity costs $0.09 per kilowatt-hour, how much does it cost per megajoule?

Heat Capacity and Calorimetry

9.29 Define the term *calorimetry*.

9.30 For the example of shallow water and sandy beaches, which material has a larger heat capacity or specific heat? How does a hot day at the beach provide evidence for your answer?

9.31 A metal radiator is made from 26.0 kg of iron. The specific heat of iron is 0.449 J g^{-1} $°C^{-1}$. How much heat must be supplied to the radiator to raise its temperature from 25.0 to 55.0°C?

9.32 The material typically used to heat metal radiators is water. If a boiler generates water at 79.5°C, what mass of water was needed to provide the heat required in the previous problem? Water has a specific heat of 4.184 J g^{-1} $°C^{-1}$.

9.33 Copper wires used to transport electrical current heat up because of the resistance in the wire. If a 140-g wire gains 280 J of heat, what is the change in temperature in the wire? Copper has a specific heat of 0.384 J g^{-1} $°C^{-1}$.

9.34 A copper nail and an iron nail of the same mass and initially at the same room temperature are both put into a vessel containing boiling water. Which one would you expect to reach 100°C first? Why?

9.35 A piece of titanium metal with a mass of 20.8 g is heated in boiling water to 99.5°C and then dropped into a coffee cup calorimeter containing 75.0 g of water at 21.7°C. When thermal equilibrium is reached, the final temperature is 24.3°C. Calculate the specific heat capacity of titanium.

9.36 Define the term *calibration*.

9.37 A calorimeter contained 75.0 g of water at 16.95°C. A 93.3-g sample of iron at 65.58°C was placed in it, giving a final temperature of 19.68°C for the system. Calculate the heat capacity of the calorimeter. Specific heats are 4.184 J g^{-1} $°C^{-1}$ for H_2O and 0.444 J g^{-1} $°C^{-1}$ for Fe.

9.38 The energy densities of various types of coal are listed below:

Anthracite 35 kJ/g	Subbituminous 31 kJ/g
Bituminous 28 kJ/g	Lignite 26 kJ/g

An unknown sample of one of these coals is burned in an apparatus with a calorimeter constant of 1.3 kJ/°C. When a 0.367-g sample is used, the temperature change is 8.75°C. Which type of coal is the sample?

9.39 How much thermal energy is required to heat all of the water in a swimming pool by 1°C if the dimensions are 4 ft deep by 20 ft wide by 75 ft long? Report your result in megajoules.

9.40 How does the specific heat of water explain why cities on the coast of large bodies of water tend to be cooler in the summer than cities several miles inland?

Enthalpy

9.41 Under what conditions does the enthalpy change equal the heat of a process?

9.42 Why is enthalpy generally more useful than internal energy in the thermodynamics of real world systems?

9.43 Define the terms *exothermic* and *endothermic*.

9.44 List at least two phase changes that are exothermic processes.

9.45 What happens to the temperature of a material as it undergoes an endothermic phase change? If heat is added, how can the temperature behave in this manner?

9.46 The heat of fusion of pure silicon is 43.4 kJ/mol. How much energy would be needed to melt a 5.24-g sample of silicon at its melting point of 1693 K?

9.47 If 14.8 kJ of heat is given off when 1.6 g of HCl condenses from vapor to liquid, what is ΔH_{cond} for this substance?

9.48 Calculate the energy required to convert 1.70 g of ice originally at −12.0°C into steam at 105°C.

9.49 ΔH_{vap} = 31.3 kJ/mol for acetone. If 1.40 kg of water were vaporized to steam in a boiler, how much acetone (in kg) would need to be vaporized to use the same amount of heat?

9.50 When a 13.0-g sample of NaOH(s) dissolves in 400.0 mL of water in a coffee cup calorimeter, the temperature of the water changes from 22.6°C to 30.7°C. Assuming that the specific heat capacity of the solution is the same as for water, calculate **(a)** the heat transfer from system to surroundings and **(b)** ΔH for the reaction

$$NaOH(s) \rightarrow Na^+(aq) + OH^-(aq)$$

9.51 Define the term *formation reaction*.

9.52 Write the formation reaction for each of the following substances. **(a)** $CH_4(g)$, **(b)** $C_3H_8(\ell)$, **(c)** HCl(g), **(d)** $C_6H_{12}O_6(s)$, **(e)** NaF(s)

9.53 Explain why each of the following chemical equations is *not* a correct formation reaction:

(a) $4\ Al(s) + 3\ O_2(g) \rightarrow 2\ Al_2O_3(s)$

(b) $N_2(g) + \frac{3}{2}H_2(g) \rightarrow NH_3(g)$

(c) $2\ Na(s) + O(g) \rightarrow Na_2O(s)$

Hess's Law and Heats of Reaction

9.54 Which of the following are state functions? **(a)** the volume of a balloon, **(b)** the time it takes to drive from your home to your college or university, **(c)** the temperature of the water in a coffee cup, **(d)** the potential energy of a ball held in your hand

9.55 Using these reactions, find the standard enthalpy change for the formation of 1 mol of PbO(s) from lead metal and oxygen gas:

$PbO(s) + C(graphite) \rightarrow Pb(s) + CO(g) \quad \Delta H° = 106.8$ kJ

$2\ C(graphite) + O_2(g) \rightarrow 2\ CO(g) \quad \Delta H° = -221.0$ kJ

If 250 g of lead reacts with oxygen to form lead(II) oxide, what quantity of thermal energy (in kJ) is absorbed or evolved?

9.56 The phase change between graphite and diamond is difficult to observe directly. Both substances can be burned, however. From these equations, calculate $\Delta H°$ for the conversion of diamond into graphite:

$C(s, graphite) + O_2(g) \rightarrow CO_2(g) \quad \Delta H° = -393.51$ kJ

$C(s, diamond) + O_2(g) \rightarrow CO_2(g) \quad \Delta H° = -395.94$ kJ

9.57 Hydrogen gas will react with either acetylene or ethylene gas. The thermochemical equations for these reactions are provided below. Write the thermochemical equation for the conversion of acetylene into ethylene by hydrogen gas.

$C_2H_2(g) + 2\ H_2(g) \rightarrow C_2H_6 \quad \Delta H° = -311$ kJ

$C_2H_4(g) + H_2(g) \rightarrow C_2H_6 \quad \Delta H° = -136$ kJ

9.58 Using heats of formation tabulated in Appendix E, calculate the heats of reaction for the following.

(a) $C_2H_2(g) + \frac{5}{2}O_2(g) \rightarrow 2\ CO_2(g) + H_2O(\ell)$

(b) $PCl_3(g) + Cl_2(g) \rightarrow PCl_5(g)$

(c) $C_2H_4(g) + H_2O(g) \rightarrow C_2H_5OH(g)$

(d) $Fe_2O_3(s) + 2\ Al(s) \rightarrow Al_2O_3(s) + 2\ Fe(\ell)$

9.59 The heat of combustion of butane is −2877 kJ/mol. Use this value to find the heat of formation of butane. (You may also need to use additional thermochemical data found in Appendix E.)

9.60 When a chemical bond breaks, is energy absorbed or released?

9.61 When a reaction is exothermic, is the sum of bond energies of products or of reactants greater?

Energy and Stoichiometry

9.62 For the reaction $C_2H_2(g) + 2\ H_2(g) \rightarrow C_2H_6$, $\Delta H° = -136$ kJ, what are the ratios that can be defined between moles of substances and energy?

9.63 For the reaction $N_2(g) + O_2(g) \rightarrow 2\ NO(g)$, $\Delta H° = 180.5$ kJ, how much energy is needed to generate 35 moles of NO(g)?

9.64 Nitroglycerine, $C_3H_5(NO_3)_3(\ell)$, is an explosive most often used in mine or quarry blasting. It is a powerful explosive because four gases (N_2, O_2, CO_2, and steam) are formed when nitroglycerine is detonated. In addition, 6.26 kJ of heat is given off per gram of nitroglycerine detonated. **(a)** Write a balanced thermochemical equation for the reaction. **(b)** What is ΔH when 4.65 mol of products is formed?

9.65 Silane, SiH_4, burns according to the reaction, $SiH_4 + 2\ O_2 \rightarrow SiO_2 + 2\ H_2O$, with $\Delta H° = -1429$ kJ. How much energy is released if 15.7 g of silane is burned?

9.66 Sulfur trioxide can be removed from the exhaust gases of power plants by reaction with lime according to the equation $CaO(s) + SO_3(g) \rightarrow CaSO_4(s)$, with $\Delta H° = -886$ kJ. If 240 kg of SO_3 is to be removed, how much heat is released?

9.67 Reactions of hydrocarbons are often studied in the petroleum industry. One such reaction is $2\ C_3H_8(g) \rightarrow C_6H_6(\ell) + 5\ H_2(g)$, with $\Delta H° = 698$ kJ. If 35 L of propane at 25°C and 0.97 atm is to be reacted, how much heat must be supplied?

9.68 In principle, ozone could be consumed in a reaction with lead and carbon with the thermochemical equation $Pb(s) + C(s) + O_3(g) \rightarrow PbCO_3(s)$ $\Delta H° = -841.0$ kJ. How much energy would be released if 110 g of ozone reacts with excess lead and carbon?

9.69 When 0.0157 g of a compound with a heat of combustion of −37.6 kJ/mole is burned in a calorimeter, 18.5 J of heat is released. What is the molar mass of the compound?

9.70 Define the term *energy density*.

9.71 Why is energy density so important in the transportation of fuels?

9.72 What are some features of petroleum that make it such an attractive fuel?

INSIGHT INTO Power Distribution and the Electrical Grid

9.73 How are the roles of transmission substations and distribution substations in the electrical grid similar? How are they different?

9.74 Residential electric service in the United States generally operates at 120 V, but transmission substations feed power onto the grid at 110 kV or higher. What advantage is realized by transmitting electricity at such a high voltage?

9.75 In recent years, the notion of a "smart grid" has emerged. Do a web search and research the smart grid concept. How would the smart grid differ from the traditional grid?

9.76 Although it can be a nuisance when a laptop computer freezes up and needs to be rebooted, we accept that as somewhat inevitable. But clearly the need to occasionally reboot the control system for the power grid would not be acceptable. Use the web to research ways that engineers ensure the reliability of crucial systems like the control infrastructure of the grid, and write a paragraph summarizing the main strategies employed.

Additional Problems

9.77 The figure below shows a "self-cooling" beverage can.

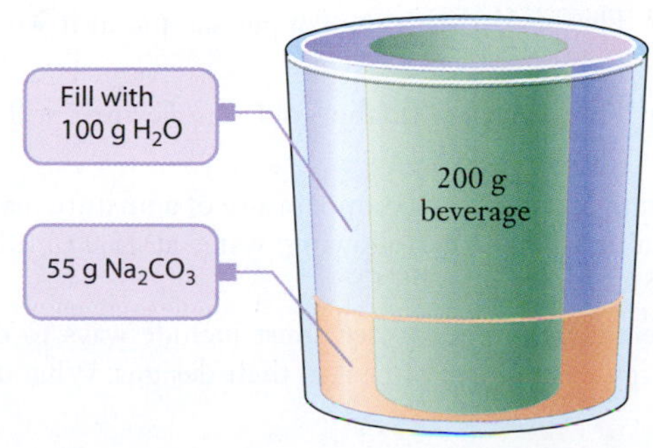

The can is equipped with an outer jacket containing sodium carbonate (Na_2CO_3), which dissolves in water rapidly and endothermically:

$$Na_2CO_3(s) \rightarrow 2\ Na^+(aq) + CO_3^{2-}(aq) \qquad \Delta H° = 67.7\ kJ$$

The user adds water to the outer jacket, and the heat absorbed in the chemical reaction chills the drink. The can contains 200 g of drink, the jacket contains 55 g of Na_2CO_3, and 100 g of water is to be added. If the initial temperatures of the can and the water are both 32°C on a summer day, what is the coldest temperature that the drink can reach? The can itself has a heat capacity of 40 J/°C. Assume that the Na_2CO_3 solution and the drink both have the same heat capacity as pure water, 4.184 J g^{-1} °C^{-1}. (HINT: Treat this like a calorimetry problem.)

9.78 You make some iced tea by dropping 134 g of ice into 500.0 mL of warm tea in an insulated pitcher. If the tea is initially at 20.0°C and the ice cubes are initially at 0.0°C, how many grams of ice will still be present when the contents of the pitcher reach a final temperature? The tea is mostly water, so assume that it has the same density (1.0 g/mL), molar mass, heat capacity (75.3 J K^{-1} mol^{-1}), and heat of fusion (6.0 kJ/mol) as pure water. The heat capacity of ice is 37.7 J K^{-1} mol^{-1}.

9.79 A student performing a calorimetry experiment combined 100.0 mL of 0.50 *M* HCl and 100.0 mL of 0.50 *M* NaOH in a Styrofoam™ cup calorimeter. Both solutions were initially at 20.0°C, but when the two were mixed, the temperature rose to 23.2°C.

(a) Suppose the experiment is repeated in the same calorimeter but this time using 200 mL of 0.50 *M* HCl and 200.0 mL of 0.50 *M* NaOH. Will the ΔT observed be greater than, less than, or equal to that in the first experiment, and why?

(b) Suppose that the experiment is repeated once again in the same calorimeter, this time using 100 mL of 1.00 *M* HCl and 100.0 mL of 1.00 *M* NaOH. Will the ΔT observed be greater than, less than, or equal to that in the first experiment, and why?

9.80 The specific heat of gold is 0.13 J g^{-1} K^{-1}, and that of copper is 0.39 J g^{-1} K^{-1}. Suppose that we heat both a 25-g sample of gold and a 25-g sample of copper to 80°C and then drop them into identical beakers containing 100 mL of cold water at 10°C.

When each beaker reaches thermal equilibrium, which of the following will be true, and why? You should not need to do any calculations here.

(i) Both beakers will be at the same temperature.

(ii) The beaker with the copper sample in it will be at a higher temperature.

(iii) The beaker with the gold sample in it will be at a higher temperature.

9.81 What will be the final temperature of a mixture made from equal masses of the following: water at 25.0°C, ethanol at 35.5°C, and iron at 95°C?

9.82 Chemical engineers often must include ways to dissipate energy in the form of heat in their designs. What does this fact say about the enthalpy change of chemical reactions that are being used?

9.83 A sample of natural gas is 80.0% CH_4 and 20.0% C_2H_6 by mass. What is the heat from the combustion of 1.00 g of this mixture? Assume the products are $CO_2(g)$ and $H_2O(\ell)$.

9.84 Many engineering designs must incorporate ways to dissipate energy in the form of heat. Water evaporators are common for this task. **(a)** What property of water makes it a good material for evaporators? **(b)** If an application could not use water, but instead was forced to use a material with a value for the property in **(a)** that was one half that of water, what changes would need to be made in the design?

9.85 You want to heat the air in your house with natural gas (CH_4). Assume your house has 275 m^2 (about 2800 ft^2) of floor area and that the ceilings are 2.50 m from the floors. The air in the house has a molar heat capacity of 29.1 J mol^{-1} K^{-1}. (The number of moles of air in the house may be found by assuming that the average molar mass of air is 28.9 g/mol and that the density of air at these temperatures is 1.22 g/L.) What mass of methane do you have to burn to heat the air from 15.0 to 22.0°C?

9.86 The curing of concrete liberates energy as heat. **(a)** What does this observation suggest happens in terms of chemical bonds as concrete cures? **(b)** What type of strategies might a civil engineer employ in designs and construction specifications to mitigate heat expansion of concrete as it cures?

9.87 An engineer is designing a product in which a copper wire will carry large amounts of electricity. The resistive heating of a 65-g copper wire is expected to add 580 J of heat energy during a 10-minute operating cycle. The specific heat of copper is 0.385 J g^{-1} °C^{-1}, the density is 8.94 g/cm^3, and the coefficient of thermal expansion is 16.6 μm m^{-1} K^{-1}. **(a)** What is the temperature increase of the wire? **(b)** What is the initial length of the wire, assuming it is a cylinder and its radius is 0.080 cm? **(c)** By what percentage does the length increase because of the temperature increase? **(d)** Do you think the engineer should be considering this expansion in the design?

9.88 Price spikes in gasoline in 2008 led to renewed interest in coal gasification projects, in which coal is converted to gasoline. Looking at the relative energy density of gasoline and coal in Table 9.4, which is more likely required in an engineering design for this project: the ability to input heat or the need to dissipate heat? Explain your reasoning.

9.89 A substance has the following properties:

$\Delta H_{fusion} = 10.0$ kJ/mol $\qquad \Delta H_{vaporization} = 20.0$ kJ/mol

$C_p(\text{solid}) = 30$ J/mol/K $\qquad C_p(\text{liquid}) = 60$ J/mol/K

$C_p(\text{gas}) = 30$ J/mol/K

(a) Which of the four graphs on the following page would be most consistent with these data? **(b)** If the melting point of the substance is 80°C, how much heat would be required to convert 1.0 mol of it from solid at 20°C into liquid at 120°C?

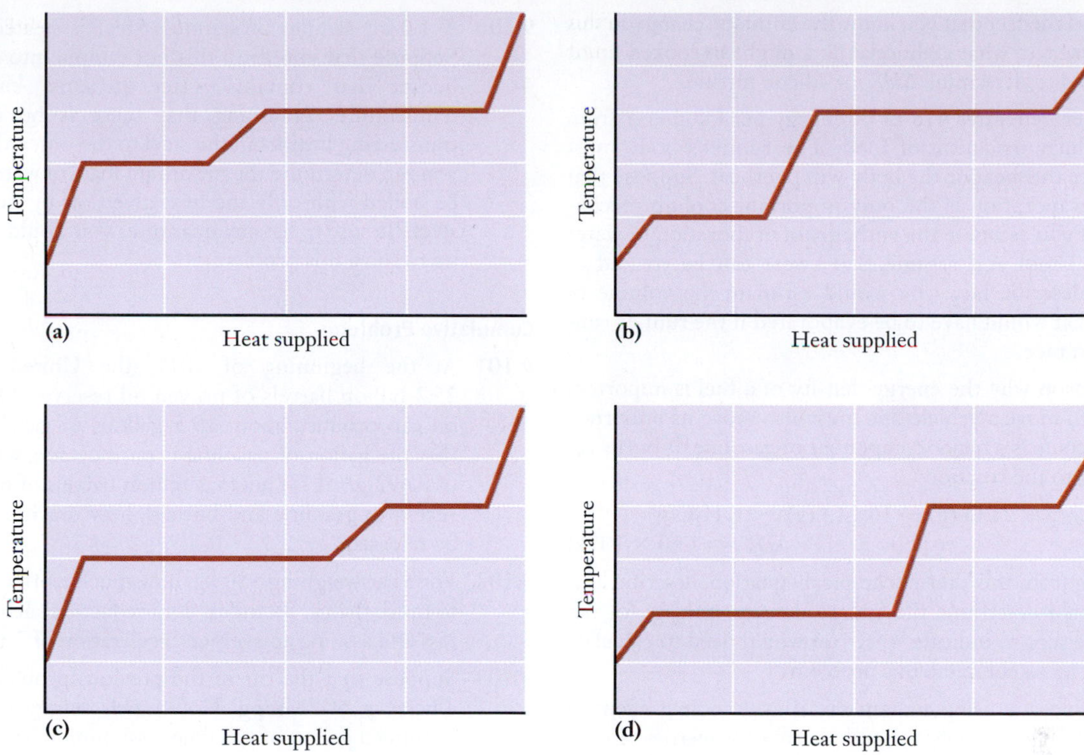

9.90 Most first aid "cold packs" are based on the endothermic dissolution of ammonium nitrate in water:

$$NH_4NO_3(s) \rightarrow NH_4^+(aq) + NO_3^-(aq) \qquad \Delta H° = 25.69 \text{ kJ}$$

A particular cold pack contains 50.0 g of NH_4NO_3 and 125.0 g of water. When the pack is squeezed, the NH_4NO_3 dissolves in the water. If the pack and its contents are initially at 24.0°C, what is the lowest temperature that this bag could reach? (Assume that the ammonium nitrate solution has a specific heat of 4.25 J g^{-1} K^{-1}, and that the heat capacity of the bag itself is small enough to be neglected.)

FOCUS ON PROBLEM SOLVING EXERCISES

9.91 Suppose that the working fluid inside an industrial refrigerator absorbs 680 J of energy for every gram of material that vaporizes in the evaporator. The refrigerator unit uses this energy flow as part of a cyclic system to keep foods cold. A new pallet of fruit with a mass of 500 kg is placed in the refrigerator. Assume that the specific heat of the fruit is the same as that of pure water because the fruit is mostly water. Describe how you would determine the mass of the working fluid that would have to be evaporated to lower the temperature of the fruit by 15°C. List any information you would have to measure or look up.

9.92 Hydrogen combines with oxygen in fuel cells according to the thermochemical equation

$$2\ H_2(g) + O_2(g) \rightarrow 2\ H_2O(g) \qquad \Delta H° = -571.7 \text{ kJ}$$

Suppose that you are working with a firm that is using hydrogen fuel cells to power satellites. The satellite requires 4.0×10^5 kJ of energy during its useful lifetime to stabilize its orbit. Describe how you would determine the mass of hydrogen you would need in your fuel cells for this particular satellite.

9.93 The chemical reaction $BBr_3(g) + BCl_3(g) \rightarrow BBr_2Cl(g) + BCl_2Br(g)$, has an enthalpy change very close to zero. Using Lewis structures of the molecules, all of which have a central boron atom, provide a molecular-level description of why $\Delta H°$ for this reaction might be very small.

9.94 Two baking sheets are made of different metals. You purchase both and bake a dozen cookies on each sheet at the same time in your oven. You observe that after 9 minutes, the cookies on one sheet are slightly burned on the bottom, whereas those on the other sheet are fine. (You are curious and you vary the conditions so you know the result is not caused by the oven.) **(a)** How can you use this observation to infer something about the specific heat of the materials in the baking sheets? **(b)** What is the mathematical reasoning (equation) that you need to support your conclusion?

9.95 Silicon nitride, Si_3N_4, has physical, chemical, and mechanical properties that make it a useful industrial material. For a particular engineering project, it is crucial that you know the heat of formation of this substance. A clever experiment allows direct determination of the $\Delta H°$ of the following reaction:

$$3\ CO_2(g) + Si_3N_4(s) \rightarrow 3\ SiO_2(s) + 2\ N_2(g) + 3\ C(s)$$

Based on the fact that you know the enthalpy change in this reaction, state what additional data might be looked up or measured to determine ΔH_f° for silicon nitride.

9.96 A runner generates 418 kJ of energy per kilometer from the cellular oxidation of food. The runner's body must dissipate this heat or the body will overheat. Suppose that sweat evaporation is the only important cooling mechanism. If you estimate the enthalpy of evaporation of water as 44 kJ/mol and assume that sweat can be treated as water, describe how you would estimate the volume of sweat that would have to be evaporated if the runner runs a 10-km race.

9.97 One reason why the energy density of a fuel is important is that to move a vehicle one must also move its unburned fuel. Octane is a major component of gasoline. It burns according to the reaction

$$2\ C_8H_{18}(\ell) + 25\ O_2(g) \rightarrow 16\ CO_2(g) + 18\ H_2O(g) \qquad \Delta H^\circ = -1.10 \times 10^4\ \text{kJ}$$

Starting from this thermochemical equation, describe how you would determine the energy density, in kJ/g, for octane. Be sure to indicate what you would need to calculate or look up to complete this problem.

9.98 An engineer is using sodium metal as a cooling agent in a design because it has useful thermal properties. Looking up the heat capacity, the engineer finds a value of 28.2 J mol^{-1} °C^{-1}. Carelessly, he wrote this number down without units. As a result, it was later taken as specific heat. **(a)** What would be the difference between these two values? **(b)** Would the engineer overestimate the ability of sodium to remove heat from the system or underestimate it because of this error? Be sure to explain your reasoning.

9.99 In passive solar heating, the goal is to absorb heat from the sun during the day and release it during the night. Which material would be better for this application: one with a high heat capacity or one with a low heat capacity? Explain.

9.100 A 1.0-kg sample of stainless steel is heated to 400°C. Suppose that you drop this hot sample into an insulated bucket that contains water at some known initial temperature. Assuming that there is no difficulty in transferring heat from the steel to the water, describe how you can determine the maximum mass of water that could be boiled with only the heat given off by this sample of steel. Be sure to list any quantities you would need to look up to solve this problem.

Cumulative Problems

9.101 At the beginning of 2011, the United States had 25.2 billion barrels of proven oil reserves. One barrel of oil can produce about 19.5 gallons of gasoline. Assume that the gallon of gasoline is pure octane, with a density of 0.692 g/mL. If all 25.2 billion barrels of oil were converted to gasoline and burned, how much energy would be released?

9.102 For a car weighing 980 kg, how much work must be done to move the car 24 miles? Ignore factors such as frictional loss and assume an average acceleration of 2.3 m/s^2.

9.103 Suppose that the car in the previous problem has a fuel efficiency of 24 mpg. How much energy is released in burning a gallon of gasoline (assuming that all of it is octane)? Based on this calculation and the work required to move the car those 24 miles, what percentage of the energy released in the combustion is wasted (doesn't directly contribute to the work of moving the car)? How do the assumptions you make in carrying out these calculations affect the value you obtain?

9.104 Suppose that there is 2.43 mol of nitrogen gas in an insulated, sealed 31.7-L container initially at 285 K. The specific heat of nitrogen gas is 1.04 kJ kg^{-1} K^{-1} (note units). If a 5.44-kg block of iron at 755 K is placed in this container and it is sealed again (with no loss of nitrogen), what is the final pressure of the nitrogen gas?

10

Entropy and the Second Law of Thermodynamics

Curbside recycling programs often collect "comingled" materials, as seen here at a Milwaukee site. Plastics, which make up about 85% of this pile, must be separated and sorted for recycling. *Thomas A. Holme*

CHAPTER OUTLINE

In our discussions of chemical bonding, we introduced the idea that bonds form because doing so reduces the overall energy of the collection of atoms involved. We've seen many examples of chemical reactions, such as combustion and explosions, which also reduce the overall energy of the atoms and molecules involved. But if you try for just a moment, you should also be able to think of many common chemical and physical processes in which the energy of the system clearly increases. Ice cubes melt. The batteries in your laptop or cell phone are recharged. At least some endothermic chemical reactions occur regularly. As each of these cases shows, the energy of a system does not always decrease, despite our intuitive sense of a preference for minimizing energy. So how can we understand and predict which changes nature will actually favor? We will need to extend our understanding by introducing the second law of thermodynamics and exploring its ramifications. Although applications abound in virtually all fields of science and engineering, the impact of thermodynamics on our understanding of chemical reactions has been especially profound. We'll explore the implications of the second law by looking into the recycling of plastics.

Chapter Objectives

After mastering this chapter, you should be able to

- describe the scientific and economic obstacles to more widespread recycling of plastics.
- explain the concept of entropy in your own words.
- deduce the sign of ΔS for many chemical reactions by examining the physical state of the reactants and products.
- state the second law of thermodynamics in words and equations and use it to predict spontaneity.
- state the third law of thermodynamics.
- use tabulated data to calculate the entropy change in a chemical reaction.
- derive the relationship between the free energy change of a system and the entropy change of the universe.
- use tabulated data to calculate the free energy change in a chemical reaction.
- explain the role of temperature in determining whether a reaction is spontaneous.
- use tabulated data to determine the temperature range for which a reaction will be spontaneous.

INSIGHT INTO

10-1 Recycling of Plastics

Standard plastic soft drink bottles are made of poly(ethylene terephthalate), or PET. In the industrial-scale synthesis of PET, the usual starting materials are dimethyl terephthalate and ethylene glycol (**Figure 10.1**). These compounds react to form bis-(2-hydroxyethyl) terephthalate (BHET) and methanol. The methanol boils off at the reaction temperature (typically around 210°C), leaving fairly pure BHET. Then, the BHET is heated further to around 270°C, where it undergoes a condensation reaction to form PET polymer. Ethylene glycol is a byproduct in this second step and can thus be reused within the plant to produce more BHET.

Condensation reactions were introduced in Section 8-6.

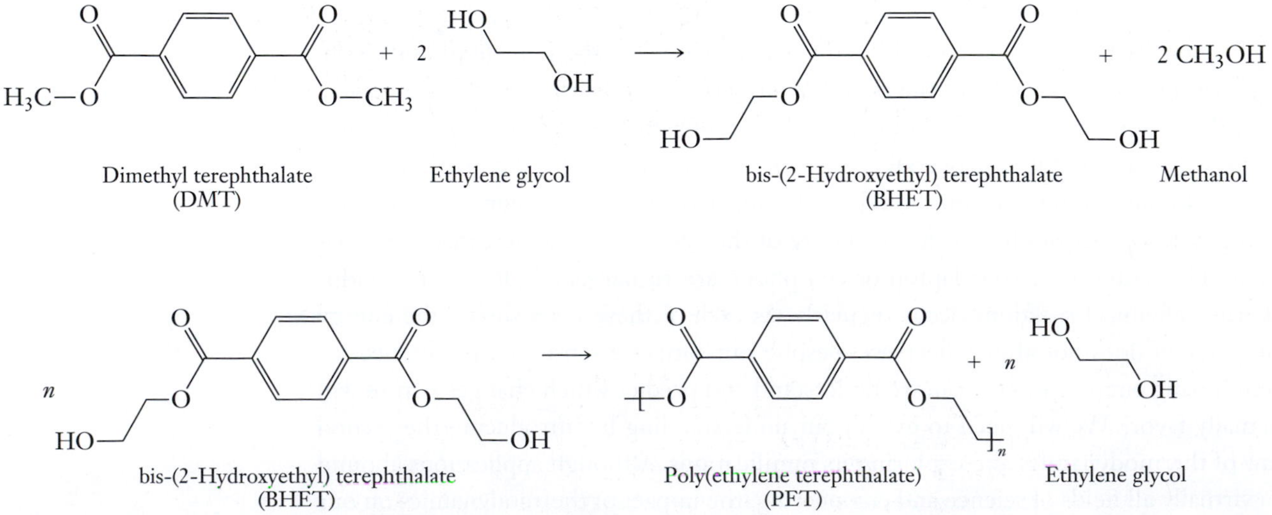

Dimethyl terephthalate (DMT) + 2 Ethylene glycol ⟶ bis-(2-Hydroxyethyl) terephthalate (BHET) + 2 CH_3OH Methanol

n bis-(2-Hydroxyethyl) terephthalate (BHET) ⟶ Poly(ethylene terephthalate) (PET) + n Ethylene glycol

Figure 10.1 ■ Steps in the industrial synthesis of PET are illustrated. Typical values of n in the polymer formula are 130–150, giving a molar mass of around 25,000 for the polymer.

The resulting polymer can be melted, blown, and molded into bottles of the desired shape. These bottles are then filled, capped, shipped, and sold. You pick up your soda, perhaps from a vending machine on your way to class. Once the bottle is empty, you toss it into the recycling bin and feel good that you've done your part to help protect the environment. Chances are you may never have thought much about what happens to the bottle from there.

Typically, the contents of the recycling bin are sold to a reclaimer—a business specializing in processing plastics. Usually, the material from the bin must be sorted into different types of plastics, and any other materials that may have been thrown into the bin are discarded. Some of this sorting is done by hand, and some takes advantage of differences in density among the various polymers that might be present. The plastics are then crushed to reduce their volume before being shipped for further processing. ◀ⓘ The next step is called reclamation, in which the sorted and compressed plastics are processed into a useable form. In most reclamation processes, the plastic is first chopped into small, uniform-sized flakes. These flakes are washed and dried, then melted and extruded into spaghetti-like strands. These are then cut into smaller pellets, which are sold to manufacturers for use in new products. The most important uses for recycled PET include fiberfill for sleeping bags and coats, fleece fabrics for outdoor wear, carpeting, and industrial strapping.

Some of the crushing and sorting is now done automatically in "reverse vending machines" designed to collect bottles for recycling. ⓘ

You may have noticed that one thing does not appear on that list of uses: new drink bottles. Although such bottles are the dominant source of PET for recycling, only very limited amounts of recycled PET are used to make new bottles. Thus the recycling of PET is far from being a "closed loop" process; large amounts of virgin plastic continue to be used in bottling despite increased collection of used bottles at the consumer level. Why is this? The simplest and shortest answer is economics: bottles can be made from virgin plastic at a lower overall cost. Several factors contribute to this. In many cases, there are legal restrictions on the use of recycled materials for food and beverage containers, due to concerns over possible contamination. Satisfying these regulations adds cost to the overall equation. Degradation of the plastic during repeated recycling processes is another concern. The average chain length of the polymer molecules tends to be somewhat lower after recycling. So if bottles were made from 100% recycled PET, they might have to be thicker and heavier. ◀ⓘ Although progress is being made in increasing the recycled content of drink bottles, most U.S. bottles still contain at least 90% virgin plastic.

Many manufacturers have introduced bottles with new rounder designs that allow the use of thinner plastic. ⓘ

One possible way to achieve a closed loop in which plastic bottles could be recycled back into plastic bottles might be to convert the polymer into monomers and then repolymerize the monomers to produce new plastic. Under what circumstances might such a scheme be feasible? Before we can explore that type of question, we will first need to learn more about thermodynamics.

10-2 Spontaneity

Nature's Arrow

The idea of time travel drives the plot in many science fiction stories. The prospect of moving forward or backward in time and existing in some other era appeals to our imagination in a way that provides fertile ground for authors. But our actual experience is that time marches inexorably from the past toward the future and that this direction is not reversible. In a sense, time is an arrow that points in the direction in which nature is headed. We've seen that large hydrocarbon molecules, such as those in gasoline, can react readily with oxygen to produce carbon dioxide and water. But your experience also tells you that the reverse reaction doesn't happen; water vapor and carbon dioxide are always present in the air, but they never react to produce gasoline. Nature clearly "knows" the correct direction for this process. This sense of the direction of life and our experience of the universe is an important intuition to carry into this chapter. But what gives nature this direction? And how can we convert our intuition into a useful quantitative model for predicting which chemical reactions will actually occur? We'll try to answer these questions by imparting a bit of mathematical rigor to our observations.

Spontaneous Processes

A more formal way of expressing the directionality of nature is to note that our intuition is predicated on the fact that some things "just happen," but others do not. Some processes occur without any outside intervention, and we say that such a process is spontaneous. From a thermodynamic perspective, then, a **spontaneous process** is one that takes place without continuous intervention. The distinction between spontaneous and nonspontaneous reactions may seem obvious, but we'll see that it is not always so.

Some spontaneous processes take place over geological timescales—the formation of petroleum used for plastics feedstocks, for example.

Students often misinterpret the word spontaneous as indicating that a process or reactions will take place quickly. But note that our actual definition does not refer to the speed of the process at all. Some spontaneous processes are very fast, but others occur only on extremely long timescales. We understand that the chemical compounds in some waste materials, like paper, may spontaneously react to decay over time. (This process can be more complicated than a simple chemical reaction, though, because of the involvement of bacteria.) But some spontaneous reactions are so slow that we have a hard time observing them at all. The combustion of diamond is thermodynamically spontaneous, yet we think of diamonds as lasting forever. Other reactions occur quickly once they start, but they don't just start on their own. Gasoline, for example, can sit more or less indefinitely in a can in the garage, in contact with oxygen in the air. Nonetheless, no reaction is observed. Yet upon being mixed with air in the cylinder of your car and ignited by the spark plug, the reaction proceeds until virtually all the gasoline is burned. Is this reaction spontaneous? The answer is yes. Even though the reaction needs a flame or spark to initiate it, once it begins, the reaction continues without any further intervention. This example emphasizes the importance of the phrase "continuous intervention" in our definition. A useful analogy is that of a rock perched precariously on a cliff. If it is nudged over the edge, it proceeds to the bottom. It does not stop midway down, unless, of course, it's a prop in a Roadrunner cartoon!

We mentioned PMMA in Section 7-1 as having been used as one of the first bone cements.

The reactions used to produce many polymers behave much like the combustion of gasoline. Once initiated, the reaction is usually spontaneous and can proceed without further intervention. The production of poly(methyl methacrylate)—Plexiglas, or PMMA—is a good example:

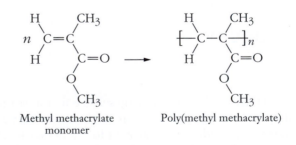

$$n\ H_2C{=}C(CH_3)(C{=}O)OCH_3 \longrightarrow {-}[CH_2{-}C(CH_3)((C{=}O)OCH_3)]_n{-}$$

Methyl methacrylate monomer — Poly(methyl methacrylate)

This reaction occurs via a free radical process, like that described in Section 2-8 for polyethylene. A small trace of an initiator is needed to start the reaction, and then it proceeds until virtually all of the available monomer has been converted into polymer.

But suppose that we wanted to convert the polymer back into monomer. In that case, the necessary reaction is the reverse of the polymerization, and it is not a thermodynamically spontaneous process at ordinary temperatures. We could still drive the reaction backward to produce methyl methacrylate monomer. But we would need to maintain a high temperature, providing enough energy to allow the molecules to go against nature's preferred direction. So what is the role of energy in the directionality of nature?

Enthalpy and Spontaneity

Recall from Chapter 9 that the enthalpy change in a chemical reaction is equal to the heat flow at constant pressure:

$$\Delta H = q_{\text{p}}$$

When ΔH is negative, the reaction is exothermic, whereas a positive value of ΔH points to an endothermic reaction. What can we say about a reaction's spontaneity based on its enthalpy change? If we were to stop and list spontaneous processes that we observe around us and then determine whether those processes are exothermic or endothermic, chances are that a majority would be exothermic. This implies that there is some relationship between enthalpy and spontaneity. The relationship is not exclusive, however. If you think for a moment you should be able to point out some endothermic reactions that obviously occur spontaneously. The melting of an ice cube at room temperature is one simple example. So at this point we might conclude that exothermic reactions seem to be preferred in some way. But clearly there must be things other than energy or enthalpy at work in determining whether or not a process is spontaneous. To develop a way to predict the spontaneity of a reaction, we must first introduce an additional thermodynamic state function—entropy.

10-3 Entropy

As we have just seen, the flow of energy as heat does not indicate whether or not a process will occur spontaneously. So we must also consider another thermodynamic state function, called **entropy (S)**. Historically, entropy was first introduced in considering the efficiency of steam engines, which was an important research topic for scientists and engineers in the 19th century. **Figure 10.2** illustrates the Carnot cycle, an important model system in which a gas goes through a cyclic change in pressure and volume. The steps labeled as 1 and 3 are isothermal processes, meaning they occur at constant temperature.

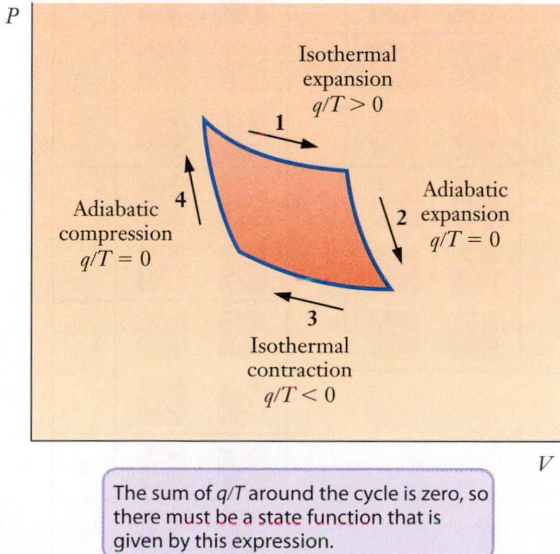

Figure 10.2 ■ In the Carnot cycle, an ideal gas undergoes a series of four processes. Two of these (labeled 1 and 3 in the figure) are isothermal, which means they occur at constant temperature. The other two steps (2 and 4) are adiabatic, meaning that $q = 0$ for those parts of the cycle. Carnot showed that the sum of the quantity q/T for the entire cycle is equal to zero. Because the cycle begins and ends with the system in the same state, this means that there must be a state function equal to q/T. We call this state function entropy.

In Step 1, the gas expands at constant temperature, a process for which q must be positive. In Step 3 the gas is compressed at constant temperature, a process for which q must be negative. It can be shown that q/T for Step 1 and q/T for Step 3 are equal in magnitude and opposite in sign. Steps 2 and 4 are adiabatic processes, in which no heat is exchanged. That means that for Steps 2 and Step 4, $q = 0$, and therefore q/T for those steps must be zero. Combining the four steps, we see that the sum of q/T around the closed path must be zero. It follows that q/T must define a new state function, and we call this new state function entropy. We'll soon see that the changes in the entropy of a system and its surroundings allow us to predict whether or not a process is spontaneous. What is entropy and how can it help us understand the production or recycling of polymers?

A state function does not depend on the system's history. So there can be no change in any state function for a process where the initial and final states are the same.

Probability and Spontaneous Change

We can observe a pattern in many changes that occur in everyday life that are analogous to events at a molecular level. For a familiar example, let's think about autumn. Although the falling leaves may be welcome as a sign of cooler temperatures, they also mean an added chore—raking the leaves into piles. Why can't the leaves simply fall in a pile to begin with? Such an event goes against our intuition because it is so unlikely that we know we'll never see it happen. This macroscopic example with perhaps thousands of leaves serves as a reasonable analogy for molecular systems with Avogadro's number of particles. Let's look at the concept of mathematical probability to solidify our understanding.

The example of leaves not falling in a pile, though perhaps obvious, is somewhat challenging to describe in mathematical terms. To establish a foundation in ideas of probability, let's think instead about rolling dice. If you take just one die and roll it, what is the chance that the roll will be a four? With six possible outcomes the chance is 1 in 6. For two dice, what is the chance the roll will be a pair of fours? This time the counting is a bit more involved, but we can quickly see from **Figure 10.3** that the

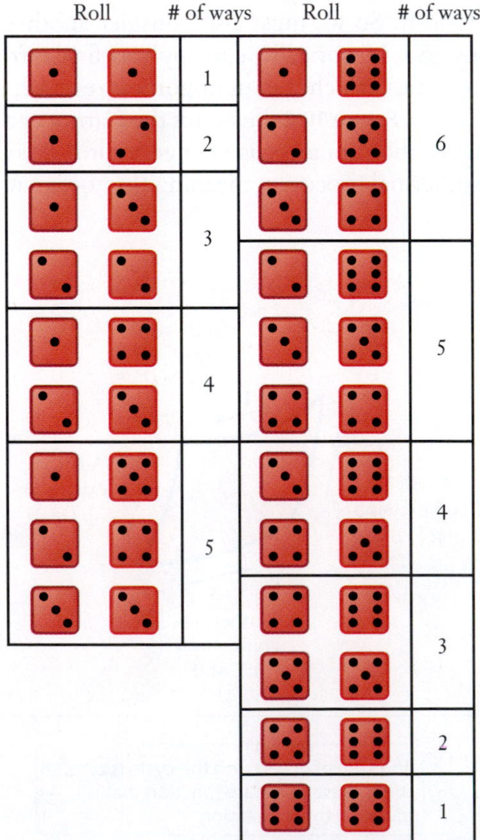

Figure 10.3 ■ The probability of rolling a given total value on a pair of dice depends on the number of different combinations that produce that total. The least likely rolls are 2 and 12, for example, because there is only one possible combination that gives each of those totals. The most likely total is seven because there are six different rolls that add up to that number. (Note that for rolls in which the two individual dice show different values, two possibilities exist. For a total roll of three, for example, the two combinations would be 1, 2 and 2, 1.)

chances are 1 in 36. If a third die is added, the chances of rolling three fours in one throw are 1 in 216. We can see the relationship that is developing for rolling all fours. There is only one way to achieve it, and the probability of that outcome grows smaller according to the following relationship:

$$\text{Probability} = \left(\frac{1}{6}\right)^N, \text{ where } N \text{ is the number of dice being thrown}$$

We should note that this relationship applies for the case at hand, but it is not general. The factor of 1 in the numerator is present because we are looking for a single specific roll (of a number four) on each die and the 6 in the denominator is there because there are six possible rolls for each die. With this relationship, however, we could easily predict that the chances of rolling the same number with five dice in one roll are 1 in 1296. (Note that the chances of rolling a *specified* number on all five dice—say all fours—are 1 in 7776. But if we do not specify in advance which of the six possible numbers we want on all five dice, then there will be six possible outcomes instead of just one.) Our experience with rolling dice is that we expect to have some random assortment of numbers present when five dice are rolled. Why? There are very many ways to obtain a "random" roll. Such a roll occurs far more often precisely because it is more probable.

Our development of the mathematics of probability has two important features. First, it shows that to obtain the probability of a collection of events based on the probability of an individual event we must multiply. This observation becomes important when we consider just how many molecules are involved when we observe something in nature or in the laboratory. Second, the number of ways to make an ordered observation (like all dice turning up four) is smaller than the number of ways to make a more random observation (no particular pattern present in the dice). When we apply these observations to a collection of molecules with $\sim 10^{23}$ particles present, the chances for highly specified arrangements become phenomenally small. To start addressing numbers with more chemical relevance, imagine rolling Avogadro's number of dice. The probability of all of them coming up four is $\left(\frac{1}{6}\right)^{6.02 \times 10^{23}}$. That number is unimaginably small. If we used all the zeros after the decimal point to replace *all* the letters in *all* the books on the planet, we would still have zeros left over!

Definition of Entropy

For large numbers of particles, then, probability favors random arrangements. Using this insight, we can tentatively define entropy as a measure of the randomness or disorder of a system. However, we still have to establish a definition that can be used quantitatively and from a molecular perspective. To do this, we turn to a branch of physical chemistry called **statistical mechanics**, or statistical thermodynamics, where we find a subtle addition to the definition. The probability of events that must be counted is not the number of ways particles can be arranged physically but rather the number of ways in which particles can achieve the same energy. (These two probabilities are often correlated with one another.)

If we recall the Maxwell-Boltzmann distribution of molecular speeds (see Section 5-6), we know that in any gas at room temperature, some particles must move slowly and others quite rapidly. We cannot, however, say precisely which particle is moving very fast or which particle is moving more slowly (**Figure 10.4**). There are a large number of different ways, with different particles assuming the various required speeds, that the sample can have the same total energy and hence the same temperature. In statistical mechanics, the way by which the collection of particles assumes a given energy is associated with a concept called a **microstate**. The number of microstates for a given energy is commonly designated by the uppercase Greek letter omega (Ω), and the entropy (S) of a system is related to the number of microstates by the equation

$$S = k_B \ln \Omega \tag{10.1}$$

Figure 10.4 ■ The Maxwell-Boltzmann distribution tells us the overall collection of molecular speeds but does not specify the speed of any individual particle. Energy exchange during molecular collisions can change the speed of individual molecules without disrupting the overall distribution.

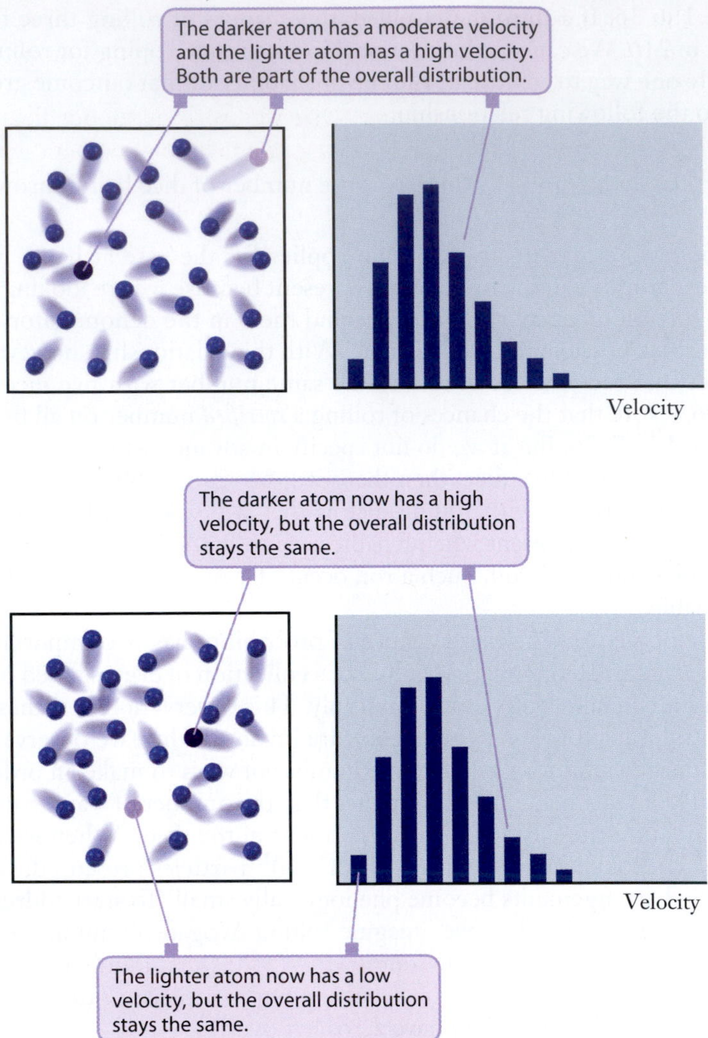

The darker atom has a moderate velocity and the lighter atom has a high velocity. Both are part of the overall distribution.
Velocity
The darker atom now has a high velocity, but the overall distribution stays the same.
Velocity
The lighter atom now has a low velocity, but the overall distribution stays the same.

Here k_B is a numerical constant called the Boltzmann constant. It is not easy to have an intuition about the number of microstates of a system, so this equation is hard to use directly at this stage of your study of chemistry. We'll soon see that we won't need to use it. It is important, however, to realize that as a system becomes "more random," the value of Ω will increase. So, the entropy of a system increases as the system moves toward more random distributions of the particles it contains because such randomness increases the number of microstates.

Judging Entropy Changes in Processes

Although the concept of a microstate is abstract, we can still assert that certain types of changes will lead to increases in entropy (because there are more available microstates). Let's see why this is so. First, consider the melting of a solid to form a liquid. As a solid, the particles are held in place rigidly, so the number of ways they can have a specific energy is limited. When the liquid forms, the movement of particles relative to each other presents a much greater number of ways to achieve a specific energy, so the number of microstates increases and so does the entropy. Similar reasoning can be applied to boiling, when molecules originally confined near each other in a liquid become much more randomly distributed in the gas phase. The increase in random molecular motion corresponds to more microstates, so entropy increases.

The entropy of 1 mole of gas is generally very much greater than that of 1 mole of liquid or solid.

Another possible way to increase the entropy of a system is to increase the number of particles present. Thus, a chemical reaction that generates 2 moles of gas where only 1 mole was present initially will increase the entropy.

Entropy also varies with temperature. One way to think about this is to begin by considering a sample of molecules at some extremely low temperature. In such a sample, it would be very unlikely to have molecules moving at high speeds because they would account for too large a percentage of the available energy. So the speeds of individual molecules would be constrained by the low total energy available. If the system were heated to a higher temperature, though, then a few of the molecules could move at high speeds because there is more total energy available. We have only considered a small portion of the distribution of speeds of the molecules, but already we can see that a hotter system has more ways to distribute its energy. This type of reasoning extends to the whole distribution of speeds, and the important result is that *heating a system increases its entropy*.

What are the implications of entropy for polymer synthesis or recycling? When a polymer is formed, a large number of monomers are converted into a single giant molecule. In most cases, this will lead to a decrease in the entropy of the system because there are more possible ways to arrange the unreacted monomers. (Note that in many polymerization reactions, other small molecules, such as water, may be formed as byproducts. In such cases, the sign of the entropy change may not be obvious.) The fact that polymerization reactions are still spontaneous under appropriate conditions tells us that entropy of the system alone is not the only important consideration. Other factors, such as energy, must favor the formation of the polymer. What about the role of entropy in recycling? As plastics are recycled, there is a possibility that the long polymer chains may be broken. From the viewpoint of entropy, this should be a favorable process. Breaking the chains gives a smaller average molecular size, and the same system of atoms will have more microstates available if more individual molecules are present. Reducing the polymer chain length tends to weaken the mechanical properties of the plastic, though, because the shorter chains do not interact with one another as strongly. So entropy provides a challenge to the recycling process. To recycle polymers without a steady loss in the quality of the material, we need to overcome nature's preference for increasing entropy. And as we'll see by the end of this chapter, this leads very directly to real economic obstacles to recycling.

10-4 The Second Law of Thermodynamics

In Chapter 9, we stressed the importance of thermodynamics in terms of the way human society uses energy. At that time, we noticed that whenever there is an attempt to convert energy from one form into another, some energy is lost or wasted. In other words, not all of the energy potentially available is directed into the desired process. How does this fact arise from thermodynamics? Entropy provides the key to understanding that the loss of useful energy is inevitable.

The Second Law

In considering the energy economy, we alluded to the second law in conjunction with the notion that it is impossible to convert heat completely to work. That is one way to express the **second law of thermodynamics.** Now let's try to understand why this is true. First, consider heat. Heat flows due to random collisions of molecules, and an increase in temperature increases the random motions of molecules. Work, by contrast, requires moving a mass some distance. To yield a net movement, there must be a direction associated with a motion, and that direction implies that there is an order to the motion. Converting heat into work, therefore, is a process that moves from

There are several equivalent ways to state the second law, but all lead to the same interpretations.

random motions toward more ordered ones. We have just seen how this type of change goes against nature's tendency to favor a more probable state (the more random one). How can we connect these ideas with entropy?

To make this connection, we must be careful to realize that changes in the universe involve both the system and its surroundings. If we focus on the system alone, we cannot understand how order is created at all. Yet the synthesis of polymers shows that it does happen, as do everyday situations such as the growth of plants, animals, and people. To express the second law of thermodynamics in terms of entropy, we must focus on the total change in entropy for the universe, ΔS_u:

$$\Delta S_u = \Delta S_{sys} + \Delta S_{surr} \tag{10.2}$$

Because nature always tends to proceed toward a more probable state, we can assert an equivalent form of the second law of thermodynamics: *In any spontaneous process, the total entropy change of the universe is positive,* ($\Delta S_u > 0$). That this statement of the second law is equivalent to our original version is not at all obvious. But remember, energy that is not converted into work (a process that would decrease entropy) is transferred to the surroundings as heat. Thus the entropy of the surroundings increases, and the total entropy change in the system and surroundings is positive.

Implications and Applications

The implications of this expression of the second law are far-reaching for calculating and predicting the outcome of chemical reactions and other processes we might wish to study. We can focus on these implications qualitatively first, by considering a polymerization reaction from the perspective of thermodynamics. Later, we will develop a quantitative approach.

Let's return to the polymerization of methyl methacrylate to form PMMA. The monomer and polymer structures are shown on page 282. Looking at both structures, we see that most of the chemical bonds are unchanged during this reaction. The only exception is the C═C double bond in the monomer, which is converted into two C—C single bonds in the polymer. From our knowledge of bond energies, then, we can say that this reaction must be exothermic. Two C—C single bonds are stronger than a C═C double bond. Because the reaction converts a large number of monomer molecules into a single polymer molecule, we can also predict that the entropy change for the system must be negative. So why is the reaction spontaneous? The fact that the reaction is exothermic means that heat must be released from the system. That same heat must flow into the surroundings. This will lead to an increase in the entropy of the surroundings. As long as that increase in the entropy of the surroundings is larger than the decrease in the entropy of the system, the overall change in entropy for the entire process can still be positive.

We can take this reasoning a little further to begin to understand the role of temperature in determining the spontaneity of a process. The surroundings will absorb an amount of heat equal to $-\Delta H$. But the surroundings represent a very large reservoir, so this heat will not produce a measurable temperature change. This means that the entropy change for the surroundings is given by

$$\Delta S_{surr} = -\frac{\Delta H}{T}$$

The entropy change for the system is just ΔS, and although we don't know its value, we do know that it will be negative. The criterion for a spontaneous polymerization is

$$\Delta S_u = \Delta S + \Delta S_{surr} > 0$$

This will be true as long as the absolute value of ΔS_{surr} is greater than that of ΔS. (Remember that ΔS is negative and ΔS_{surr} is positive.) The magnitudes of ΔS and ΔH are essentially independent of the temperature, but ΔS_{surr} will decrease as the temperature increases. So at some sufficiently high temperature, ΔS_u will no longer be positive, and the reaction will cease to be spontaneous.

This same argument points to a possible route to depolymerization, which might be useful in recycling. Suppose that we raise the temperature high enough that ΔS_u for the polymerization reaction becomes negative. That must mean that ΔS_u for the reverse reaction in which polymer is converted back into monomer must become positive. So if we heat the polymer above some threshold temperature, we should be able to regenerate methyl methacrylate monomer. When PMMA is heated above about 400°C, it is converted into monomer with a very high efficiency. This process, called *thermolysis*, is one example of what is often called *advanced recycling* or *feedstock recycling*. Because the recovered monomer can be purified by distillation or other means, it can then be repolymerized to produce PMMA that is indistinguishable from virgin material. Thermolysis is not practical for most plastics, though, because the monomers themselves often break down or undergo other undesirable reactions at the high temperatures that are required. In the particular case of PMMA, thermolysis is used mainly within manufacturing plants to reclaim scraps that are left behind in the production of items such as automobile taillight lenses.

Purification involves separating the monomer from other substances. Distillation is a common method for separation of chemical mixtures.

10-5 The Third Law of Thermodynamics

Thus far, we have taken a purely qualitative approach to entropy changes and have not attempted to find numerical values for ΔS. To move toward a quantitative view, though, all we really need is to define some reference point with a fixed value of entropy. Then, as long as we can calculate entropy changes, we should be able to obtain values of interest. When we seek to calculate entropy changes for chemical reactions, the most convenient method comes from the **third law of thermodynamics**, which says that *the entropy of a perfect crystal of any pure substance approaches zero as the temperature approaches absolute zero*. An additional implication of the third law is that it is impossible to attain a temperature of absolute zero, although scientists have come very close to that value. Because all substances can be cooled to temperatures near zero (at least in principle), it is possible to evaluate the entropy of 1 mole of any given chemical substance under standard conditions, by determining the change in entropy from 0 K to 298 K at a pressure of 1 atm. This process yields the **standard molar entropy ($S°$)** **Table 10.1** provides $S°$ values for some substances. A more complete list is given in Appendix E.

Table 10.1

Standard molar entropies ($S°$) for selected substances. A much larger listing appears in Appendix E. Values for many compounds can also be found online in the NIST Chemistry WebBook at http://webbook.nist.gov/chemistry.

Compound	$S°$ (J mol^{-1} K^{-1})	Compound	$S°$ (J mol^{-1} K^{-1})
$H_2(g)$	130.6	$CO_2(g)$	213.6
$O_2(g)$	205.0	$C_4H_{10}(g)$	310.03
$H_2O(\ell)$	69.91	$CH_4(g)$	186.2
$H_2O(g)$	188.7	$C_2H_4(g)$	219.5
$NH_3(g)$	192.3	$C_3H_3N(\ell)$	178.91

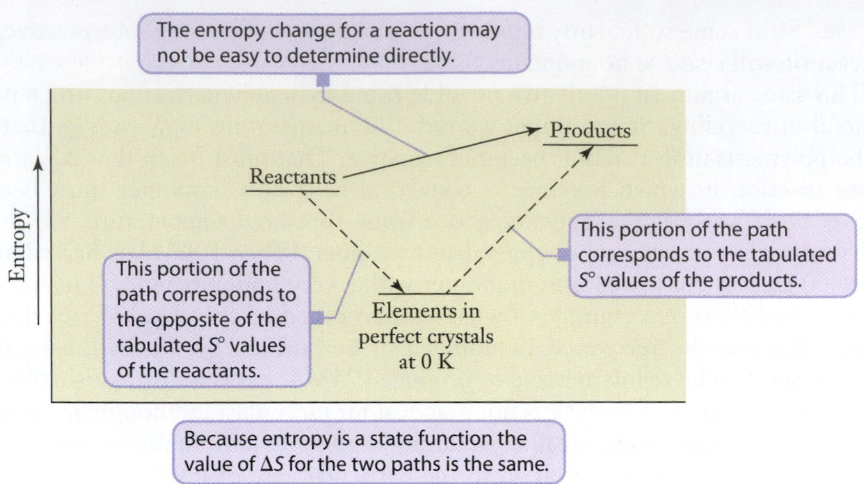
The entropy change for a reaction may not be easy to determine directly.
Products
Reactants
Entropy
This portion of the path corresponds to the tabulated S° values of the products.
This portion of the path corresponds to the opposite of the tabulated S° values of the reactants.
Elements in perfect crystals at 0 K
Because entropy is a state function the value of ΔS for the two paths is the same.

Figure 10.5 ■ Entropy is a state function, so the value of ΔS must be independent of the path taken from reactants to products. When working with tabulated standard molar entropies, we implicitly choose a path in which reactants are converted to pure perfect crystals of elements at 0 K, followed by the reaction of those elements to form the desired products. Such a path is obviously not feasible. Nevertheless, it allows us to obtain accurate values of entropy changes.

Because entropy is a state function and because the third law allows us to obtain a value for the standard molar entropy of any substance, we can derive a useful equation for the entropy change in a reaction. **Figure 10.5** shows how the entropy change in a reaction may be determined by a method that is reminiscent of the way we used heats of formation and Hess's law in Chapter 9:

$$\Delta S° = \sum_i \nu_i S°(\text{product})_i - \sum_j \nu_j S°(\text{reactant})_j \qquad (10.3)$$

As in Equation 9.12, we designate the stoichiometric coefficient by the Greek letter ν, and the subscripts i and j refer to individual product and reactant species. In practice, this equation is used much like Hess's law, as shown in Example Problem 10.1. One subtle difference between this and Equation 9.12 is that here we use *absolute* entropy values ($S°$) in contrast to the enthalpy *changes* for formation reactions ($\Delta H_f°$).

EXAMPLE PROBLEM 10.1

Polymerization reactions are complicated somewhat because they involve very large numbers of molecules. But we can demonstrate the general features of the thermodynamics of polymerization by considering a much smaller model system. Instead of considering the formation of polyethylene, for example, we can begin with the following reaction in which two ethylene molecules combine with hydrogen to form butane:

$$2\ C_2H_4(g) + H_2(g) \rightarrow C_4H_{10}(g)$$

Use data from Table 10.1 to calculate $\Delta S°$ for this reaction.

Strategy Any time we are asked to calculate the standard entropy change for a reaction, our first thought should be to look up values for standard molar entropy and use them in Equation 10.3. The two main things we need to be careful about are (1) to watch the state of the substances (in this case all are gases) and (2) to make sure we don't forget to include the stoichiometric coefficients in our calculations. Unlike heats of formation, the standard molar entropy of an element in its standard state is *not* zero, so we need to be sure to include everything appearing in the equation.

Solution

$$\begin{aligned}\Delta S° &= S°[C_4H_{10}(g)] - 2S°[C_2H_4(g)] - S°[H_2(g)]\\ &= (310.03\ \text{J K}^{-1}) - 2(219.5\ \text{J K}^{-1}) - (130.6\ \text{J K}^{-1}) = -259.6\ \text{J K}^{-1}\end{aligned}$$

Analyze Your Answer You probably don't have an intuition for the size of an entropy change, but at least we can think about the sign. Does it make sense that this reaction has a negative change of entropy? The answer is yes, because we have decreased the amount of gas present as a result of this reaction: 3 moles of gaseous reactants are consumed, and only 1 mole of gaseous product is formed.

Discussion Can this reaction be spontaneous? Yes, as it turns out. The explanation lies in the change in entropy of the surroundings. Because it forms a number of strong C—H and C—C bonds, this reaction is exothermic. The release of heat will increase the entropy of the surroundings. This factor must be large enough to compensate for the decrease in entropy of the system itself.

Check Your Understanding Acrylonitrile (C_3H_3N) is an important monomer for the production of many acrylic fibers. It can be synthesized from propene and ammonia according to the following reaction:

$$2\ C_3H_6(g) + 2\ NH_3(g) + 3\ O_2(g) \rightarrow 2\ C_3H_3N(\ell) + 6\ H_2O(g)$$

Given that $\Delta S°$ for this reaction is -43.22 J/mol, use data from Table 10.1 to calculate the standard molar entropy ($S°$) of $C_3H_6(g)$.

Example Problem 10.1 shows one of the dangers of attempting to use ΔS to predict spontaneity. The need to consider both the system *and* surroundings creates additional work, and it also allows drawing an incorrect conclusion if we forget to include the surroundings. As you might guess, having to account for the surroundings is rarely convenient. After all, by definition, the system is what we are really interested in! Ideally we would be able to use a state function that predicts spontaneity with a single calculation for the system alone. Fortunately, such a state function exists, and so we now proceed to introduce the concept of free energy.

10-6 Gibbs Free Energy

In many ways, thermodynamics is one of the more powerful examples of the application of mathematics to science. If you continue in engineering, there is a good chance that you will eventually take a full course in thermodynamics in which you will develop a much more rigorous mathematical perspective. For now, we simply will introduce the fruits of the mathematical efforts of one of the scientists who invented thermodynamics, J. Willard Gibbs. Gibbs, motivated by an interest in predicting spontaneous processes, defined a new function that was ultimately called the **Gibbs free energy**, G:

An alternative function called the Helmholtz free energy is useful for constant volume conditions. This is less common in engineering applications, so we will not consider it.

$$G = H - TS$$

He realized that changes in this function could predict whether or not a process is spontaneous under conditions of constant pressure and temperature. This constraint is not that demanding because many laboratory processes occur under these conditions (or approximately so). Now, if we focus on the change in the Gibbs free energy for a process at constant temperature, we have the following result:

$$\Delta G = \Delta H - T\Delta S \quad (10.4)$$

From a practical viewpoint, this could be the most important equation in this chapter.

Free Energy and Spontaneous Change

How is this change in free energy connected to the spontaneity of a process? We have already established that the total entropy of the system plus its surroundings must increase for a spontaneous process. We can write that total entropy change as follows:

$$\Delta S_u = \Delta S_{sys} + \Delta S_{surr}$$

We usually refer to the entropy change in the system simply as ΔS, so we'll drop the "sys" subscript:

$$\Delta S_u = \Delta S + \Delta S_{surr}$$

The entropy change of the surroundings is due to the flow of heat, and because T is constant, it can be shown that ΔS_{surr} is just q_{surr}/T. And q_{surr} is $-q_{sys}$. For a process at constant pressure, q_{sys} is ΔH. Combining these ideas, we have now established that

$$\Delta S_{surr} = -\frac{\Delta H}{T}$$

We can insert that into our equation above for ΔS_u:

$$\Delta S_u = \Delta S - \frac{\Delta H}{T}$$

If we multiply both sides of this by T, we should begin to see a connection to ΔG:

$$T\Delta S_u = T\Delta S - \Delta H$$

Comparing this with

$$\Delta G = \Delta H - T\Delta S$$

we can see that

$$\Delta G = -T\Delta S_u$$

In all of these equations, T is the absolute temperature, so its value is always positive. That means that the last equation ensures that the sign of ΔG will always be opposite that of ΔS_u. So if ΔS_u is positive—as it must be for any spontaneous process—then ΔG must be negative.

All of this shows why chemists find ΔG such a useful thermodynamic quantity. First, it is a state function of the system, so it can be calculated fairly easily. Second, the sign of ΔG is sufficient to tell us whether or not a process will be spontaneous.

We can also use Equation 10.4 to help formalize our understanding of the roles of ΔH and ΔS in determining the spontaneity of a given reaction. We have already argued that exothermic reactions, with $\Delta H < 0$, seem to be preferred over endothermic ones and also that reactions where ΔS is positive seem to be preferred. Equation 10.4 shows us that if ΔH is negative and ΔS is positive, then ΔG will *always* be negative. But not all spontaneous processes fit this specific pattern. **Table 10.2**

Table ■ 10.2

The four possible combinations for the signs of ΔH and ΔS

Sign of ΔH	Sign of ΔS	Implications for Spontaneity
−	+	Spontaneous at all temperatures
+	−	Never spontaneous
−	−	Spontaneous only at low temperatures
+	+	Spontaneous only at high temperatures

shows the four possible combinations of signs for ΔH and ΔS. If a process is exothermic, but entropy decreases, we see that the sign on ΔG depends on temperature. As T increases, the relative importance of the $-T\Delta S$ term also increases, so such processes will be spontaneous only at lower temperatures where the ΔH term is dominant. Processes that occur spontaneously only at lower temperatures are sometimes said to be **enthalpy driven** because the enthalpy term is responsible for the negative value of ΔG. ◀ⓘ For endothermic processes, if the entropy of the system decreases, the sign on ΔG will always be positive, and the process is never spontaneous. An endothermic process that increases the entropy, however, may be spontaneous at high temperatures, where the $-T\Delta S$ term becomes larger than the ΔH term of Equation 10.4. These processes are said to be **entropy driven**. The reasoning associated with Table 10.2 can be used to understand the nature of phase changes, as noted in Example Problem 10.2.

The word "driven" is used here to imply that either the entropy or enthalpy term is dominant in determining the sign of ΔG. There is no force due to enthalpy or entropy responsible for driving spontaneous changes. ⓘ

EXAMPLE PROBLEM 10.2

Use the signs of ΔH and ΔS to explain why ice spontaneously melts at room temperature but not outside on a freezing winter day.

Strategy This problem calls for the same type of reasoning used to construct Table 10.2. We must determine whether the process is endothermic or exothermic and whether it increases or decreases the entropy of the system. Then, by considering the signs of ΔH and ΔS in conjunction with Equation 10.4, we can attempt to explain the behavior described.

Solution Melting is an endothermic process ($\Delta H > 0$) because we must heat the system to effect the change. It is also a process that increases entropy because molecules formerly held in place in a solid have greater freedom of motion in a liquid and are therefore less ordered. Thus both ΔH and ΔS have positive values, and ΔG will be negative at high temperatures, where the $T\Delta S$ term is larger. So melting tends to occur at higher temperatures. At least for water, we know that room temperature is sufficient to make melting spontaneous. At low temperatures, the ΔH term is more important, so on a freezing day, the sign of ΔG is positive. Melting is not spontaneous and therefore is not observed. At the freezing point, ΔG is equal to zero, and ice and water can coexist in any proportions.

Discussion The result that ice melts if it is warm but not if it is cold is intuitively obvious. But this problem emphasizes the importance of gaining a qualitative understanding of entropy changes.

Check Your Understanding At what temperatures would you expect a gas to condense? Explain why the free energy change is negative for this process at these temperatures.

If the actual values of ΔH and ΔS are known, then the same type of argument used in Example Problem 10.2 can also be extended to obtain quantitative information, as shown in Example Problem 10.3.

EXAMPLE PROBLEM 10.3

The heat of fusion of crystalline polyethylene is approximately 7.7 kJ/mol, and the corresponding entropy change for melting is 19 J mol^{-1} K^{-1}. Use these data to estimate the melting point of polyethylene.

Strategy Because ΔH and ΔS are both positive, we know that ΔG must be positive at low temperatures and negative at higher temperatures. The melting point marks the dividing line between low and high temperatures. So at the melting point itself, ΔG

must be equal to zero. We can therefore start with Equation 10.4 and set $\Delta G = 0$. That leaves T as the only unknown, so we can solve for the desired melting point.

Solution We begin with Equation 10.4 and set $\Delta G = 0$. (Adding a subscript "m" to the temperature will help remind us that this equation is only valid at the melting temperature.)

$$\Delta G = \Delta H - T_m \Delta S = 0$$

Rearranging this gives us

$$\Delta H = T_m \Delta S$$

Because ΔH and ΔS are known, solving for T_m is simple. (We do need to take care in handling the units, though, because ΔH is in kJ and ΔS is in J.)

$$T_m = \frac{\Delta H}{\Delta S} = \frac{7700 \text{ J mol}^{-1}}{19 \text{ J mol}^{-1} \text{ K}^{-1}} = 410 \text{ K}$$

Analyze Your Answer The result seems reasonable: 410 K is 140°C, or 280°F, which is a plausible melting point for a plastic. Note that if we had been sloppy with units and forgotten to convert ΔH from kJ/mol into J/mol, we would have gotten an answer of 0.4 K. Such an extremely low temperature should then have tipped us off to the mistake.

Discussion In our calculation, we have assumed that ΔH and ΔS do not change with temperature. This is often a reasonable assumption, but it is not strictly true. So the value we find should be considered as an estimate of the actual melting point. You may recall from our earlier discussions of polymers that the molecules in a given polymer usually do not all have the same chain length. Thus there is a bit of difficulty in specifying quantities per mole. The values used here were actually measured per mole of monomer unit.

Check Your Understanding Poly(tetrafluoroethylene) melts at approximately 327°C. If the heat of fusion is $\Delta H_{fus} = 5.28 \text{ kJ mol}^{-1}$, what is the molar entropy change of fusion (ΔS_{fus})?

Free Energy and Work

By now, you are probably wondering about the name. Why is it called *free energy*? The change in Gibbs free energy can be shown to be equal to the maximum useful work that can be done by the system

$$\Delta G = -w_{max} \tag{10.5}$$

We must include a minus sign in this expression to be consistent with the convention that w is the work done *on the system*. This relationship suggests that ΔG tells how much energy is "free" or available to do something useful. For an engineer interested in making practical use of a chemical reaction, the implications should be clear.

Keep in mind that work is not a state function. So the maximum work will be realized only if we carry out the process by a specific path. In this case, the requirement is that the change is carried out along a **reversible** path. This means that the system is near equilibrium, and a small incremental change in a variable will bring the system back to its initial state. Maximum work is possible only for reversible processes. Systems that are far from equilibrium usually undergo irreversible changes.

In an **irreversible** change, a small incremental change of any variable does not restore the initial state. The amount of work available in an irreversible change is always less than the maximum work. So, although the free energy change can be used to establish an upper bound to the amount of work obtained from a given process, the actual work produced in any real application may be considerably less. Reactant mixtures such as those for combustion reactions are generally very far from equilibrium. Systems that are far from equilibrium often change rapidly, and rapid changes tend to be irreversible.

Calculations for irreversible changes are challenging, and we will not attempt any such exercises.

To use these free energy concepts quantitatively, of course, we still need a simple method for obtaining accurate values of free energy changes.

10-7 Free Energy and Chemical Reactions

Free energy is a state function, and so the value of the free energy of a system depends on specific variables such as concentration or pressure. As usual, to provide consistent comparison, we define a standard state as 1 atmosphere of pressure and concentrations of solutions of 1 M. The free energy change under these conditions is the **standard Gibbs free energy change**, $\Delta G°$. Although it is feasible to calculate this value from $\Delta H°$ and $S°$ at a given temperature using Equation 10.4, the most convenient means for calculating the change in free energy for many reactions is to use a formulation analogous to Hess's law for enthalpy changes:

Hess's law relating enthalpy changes to heats of formation was given in Equation 9.12, page 272.

$$\Delta G° = \sum_i \nu_i \Delta G_f°(\text{product})_i - \sum_j \nu_j \Delta G_f°(\text{reactant})_j \qquad (10.6)$$

Again, in Equation 10.6 we use the concept of the formation reaction from Chapter 9. Tabulated values of free energies of formation for a few substances are provided in **Table 10.3**, and a more extensive list appears in Appendix E. Note that $\Delta G_f°$ is zero for elements in their standard states, for the same reason that their heats of formation are zero. Because the formation reaction uses elements in their standard states to define the reactants, a formation reaction of an element in its standard state would have the same chemical species as both reactant and product, and this process would clearly have no change in any thermodynamic state function. We can use Equation 10.6 to calculate standard free energy changes, as demonstrated in Example Problem 10.4.

Table 10.3

Values of the free energy change of formation, $\Delta G_f°$, for selected compounds. A much larger list appears in Appendix E.

Substance	$\Delta G_f°$ (kJ mol^{-1})	Substance	$\Delta G_f°$ (kJ mol^{-1})
$H_2(g)$	0	$CO_2(g)$	−394.4
$O_2(g)$	0	$C_4H_{10}(g)$	−15.71
$H_2O(\ell)$	−237.2	$CH_4(g)$	−50.75
$H_2O(g)$	−228.6	$C_2H_4(g)$	68.12
$NH_3(g)$	−16.5	$C_3H_6(g)$	62.75

EXAMPLE PROBLEM 10.4

In Example Problem 10.1, we considered an addition reaction involving two ethylene molecules and found that the entropy change was negative. We suggested at the time that this reaction would still be spontaneous because it is strongly exothermic. Confirm this by calculating the standard free energy change for the same reaction using values from Table 10.3.

$$2\ C_2H_4(g) + H_2(g) \rightarrow C_4H_{10}(g)$$

Strategy Any problem that requests a calculation of a state function using values from a table generally means we'll be using a formulation like Equation 10.6. Free energy is the third such state function we have encountered that can be treated in this way. If we're careful to note the states of the molecules present and watch the stoichiometric coefficients, this type of problem can be solved readily.

Solution

$$\Delta G° = \Delta G_f°[C_4H_{10}(g)] - 2\Delta G_f°[C_2H_4(g)] - \Delta G_f°[H_2(g)]$$
$$= (-15.71\ \text{kJ}) - 2(68.12\ \text{kJ}) - (0) = -151.95\ \text{kJ}$$

Analyze Your Answer This value is negative, indicating that this reaction would be spontaneous under standard conditions, as we had said. Many reactions have free energy changes on the order of 10^2 kJ/mol, so our result seems plausible.

Discussion It's worth pointing out again, though, that this does *not* mean that the reaction will occur readily if we mix the reactants. Simply because a reaction has a negative value for $\Delta G°$ does not mean we will observe a rapid conversion. In some cases, the rate of the reaction is so slow that, despite the thermodynamic indication of spontaneity, the reaction is not observed. Thermodynamics does not tell us how rapidly a spontaneous process will take place.

Check Your Understanding For the reaction shown below, $\Delta G° = -1092.3$ kJ. Find $\Delta G_f°$ for liquid acrylonitrile (C_3H_3N).

$$2\ C_3H_6(g) + 2\ NH_3(g) + 3\ O_2(g) \rightarrow 2\ C_3H_3N(\ell) + 6\ H_2O(g)$$

Implications of $\Delta G°$ for a Reaction

Now that we have a method for calculating the standard free energy change, what does this value tell us? One answer is that the free energy change tells us the maximum useful work that can be obtained from a reaction. But the free energy change also has very important implications for the many chemical reactions that can be made to run in either direction. Earlier in this chapter, we considered the thermal depolymerization (or thermolysis) of PMMA as a possible alternative to mechanical recycling. Now, we can use free energy changes to explore that idea quantitatively. The polymerization of methyl methacrylate has $\Delta H° = -56$ kJ and $\Delta S° = -117$ J/K. We can use these values in Equation 10.4 to find $\Delta G° = -21$ kJ. (We use the standard temperature of 298 K, which gives us the standard free energy change.) The negative value tells us that the formation of the polymer is spontaneous at 298 K. This, of course, means that the reverse reaction in which PMMA is broken down into monomers cannot be spontaneous at this temperature. Because the depolymerization amounts to running the polymerization reaction backward, we can conclude that it has $\Delta G° = +21$ kJ.

Polymerization: methyl methacrylate monomer → PMMA polymer

$$\Delta G° = -21 \text{ kJ}$$

Depolymerization: PMMA polymer → methyl methacrylate monomer

$$\Delta G° = +21 \text{ kJ}$$

The fact that the numerical values of these free energy changes are relatively small hints at the fact that a relatively small shift in temperature might flip the signs, making depolymerization the thermodynamically preferred process. We can quantify this idea, too. Using the same sort of calculation as we did in Example Problem 10.3, we can find the temperature at which $\Delta G°$ will equal zero:

$$T = \frac{\Delta H}{\Delta S} = \frac{-56 \text{ kJ}}{-0.117 \text{ kJ K}^{-1}} = 480 \text{ K}$$

Above this temperature, entropy wins out over enthalpy, and the reaction will be driven backward, toward the monomer. The fact that this temperature threshold is reasonably low is one reason why depolymerization is a viable recycling strategy for PMMA. Most polymerization reactions have $\Delta S°$ values around −100 J/K, but many are more strongly exothermic than that for PMMA. Looking at the equation above, we can see that this leads to a higher temperature requirement for depolymerization. So thermolysis is less feasible for two reasons. First, the need for higher temperatures implies higher cost. Second, the higher temperatures increase the likelihood that additional chemical reactions, such as breakdown of monomers into other compounds, would compete with thermolysis.

10-8 The Economics of Recycling

According to the Container Recycling Institute, about 50% of all aluminum beverage cans sold in the United States in 2010 were recycled. But in the same year, only about 29% of PET bottles were recycled. It seems fair to assume that individual consumers are no more anxious to recycle aluminum cans than plastic bottles. So there must be a real cause for the difference between those two numbers, and that cause lies in economics. When it comes right down to it, the business of recycling is all about trying to sell your trash. And as you might guess, selling trash can be a difficult business. In this section we'll take a look at some of the factors that make aluminum recycling so much more attractive than plastic recycling.

Recycling rates vary significantly from place to place, and are significantly higher in states with deposit laws. The numbers cited here are national averages.

We'll begin by looking at the recycling of aluminum, which now accounts for nearly all beverage cans sold in the United States. The chemistry of aluminum makes it an excellent target for recycling. Recall that in Chapter 1 we pointed out how difficult it is to obtain pure aluminum from ores like bauxite. Aluminum reacts readily with oxygen, forming strong chemical bonds that are not easily disrupted. Extracting aluminum from its ores requires very high temperatures and therefore consumes a lot of energy. When aluminum cans are recycled, however, the paint and other coatings used can be removed relatively easily, allowing the underlying aluminum to be melted down and reprocessed into a new can. Current industry estimates are that making four new cans from recycled aluminum uses the same amount of energy as producing one can from raw aluminum ore. This provides strong economic incentives throughout the entire process. First, the cost savings mean that the beverage industry has a strong motivation to encourage recycling and to seek recycled aluminum to produce new cans. This, in turn, means that communities and private companies in the recycling business are assured of finding a good market for any aluminum cans they can collect. This is an important concern because there will always be costs for collecting

Aluminum foil made entirely from recycled aluminum has also been introduced recently.

recyclables. Other positive factors also exist in the recycling equation for aluminum. The fact that virtually all beverage cans in use are made of aluminum means there is no need to separate collected cans into different types. And because aluminum cans are so easily crushed, the collected material can be compressed so that storage and transportation are easier and less expensive. All in all, aluminum is an ideal candidate for recycling.

Now let's contrast that with plastic recycling. We can begin by thinking about the production costs of virgin polymers because these costs will set the standard against which the price of recycled plastics will be measured. Feedstocks for virtually all commercial polymerization reactions have their roots in petroleum. Therefore, the cost of raw materials for synthesizing most plastics is linked to the price of oil. Oil is a complex mixture of compounds, and our new understanding of entropy should make clear that such a mixture is unlikely to separate spontaneously into its various components. Various separation and purification methods are used to obtain the needed monomer molecules. Most of these schemes involve cracking and distillation, which require heating the crude oil until various components boil off and can be reclaimed from the vapor phase. This need for heating means that there is also an energy cost for producing the feedstock for polymerization. Once we have a supply of the appropriate monomer (and any other reagents that may be needed), we will have to pay to transport them to the plant where the polymer will be produced. That cost can be minimized, of course, if the plant where the polymer will be made is located close to a refinery where the oil is processed.

Many polymerization reactions are spontaneous under ordinary conditions, but as we've learned, that doesn't necessarily mean that those reactions are fast. If we are in the business of making plastics, we will probably not be satisfied if our polymer production requires days or even years. Most reactions run faster when heated, though, as we'll see when we examine chemical kinetics in Chapter 11. So, we will probably want to carry out our polymerization at higher temperatures. This will add a further energy cost to the bill for producing our polymer.

So how will this compare to the cost of recycling plastics? Just as for aluminum cans, there will be some costs for collecting the bottles to be recycled. But the fact that plastic bottles are made from a wide array of polymers offers some added complications here. The various types of polymers usually must be separated before they can be processed. Consumers can be encouraged to do some of this separation based on the recycling codes commonly found on bottles (see **Table 10.4**). But even different colors of the same type of plastic may be incompatible, and inevitably there will be some mixing of bottle types in the collection bins. So the recycler must expect to have to sort the materials before they can be sold. This is most often done by hand, although sometimes flotation and separation based on density may be possible. No matter the method, though, this separation adds cost to the overall recycling effort. Once bottles of a particular type have been separated, they are ready to be processed. This would typically include crushing, washing, and melting the plastic before forming it into pellets to be resold. Again, each of those steps adds some cost.

Although consumers may toss a bottle into a recycling bin out of a sense of doing a good deed, a company that might be interested in buying the recycled plastic pellets is more likely to have its eye on the bottom line. For the resulting recycled pellets to be attractive, they will have to be priced competitively with virgin polymer. And here things are not nearly as favorable as for aluminum. The production of most plastics from raw feedstocks is generally less expensive than the processing of aluminum from its ore. So, in economic terms, the standard to which recycled plastic will be held is more demanding. There is also an important difference in the ability of recycling to produce high-quality materials. Once the aluminum has been reclaimed from a recycled can and melted down, it is indistinguishable from new aluminum freshly extracted from ore. But for plastics, the recycling process leads to degradation of the polymer. This is really just an effect of the second law: the entropy of a polymer molecule will generally increase if the long chain is broken into shorter pieces. Chain

Table ■ 10.4

Symbols, structures, sources, and uses for various recycled plastics

Symbol	Polymer	Repeat Unit	Sources	Recycled Products
1 PETE	Polyethylene terephthalate		Soda bottles, peanut butter jars, vegetable oil bottles	Fiber, tote bags, clothing, film and sheet, food and beverage containers, carpet, fleece wear
2 HDPE	High-density polyethylene		Milk and water jugs, juice and bleach bottles	Bottles for laundry detergent, shampoo, and motor oil; pipe, buckets, crates, flower pots, garden edging, film and sheet, recycling bins, benches, dog houses, plastic lumber
3 PVC	Polyvinyl chloride	Cl	Detergent / cleanser bottles, pipes	Packaging, loose-leaf binders, decking, paneling, gutters, mud flaps, film and sheet, floor tiles and mats, resilient flooring, electrical boxes, cables, traffic cones, garden hose
4 LDPE	Low-density polyethylene		Six-pack rings, bread bags, sandwich bags	Shipping envelopes, garbage can liners, floor tile, furniture, film and sheet, compost bins, paneling, trash cans, landscape timber, lumber
5 PP	Polypropylene	CH_3	Margarine tubs, straws, screw-on lids	Automobile battery cases, signal lights, battery cables, brooms, brushes, ice scrapers, oil funnels, bicycle racks, rakes, bins, pallets, sheeting, trays
6 PS	Polystyrene		Styrofoam, packing peanuts, egg cartons, foam cups	Thermometers, light switch plates, thermal insulation, egg cartons, vents, desk trays, rulers, license plate frames, foam packing, foam plates, cups, utensils
7 Other	Miscellaneous and multilayer plastics	N/A	Squeezable ketchup and syrup bottles	Bottles, plastic lumber

lengths in recycled plastics invariably are shorter than in virgin materials, and this can raise concerns as to whether the recycled material offers sufficient strength to satisfy design requirements. On balance, then, many businesses find it less expensive to use virgin plastics rather than recycled polymers.

Recall that ordinary samples contain a range of polymer chain lengths. The degradation associated with recycling shifts this distribution toward shorter chains.

Business decisions are driven by economics, of course, so changes in the costs associated with producing new plastic or recycling old plastic could shift that balance. One factor that could alter the economics of recycling plastics would be changes in public policy that might force businesses to consider environmental costs in their economic models. Recent studies suggest that only about half of the plastic produced each year can be accounted for as having been recycled or buried in landfills. Up to 150 million tons of plastic per year are not accounted for. Much of this plastic ends up in the oceans, where the effects on marine ecosystems can be significant and harmful. Ocean currents form large gyres, or vortex-like flows over distances of many kilometers. These gyres tend to concentrate the plastics that enter the ocean into relatively small regions, as shown in **Figure 10.6**. Under the mechanical pounding of ocean waves, plastics often break down into very small particles, with sizes on the order of a few microns. Small sea animals mistake these small bits of plastic for plankton and eat them. Once the plastic is consumed, it works its way up the food chain toward larger animals. The resulting accumulation of chemical substances in the affected animals can be substantial. Seabirds that ingest plastic waste, for example, have been found to have concentrations of polychlorinated biphenyls, or PCBs, in their bodies that are three times higher than in birds that have not eaten plastic.

Because oceans and the marine ecosystems they support are a shared global resource, ecological issues related to them result in international discussions. Such discussions are often time consuming, and action is more deliberate than might occur within just a single country. Nonetheless, it's entirely conceivable that the long-term economics of plastic production and recycling will one day include international-level policy directives. Engineers and scientists will be called upon to help limit the damages that result from plastics finding their way into the ocean, and the economic considerations of plastics recycling may become more strongly aligned with the environmental arguments that often come to mind when recycling programs are proposed.

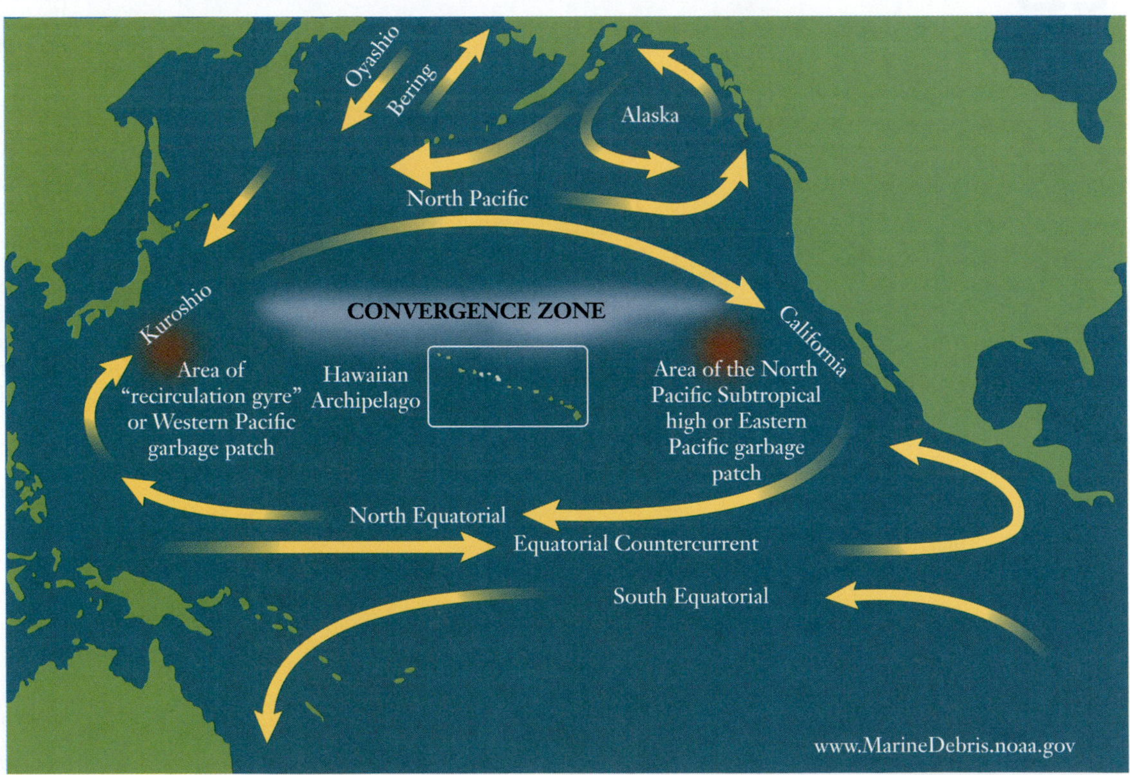

Figure 10.6 ■ This illustration shows a simplified version of currents in the Pacific Ocean. The areas shown in red have been referred to as "garbage patches" because large amounts of debris tend to accumulate there. ◀ⓘ The actual size and location of these garbage patches varies with the seasons and with global weather patterns.

Similar garbage patches have recently been observed in the Great Lakes. ⓘ

FOCUS ON PROBLEM SOLVING

Question Suppose that you need to know the melting point of an oil for the design of a microwaveable food package. You have no sample of the oil, so you can't measure the melting point. But you can find tabulated thermodynamic data for the oil in both solid and liquid phases. What specific values would you need to look up in the table, and how would you use them to determine the melting point of the oil?

Strategy This question asks us to think about how a phase change is described thermodynamically, and how we can use tabulated data along with the equations to get the melting point. It will help if we think of melting the oil as a simple reaction:

$$\text{Oil(s)} \rightarrow \text{Oil}(\ell)$$

Solution From Example Problem 10.3, we know that the melting point can be estimated using the ratio of $\Delta H/\Delta S$. If we have data for both the solid and the liquid forms of the oil, we can determine the enthalpy of fusion (ΔH_{fus}) by subtracting the heat of formation of the solid from that of the liquid. Similarly, we can get the entropy change for fusion (ΔS_{fus}) by subtracting the absolute entropy of the solid from that of the liquid. Taking the ratio of these two numbers will tell us the approximate melting temperature.

SUMMARY

Thermodynamics can provide more information than just the amount of energy released or absorbed in a chemical or physical process, as discussed in Chapter 9. By defining additional state functions, we can also determine the spontaneous direction of change of any system—a very powerful tool for understanding chemistry and an important factor in many engineering designs.

To predict spontaneity, we introduced two new concepts and state functions: entropy and free energy. We can define entropy in either of two ways: as the ratio of the heat flow to the temperature or as a measure of the number of ways that a system can have the same energy. This latter definition, for practical purposes, is a measure of the extent of the randomness of the system at the atomic and molecular level. A more random system will have more ways to distribute its energy, so entropy increases with the extent of randomness.

The second law of thermodynamics tells us that spontaneous changes always increase the entropy of the universe. But calculations that must consider the universe are rarely practical. So we define another state function, called Gibbs free energy, so that its change predicts spontaneity. The change in Gibbs free energy, given by $\Delta G = \Delta H - T\Delta S$, is always negative for a spontaneous process. Both entropy changes and free energy changes can be calculated for many chemical and physical processes by using tabulated thermodynamic data.

KEY TERMS

enthalpy driven (10-6)
entropy (S) (10-3)
entropy driven (10-6)
Gibbs free energy (G) (10-6)
irreversible (10-6)
microstate (10-3)
reversible (10-6)
second law of thermodynamics (10-4)
spontaneous process (10-2)
standard Gibbs free energy change ($\Delta G°$) (10-7)
standard molar entropy ($S°$) (10-5)
statistical mechanics (10-3)
third law of thermodynamics (10-5)

PROBLEMS AND EXERCISES

INSIGHT INTO Recycling of Plastics

10.1 "Reduce, reuse, recycle" is a common slogan among environmentalists, and the order of the three words indicates their perceived relative benefits. Why is recycling the least desirable of these three approaches to waste reduction?

10.2 Is the recycling of most plastics primarily a chemical or a physical process? Explain and defend your choice.

10.3 List some consumer products made from recycled PET.

10.4 Why is recycled PET rarely used to make new soft drink bottles?

10.5 Use the web to research a company that specializes in the recycling of plastics. Does the material on their website emphasize environmental, scientific, or economic concerns? Write a brief essay on the company's positions, explaining how they fit with the ideas expressed in this chapter.

10.6 Use the web to learn how many pounds of plastics are recycled in your area each year. How has this value changed during the past decade?

Spontaneity

10.7 On the basis of your experience, predict which of the following reactions are spontaneous.

(a) $CO_2(s) \rightarrow CO_2(g)$ at 25°C

(b) $NaCl(s) \rightarrow NaCl(\ell)$ at 25°C

(c) $2\ NaCl(s) \rightarrow 2\ Na(s) + Cl_2(g)$

(d) $CO_2(g) \rightarrow C(s) + O_2(g)$

10.8 In the thermodynamic definition of a spontaneous process, why is it important that the phrase "continuous intervention" be used rather than just "intervention?"

10.9 If the combustion of butane is spontaneous, how can you carry a butane lighter safely in your pocket or purse?

10.10 Identify each of the processes listed as spontaneous or nonspontaneous. For each nonspontaneous process, describe the corresponding spontaneous process in the opposite direction.

(a) A group of cheerleaders builds a human pyramid.

(b) Table salt dissolves in water.

(c) A cup of cold coffee in a room becomes steaming hot.

(d) Water molecules in the air are converted to hydrogen and oxygen gases.

(e) A person peels an orange, and you smell it from across the room.

10.11 Identify each of the processes listed as spontaneous or nonspontaneous. For each nonspontaneous process, describe the corresponding spontaneous process in the opposite direction.

(a) Oxygen molecules dissociate to form oxygen atoms.

(b) A tray of water is placed in the sun on a warm day and freezes.

(c) A solution of salt water forms a layer of acid on top of a layer of base.

(d) Silver nitrate is added to a solution of sodium chloride and a precipitate forms.

(e) Sulfuric acid sitting in a beaker turns into water by giving off gaseous SO_3.

10.12 Athletic trainers use instant ice packs that can be cooled quickly on demand. Squeezing the pack breaks an inner container, allowing two components to mix and react. This reaction makes the pack become cold. Describe the heat flow for this spontaneous process.

10.13 Are any of the following exothermic processes not spontaneous under any circumstances?

(a) Snow forms from liquid water.

(b) Liquid water condenses from water vapor.

(c) Fossil fuels burn to form carbon dioxide and water.

(d) Monomers react to form a polymer.

10.14 Enthalpy changes often help predict whether or not a process will be spontaneous. What type of reaction is more likely to be spontaneous: an exothermic or an endothermic one? Provide two examples that support your assertion and one counterexample.

10.15 When a fossil fuel burns, is that fossil fuel the system? Explain your answer.

10.16 Murphy's law is a whimsical rule that says that anything that can go wrong will go wrong. But in an article in the *Journal of Chemical Education*, Frank Lambert writes, "Murphy's law is a fraud." He also writes, "The second law of thermodynamics is time's arrow, but chemical kinetics is its clock." Read Lambert's article (*J. Chem. Ed.*, 74(8), 1997, p. 947), and write an essay explaining, in the context of the latter quotation, why Lambert claims that Murphy's law is a fraud. (For more of Professor Lambert's unique insights into thermodynamics, see his website at http://secondlaw.oxy.edu/)

10.17

Humpty Dumpty sat on a wall,
Humpty Dumpty had a great fall.
All the King's horses and all the King's men
Couldn't put Humpty together again.

In Lewis Carroll's *Through the Looking Glass*, Alice encounters Humpty Dumpty, a human-sized egg sitting on a wall. Alice, who is familiar with the nursery rhyme, asks anxiously, "Don't you think you'd be safer on the ground? That wall is so narrow." Humpty, an egg with an attitude, growls, "Of course I don't think so. Why if I ever *did* fall off—which there's no chance of—but if I did . . . the King has promised me—with his very own mouth—(that) they'd pick me up again in a minute, they would!"

Write a paragraph in the voice of seven-and-a-half-year-old Alice, explaining to Humpty in the context of this chapter **(a)** the probability that Humpty will fall off the wall and **(b)** the probability that the King's horses and men will be able to put him back together again.

10.18 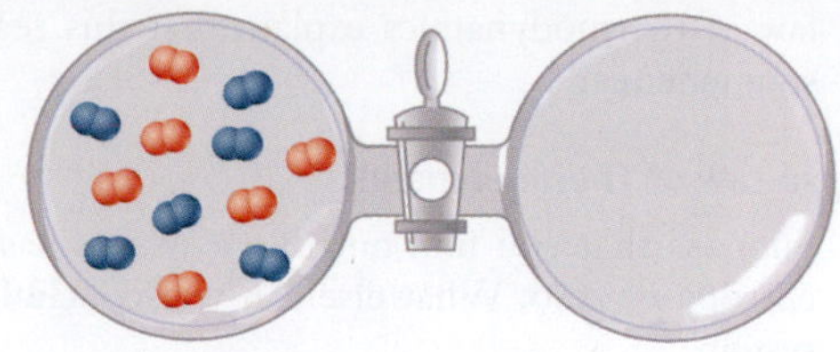

The vessel on the left contains a mixture of oxygen and nitrogen at atmospheric pressure. The vessel on the right is evacuated.

(a) Describe what will happen when the stopcock is opened.

(b) If you could see the individual molecules, what would you observe after a period of time has passed?

(c) Explain your answers to **(a)** and **(b)** in terms of probabilities.

(d) What is the probability that at any one moment all the oxygen molecules will be in one vessel and all the nitrogen molecules will be in the other? Explain.

Entropy

10.19 What observation about the Carnot engine led Carnot to propose the existence of a new state function?

10.20 Some games include dice with more than six sides. If you roll two eight-sided dice, with faces numbered one through eight, what is the probability of rolling two eights? What is the most probable roll?

10.21 How does probability relate to spontaneity?

10.22 Define the concept of a microstate. How is this concept related to the order or disorder of a system?

10.23 For each pair of items, tell which has the higher entropy and explain why.

(a) Item 1, a sample of solid CO_2 at $-78°C$, or item 2, CO_2 vapor at 0°C

(b) Item 1, solid sugar, or item 2, the same sugar dissolved in a cup of tea

(c) Item 1, a 100-mL sample of pure water and a 100-mL sample of pure alcohol, or item 2, the same samples of water and alcohol after they have been poured together and stirred

10.24 When ice melts, its volume decreases. Despite this fact, the entropy of the system increases. Explain **(a)** why the entropy increases and **(b)** why under most circumstances, a decrease in volume results in an entropy decrease.

10.25 If a sample of air were separated into nitrogen and oxygen molecules (ignoring other gases present), what would be the sign of ΔS for this process? Explain your answer.

10.26 For each process, tell whether the entropy change of the system is positive or negative. **(a)** A glassblower heats glass (the system) to its softening temperature. **(b)** A teaspoon of sugar dissolves in a cup of coffee. (The system consists of both sugar and coffee.) **(c)** Calcium carbonate precipitates out of water in a cave to form stalactites and stalagmites. (Consider only the calcium carbonate to be the system.)

10.27 Without doing a calculation, predict whether the entropy change will be positive or negative when each of the following reactions occurs in the direction it is written.

(a) $CH_3OH(\ell) + 3/2\ O_2(g) \rightarrow CO_2(g) + 2\ H_2O(g)$

(b) $Br_2(\ell) + H_2(g) \rightarrow 2\ HBr(g)$

(c) $Na(s) + 1/2\ F_2(g) \rightarrow NaF(s)$

(d) $CO_2(g) + 2\ H_2(g) \rightarrow CH_3OH(\ell)$

(e) $2\ NH_3(g) \rightarrow N_2(g) + 3\ H_2(g)$

10.28 For the following chemical reactions, predict the sign of ΔS for the system. (Note that this should not require any detailed calculations.)

(a) $Fe(s) + 2\ HCl(g) \rightarrow FeCl_2(s) + H_2(g)$

(b) $3\ NO_2(g) + H_2O(\ell) \rightarrow 2\ HNO_3(\ell) + NO(g)$

(c) $2\ K(s) + Cl_2(g) \rightarrow 2\ KCl(s)$

(d) $Cl_2(g) + 2\ NO(g) \rightarrow 2\ NOCl(g)$

(e) $SiCl_4(g) \rightarrow Si(s) + 2\ Cl_2(g)$

10.29 In many ways, a leaf is an example of exquisite order. So how can it form spontaneously in nature? What natural process shows that the order found in a leaf is only temporary?

The next four questions relate to the following paragraph (Frank L. Lambert, *Journal of Chemical Education*, 76(10), 1999, 1385).

"The movement of macro objects from one location to another by an external agent involves no change in the objects' physical (thermodynamic) entropy. The agent of movement undergoes a thermodynamic entropy increase in the process."

10.30 A student opens a stack of new playing cards and shuffles them. In light of the paragraph above, have the cards increased in entropy? Explain your answer in terms of thermodynamics. Explain why the agent (the shuffler) undergoes an increase in entropy.

10.31 An explosion brings down an old building, leaving behind a pile of rubble. Does this cause a thermodynamic entropy increase? If so, where? Write a paragraph explaining your reasoning.

10.32 Write two examples of your own that illustrate the concept in the paragraph above.

10.33 According to Lambert, leaves lying in the yard and playing cards that are in disarray on a table have not undergone an increase in their thermodynamic entropy. Suggest another reason why leaves and playing cards may not be a good analogy for the entropy of a system containing, for example, only H_2O molecules or only O_2 molecules.

10.34 A researcher heats a sample of water in a closed vessel until it boils.

(a) Does the entropy of the water increase?

(b) Has the randomness of the molecules increased? (In other words, are there more physical positions that the molecules can occupy?)

(c) What else has increased that affects the entropy of the system?

The researcher now heats the water vapor from 400 K to 500 K, keeping the volume constant.

(d) Does the entropy of the system increase?

(e) Has the randomness of the molecules increased? (In other words, are there more physical positions that the molecules can occupy?)

(f) Why has an increase in temperature of the gas at constant volume caused an increase in entropy?

(To delve more deeply into the concept of entropy, read John P. Lowe's article "Entropy: Conceptual Disorder" in the *Journal of Chemical Education*, 65(5), 1988, 403–406.)

The Second Law of Thermodynamics

10.35 What happens to the entropy of the universe during a spontaneous process?

10.36 Why do we need to consider the surroundings of a system when applying the second law of thermodynamics?

10.37 One statement of the second law of thermodynamics is that heat cannot be turned completely into work. Another is that the entropy of the universe always increases. How are these two statements related?

10.38 According to the second law of thermodynamics, how does the sign of ΔS_u relate to the concept that some energy is wasted or lost to the surroundings when we attempt to convert heat into work?

10.39 How does the second law of thermodynamics explain a spontaneous change in a system that becomes more ordered when that process is exothermic?

10.40 Some say that the job of an engineer is to fight nature and the tendencies of entropy. **(a)** Does this statement seem accurate in any way? **(b)** How can any engineering design create order without violating the second law of thermodynamics?

10.41 When a reaction is exothermic, how does that influence ΔS of the system? Of the surroundings?

10.42 Which reaction occurs with the greater increase in entropy? Explain your reasoning.

(a) $2\ H_2O(\ell) \rightarrow 2\ H_2(g) + O_2(g)$

(b) $C(s) + O_2(g) \rightarrow CO_2(g)$

10.43 Which reaction occurs with the greater increase in entropy? Explain your reasoning.

(a) $2\ NO(g) \rightarrow N_2(g) + O_2(g)$

(b) $Br_2(g) + Cl_2(g) \rightarrow 2\ BrCl(g)$

10.44 Methanol is burned as fuel in some race cars. This makes it clear that the reaction is spontaneous once methanol is ignited. Yet the entropy change for the reaction $2\ CH_3OH(\ell) + 3\ O_2(g) \rightarrow 2\ CO_2(g) + 4\ H_2O(\ell)$ is negative. Why doesn't this violate the second law of thermodynamics?

10.45 Limestone is predominantly $CaCO_3$, which can undergo the reaction $CaCO_3(s) \rightarrow CaO(s) + CO_2(g)$. We know from experience that this reaction is not spontaneous, yet ΔS for the reaction is positive. How can the second law of thermodynamics explain that this reaction is not spontaneous?

The Third Law of Thermodynamics

10.46 Suppose that you find out that a system has an absolute entropy of zero. What else can you conclude about that system?

10.47 Use tabulated thermodynamic data to calculate the standard entropy change of each of the reactions listed below.

(a) $Fe(s) + 2\ HCl(g) \rightarrow FeCl_2(s) + H_2(g)$

(b) $3\ NO_2(g) + H_2O(\ell) \rightarrow 2\ HNO_3(\ell) + NO(g)$

(c) $2\ K(s) + Cl_2(g) \rightarrow 2\ KCl(s)$

(d) $Cl_2(g) + 2\ NO(g) \rightarrow 2\ NOCl(g)$

(e) $SiCl_4(g) \rightarrow Si(s) + 2\ Cl_2(g)$

10.48 If you scan the values for $S°$ in Appendix E, you will see that some aqueous ions have values that are less than zero. The third law of thermodynamics states that for a pure substance the entropy goes to zero only at 0 K. Use your understanding of the solvation of ions in water to explain how a negative value of $S°$ can arise for aqueous species.

10.49 Calculate $\Delta S°$ for the dissolution of magnesium chloride: $MgCl_2(s) \rightarrow Mg^{2+}(aq) + 2\ Cl^-(aq)$. Use your understanding of the solvation of ions at the molecular level to explain the sign of $\Delta S°$.

10.50 Calculate the standard entropy change for the reaction $CO_2(g) + 2\ H_2O(\ell) \rightarrow CH_4(g) + 2\ O_2(g)$. What does the sign of $\Delta S°$ say about the spontaneity of this reaction?

10.51 Through photosynthesis, plants build molecules of sugar containing several carbon atoms from carbon dioxide. In the process, entropy is decreased. The reaction of CO_2 with formic acid to form oxalic acid provides a simple example of a reaction in which the number of carbon atoms in a compound increases:

$$CO_2(aq) + HCOOH(aq) \rightarrow H_2C_2O_4(aq)$$

(a) Calculate the standard entropy change for this reaction and discuss the sign of $\Delta S°$.

(b) How do plants carry out reactions that increase the number of carbon atoms in a sugar, given the changes in entropy for reactions like this?

10.52 Find websites describing two different attempts to reach the coldest temperature on record. What features do these experiments have in common?

10.53 Look up the value of the standard entropy for the following molecules: $CH_4(g)$, $C_2H_5OH(\ell)$, $H_2C_2O_4(s)$. Rank the compounds in order of increasing entropy and then explain why this ranking makes sense.

10.54 Look up the value of the standard entropy for the following molecules: $SiO_2(s)$, $NH_3(g)$, $C_2H_6(g)$. Rank the compounds in order of increasing entropy and then explain why this ranking makes sense.

10.55 A beaker of water at 40°C (on the left in the drawing) and a beaker of ice water at 0°C are placed side by side in an insulated container. After some time has passed, the temperature of the water in the beaker on the left is 30°C and the temperature of the ice water is still 0°C.

Describe what is happening in each beaker **(a)** on the molecular level and **(b)** in terms of the second law of thermodynamics.

Gibbs Free Energy

10.56 Describe why it is easier to use ΔG to determine the spontaneity of a process rather than ΔS_u.

10.57 Under what conditions does ΔG allow us to predict whether a process is spontaneous?

10.58 There is another free energy state function, the Helmholtz free energy (A), defined as $A = E - TS$. Comparing this to the definition of G, we see that internal energy has replaced enthalpy in the definition. Under what conditions would this free energy tell us whether or not a process is spontaneous?

10.59 Calculate $\Delta G°$ at 45°C for reactions for which

(a) $\Delta H° = 293$ kJ; $\Delta S° = -695$ J/K

(b) $\Delta H° = -1137$ kJ; $\Delta S° = 0.496$ kJ/K

(c) $\Delta H° = -86.6$ kJ; $\Delta S° = -382$ J/K

10.60 Discuss the effect of temperature change on the spontaneity of the following reactions at 1 atm.

(a) $Al_2O_3(s) + 2\,Fe(s) \rightarrow 2\,Al(s) + Fe_2O_3(s)$
$\Delta H° = +851.5$ kJ; $\Delta S° = +38.5$ J/K

(b) $N_2H_4(\ell) \rightarrow N_2(g) + 2\,H_2(g)$
$\Delta H° = -50.6$ kJ; $\Delta S° = 0.3315$ kJ/K

(c) $SO_3(g) \rightarrow SO_2(g) + \frac{1}{2}O_2(g)$
$\Delta H° = 98.9$ kJ; $\Delta S° = +0.0939$ kJ/K

10.61 The reaction

$$CO_2(g) + H_2(g) \rightarrow CO(g) + H_2O(g)$$

is not spontaneous at room temperature but becomes spontaneous at a much higher temperature. What can you conclude from this about the signs of $\Delta H°$ and $\Delta S°$, assuming that the enthalpy and entropy changes are not greatly affected by the temperature change? Explain your reasoning.

10.62 Why is the free energy change of a system equal to the *maximum* work rather than just the work?

10.63 Distinguish between a reversible and an irreversible process.

10.64 For the reaction $NO(g) + NO_2(g) \rightarrow N_2O_3(g)$, use tabulated thermodynamic data to calculate $\Delta H°$ and $\Delta S°$. Then use those values to answer the following questions.

(a) Is this reaction spontaneous at 25°C? Explain your answer.

(b) If the reaction is not spontaneous at 25°C, will it become spontaneous at higher temperatures or lower temperatures?

(c) To show that your prediction is accurate, choose a temperature that corresponds to your prediction in part **(b)** and calculate ΔG. (Assume that both enthalpy and entropy are independent of temperature.)

10.65 The combustion of acetylene is used in welder's torches because it produces a very hot flame:

$$C_2H_2(g) + \frac{5}{2}O_2(g) \rightarrow 2CO_2(g) + H_2O(g) \qquad \Delta H° = -1255.5 \text{ kJ}$$

(a) Use data in Appendix E to calculate $\Delta S°$ for this reaction.

(b) Calculate $\Delta G°$ and show that the reaction is spontaneous at 25°C.

(c) Is there any temperature range in which this reaction is not spontaneous?

(d) Do you think you could use Equation 10.4 to calculate such a temperature range reliably? Explain your answer.

10.66 Natural gas (methane) is being used in experimental vehicles as a clean-burning fuel.

(a) Write the equation for the combustion of $CH_4(g)$, assuming that all reactants and products are in the gas phase.

(b) Use data from Appendix E to calculate $\Delta S°$ for this reaction.

(c) Calculate $\Delta G°$ and show that the reaction is spontaneous at 25°C.

10.67 Silicon forms a series of compounds that is analogous to the alkanes and has the general formula Si_nH_{2n+2}. The first of these compounds is silane, SiH_4, which is used in the electronics industry to produce thin ultrapure silicon films. $SiH_4(g)$ is somewhat difficult to work with because it is *pyrophoric* at room temperature—meaning that it bursts into flame spontaneously.

(a) Write an equation for the combustion of $SiH_4(g)$. (The reaction is analogous to hydrocarbon combustion, and SiO_2 is a solid under standard conditions. Assume the water produced will be a gas.)

(b) Use the data from Appendix E to calculate $\Delta S°$ for this reaction.

(c) Calculate $\Delta G°$ and show that the reaction is spontaneous at 25°C.

(d) Compare $\Delta G°$ for this reaction to the combustion of methane. (See the previous problem.) Are the reactions in these two exercises enthalpy or entropy driven? Explain.

Free Energy and Chemical Reactions

10.68 Explain why $\Delta G_f°$ of $O_2(g)$ is zero.

10.69 Using tabulated thermodynamic data, calculate $\Delta G°$ for these reactions.

(a) $Fe(s) + 2\ HCl(g) \rightarrow FeCl_2(s) + H_2(g)$

(b) $3\ NO_2(g) + H_2O(\ell) \rightarrow 2\ HNO_3(\ell) + NO(g)$

(c) $2\ K(s) + Cl_2(g) \rightarrow 2\ KCl(s)$

(d) $Cl_2(g) + 2\ NO(g) \rightarrow 2\ NOCl(g)$

(e) $SiCl_4(g) \rightarrow Si(s) + 2\ Cl_2(g)$

10.70 Using tabulated thermodynamic data, calculate $\Delta G°$ for these reactions.

(a) $Mg_3N_2(s) + 6\ H_2O(\ell) \rightarrow 2\ NH_3(g) + 3\ Mg(OH)_2(s)$

(b) $4\ CH_3NH_2(g) + 9\ O_2(g) \rightarrow 4\ CO_2(g) + 10\ H_2O(\ell) + 2\ N_2(g)$

(c) $Fe_3O_4(s) + 4\ CO(g) \rightarrow 3\ Fe(s) + 4\ CO_2(g)$

(d) $P_4O_{10}(s) + 6\ H_2O(\ell) \rightarrow 4\ H_3PO_3(aq)$

10.71 Calculate $\Delta G°$ for the dissolution of both sodium chloride and silver chloride using data from Appendix E. Explain how the values you obtain relate to the solubility rules for these substances.

10.72 Phosphorus exists in multiple solid phases, including two known as red phosphorus and white phosphorus. **(a)** Based on their respective heats of formation, which form of phosphorus is defined as the standard state? **(b)** Now consider the phase transition between white and red phosphorous: $P_4(s, white) \rightarrow 4\ P(s, red)$. Use data from Appendix E to determine which form of phosphorous is actually more stable at 25°C. (Your result should reveal that phosphorus is an exception to the usual convention for defining the standard state.) **(c)** Is the same form of the solid more stable at all temperatures? If not, what temperatures are needed to make the other form more stable?

10.73 The normal melting point of benzene, C_6H_6, is 5.5°C. For the process of melting, what is the sign of each of the following? **(a)** $\Delta H°$, **(b)** $\Delta S°$, **(c)** $\Delta G°$ at 5.5°C, **(d)** $\Delta G°$ at 0.0°C, **(e)** $\Delta G°$ at 25.0°C

10.74 Calculate $\Delta G°$ for the complete combustion of 1 mole of the following fossil fuels: methane (CH_4), ethane (C_2H_6), propane (C_3H_8), and *n*-butane (C_4H_{10}). Identify any trends that are apparent from these calculations.

10.75 Estimate the temperature range over which each of the following reactions is spontaneous.

(a) $2\ Al(s) + 3\ Cl_2(g) \rightarrow 2\ AlCl_3(s)$

(b) $2\ NOCl(g) \rightarrow 2\ NO(g) + Cl_2(g)$

(c) $4\ NO(g) + 6\ H_2O(g) \rightarrow 4\ NH_3(g) + 5\ O_2(g)$

(d) $2\ PH_3(g) \rightarrow 3\ H_2(g) + 2\ P(g)$

10.76 Recall that incomplete combustion of fossil fuels occurs when too little oxygen is present and results in the production of carbon monoxide rather than carbon dioxide. Water is the other product in each case.

(a) Write balanced chemical equations for the complete and incomplete combustion of propane.

(b) Using these equations, predict which will have the larger change in entropy.

(c) Use tabulated thermodynamic data to calculate $\Delta G°$ for each reaction.

(d) Based on these results, predict the sign and value of $\Delta G°$ for the combustion of carbon monoxide to form carbon dioxide.

INSIGHT INTO the Economics of Recycling

10.77 During polymerization, the system usually becomes more ordered as monomers link together. Could an endothermic polymerization reaction ever occur spontaneously? Explain.

10.78 If a particular polymerization reaction happened to be endothermic, how could it ever take place? Where would the increase in the entropy of the universe have to arise?

10.79 When polymers are recycled, the ends of the long-chain polymer molecules tend to break off, and this process eventually results in a degradation of physical properties, rendering the recycled polymer unusable. Explain why the breaking off of the ends of the polymer molecules is favorable from the standpoint of the entropy of the system.

10.80 The recycling of polymers represents only one industrial process that allows creating order in one location by creating greater disorder at some other location, often at a power plant. List three other industrial processes that must create disorder in the surroundings to generate the desired material.

Additional Problems

10.81 Diethyl ether is a liquid at normal temperature and pressure, and it boils at 35°C. Given that ΔH is 26.0 kJ/mol for the vaporization of diethyl ether, find its molar entropy change for vaporization.

10.82 Calculate the entropy change, $\Delta S°$, for the vaporization of ethanol, C_2H_5OH, at the boiling point of 78.3°C. The heat of vaporization of the alcohol is 39.3 kJ/mol.

$$C_2H_5OH(\ell) \rightarrow C_2H_5OH(g) \qquad \Delta S° = ?$$

10.83 Gallium metal has a melting point of 29.8°C. Use the information below to calculate the boiling point of gallium in °C.

Substance	$\Delta H_f°$ (kJ mol^{-1})	$\Delta G_f°$ (kJ mol^{-1})	$S°$ (J mol^{-1}K^{-1})
Ga(s)	0	0	40.83
Ga(ℓ)	5.578	0.0888	59.25
Ga(g)	271.96	233.76	169.03

For a metal, gallium has a very low melting point.

10.84 Methane can be produced from CO and H_2. The process might be done in two steps, as shown below, with each step carried out in a separate reaction vessel within the production plant.

Reaction #1 $CO(g) + 2\ H_2(g) \rightarrow CH_3OH(\ell)$

$$\Delta S° = -332 \text{ J/K}$$

Reaction #2 $CH_3OH(\ell) \rightarrow CH_4(g) + \frac{1}{2} O_2(g)$

$$\Delta S° = +162 \text{ J/K}$$

Substance	$\Delta H_f°$ (kJ mol^{-1})	$\Delta G_f°$ (kJ mol^{-1})	$S°$ (J mol^{-1}K^{-1})
$CO(g)$	−110.5		197.674
$CH_3OH(\ell)$	−238.7	−166.4	126.8
$CH_4(g)$	−74.8		186.2

NOTE: You should be able to work this problem *without* using any additional tabulated data.

(a) Calculate $\Delta H°$ for reaction #1.

(b) Calculate $\Delta G_f°$ for CO(g).

(c) Calculate $S°$ for $O_2(g)$.

(d) At what temperatures is reaction #1 spontaneous?

(e) Suggest a reason why these two steps would need to be carried out separately.

10.85 Iodine is not very soluble in water, but it dissolves readily in a solution containing iodide ions by the following reaction:

$$I_2(aq) + I^-(aq) \rightarrow I_3^-(aq)$$

The following graph shows the results of a study of the temperature dependence of $\Delta G°$ for this reaction. (The solid line is a best fit to the actual data points.) Notice that the quantity on the y axis is $\Delta G°/T$, not just $\Delta G°$. Additional data relevant to this reaction are also given in the table that follows the graph.

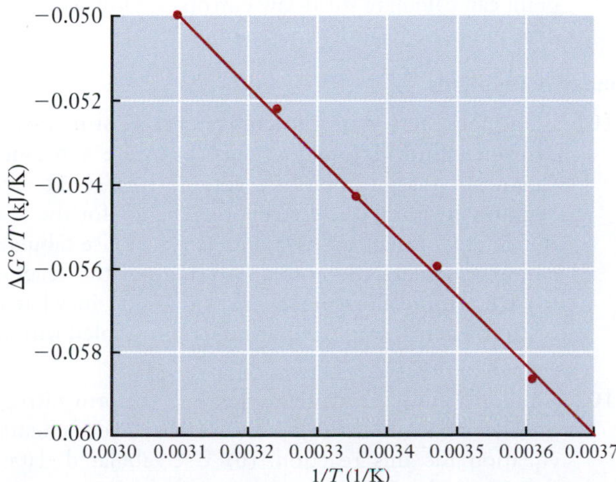

Substance	$\Delta G_f°$ (kJ mol^{-1})
$I_2(aq)$	16.37
$I^-(aq)$	−51.57
$I_3^-(aq)$	−51.4

(a) Calculate $\Delta G°$ for this reaction at 298 K. (DO NOT read this value off the graph. Use the data given to calculate a more accurate value.)

(b) Determine $\Delta H°$ for this reaction. Assume that $\Delta H°$ is independent of T. (HINT: You will need to use the graph provided to find $\Delta H°$. It may help if you realize that the graph is a straight line and then try to write an equation for that line.)

10.86 The enthalpy of vaporization for water is 40.65 kJ mol^{-1}. As a design engineer for a project in a desert climate, you are exploring the option of using evaporative cooling. **(a)** If the air has an average volumetric heat capacity of 0.00130 J cm^{-3} K^{-1}, what is the minimum mass of water that would need to evaporate in order to cool a 5 m × 5 m room with a 3 m ceiling by 5°F using this method? **(b)** Is this a spontaneous or nonspontaneous process?

10.87 Determine whether each of the following statements is true or false. If false, modify to make the statement true.

(a) An exothermic reaction is spontaneous.

(b) When $\Delta G°$ is positive, the reaction cannot occur under any conditions.

(c) $\Delta S°$ is positive for a reaction in which there is an increase in the number of moles.

(d) If $\Delta H°$ and $\Delta S°$ are both negative, $\Delta G°$ will be negative.

10.88 Nickel metal reacts with carbon monoxide to form tetracarbonyl nickel, $Ni(CO)_4$:

$$Ni(s) + 4\ CO(g) \rightarrow Ni(CO)_4(g)$$

This reaction is exploited in the Mond process in order to separate pure nickel from other metals. The reaction above separates nickel from impurities by dissolving it into the gas phase. Conditions are then changed so that the reaction runs in the opposite direction to recover the purified metal.

(a) Predict the signs of $\Delta H°$ and $\Delta S°$ for the reaction as written above. (Note that bonds are formed but none are broken.)

(b) Use tabulated thermodynamic data to calculate $\Delta H°$, $\Delta S°$, and $\Delta G°$ for the reaction.

(c) Find the range of temperatures at which this reaction is spontaneous in the forward direction.

10.89 Polyethylene has a heat capacity of 2.3027 J g^{-1} °C^{-1}. You need to decide if 1.0 ounce of polyethylene can be used to package a material that will be releasing heat when in use. Consumer safety specifications indicate that the maximum allowable temperature for the polyethylene is 45°C; it can be assumed that the plastic is initially at room temperature. **(a)** What temperature will the polyethylene reach if the product generates 1500 J of heat and all of this energy is absorbed by the plastic package? **(b)** Is this a realistic estimate of the temperature that the polyethylene packaging would reach? Explain your answer. **(c)** What is the enthalpy change of the polyethylene? **(d)** Estimate the entropy change of the polyethylene. (You will need to assume that the temperature of the plastic is constant.)

10.90 A key component in many chemical engineering designs is the separation of mixtures of chemicals. **(a)** What happens to the entropy of the system when a chemical mixture is separated? **(b)** Are designs for chemical separation more likely to rely on spontaneous or nonspontaneous processes?

10.91 The reaction shown below is involved in the refining of iron. (The table that follows provides all of the thermodynamic data you should need for this problem.)

$$2\ Fe_2O_3(s) + 3\ C(s,\ graphite) \rightarrow 4\ Fe(s) + 3\ CO_2(g)$$

Compound	ΔH_f° (kJ mol^{-1})	S° (J mol^{-1} K^{-1})	ΔG_f° (kJ mol^{-1})
Fe_2O_3(s)	−824.2	?	−742.2
C(s, graphite)	0	5.740	0
Fe(s)	0	27.3	0
CO_2(g)	−393.5	213.6	−394.4

(a) Find ΔH° for the reaction.

(b) ΔS° for the reaction above is 557.98 J/K. Find S° for Fe_2O_3(s).

(c) Calculate ΔG° for the reaction at the standard temperature of 298 K. (There are two ways that you could do this.)

(d) At what temperatures would this reaction be spontaneous?

10.92 Using only the data given below, determine ΔG° for the following reaction:

$$NO(g) + O(g) \rightarrow NO_2(g)$$

(Remember that ΔG is a state function, just like ΔH.)

$$2\ O_3(g) \rightarrow 3\ O_2(g) \qquad \Delta G^\circ = -326\ kJ$$

$$O_2(g) \rightarrow 2\ O(g) \qquad \Delta G^\circ = 463.6\ kJ$$

$$NO(g) + O_3(g) \rightarrow NO_2(g) + O_2(g) \qquad \Delta G^\circ = -198.3\ kJ$$

10.93 The graph below shows ΔG° as a function of temperature for the synthesis of ammonia from nitrogen and hydrogen.

$$N_2(g) + 3\ H_2(g) \rightarrow 2\ NH_3(g)$$

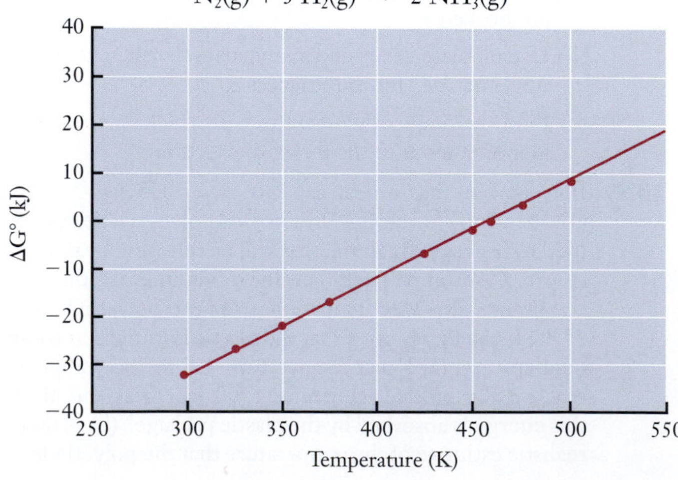

(a) Use the graph to estimate ΔS° for the ammonia synthesis reaction.

(b) Given that the standard free energy change of formation for ammonia (ΔG_f°) is −16.50 kJ/mol, estimate ΔH° for the ammonia synthesis reaction.

FOCUS ON PROBLEM SOLVING EXERCISES

10.94 Suppose that you need to know the heat of formation of 2-pentene, C_5H_{10}, but the tables you have do not provide the value. You have a sample of the chemical. What could you do to determine the heat of formation?

10.95 Suppose that you need to know the heat of formation of cyclohexane, C_6H_{12}, but the tables you have do not provide the value. You have a sample of the chemical. What could you do to determine the heat of formation?

10.96 You have a table of thermodynamic variables that includes heats of formation and standard entropies but not free energies of formation. How could you use the information you have to estimate the free energy of formation of a substance that is listed in your table?

10.97 You have a table of thermodynamic variables that includes standard entropies and free energies of formation but not heats of formation. How could you use the information you have to estimate the heat of formation of a substance that is listed in your table?

10.98 From a series of hydrocarbons containing only single carbon–carbon bonds, how could you identify a trend for the heat of formation for successive carbon–carbon bond formation?

10.99 Thermodynamics provides a way to interpret everyday occurrences. If you live in northern climates, one common experience is that during early winter, snow falls but then melts when it hits the ground. Both the formation and the melting happen spontaneously. How can thermodynamics explain both of these seemingly opposed events?

10.100 Suppose that you are designing a chemical reactor and you need to know the heat of vaporization for a solvent. You don't have any thermodynamic tables handy, but you know the boiling point of the liquid and your CAD program can calculate third-law entropies. How would you determine the heat of vaporization?

Cumulative Problems

10.101 Fluorine reacts with liquid water to form gaseous hydrogen fluoride and oxygen. **(a)** Write a balanced chemical equation for this reaction. **(b)** Use tabulated data to determine the free energy change for the reaction and comment on its spontaneity. **(c)** Use tabulated data to calculate the enthalpy change of the reaction. **(d)** Determine how much heat flows and in what direction when 34.5 g of fluorine gas is bubbled through excess water.

10.102 Ammonia can react with oxygen gas to form nitrogen dioxide and water. **(a)** Write a balanced chemical equation for this reaction. **(b)** Use tabulated data to determine the free energy change for the reaction and comment on its spontaneity. **(c)** Use tabulated data to calculate the enthalpy change of the reaction. **(d)** Determine how much heat flows and in what direction when 11.4 g of ammonia gas is burned in excess oxygen.

10.103 Consider the following thermodynamic data for oxides of manganese.

Substance	$\Delta H_f°$ (kJ mol^{-1})	$\Delta G_f°$ (kJ mol^{-1})	$S°$ (J mol^{-1}K^{-1})
MnO	−385.2	−362.9	59.71
MnO_2	−520.0	−465.2	53.05
Mn_2O_3	−959.0	−881.2	110.5
Mn_3O_4	−1388	−1283	155.6

(a) What is the correct chemical nomenclature for each of the first three oxides? **(b)** Write and balance chemical equations for the conversion of each of these oxides into Mn_3O_4. **(c)** Based on the free energy changes of these reactions, which oxide is the most stable at room temperature?

10.104 **(a)** When a chemical bond forms, what happens to the entropy of the system? **(b)** Thermodynamically, what allows for any bond formation to occur? **(c)** What do your answers to parts **(a)** and **(b)** suggest must be true about the formation of chemical bonds for the octet rule to hold?

11 Chemical Kinetics

CHAPTER OUTLINE

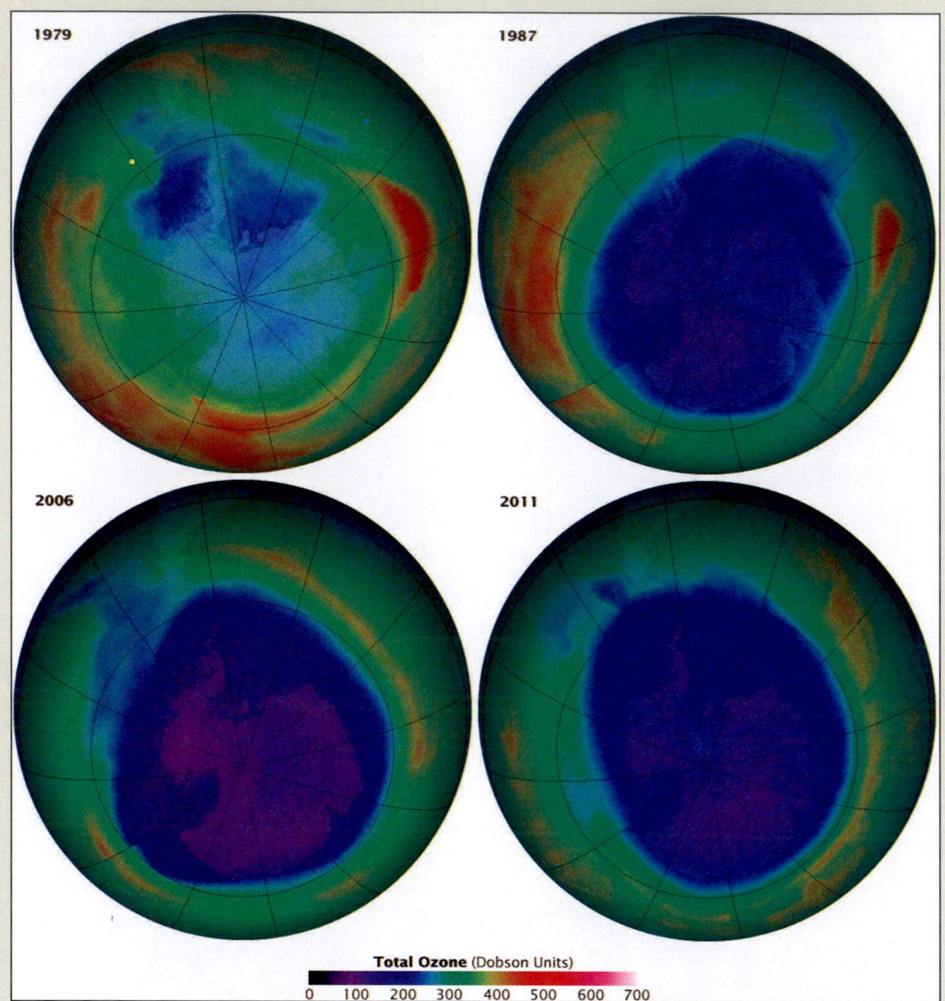

The figure shows Antarctic ozone levels as measured on September 16 in the years 1979, 1987, 2006, and 2011. Blue areas correspond to the ozone hole, where atmospheric ozone concentration is depleted. The loss of ozone between 1979 and 1987 is dramatic. In 1987 the Montreal Protocol went into effect, limiting the production and use of chlorofluorocarbon (CFC) refrigerants. Since that time ozone levels have stabilized, and may be recovering. *NASA animation by Robert Simmon, using imagery from the Ozone Hole Watch.*

When we considered thermodynamics, we concentrated on what should *eventually* happen to a chemical system. But we have not yet given any thought to how long it might take to reach this eventual outcome or exactly *how* a collection of reactants might be transformed into products. Fundamentally, chemistry is about change. So if we want to understand chemistry, we will need to examine both the process of change and the speed with which it occurs. On the Fourth of July, it is exhilarating to watch the explosive oxidation reactions of fireworks, but a holiday event to watch iron rust would probably not draw a crowd. What factors determine whether a reaction proceeds with explosive speed or at an imperceptible rate? Could we learn to manipulate these factors to force a reaction to occur at the speed we choose? The area of chemistry dealing with these questions of "How?" or "How fast?" is called chemical kinetics, and we begin with a glimpse into stratospheric ozone and its depletion.

Chapter Objectives

After mastering this chapter, you should be able to

- explain the role of chemical kinetics in the formation and destruction of ozone in the atmosphere.
- define the rate of a chemical reaction and express the rate in terms of the concentrations of individual reactants or products.
- use the method of initial rates to determine rate laws from experimental data.
- use graphical methods to determine rate laws from experimental data.
- explain the difference between elementary reactions and multistep reactions.
- find the rate law predicted for a particular reaction mechanism.
- use a molecular perspective to explain the significance of the terms in the Arrhenius equation.
- calculate the activation energy for a reaction from experimental data.
- explain the role of a catalyst in the design of practical chemical reactions.

11-1 Ozone Depletion

Walk outside right after a summer thunderstorm and you will probably notice a pleasantly fresh smell. The source of that enjoyable scent is ozone (O_3), the less common allotrope of oxygen. Though its pungent odor can be pleasing in very low concentration, it becomes unpleasant and even toxic in concentrations greater than about one part per million, causing headache and difficulty in breathing. In a thunderstorm, energy from lightning helps drive reactions that produce ozone from ordinary oxygen. But ozone can also be formed through reactions of various compounds in the exhaust gases from automobiles and industrial processes. Weather forecasts for many urban areas now include ozone alerts on days when the ozone level in the air is expected to rise above about 0.1 ppm. Ozone is a major air pollutant in the troposphere (the atmosphere at ground level—see **Figure 11.1**), and yet a frequent topic in the news is concern about the depletion of the stratospheric ozone layer. Why is ozone a problem in the troposphere but a necessity in the stratosphere? What is the chemistry behind these issues?

The smell of ozone after a thunderstorm disappears quickly, not only because it diffuses through the air (and because your olfactory nerves dull) but also because ozone is unstable and decomposes to form ordinary oxygen. The decomposition reaction has a deceptively simple overall equation: $2\ O_3 \rightarrow 3\ O_2$. That ozone is reactive and cannot exist for long at the earth's surface suggests two important facts. First, O_2 is the more stable of the two allotropes of oxygen. So thermodynamics must favor the decomposition of ozone. Second, for the ozone layer to exist, the upper atmosphere must have some conditions that allow the formation of the less stable allotrope, O_3, in significant concentration. The accumulation of ozone in the stratosphere is governed by the way these conditions influence the rates of chemical reactions and therefore provides our initial insight into the role of kinetics in chemistry.

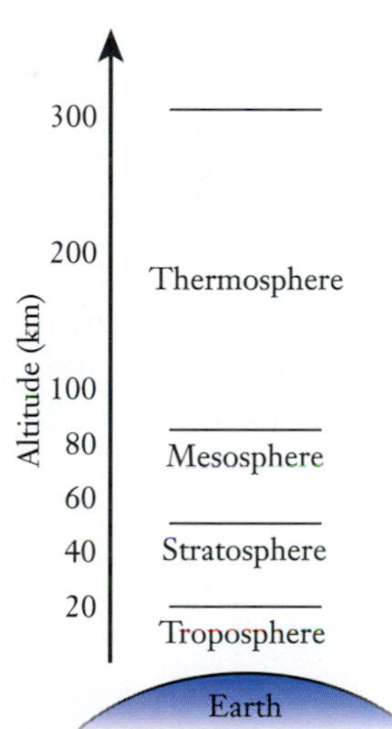

Figure 11.1 ■ Several layers of the atmosphere are identified in the figure. The ozone layer is in the stratosphere at an altitude of about 30 km.

Figure 11.2 ■ The Chapman cycle for the formation and destruction of ozone in the stratosphere.

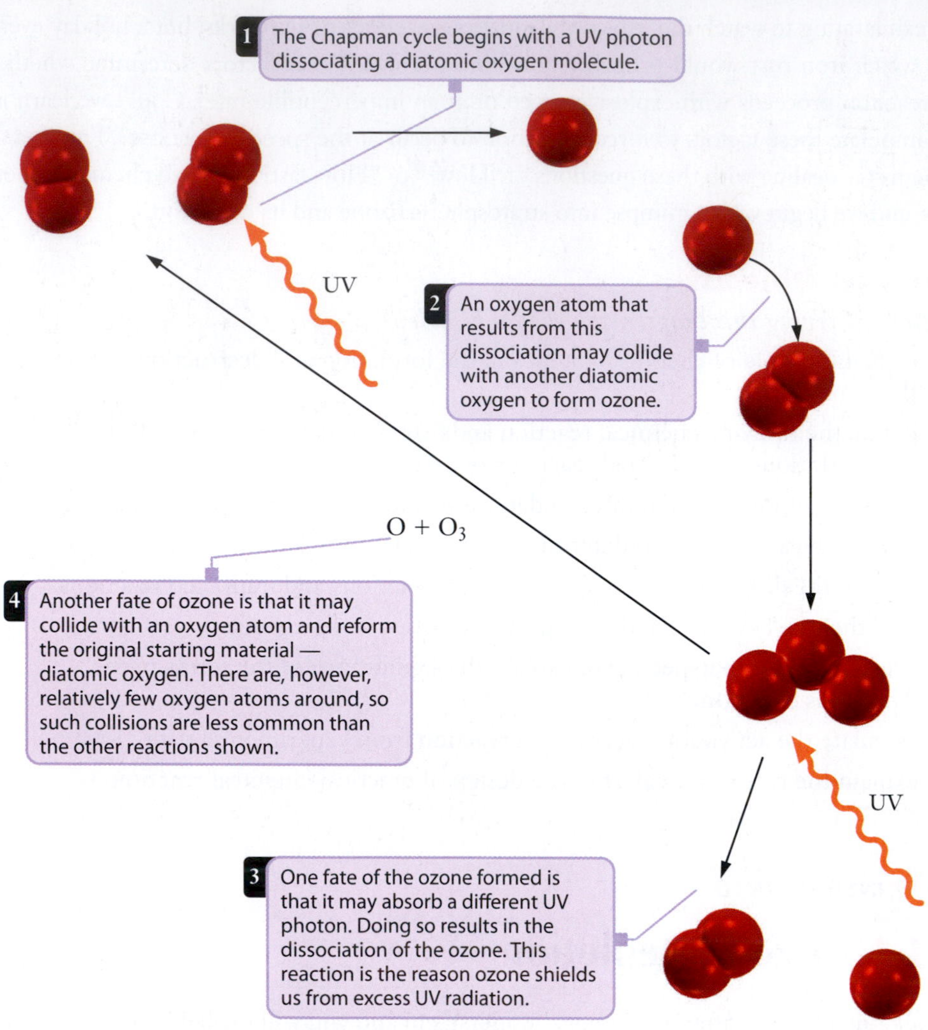

1 The Chapman cycle begins with a UV photon dissociating a diatomic oxygen molecule.
UV
2 An oxygen atom that results from this dissociation may collide with another diatomic oxygen to form ozone.
O + O_3
4 Another fate of ozone is that it may collide with an oxygen atom and reform the original starting material — diatomic oxygen. There are, however, relatively few oxygen atoms around, so such collisions are less common than the other reactions shown.
UV
3 One fate of the ozone formed is that it may absorb a different UV photon. Doing so results in the dissociation of the ozone. This reaction is the reason ozone shields us from excess UV radiation.

Sydney Chapman was a mathematician. He devised the idea of the Chapman cycle while working on a mathematical theory of gases.

In 1903, Sydney Chapman first explained the chemistry of the formation of ozone in the upper atmosphere when he proposed what is now known as the **Chapman cycle.** Although Chapman's original proposal was quite speculative, modern measurements of a wide array of quantities support his hypothesis. The Chapman cycle begins and ends with diatomic oxygen, the more stable allotrope, as shown in **Figure 11.2.** The first step is the photochemical dissociation of O_2 to form oxygen atoms, which may then react with O_2 molecules to form ozone, O_3. Mechanisms also exist for ozone decomposition, so that, even in the stratosphere, ozone is not stable. So why does the ozone layer exist? To answer this, we need to consider the rates at which ozone is produced and consumed. If ozone is produced faster than it is consumed, its concentration will increase. But if ozone is destroyed faster than it is formed, the ozone level will decrease. The existence of the ozone layer, then, depends on this balance between the rates at which ozone is produced and destroyed. To understand ozone depletion, we will need to understand reaction rates.

The ozone layer would probably not make headlines were it not at risk. During the past few decades, scientists have been measuring a decrease in ozone concentration that returns during the early spring months, particularly over Antarctica and (less dramatically) over North America. Evidence of this effect is presented graphically in **Figure 11.3.** The decrease in stratospheric ozone concentration from that of 30 years ago is what is commonly referred to as the ozone hole.

Other measurements make clear that the depletion of ozone over polar areas of the planet is a seasonal process.

How does the ozone hole arise? We must consider the role of species other than the allotropes of oxygen in the Chapman cycle. There is evidence that chlorine and

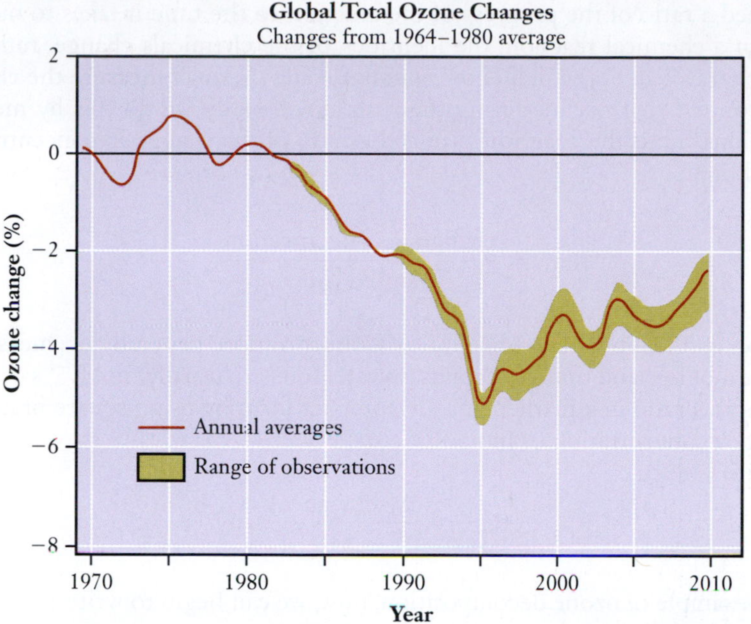

Figure 11.3 ■ The data in this graph clearly show a decrease in the overall levels of atmospheric ozone from the late seventies through the early nineties. The increase in ozone levels in recent years is due mainly to the impact of the Montreal Protocol, which curtailed the use of ozone-damaging substances. Source: WMO [World Meteorological Organization] *Scientific Assessment of Ozone Depletion: 2006*, Global Ozone Research and Monitoring Project, Report No. 50, Geneva, 2007.

bromine in the stratosphere lead to a decrease in the amount of ozone present. How does this occur? What factors influence whether or not the decrease occurs? And what might be done to combat the loss of ozone? These questions all point to aspects of chemical kinetics that we will address in this chapter. To begin this study, we focus first on the concepts of rate of reaction and the ways in which reaction rate can be measured.

11-2 Rates of Chemical Reactions

We suggested that the rates of different chemical reactions govern ozone levels in the atmosphere, and this seems sensible. If ozone—or anything else—is produced faster than it is consumed, it will accumulate. To discuss these ideas quantitatively, we must first raise two fundamental issues in chemical kinetics. How do we define the rate of a reaction? And how can we measure it?

Concept of Rate and Rates of Reaction

Aside from our interest in the rate of destruction of ozone, we have some general experience with rates. If we travel 55 miles from home and it takes us an hour to go that distance, we say that our average speed is 55 miles per hour:

$$\text{Average speed} = \frac{\text{distance traveled}}{\text{time elapsed}}$$

In this ratio, there is an indication of progress (distance traveled) related to the time required to make that progress. Because the distance measured was from zero miles (at home) to 55 miles from home, the change in distance from home can be symbolized by Δd, where Δ is the symbol for change. The time changed from $t = 0$ (at home) to $t = 1$ hour at our destination. So the time elapsed can be represented by Δt.

$$\text{Average speed} = \frac{\Delta d}{\Delta t} = \frac{55 \text{ miles}}{1 \text{ hour}}$$

How do we translate this ratio into one that is meaningful for chemical reactions?

We need a ratio of the progress of the reaction to the time it takes to make that progress. In a chemical reaction, the identities of the chemicals change, rather than the distance traveled. So, to define the chemical rate, we must measure the change in chemical content that occurs during the reaction. Usually we do this by measuring concentrations. Thus the **reaction rate** is the ratio of the change in concentration to the elapsed time:

$$\text{Rate} = \frac{\text{change in concentration}}{\text{elapsed time}} \tag{11.1}$$

The timescales of interest in chemical kinetics can vary over a very wide range, from picoseconds to years.

Notice the units implied by this ratio. Concentration generally is measured in molarity or mol L^{-1} and time in seconds, so rate units are usually mol L^{-1} s^{-1}. We can express this ratio in a mathematically compact form by using square brackets, [ ], to designate a concentration or molarity:

$$\text{Rate} = \frac{\Delta[\text{substance}]}{\Delta t}$$

Using our example of ozone decomposition, now, we can begin to write an expression for the rate of reaction

$$\text{Rate} = \frac{\Delta[O_3]}{\Delta t}$$

Stoichiometry and Rate

The overall reaction for the destruction of ozone was mentioned in the opening "Insight" section:

$$2\ O_3 \rightarrow 3\ O_2$$

As we saw above, the rate of this reaction can be given by the rate at which ozone decomposes:

$$\text{Rate} = \frac{\Delta[O_3]}{\Delta t}$$

But suppose that we had monitored the increase in O_2 concentration over time, instead of the decrease in ozone concentration. The change in concentration we'd measure would be different, yet it must also be determined by the rate of the same reaction. As the reaction proceeds, oxygen concentration increases more quickly than ozone concentration decreases because three oxygen molecules form for every two ozone molecules destroyed. Which rate should we measure, the rate of O_2 production or the rate of disappearance of O_3? The answer is either one, but to assure a consistent result, we must take into account the stoichiometry of the reaction.

If we are measuring the increase in the concentration of the product, the rate of a reaction is naturally a positive number. When we observe a *decrease* in the concentration of a reactant, we must include a minus sign in the rate statement to obtain a positive value for the rate. Thus rate might be defined as either

$$\text{Rate} = \frac{\Delta[\text{product}]}{\Delta t}$$

or

$$\text{Rate} = -\frac{\Delta[\text{reactant}]}{\Delta t}$$

But this still does not account for the fact that the number of O_2 molecules produced in a given time is greater than the number of O_3 molecules consumed. To account for this, we also include the stoichiometric coefficient in the denominator of the rate expression:

$$\text{Rate} = \frac{\Delta[\text{product}]}{\nu_{\text{prod}}\,\Delta t} \tag{11.2.a}$$

or

$$\text{Rate} = -\frac{\Delta[\text{reactant}]}{\nu_{\text{react}}\,\Delta t} \tag{11.2.b}$$

Here ν_{prod} is the stoichiometric coefficient of the product being measured, and ν_{react} is the coefficient of the reactant. Accounting for stoichiometry in this way ensures that the same reaction rate is obtained using either the rate of disappearance of reactants or the rate of appearance of products. Example Problem 11.1 illustrates the connection between stoichiometry and rates.

We also used the symbol ν to represent stoichiometric coefficients in several thermodynamic equations.

EXAMPLE PROBLEM 11.1

The conversion of ozone to oxygen, $2\ O_3 \rightarrow 3\ O_2$, was studied in an experiment, and the rate of ozone consumption was measured as 2.5×10^{-5} mol L^{-1} s^{-1}. What was the rate of O_2 production in this experiment?

Strategy Use the stoichiometry of the reaction to relate the respective rate expressions. Because the rate of the reaction is the same whichever quantity is measured, we can set the expressions for the rate of oxygen appearing and ozone disappearing equal to each other and solve for the unknown quantity.

Solution Begin by writing the expressions for the rate of reaction in terms of both O_2 and O_3 concentration:

$$\text{Rate} = -\frac{\Delta[O_3]}{2\Delta t} = \frac{\Delta[O_2]}{3\Delta t}$$

Remember that the change in concentration of the reactant is a *negative* quantity and substitute -2.5×10^{-5} mol L^{-1} s^{-1} for $\dfrac{\Delta[O_3]}{\Delta t}$:

$$\frac{2.5 \times 10^{-5}\text{mol L}^{-1}\text{s}^{-1}}{2} = \frac{\Delta[O_2]}{3\Delta t}$$

Solve for the rate of O_2 formation:

$$\frac{\Delta[O_2]}{\Delta \text{t}} = \frac{3\ (2.5 \times 10^{-5}\text{ mol L}^{-1}\text{s}^{-1})}{2}$$

$$= 3.8 \times 10^{-5}\text{ mol L}^{-1}\text{ s}^{-1}$$

Discussion For some students, the most troubling part of this mathematical description of rate is the minus sign when reactant concentrations are used. By recalling our standard definition for Δ, however, we can see the need for this sign more explicitly. In this case, we are interested in $\Delta[O_3]$, which is $[O_3]_{\text{later}} - [O_3]_{\text{initial}}$. The initial concentration of a reactant is larger than its concentration at some later time, so $\Delta[O_3]$ is negative. We need the minus sign in the equation so that the rate is not a negative number.

Check Your Understanding Like ozone, dinitrogen pentoxide decomposes to form O_2 (along with nitrogen dioxide) by the reaction $2\ N_2O_5 \rightarrow 4\ NO_2 + O_2$. If the rate of disappearance of N_2O_5 is 4.0×10^{-6} mol L^{-1} s^{-1}, what is the rate of appearance of each product?

Average Rate and Instantaneous Rate

Let's think about a simple experiment like the one shown in **Figure 11.4**. Take a glass and place a candle inside. Light the candle, let it burn for a few moments, and cover the glass with a saucer. Observe the flame. What happens? The brightly burning flame gradually decreases in size until it goes out because the concentration of one reactant, oxygen, has decreased. During this experiment, the rate of the combustion reaction decreases until the reaction stops. What is the rate of the reaction? Is it the rate at which oxygen is consumed in the brightly burning flame or in the nearly extinguished one? This simple experiment shows that the rate we observe depends on when or how we make our observations.

Figure 11.5 shows a graph of the concentration of ozone as a function of time in a laboratory experiment. Looking closely at this figure, we can see the slight difference between the average reaction rate and the instantaneous reaction rate. The difference between these two methods of defining rate lies in the amount of time needed to make the observation. For the **average rate**, two concentrations are measured at times separated by a finite difference, and the slope of the line between them gives the rate. The **instantaneous rate** refers to the rate at a single moment, and it is given by the slope of a line tangent to the curve defined by the change in concentration versus time. As Figure 11.5 shows, these two slopes can differ. (Returning to the speed analogy, your instantaneous speed at any moment during a long drive might be very different from your average speed during the same trip. You may accelerate to pass another car or come to a stop in traffic.) In most cases, we prefer to work with instantaneous rates in kinetics. One commonly measured rate is the initial rate: the instantaneous rate of the reaction just as it begins. In Figure 11.5, the tangent line and the concentration curve coincide for at least a brief period at the beginning of the reaction, so the initial instantaneous rate is often relatively easy to measure.

If you are familiar with calculus, you'll realize that the instantaneous rate is the derivative of concentration with respect to time.

To determine rates of reaction in complicated systems such as the stratosphere, we typically try to measure the rates of the same reactions independently in

Photos: Thomas A. Holme

Figure 11.4 ■ When a candle burns in a closed container, the flame will diminish and eventually go out. As the amount of oxygen present decreases, the rate of combustion will also decrease. Eventually, the rate of combustion is no longer sufficient to sustain the flame even though there is still some oxygen present in the vessel.

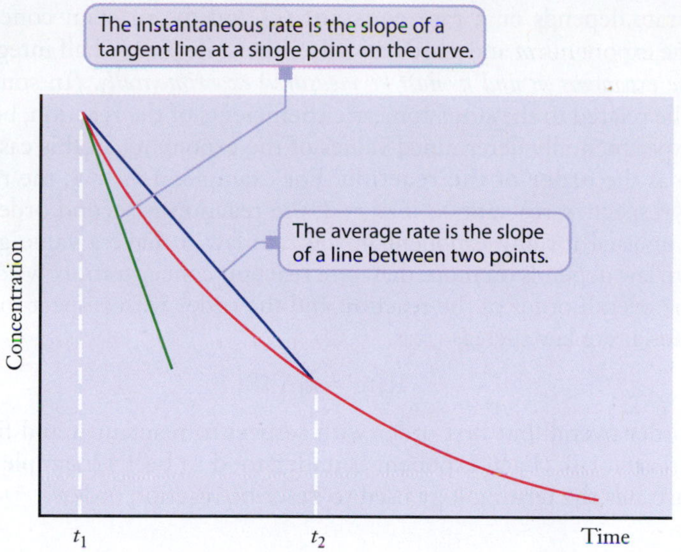

The instantaneous rate is the slope of a tangent line at a single point on the curve.
The average rate is the slope of a line between two points.
Concentration
t_1
t_2
Time

Figure 11.5 ■ The distinction between the instantaneous rate of reaction and the average rate measured over some period of time is illustrated here. The slope of the green line to the left gives the instantaneous rate at time t_1. The slope of the blue line to the right gives the average rate for the period from time t_1 to time t_2.

the laboratory. Thus to consider the reactions of the Chapman cycle, a chemist would have to devise an experiment that measures the rate of reaction of some very reactive species, such as oxygen atoms. The reactive species is first created (maybe by using a pulse of laser light to break the bond in O_2), and then its rapid consumption is monitored. This can be a difficult measurement because the reaction may proceed extremely quickly. Reactive species such as oxygen atoms react with most chemicals in the reaction container, so great care must be taken to isolate the reacting species to ensure that only the reaction of interest can occur. Once concentrations of reacting species are measured, an equation can be formulated to describe how the rate depends on these concentrations.

11-3 Rate Laws and the Concentration Dependence of Rates

We saw in Figure 11.4 that as O_2 is consumed, the rate of combustion decreases. The rate of a chemical reaction depends on a number of factors. One of these factors, the concentration of the reacting species, can help us understand why there is an ozone layer in the upper atmosphere, despite the fact that O_2 is the more stable of the two oxygen allotropes. To approach this issue for the stratosphere, let's first investigate how reaction rates measured in a laboratory depend on concentration.

The Rate Law

Observations of many chemical reactions show that the dependence of reaction rate on concentration often follows relatively simple mathematical relationships. This behavior can be summarized in a mathematical equation known as the **rate law**. There are two useful forms of the rate law, and we begin with the **differential rate law** (the name and the relationship are derived from calculus). For a reaction between substances X and Y, the rate of reaction can usually be described by an equation of the form

$$\text{Rate} = k[\text{X}]^m[\text{Y}]^n \qquad (11.3)$$

where the rate depends on a **rate constant *(k)*** and the reactant concentrations [X] and [Y]. The exponents m and n are typically either integers or half integers. *The actual values of the exponents m and n must be measured experimentally.* (In some cases, these values can be related to the stoichiometric coefficients of the reaction, but not always!)

The experimentally determined values of the exponents, in this case m and n, are referred to as the **order of the reaction**. For example, if $m = 1$, the reaction is first order with respect to reactant X; if $m = 2$, the reaction is second order with respect to X. It is unusual for any exponent in the rate law to have a value greater than 2. When a rate law depends on more than one reactant concentration, we can distinguish between the overall order of the reaction and the order with respect to individual reactants. Thus a rate law such as

$$\text{Rate} = k[\text{A}][\text{B}]$$

is second order overall but first order with respect to reactant A and first order with respect to reactant B. (Each exponent is understood to be 1.) Example Problem 11.2 further illustrates the terminology used to describe reaction orders.

EXAMPLE PROBLEM 11.2

In the following rate laws, determine the orders with respect to each substance and the overall order of the reaction. (a) Rate $= k[\text{A}]^2[\text{B}]$, (b) Rate $= k[\text{A}][\text{B}]^{1/2}$

Strategy The order with respect to each individual species is the exponent, and the overall order is the sum of the individual orders.

Solution

(a) The order with respect to A is 2 and with respect to B is 1. The overall order is 3.
(b) The order with respect to A is 1 and with respect to B is 1/2. The overall order is 3/2.

Discussion It is important to remember that the order of the reaction is determined experimentally. Although this problem focuses on the mathematical detail of the definition of reaction order, we must keep in mind that reaction order and stoichiometry are not directly related.

Check Your Understanding The following reactions were studied in the laboratory, and their respective rate laws were determined. Find the orders with respect to each reactant and the overall order for each of the following reactions.

(a) $H_2(g) + Br_2(g) \rightarrow 2\ HBr$; rate $= k[H_2][Br_2]$
(b) $2\ N_2O_5 \rightarrow 4\ NO_2 + O_2$; rate $= k[N_2O_5]$

The rate constant gives us valuable information about the kinetics of a chemical reaction. Perhaps most importantly, the magnitude of the rate constant tells us whether or not a reaction proceeds quickly. If the rate constant is small, the reaction is likely to proceed slowly. By contrast, a large rate constant indicates a rapid reaction. Rate constants for chemical reactions range over many orders of magnitude. We should also point out that the value of the rate constant depends on the temperature of the reaction. (We will consider temperature dependence further in Section 11-5, but it may take some time to get used to the idea of a "constant" that is a function of a variable.) The temperature dependence of the rate of a reaction is described in terms of the rate constant, as we'll see in Section 11-5.

The units of a rate constant depend on the overall order of the reaction. We have already established that rate has the units of mol L^{-1} s^{-1} when concentrations are given in mol L^{-1}. The units of the rate constant must be chosen so that units match on both sides of the rate law. You should be able to verify that the units for the rate constant of a first-order reaction must be s^{-1}, whereas for second-order reactions, the rate constant has the units L mol^{-1} s^{-1}.

Not all rate laws are as simple as those mentioned thus far, all of which have included only concentrations of reactants. One reaction for the destruction of stratospheric O_3, for example, has a concentration dependence on the product, O_2. Increased O_2 concentration, however, slows the reaction rather than increasing it. This observation requires a negative exponent for oxygen in the rate law. Thus the rate law for $2\ O_3 \rightarrow 3\ O_2$ is given by

$$\text{Rate} = k\frac{[O_3]^2}{[O_2]}$$

or

$$\text{Rate} = k[O_3]^2[O_2]^{-1}$$

It is also not uncommon to find orders of 1/2 or −1/2 in rate laws for some reactions.

Determination of the Rate Law

As noted earlier, a rate law must be determined experimentally. There are two common ways to do this. One is to use a series of graphs to compare data to various possible rate laws, and that method is emphasized in Section 11-4. First we will look at the other approach, which is to measure the initial rate of the reaction while adjusting the concentrations of the various reactants.

To see the logic needed to determine a rate law, let's think about a simple reaction in which some substance we'll label as A is the only reactant. The rate for such a reaction should be given by a rate constant times the concentration of A raised to some power.

$$\text{Rate} = k[A]^n$$

Keep in mind that the order of reaction (n) is usually an integer and is rarely greater than two. That gives us three likely orders: 0, 1, and 2. In each of these cases, if the concentration of A is doubled, the rate will change in a simple and predictable way.

1. If $n = 0$, doubling the concentration of A does not change the rate at all because any quantity raised to the 0 power is 1.
2. If $n = 1$, doubling the concentration of A *doubles* the rate.
3. If $n = 2$, doubling the concentration of A increases the rate by a *factor of 4* (because $2^2 = 4$).

Other orders of reaction are also possible and can give more complicated relationships. But these three cases cover the great majority of the reactions we will encounter. Example Problem 11.3 shows how we can use the method of initial rates to determine the rate law for a reaction involving only one reactant.

EXAMPLE PROBLEM 11.3

Earlier in the chapter we mentioned the decomposition of N_2O_5:

$$2\ N_2O_5(g) \rightarrow 4\ NO_2(g) + O_2(g)$$

Consider the following data for the kinetics of this reaction:

Experiment	Initial $[N_2O_5]$ (mol L^{-1})	Initial Rate of Reaction (mol L^{-1} s^{-1})
1	3.0×10^{-3}	9.0×10^{-7}
2	9.0×10^{-3}	2.7×10^{-6}

Determine the rate law and rate constant for this reaction at the temperature of these experiments.

Strategy To establish the rate law, determine the order with respect to the reactant by noticing how the rate changes with changing concentration. Write the rate law and calculate k by substituting the concentration and rate from one of the experiments in the rate law.

Solution Looking at the data, we see that the initial concentration triples from the first to the second experiment:

$$3(3.0 \times 10^{-3}) = 9.0 \times 10^{-3} \text{ mol L}^{-1}$$

And, the rate also triples:

$$3(9.0 \times 10^{-7}) = 2.7 \times 10^{-6} \text{ mol L}^{-1} \text{ s}^{-1}$$

Because tripling the concentration triples the rate, the reaction must be first order with respect to N_2O_5:

$$\text{Rate} = k[N_2O_5]$$

Now use the values from either experiment to determine k. Using Experiment #1 from the table,

$$9.0 \times 10^{-7} \text{ mol L}^{-1} \text{ s}^{-1} = k(3.0 \times 10^{-3} \text{ mol L}^{-1})$$

Therefore, $k = 3.0 \times 10^{-4} \text{ s}^{-1}$.

Analyze Your Answer This is a problem where we have little initial intuition, but we can still take a quick look at the numbers and see if they are reasonable by focusing on the orders of magnitude. Looking at only the powers of 10, the rates measured and the concentrations used in the experiments have differences of either 10^3 or 10^4. Getting a value for k in this range, therefore, makes sense.

Discussion The absolute size of the initial concentration does not matter when we carry out this type of experiment. Neither does the choice to double or triple that initial concentration. We could equally well have chosen to cut the initial concentration in half and then measured the new rate. The logic used to determine the rate law remains the same. If we were doing the experiment, we would probably choose to change the concentration by a factor of 2 or 3 rather than by some arbitrary amount, but even that is not really necessary. It just makes it easier to recognize the expected effects for each possible order of reaction.

Check Your Understanding Would the result change if we used Experiment 2 to determine the rate constant?

Now let's confront the situation in which two reactants are used and we need to determine the order with respect to each of them. In this case, our rate law will have three unknowns: the rate constant and the two reaction orders. To determine values for these three parameters, we must carry out at least three experiments. (This follows from the idea that at least three equations are needed to solve for three unknowns.) When working with a system that depends on several variables, it is always a good idea to try to separate the influence of one variable from the others. In this case, we can do that by holding one reactant concentration constant while changing the other to determine its effect on the rate as shown in Example Problem 11.4.

EXAMPLE PROBLEM 11.4

The study of the kinetics of real systems can be complicated. For example, there are several ways by which O_3 can be converted to O_2. One such reaction is

$$NO_2 + O_3 \rightarrow NO_3 + O_2$$

Three experiments were run at the same temperature, and the following data obtained:

Experiment	Initial $[NO_2]$ (mol L^{-1})	Initial $[O_3]$ (mol L^{-1})	Initial Rate of Reaction (mol L^{-1} s^{-1})
1	2.3×10^{-5}	3.0×10^{-5}	1.0×10^{-5}
2	4.6×10^{-5}	3.0×10^{-5}	2.1×10^{-5}
3	4.6×10^{-5}	6.0×10^{-5}	4.2×10^{-5}

Determine the rate law and rate constant for this reaction.

Strategy To determine the order with respect to each of the reactants, look for pairs of experiments that differ only in the concentration of one reactant and compare their rates. Any difference in rate *must* be due to the effect of the reactant whose concentration changed. Write the rate law and use it to determine k.

Solution

Order with respect to NO_2

From Experiments 1 and 2, we see that, if we double $[NO_2]$ while holding $[O_3]$ constant, we double the rate (within experimental error). Thus the order in NO_2 is 1.

Because rates are determined experimentally, the change in rate may not be an exact multiple of an integer.

Order with respect to O_3

From Experiments 2 and 3, we see that if we double $[O_3]$ while holding $[NO_2]$ constant, we double the rate. Thus the order in O_3 must also be 1. The rate law is

$$\text{Rate} = k[NO_2][O_3]$$

Evaluate k, using data from any of the three experiments. Here we will use Experiment 2:

$$\text{Rate} = k[NO_2][O_3]$$

$$2.1 \times 10^{-5} \text{ mol L}^{-1} \text{ s}^{-1} = k(4.6 \times 10^{-5} \text{ mol L}^{-1})(3.0 \times 10^{-5} \text{ mol L}^{-1})$$

$$k = 1.5 \times 10^{4} \text{ L mol}^{-1} \text{ s}^{-1}$$

Analyze Your Answer Once again, we can take a quick look to see if the numbers are reasonable by focusing on the powers of 10. We have two factors of 10^{-5} (the concentrations) on the right-hand side of the equation and one factor of 10^{-5} (the rate) on the left. Getting a value for k that roughly offsets one of the concentrations makes sense.

Discussion The key to solving this type of problem lies in recognizing which concentrations change and which stay the same. Sometimes when rates are reported using scientific notation, it is easy to miss a change in the exponent. These types of organizational details are the only added complication of this type of problem relative to Example Problem 11.3.

Check Your Understanding Why are the units for k in this example different from the units in Example Problem 11.3?

11-4 Integrated Rate Laws

The rate law determined by the method described in the previous section allows us to predict the rate of a reaction for a given set of concentrations. But because the concentrations of the reacting substances will change with time, the rate law does not let us easily predict the concentrations or rate at some later time. To do that, we need to formulate an equation that tells us explicitly how the concentrations will change as a function of

The differential and integrated forms of the rate law are both equally valid. The integrated rate law is useful if we want to consider concentrations as a function of time.

time. This new equation, called the **integrated rate law**, can be derived from the rate law itself. (See MathConnections box below.) The reasoning behind the integrated rate law is this: if we know all of the concentrations in a reacting system at any given time, and we know how the reaction rate changes with time, then we should be able to predict what the concentrations will be at some later time. This is similar in concept to a physics problem where you might calculate the position of a projectile from its initial position and velocity.

The form of the integrated rate law depends on the order of reaction. If we know the integrated rate laws for a few common reaction orders, we can use them as models for comparison with data. As we'll see below, this provides a useful alternative way of determining the rate law for a reaction.

MathConnections

You may have already learned about integrals in calculus. If so, you could probably guess from the name that the integrated rate law is found by integrating the rate law. Let's take a look at how that's done. (If you have not learned about integrals yet, don't worry. You can work with the integrated rate law without knowing how to derive it.) The simplest case to start with is a zero-order reaction with a single reactant we'll again call A. For such a case, the rate law is just

$$\text{Rate} = k[\text{A}]^0 = k$$

Now remember that the rate is the change in concentration over the change in time. We've been writing this in terms of macroscopic changes, using the "Δ" notation:

$$\text{Rate} = -\frac{\Delta[A]}{\Delta t} = k$$

But if we simply let the changes become infinitesimal, we can convert from deltas to differentials:

$$-\frac{d[A]}{dt} = k$$

We can rearrange this to put one of the differential terms on each side:

$$d[\text{A}] = -kdt$$

Now we can take the integral of both sides to eliminate the differentials. We will start at the beginning of the reaction, when $t = 0$ and $[\text{A}] = [\text{A}]_0$, and integrate from there to some later time when the corresponding values are just t and $[\text{A}]_t$:

$$\int_{[\text{A}]_0}^{[\text{A}]_t} d[\text{A}] = -k\int_0^t dt$$

Both of these integrals are simple to evaluate, and we get the following result:

$$[\text{A}]_t - [\text{A}]_0 = -kt$$

This is the integrated rate law for the zero-order reaction.

If you are familiar with integration, you should be able to follow the same procedure to derive the integrated rate laws for first-order and second-order kinetics.

Zero-Order Integrated Rate Law

Zero-order rate laws are relatively rare.

The simplest rate law is for a **zero-order reaction**:

$$\text{Rate} = k[\text{A}]^0 = k$$

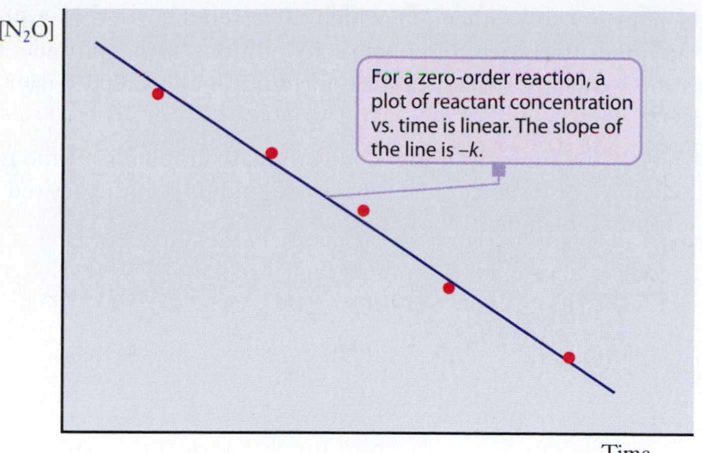
[N_2O]
For a zero-order reaction, a plot of reactant concentration vs. time is linear. The slope of the line is $-k$.
Time

Figure 11.6 ■ The decomposition of N_2O over gold is a zero-order process, for which the reactant concentration decreases linearly as a function of time.

In such instances, the rate of the reaction will not change as the reactants are consumed. The integrated rate law for this type of kinetics is given by

$$[A]_t = [A]_0 - kt \tag{11.4a}$$

Rearrange this equation slightly and compare it to the equation for a straight line with the form $y = mx + b$:

$$[A]_t = -kt + [A]_0 \tag{11.4b}$$

$$y = mx + b$$

Because the forms of these expressions match, we can see that if we plot [A] (on the y axis) as a function of t (on the x axis), we will get a straight line. We can also see that the slope of that line (m) must be equal to $-k$ and the y intercept (b) must be equal to $[A]_0$, the initial concentration of reactant A. Equation 11.4 provides us a model of the behavior expected for a system obeying a zero-order rate law. To test this model, we simply need to compare it with data for a particular reaction. So we could measure the concentration of reactant A as a function of time and then plot [A] versus t. If the plot is linear, we could conclude that we were studying a zero-order reaction. The catalytic destruction of N_2O in the presence of gold is an example of this type of kinetics. A graphical analysis of the reaction is shown in **Figure 11.6**.

Catalysis and other catalytic reactions are discussed in Section 11-7.

First-Order Integrated Rate Law

For a **first-order reaction** with a single reactant, A, the integrated rate law is given by the equations

$$\ln \frac{[A]_t}{[A]_0} = -kt \tag{11.5}$$

or

$$[A]_t = [A]_0 e^{-kt} \tag{11.6}$$

$[A]_0$ represents the initial concentration of the reactant at time $t = 0$. With these equations, if we know the rate constant and the initial concentration, we can predict the concentration at any subsequent time, as shown in Example Problem 11.5.

EXAMPLE PROBLEM 11.5

The photodissociation of ozone by ultraviolet light in the upper atmosphere is a first-order reaction with a rate constant of 1.0×10^{-5} s^{-1} at 10 km above the planet's surface:

$$O_3 + h\nu \rightarrow O + O_2$$

Consider a laboratory experiment in which a vessel of ozone is exposed to UV radiation at an intensity chosen to mimic the conditions at that altitude. If the initial O_3 concentration is 5.0 mM, what will the concentration be after 1.0 day?

Strategy Convert the time given to seconds to match the units of the rate constant and then use the first-order integrated rate law to find $[O_3]$ at the desired time. (Note that mM is just millimolar, or 10^{-3} M.)

Solution

$$1.0 \text{ day}\left(\frac{24 \text{ h}}{1 \text{ day}}\right)\left(\frac{60 \text{ min}}{1 \text{ h}}\right)\left(\frac{60 \text{ s}}{1 \text{ min}}\right) = 8.6 \times 10^4 \text{ s}$$

$$[O_3] = [O_3]_0 e^{-kt}$$

$$= 5.0 \text{ mM } (e^{-(1.0\times10^{-5}\text{ s}^{-1})(8.6\times10^4\text{ s})})$$

$$= 2.1 \text{ mM}$$

Discussion This problem reminds us of the importance of using consistent units. The argument of an exponent can never have units, so when we put in values for k and t, they must have the same time unit (seconds or minutes, for example).

Check Your Understanding In the upper stratosphere, at an altitude of about 40 km, this same reaction proceeds with a rate constant of 1.0×10^{-3} s^{-1}. (The difference is caused by the higher intensity of UV radiation at higher altitude.) If an experiment were to simulate these conditions, how long would it take for an initial ozone concentration of 5.0 mM to fall to 2.1 mM?

Figure 11.7 shows a plot of $[O_3]$ versus time for the experiment described in Example Problem 11.5.

A quick glance at Figures 11.6 and 11.7 shows a glaring difference between the two plots of concentration versus time: one is linear and the other is clearly not. The curve shown for the first-order reaction in Figure 11.7 is an example of exponential decay. You are probably familiar with this type of curve from math classes, and you will see many other physical phenomena that show this kind of behavior as you continue your engineering studies.

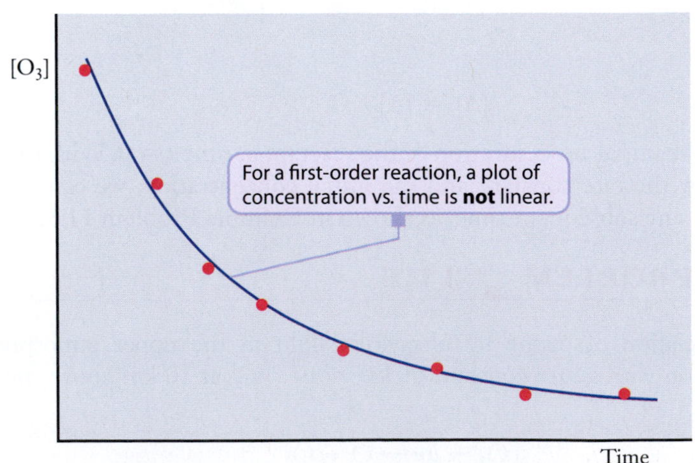
[O3]
For a first-order reaction, a plot of concentration vs. time is **not** linear.
Time

Figure 11.7 ■ This plot of $[O_3]$ vs. time shows that the dissociation of ozone does not follow a zero-order rate law because the graph is clearly nonlinear.

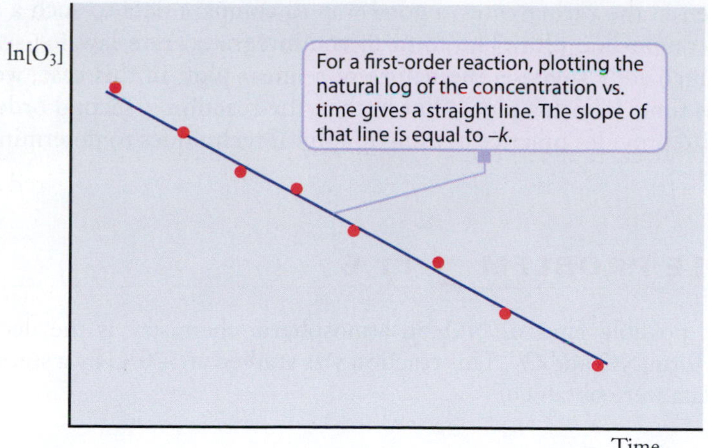
ln[O_3]
For a first-order reaction, plotting the natural log of the concentration vs. time gives a straight line. The slope of that line is equal to –k.
Time

Figure 11.8 ■ The first-order integrated rate law predicts that a plot of the natural logarithm of reactant concentration versus time should be linear. Here we see that this is the case for the dissociation of ozone.

Suppose that we had measured the data shown in Figure 11.7, but we did not yet know the order of the reaction. The fact that the plot of concentration versus time is not linear tells us that the reaction is *not* zero order; it doesn't fit the model for that rate law. Next we might ask if the reaction is first order. We can answer this question by comparing the data to the model of first-order behavior. One way to do this is to attempt to fit an exponential decay (Equation 11.6) to the data. This might be done using a graphing calculator, a spreadsheet, or some other tool that can fit nonlinear functions. If we are examining the data visually, though, a curved plot is not as easy to judge as a straight line, so we might want to manipulate the data so that a straight line results. This requires only a little algebra to arrange the integrated rate law into a linear form:

Transforming data to give a linear plot is a convenient way to compare the data with a model.

$$\ln[\mathrm{A}] = -kt + \ln[\mathrm{A}]_0$$

Now we have an equation that provides a straight line when we plot ln[A] versus time. **Figure 11.8** shows the same data as Figure 11.7, manipulated to give such a straight-line plot.

At this point, we can see that (1) if a plot of [A] versus time is linear, the reaction is zero order; and (2) if a plot of ln[A] versus time is linear, the reaction is first order. From the form of the integrated rate law equations for these cases, you should also be able to see that the slope of the linear plot must be equal to $-k$. So we can find both the rate constant and the reaction order from our graph. Next, we will examine the correct model for second-order kinetics.

The quantity ln[A] has no units, so the units of the slope—and of the first-order rate constant—are simply 1/time.

Second-Order Integrated Rate Law

For **second-order reactions**, the mathematical form of the integrated rate law is quite different from the one we used for first-order reactions. For the consumption of reactant A, the integrated rate law is

$$\frac{1}{[\mathrm{A}]_t} - \frac{1}{[\mathrm{A}]_0} = kt \tag{11.7}$$

As before, $[\mathrm{A}]_t$ is the concentration of X at time t and $[\mathrm{A}]_0$ is the initial concentration. If we know the value of k, we can use this equation and the same strategies used in Example Problem 11.5 to consider second-order reactions. This equation provides us with a mathematical model for a second-order reaction as a function of time. And as we've seen in the earlier cases, a good way to compare data to such a model is to rely on a straight-line plot. The form of the integrated rate law for second-order kinetics immediately suggests the nature of a linear plot; in this case, we must plot 1/[A] versus time. If such a plot is linear, then the reaction is second order. Example Problem 11.6 provides practice at using graphical techniques to determine the order of reactions.

EXAMPLE PROBLEM 11.6

Among the possible fates of NO_2 in atmospheric chemistry is the decomposition reaction to form NO and O_2. This reaction was studied at 370°C by a student, and the following data were obtained:

Time (s)	$[NO_2]$ (mol L^{-1})
0.0	3.00×10^{-1}
5.0	1.97×10^{-2}
10.0	1.00×10^{-2}
15.0	7.00×10^{-3}
20.0	5.20×10^{-3}
25.0	4.10×10^{-3}
30.0	3.50×10^{-3}

Based on these data, determine the order of this reaction and the rate constant.

Strategy The available data are the reactant concentration as a function of time for a single experiment, so we will need to use graphical techniques to determine the order of the reaction. There are three possibilities we can explore using the integrated rate laws we've examined. The reaction could be zero order, first order, or second order with respect to NO_2. We will need to manipulate and plot the data in various ways to determine whether there is a good fit with any of these models. (Other orders are also possible, so we should be aware that all three tests could conceivably fail.) With a spreadsheet or a graphing calculator, such manipulation of data is easy. For this example, first we will calculate all of the data needed for all three plots and then make the appropriate graphs to find the linear relationship and determine the rate law.

Solution The three versions of concentration we need to plot versus time are $[NO_2]$ (zero-order model), $\ln[NO_2]$ (first-order model), and $1/[NO_2]$ (second-order model). The following table gives the data we will need to plot:

Time	$[NO_2]$	$\ln[NO_2]$	$1/[NO_2]$
0.0	3.00×10^{-1}	−1.20	3.33
5.0	1.97×10^{-2}	−3.93	50.8
10.0	1.00×10^{-2}	−4.61	100.
15.0	7.00×10^{-3}	−4.96	140.
20.0	5.20×10^{-3}	−5.26	190.
25.0	4.10×10^{-3}	−5.50	240.
30.0	3.50×10^{-3}	−5.65	290.

First, plot concentration versus time to determine whether the reaction is zero order:

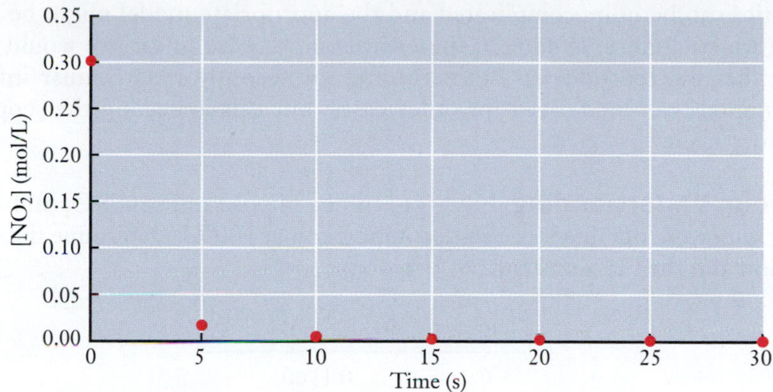

This plot is clearly not linear, so we must proceed to the other possibilities. To check for first-order kinetics, we need to plot $\ln[NO_2]$ versus time:

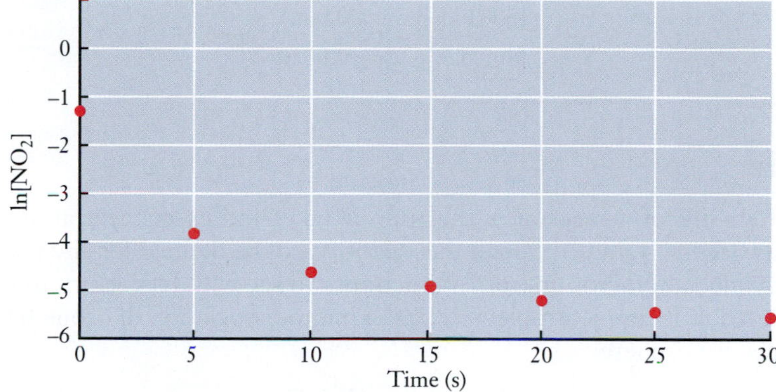

Just like the plot for zero-order kinetics, this plot is not linear, so the reaction is not first order. Our final option is to plot $1/[NO_2]$ versus time to see if the reaction is second order:

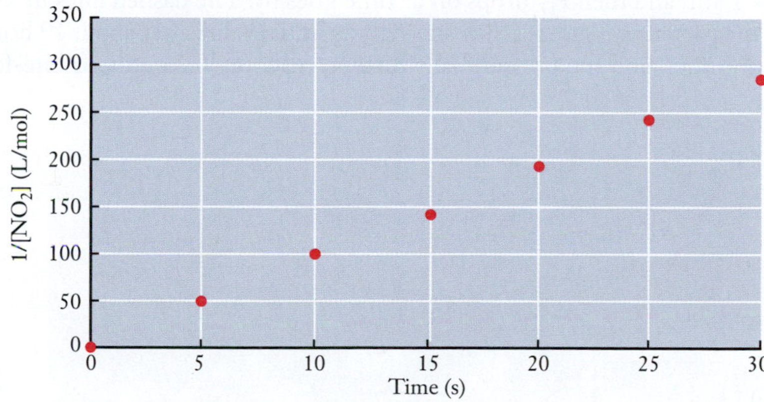

This plot is linear, so the reaction must be second order. To determine the rate constant, we need to determine the slope of the line. Estimating from the graph, the line crosses 50 at 5 seconds and 200 at 21 seconds. Calculating the slope (rise/run) from those numbers, we have 150/16, which is 9.4 L mol^{-1} s^{-1}. (The units are also given by the rise/run concept from the graph.)

Linear regression functions in a calculator or spreadsheet can determine the slope of a best-fit line.

Discussion Because this is an introductory text, we have chosen our examples so that they will fit one of these three simple models. In some cases, however, reaction kinetics can be quite complicated and the appropriate model might be none of the three we tried here. If none of these three plots were linear, we would have to conclude that the reaction was not zero, first, or second order. Similar integrated rate law models can be derived for other cases, but this is beyond the scope of an introductory class.

Check Your Understanding $N_2O_5(g)$ can decompose to form NO_2 and O_2. The following data were obtained for this decomposition at 300°C. Determine the order of the reaction and the rate constant.

Time (s)	$[N_2O_5]$ (mol L^{-1})
0.0	0.1500
300.0	0.0729
600.0	0.0316
900.0	0.0203
1200.0	0.0108
1500.0	0.0051
1800.0	0.0020

Half-Life

The **half-life** of a reactant is the time it takes for its concentration to fall to one-half its original value. Although this quantity can be defined for any reaction, it is particularly meaningful for first-order reactions. To see why, let's return to the system we considered in Example Problem 11.5, the photodissociation of ozone by UV light in the upper atmosphere:

$$O_3 + h\nu \rightarrow O + O_2$$

We said that this is a first-order reaction with a rate constant of $1.0 \times 10^{-5}\ s^{-1}$ at an altitude of 10 kilometers above the earth's surface. Suppose that we designed a laboratory experiment in which the temperature and light levels at that altitude could be reproduced. **Figure 11.9** shows the type of results we might obtain. The experiment begins with $P_{O_3} = 1$ atm and then P_{O_3} drops off as time goes by. The dashed lines in the figure show that the partial pressure reaches one-half its initial value after about 19 hours. The reaction continues, and the pressure falls further, reaching 0.25 atm, or one-fourth of

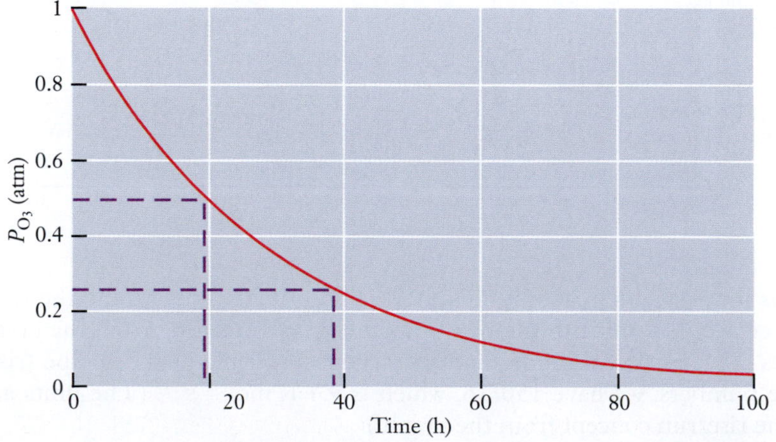

Figure 11.9 ■ Ozone pressure as a function of time in an experiment designed to model the destruction of ozone in the stratosphere. The ozone pressure is reduced by 1/2 every 19 hours.

its initial value after about 38 hours. In the first 19-hour period, the ozone pressure decreased by one-half its value, from 1 to 0.5. Between about 19 hours and 38 hours on the graph, the pressure again decreased by one-half, this time falling from 0.5 to 0.25. For the particular conditions of this experiment, the ozone pressure decreases by a factor of 2 every 19 hours. So the half-life is constant, no matter how much ozone we start with.

We can obtain a mathematical expression for the half-life of a first-order reaction by substituting in the integrated rate law (Equation 11.5). By definition, when the reaction has been proceeding for one half-life ($t_{1/2}$), the concentration of the reactant must be $[A] = \frac{1}{2}[A]_0$. Thus we have

$$\ln\left(\frac{\frac{1}{2}[A]_0}{[A]_0}\right) = \ln\left(\frac{[A]_0}{2[A]_0}\right) = -kt_{1/2}$$

which simplifies to

$$\ln\frac{1}{2} = -kt_{1/2}$$

$$\ln 1 - \ln 2 = -kt_{1/2}$$

or

$$\ln 2 = kt_{1/2}$$

Solving this equation for the half-life gives

$$t_{1/2} = \frac{\ln 2}{k} = \frac{0.693}{k} \tag{11.8}$$

Equation 11.8 relates the half-life of *any first-order reaction* to its rate constant. Because k does not depend on the amount of substance present, neither does $t_{1/2}$. The half-life is most often used to describe the kinetics of nuclear decay. All radioactive decay processes follow first-order kinetics, and people working with nuclear processes typically use half-lives rather than rate constants to discuss these systems. According to Equation 11.8, we can easily convert the half-life to the rate constant (or vice versa), so knowing either of these values gives us the same amount of information about the reaction kinetics. (For nuclear decay, the half-life is also independent of temperature.) Example Problem 11.7 demonstrates the link between the rate constant and the half-life.

Radioactive decay and its kinetics are discussed in detail in Chapter 14.

EXAMPLE PROBLEM 11.7

The rate of the photodissociation of ozone in the example shown in Figure 11.9 may seem slow. But it is actually tremendously faster than what we would see in the absence of ultraviolet light. The rate constant, k, for the thermal decomposition of ozone in the dark at 25°C is just $3 \times 10^{-26}\ s^{-1}$. What is the half-life of ozone under these conditions?

Strategy We can see from the units of the rate constant that this is a first-order reaction. So we can use the equation for the first-order half-life.

Solution

$$t_{1/2} = \frac{0.693}{k} = \frac{0.693}{3 \times 10^{-26}\ s^{-1}} = 2 \times 10^{25}\ s$$

Analyze Your Answer A number this large may seem disconcerting. But if we stop to realize that the rate constant and the half-life are inversely related to each other, such a large number makes sense. If the rate constant is small, the half-life must be long.

Discussion Comparing this result with the half-life of 19 hours (or about 68,000 s) we saw above shows the importance of UV photons in the photodissociation reaction. In the absence of light, the rate constant is smaller by more than 20 orders of magnitude, making the half-life tremendously long.

Check Your Understanding The half-life of the photochemical decomposition of N_2O_5 in the stratosphere (at 40 km) is roughly 43,000 s. What is the rate constant for the reaction?

11-5 Temperature and Kinetics

The most widely recognized causes of ozone depletion in the stratosphere were chemicals that were widely used in refrigerators for many years—chlorofluorocarbons (CFCs). Freon-12® is the registered trademark for CCl_2CF_2, the particular chlorofluorocarbon that was most widely used in refrigerators and air conditioners.

Have you ever asked yourself why you need a refrigerator? Aside from keeping your favorite beverage refreshingly cold, what advantages does the cooled storage space provide? The answer lies in the temperature dependence of the rates of chemical reactions. Without a refrigerator, food would spoil more quickly. Maintaining healthful foodstuffs is a critically important use for refrigeration. It works because at cooler temperatures, chemical reactions—including those that allow the growth of dangerous bacteria in food—are slower. You can measure this temperature effect with a simple experiment. Set up one sample of milk at room temperature and another sample in the refrigerator. Determine the time it takes the milk to smell sour. It is pretty obvious how this will turn out: the room temperature milk will sour much faster. Why does temperature affect reaction kinetics, and how can we quantify that effect?

Temperature Effects and Molecules That React

We can investigate the role of temperature in kinetics by referring to the most useful model for describing the motion of molecules—the kinetic-molecular theory. Let's consider the events that lead to a chemical reaction between two gas-phase molecules. (The same ideas apply to reactions involving solids and liquids, too. The gas-phase case is just easier for us to discuss.)

We discussed the kinetic-molecular theory in Section 5-6.

The kinetic-molecular theory says that molecules interact with one another only through collisions. So, for two molecules to react, first they must collide. We should not expect that all collisions between molecules result in chemical reactions, though. Our experience suggests that the air surrounding us is stable, even though the N_2 and O_2 molecules are constantly colliding. So when nitrogen and oxygen collide, the probability of reaction must be very low. If collisions alone cannot explain reactivity, what additional variables come to mind? One possibility is the kinetic energy of the colliding molecules. Slow moving molecules collide with gentle impacts that do not usually result in reaction, whereas fast moving molecules seem more likely to react.

Recall from Section 5-6 that the Boltzmann distribution describes the speeds of molecules. **Figure 11.10** shows the speed distributions for gas molecules at two temperatures. At the higher temperature, a larger fraction of the molecules moves at high speeds. Thus at higher temperatures, reactions tend to proceed more quickly because more of the collisions between reactants have the high energies needed to bring about reaction.

Why do reactions require high-energy collisions? Chemical reactions involve the breaking and forming of chemical bonds. From thermodynamics, we know that breaking bonds requires energy input and forming bonds liberates energy. Intuition suggests

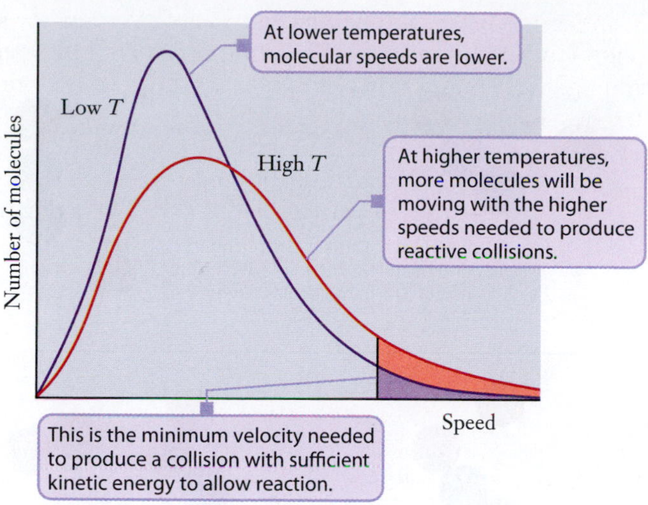
At lower temperatures, molecular speeds are lower.
Low T
High T
At higher temperatures, more molecules will be moving with the higher speeds needed to produce reactive collisions.
Number of molecules
Speed
This is the minimum velocity needed to produce a collision with sufficient kinetic energy to allow reaction.

Figure 11.10 ■ The two curves show the Maxwell-Boltzmann distribution of molecular speeds for two temperatures. The shaded areas represent the number of molecules traveling fast enough to collide with sufficient energy to react. Note that there are many more fast moving molecules at higher temperature.

that before new bonds can form, old ones must break, or at least begin to break. So, many chemical reactions need energy input at the outset to start breaking bonds in reactants. **Figure 11.11** provides one way to visualize this requirement. Standard thermodynamic data allow us to predict whether a reaction is exothermic or endothermic, as we saw in Chapter 9. But to get from reactants to products, the reaction must first overcome an energy threshold, called the **activation energy (E_a)** or activation barrier. Collisions of fast-moving particles provide sufficient kinetic energy to overcome the activation barrier. The larger this barrier, the more kinetic energy will be needed. So the activation energy is what dictates how energetic a collision must be to lead to reaction.

Collision geometry also plays a role in whether a collision between fast-moving molecules is effective. As you can see in **Figure 11.12**, a collision between a molecule of N_2O and an oxygen atom may or may not lead to a reaction, depending on the geometry of the collision. At the moment of an effective collision, both bond breaking and bond formation are occurring. For an instant, as bond rearrangement is occurring, an unstable intermediate species, called an **activated complex**, exists

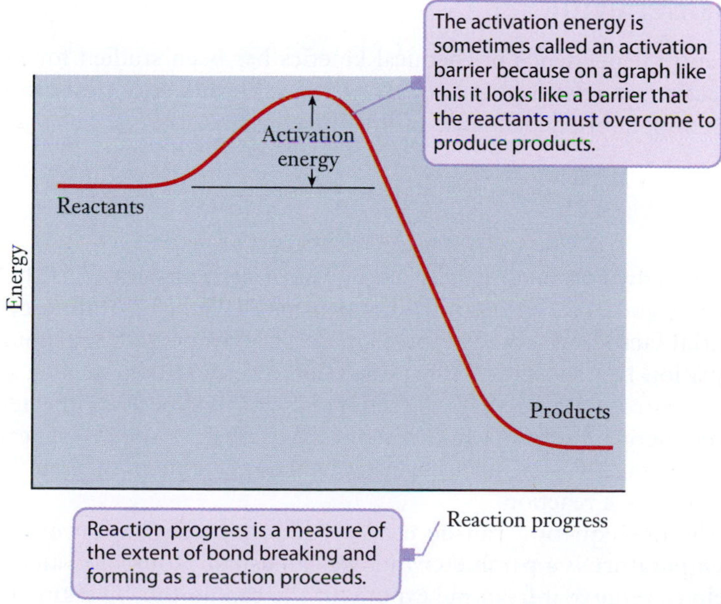
The activation energy is sometimes called an activation barrier because on a graph like this it looks like a barrier that the reactants must overcome to produce products.
Activation energy
Reactants
Energy
Products
Reaction progress
Reaction progress is a measure of the extent of bond breaking and forming as a reaction proceeds.

Figure 11.11 ■ Potential energy plot for an exothermic chemical reaction. To progress from reactants to products, the molecules must collide with enough energy to pass over the activation barrier.

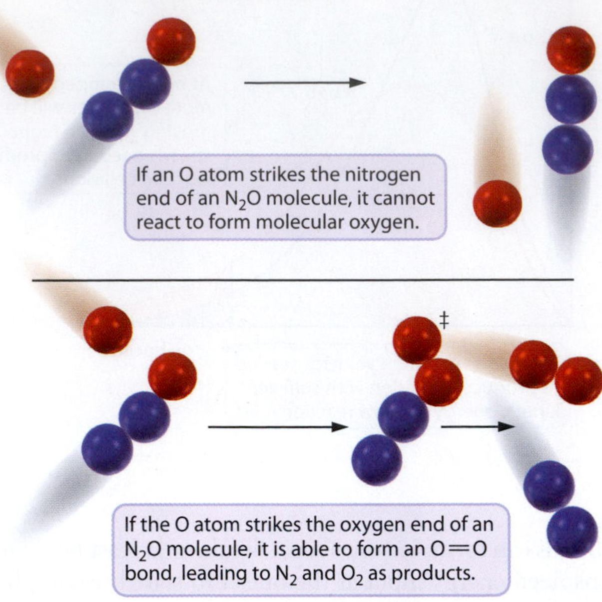

If an O atom strikes the nitrogen end of an N_2O molecule, it cannot react to form molecular oxygen.
‡
If the O atom strikes the oxygen end of an N_2O molecule, it is able to form an O=O bond, leading to N_2 and O_2 as products.

Figure 11.12 ■ Geometric factors can also be important in determining whether a molecular collision is reactive. For the oxygen atom to react with N_2O, it must strike the oxygen end of the molecule. The importance of collision geometry depends on the shapes of the molecules in the reaction.

in the reaction mixture. (This activated complex is denoted ‡ in Figure 11.12.) The activated complex represents the highest energy point along the route from reactants to products. Because it is highly unstable, the activated complex may have a lifetime as short as 10^{-15} s. Formation of an activated complex and its subsequent decomposition into reaction products depends on both the energy and the geometry of the collision.

More specifically, the activated complex is the highest energy point along the minimum energy path to products.

Arrhenius Behavior

The temperature dependence of chemical kinetics has been studied for many years. For most reactions, a relationship known as the **Arrhenius equation** can be used to describe the temperature dependence of the rate constant, k:

Svante Arrhenius also studied acids and bases. The notion that acids form H^+ and bases form OH^- in water is attributed to him.

$$k = Ae^{-E_a/RT} \tag{11.9}$$

Here E_a is the activation energy, R is the universal gas constant, T is the temperature (in kelvins), and A is a proportionality constant called the **frequency factor** or **preexponential factor.** We noted earlier that a large activation energy should hinder a reaction. Equation 11.9 shows this effect: As E_a increases, k will be smaller, and smaller rate constants correspond to slower reactions. Note that because the temperature and activation energy appear in the exponent, the rate constant will be very sensitive to these parameters. That's why fairly small changes in temperature can have drastic effects on the rate of a reaction.

The Arrhenius equation can be used to determine activation energy experimentally. Temperature is a parameter that we can usually control in an experiment, so it may help to remove it from the exponent. We can do this by taking the natural

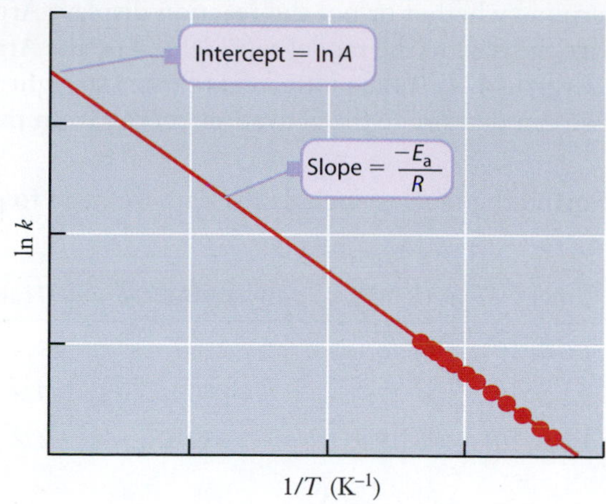

Figure 11.13 ■ According to the Arrhenius equation, a plot of ln k vs. $1/T$ should give a straight line with a slope equal to $-E_a/R$. Such a plot is frequently used to determine the activation energy of a reaction.

The quantity ln k is considered unitless since the logarithm of a unit has no physical meaning. So the units of the slope of this graph will be kelvins.

logarithm of both sides of Equation 11.9:

$$\ln k = \frac{-E_a}{RT} + \ln A$$

Rearranging, we get

$$\ln k = \frac{-E_a}{R}\left(\frac{1}{T}\right) + \ln A$$

This shows us how we can determine values for E_a or A. We will need to measure k for at least two temperatures, but we already know ways to do that. The rearranged equation is in the familiar form of a line, $y = mx + b$, so a plot of ln k versus $1/T$ should be a straight line. The line's slope is $-E_a/R$, where R is the gas constant, whose value is known. So we can determine the activation energy from the slope, as shown in **Figure 11.13** and Example Problem 11.8. This type of problem is easily attacked with a graphing calculator or a spreadsheet, and there are several such problems at the end of this chapter.

When working with the Arrhenius equation, the most commonly used form for R is 8.314 J K^{-1} mol^{-1}. It is essential that all units are chosen so that the quantity (E_a/RT) is unitless.

EXAMPLE PROBLEM 11.8

In the troposphere, ozone can be converted to O_2 by the following reaction with hydroxyl radicals:

$$HO\cdot(g) + O_3(g) \rightarrow HO_2\cdot(g) + O_2(g)$$

The following values for the rate constant, k, for this reaction were measured in experiments at various temperatures:

k (L mol^{-1} s^{-1})	Temperature (K)
1.0×10^7	220
5.1×10^7	340
1.1×10^8	450

(a) Does this reaction display Arrhenius behavior? (b) Estimate the activation energy from these data.

Strategy To determine whether or not the reaction displays Arrhenius behavior, we will need to compare the data to the model represented by the Arrhenius equation. To do this, we plot ln k versus $1/T$. If the plotted data form a straight line, the slope of the line is $-E_a/R$. So, we can determine the activation energy from that slope.

Solution First, expand the table to include the data we need to plot:

k (L mol^{-1} s^{-1})	ln k	Temperature (K)	$1/T$ (K^{-1})
1.0×10^7	16.1	220	4.5×10^{-3}
5.1×10^7	17.7	340	2.9×10^{-3}
1.1×10^8	18.5	450	2.2×10^{-3}

Now plot ln k (y axis) versus $1/T$ (x axis):

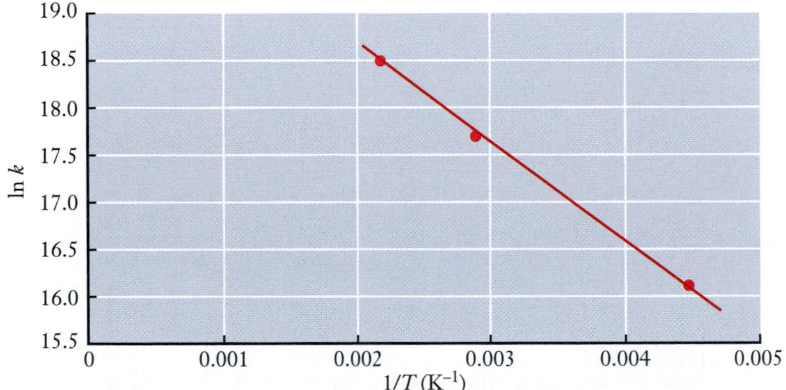

The data do appear to fall on a straight line. So our answer to question (a) is that yes, the reaction does exhibit Arrhenius behavior. (It would be better if we had more than three data points on which to base this conclusion.) To determine the slope of the straight line, we could use the curve-fitting routine built into a graphing calculator or spreadsheet program. Or we can determine the slope from two points on the line. If our straight line for these data includes the first and last points, which the graph suggests as a reasonable estimate, we can use data from our expanded table directly:

$$\text{Slope} = \frac{\text{Rise}}{\text{Run}} = \frac{\Delta \ln k}{\Delta\left(\frac{1}{T}\right)} = \frac{18.5-16.1}{(2.2 \times 10^{-3}-4.5 \times 10^{-3})\ \text{K}^{-1}}$$

$$= -1.04 \times 10^3\ \text{K} = -\frac{E_a}{R}$$

We can insert $R = 8.314\ \text{J K}^{-1}\ \text{mol}^{-1}$ and solve for E_a:

$$E_a = (1.04 \times 10^3\ \text{K})\ (8.314\ \text{J K}^{-1}\ \text{mol}^{-1})$$
$$= 8.7 \times 10^3\ \text{J mol}^{-1} = 8.7\ \text{kJ mol}^{-1}$$

Analyze Your Answer To assess this answer, we need some familiarity with the magnitude of activation energies. Typical E_a values are on the order of tens of kJ mol^{-1}, so this is somewhat small. But free radicals are very reactive, and that is consistent with a small activation energy.

Discussion Often the best-fit line for experimental data does not explicitly cross through any of the individual data points. Graphing calculators or spreadsheet programs provide the best and simplest way to determine the slope of such a line. But it is important to keep in mind that the slope calculation is only meaningful if the straight-line model provides a reasonable fit to the data. It is always a good idea to look at the data and the best-fit line on the same graph to verify that you are satisfied with the quality of the fit. If the data do not suggest a linear relationship, then we should not put our confidence in the meaning of a slope or other parameter taken from a linear fit.

Check Your Understanding Once generated, $HO_2\cdot$ may react with carbon monoxide to reform the hydroxyl radical and carbon dioxide. This reaction plays a role in ozone pollution in major cities. Assume that an experiment to determine the rate constant of this reaction provides the data tabulated here. What is the activation energy of this reaction?

k (L mol^{-1} s^{-1})	Temperature (K)
5.9	301
4.7×10^4	650

For most students, the most challenging mental connections that are forged in general chemistry are those between bulk macroscopic properties and molecular-level origins of those bulk properties. The Arrhenius equation provides us with a good opportunity to focus on this type of mental connection:

$$k = Ae^{-E_a/RT}$$

There are two important factors in this equation—the argument of the exponent and the frequency factor, A. **Figure 11.14** shows how these two components of the equation arise from considerations at the molecular level. The figure shows a reaction in which two molecules of ClO are converted to Cl_2 and O_2:

$$2\,ClO \rightarrow Cl_2 + O_2$$

In keeping with our usual color code, the figure shows chlorine atoms in green and oxygen atoms in red. This particular reaction was chosen because it is related to the atmospheric chemistry of ozone. But the same arguments can be applied to any other reaction, too.

Perhaps the effect of temperature on the Boltzmann distribution is familiar by now, but we show it for two separate cases—low temperature (2) and higher temperature (3). We have noted previously that Boltzmann distributions imply that more molecules can react at higher temperatures. In the final three illustrations, we depict the connection between the molecular velocities that are incorporated in (2) and (3) and reactivity.

Drawing (4) shows that the slow-moving molecules have insufficient energy to react; the collision energy does not exceed the activation energy. In (5), however, the high-velocity molecules have large collision energies—in excess of the activation energy—and reaction occurs. We could just as easily have drawn this effective collision originating from the higher temperature Boltzmann distribution, where even more molecules will undergo these high-energy collisions. The accompanying MathConnections section examines how the equation makes these ideas quantitative.

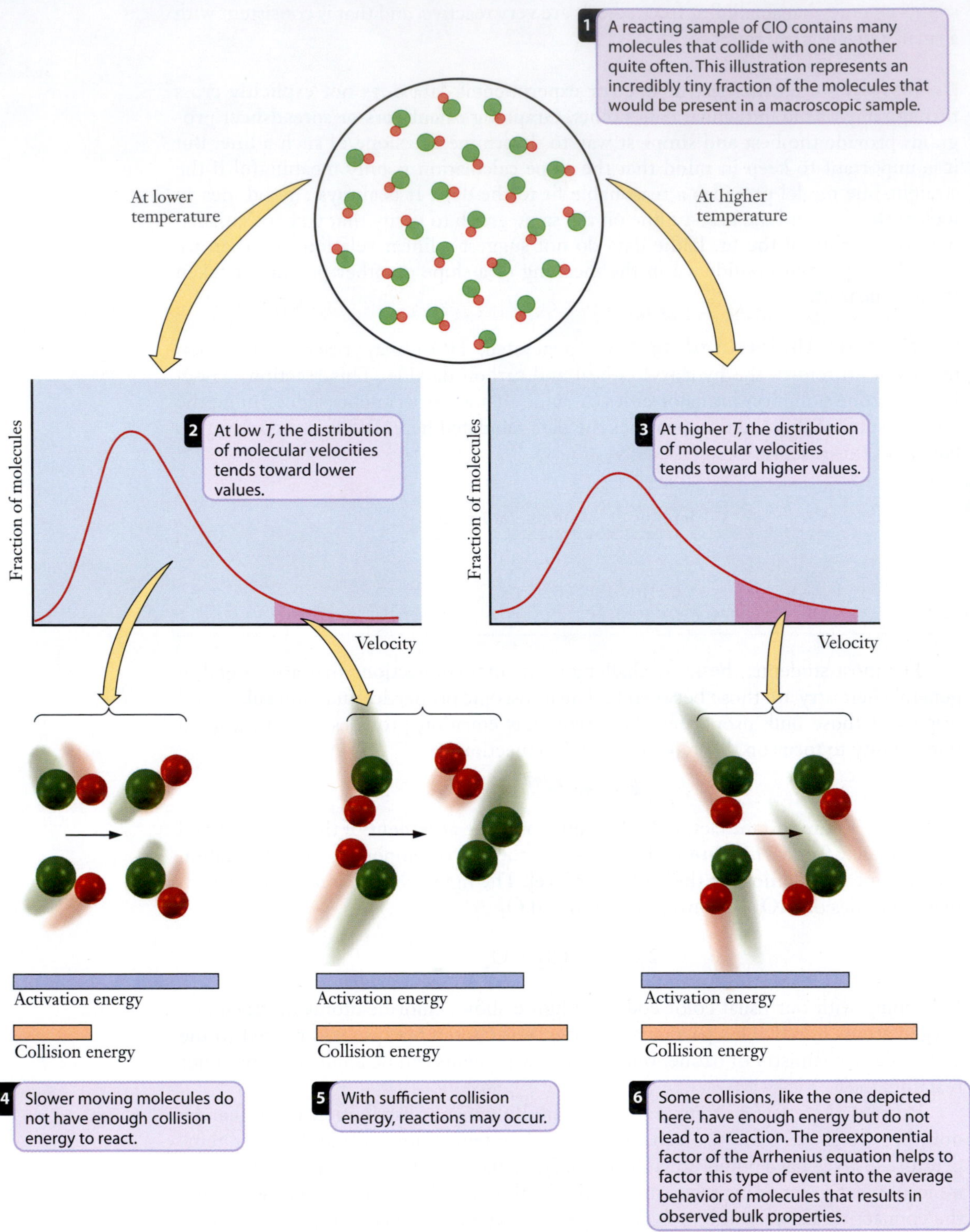

1 A reacting sample of ClO contains many molecules that collide with one another quite often. This illustration represents an incredibly tiny fraction of the molecules that would be present in a macroscopic sample.
At lower temperature
At higher temperature
Fraction of molecules
2 At low T, the distribution of molecular velocities tends toward lower values.
Velocity
Fraction of molecules
3 At higher T, the distribution of molecular velocities tends toward higher values.
Velocity
Activation energy
Collision energy
Activation energy
Collision energy
Activation energy
Collision energy
4 Slower moving molecules do not have enough collision energy to react.
5 With sufficient collision energy, reactions may occur.
6 Some collisions, like the one depicted here, have enough energy but do not lead to a reaction. The preexponential factor of the Arrhenius equation helps to factor this type of event into the average behavior of molecules that results in observed bulk properties.

Figure 11.14 ■ The connection between temperature, collision energy, and reaction rate is developed here. The activation energy establishes a minimum threshold that must be overcome before a collision can lead to reaction. The number of energetic molecules increases with temperature, according to the Boltzmann distribution.

MathConnections

Figure 11.14 shows how important it is for the collision energy to exceed the activation barrier, but how is this concept expressed in the Arrhenius equation? The argument of the exponential function is the ratio of these two values. Recall from Section 5-6 where we introduced gas kinetic theory that the average velocity (the root-mean-square velocity, to be more precise) is given by the equation

$$\left[v_{\text{rms}} = \sqrt{\frac{3RT}{MM}}\right]$$

MM is the molar mass of the gas. If we square both sides of this equation, we can isolate *RT*:

$$\left[RT = \frac{1}{3}MMv_{\text{rms}}^2\right]$$

Noting that the molar mass is the mass of a molecule (m) times Avogadro's number (N_A), we can see that

$$RT = \frac{N_A}{3}mv_{\text{rms}}^2$$

Because kinetic energy is expressed as $\frac{1}{2}mv^2$, *RT* is directly related to the average kinetic energy:

$$RT = \frac{2N_A}{3}\left(\frac{1}{2}mv_{\text{rms}}^2\right) = \frac{2\,N_A}{3}\,\text{KE}_{\text{avg}}$$

Thus the macroscopic Arrhenius equation is based on the average kinetic energy of the molecules in a sample. The Boltzmann distribution can be evaluated to quantify that average behavior, and the connection between macroscopic and microscopic is made. The Arrhenius equation contains a ratio of

$$\frac{\text{Activation energy}}{\text{Kinetic energy}}$$

When we change the kinetic energy part of this ratio by changing the temperature, the activation energy remains the same.

The frequency factor, *A*, accounts for the collision frequency, which will depend on the sizes of the molecules involved. It also reflects the fact that even at high temperatures and high collision energies, not all collisions result in reaction, due to geometric effects like those illustrated in Figure 11.12. For the reaction considered in Figure 11.14, an effective collision might require that the oxygen ends of the two ClO molecules collide directly with one another, for example, to allow the O═O bond to begin to form.

11-6 Reaction Mechanisms

Earlier, we noted the net reaction of the Chapman cycle: $3\ O_2 \rightarrow 3\ O_2$. In some ways, it appears pointless to talk about the kinetics of a reaction that has no net change. But as we have already seen, the importance of the Chapman cycle lies in the details. Without the Chapman cycle, the upper atmosphere would not have an ozone layer, and the quantity of harmful UV radiation reaching the surface of the planet would be significantly greater. This natural cycle points to the importance of the mechanism of the reaction. A **reaction mechanism** is a collection of one or more molecular steps that account for the way reactants become products. If the overall equation is like a parts list and a finished product, then the mechanism is like the assembly instructions. In many cases, including the Chapman cycle, if we want to understand the kinetics of the process we must consider the reaction mechanism. What is the stepwise process of bond breaking and bond formation behind the overall reaction?

The study of reaction mechanisms on a molecular level is referred to as chemical dynamics.

Elementary Steps and Reaction Mechanisms

We have already reasoned that chemical reactions occur when reactants collide. Based on the kinetic-molecular theory, collisions between two particles should be rather common. Three-particle collisions are much less likely because they require that three randomly moving species arrive at the same place at the same time. And four-particle collisions are virtually unheard of. However, when we look at the stoichiometric coefficients of chemical equations, they seem to imply that collisions can occur among many particles: balanced equations often involve stoichiometric coefficients far greater than 2 or 3. To understand reaction mechanisms, we need to distinguish between the overall stoichiometry of a reaction and the steps in the reaction mechanism.

The individual steps in a reaction mechanism are called **elementary steps**. Unlike the overall stoichiometric equation, the coefficients for reactants in an elementary step *do* provide the exponents in the rate law for that step. According to our reasoning above, only three types of elementary steps are likely to occur, those involving one, two, or three molecules. Steps with one reactant are called **unimolecular**, and those with two and three reactants are called **bimolecular** and **termolecular**, respectively. The molecularity tells us the overall order of the rate law for the elementary step, as summarized in **Table 11.1**.

It is very important to remember that we can use the stoichiometric coefficients as the reaction orders only *when we are considering an elementary step!*

Let's consider a particular reaction mechanism so that we can define some other terms. The destruction of ozone by chlorine radicals is expressed by the net reaction, $2\ O_3 \rightarrow 3\ O_2$. The accepted mechanism for this reaction is

$$Cl\cdot + O_3 \rightarrow ClO\cdot + O_2$$

$$ClO\cdot + O_3 \rightarrow Cl\cdot + 2\ O_2$$

$$\text{Net:} \quad 2\ O_3 \rightarrow 3\ O_2$$

Notice that the equations for elementary steps look just like those for any other chemical reaction. So it is important to remember that when we write a reaction

Table ■ 11.1

Summary of molecularity of elementary reactions

Type of Elementary Reaction	Molecularity	Rate Law
$A \rightarrow$ products	Unimolecular	Rate = $k[A]$
$A + B \rightarrow$ products	Bimolecular	Rate = $k[A][B]$
$2\ A \rightarrow$ products		Rate = $k[A]^2$
$A + B + C \rightarrow$ products	Termolecular	Rate = $k[A][B][C]$
$2\ A + B \rightarrow$ products		Rate = $k[A]^2[B]$
$A + B + C + D \rightarrow$ products	Not observed	

mechanism, each equation is always an elementary step. In the mechanism above, you can see two important characteristics of reaction mechanisms:

1. The ClO· generated in the first reaction is consumed in the second reaction. A chemical species that is generated in one step and consumed in a later step is called an intermediate or a **reactive intermediate**. Many mechanisms involve one or more intermediates.
2. When the steps of the mechanism are properly summed, they give the observed stoichiometry of the overall reaction.

An intermediate in the overall reaction will be a product in one elementary step and a reactant in another.

Because of the enormous number of collisions occurring at any moment in a gas-phase reaction and the rapid formation and decomposition of reactive intermediates, reaction mechanisms can never be known with certainty. However, by observing reaction rates and comparing proposed mechanisms with experimental rates and with the overall stoichiometry, we can propose a mechanism that fits available data. If more than one mechanism seems plausible, we may be able to design additional experiments to try choosing between them. Example Problem 11.9 provides some practice in identifying valid mechanisms and the terms associated with them.

EXAMPLE PROBLEM 11.9

The decomposition of N_2O_5 is given by the equation

$$2\ N_2O_5(g) \rightarrow 4\ NO_2(g) + O_2(g)$$

The following mechanism is proposed for this reaction:

$$N_2O_5 \rightarrow NO_2 + NO_3$$

$$NO_2 + NO_3 \rightarrow NO_2 + NO + O_2$$

$$NO + NO_3 \rightarrow 2\ NO_2$$

(a) Does this mechanism as written provide the correct stoichiometry? If not how does it need adjustment? (b) Identify all intermediates in the mechanism. (c) Identify the molecularity of each step in the mechanism.

Strategy Consider the definitions for reaction mechanism and see how this example of a mechanism illustrates those definitions.

Solution

(a) A simple sum of the mechanism yields

$$N_2O_5 + NO_3 \rightarrow O_2 + 3\ NO_2$$

This is not the desired stoichiometry, but if we multiply the first step by 2, we have

$$2\ N_2O_5 \rightarrow 2\ NO_2 + 2\ NO_3$$

Adding that to steps 2 and 3,

$$2\ N_2O_5 \rightarrow 2\ NO_2 + 2\ NO_3$$

$$NO_2 + NO_3 \rightarrow NO_2 + NO + O_2$$

$$NO + NO_3 \rightarrow 2\ NO_2$$

yields

$$2\ N_2O_5 \rightarrow O_2 + 4\ NO_2$$

So the mechanism *does* provide the correct stoichiometry, as long as we realize that some steps may need to occur more often than others.

(b) The intermediates in this case are NO_3 and NO.
(c) The first step is unimolecular, whereas steps two and three are bimolecular.

Discussion This type of problem solving is really performed in reverse. We knew what the target reaction had to be (the end of the problem) and had to go back to other data given to solve the problem. Reaction mechanisms certainly aren't the only time we use this problem-solving strategy in general chemistry. Stop for a moment and try to think of another type of chemistry problem where working the problem in reverse makes the most sense.

Check Your Understanding The following mechanism was once proposed for the decomposition of N_2O_5:

$$N_2O_5 \rightarrow N_2O_3 + O_2$$

$$N_2O_3 \rightarrow NO_2 + NO$$

$$NO + N_2O_5 \rightarrow 3\ NO_2$$

(a) Is this mechanism plausible? (b) Identify the intermediates in the reaction. (c) Determine the molecularity of each step.

Mechanisms and Rate: The Rate-Determining Step

For an elementary step, the orders of reaction are given by the coefficients in the balanced equation.

Once we have proposed a mechanism, it is easy for us to write the rate law for each elementary step in that mechanism. What we can measure experimentally, however, is the rate law for the overall reaction. How can we relate the observed overall rate law to the rate laws for the individual elementary steps in a mechanism? The answer lies in the fact that most mechanisms include one step that is much slower than all the others. This slow step is called the **rate-determining step** and it dictates the rate law. Why does the slowest step assume this role?

Consider this analogy: Suppose you eat lunch at a very popular cafeteria where the food is great and the prices are low. The only problem is that the service is a little slow. As the line moves along, it takes each person 5 minutes to get food and then another 30 seconds to pay. If we watch the cash register in this cafeteria, we can expect to see one customer walking away with food every 5 minutes—the rate of the slow step in the "mechanism" for getting lunch. We can take this analogy one step further. Suppose that the owner is concerned that the line is too long and is driving customers away. To speed things up, he buys a new cash register that cuts the paying time down to just 15 seconds per customer. Much to his surprise, he finds that one customer still walks away from the register every 5 minutes! Speeding up the fast part of the process has no impact on the overall rate because that rate is set by the slow step. That's why we call the slow step in a reaction mechanism "rate determining." In a chemical reaction, the fast steps may be several orders of magnitude faster than the slow step.

In the formation of ozone through the Chapman cycle, the slow step is the last one, the reaction of ozone with atomic oxygen:

$$O + O_3 \rightarrow 2\ O_2$$

It is because this step is rate determining for the cyclic reaction that there is a protective ozone layer in the stratosphere. Once oxygen atoms combine to form ozone, the O_3 survives long enough to absorb UV radiation, and that absorption prevents the UV light from reaching the surface of the planet. If the last step were not slow, the ozone created earlier in the cycle would be destroyed chemically before it encountered a photon of UV light. Without the ozone to absorb it, that light would pass through the stratosphere to harm living organisms in the troposphere.

11-7 Catalysis

We noted from the outset of this chapter that environmental concerns about ozone depletion arise from the increase in the rate of destruction of ozone in the stratosphere. This is a process apparently caused by man-made chemicals. Specifically, chlorofluorocarbons used as refrigerants have been cited as a cause of ozone depletion. Yet a look at the Chapman cycle does not show any obvious role for CFCs. So how can these molecules accelerate the destruction of ozone if they do not even appear in the underlying chemical equations? Catalysis is a process in which a reaction rate is influenced by the presence of substances that are neither reactants nor products in the overall equation. A **catalyst** is a substance that increases the rate of the reaction but is neither created nor destroyed in the process. How can CFCs catalyze ozone depletion?

Homogeneous and Heterogeneous Catalysts

Catalysts can be divided into two broad categories. **Homogeneous catalysts** are those that are in the same phase as the reacting substances, whereas **heterogeneous catalysts** are in a different phase from the reacting species. For gas-phase reactions, the heterogeneous catalyst typically is a solid surface. Both types of catalysis are important in atmospheric processes.

The catalytic destruction of ozone in the stratosphere involves reactions between gases there, so it is an example of homogeneous catalysis. The most important catalyst for this process is chlorine. Much of the chlorine present in the stratosphere comes from CFC molecules that were released in the troposphere and slowly migrated to the stratosphere. (Because they are very unreactive at ground level, nearly all CFCs that are released into the atmosphere eventually find their way to the stratosphere.) Upon absorption of UV light, the CFCs initiate a catalyzed reaction mechanism:

Dr. Susan Solomon's work at the National Oceanic and Atmospheric Administration shows that these reactions occur on the surface of ice particles.

Step 1: $CF_2Cl_2 + h\nu \rightarrow CF_2Cl + Cl\cdot$

Step 2: $Cl\cdot + O_3 \rightarrow ClO\cdot + O_2$

Step 3: $ClO\cdot + O_3 \rightarrow Cl\cdot + 2\ O_2$

We encountered the second and third reactions earlier in our discussion of mechanisms. The chlorine atoms serve as a catalyst for ozone decomposition because (a) they are not part of the reaction stoichiometry, (b) they are not consumed by the reaction, and (c) they increase the rate of the net reaction. Because the chlorine atom is regenerated in Step 3, it can catalyze another reaction. An individual chlorine atom may cycle through this mechanism often enough to destroy 100,000 or more ozone molecules. Because of this, the United States banned the use of CFCs as propellants in aerosol cans in 1978. The Montreal Protocol, an international treaty, calls for an end to CFC production. Non-CFC refrigerants and propellants are now in use in new equipment, although old refrigerators and air conditioners still contain CFCs that may one day be released to the environment.

Heterogeneous catalysis has a role in the atmospheric chemistry of ozone in the troposphere as well. Catalytic converters in automobiles are filled with a porous ceramic material, which provides a surface that catalyzes the removal of CO and NO_x

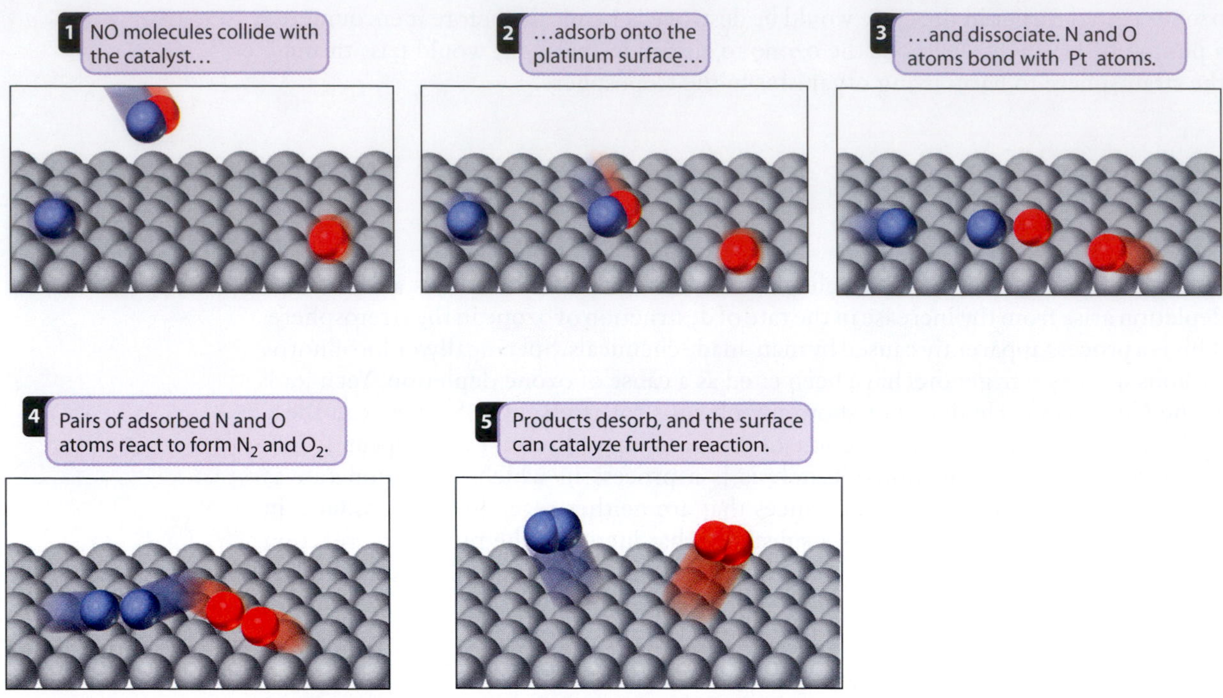

Figure 11.15 ■ A schematic model of the steps involved in the catalytic removal of NO from exhaust gas.

(nitrogen oxides) from the exhaust. (Nitrogen oxides initiate the formation of ozone and other lung irritants in photochemical smog. We will examine this process in detail in Section 11-8.) The process by which catalytic converters operate is shown in **Figure 11.15**. The types of steps shown there are found in most examples of heterogeneous catalysis.

1. The reactive species adsorb, or stick, onto the surface of the catalyst. (When a molecule is *adsorbed*, it sticks to the surface. When a molecule or substance is *absorbed*, it penetrates into the bulk of a material, like water into a sponge. Desorption is the opposite of adsorption.)
2. Species migrate on the surface until they encounter each other.
3. The reaction takes place on the surface.
4. The products desorb from the catalyst surface.

Molecular Perspective of Catalysis

We noted earlier that the activation energy of a chemical reaction plays a critical role in its kinetics. Catalysts increase the rate of reaction by providing a new reaction pathway that lowers the activation energy. **Figure 11.16** illustrates this concept for the conversion of two ozone molecules to three oxygen molecules. Despite the fact that the addition of chlorine atoms requires an extra step relative to direct collisions of two ozone molecules, the net energy required for the reaction is lower, so the catalyzed reaction is faster.

We saw in Section 11-5 that the energy needed to overcome the activation barrier is normally acquired from collisions of fast-moving particles. When a catalyst is introduced, the amount of energy needed decreases, and a larger fraction of the molecules have speeds that are high enough to provide the needed energy. So the direct

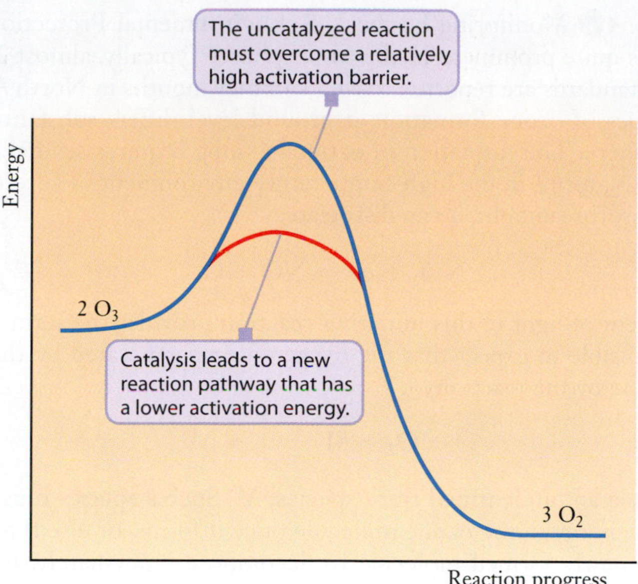
The uncatalyzed reaction must overcome a relatively high activation barrier.
Energy
2 O$_3$
Catalysis leads to a new reaction pathway that has a lower activation energy.
3 O$_2$
Reaction progress

Figure 11.16 ■ Catalysts increase the rate of reaction by providing an alternative reaction pathway with lower activation energy.

effect of a catalyst at the molecular level is to lower the activation energy, and the net result is an increased rate of reaction.

Catalysis and Process Engineering

One of the most prominent challenges of modern chemistry and materials science is the rational design of catalysts for industrial processes. Catalysts are important in industries such as petroleum refining, where billions of dollars are at stake, so the incentives for research in these areas are great. As in any design process, we can begin by listing some desirable features for a catalyst. It should have sufficient longevity or durability to last for many reaction cycles. Although the definition of a catalyst might seem to imply that the catalyst should last indefinitely, usually this is not true in practice. Many catalysts suffer from chemical or mechanical degradation and eventually become useless. A good catalyst should also have a high "turnover number." This means that it can generate large amounts of products in short times. (**Turnover number** is defined as the number of molecules that can react per catalyst binding site per unit time.) Another crucial property of catalysts is selectivity. We often want the catalyst to speed up one, and only one, reaction to minimize the formation of unwanted by-products. A highly selective catalyst will accomplish this goal. Historically, catalysts have often been chosen by trial and error. But as our understanding of reaction mechanisms, molecular structure, and materials properties has increased, many researchers feel it is now becoming possible to apply standard engineering design principles to improve catalysts.

11-8 Tropospheric Ozone

Ozone in the stratosphere helps make life possible by blocking large amounts of harmful ultraviolet radiation. In the troposphere, however, it is a major lung irritant. It causes localized problems when it is formed from O_2 by an electrical discharge (such as lightning or the electrostatic mechanism in a photocopy machine) or by ultraviolet light. As the major constituent of **photochemical smog**, ozone contributes to health problems

We first discussed photochemical smog in Section 5-1; ideas from this chapter allow us to provide a deeper explanation.

in urban areas. Monitoring by the U.S. Environmental Protection Agency suggests that O_3 is quite prominent in the troposphere. Typically, almost 2000 violations of air quality standards are reported during summer months in North America.

The kinetics of ozone formation at ground level differs substantially from that in the stratosphere. The formation of ozone in smog requires several steps. The key compound, NO_2, forms in the high-temperature environment of automobile engines. When NO_2 absorbs sunlight, it can dissociate:

$$NO_2 + h\nu \rightarrow NO + O \qquad (11.10.a)$$

The involvement of light in this initiating reaction provides the term photochemical smog. It is plausible to expect that the oxygen atoms generated by this reaction can then form ozone by the reaction

$$O + O_2 + M \rightarrow O_3 + M \qquad (11.10.b)$$

Here we include an unidentified third species, M. Such a species must be present to remove excess energy in the ozone molecule once it forms, or else that excess energy will cause the newly formed molecule to decompose immediately. Even if reaction 11.10.b occurs, the ozone that forms could be consumed quickly by reaction with NO:

$$O_3 + NO \rightarrow NO_2 + O_2 \qquad (11.10.c)$$

Just as in the stratosphere, the balance between the rates at which ozone is formed and destroyed will determine the ozone level. The rate constants in **Table 11.2** show that reaction 11.10.c is much faster than reaction 11.10.b. Thus if no other chemicals present react with NO, ozone is not likely to form in the troposphere.

The chemicals that most often give rise to the consumption of NO are volatile organic compounds (VOCs). Common VOCs are industrial solvents such as pentane, hexane, and benzene (**Table 11.3**). As a group, these are known as petroleum distillates because they are obtained by distilling crude oil. We can designate this class of chemicals as RH, where R can be any organic radical, such as $—CH_3$, $—C_2H_5$, etc.

Product ingredient lists sometimes use the general term "petroleum distillates" rather than actual chemical names such as hexane.

Table ■ 11.2

Rate constants for some reactions in atmospheric chemistry (Data adapted from NBS technical Note 866, "Chemical Kinetic and Photochemical Data for Modeling Atmospheric Chemistry," U.S. Department of Commerce, National Bureau of Standards, 1975.)

Reaction (at 300 K unless indicated)	Rate Constant
$O + O_2 + M \rightarrow O_3 + M$	5.0×10^5 L^2 mol^{-2} s^{-1}
$O_3 + NO \rightarrow NO_2 + O_2$	1.0×10^7 L mol^{-1} s^{-1}
$O_3 + NO_2 \rightarrow NO_3 + O_2$	3.0×10^4 L mol^{-1} s^{-1}
$O_3 + OH \rightarrow OOH + O_2$	4.8×10^7 L mol^{-1} s^{-1}
$O_3 + Cl \rightarrow OCl + O_2$	1.1×10^{10} L mol^{-1} s^{-1}
$O + N_2O \rightarrow N_2 + O_2$ (1200 K)	8.1×10^5 L mol^{-1} s^{-1}
$NO + CH_3O_2 \rightarrow CH_3O + NO_2$	3.8×10^8 L mol^{-1} s^{-1}
$ClO + ClO \rightarrow Cl_2 + O_2$	1.4×10^7 L mol^{-1} s^{-1}

Table ■ 11.3

Examples of volatile organic chemicals and their sources

VOC	Source
Petroleum distillates (pentane, hexane, benzene)	Gasoline spills and evaporation
Terpenes	Emitted from live plants (e.g., odors of trees)
Alcohols and aldehydes	Solvents (for paint thinners, etc)
Plasticizers	Outgassing from new rugs, upholstery, electronic equipment

The role of these RH species in smog formation begins when they react with HO· radicals:

$$RH + HO\cdot \rightarrow R\cdot + H_2O \quad (11.10.d)$$

We have designated radical species by a dot representing an unpaired electron. The product radical, R·, can react with oxygen in a fast reaction:

$$R\cdot + O_2 + M \rightarrow RO_2\cdot + M \quad (11.10.e)$$

The species $RO_2\cdot$ is called an alkylperoxy radical. It removes NO and thereby increases the lifetime of any ozone that is formed. The reaction produces NO_2 and another radical, RO·:

$$RO_2\cdot + NO \rightarrow RO\cdot + NO_2 \quad (11.10.f)$$

Representative rate constants for these reactions are included in Table 11.2. Note the wide range of rate constant values. We can see that the reactions to remove NO are roughly 40 times faster than the reaction of NO with ozone. The result is that if NO_2 (the initiator) and VOCs are both present, then with adequate sunlight, smog may form and ozone can form within that smog.

Modern gas caps and pump nozzles are designed to minimize the emission of VOCs and reduce smog formation.

These few reactions do not completely describe the chemical processes that form smog. Additional reactions not shown here produce other substances that are strong lung irritants. But even with the few reactions we have examined here, we can begin to appreciate how complicated the study of atmospheric chemistry is. Many pollutants are present at concentrations of parts per million or less, and their concentrations may vary with the seasons, the weather, or even the time of day. Yet they result in real threats to human health.

How can atmospheric chemists hope to unravel such a complex problem? The usual approach is to construct a model that incorporates as many of the relevant reactions as possible. Scientists then use computers to solve simultaneous rate law equations for the connected processes. One commonly used model, known as the "urban airshed model," incorporates 86 chemical reactions among 36 different species. The role of individual reactions can be explored by adjusting their rate constants or even by removing them from the model entirely. Several of the species involved are those we have discussed in this chapter, and the major pollutant that results is ground-level ozone. Among all of the reactions included in the mechanism, ozone is produced in only one of them, and yet it is the key product of the entire mechanism. Of course, uncertainty in the values of any of the rate constants used in this model can shift the expected concentrations of many substances, sometimes dramatically. Our brief introduction to chemical kinetics cannot provide us with the tools needed to investigate such a complicated process any further. But it does provide some insight into the roles of both chemistry and mathematical modeling in understanding and confronting an important societal problem.

FOCUS ON PROBLEM SOLVING

Question Suppose you perform kinetics experiments on two different reactions (A and B) and find that they both have the same rate constant at room temperature (23°C). Computer simulations suggest that the activation energy of reaction A should be roughly twice as large as that of reaction B. Which reaction will have the higher rate constant at 50°C?

Strategy This question involves rates and temperature, and we know that the connection between those is given by the Arrhenius equation. We do not have enough information to calculate the rates at 50°C and compare them. So we will need to think through the implications of the Arrhenius equation conceptually rather than numerically. In cases like this, graphical methods can often be useful. Recall that a graph of ln k vs. $1/T$ is a straight line, with a slope related to E_a. If we can imagine what that graph would look like for each reaction, we will be able to answer the question.

Solution To visualize the graph, we need to use the two pieces of information that we've been given. First, the rate constants of the two reactions are the same at 23°C, so the lines we get when we plot ln k vs. $1/T$ must intersect at that temperature. Second, one reaction has a higher activation energy than the other, which means it will have a greater slope. That is enough to let us sketch the graphs:

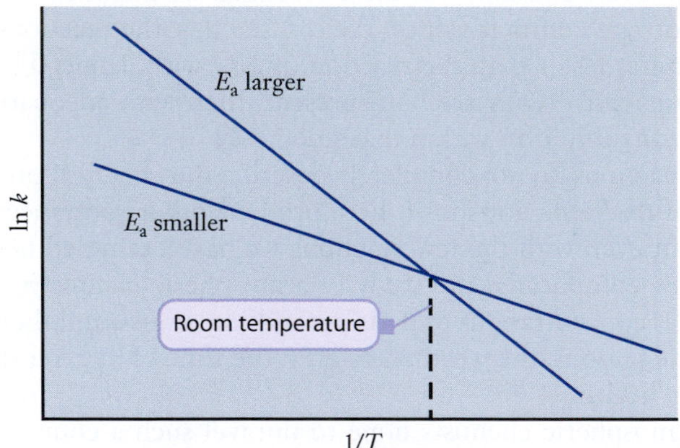

To answer the question, we need to understand what happens at higher temperatures. Because we are plotting $1/T$, temperature will increase from right to left on the graph. If we look to the left of the intersection point, we can see that the reaction with the larger activation energy will have a higher rate constant as the temperature increases. This result might be counterintuitive, but by using the data we had and constructing a graphical interpretation, we can reach the correct conclusion.

SUMMARY

Because chemical reactions are examples of change, it is only natural that understanding the rate of the change can be important. The study of chemical kinetics includes measurements of rates of reactions, as well as studies of the detailed mechanisms by which they occur.

The rate of a chemical reaction is expressed in terms of the changes in concentration per unit time, and a careful definition must account for the stoichiometry of the reaction. Reaction rates change as a reaction proceeds, so we also need to distinguish between average and instantaneous rates.

A number of factors influence the rate of a chemical reaction. The concentration of reactants present is one critical factor, and the experimental determination of this effect results in an expression referred to as the rate law. We can work with rate laws in either of two forms: the differential rate law and the integrated rate law. The differential rate law tells us the rate if we know the appropriate concentrations, whereas the integrated rate law predicts the concentration of a reactant as a function of time.

Temperature also affects the rate of chemical reactions. An increase in temperature generally results in an increased rate of reaction. This can be understood by noting that most reactions, viewed from the molecular perspective, require some energy input to get them started. Increasing the temperature creates more energetic molecular collisions, and the rate increases. Experimental studies of reaction rates as a function of temperature provide the data needed to measure the activation energy.

For many reactions, the molecular-level understanding is more complicated than asking if the reactants collide with enough energy to overcome the activation barrier. Many reactions proceed through reaction mechanisms that may comprise several elementary steps. Mechanisms are difficult to prove experimentally, but any mechanism must be consistent with experimental evidence for it to be useful as a model that explains how a reaction proceeds. Understanding reaction mechanisms helps us explain catalysis. A catalyst provides an alternative mechanism for a reaction. It lowers the activation energy and increases the rate.

KEY TERMS

activated complex (11-5)
activation energy (E_a) (11-5)
Arrhenius equation (11-5)
average rate (11-2)
bimolecular (11-6)
catalyst (11-7)
Chapman cycle (11-1)
differential rate law (11-3)
elementary steps (11-6)
first-order reaction (11-4)
frequency factor (11-5)
half-life (11-4)
heterogeneous catalyst (11-7)
homogeneous catalyst (11-7)
instantaneous rate (11-2)
integrated rate law (11-4)
order of reaction (11-3)
photochemical smog (11-8)
preexponentical factor (11-5)
rate constant (k) (11-3)
rate law (11-3)
rate-determining step (11-6)
reaction mechanism (11-6)
reaction rate (11-2)
reactive intermediate (11-6)
second-order reaction (11-4)
termolecular (11-6)
turnover number (11-7)
unimolecular (11-6)
zero-order reaction (11-4)

PROBLEMS AND EXERCISES

INSIGHT INTO Ozone Depletion

11.1 What is an allotrope?

11.2 Describe what would happen to the ozone molecules if a small amount of ozone were generated and released in a room.

11.3 In what region of the atmosphere is ozone considered a pollutant? In what region is it considered beneficial?

11.4 What are the steps in the Chapman cycle? Explain the cycle's importance.

11.5 What is the net chemical reaction associated with the Chapman cycle?

11.6 At what points in the Chapman cycle do photochemical reactions take place?

11.7 Which photochemical reaction in the Chapman cycle has the highest energy input via light?

11.8 Describe what is meant by the term *ozone hole*. Would you say that the average nonscientist pictures this issue correctly based on the phrase?

11.9 Is the ozone hole permanent?

Rates of Chemical Reactions

11.10 For each of the following, suggest appropriate rate units.

(a) Driving from one place to another
(b) Drying dishes by hand
(c) The beating wings of a mosquito
(d) Eyes blinking

11.11 For each of the following, suggest an appropriate rate unit.

(a) Heart beating
(b) Videotape rewinding
(c) Automobile wheels rotating
(d) Gas evolving in a very fast chemical reaction

11.12 Rank the following in order of increasing reaction rate.

(a) Dynamite exploding
(b) Iron rusting
(c) Paper burning

11.13 Distinguish between instantaneous rate and average rate. In each of the following situations, is the rate measured the instantaneous rate or the average rate?

(a) In a hot dog eating contest, it took the winner only 4 minutes to eat 20 hot dogs, so he ate 5 hot dogs per minute.

(b) At minute 1.0, the winner was eating 6 hot dogs per minute, but at minute 3.0, he was down to only 3 hot dogs per minute.

11.14 Candle wax is a mixture of hydrocarbons. In the reaction of oxygen with candle wax in Figure 11.4, the rate of consumption of oxygen decreased with time after the vessel was covered, and eventually the flame went out. From the perspective of the kinetic-molecular theory, describe what is happening in the vessel.

11.15 In the description of the candle in Figure 11.4, we mentioned the consumption of oxygen. Assuming that candle wax is a mixture of hydrocarbons with the general formula C_nH_{2n+2}, what other variables could be measured besides the concentration of oxygen to determine the rate of the reaction?

11.16 The reaction for the Haber process, the industrial production of ammonia, is

$$N_2(g) + 3\,H_2(g) \rightarrow 2\,NH_3(g)$$

Assume that under certain laboratory conditions ammonia is produced at the rate of 6.29×10^{-5} mol L^{-1} s^{-1}. At what rate is nitrogen consumed? At what rate is hydrogen consumed?

11.17 Ammonia can react with oxygen to produce nitric oxide and water:

$$4\,NH_3(g) + 5\,O_2(g) \rightarrow 4\,NO(g) + 6\,H_2O(g)$$

If the rate at which ammonia is consumed in a laboratory experiment is 4.23×10^{-4} mol L^{-1} s^{-1}, at what rate is oxygen consumed? At what rate is NO produced? At what rate is water vapor produced?

11.18 The following data were obtained in the decomposition of $H_2O_2(aq)$ to $O_2(g)$ and $H_2O(\ell)$. The rate at which oxygen gas was produced was measured. (No oxygen was present initially.)

Time (min)	Volume of O_2 (mL)
3.3	10.4
6.9	17.3

(a) Calculate the average rate in mL/min for the first 3.3 minutes.

(b) Calculate the average rate in mL/min for the first 6.9 minutes.

11.19 A gas, AB, decomposes and the volume of B_2 produced is measured as a function of time. The data obtained are as follows:

Time (min)	0	8.3	15.4	19.0
Volume (L)	0	4.2	8.6	11.5

What is the average rate of production of B_2 for the first 8.3 min? For the first 19 min?

11.20 Experimental data are listed here for the reaction $A \rightarrow 2\,B$:

Time (s)	[B] (mol/L)
0.00	0.000
10.0	0.326
20.0	0.572
30.0	0.750
40.0	0.890

(a) Prepare a graph from these data, connect the points with a smooth line, and calculate the rate of change of [B] for each 10-s interval from 0.0 to 40.0 s. Does the rate of change decrease from one time interval to the next? Suggest a reason for this result.

(b) How is the rate of change of [A] related to the rate of change of [B] in each time interval? Calculate the rate of change of [A] for the time interval from 10.0 to 20.0 s.

(c) What is the instantaneous rate, $\Delta[B]/\Delta t$, when [B] = 0.750 mol/L?

11.21 Azomethane, CH_3NNCH_3, is not a stable compound, and once generated, it decomposes. The rate of decomposition was measured by monitoring the partial pressure of azomethane, in torr:

Time (min)	0	15	30	48	75
Pressure (torr)	36.2	30.0	24.9	19.3	13.1

Plot the data and determine the instantaneous rate of decomposition of azomethane at t = 20 min.

Rate Laws and the Concentration Dependence on Rates

11.22 Which of the following would be likely to have a very large rate constant? Which would have a very small rate constant?

(a) Dynamite exploding

(b) Iron rusting

(c) Paper burning

11.23 A reaction has the experimental rate equation Rate = $k[A]^2$. How will the rate change if the concentration of A is tripled? If the concentration of A is halved?

11.24 Second-order rate constants used in modeling atmospheric chemistry are commonly reported in units of cm^3 $molecule^{-1}$ s^{-1}. Convert the following rate constants to $L\ mol^{-1}\ s^{-1}$:

(a) 3.5×10^{-14} cm^3 $molecule^{-1}$ s^{-1}

(b) 7.1×10^{-18} cm^3 $molecule^{-1}$ s^{-1}

(c) 6.1×10^{-30} cm^3 $molecule^{-1}$ s^{-1}

11.25 For each of the rate laws below, what is the order of the reaction with respect to the hypothetical substances X, Y, and Z? What is the overall order?

(a) Rate = $k[X][Y][Z]$, (b) Rate = $k[X]^2[Y]^{1/2}[Z]$,
(c) Rate = $k[X]^{1.5}[Y]^{-1}$, (d) Rate = $k[X]/[Y]^2$

11.26 The reaction of $CO(g) + NO_2(g)$ is second-order in NO_2 and zero-order in CO at temperatures less than 500 K.

(a) Write the rate law for the reaction.

(b) How will the reaction rate change if the NO_2 concentration is halved?

(c) How will the reaction rate change if the concentration of CO is doubled?

11.27 Show that if the units of rate are mol L^{-1} s^{-1}, then the units of the rate constant in the following second-order reaction are L mol^{-1} s^{-1}:

$H_2(g) + Br_2(g) \rightarrow 2\ HBr$ rate = $k\ [H_2][Br_2]$

11.28 One reaction that destroys O_3 molecules in the stratosphere is

$$NO + O_3 \rightarrow NO_2 + O_2$$

When this reaction was studied in the laboratory, it was found to be first order with respect to both NO and O_3, with a rate constant of 1.9×10^4 L mol^{-1} s^{-1}. If [NO] = 1.2×10^{-5} mol L^{-1} and $[O_3] = 2.0 \times 10^{-5}$ mol L^{-1}, what is the rate of this reaction?

11.29 The hypothetical reaction, A + B → C, has the rate law

$$\text{Rate} = k[A]^x[B]^y$$

When [A] is doubled and [B] is held constant, the rate doubles. But the rate increases fourfold when [B] is doubled and [A] is held constant. What are the values of x and y?

11.30 The rate of the decomposition of hydrogen peroxide, H_2O_2, depends on the concentration of iodide ion present. The rate of decomposition was measured at constant temperature and pressure for various concentrations of H_2O_2 and of KI. The data appear below. Determine the order of reaction for each substance, write the rate law, and evaluate the rate constant.

Rate (mL min^{-1})	$[H_2O_2]$ (mol L^{-1})	[KI] (mol L^{-1})
0.090	0.15	0.033
0.178	0.30	0.033
0.184	0.15	0.066

11.31 Give the order with respect to each reactant and the overall order for the hypothetical reaction

$$A + B + C \rightarrow D + E$$

which obeys the rate law Rate = $k[A][B]^2$.

11.32 The following experimental data were obtained for the reaction

$$2\ A + 3\ B \rightarrow C + 2\ D$$

[A] (mol L^{-1})	[B] (mol L^{-1})	Rate = $\Delta[C]/\Delta t$ (mol L^{-1} s^{-1})
0.127	0.15	0.033
0.127	0.30	0.132
0.255	0.15	0.066

Determine the reaction order for each reactant and the value of the rate constant.

11.33 The following experimental data were obtained for the reaction of NH_4^+ and NO_2^- in acidic solution.

$$NH_4^+(aq) + NO_2^-(aq) \rightarrow N_2(g) + 2\ H_2O(\ell)$$

$[NH_4^+]$ (mol L^{-1})	$[NO_2^-]$ (mol L^{-1})	Rate = $\Delta[N_2]/\Delta t$ (mol L^{-1} s^{-1})
0.0092	0.098	3.33×10^{-7}
0.0092	0.049	1.66×10^{-7}
0.0488	0.196	3.51×10^{-6}
0.0249	0.196	1.80×10^{-6}

Determine the rate law for this reaction and calculate the rate constant.

11.34 Rate data were obtained at 25°C for the following reaction. What is the rate law expression for this reaction?

$$A + 2\ B \rightarrow C + 2\ D$$

Expt.	Initial [A] (mol L^{-1})	Initial [B] (mol L^{-1})	Initial Rate of Formation of C (mol L^{-1} min^{-1})
1	0.10	0.10	3.0×10^{-4}
2	0.30	0.30	9.0×10^{-4}
3	0.10	0.30	3.0×10^{-4}
4	0.20	0.40	6.0×10^{-4}

11.35 For the reaction

$$2\ NO(g) + 2\ H_2(g) \rightarrow N_2(g) + 2\ H_2O(g)$$

at 1100°C, the following data have been obtained:

[NO] (mol L^{-1})	$[H_2]$ (mol L^{-1})	Rate = $\Delta[N_2]/\Delta t$ (mol L^{-1} s^{-1})
5.0×10^{-3}	0.32	0.012
1.0×10^{-2}	0.32	0.048
1.0×10^{-2}	0.64	0.096

Derive a rate law for the reaction and determine the value of the rate constant.

11.36 The reaction

$$NO(g) + O_2(g) \rightarrow NO_2(g) + O(g)$$

plays a role in the formation of nitrogen dioxide in automobile engines. Suppose that a series of experiments measured the rate of this reaction at 500 K and produced the following data:

[NO] (mol L^{-1})	$[O_2]$ (mol L^{-1})	Rate = $-\Delta[NO]/\Delta t$ (mol L^{-1} s^{-1})
0.002	0.005	8.0×10^{-17}
0.002	0.010	1.6×10^{-16}
0.006	0.005	2.4×10^{-16}

Derive a rate law for the reaction and determine the value of the rate constant.

11.37 In a heterogeneous system such as wood burning in oxygen, the surface area of the solid can be a factor in the rate of the reaction. Increased surface area of the wood means increased collisions with oxygen molecules. To understand how surface increases when dividing a solid, do the following:

(a) Draw a cube that is 2 cm on a side. What is its total surface area?

(b) Divide each face of the cube into four equal sections. If you were able to take the cube apart on the lines you have just drawn, how many cubes would you have?

(c) What are the dimensions of the new cubes?

(d) What is the total surface area of all the new cubes? What will happen if you divide the cubes again?

(e) Now consider burning a log or burning all the toothpicks that could be made from the log. What effect would surface area have on the rate that each burns? Would you rather try to start a campfire with a log or with toothpicks?

11.38 In wheat-growing areas, such as the plains of the central United States and Canada, harvested wheat is stored in tall grain elevators that are visible for miles in the flat prairie. The wheat is dumped from trucks into a lower part of the elevator and then moved up the elevators into storage areas. The combustible dust produced as the wheat grains rub together has caused grain elevator explosions. Use the cube analogy in problem 11.37 above to explain why wheat dust dispersed in the air in a grain elevator can react explosively if sparked, whereas whole wheat grains do not.

Integrated Rate Laws

11.39 The decomposition of N_2O_5 in solution in carbon tetrachloride is a first-order reaction:

$$2\ N_2O_5 \rightarrow 4\ NO_2 + O_2$$

The rate constant at a given temperature is found to be $5.25 \times 10^{-4}\ s^{-1}$. If the initial concentration of N_2O_5 is 0.200 M, what is its concentration after exactly 10 minutes have passed?

11.40 In Exercise 11.39, if the initial concentration of N_2O_5 is 0.100 M, how long will it take for the concentration to drop to 0.0100 times its original value?

11.41 For a drug to be effective in treating an illness, its levels in the bloodstream must be maintained for a period of time. One way to measure the level of a drug in the body is to measure its rate of appearance in the urine. The rate of excretion of penicillin is first order, with a half-life of about 30 min. If a person receives an injection of 25 mg of penicillin at $t = 0$, how much penicillin remains in the body after 3 hours?

11.42 Amoxicillin is an antibiotic packaged as a powder. When it is used to treat babies and small animals, the pharmacist or veterinarian must suspend it in water, so that it can be administered orally with a medicine dropper. The label says to dispose of unused suspension after 14 days. It also points out that refrigeration is required. In the context of this chapter, what is implied in the latter two statements?

11.43 As with any drug, aspirin (acetylsalicylic acid) must remain in the bloodstream long enough to be effective. Assume that the removal of aspirin from the bloodstream into the urine is a first-order reaction, with a half-life of about 3 hours. The instructions on an aspirin bottle say to take 1 or 2 tablets every 4 hours. If a person takes 2 aspirin tablets, how much aspirin remains in the bloodstream when it is time for the second dose? (A standard tablet contains 325 mg of aspirin.)

11.44 A possible reaction for the degradation of the pesticide DDT to a less harmful compound was simulated in the laboratory. The reaction was found to be first order, with $k = 4.0 \times 10^{-8}\ s^{-1}$ at 25°C. What is the half-life for the degradation of DDT in this experiment, in years?

11.45 The initial concentration of the reactant in a first-order reaction A $\rightarrow$ products is 0.64 mol/L and the half-life is 30.0 s.

(a) Calculate the concentration of the reactant exactly 60 s after initiation of the reaction.

(b) How long would it take for the concentration of the reactant to drop to one-eighth its initial value?

(c) How long would it take for the concentration of the reactant to drop to 0.040 mol/L?

11.46 A substance undergoes first-order decomposition. After 40.0 min at 500°C, only 12.5% of the original sample remains. What is the half-life of the decomposition? If the original sample weighed 243 g, how much would remain after 2.00 h?

11.47 Show that the half-life of a second-order reaction is given by

$$t_{1/2} = \frac{1}{k[A]_0}$$

In what fundamental way does the half-life of a second-order reaction differ from that of a first-order reaction?

11.48 The following data were collected for the decomposition of N_2O_5:

Time, t (min)	$[N_2O_5]$ (mol L^{-1})
0	0.200
5	0.171
10	0.146
15	0.125
20	0.106
25	0.0909
30	0.0777
35	0.0664
40	0.0570

(a) Use appropriate graphs to determine the rate constant for this reaction.

(b) Find the half-life of the reaction.

11.49 The rate of photodecomposition of the herbicide picloram in aqueous systems was determined by exposure to sunlight for a number of days. One such experiment produced the following results. (Data from R.T. Hedlun and C.R. Youngson, "The Rates of Photodecomposition of Picloram in Aqueous Systems," *Fate of Organic Pesticides in the Aquatic Environment*, Advances in Chemistry Series, #111, American Chemical Society (1972), 159–172.)

Exposure Time, t (days)	[Picloram] (mol L^{-1})
0	4.14×10^{-6}
7	3.70×10^{-6}
14	3.31×10^{-6}
21	2.94×10^{-6}
28	2.61×10^{-6}
35	2.30×10^{-6}
42	2.05×10^{-6}
49	1.82×10^{-6}
56	1.65×10^{-6}

Determine the order of reaction, the rate constant, and the half-life for the photodecomposition of picloram.

11.50 The rate of decomposition of SO_2Cl_2 according to the reaction

$$SO_2Cl_2(g) \rightarrow SO_2(g) + Cl_2(g)$$

can be followed by monitoring the total pressure in the reaction vessel. Consider the following data:

Time, t (s)	Total Pressure (torr)
0.0	491.7
185.3	549.6
242.8	566.6
304.5	584.1
362.7	599.9
429.5	617.2
509.7	637.0
606.3	659.5

What must you do to convert these total pressures into changes in the pressure of the SO_2Cl_2? Manipulate the data as needed and then use a graphing calculator or spreadsheet to plot the data and determine the order of reaction and the rate constant.

11.51 Peroxyacetyl nitrate (PAN) has the chemical formula $C_2H_3NO_5$ and is an important lung irritant in photochemical smog. An experiment to determine the decomposition kinetics of PAN gave the data below. Determine the order of reaction and calculate the rate constant for the decomposition of PAN.

Time, t (min)	Partial Pressure of PAN (torr)
0.0	2.00×10^{-3}
10.0	1.61×10^{-3}
20.0	1.30×10^{-3}
30.0	1.04×10^{-3}
40.0	8.41×10^{-4}
50.0	6.77×10^{-4}
60.0	5.45×10^{-4}

11.52 The following series of pictures represents the progress of a reaction in which A_2 molecules dissociate into atoms: $A_2 \rightarrow 2\ A$. Each picture represents a "snapshot" of the reaction mixture at the indicated time.

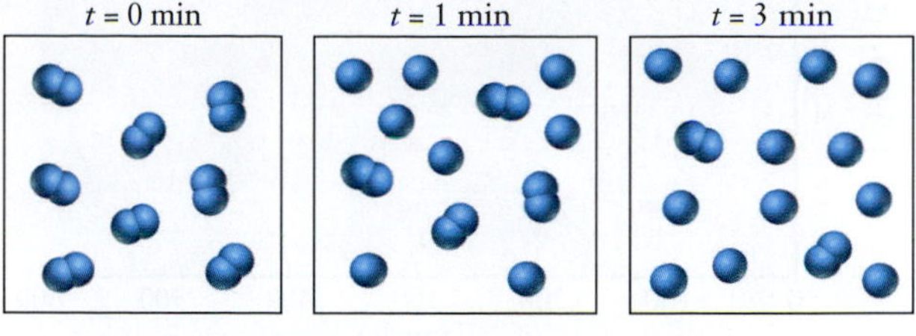

Find the rate law for this reaction, including the numerical value of the rate constant. Assume that any concentration terms in the rate law are expressed in units of mol L^{-1}.

11.53 Hydrogen peroxide (H_2O_2) decomposes into water and oxygen:

$$H_2O_2(aq) \rightarrow H_2O(\ell) + \frac{1}{2}O_2(g)$$

Ordinarily this reaction proceeds rather slowly, but in the presence of some iodide ions (I^-), the decomposition is much faster. The decomposition in the presence of iodide was studied at 20°C, and the data were plotted in various ways. Use the graphs below, where concentrations are in moles/liter and time is in seconds, to answer the questions that follow.

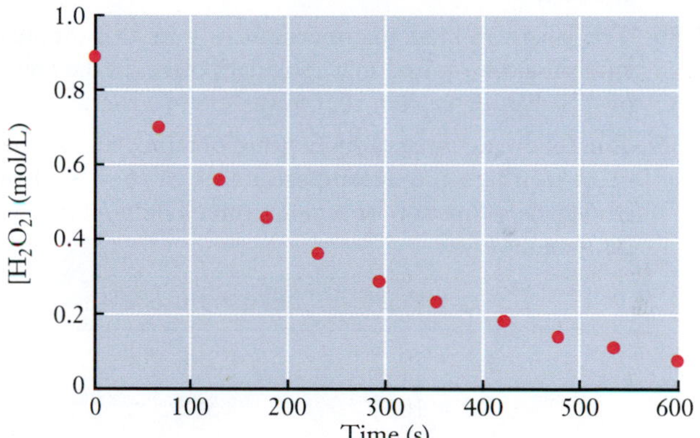

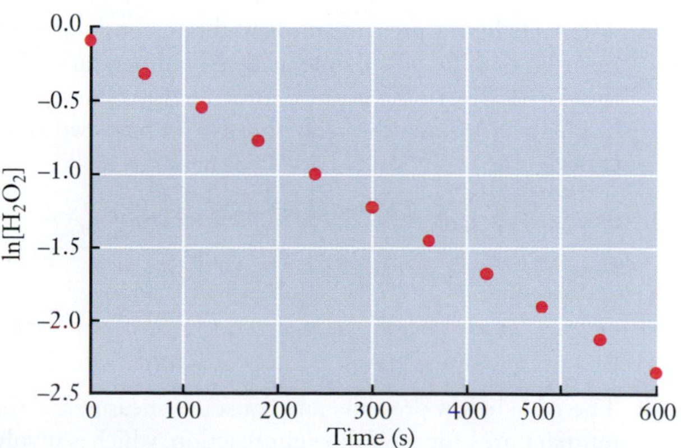

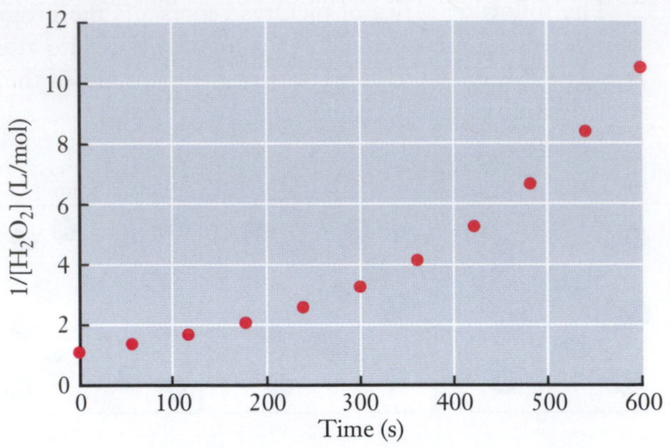

(a) What is the order of reaction for the decomposition of hydrogen peroxide?

(b) Find the numerical value of the rate constant at 20°C, including the correct units.

(c) Obtain an estimate of the initial rate of reaction in the experiment that produced the graphs (i.e., the rate at $t = 0$ in the graphs).

Temperature and Kinetics

11.54 Use the kinetic-molecular theory to explain why an increase in temperature increases reaction rate.

11.55 The activation energy for the reaction in which CO_2 decomposes to CO and a free radical oxygen atom, O, has an activation energy of 460 kJ mol^{-1}. The frequency factor is 2×10^{11} s^{-1}. What is the rate constant for this reaction at 298 K?

11.56 The labels on most pharmaceuticals state that the medicine should be stored in a cool, dark place. In the context of this chapter, explain why this is sound advice.

11.57 The following rate constants were obtained in an experiment in which the decomposition of gaseous N_2O_5 was studied as a function of temperature. The products were NO_2 and NO_3.

k (s^{-1})	Temperature (K)
3.5×10^{-5}	298
2.2×10^{-4}	308
6.8×10^{-4}	318
3.1×10^{-3}	328

Determine E_a for this reaction in kJ/mol.

11.58 The table below presents measured rate constants for the reaction of NO with ozone at three temperatures. From these data, determine the activation energy of the reaction in kJ/mol. (Assume the temperatures all have two significant figures.)

k (L mol^{-1} s^{-1})	Temperature (K)
1.3×10^{6}	200
4.4×10^{6}	250
9.9×10^{6}	300

11.59 The table below presents rate constants measured at three temperatures for the following reaction, which is involved in the production of nitrogen oxides in internal combustion engines. (Assume the temperatures all have two significant figures.)

$$O(g) + N_2(g) \rightarrow NO(g) + N(g)$$

k (L mol^{-1} s^{-1})	Temperature (K)
4.4×10^{2}	2000
2.5×10^{5}	3000
5.9×10^{6}	4000

Determine the activation energy of the reaction in kJ/mol.

11.60 Many reactions occur in the formation of photochemical smog, including the reaction of ozone with various volatile organic chemicals, or VOCs. The following table shows the Arrhenius expression for four such reactions.

Volatile Organic Chemical	Arrhenius Expression
Allene	9.6×10^{5} exp $(-2750/T)$
Propane	6.6×10^{6} exp $(-1970/T)$
1-Butene	1.7×10^{6} exp $(-1696/T)$
cis-2-Butene	1.9×10^{6} exp $(-956/T)$

(a) Which reaction is likely to be the fastest at 310 K?

(b) Which reaction has the lowest activation energy?

(c) Which has the highest activation energy?

11.61 Bacteria cause milk to go sour by generating lactic acid. Devise an experiment that could measure the activation energy for the production of lactic acid by bacteria in milk. Describe how your experiment will provide the information you need to determine this value. What assumptions must be made about this reaction?

Reaction Mechanisms

11.62 Define the term *intermediate* in the context of reaction mechanisms.

11.63 Can a reaction mechanism ever be proven correct? Can it be proven incorrect?

11.64 What is the rate law for each of the following elementary reactions?

(a) $Cl(g) + ICl(g) \rightarrow I(g) + Cl_2(g)$

(b) $O(g) + O_3(g) \rightarrow 2\ O_2(g)$

(c) $2\ NO_2(g) \rightarrow N_2O_4(g)$

11.65 Describe how the Chapman cycle is a reaction mechanism. What is the molecularity of each reaction in the Chapman cycle?

11.66 What are the reaction intermediates in the Chapman cycle?

11.67 Propane is used as a household fuel in areas where natural gas is not available. The combustion of propane is given by the equation

$$C_3H_8 + 5\ O_2 \rightarrow 3\ CO_2 + 4\ H_2O$$

Does this equation represent a probable mechanism for the reaction? Defend your answer.

11.68 Five people are responsible for cleaning up after a very large banquet. Two are clearing the tables, while one washes the dishes and two dry. Which step is likely to be rate determining?

11.69 The following mechanism is proposed for a reaction:

$$NO + Br_2 \rightarrow NOBr_2 \qquad \text{(slow)}$$
$$NOBr_2 + NO \rightarrow 2\ NOBr \qquad \text{(fast)}$$

(a) Write the overall equation for the reaction.

(b) What is the rate-determining step?

(c) What is the intermediate in this reaction?

(d) What is the molecularity of each step of the reaction?

(e) Write the rate expression for each step.

11.70 HBr is oxidized in the following reaction:

$$4\ HBr(g) + O_2(g) \rightarrow 2\ H_2O(g) + 2\ Br_2(g)$$

A proposed mechanism is

$$HBr + O_2 \rightarrow HOOBr \qquad \text{(slow)}$$
$$HOOBr + HBr \rightarrow 2\ HOBr \qquad \text{(fast)}$$
$$HOBr + HBr \rightarrow H_2O + Br_2 \qquad \text{(fast)}$$

(a) Show that this mechanism can account for the correct stoichiometry.

(b) Identify all intermediates in this mechanism.

(c) What is the molecularity of each elementary step?

(d) Write the rate expression for each elementary step.

(e) Identify the rate-determining step.

11.71 The reaction of $NO_2(g)$ and CO(g) is thought to occur in two steps:

Step 1: $NO_2(g) + NO_2(g) \rightarrow NO(g) + NO_3(g)$ (slow)
Step 2: $NO_3(g) + CO(g) \rightarrow NO_2(g) + CO_2(g)$ (fast)

(a) Show that the elementary steps add up to give the overall stoichiometric equation.

(b) What is the molecularity of each step?

(c) For this mechanism to be consistent with kinetic data, what must be the experimental rate equation?

(d) Identify any intermediates in this reaction.

Catalysis

11.72 If a textbook defined a catalyst as "a substance that increases the rate of a reaction," would that definition be adequate?

11.73 The word *catalyst* has been adopted outside of science. In sports, a reserve who enters the game might be described as the catalyst for a comeback victory. Discuss how this use of the word is similar to the scientific use of the word. How is it different from the scientific use of the word?

11.74 In the process by which CFCs contribute to the depletion of stratospheric ozone, is light a catalyst? Why or why not?

11.75 What distinguishes homogeneous and heterogeneous catalysis?

11.76 Define the terms *adsorbed* and *desorbed.* Do the processes to which these terms refer occur only in heterogeneous catalysis?

11.77 Based on the kinetic theory of matter, what would the action of a catalyst do to a reaction that is the reverse of some reaction that we say is catalyzed?

11.78 Draw a hypothetical activation energy diagram for an uncatalyzed reaction. On the same set of coordinates, draw a similar graph for the catalyzed reaction. Why does reaction rate increase with the use of the catalyst?

11.79 In Chapter 3, we discussed the conversion of biomass into biofuels. One important area of research associated with biofuels is the identification and development of suitable catalysts to increase the rate at which fuels can be produced. Do a web search to find an article describing biofuel catalysts. Then, write one or two sentences describing the reactions being catalyzed, and identify the catalyst as homogeneous or heterogeneous.

11.80 The label on a bottle of 3% (by volume) hydrogen peroxide, H_2O_2, purchased at a grocery store, states that the solution should be stored in a cool, dark place. H_2O_2 decomposes slowly over time, and the rate of decomposition increases with an increase in temperature and in the presence of light. However, the rate of decomposition increases dramatically if a small amount of powdered MnO_2 is added to the solution. The decomposition products are H_2O and O_2. MnO_2 is not consumed in the reaction.

(a) Write the equation for the decomposition of H_2O_2.

(b) What role does MnO_2 play?

(c) In the chemistry lab, a student substituted a chunk of MnO_2 for the powdered compound. The reaction rate was not appreciably increased. What is one possible explanation for this observation?

(d) Is MnO_2 part of the stoichiometry of the decomposition of H_2O_2?

INSIGHT INTO Tropospheric Ozone

11.81 Why is smog appropriately labeled *photochemical* smog?

11.82 When oxygen molecules, O_2, react with oxygen atoms to form ozone, a third molecule must be present.

(a) What is the molecularity of this elementary reaction?

(b) Would you expect this reaction to be fast or slow? Explain your reasoning.

(c) Why is the third molecule required?

11.83 What is a VOC? What role do VOCs play in photochemical smog?

11.84 Why does the formation of ozone in smog depend on having *less* NO(g) present?

11.85 Many urban centers in North America do not meet air quality standards for ozone levels. What factors associated with living in a city contribute most strongly to the formation of potentially dangerous levels of ozone?

11.86 Ozone is not stable at pressures typically found in the troposphere. How might this contribute to the variation of ozone levels over the course of each day?

11.87 Why do ozone concentrations lag in time relative to other pollutants in photochemical smog?

11.88 Reducing ozone formation in photochemical smog would likely require action at both the chemical and societal levels.

(a) Propose strategies at the chemical level—i.e., which chemicals would need to be accentuated or eliminated?

(b) Propose strategies at the societal level—i.e., what changes would the average person need to make to accomplish the changes you suggested at the chemical level?

11.89 Find a website that tells you whether you live in an area that has unhealthy levels of ozone.

(a) What is the URL of the site? Does this site strike you as a reliable source of information?

(b) Does your area have unhealthy air?

11.90 Use the web to find out what it means if the U.S. Environmental Protection Agency lists a city as a nonattainment area.

11.91 Western Michigan can experience higher than normal ozone levels in the summer because prevailing southwest winds carry industrial pollution in the form of VOCs from Milwaukee and Chicago. (For maps, visit the EPA Region 5 website at http://www.epa.gov/region5)

(a) Rewrite equations 11.10.d−f, using hexane, C_6H_{14}, as the VOC pollutant. **(b)** Explain how the NO_2 produced in the reactions you wrote in (a) can result in tropospheric ozone production in Western Michigan.

11.92 Along the Front Range of Colorado, the greatest contributor to Denver's "brown cloud" is vehicle traffic. Motor vehicles emit VOCs while they stir up the dust on the highways, increasing particulates. (Visit the Regional Air Quality Control's website at http://www.raqc.org/ for more information.)

(a) In 2007, Denver exceeded the EPA's 8-hour ozone standard, which at that time was 0.08 ppm. Imagine that you are responsible for helping Denver (or your community) decrease ozone concentrations. Using the information in Table 11.3, list four things you would tell Denver residents to do around the house that would decrease emissions of VOCs and thus decrease ozone formation.

(b) One of the suggestions given to residents of the Denver area to help decrease ozone formation is to fill their car's gas tank after dark. Explain why this could be effective.

Additional Problems

11.93 According to an article published in 1966 in the *Journal of Chemical Education*, xenon reacts photochemically with fluorine at room temperature (298 K) to produce XeF_2. A portion of a table showing ratios of gases in the reaction vessel for four experiments is reproduced below. (Assume that atmospheric pressure is exactly 1 atm.) The reaction is as shown:

$$Xe(g) + F_2(g) \rightarrow XeF_2(s)$$

Experiment	1	2	3	4
Volume of Xe, in mL at atmospheric pressure	300.	350.	400.	450.
Volume of F_2, in mL at atmospheric pressure	271	260	262	183

(Data from John H. Holloway, *J. Chem. Ed.*, **43**, 203 (1966).)

(a) Draw the Lewis structure of XeF_2 and predict the molecule's shape.

(b) Calculate the partial pressure of each gas in Experiment 3 if the amounts shown are combined such that the total pressure remains 1 atm.

(c) Calculate the theoretical yield of XeF_2 in Experiment 3.

(d) After the reactants were introduced, the bulb was placed in sunlight. XeF_2 began crystallizing on the inner surface of the bulb about 24 hours later. Compare the rate of this reaction to the rate of precipitation of CaF_2 from solutions of $Ca^{2+}(aq)$ and $F^-(aq)$ in the laboratory.

11.94 In the previous problem, the reaction to produce XeF_2 occurs photochemically at room temperature. However, to produce $XeF_4(s)$, higher temperature and pressure are required. Production of XeF_6 requires even higher pressures. Discuss the differences in reaction conditions for these three reactions in terms of activation energy and collision theory.

11.95 On a particular day, the ozone level in Milwaukee exceeded the EPA's 1-hour standard of 0.12 ppm by 10 ppb. How many ozone molecules would be present in 1 liter of air at the detection site?

11.96 The EPA's 8-hour ozone standard for communities is 0.075 ppm. On a certain day, the concentration of ozone in the atmosphere in Denver was 1.5×10^{-3} mol/L. Was Denver in compliance with EPA regulations that day?

11.97 The following is a thought experiment. Imagine that you put a little water in a test tube and add some NaF crystals. Immediately after you add NaF, you observe that the crystals begin dissolving. The quantity of solid NaF decreases, but before long, it appears that no more NaF is dissolving. The solution is saturated.

(a) The equation for the dissolution of NaF in water is $NaF(s) \rightarrow Na^+(aq) + F^-(aq)$. As NaF dissolves, what do you think happens to the rate of dissolution? Describe what is occurring on the molecular level.

(b) Assume that the reverse reaction, $Na^+(aq) + F^-(aq) \rightarrow NaF(s)$, also occurs as the crystal dissolves. In other words, both dissolution and precipitation are taking place. When it appears that there is no more change in the quantity of NaF dissolving (the solution is saturated), what has happened to the rates of the forward and reverse reactions? Explain your answer.

11.98 The following statements relate to the reaction for the formation of HI:

$$H_2(g) + I_2(g) \rightarrow 2\ HI(g) \qquad \text{Rate} = k[H_2][I_2]$$

Determine which of the following statements are true. If a statement is false, indicate why it is incorrect.

(a) The reaction must occur in a single step.

(b) This is a second-order reaction overall.

(c) Raising the temperature will cause the value of k to decrease.

(d) Raising the temperature lowers the activation energy for this reaction.

(e) If the concentrations of both reactants are doubled, the rate will double.

(f) Adding a catalyst in the reaction will cause the initial rate to increase.

11.99 The Haber process, which produces ammonia from nitrogen and hydrogen, is one of the world's most important industrial chemical reactions:

$$N_2(g) + 3\ H_2(g) \rightarrow 2\ NH_3(g)$$

For this reaction, both $\Delta H°$ and $\Delta S°$ are negative, meaning that product formation is thermodynamically favored by low temperatures. Yet the industrial reaction is commonly carried out at temperatures between 400 and 600°C. These high temperatures require a significant input of energy; some estimates indicate that the industrial production of ammonia is responsible for 1% of the world's total energy consumption. Suggest a reason why these high temperatures are employed.

11.100 Experiments show that the reaction of nitrogen dioxide with fluorine,

$$2\ NO_2(g)+F_2(g) \rightarrow 2\ FNO_2(g)$$

has the rate law

$$\text{Rate} = k[NO_2][F_2]$$

The reaction is thought to occur in two steps.

Step 1: $NO_2(g) + F_2(g) \rightarrow FNO_2(g) + F(g)$

Step 2: $NO_2(g) + F(g) \rightarrow FNO_2(g)$

(a) Show that the sum of this sequence of reactions gives the balanced equation for the overall reaction.

(b) Which step is rate determining?

11.101 Substances that poison a catalyst pose a major concern for many engineering designs, including catalytic converters. One design concept is to add materials that react with potential poisons before they reach the catalyst. Among the commonly encountered catalyst poisons are silicon and phosphorus, which typically form phosphate or silicate ions in the oxidizing environment of an engine. Group 2 elements are added to the catalyst to react with these contaminants before they reach the working portion of the catalytic converter. If estimates show that a catalytic converter will be exposed to 625 g of silicon during its lifetime, what mass of beryllium would need to be included in the design?

11.102 Chemical engineering design often uses mol% as a measure of concentration. A combustion reactor has an inlet flow of 5.50 mol% propane in air. Analysis of the product stream from the reactor finds 6.25 mol% CO_2. Does this measurement indicate complete combustion? What other species might you suggest monitoring in order to determine if the reaction kinetics are working in a predictable way.

11.103 The flashing of fireflies is the result of a chemical reaction, and the rate of flashing can be described by the Arrhenius equation. A certain batch of fireflies were observed to flash at a rate of 16.0 times per minute at 25°C and at a rate of 5.5 times per minute at 15°C. Use these data to find the apparent activation energy for the reaction that causes the flies to flash.

FOCUS ON PROBLEM SOLVING EXERCISES

11.104 Suppose that you are studying a reaction and need to determine its rate law. Explain what you would need to measure in order to accomplish this in a single experiment, and how you could use graphical methods to get from the experimental data to a complete rate law.

11.105 You measure an archeological specimen and find that the count rate from carbon-14 is x% of that seen in living organisms. How does this information allow you to establish the age of the artifact? What would you have to know or look up?

11.106 At room temperature, it takes about 3 days for a container of milk to turn sour. An identical container stored in a refrigerator turns sour only after 11 days. How could you use this information to estimate the activation energy of the chemical reaction that turns milk sour? What information would have to be looked up or estimated?

11.107 You are designing a self-heating food package. The chemicals you are using seem to be used up too quickly. What experiments could be run to devise conditions that would allow the heating pack to last longer?

Cumulative Problems

11.108 Use the web to look up the rates of reaction for different alkali metals with water. How can you explain the trend in these rates in terms of the concepts presented in this chapter?

11.109 Fluorine often reacts explosively. What does this fact suggest about fluorine reactions at the molecular level?

11.110 In Chapter 4, we noted that compressibility is an important factor for fuels, and that some hydrocarbons are more compressible than others. If a fuel is not very compressible, what can we infer about the activation energy for its combustion reaction?

11.111 A free radical is often used as a catalyst for polymerization reactions. Such reactions normally need to be carried out at higher temperatures. Explain this fact from the molecular perspective in terms of the activation energy of the reactants.

11.112 When formic acid is heated, it decomposes to hydrogen and carbon dioxide in a first-order decay:

$$HCOOH(g) \rightarrow CO_2(g) + H_2(g)$$

The rate of reaction is monitored by measuring the total pressure in the reaction container.

Time (s)	Pressure (torr)
0	220
50	324
100	379
150	408
200	423
250	431
300	435

Calculate the rate constant and half-life in seconds for the reaction. At the start of the reaction (time = 0), only formic acid is present. (HINT: Find the partial pressure of formic acid using Dalton's law of partial pressure and the reaction stoichiometry to find P_{HCOOH} at each time.)

12 Chemical Equilibrium

CHAPTER OUTLINE

The Natchez Trace Parkway Bridge, in Williamson County, Tennessee, is a concrete double arch bridge. The unique open design of the bridge places strict strength requirements on the concrete used. The weight of the bridge is concentrated at the crown of each arch. *National Park Service*

In our study of kinetics, we saw that the rate of a reaction changes over time, invariably slowing down as the reaction progresses. If we were to observe many reactions, this behavior would be evident each time. What does this imply? Eventually, the net rate of any chemical reaction approaches zero. When the net rate goes to zero, the reaction has reached a state of equilibrium. The concept of chemical equilibrium is critically important, and we will briefly look at several aspects of equilibrium in this chapter. Let's begin by considering a very familiar building material—concrete.

Chapter Objectives

After mastering this chapter, you should be able to

- list chemical reactions important in the production and weathering of concrete.
- explain that equilibrium is dynamic, and that at equilibrium the forward and backward reaction rates are equal.
- write the equilibrium constant expression for any reversible reaction.
- calculate equilibrium constants from experimental data.
- calculate equilibrium composition from initial data and the numerical value of the equilibrium constant.

- calculate molar solubility from K_{sp} or vice versa.
- write equilibrium constants for the dissociation of weak acids and weak bases and use them to calculate pH or the degree of ionization.
- use LeChatelier's principle to explain the response of an equilibrium system to applied stresses.
- calculate the new equilibrium composition of a system after an applied stress.
- explain the importance of both kinetic and equilibrium considerations in the design of industrial chemical processes.

12-1 Concrete Production and Weathering

There are few building materials that are more ubiquitous than concrete. When people think about concrete, usually they associate it with traits like durability and strength. You may never have associated concrete with chemistry, but there are actually a number of complex chemical systems that must be considered in many engineering designs that use concrete. Some of the most interesting chemistry occurs in the weathering of concrete due to prolonged exposure to the environment. Before we can consider weathering of concrete, we will need to look at its production.

Traditionally concrete has been composed of cement, water, and aggregate. Modern concrete preparation includes the use of additional components called **admixtures** to help manipulate the concrete into having desired properties, either as it is being poured or in long-term usage. From a chemical perspective, the most interesting reactions arise in the preparation of the cement and the admixtures.

Most concrete uses **Portland cement.** Portland cement starts with the production of CaO from limestone, which is mostly calcium carbonate, $CaCO_3$:

The name Portland cement *was first used in an 1824 patent in England by Joseph Aspdin, who thought the material looked like a type of limestone quarried near Portland, England.*

$$CaCO_3 \rightarrow CaO + CO_2$$

This step in the production of cement for concrete accounts for an estimated 5% of the total amount of CO_2 released into the atmosphere annually.

In addition to CaO, cement includes oxides of silicon and aluminum. These combined materials are then hydrated (have water added) when the concrete is mixed. The resulting mixture may have a fairly wide range of composition, so the equations that represent the chemistry are often written to include a variable, x. This variable may assume different values depending on the composition of a particular concrete, but it doesn't fundamentally alter the nature of the chemical reaction. Three representative reactions for the hydration process are

$$3\ CaO{\cdot}Al_2O_3 + 6\ H_2O \rightarrow Ca_3Al_2(OH)_{12}$$

$$2\ CaO{\cdot}SiO_2 + x\ H_2O \rightarrow Ca_2SiO_4{\cdot}xH_2O$$

$$3\ CaO + SiO_2 + (x + 1)H_2O \rightarrow Ca_2SiO_4{\cdot}xH_2O + Ca(OH)_2$$

If you count the chemical bonds in the reactants and products of these reactions, it should not be surprising that they release energy as heat, because the net result of the process is the formation of additional chemical bonds. **Figure 12.1** shows a graph of the energy liberated by the hydration of concrete as a function of time. Note that the reaction is fairly slow: the time scale in the graph is more than a month. This helps explain why engineers who use concrete need to specify curing times to achieve the desired characteristics of the concrete for their designs.

Recently, the use of fly ash to partially replace Portland cement has become increasingly common. Fly ash is generated when coal is burned in power plants. Minerals that are present in the coal react with oxygen at high temperatures and produce a residue known as fly ash. The average composition of fly ash is similar to that of Portland

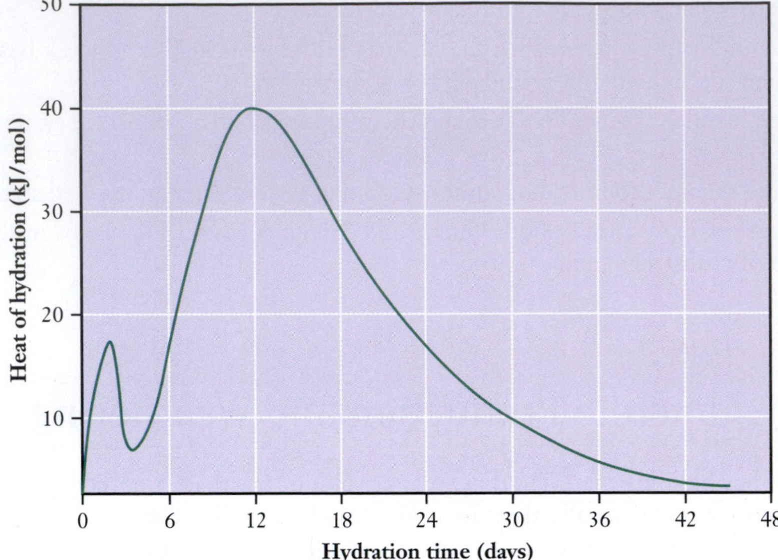

Figure 12.1 ■ The hydration reaction for concrete formation is exothermic and takes place over relatively long times. After a rapid temperature rise, called the preinduction period, the main energy release occurs about a week after the original mixing of the concrete.

cement, with the four main components being SiO_2, Al_2O_3, Fe_2O_3, and CaO. Fly ash typically consists of small spherical particles, and the incorporation of these particles can improve the strength of concrete. Inclusion of fly ash in concrete also provides a use for a material that might otherwise be considered waste. The use of fly ash is increasingly suggested as a more environmentally friendly way to manufacture concrete cement.

From the perspective of design considerations, modern engineers have notably more options in the specifications of concrete because of the use of admixtures. **Table 12.1** indicates uses for admixtures and common chemicals that provide the desired characteristics. Water reducers are used to lower the amount of water in the concrete while not reducing the ability to work with the material. Air entraining admixtures improve durability of cement by stabilizing small air bubbles within the cement portion of the concrete, particularly when concrete is exposed to water during freeze-thaw cycles. Other forms of admixtures combat the effects of moisture via waterproofing. Finally, some admixtures are useful as either accelerators of the hardening process

Table ■ 12.1

Chemical composition of selected admixtures

Function	Compound	Origin
Water reduction	Lignosulfonate	Wood/pulp byproduct
Water reduction	Hydroxycarboxylic acids	Chemical production
Air entrainment	Abietic and pimeric acids	Wood resins
Air entrainment	Alkyl-aryl sulphonates	Industrial detergents
Waterproofing	Fatty acids	Vegetable and animal fats
Acceleration	Calcium chloride	Chemical production
Acceleration	Calcium formate	Chemical production byproduct
Acceleration	Triethanolamine	Chemical production
Retardation	Borates	Borax mining
Retardation	Magnesium salts	Chemical production

Figure 12.2 ■ Treatment of concrete with phenolphthalein reveals the chemical effects of weathering. The pink color shows that the bulk of this sample remains basic, even after extended exposure to the elements. The colorless area at the top, which is the surface that was actually exposed, is no longer basic. Reaction with atmospheric CO_2 has neutralized the hydroxide in this surface region.

or retarders of that process. Once in the concrete, any of these admixtures can play a role in how it functions in the environment.

Most retarders also serve as water-reducing admixtures.

Weathering of concrete has several components, including the freeze-thaw cycle just noted as being partially mitigated by air entraining admixtures. We will focus on those components whose effects are primarily chemical from the start.

A key chemical reaction related to the aging of concrete is carbonation. In carbonation, CO_2 from the air diffuses into the concrete. Once there, it can react with calcium hydroxide in a two-step process represented by the following reactions:

$$Ca(OH)_2(s) \rightarrow Ca^{2+}(aq) + 2\ OH^-(aq)$$

$$Ca^{2+}(aq) + 2\ OH^-(aq) + CO_2(g) \rightarrow CaCO_3(s) + H_2O(\ell)$$

As we have noted earlier, the presence of hydroxide ions means that a solution is basic. So it is possible to observe the carbonation of concrete using an acid–base indicator. **Figure 12.2** shows a cross section of a piece of concrete that has been treated with an indicator called phenolphthalein, which turns pinks in the presence of a base. The bulk of this sample appears pink, confirming that the interior of the concrete remains basic. But the top 1/8 inch, which is the surface that was actually exposed, is not colored at all. This indicates that CO_2 from the environment has reacted with and neutralized the hydroxide originally present in the concrete. As is true in the environmental exposure of most engineering designs, the reactions that take place in concrete have long times in which to occur, and as a result, they are capable of reaching chemical equilibrium.

We introduced the concepts of acids and bases in Section 3-3 and will explore them further in this chapter.

12-2 Chemical Equilibrium

Equilibria in applied chemical systems such as concrete are unavoidably complex. But we can still understand much of the chemistry of these complex equilibria in terms of a few underlying principles. So as we begin our exploration of equilibria, we will use simpler systems to allow us to focus on those principles.

Forward and Reverse Reactions

If you place a glass of water on a table and leave it overnight, in the morning you may find an empty glass, or at least find that the water level has decreased **(Figure 12.3)**. This won't surprise you, because you know that the missing water has evaporated.

0 1

1 1

1 7

Figure 12.3 ■ This photo sequence shows the water level in two glasses over the course of time. The glass on the left is covered, whereas that on the right is open to the air. In the left-hand photo, the glasses are filled to the same level. In the second photo, taken 10 days later, the water level in the right-hand glass is visibly lower, whereas that in the left-hand glass remains unchanged. The trend continues in the final photo.

Try the experiment again, this time covering the glass, and you'll find that all of the water stays in the glass. In a closed system—like the covered glass—a dynamic equilibrium is established between liquid water and water vapor **(Figure 12.4)**. At equilibrium, the rate at which water molecules leave the surface of the liquid equals the rate at which they return to the surface. The rate of evaporation will be determined by the distribution of molecular energies in the liquid. Only those molecules with sufficient kinetic energy can overcome the intermolecular forces in the liquid and escape into the vapor phase. Assuming that the water is at a constant temperature, the rate of evaporation will be constant. What about the rate of condensation? Each time a gas phase molecule strikes the surface of the liquid, it has some probability of sticking to

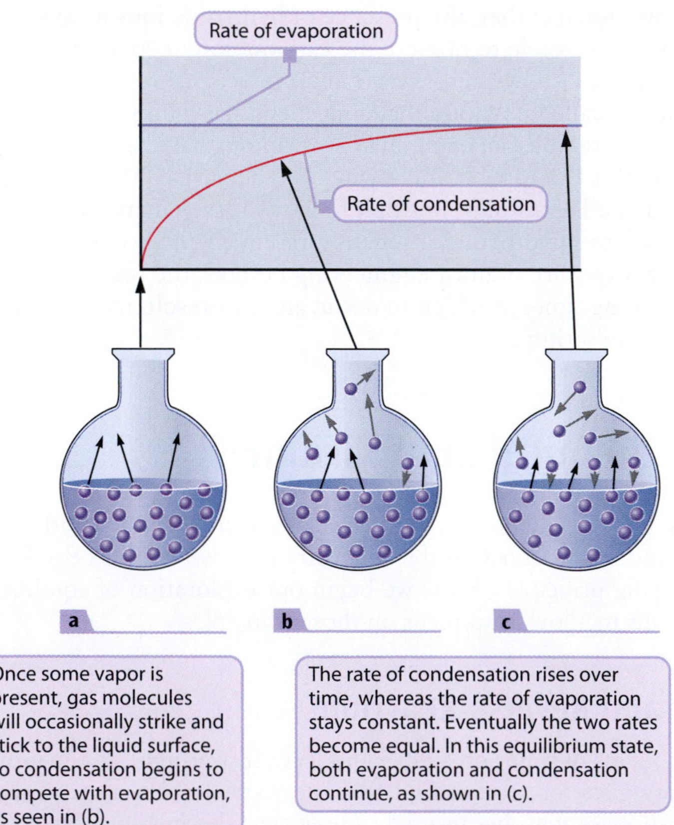
Rate of evaporation
Rate of condensation
a
b
c
Once some vapor is present, gas molecules will occasionally strike and stick to the liquid surface, so condensation begins to compete with evaporation, as seen in (b).
The rate of condensation rises over time, whereas the rate of evaporation stays constant. Eventually the two rates become equal. In this equilibrium state, both evaporation and condensation continue, as shown in (c).

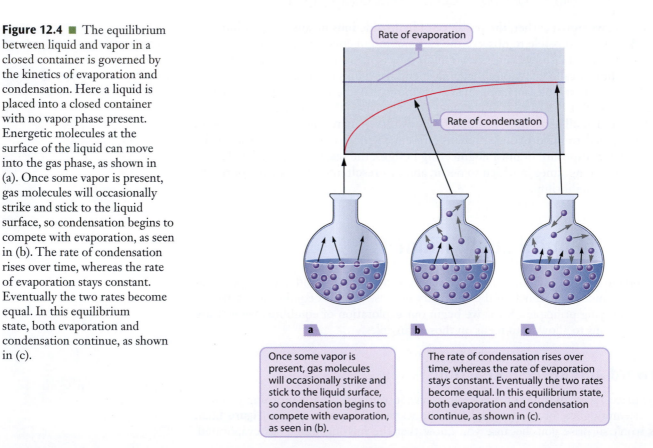

Figure 12.4 ■ The equilibrium between liquid and vapor in a closed container is governed by the kinetics of evaporation and condensation. Here a liquid is placed into a closed container with no vapor phase present. Energetic molecules at the surface of the liquid can move into the gas phase, as shown in (a). Once some vapor is present, gas molecules will occasionally strike and stick to the liquid surface, so condensation begins to compete with evaporation, as seen in (b). The rate of condensation rises over time, whereas the rate of evaporation stays constant. Eventually the two rates become equal. In this equilibrium state, both evaporation and condensation continue, as shown in (c).

that surface and returning to the liquid phase. The rate at which gas molecules strike the surface of the liquid is proportional to the number of gas molecules present, or to the gas pressure. If only liquid is present initially, the rate of condensation will be zero. But because evaporation is occurring, after a short time there will be some vapor in the container. That means condensation will begin. As long as the (constant) rate of evaporation is greater than the rate of condensation, the net effect will be to increase the pressure of the vapor phase. That, in turn, will increase the rate of condensation. Eventually the two rates will become equal, and after that point, no net change in the amounts of liquid or vapor will be seen.

We call this a **dynamic equilibrium** because the evaporation and condensation do not stop at equilibrium. The rates of these competing processes become equal to one another, but they do *not* go to zero. So on the microscopic level, individual molecules continue to move from the liquid to the vapor phase and back again. But from a macroscopic perspective, the observable amounts of liquid and vapor no longer change.

We introduced the idea of dynamic equilibrium when discussing vapor pressure in Section 8-5.

Turning to chemical reactions, the manner in which we have written chemical equations thus far has been to consider only the *forward* direction of the reaction. In the *reverse* reaction, the substances we've been calling products react to form reactants. In some cases, the reverse reaction may occur to a large degree and in other cases to an infinitesimal and immeasurable degree, but in principle it can occur in *every* chemical reaction. This means that in a closed system, any chemical reaction will eventually reach a dynamic equilibrium analogous to that we just described for evaporation.

At the instant we initiate a chemical reaction between two reactants, there are no products present **(Figure 12.5)**. We have already seen that the rate of a chemical reaction depends on the concentrations of reactants, so initially the rate of the reverse reaction must be zero because the concentrations of its reactants—which are the products of the forward reaction—are zero. Thus when we first mix the reactants, the rate of the forward reaction is greater than that of the reverse reaction. Over time, the concentrations of the reactants decrease, and those of the products increase. These changes in concentration are accompanied by changes in rate: the forward reaction slows down and the reverse reaction speeds up. Ultimately, chemical equilibrium is reached when the two rates become equal, and the observable concentrations of both reactants and products become constant. In any chemical system at equilibrium, *the rate of the forward reaction equals the rate of the reverse reaction.*

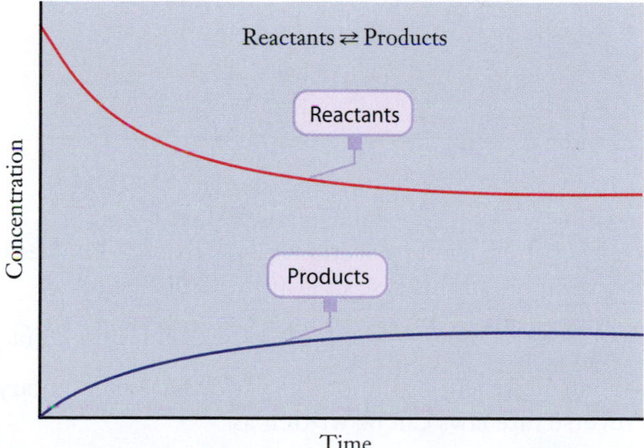

Figure 12.5 ■ The graph depicts the approach of a chemical system to equilibrium. Initially only reactants are present; the forward reaction proceeds and forms products. As the product concentrations increase, the reverse reaction becomes increasingly important. Eventually the rates of the forward and reverse reactions become equal, and the measured concentrations of reactants and products all become constant at equilibrium.

This graph implies that the reaction does not proceed very far toward products. We'll see next how this type of observation can be quantified.

Mathematical Relationships

The rate laws of the forward and reverse reactions suggest that a mathematical relationship can be written to describe equilibrium. For any reaction involving reactants R and products P, we write the chemical equation at equilibrium with two arrows, one pointing in each direction, to emphasize the dynamic character of the process:

$$R \rightleftarrows P$$

For simplicity, we will assume that this reaction proceeds in each direction through a single elementary step. (See Section 11-6. The result of our derivation here does not actually depend on this assumption, but it is easier to understand this way.) We can write the rate law for the forward direction of this reaction as

$$\text{Rate}_{\text{for}} = k_{\text{for}}[\text{R}]$$

Similarly, for the reverse reaction, we can write

$$\text{Rate}_{\text{rev}} = k_{\text{rev}}[\text{P}]$$

We know that these two rates must be equal at equilibrium, so

$$\text{Rate}_{\text{for}} = \text{Rate}_{\text{rev}}$$

Therefore,

$$k_{\text{for}}[\text{R}]_{\text{eq}} = k_{\text{rev}}[\text{P}]_{\text{eq}}$$

Because the rates are equal only if the system is at equilibrium, we have explicitly designated the concentrations as *equilibrium* concentrations here by adding a subscript "eq." We can manipulate this equation to obtain

$$\frac{k_{\text{for}}}{k_{\text{rev}}} = \frac{[\text{P}]_{\text{eq}}}{[\text{R}]_{\text{eq}}}$$

Both k_{for} and k_{rev} are rate constants, so as long as temperature is constant, the left-hand side of this equation is also constant. This means that at a given temperature, the ratio $[\text{P}]_{\text{eq}}/[\text{R}]_{\text{eq}}$ is a constant. (Remember that rate constants vary with temperature, so this ratio may also change as a function of temperature.) This expression will be at the heart of our mathematical treatment of equilibrium in Section 12-3.

To make this idea more concrete, consider what happens when salicylic acid, $C_6H_4(OH)COOH$, a water-reducing admixture component, dissolves in water.

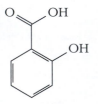

salicylic acid

As we will see in greater detail in Section 12-7, an acid dissolves in water to form hydrogen ions and an anion, in this case $C_6H_4(OH)COO^-$:

$$C_6H_4(OH)COOH\ (aq) \rightleftarrows H^+(aq) + C_6H_4(OH)COO^-(aq)$$

If we assume that the reactions in both directions are elementary rate processes, the forward and reverse rate laws can be written as

$$\text{Rate}_{\text{for}} = k_{\text{for}}[C_6H_4(OH)COOH]$$

$$\text{Rate}_{\text{rev}} = k_{\text{rev}}[H^+][C_6H_4(OH)COO^-]$$

Setting these rates equal to one another at equilibrium gives us

$$k_{\text{for}}[C_6H_4(OH)COOH] = k_{\text{rev}}[H^+][C_6H_4(OH)COO^-]$$

We can rearrange this so that the two rate constants are on one side and the concentrations are on the other:

$$\frac{k_{\text{for}}}{k_{\text{rev}}} = \frac{[H^+]_{eq}[C_6H_4(OH)COO^-]_{eq}}{[C_6H_4(OH)COOH]_{eq}} = \text{constant}$$

12-3 Equilibrium Constants

We noted earlier that Portland cement includes calcium oxide that is mixed with other materials. The decomposition of limestone, $CaCO_3$, noted in Section 12-1, occurs in a kiln via the following reaction:

$$CaCO_3(s) \rightarrow CaO(s) + CO_2(g)$$

The extent of this reaction and therefore the amount of CaO recovered for use in the cement is an example of an equilibrium process. For this reaction to be used in an industrial process such as the manufacture of cement, it is important to know how much CaO is available. And because of the environmental concerns associated with increased atmospheric CO_2 level, the amount of this gas that is released is also relevant.

The need for quantitative answers to questions like this requires a mathematical model for equilibrium. In the previous section, we saw that a relationship exists between concentrations of reactants and products. This relationship, which was first proposed by Cato Maximilian Guldberg and Peter Waage in 1864, is often called the law of mass action.

The Equilibrium (Mass Action) Expression

For a general chemical reaction:

$$a\,A + b\,B \rightleftarrows c\,C + d\,D$$

we can define a ratio of concentrations, whether or not at equilibrium:

$$Q = \frac{[C]^c[D]^d}{[A]^a[B]^b} \tag{12.1}$$

We refer to this ratio as the **reaction quotient (*Q*)**. Product concentrations appear in the numerator of the expression, and reactants appear in the denominator. Each concentration is raised to a power corresponding to the stoichiometric coefficient of that species in the balanced chemical equation for the equilibrium. At equilibrium, however, this ratio becomes the **equilibrium expression**, and the corresponding value of *Q* is called the **equilibrium constant (*K*)**:

$$K = \frac{[C]^c_{\text{eq}}[D]^d_{\text{eq}}}{[A]^a_{\text{eq}}[B]^b_{\text{eq}}} \tag{12.2}$$

Here we use the subscript "eq" to emphasize that the system is at equilibrium. This subscript is often omitted, and the use of K rather than Q implies equilibrium.

To determine the value of *K* for a particular reaction, we must measure the concentrations of the reactants and products at equilibrium. Note that, unlike rate expressions in chemical kinetics, *the expression for the equilibrium constant is always based directly on the reaction stoichiometry*. Example Problem 12.1 shows how to write an equilibrium expression.

EXAMPLE PROBLEM 12.1

Concrete used in smokestacks has to be designed to withstand sometimes caustic conditions. Sulfur oxides are produced in some cases, for example, and they would establish equilibrium if they did not disperse:

$$2\ SO_2(g) + O_2(g) \rightleftarrows 2\ SO_3(g)$$

Write the equilibrium expression for this reaction.

Strategy We use the definition of the equilibrium expression. Write the products in the numerator and the reactants in the denominator. The exponent for each concentration term is given by the corresponding stoichiometric coefficient from the chemical equation.

Solution

$$K = \frac{[SO_3]^2}{[SO_2]^2[O_2]}$$

Analyze Your Answer The only way we can assess an answer like this is to make sure that we used the definition of the equilibrium constant correctly. We have the product (SO_3) in the numerator and the reactants (SO_2 and O_2) in the denominator, which is correct, and we have exponents to match the stoichiometric coefficients. So our answer is correct.

Check Your Understanding Nitrogen oxides in the atmosphere may acidify rainwater. Write the equilibrium expression for the following reaction in which gaseous nitrous acid forms in the atmosphere:

$$NO(g) + NO_2(g) + H_2O(g) \rightleftarrows 2\ HNO_2(g)$$

Gas Phase Equilibria: K_p vs. K_c

Many equilibrium reactions, including that in the previous example, take place in the gas phase. The equilibrium expression we wrote in Example Problem 12.1 includes the molar concentrations of three gases. Although there is nothing wrong with such concentrations, it is often easier to describe gases using partial pressures rather than molar concentrations. Hence an alternative definition of the equilibrium constant in which partial pressures replace molarities is sometimes used for gas-phase reactions. The symbol K_p is used to indicate such an equilibrium constant. Let's return to the general equilibrium we used to develop Equations 12.1 and 12.2 and modify it slightly to show all species as gases:

$$a\ A(g) + b\ B(g) \rightleftarrows c\ C(g) + d\ D(g)$$

From this we can write an expression for K_p:

$$K_p = \frac{(P_C)^c_{eq}(P_D)^d_{eq}}{(P_A)^a_{eq}(P_B)^b_{eq}} \qquad (12.3)$$

Because we now have two alternative definitions for equilibrium constants, it is appropriate to ask whether they will lead to the same values. To answer that question, we need to relate the molar concentration of a gas to its partial pressure. That connection can be made through the ideal gas law and more specifically by Equation 5.5 (page 124), which defines the partial pressure of a gas i:

$$P_i = \frac{n_iRT}{V}$$

The molar concentration of this gas is just (n_i/V). Switching from "i" to "A" to match our generalized reaction, we can write

$$P_A = \frac{n_A}{V}RT = [A]RT$$

Similar expressions can be written for the remaining gases in our reaction (B, C, and D). Inserting all of these into Equation 12.3,

$$K_p = \frac{([C]_{eq}RT)^c([D]_{eq}RT)^d}{([A]_{eq}RT)^a([B]_{eq}RT)^b}$$

$$= \frac{[C]^c_{eq}[D]^d_{eq}}{[A]^a_{eq}[B]^b_{eq}} \times RT^{(c+d-a-b)}$$

$$= K_c \times RT^{(c+d-a-b)}$$

Here the subscript "c" on the last K is added to emphasize that this is the equilibrium constant written in molar concentrations. It is immediately clear that unless $RT^{(c+d-a-b)} = 1$, K_p and K_c will not be equal. The exponent $(c + d - a - b)$ in that term represents the change in the number of moles of gas from the reactants to the products. Analogously to the way we have written the changes in various thermodynamic quantities, we can write this as Δn_{gas}, defined as the number of moles of gas on the product side minus the number of moles of gas on the reactant side. Using this definition, we can write

$$K_p = K_c \times RT^{(\Delta n_{gas})} \tag{12.4}$$

The two equilibrium constants, K_p and K_c, will be equal only if $\Delta n_{gas} = 0$. When working with gas phase equilibrium constants, therefore, it is always important to be certain that you know whether the value you look up is K_p or K_c. For simplicity, *all equilibrium constants in this text will be based on molar concentrations, not pressures.* Accordingly, we will not use the subscript "c" when writing these K values.

Homogeneous and Heterogeneous Equilibria

The equilibria we have examined so far are all examples of **homogeneous equilibria,** in which reactants and products are in the same phase, either gaseous or aqueous. If any of the reactants or products is in the solid or liquid phase, though, we will need to adjust our formulation of the equilibrium expression slightly. The concentration of a pure solid or liquid will not change as the reaction proceeds. Indeed, the phrase "concentration of a pure solid or liquid" should strike you as a bit odd; unlike gases or solutions, solids and liquids do not have variable concentrations. For example, suppose that the decomposition of calcium carbonate that occurs in the making of Portland cement were carried out in a closed system (**Figure 12.6**):

$$CaCO_3(s) \rightleftarrows CaO(s) + CO_2(g)$$

Once equilibrium has been established at a particular temperature, the concentration of gaseous CO_2 can be expressed in mol/L. If the system is heated, more CaO and CO_2 form, and, if the volume of the system is held constant, the concentration of CO_2 in mol/L increases as equilibrium is established at a higher temperature. But the concentrations of the solids do not change because, as the numbers of moles of each changes, so do their volumes. The densities of solids change very little with a change in temperature, and the ratio of the number of moles (and mass) of $CaCO_3$, for example, to the volume of $CaCO_3$ remains constant. A solid's concentration does not change, nor does that of a liquid. This fact has led to a different way to describe the equilibrium expression for **heterogeneous equilibrium** in a multiphase system. Because their concentrations are constant, reactants and products that are either liquid or solid do not appear in the expression, so the equilibrium expression for the system above includes only the carbon dioxide concentration:

$$K = [CO_2]$$

In the Example Problem 12.2, you can see how this idea can be applied to precipitation reactions.

Figure 12.6 ■ Because solids do not have variable concentrations, the equilibrium constant for this reaction is just the concentration of CO_2. This means that the CO_2 pressure at a given temperature is independent of the amounts of $CaCO_3$ and CaO present. Here the CO_2 pressure is the same over two different mixtures of the two solids.

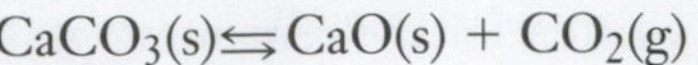

$$CaCO_3(s) \leftrightarrows CaO(s) + CO_2(g)$$

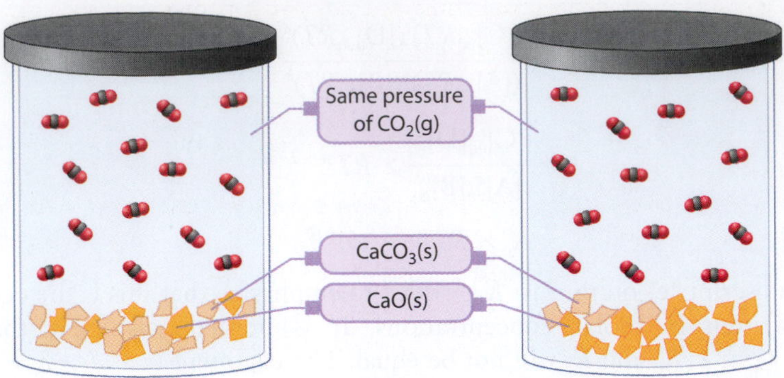

EXAMPLE PROBLEM 12.2

Calcium hydroxide will precipitate from solution by the following equilibrium:

$$Ca^{2+}(aq) + 2\ OH^-(aq) \rightleftarrows Ca(OH)_2(s)$$

Write the equilibrium expression for this reaction.

Strategy Write the equilibrium expression as before but do not include a term for the concentration of the solid calcium hydroxide product.

Solution

$$K = \frac{1}{[Ca^{2+}][OH^-]^2}$$

Analyze Your Answer This answer may seem troublesome because there are no concentration terms in the numerator. But when we look at the reaction, we see that the only product is a solid. Because pure substances do not appear in the equilibrium constant expression, we are left with just a "1" in the numerator, and the answer is correct.

Check Your Understanding Write the chemical equation for the precipitation of copper(II) phosphate from a solution containing copper(II) and phosphate ions and then write the corresponding equilibrium expression.

Numerical Importance of the Equilibrium Expression

Being able to write an expression relating equilibrium concentrations in chemical reactions provides a powerful tool for understanding chemical equilibrium. We will soon see how to evaluate these expressions for several classes of reactions. But first, let's look at what we can learn from the numerical values of equilibrium constants. For example, we could ask the question, "Which calcium compound is more likely to form if more than one anion is present?" How can we use equilibrium constants to answer this type of question?

Equilibrium constants range over many orders of magnitude. Values as small as 10^{-99} are known, as are values as high as 10^{99} or greater. What does the size of the equilibrium constant tell us about the reaction? If we look at the equilibrium expression for a reaction as an algebraic relationship, the answer will become apparent:

Sometimes, the power of 10 in an equilibrium constant exceeds 99, causing errors on some calculators. Such cases can be handled by algebraically manipulating the exponents.

$$K = \frac{[\text{products}]}{[\text{reactants}]}$$

Concentrations of products are in the numerator, and concentrations of reactants are in the denominator. For a very large value of K, the product concentrations must be large and the reactant concentrations must be small. By contrast, for the value of K to be small, concentrations of the reactants must be large and those of products must be small.

$$K = \frac{\text{[products]}}{\text{[reactants]}} = \text{large value} \qquad K = \frac{\text{[products]}}{\text{[reactants]}} = \text{small value}$$

Thus the value of K can tell us the direction in which a chemical reaction is likely to proceed. For large values of K, products are favored, whereas small values of K (meaning $K << 1$) mean that at equilibrium most of the reactants will remain intact. If K is on the order of 1, the equilibrium has significant concentrations of both reactants and products. Example Problem 12.3 provides some practice in interpreting the values of equilibrium constants.

EXAMPLE PROBLEM 12.3

In Example Problem 12.2, we saw that hydroxide ions precipitate with calcium. Magnesium ions show similar behavior. The two pertinent equilibria are

$$Ca^{2+}(aq) + 2\ OH^{-}(aq) \rightleftarrows Ca(OH)_2(s) \quad K = 1.3 \times 10^5$$

$$Mg^{2+}(aq) + 2\ OH^{-}(aq) \rightleftarrows Mg(OH)_2(s) \quad K = 6.7 \times 10^{10}$$

Which ion is more likely to precipitate hydroxide from a solution, assuming roughly equal concentrations of calcium and magnesium ions?

Strategy The likely extent of each reaction can be predicted by the size of the equilibrium constant. By comparing the two numbers, we can see which cation is more likely to precipitate the hydroxide.

Solution Both equilibrium constants are much greater than 1, but the value for the reaction with magnesium ions is significantly larger than that for calcium. Thus magnesium hydroxide has a much greater tendency to form and will be the more likely product.

Check Your Understanding Manganese(II) hydroxide precipitates according to the equation

$$Mn^{2+}(aq) + 2\ OH^{-}(aq) \rightleftarrows Mn(OH)_2(s) \quad K = 2.17 \times 10^{13}$$

Relative to calcium and magnesium ions, how would manganese(II) ions rate in terms of their ability to precipitate hydroxide ions?

Mathematical Manipulation of Equilibrium Constants

When we work with precipitation reactions like those in the previous example, we need to look up values for the equilibrium constants. But the values we find in a reference book typically refer to the opposite process—solubility. So we will need to know how an equilibrium constant changes if we modify the chemical equation to which it refers.

Reversing the Chemical Equation

When we write chemical equations for reactions that go to completion, the choice of reactants and products seems logically obvious. But for an equilibrium reaction, the fact that both forward and backward reactions are important makes the choice seem much more arbitrary. We can write equations for equilibria in either direction. Neither choice is wrong, provided we also realize that the equilibrium constant and expression we use must match the equation we choose. Consider the precipitation of calcium hydroxide, which we've seen in the previous example problems. We can write the equilibrium expressions for both forms of the chemical equation:

$$Ca^{2+}(aq) + 2\ OH^-(aq) \rightleftarrows Ca(OH)_2(s) \quad K = \frac{1}{[Ca^{2+}][OH^-]^2}$$

$$Ca(OH)_2(s) \rightleftarrows Ca^{2+}(aq) + 2\ OH^-(aq) \quad K' = [Ca^{2+}][OH^-]^2$$

From these two expressions, we can see that the equilibrium expressions are inverses of one another. Thus when we reverse a chemical equation by switching reactants and products, we invert the equilibrium constant expression and its value:

$$K' = \frac{1}{K}$$

Adjusting the Stoichiometry of the Chemical Reaction

The stoichiometric coefficients play a direct role in the equilibrium expression, and if adjustments are made in the stoichiometry, they will affect the value of the equilibrium constant. For example, a reaction called the Andrussow process, which produces HCN for use in the polymer industry, can be written either as

$$2\ NH_3(g) + 2\ CH_4(g) + 3\ O_2(g) \rightleftarrows 2\ HCN(g) + 6\ H_2O(g)$$

or

$$NH_3(g) + CH_4(g) + \frac{3}{2}\ O_2(g) \rightleftarrows HCN(g) + 3\ H_2O(g)$$

How is the equilibrium expression affected by this change in the stoichiometry? The two expressions are

$$K_1 = \frac{[HCN]^2[H_2O]^6}{[NH_3]^2[CH_4]^2[O_2]^3} \text{ and } K_2 = \frac{[HCN][H_2O]^3}{[NH_3][CH_4][O_2]^{3/2}}$$

As you can see, K_2 is the square root of K_1. Thus dividing by two in the stoichiometry leads to a division by 2 in the *exponents* of the equilibrium expression. It's clear that we must look carefully at the way a chemical equation is written when we consider the value of the equilibrium constant. Example Problem 12.4 explores this idea further.

EXAMPLE PROBLEM 12.4

Ammonia is an important starting material for several industrial processes, including the production of fertilizers, polymers, and admixture components for cement. Thus the production of ammonia from nitrogen and hydrogen is one of the world's most important industrial reactions:

$$N_2(g) + 3\ H_2(g) \rightleftarrows 2\ NH_3(g)$$

Write equilibrium expressions for (a) the reaction as written; (b) the reverse reaction; and (c) the reaction as written but with all coefficients in the equation halved.

Strategy Write the equilibrium expression for the reaction as written and manipulate the expression as needed to match the other conditions. If we are not confident with these manipulations, we can always write out the specified versions of the equilibrium equations and use them to obtain the needed equilibrium expressions.

Solution

(a) $K = \frac{[NH_3]^2}{[N_2][H_2]^3}$

(b) $K' = \frac{1}{K} = \frac{[N_2][H_2]^3}{[NH_3]^2}$

(c) $K'' = K^{1/2} = \frac{[NH_3]}{[N_2]^{1/2}[H_2]^{3/2}}$

Check Your Understanding Consider the following reaction in a closed system:

$$2\ NH_3(g) + \frac{5}{2}O_2(g) \rightleftarrows 2\ NO(g) + 3\ H_2O(g)$$

Write the equilibrium expressions for (a) the reaction as written; (b) the reverse reaction; and (c) the reaction as written but with all coefficients in the equation doubled.

Equilibrium Constants for a Series of Reactions

In complicated chemical systems, a product of one chemical reaction can be a reactant in a subsequent reaction. Because of the nature of the equilibrium expression, if we know the equilibrium constant for the two separate reactions, we can determine an overall value for the combination of the two. Consider, for example, the stepwise reaction of phosphate ions with hydrogen ions:

Phosphates are occasionally used in admixtures for concrete.

$$PO_4^{3-}(aq) + H^+(aq) \rightleftarrows HPO_4^{2-}(aq) \qquad K_1 = \frac{[HPO_4^{2-}]}{[PO_4^{3-}][H^+]}$$

$$HPO_4^{2-}(aq) + H^+(aq) \rightleftarrows H_2PO_4^-(aq) \qquad K_2 = \frac{[H_2PO_4^-]}{[HPO_4^{2-}][H^+]}$$

$$PO_4^{3-}(aq) + 2\ H^+(aq) \rightleftarrows H_2PO_4^-(aq) \qquad K_3 = \frac{[H_2PO_4^-]}{[PO_4^{3-}][H^+]^2}$$

The third equation is the sum of the first two equations. How are their equilibrium constants related? To obtain the chemical equation associated with K_3, we added the chemical equations for K_1 and K_2. To get the equilibrium constant for equation 3, however, we must multiply the expressions:

$$K_3 = K_1 \times K_2$$

$$K_3 = \frac{[HPO_4^{2-}]}{[PO_4^{3-}][H^+]} \times \frac{[H_2PO_4^-]}{[HPO_4^{2-}][H^+]} = \frac{[H_2PO_4^-]}{[PO_4^{3-}][H^+]^2}$$

This relationship holds true any time we add chemical equations to obtain a new equation: multiply the equilibrium constants to find the equilibrium constant for the reaction of interest. Example Problem 12.5 provides more practice with this idea.

The dependence of K on stoichiometry is a special case of this relationship. Doubling all coefficients leads to squaring the equilibrium constant.

EXAMPLE PROBLEM 12.5

Given the equilibria

$$CO_2(g) \rightleftarrows CO(g) + \frac{1}{2}O_2(g)$$

$$H_2(g) + \frac{1}{2}O_2(g) \rightleftarrows H_2O(g)$$

determine the equilibrium expression for the sum of the two reactions.

Strategy Write the equilibrium expressions for the two reactions. Multiply them to obtain the equilibrium expression for the sum of the two reactions. Check your answer by adding the two reactions and comparing the sum to the equilibrium expression you determined.

Solution

$$CO_2(g) \rightleftarrows CO(g) + \frac{1}{2}O_2(g) \quad K_1 = \frac{[CO][O_2]^{1/2}}{[CO_2]}$$

$$H_2(g) + \frac{1}{2}O_2(g) \rightleftarrows H_2O(g) \quad K_2 = \frac{[H_2O]}{[H_2][O_2]^{1/2}}$$

$$K_3 = \frac{[CO][O_2]^{1/2}}{[CO_2]} \times \frac{[H_2O]}{[H_2][O_2]^{1/2}} = \frac{[CO][H_2O]}{[CO_2][H_2]}$$

Analyze Your Answer The sum of the two reactions is

$$CO_2(g) + H_2(g) \rightleftarrows CO(g) + H_2O(g)$$

So it is fairly easy to see that the equilibrium expression we obtained above is correct.

Check Your Understanding Given the following equilibrium expressions for a pair of gas phase reactions, find the equilibrium expression for the sum of the two reactions and write its chemical equation:

$$K_1 = \frac{[NO_2]^2}{[N_2][O_2]^2}$$

$$K_2 = \frac{[NO]^2[O_2]}{[NO_2]^2}$$

Units and the Equilibrium Constant

As you've already begun to see, the equilibrium constant can be used in various mathematical expressions. Later, we'll see cases where an equilibrium constant appears as the argument in a natural log function. When a variable appears as an argument of such a function, it must be dimensionless. To produce equilibrium constants without units, we adjust the concentrations of all species by comparing them to a standard concentration of 1 M. Thus the concentration of hydroxide ion we use in the expressions we encounter in basic solutions will actually be $[OH^-] = [OH^-]_{eq}/1\ M$. This means that the numbers we use in equilibrium constant calculations carry no units, and therefore the equilibrium constant will also be unitless. Because we choose 1 M as the standard concentration, the manipulation has no numerical consequences. It does, however, help avoid confusion later when we use the equilibrium constant in mathematical equations.

12-4 Equilibrium Concentrations

Often, equilibrium concentrations are measured in lab experiments to determine the value of equilibrium constants. But other situations call for the prediction of equilibrium concentrations based on known equilibrium constants. We can look up the equilibrium constant for the dissociation of salicylic acid in water, but how does this tell us the concentrations of the various ions in a given solution? We can address this type of question by carrying out calculations using the equilibrium constant expression and the stoichiometry of the chemical reaction it describes.

Predictions of equilibrium concentrations might be useful in the early stages of designing a process.

Calculations based on chemical equilibria tend to be somewhat complicated and often involve a large amount of data. So it is a good idea to approach these problems systematically. We'll illustrate how to handle some specific types of problems throughout this chapter. But first, it is worth noting three basic features of the strategy we'll employ in *any* equilibrium calculation:

- Write a balanced chemical equation for the relevant equilibrium or equilibria.
- Write the corresponding equilibrium expression or expressions.
- Create a table of concentrations for all of the reacting species.

The table we set up in the third step will generally contain a column for each species involved in the equilibrium. The first row contains the initial concentrations, the second row contains the changes in these concentrations that occur as the system comes to equilibrium, and the third row contains the final equilibrium concentrations. Much of this information will not be known at the outset of the problem. But the table provides a convenient way of organizing all of the relevant information. As we work through a problem, more and more of the cells in this table will be filled in.

Equilibrium Concentrations from Initial Concentrations

A batch reaction is one that is carried out in a closed container, with no material added or removed during the reaction. If a process is run in such a batch mode, how will the equilibrium concentrations of reactants and products be related to the initial concentrations? We can find answers to questions like this, and doing so represents an important starting point in the design of many industrial processes. To get started with this type of problem, Example Problem 12.6 looks at an algebraically simple example involving hydrogen and iodine. This will allow us to illustrate the use of the problem-solving framework that we will apply for equilibrium calculations. Once we've mastered that framework, we will also be able to apply it to more complicated systems, such as those responsible for the weathering of concrete.

Industrial reactions are often carried out in flow mode, where a stream of reactants and products is constantly pumped through a reaction vessel.

EXAMPLE PROBLEM 12.6

When hydrogen gas reacts with iodine gas at elevated temperatures, the following equilibrium is established:

$$H_2(g) + I_2(g) \rightleftarrows 2\,HI(g)$$

A student measured the equilibrium constant as 59.3 at 400°C. If one trial begins with a mixture that includes 0.050 M hydrogen and 0.050 M iodine, what will be the equilibrium concentrations of reactants and products?

Strategy It may help to begin by summarizing the information from the problem.

Given: $[H_2]_{initial} = [I_2]_{initial} = 0.050$ M; $K = 59.3$ Find: $[H_2]_{eq}$, $[I_2]_{eq}$, and $[HI]_{eq}$

We'll use (and clarify) the three-point strategy outlined above. The chemical equation for the equilibrium is given in the problem. From that equation, we can write the equilibrium expression easily enough:

$$K = \frac{[HI]^2}{[H_2][I_2]} = 59.3$$

Next, we construct the table of concentrations, with places to enter the initial concentrations, the changes in concentration, and the final (equilibrium) concentrations for the three species. So far, only the given initial concentrations are known. (Because only H_2 and I_2 are present initially, the initial concentration of HI is zero.)

	H_2	I_2	HI
Initial Concentration	0.050 M	0.050 M	0 M
Change in Concentration			
Final Concentration			

The rest of the table will be filled in as we work our way through the solution.

At this point, the problem seems daunting; we have three unknown equilibrium concentrations, and only one real equation to work with—the equilibrium expression. We will need to use the stoichiometry of the reaction to relate the changes in concentration to one another.

Solution Because only H_2 and I_2 are present initially, we know that their concentrations must decrease (and the HI concentration must increase) as the system goes to equilibrium. We can also see from the balanced equation that each H_2 molecule must react with one I_2 molecule to form two HI molecules. This gives us the relationship we need to express the changes in concentration in terms of a single variable. Let's call the change in $[H_2]$ "$-x$." (The negative sign shows that the concentration decreases.) The 1:1 mole ratio between H_2 and I_2 means that the change in $[I_2]$ must also be $-x$, and the 1:2 mole ratio between H_2 and HI means that the change in [HI] must be $+2x$. This gives us the second row in the table.

	H_2	I_2	HI
Initial Concentration	0.050 M	0.050 M	0 M
Change in Concentration	$-x$	$-x$	$+2x$
Final Concentration			

Next, we can fill in the third row. To do this, we use the simple relationship that the concentration of each species at equilibrium will be its initial concentration plus the change in concentration. So we just need to sum each column to get expressions for the third row.

	H_2	I_2	HI
Initial Concentration	0.050 M	0.050 M	0 M
Change in Concentration	$-x$	$-x$	$+2x$
Final Concentration	$0.050 - x$	$0.050 - x$	$2x$

Of course, our aim is to find actual values for those three equilibrium concentrations, so we still need to find a way to solve for the variable (x). To do this, we return to the equilibrium expression. If we insert the expressions from the third row of our table into the equilibrium expression, we'll get the following equation:

$$K = \frac{[HI]^2}{[H_2][I_2]} = \frac{(2x)^2}{(0.050 - x)(0.050 - x)} = 59.3$$

or

$$\frac{(2x)^2}{(0.050 - x)^2} = 59.3$$

This equation has only one unknown, so we should be able to solve it for x and then use that result to obtain values for the final concentrations. We could multiply the denominator out and solve using the quadratic formula. Or we might save a little algebra by realizing that the left-hand side is a perfect square. This lets us take the square root of both sides to get a linear equation, which we can solve more easily:

$$\frac{2x}{0.050 - x} = 7.70$$

$$9.70x = 0.39$$

$$x = 0.040$$

Finally, we use this value of x to calculate all three equilibrium concentrations:

$$[H_2] = [I_2] = 0.050 - 0.040 = 0.010 \text{ M}$$

$$[HI] = 2x = 0.080 \text{ M}$$

Analyze Your Answer We can compare this result to the initial amounts to make sure it seems reasonable. First, we know that x can't be bigger than 0.05 because that would make $[H_2]$ and $[I_2]$ negative. Looking more closely, we can also see that because K is noticeably greater than one, we expect most of the original H_2 and I_2 to end up on the product side as HI. This is consistent with our answer.

Check Your Understanding The same student measures the equilibrium constant for this reaction at a higher temperature and obtains a value of $K = 51.4$. An experiment is started with 0.010 M each of hydrogen and iodine. Predict the equilibrium concentrations of all species.

Two specific factors helped us to simplify the calculations in Example Problem 12.6. First, the stoichiometry of the reaction in which 1 mole of each of the two reactants yielded 2 moles of products contributed to getting perfect squares in both the numerator and the denominator. Second, we started with the same initial concentrations of each reactant. Had those initial amounts been different from each other, we would not have had a perfect square in the denominator when we filled in the equilibrium expression. When we are not so fortunate in our reaction of interest, we may have to use the quadratic formula to solve the resulting algebraic equation. Although this may be a minor nuisance, it isn't really a major hurdle. We consider this situation in Example Problem 12.7.

EXAMPLE PROBLEM 12.7

The equilibrium constant for the reaction of chlorine gas with phosphorus trichloride to form phosphorus pentachloride is 33 at 250°C. If an experiment is initiated with concentrations of 0.050 M PCl_3 and 0.015 M Cl_2, what are the equilibrium concentrations of all three gases?

$$Cl_2(g) + PCl_3(g) \rightleftarrows PCl_5(g)$$

Strategy The approach here is the same as in Example Problem 12.6. The only change occurs when we need to solve for x after we set up the equilibrium constant expression. The stoichiometry and the initial amounts in this case will force us to use the quadratic formula to solve for x.

Solution Construct a table of concentrations, as we did in Example Problem 12.6. We'll insert the initial concentrations and the changes in concentration. Because the stoichiometry of this reaction is 1:1:1, the changes can be written as $-x$ for both Cl_2 and PCl_3 and $+x$ for PCl_5.

	Cl_2	PCl_3	PCl_5
Initial Concentration	0.015 M	0.050 M	0 M
Change in Concentration	$-x$	$-x$	$+x$
Final Concentration	$0.015 - x$	$0.050 - x$	x

Substitute in the equilibrium expression:

$$K = \frac{[PCl_5]}{[Cl_2][PCl_3]} = 33$$

$$\frac{x}{(0.015 - x)(0.050 - x)} = 33$$

To solve this, we must use the quadratic formula, so we first expand it:

$$33x^2 - 3.145x + 0.025 = 0$$

Using the quadratic formula,

$$x = \frac{-b \pm \sqrt{b^2 - 4ac}}{2a}$$

$$= \frac{3.145 \pm \sqrt{9.89 - 3.3}}{66}$$

$$x = 0.087 \text{ or } x = 0.0088$$

It may seem disconcerting to get two numerical solutions to a problem like this. How do we decide which root is correct? The resulting equilibrium concentrations must be physically reasonable. Among other things, that means that none of them can be negative. But if we chose the first value, $x = 0.087$, we would get negative concentrations for both Cl_2 and PCl_3. So the correct root must be the second one, $x = 0.0088$, which gives us the following results:

$$[PCl_5] = x = 0.0088 \text{ M}$$

$$[Cl_2] = 0.015 - x = 0.015 - 0.0088 = 0.006 \text{ M}$$

$$[PCl_3] = 0.050 - x = 0.050 - 0.0088 = 0.041 \text{ M}$$

Analyze Your Answer One check we can always do for this type of equilibrium problem is to put the final values back into the equilibrium expression and make sure that they produce a reasonable value for K. In this case, that gives $K = (0.0088)/(0.006)(0.041) = 36$. This number is not quite 33, but because all of our data have only two significant figures, we should not expect an exact match. Being this close to the original value of K is reassuring and means that we have probably done the arithmetic correctly.

Check Your Understanding This reaction can also be considered as the decomposition of $PCl_5(g)$, for which the equilibrium constant is 0.030 (at the same temperature). If you begin an experiment with 0.040 M $PCl_5(g)$, what are the expected equilibrium concentrations of all species?

Mathematical Techniques for Equilibrium Calculations

The general approach we have put forward in these example problems will work for any equilibrium calculation. But we should point out that these examples were chosen at least in part because they can be solved with relative ease. Even our more complicated case required nothing more than using the quadratic formula. But not all equilibria lend themselves to such simple solutions. Consider the Andrussow process for the production of HCN, which we mentioned earlier. The overall equation is shown below. Because all species are gases under the high-temperature conditions needed for reaction, five concentration terms appear in the equilibrium expression, along with exponents as high as six:

$$2\ NH_3 + 2\ CH_4 + 3\ O_2 \rightleftarrows 2\ HCN + 6\ H_2O$$

$$K = \frac{[HCN]^2[H_2O]^6}{[NH_3]^2[CH_4]^2[O_2]^3}$$

If we were to construct a problem similar to those we solved above but based on this equilibrium, we would be faced with solving an equation that included terms in x^8.

Clearly this would present challenges well beyond solving a quadratic equation. Still, modern software does offer hope for solving such equations. Maple®, Mathematica®, and other packages provide tools for finding solutions to equations of this sort. We will not spend our time on such calculations here, though, because we would rather focus on the underlying chemical concepts.

Being able to determine the equilibrium concentration of a desired reactant provides a useful starting point for exploring industrial processes like the production of hydrogen cyanide, but the equilibrium yield of a product may not be sufficient to make a process economically viable. In such cases, a process engineer needs to find ways to manipulate the equilibrium and allow higher levels of production or greater efficiency. How can equilibria be manipulated? Our next section addresses this question.

12-5 LeChatelier's Principle

For the production of chemical commodities, chemical and industrial engineers choose reactions whose equilibrium constants favor products whenever possible. Often, this type of choice simply is not available. For example, equilibrium between H_2, N_2, and NH_3 in the production of ammonia does not favor the formation of ammonia. Nevertheless, the manufacture of ammonia is commercially viable, and it is a major component of the chemical industry—chiefly because of the importance of ammonia in fertilizers and other agricultural chemicals. What are the variables that can be controlled in chemical plants, and how do they influence equilibria?

LeChatelier's principle summarizes the ways that any equilibrium system responds to changes. *When a system at equilibrium is stressed, it responds by reestablishing equilibrium to reduce the applied stress.* There are three common means for introducing such stresses in chemical equilibria: changes in concentrations, changes in pressure, and changes in temperature. We will look at each of these in turn and will think about them in the context of a single simple equilibrium system: the dimerization of NO_2 to form N_2O_4.

Frenchman Henri Louis LeChatelier came from a family of architects and engineers. He carried out his work in chemistry from the 1880s to the early 1900s.

Effect of a Change in Concentration of Reactant or Product on Equilibrium

Chemical equilibrium results in a balance among the concentrations of the various species involved. So the most direct way to disturb a chemical system at equilibrium is to change the concentration of one or more reactants or products. Suppose that a closed flask contains an equilibrium mixture of NO_2 and N_2O_4 gases. If we somehow inject additional NO_2 into the system, how will the equilibrium respond? The plot in **Figure 12.7** shows the expected effect: additional N_2O_4 must form to reduce the stress applied to the equilibrium.

But how does this response offset the applied stress? When we began our discussion of the equilibrium constant in Section 12-3, we first defined the reaction quotient, Q, and then considered the equilibrium constant as a special case of this ratio. Thus at equilibrium, we have the relationship, $Q = K$. But if we change the concentrations from their equilibrium values, Q will no longer be equal to K. Immediately following an increase in the concentration of a reactant, as in Figure 12.7, Q must become smaller than K. So to establish a new equilibrium, the value of the ratio must increase. Thus the system must respond by increasing the concentrations in the numerator and decreasing those in the denominator. Because products are in the numerator, this means that when $Q < K$, the reaction must shift toward products to reach equilibrium. Similar reasoning suggests that when $Q > K$, the reaction must shift toward reactants. **Table 12.2** summarizes the way that an equilibrium system responds to the four possible types of concentration changes, and Example Problem 12.8 explores this further.

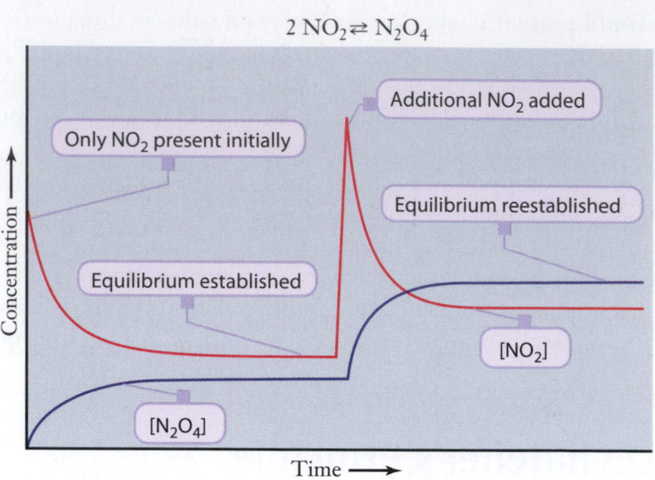

Figure 12.7 ■ The graph illustrates the response of an equilibrium to an increase in reactant concentration. Initially, only NO_2 is present, and the formation of N_2O_4 establishes equilibrium. Midway through the time in the plot, a large amount of NO_2 is added to the system, perturbing the equilibrium. The system responds by forming additional N_2O_4 and establishing a new equilibrium.

Table ■ 12.2

The effects of concentration changes on equilibrium can be rationalized by considering the reaction quotient, Q, and comparing it to the equilibrium constant, K.

Type of Concentration Change	Resulting Change in Q	Response of System
[Products] increased	$Q > K$	More reactants formed
[Products] decreased	$Q < K$	More products formed
[Reactants] increased	$Q < K$	More products formed
[Reactants] decreased	$Q > K$	More reactants formed

EXAMPLE PROBLEM 12.8

Acetic acid, CH_3COOH, has the peculiar property of being a retarding agent for concrete at low concentrations and an accelerator at high concentrations. A solution of acetic acid in equilibrium with hydrogen ions and acetate ions, CH_3COO^-, can be disturbed in several ways. Predict the change in the reaction quotient, Q, when (a) sodium acetate is added to the system, (b) additional acetic acid is added to the system, and (c) sodium hydroxide is added to the system. Then use those predictions to explain how the equilibrium shifts in response to each stress.

Strategy In each case, determine what change the added substance induces in the equilibrium concentrations. By noting whether the concentration change affects the numerator or denominator in the equilibrium expression, we can then judge whether the resulting reaction quotient is larger than or smaller than K. Comparing the expected Q with K, we can predict the direction that equilibrium will move to offset the stress.

Solution The equilibrium is

$$CH_3COOH(aq) \rightleftarrows H^+(aq) + CH_3COO^-(aq)$$

(a) Sodium acetate is added and will dissociate into Na^+ and CH_3COO^-. The sodium ion is not involved in this equilibrium, but by adding CH_3COO^-, we increase the concentration of a product. This in turn increases the value of Q, so $Q > K$. Equilibrium must shift toward reactants, so additional CH_3COOH will form.

(b) Additional acetic acid is added. Because CH_3COOH is a reactant, the increase in concentration of the acid decreases the value of Q. Q becomes less than K, so equilibrium shifts toward products. Some of the added CH_3COOH will dissociate, and the concentration of both H^+ and CH_3COO^- will increase.

(c) Sodium hydroxide is added. Again the sodium ion does not play a significant role in this perturbation, and at first glance, it might seem like the hydroxide ion doesn't either. But if you use a little bit of the chemical intuition you've developed by this point in the course, you should realize that OH^- and H^+ ions will react with one another to form water. This will decrease the concentration of H^+. Because the concentration of a product is reduced, the size of the numerator decreases, making $Q < K$. Thus the equilibrium shifts to form more products.

Check Your Understanding Describe two ways you could make this equilibrium shift toward reactants that haven't already been discussed in this problem. Explain your reasoning.

Effect of a Change in Pressure on Equilibrium When Gases Are Present

Several aspects of the production and weathering of concrete involve gases. But let's consider a much more familiar example to illustrate the effect of pressure on a chemical system at equilibrium. Think about carbon dioxide dissolving in water. A chemical reaction occurs, and equilibrium is established:

$$CO_2(g) + H_2O(\ell) \rightleftarrows H_2CO_3(aq)$$

You might even have an example on your desk as you read this text—your can of soda. The carbonation of soft drinks results from dissolving gaseous CO_2 into water at high pressures. What happens when you open a soda can? The volume available to the carbon dioxide gas becomes much larger, so the pressure decreases. As pressure decreases, the soft drink fizzes, showing that carbon dioxide is leaving the solution. How does this observation conform to LeChatelier's principle?

LeChatelier's principle states that the equilibrium will shift in a way that tends to offset the applied stress. In this case, we have reduced the pressure of CO_2 above the soda. How could the system try to increase this pressure? By having more dissolved gas leave the solution (the soda) and enter the gas phase, thereby contributing to the pressure. When the soft drink fizzes, that is precisely what the system is doing—shifting in the direction that produces more gas. Because the container is open to the atmosphere, the CO_2 pressure never really rises, and gas continues to be released until the soda goes flat.

This effect can be generalized for any chemical reaction that includes gases. If the number of moles of gases differs between reactants and products, a shift in pressure (due to a volume change) will result in a change in the equilibrium position. Carbon dioxide dissolving in water is an example of this because, as the stoichiometry shows, there are more moles of gas phase reactants (1 mole of CO_2) than there are moles of gas phase products (no moles of gas). If the total number of moles of gaseous reactants equals the total number of moles of gaseous products, inducing a pressure change produces no change in the equilibrium position.

Changing the pressure by adding an inert gas does not affect the equilibrium because it does not change the partial pressures of the gases in the reaction.

Now let's return to the equilibrium between NO_2 and N_2O_4: 2 moles of gaseous NO_2 combine to form 1 mole of gaseous N_2O_4. So if pressure is applied, the system responds by producing more N_2O_4. This is illustrated in **Figure 12.8**, and Example Problem 12.9 extends the same idea to other systems.

Figure 12.8 ■ The effect of a pressure increase on the NO_2/N_2O_4 equilibrium is illustrated. A system is initially equilibrated in a volume of 10 L at a total pressure of 1.0 atm. Decreasing the volume to 2 L would cause an immediate pressure increase to 5.0 atm. The equilibrium responds by producing more N_2O_4, which reduces the total number of moles of gas present and reduces the extent of the pressure increase. The partial pressure of NO_2 in the 2-L volume will still be greater than that in the original 10-L volume, so the gas would appear a darker brown color. (The actual final pressure of 4.6 atm was calculated using the numerical value of the equilibrium constant.)

EXAMPLE PROBLEM 12.9

HCN is an important but potentially dangerous industrial chemical. It can be produced in several ways, including by the reactions shown below. Assuming that each system is initially at equilibrium, predict the direction in which the reaction will go to respond to the indicated stress.

(a) $NH_3(g) + CH_4(g) \rightleftarrows HCN(g) + 3\ H_2(g)$; pressure is increased
(b) $2\ NH_3(g) + 2\ CH_4(g) + 3\ O_2(g) \rightleftarrows 2\ HCN(g) + 6\ H_2O(g)$; pressure is decreased

Strategy Count the total number of moles of gas on each side of the equation. If pressure is increased, the reaction that produces fewer moles of gas is favored; if pressure is decreased, the reaction that produces more moles of gas is favored.

Solution

(a) There are 2 moles of gas on the left and 4 moles of gas on the right. If pressure is increased, the equilibrium will move to the left.

(b) There are 7 moles of gas on the left and 8 moles of gas on the right. Decreasing pressure favors the reaction that produces more moles of gas, so the reaction to the right is favored.

Analyze Your Answer This type of question addresses your conceptual understanding, so we need a qualitative approach to check the answer. One strategy is to ask what would happen if you answered the question the opposite way. If we said equilibrium moved to the right in part (a) for example, what would that imply? It would say we have much more gas present, and more gas would increase pressure. Because the problem initially indicated an increase in pressure, this result would not oppose the applied change—so the opposite direction of our answer is clearly wrong, and our answer must make sense.

Check Your Understanding In each of the following, determine the direction in which the equilibrium will shift in response to the indicated stress.

(a) $N_2(g) + 3\ H_2(g) \rightleftarrows 2\ NH_3(g)$; pressure is decreased
(b) $CO_2(g) + H_2(g) \rightleftarrows CO(g) + H_2O(g)$; pressure is increased

Effect of a Change in Temperature on Equilibrium

In Example Problem 5.7, we pointed out that the reaction of CaO with CO_2 is a possible method for capturing carbon dioxide. In this chapter, we noted that the reverse reaction—the decomposition of $CaCO_3$ in a kiln—is the first step in concrete production. To get a sense of how temperature will affect a system, the key idea is to think of a temperature change as a flow of heat into or out of the reacting system. When put in those terms, you might recognize that the link between chemical reactions and heat comes from thermodynamics. As long as we know whether a reaction is exothermic or endothermic, we can predict how it will respond to temperature changes.

For an exothermic reaction, heat flows out of the system. We might say that this heat is a product of the reaction. We can note this explicitly in the form of a thermochemical equation:

Adding the word "heat" to an equation helps us see which direction the reaction will shift. But it should not imply that heat is a quantity that a substance can possess.

$$\text{Exothermic reaction: reactants} \rightleftharpoons \text{products} + \text{heat}$$

Once we do this, the situation begins to resemble our discussion of the addition or removal of a product from the reaction mixture. When temperature is increased, we stress the equilibrium by adding heat. According to LeChatelier's principle, the system will respond to offset the stress and absorb the excess heat, moving the equilibrium toward reactants. On the other hand, when the temperature is lowered, heat flows out from the system, and the system will respond by generating additional heat—moving toward products.

The equilibrium between NO_2 and N_2O_4 that we have been considering must be an exothermic reaction because a bond is formed between the two nitrogen atoms and no bonds are broken. Thus we can predict that formation of N_2O_4 is favored at lower temperatures, and **Figure 12.9** confirms this.

For an endothermic reaction, we can write

$$\text{Endothermic reaction: reactants} + \text{heat} \rightleftharpoons \text{products}$$

Using similar reasoning, if the temperature is increased, the endothermic system will move in the direction that absorbs heat—toward the products. If the temperature is decreased, a shift toward reactants liberates heat to offset this stress. We summarize the possible responses of chemical reactions to temperature changes in **Table 12.3**.

Unlike changes in the concentrations of reactants or products, which leave the equilibrium constant unaffected, a temperature shift also changes the *value* of the

Figure 12.9 ■ The flask in these photos contains a mixture of NO_2, which has a brownish color, and N_2O_4, which is colorless. In the left-hand photo, the flask is held in an ice bath, and the contents have a pale color. In the right-hand photo, the same flask is held in a water bath at 50°C, and the color is markedly darker. This deepening of the color indicates that more NO_2 is present at the higher temperature, as predicted by LeChatelier's principle.

Table ■ 12.3

The effects of temperature changes on a chemical system at equilibrium depend on whether the reaction is exothermic or endothermic. Unlike concentration or pressure changes, temperature changes also alter the *value* of the equilibrium constant.

Type of Reaction	Type of Temperature Change	Response of System
Exothermic	T increase	More reactants formed
Exothermic	T decrease	More products formed
Endothermic	T increase	More products formed
Endothermic	T decrease	More reactants formed

equilibrium constant. (As for rate constants, we are again confronted with a "constant" that varies as a function of temperature!) The fact that equilibrium constants vary with temperature, however, is important in the way chemical reactions occur in industrial processes. Often, high temperatures are required for kinetic reasons, but what if the equilibrium constant is then less favorable? In such cases, chemists and engineers work together to find combinations of factors that make a reaction commercially viable. Nearly every adjustment affects the reaction via LeChatelier's principle.

Effect of a Catalyst on Equilibrium

Equilibrium is established when the rate of the forward reaction equals the rate of the reverse reaction. But in general, the equilibrium position tells us nothing about the rate at which equilibrium was attained. What happens when a catalyst is added to a system at equilibrium? A catalyst increases the rates of both the forward and reverse reactions equally, and therefore does not affect the equilibrium position or concentrations.

12-6 Solubility Equilibria

Earlier we pointed out that when hydroxide ions encounter calcium ions, a precipitate forms. This reaction plays a role in how cement forms, and other precipitation reactions, such as those of phosphates or carbonates, can take place with a wide range of cations. Concrete is a very complex mixture, so even sorting out the details of all the possible precipitation reactions that may take place is a challenge for chemists and materials engineers. The ability to manipulate concrete properties based on the addition of admixtures depends on changing the nature of the chemical equilibria that arise during the hydration and setting processes. Advances in the production of new cements and concrete often follow from the study of the solubility of specific compounds and the equilibrium reactions they establish.

Solubility Product Constant

In Section 3-3, we introduced the term *solubility* to measure how much of a solute a particular solvent can dissolve. This idea is frequently translated into the solubility rules for ionic substances, which we introduced in Section 3-3 (page 69). Those rules classify both calcium hydroxide and magnesium hydroxide as insoluble. But in Example Problem 12.3, we implied that there was a measurable difference in solubility between those two compounds. So we need to refine our notion of what it means for a compound to be "insoluble." We will be more accurate if we refer to salts such as calcium hydroxide and magnesium hydroxide as **sparingly soluble**. Given enough time and a constantly refreshing solvent (in other words, *not* equilibrium conditions), even the most insoluble salt will dissolve. The dissolution of mountains by rainfall represents such a process, albeit one that requires hundreds of thousands of years. So how can we specify just how soluble an "insoluble" compound is?

Defining the Solubility Product Constant

Let's stay with calcium hydroxide and magnesium hydroxide as examples. If either of these substances is placed in water and the system is allowed to sit undisturbed, a dynamic equilibrium will be established between the salt and its constituent ions. The equilibria that describe these reactions are heterogeneous, as are those of any solubility process. So the equilibrium constant expression does not contain the concentration of the solid salt; only the product of the concentrations of the dissolved ions will appear. Because this form is universally employed for solubility equilibrium constant expressions, the constant has been named the **solubility product constant (K_{sp})**. Thus for the two salts we are considering, we can write the following expressions:

$$Ca(OH)_2(s) \rightleftarrows Ca^{2+}(aq) + 2\ OH^-(aq) \quad K_{sp} = [Ca^{2+}][OH^-]^2 = 7.9 \times 10^{-6}$$

$$Mg(OH)_2(s) \rightleftarrows Mg^{2+}(aq) + 2\ OH^-(aq) \quad K_{sp} = [Mg^{2+}][OH^-]^2 = 1.5 \times 10^{-11}$$

From the very small values of these equilibrium constants, we can see that these compounds are not very soluble.

Solubility product constants are important for a wide variety of applications and have been determined for many sparingly soluble salts. **Table 12.4** lists some common salts and their K_{sp} values. A more extensive table is included in Appendix H. Example Problem 12.10 shows how we write the solubility product expression for an ionic compound.

Table 12.4

Values for a few compounds illustrate the wide range over which solubility product constants can vary.

Salt	K_{sp}
Ag_2CO_3	8.1×10^{-12}
Ag_3PO_4	1.3×10^{-20}
AgBr	5.3×10^{-13}
AgCl	1.8×10^{-10}
AgCN	6.0×10^{-17}
$Ca_3(PO_4)_2$	2.0×10^{-33}
$CaCO_3$	4.8×10^{-9}
$FeCO_3$	3.5×10^{-11}
$Mg_3(PO_4)_2$	9.9×10^{-25}
$PbCO_3$	1.5×10^{-13}
PtS	9.9×10^{-74}
ZnS	1.1×10^{-21}

EXAMPLE PROBLEM 12.10

Write the solubility product constant expression for calcium fluoride.

Strategy Identify the formula for the salt and write the chemical equation that represents the solubility equilibrium. Write the solubility product constant expression based on the chemical equation.

Solution Because fluorine forms a 1− ion and calcium forms a 2+ ion, calcium fluoride is CaF_2. The solubility equilibrium is

$$CaF_2(s) \rightleftarrows Ca^{2+}(aq) + 2\ F^-(aq)$$

so,

$$K_{sp} = [Ca^{2+}][F^-]^2$$

Check Your Understanding Write the solubility product constant expressions for the following sparingly soluble salts. (a) manganese(II) hydroxide, (b) copper(II) sulfide, (c) copper(I) iodide, (d) aluminum sulfate

The Relationship Between K_{sp} and Molar Solubility

We have referred to the solubility of calcium salts regularly in this chapter. In Chapter 3, we defined solubility in terms of the mass of solute that dissolves in 100 g of solute, but this was just one possible choice of concentration units. **Molar solubility** is the concentration of a dissolved solid present in a saturated solution, expressed in molarity. We can convert between these two different expressions for solubility as needed. Molar solubility is easily determined from K_{sp}, as you can see in Example Problem 12.11.

EXAMPLE PROBLEM 12.11

What is the molar solubility of calcium phosphate, given that $K_{sp} = 2.0 \times 10^{-33}$? Assuming that the density of the saturated solution is 1.00 g/cm^3, what is the solubility in grams of $Ca_3(PO_4)_2$ per 100 grams of solvent?

Strategy This problem asks us to find the equilibrium concentration of ions in a saturated solution. So it is actually similar to problems we solved earlier in this chapter. Hence we will use the same general approach. To find molar solubility, begin by creating an equilibrium table for a solution with unknown equilibrium concentrations and solve the resulting equilibrium expression for the unknown. Convert the concentration units to grams of solute per 100 grams of solvent, using the molar mass of the solute and the density of the solution. (Because the solution will be very dilute, we will assume that there is no significant difference between 100 g of solution and 100 g of solvent.)

Solution The relevant equilibrium is the dissolution of calcium phosphate:

$$Ca_3(PO_4)_2(s) \rightleftarrows 3\ Ca^{2+}(aq) + 2\ PO_4^{\ 3-}(aq)$$

We will need to work with the solubility product expression for this reaction, and the value of K_{sp} is given:

$$K_{sp} = [Ca^{2+}]^3[PO_4^{\ 3-}]^2 = 2.0 \times 10^{-33}$$

We can envision the process as starting with only solid present. Dissolution of some of the solid leads to the presence of ions in the saturated solution at equilibrium. In setting up our table, we must be sure to account for the 3:2 stoichiometry. For every mole of $Ca_3(PO_4)_2(s)$ that dissolves, 3 moles of calcium ion and 2 moles of phosphate ion are released into solution.

	$Ca_3(PO_4)_2(s)$	$Ca^{2+}(aq)$	$PO_4^{\ 3-}(aq)$
Initial Concentration	Solid	0 M	0 M
Change in Concentration	Solid	$+3x$	$+2x$
Final Concentration	Solid	$3x$	$2x$

Substitute in the equilibrium expression:

$$K_{sp} = 2.0 \times 10^{-33} = [Ca^{2+}]^3[PO_4^{\ 3-}]^2 = (3x)^3(2x)^2$$

We can solve this for x:

$$2.0 \times 10^{-33} = 108x^5$$

$$x = 1.1 \times 10^{-7}$$

This means that the molar solubility of calcium phosphate is 1.1×10^{-7} M. Now convert units to find the solubility in terms of mass:

$$\text{Solubility} = \frac{1.1 \times 10^{-7}\text{mol } Ca_3(PO_4)_2}{\text{L}} \times \frac{1\text{ L}}{10^3\text{cm}^3} \times \frac{1\text{ cm}^3}{1.0\text{ g}} \times \frac{310.2\text{ g } Ca_3(PO_4)_2}{1\text{ mol } Ca_3(PO_4)_2}$$

$$= 3.5 \times 10^{-8}\text{ g solute/g } H_2O$$

Multiplying this by 100 gives us a solubility of 3.5×10^{-6} g $Ca_3(PO_4)_2$ per 100 g H_2O.

Analyze Your Answer The molar solubility we obtained is very small. Does that make sense? The value for K_{sp} is very small, so we can expect that the equilibrium will favor reactants strongly. Favoring reactants means few ions in solution, so a small solubility makes sense.

Check Your Understanding Calculate the molar solubility of silver cyanide. What does this correspond to in grams of solute per 100 grams of solvent? What assumptions must be made?

Common Ion Effect

When we look at the solubility of calcium hydroxide or magnesium hydroxide, we presume that equilibrium is established for the salt in pure water. What if additional hydroxide ions were introduced into the system by adding NaOH? How does the presence of an ion that is part of an equilibrium but is introduced into a system from another source influence the solubility of a salt? This question can be important in the manufacturing of cement, because the complex mixture may have multiple sources of some ions that are present. The answer can be deduced from LeChatelier's principle. Ions that are part of an equilibrium but are introduced from an outside source are called common ions because they are common to the solution and the solid. If we were to add a common ion (OH^- in our example) to a solution of $Mg(OH)_2$ already at equilibrium, the resulting shift would be in a direction that consumes the added ion, forming additional solid $Mg(OH)_2$. This is the **common ion effect (Figure 12.10)**. If instead we add a sparingly soluble solid to a solution in which common ions are already present, their presence will suppress solubility of the solid (because they tend to stress the system toward the production of reactants). We explore this idea in Example Problem 12.12.

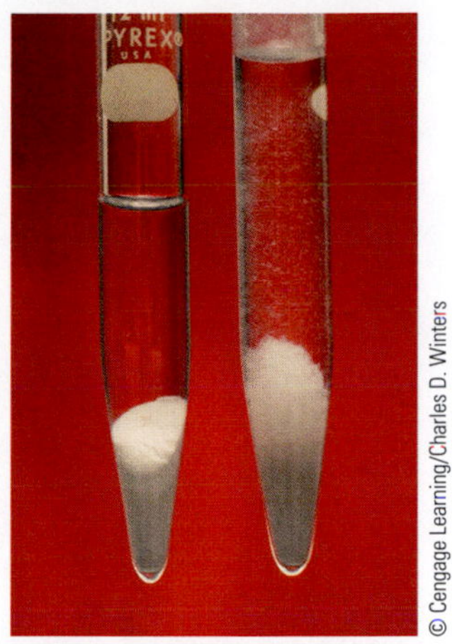

© Cengage Learning/Charles D. Winters

Figure 12.10 ■ The test tube on the left contains a saturated solution of silver acetate. Adding aqueous $AgNO_3$ causes additional solid silver acetate to precipitate due to the common ion effect, as seen in the tube on the right.

EXAMPLE PROBLEM 12.12

The K_{sp} of $Ca_3(PO_4)_2$ is 2.0×10^{-33}. Determine its molar solubility in 0.10 M $(NH_4)_3PO_4$. Compare your answer to the molar solubility of $Ca_3(PO_4)_2$ in water, which we calculated in Example Problem 12.11.

Strategy $(NH_4)_3PO_4$ is very soluble in water, so $[PO_4^{3-}]$ in 0.10 M $(NH_4)_3PO_4$ is 0.10 M. When solid $Ca_3(PO_4)_2$ is added and equilibrium is established, molar solubility can be found from the amount of $[Ca^{2+}]$ in solution. Write the reaction of interest and set up the usual equilibrium table. Because the value of K_{sp} is very small, we can use a simple approximation to solve for molar solubility.

Solution Let x = moles of $Ca_3(PO_4)_2$ that dissolve per liter (i.e., the molar solubility that we want to calculate). The initial concentration of Ca^{2+} is zero, but that of PO_4^{3-} is now 0.10 M. This gives us the following equilibrium table:

$$Ca_3(PO_4)_2(s) \rightleftarrows 3\ Ca^{2+}(aq) + 2\ PO_4^{3-}(aq)$$

	$Ca_3(PO_4)_2(s)$	$Ca^{2+}(aq)$	$PO_4^{3-}(aq)$
Initial Concentration	Solid	0 M	0.10 M
Change in Concentration	Solid	$+3x$	$+2x$
Final Concentration	Solid	$3x$	$0.10 + 2x$

Substitute in the solubility product expression:

$$K_{sp} = 2.0 \times 10^{-33} = [Ca^{2+}]^3[PO_4^{3-}]^2 = (3x)^3(0.10 + 2x)^2$$

This looks daunting: multiplying out would give us a term involving x^5. But we can take advantage of a clever approximation. We know from the last example problem that the solubility in pure water was on the order of 10^{-7} M. And in this case the presence of the common ion should decrease the solubility considerably from that level. So we know that x will be less than 10^{-7}. This means that we can neglect the $+2x$ piece of the last term in comparison to 0.10:

$$0.10 + 2x \approx 0.10$$

This simplifies the problem considerably:

$$2.0 \times 10^{-33} = (3x)^3(0.10)^2$$

$$x = 1.9 \times 10^{-11}\ M$$

In the previous example problem, we calculated the molar solubility of $Ca_3(PO_4)_2$ in water as 1.1×10^{-7} M. The presence of a relatively high concentration of phosphate ion in solution has depressed the solubility by almost four orders of magnitude, to 1.9×10^{-11} M.

Analyze Your Answer Whenever we make a simplifying approximation of the sort used here, it is important to verify that the eventual solution is consistent with the underlying assumption. Here we assumed that $2x$ was negligible compared to 0.10. Because we found that x was on the order of 10^{-11}, that assumption was certainly justifiable. We could have used appropriate software tools to find a numerical solution to the actual equation, but the added work would not have improved our result discernibly. Note that the underlying reason why this assumption could be made was the fact that the K_{sp} value was a very small number.

Check Your Understanding The solubility product of CaF_2 is 1.7×10^{-10}. Calculate its solubility in water and in 0.15 *M* NaF.

Reliability of Using Molar Concentrations

We need to be aware that the calculations we've been carrying out have been simplified in various ways to make them compatible with the level of this course. These simplifications are reasonable under many typical circumstances but can introduce significant errors under other conditions. One of the fundamental simplifications we have made is that we have been using the molarity of the ionic species in all calculations of solubility equilibrium problems. In some cases, however, the utility of molarity dwindles. When ion concentrations are high, the interactions between oppositely charged particles, even other ions that are not part of a particular equilibrium, alter ion behavior. To take these effects into account, we would use the activity, or effective concentration, of an ion rather than the molarity of the ion. Activity is not a particularly intuitive modification of the molarity of a species, and we will not undertake any calculations using it here. We must note, however, that the types of calculations we have presented here do have limitations. In industrial chemistry, the models used to understand a synthetic process often require higher accuracy than is afforded by using molarity, so computer codes frequently offer the option of using activity instead.

Activities are often obtained from molarities by multiplying by an activity coefficient, whose value is obtained experimentally.

12-7 Acids and Bases

We pointed out back in Chapter 2 that molecules containing the functional group —COOH are called carboxylic acids. And in this chapter, we have seen multiple examples illustrating the idea that these compounds do indeed act as acids in aqueous solution. We introduced a few fundamental concepts of acids and bases in Chapter 3, but the context of equilibrium allows us to explore them further. Recall that we distinguished between strong acids (or bases), which dissociate completely in solution, and weak acids (or bases), which dissociate only partially. At this point in our study of chemistry, we should realize that this partial dissociation of weak electrolytes was an example of a system reaching equilibrium. So we can use equilibrium constants to characterize the relative strengths of weak acids or bases. One common way to do this is to use the pH scale, which we will define in this section.

The Brønsted–Lowry Theory of Acids and Bases

Our initial definition of acids and bases in Section 3-3 involved the formation of hydronium ions (for acids) or hydroxide ions (for bases). This definition is attributed to Svante Arrhenius, but it must be expanded if we wish to include nonaqueous solutions, among other things. Formulated independently in 1923 by two chemists, Johannes Brønsted in Denmark and Thomas Lowry in England, the Brønsted–Lowry definition does just that. According to this definition, a **Brønsted–Lowry acid** is a proton (H^+) donor, and a base is a proton acceptor.

Brønsted received his undergraduate degree in chemical engineering in 1899.

This more general definition of an acid includes Arrhenius acids and bases, as well. So HCl, for example, should fit both the Arrhenius and Brønsted–Lowry theories. To see whether it does, let's look at what happens when HCl reacts with water. As hydrogen chloride gas dissolves in water, dipole–dipole forces come into play, and the partially positive hydrogen of the HCl molecule is attracted to the partially negative oxygen of a water molecule. The attraction is strong enough to cause the transfer of the hydrogen ion from HCl to the water molecule. The result is a hydrated proton, or hydronium ion, H_3O^+:

$$\mathrm{H{:}\ddot{\underset{\cdot\cdot}{O}}{:}\,(H)} + \mathrm{H{:}\ddot{\underset{\cdot\cdot}{Cl}}{:}} \longrightarrow \left[\mathrm{H{:}\ddot{\underset{\cdot\cdot}{O}}{:}H\ (H)}\right]^+ + \mathrm{{:}\ddot{\underset{\cdot\cdot}{Cl}}{:}^-}$$

HCl has *donated* a proton to the water molecule, which *accepted* the proton, and we can write an equation to reflect this:

$$HCl(g) + H_2O(\ell) \rightarrow H_3O^+(aq) + Cl^-(aq)$$

Remember that we use a single arrow when a reaction proceeds extremely far to the right.

In the Brønsted–Lowry theory, a base is a proton acceptor. The $OH^-(aq)$ ion produced by the ionization of NaOH(s) is a base because it can accept a proton from any proton donor:

$$OH^-(aq) + H^+(aq) \rightleftarrows H_2O(\ell)$$

The Role of Water in the Brønsted–Lowry Theory

Next, let's consider the ionization of HCN in water:

$$HCN(aq) + H_2O(\ell) \rightleftarrows H_3O^+(aq) + CN^-(aq)$$

The HCN molecule donates a proton, so it is clearly an acid. But the water molecule accepts that proton. Does this make water a base? By the Brønsted–Lowry theory, it does. But by the same theory, water can also react as an acid, as you can see in the following reaction between ammonia and water:

$$NH_3(g) + H_2O(\ell) \rightarrow NH_4^+(aq) + OH^-(aq)$$

The water molecule (an acid) donates a proton to the ammonia molecule (a base), which accepts the proton. So water can act as both acid and base, depending on the identity of the other reactant. We use the term **amphoteric** to refer to a substance that can be either an acid or a base.

The term amphiprotic *is finding increasing use and is slowly replacing the term* amphoteric.

We also must be careful to note that the reaction of ammonia with water does not go to completion. The reaction above is actually in dynamic equilibrium:

$$NH_3(aq) + H_2O(\ell) \rightleftarrows NH_4^+(aq) + OH^-(aq)$$

Because the reverse reaction also involves a donated proton (this time from the ammonium ion), both the forward and reverse reactions are acid–base reactions:

$$\underset{\text{Base}_1}{NH_3(aq)} + \underset{\text{Acid}_2}{H_2O(\ell)} \rightleftarrows \underset{\text{Acid}_1}{NH_4^+(aq)} + \underset{\text{Base}_2}{OH^-(aq)}$$

The equation contains two pairs of acids and bases: NH_4^+/NH_3 and H_2O/OH^-. These are called **conjugate acid–base pairs**. The **conjugate acid** of a base is the acid formed when the base accepts a proton. Similarly, the **conjugate base** of an acid is the base formed when the acid donates a proton. NH_3 is the conjugate base of NH_4^+, and NH_4^+ is the conjugate acid of NH_3; H_2O is the conjugate acid of OH^-, and OH^- is the conjugate base of H_2O. We can link the pairs using lines under the equation to highlight these relationships:

$$\underset{\text{Base}_1}{NH_3(aq)} + \underset{\text{Acid}_2}{H_2O(\ell)} \leftrightarrows \underset{\text{Acid}_1}{NH_4^+(aq)} + \underset{\text{Base}_2}{OH^-(aq)}$$

Example Problem 12.13 looks at another set of conjugate acid–base pairs.

EXAMPLE PROBLEM 12.13

When dissolved in water, CH_3COOH is called acetic acid. Write the equilibrium for its reaction in water and identify the conjugate acid–base pairs.

Strategy Because acetic acid is behaving as a Brønsted–Lowry acid, it will require some species to accept a proton. Here the proton acceptor is water, as is often the case. This allows us to write the equilibrium for the CH_3COOH solution. Once it is written, we identify the conjugate pairs by noting which species differ by only the addition or subtraction of a proton.

Solution

$$\underset{\text{Acid}_1}{CH_3COOH(aq)} + \underset{\text{Base}_2}{H_2O(\ell)} \leftrightarrows \underset{\text{Base}_1}{CH_3COO^-(aq)} + \underset{\text{Acid}_2}{H_3O^+(aq)}$$

Check Your Understanding Identify the conjugate acid–base pairs in the reaction between carbonate ion and water to form hydrogen carbonate and hydroxide ions.

One of the advantages of the Brønsted–Lowry theory is that it can describe acid–base reactions in a nonaqueous medium like liquid ammonia. In liquid ammonia, HCl(g) donates a proton to NH_3, forming NH_4^+ and Cl^-. No water is involved, so there are no H_3O^+ or OH^- ions formed. Ammonium and chloride ions are solvated by ammonia molecules, so instead of the usual "aq", we might use "am" to show solvation by ammonia molecules:

$$HCl(g) + NH_3(\ell) \rightarrow NH_4^+(am) + Cl^-(am)$$

Weak Acids and Bases

Many acids, including all of the hydroxycarboxylic acids that are used as water-reducing admixtures in the production of concrete, do not ionize completely in water. We learned in Section 3-3 that these compounds are weak acids. A weak acid and water react to produce a conjugate acid–base system in which the acid and base on the right are stronger than their conjugate acid and base on the left. This means that equilibrium will favor the left-hand side, and only a small percentage of the acid molecules ionize. We can write a general equation for the dissociation of any weak acid, denoted here as HA:

$$\underset{\text{Weaker acid}}{HA(aq)} + \underset{\text{Weaker base}}{H_2O(\ell)} \rightleftarrows \underset{\text{Stronger acid}}{H_3O^+(aq)} + \underset{\text{Stronger base}}{A^-(aq)}$$

The equilibrium expression for this reaction might be written as

$$K = \frac{[H_3O^+][A^-]}{[HA][H_2O]}$$

The water in the equilibrium reaction is a pure liquid, of course. So just like the pure solids in solubility equilibria, its concentration is constant. So $[H_2O]$ is dropped from the expression, and the resulting K is called an **acid ionization constant (K_a)**:

$$K_a = \frac{[H_3O^+][A^-]}{[HA]}$$

Table 12.5 lists acid ionization constants at 25°C for some common weak acids, and additional values can be found in Appendix F.

For weak bases, the ionization equilibrium can be written as

$$\underset{\text{Weaker base}}{B(aq)} + \underset{\text{Weaker acid}}{H_2O(\ell)} \rightleftarrows \underset{\text{Stronger acid}}{BH^+(aq)} + \underset{\text{Stronger base}}{OH^-(aq)}$$

Just as for weak acids, the equilibrium constant expression is written without including a concentration term for pure water. The resulting equilibrium constant is called a **base ionization constant (K_b)**:

$$K_b = \frac{[BH^+][OH^-]}{[B]}$$

Values of K_b for common weak bases can be found in Appendix G.

Table ■ 12.5

Acid ionization constants are shown for some common weak acids at 25°C. Larger values of K_a indicate stronger acids.

Name of Acid	Formula	K_a
Organic Acids		
Formic	$HCOOH$	1.8×10^{-4}
Acetic	CH_3COOH	1.8×10^{-5}
Propanoic	CH_3CH_2COOH	1.3×10^{-5}
Butanoic	$CH_3CH_2CH_2COOH$	1.5×10^{-5}
Salicylic	$C_6H_4(OH)COOH$	1.1×10^{-3}
Gluconic	$HOCH_2(CHOH)_4COOH$	2.4×10^{-4}
Heptonic	$HOCH_2(CHOH)_5COOH$	1.3×10^{-5}
Inorganic Acids		
Hydrofluoric	HF	6.3×10^{-4}
Carbonic	H_2CO_3	4.4×10^{-7}
Hydrocyanic	HCN	6.2×10^{-10}

Figure 12.11 ■ The pH scale provides an easy way to measure the relative acidity or basicity of aqueous solutions, including many common substances from our daily lives. Pure water has a pH of 7.0 at 25°C. So solutions with pH < 7 are acidic, and those with pH > 7 are basic. Because pH is measured on a logarithmic scale, it can describe H_3O^+ concentrations that vary over many orders of magnitude without using scientific notation. Note that pH values less than zero or greater than 14 are also possible.

pH	$[H_3O^+]$ mol/L	Common substances in pH range
0	10^{0}	1.0 M HCl
1	10^{-1}	HCl in the human stomach
2	10^{-2}	
3	10^{-3}	Vinegar : $CH_3COOH(aq)$
4	10^{-4}	Soft drinks
5	10^{-5}	
6	10^{-6}	Milk
7	10^{-7}	Pure water, blood
8	10^{-8}	Seawater
9	10^{-9}	
10	10^{-10}	Milk of magnesia : $Mg(OH)_2(aq)$
11	10^{-11}	Household ammonia
12	10^{-12}	
13	10^{-13}	
14	10^{-14}	1.0 M NaOH

Increasing acidity

Increasing basicity

It should be clear that these reactions of weak acids and bases are a special case of the equilibrium that we have been exploring throughout this chapter. When any acid dissolves in water, hydronium ions are always formed as one of the products. This species is so common that an additional way to describe its concentration has been devised. To avoid dealing with small numbers in scientific notation, we often use the pH scale in which pH is defined as the negative logarithm of the hydronium ion concentration:

$$\text{pH} = -\log[H_3O^+] \tag{12.5}$$

Consider a solution with a hydronium ion concentration of 0.1 M (or 10^{-1} M). This solution would have pH $= -\log(10^{-1})$; the logarithm of 10^{-1} is -1, so pH $= 1$. Applying the same logic to other hydronium ion concentrations in aqueous solution, the result is the pH scale (**Figure 12.11**). From the table you can see that if the concentration of the hydronium ion is 1×10^{-x}, the pH is x. Special laboratory instruments, called pH meters, allow us to determine the pH directly.

Many students have misconceptions about pH. For example, pH can be less than 0 or greater than 14.

If we know the K_a of an acid and its initial concentration, we can calculate the hydronium ion concentration and therefore the pH of the solution, as shown in Example Problem 12.14.

EXAMPLE PROBLEM 12.14

The K_a of acetic acid, CH_3COOH, is 1.8×10^{-5}. Calculate the pH of a 0.10 M acetic acid solution.

Strategy We begin as usual, by setting up an equilibrium table and determining the concentrations of all species. Once we know the concentration of H_3O^+, we can find the pH.

Solution The initial concentration of acetic acid is 0.10 M and those of the ions are zero, leading to the following equilibrium table:

$$CH_3COOH(aq) + H_2O(\ell) \rightleftarrows H_3O^+(aq) + CH_3COO^-(aq)$$

	$CH_3COOH(aq)$	$H_3O^+(aq)$	$CH_3COO^-(aq)$
Initial Concentration	0.10	0	0
Change in Concentration	$-x$	$+x$	$+x$
Final Concentration	$0.10 - x$	x	x

Write the equilibrium constant expression and substitute values from the table:

$$K_a = \frac{[H_3O^+][CH_3COO^-]}{[CH_3COOH]}$$

$$K_a = \frac{(x)(x)}{(0.10 - x)} = \frac{x^2}{(0.10 - x)} = 1.8 \times 10^{-5}$$

We should expect the extent of dissociation to be small for a weak acid. So we can simplify the calculation if we assume that x is small enough that $(0.10 - x) \approx 0.10$. With that assumption, we can rewrite the equation above so that we will not need to use the quadratic formula:

$$K_a \approx \frac{x^2}{0.10} = 1.8 \times 10^{-5}$$

Solve this for x:

$$x^2 = 1.8 \times 10^{-5}(0.10) = 1.8 \times 10^{-6}$$

$$x = 1.3 \times 10^{-3}\ \text{M}$$

Therefore, $[H_3O^+] = [CH_3COO^-] = 1.3 \times 10^{-3}$ M. Because we now know $[H_3O^+]$, it is easy to find the pH:

$$\text{pH} = -\log(1.3 \times 10^{-3}) = 2.87$$

Analyze Your Answer We should confirm that the assumption we made was reasonable. We assumed that $(0.10 - x) \approx 0.10$. So we can test this by subtracting the value obtained for x from the initial concentration of CH_3COOH:

$$0.10\ \text{M} - 1.3 \times 10^{-3}\ \text{M} = 0.0987 \approx 0.10\ \text{M}$$

Our initial concentration was given only to two significant figures, so our simplifying approximation is reasonably valid.

Check Your Understanding Use the approximation method shown above to determine the percentage ionization of a 0.10 M solution of salicylic acid. Is the approximation valid in this case?

Acid–base chemistry plays an important role in the weathering of concrete. Recall from Section 12-1 that CO_2 will react with $Ca(OH)_2$ to form calcium carbonate and water. At first glance, this may not look like acid–base chemistry, but CO_2 is acting as an acid in this case. If CO_2 first dissolved in water, the resulting carbonic acid clearly would react with the basic calcium hydroxide:

CO_2 is sometimes called an acid anhydride, meaning an acid "without water."

$$H_2CO_3(aq) + Ca(OH)_2(aq) \rightarrow CaCO_3(s) + 2\ H_2O(\ell)$$

Note that the products of this reaction are the same as those we saw for the carbonation of concrete—namely, calcium carbonate and water. The diffusion of CO_2

Figure 12.12 ■ Chemical weathering of concrete by exposure to atmospheric CO_2 results in carbonation. When the carbonated zone reaches the metal reinforcement (rebar, shown as the red circles in this cross section) within the concrete, it can lead to an enhanced rate of corrosion of the reinforcement—a topic taken up in more detail in Chapter 13.

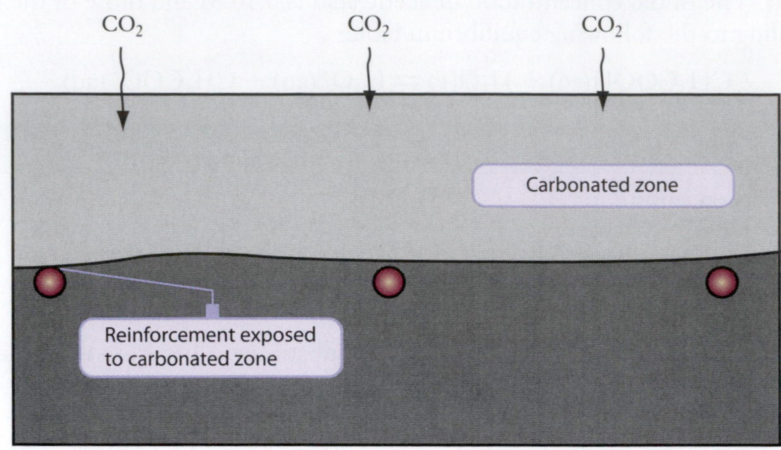

into concrete and its reaction with the cement present is referred to as the *ingress* of the carbonated zone, as depicted in **Figure 12.12.** When this zone reaches the iron reinforcements in concrete, it will accelerate the corrosion reactions that take place—a topic we will look at in more detail in the next chapter.

12-8 Free Energy and Chemical Equilibrium

We've approached equilibrium from the perspective of the kinetics of forward and reverse reactions. But because equilibrium is the eventual outcome of any closed system, it must somehow be connected to free energy, too. What is the relationship between free energy and equilibrium? *Equilibrium is a state of minimum free energy.* All systems move toward equilibrium because doing so will lower their free energy. When they reach equilibrium, the free energy no longer changes. Hence $\Delta G = 0$ at equilibrium.

Graphical Perspective

The graphs in **Figure 12.13** show free energy as a function of progress along a reaction pathway. (This is similar in some ways to the potential energy diagrams we saw in Chapter 11.) Although the vertical axis is the Gibbs free energy, keep in mind that we cannot determine an absolute value for the free energy of a system at any point. Instead, we can only measure *changes* in free energy. The purpose of the graph, therefore, is to help you understand that free energy decreases as you move toward equilibrium from either direction (reactants or products). *A chemical system tends to move spontaneously toward equilibrium.* When equilibrium is reached, the change in free energy will be zero.

Figure 12.13 also reminds us that $\Delta G°$ is the difference between the free energy of the reactants and the products in their standard states. For the reaction depicted in the right-hand panel of the figure, $\Delta G°$ is positive because $G°_{\text{products}} > G°_{\text{reactants}}$. In Chapter 10, we asserted that a reaction with $\Delta G° > 0$ will not proceed spontaneously. Now we see that this was a bit of a simplification. All chemical reactions will proceed to the point of minimum free energy. Though there is variability from one reaction to the next, the free energy minimum along the reaction progress axis tends to lie closer to the state (reactants or products) that is lower in free energy. Thus for a reaction where $\Delta G° < 0$, the equilibrium position lies closer to the products, and we tend to view such a reaction as proceeding toward products. For a reaction where $\Delta G° > 0$, on the other hand, the equilibrium position is usually achieved when a majority of reactants are still present. We tend to view these reactions as not proceeding spontaneously toward products because they do not move very far in that direction before reaching equilibrium.

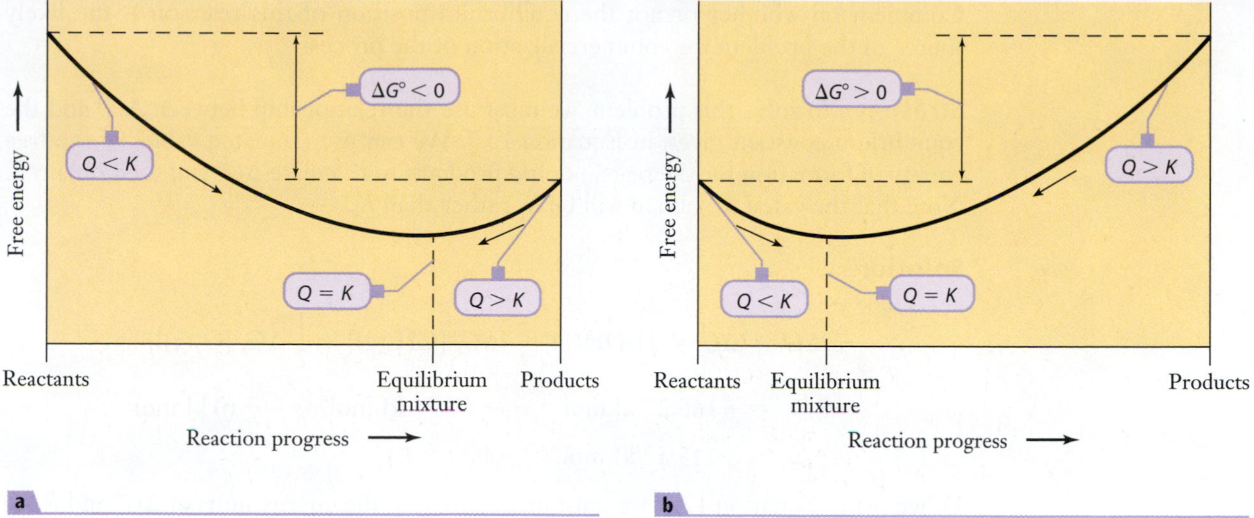

Figure 12.13 ■ Chemical reactions always proceed toward a minimum in free energy. The left-hand plot illustrates what this means for a reaction with $\Delta G^\circ < 0$. In this case, the equilibrium constant will be greater than one, and equilibrium will favor products. The right-hand figure shows a reaction with $\Delta G^\circ > 0$. Here the equilibrium constant will be less than one, and the equilibrium will favor reactants. In either case, equilibrium will occur at the point at which free energy is minimized.

Free Energy and Nonstandard Conditions

To obtain a mathematical relationship between ΔG° and the equilibrium constant, K, first we must realize that the equilibrium state does not correspond to standard conditions. We cannot expect all reactants and products to be present at one atmosphere pressure and one molar concentration for every equilibrium reaction! The derivation of the mathematical relationship that adjusts for nonstandard conditions is one you will learn if you continue to study chemistry. For now, we simply offer the expression itself:

$$\Delta G = \Delta G^\circ + RT \ln Q \qquad (12.6)$$

Here R is the universal gas constant, T is the temperature, and Q is the reaction quotient. Recall from Section 12-3, however, that Q approaches K as the system reaches equilibrium. Combining this with the fact that $\Delta G = 0$ at equilibrium means that Equation 12.6 becomes

$$0 = \Delta G^\circ + RT \ln K$$

Rearranging this, we obtain

$$\Delta G^\circ = -RT \ln K \qquad (12.7)$$

With this equation, we can calculate an equilibrium constant if we know the standard free energy change for the desired reaction. ◀ As we learned in Section 10-7, we can often obtain that from tabulated data. Example Problem 12.15 applies Equation 12.7 to the conversion of methane to methanol.

For equilibria involving gases, Equation 12.7 relates ΔG° to K_p.

EXAMPLE PROBLEM 12.15

The conversion of methane gas (CH_4) to methanol (CH_3OH), which is a liquid at room temperature, is an area in which considerable research is being done. Still, this process is not yet economically viable. Using tabulated thermodynamic data, calculate the equilibrium constant for the following reaction at 25°C:

$$CH_4(g) + \frac{1}{2} O_2(g) \rightleftharpoons CH_3OH(\ell)$$

Comment on whether or not the equilibrium position of this reaction is the likely source of the problem for commercialization of the process.

Strategy To solve this problem, we must use the relationship between $\Delta G°$ and the equilibrium constant given in Equation 12.7. We can use tabulated values of the free energy of formation for the reactants and products to calculate $\Delta G°$ and then obtain K. Note that the value we obtain will be K_p rather than K_c.

Solution

$$\begin{aligned}\Delta G° &= \Delta G_f°[CH_3OH(\ell)] - \{\Delta G_f°[CH_4(g)] + \frac{1}{2}\Delta G_f°[O_2(g)]\} \\ &= -166.27 \text{ kJ mol}^{-1} - (-50.75 \text{ kJ mol}^{-1}) - \frac{1}{2}(0 \text{ kJ mol}^{-1}) \\ &= -115.52 \text{ kJ mol}^{-1}\end{aligned}$$

When using Equation 12.7, we must make sure that the energy units of $\Delta G°$ and R are consistent. Here it is easy to write $\Delta G°$ in joules rather than kilojoules:

$$-115.52 \times 10^3 \text{ J mol}^{-1} = -(8.3145 \text{ J mol}^{-1} \text{ K}^{-1})(298.15 \text{ K}) \ln K$$

so

$$\ln K = 46.600$$

and

$$K = e^{46.600} = 1.73 \times 10^{20}$$

Discussion This reaction can be thought of as the partial oxidation of methane, and equilibrium clearly favors the products. The problem that scientists and engineers face with this process is that methane also has a strong tendency to oxidize completely to carbon dioxide and water. This issue is explored further in the "Check Your Understanding" problem.

Check Your Understanding Calculate the equilibrium constant for the complete combustion of methane:

$$CH_4(g) + 2\ O_2(g) \rightarrow CO_2(g) + 2\ H_2O(\ell)$$

Compare the equilibrium constant for the complete combustion with that for the partial oxidation calculated previously. Does this result help explain why it is so difficult to carry out the partial oxidation process?

Example Problem 12.15 shows that the equilibrium constant can be enormous if the free energy change is large and negative. Conversely, a reaction with an extremely small equilibrium constant will have a standard free energy change that is large and positive. Only if the magnitude of $\Delta G°$ is small do we find equilibrium constants near one.

12-9 Bendable Concrete

We've seen throughout this chapter that there are many chemical properties that can be used to characterize the production and aging of concrete. Even so, it may be a physical property of concrete—brittleness—that represents the biggest constraint on how it is used in engineering designs. You can see the effects of this almost anytime you walk along a concrete sidewalk or driveway: concrete cracks under stress. And although reinforcement with steel or other materials can minimize the impact of cracking, it remains an important issue in the design of concrete structures. What makes concrete brittle and can anything be done to change this? Is there a way to make bendable concrete?

The main way to distinguish between a brittle material and a more ductile one is to observe how it fails. If a brittle material is put under stress, the basic mechanism of failure is the formation and propagation of cracks. While the details of failures due to cracks are something you are likely to encounter in a later engineering course, a key characteristic as to whether a crack will lead to failure is its size. A single large crack is more likely to contribute to failure than many smaller cracks. Unfortunately, traditional concrete is a type of material that is prone to the formation of larger cracks.

In order to make concrete less brittle, therefore, the key is to alter the composition so that smaller cracks, sometimes called micro-cracks, are more likely to form. A group of materials known as "engineered cementitious composites," or ECC, shows significant promise in achieving this goal. **Figure 12.14** compares the compositions of traditional concrete and ECC, and shows that many of the same constituent materials are used in both cases. But there are also important differences. Traditional concrete includes substantial amounts of gravel, whereas ECC does not. ECC includes fly ash, where traditional concrete does not. Perhaps most importantly, ECC includes small polyvinyl alcohol (PVA) fibers. Even though these fibers are present in relatively small amounts, their inclusion significantly alters how cracks form in the material.

When put under stress, the ECC material forms many small cracks and few or no larger cracks. And such small cracks are much less likely to lead to structural failure. The strain capacity of traditional concrete is quite small, typically on the order of 0.01%, indicating that the material is very brittle. By contrast, the ECC concrete can be engineered with a strain capacity of roughly 5%. **Figure 12.15** shows a sample of ECC being tested and illustrates the remarkable flexibility the material offers.

The development of materials like ECC is an area of ongoing research, and recent work suggests that it may be possible to produce materials in which small micro-cracks may even be able to heal themselves over time. Thus while bendable concrete may sound exotic at this time, engineering designs of the future might be able to routinely take advantage of such materials. Because the improved properties are derived primarily from the inclusion of a small amount of polymer fiber in the concrete,

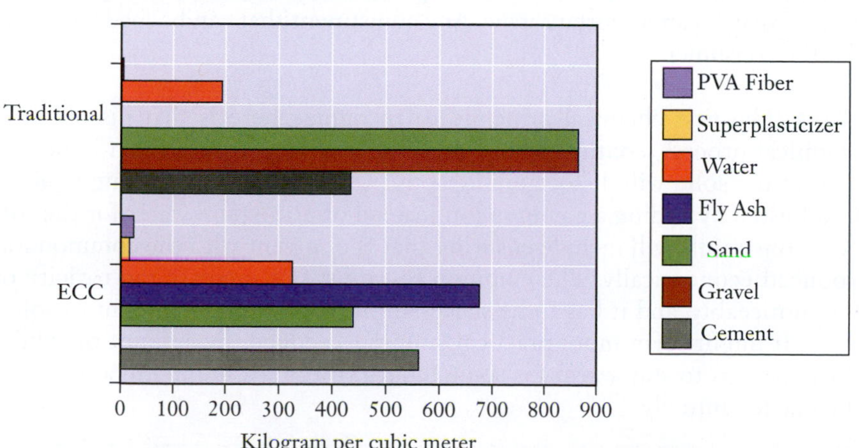

Figure 12.14 ■ Typical compositions of ECC and traditional concrete are compared. The inclusion of relatively small amounts of PVA fiber makes the ECC much more resistant to the formation of large cracks.

Nicole Casal Moore, U-M News Service and College of Engineering

Figure 12.15 ■ A sample of bendable concrete undergoes stress testing in a University of Michigan lab. The formation of many micro-cracks rather than a single large crack allows the material to bend under stress that would fracture traditional concrete. (The scale on the rulers seen in the photo is in inches.)

it is likely that properties could be adjusted by making changes in the composition of the concrete or the chemical identity of the fibers used. As we've seen throughout the text, connections between chemistry and engineering can be found in many unexpected places.

FOCUS ON PROBLEM SOLVING

Question You are designing an instant heat pack for a food application. Your design includes a nontoxic ionic salt that dissolves exothermically in water. You need to control the amount of heat released. Initial prototypes get too warm, even with a fairly small amount of salt used. What else could you do to reduce the heat released in your product? Identify any key limitations that would have to be addressed.

Strategy This is a conceptual question that requires that we identify both the chemical properties that are important and the practical ones. For a food application, for example, toxicity is an important issue. Any adjustment that can be made must still be safe for the consumer.

Solution The first chemical principle to recognize here is that dissolution, like any chemical process, continues until equilibrium is reached. For an ionic salt, we expect that the solid will dissociate into ions in solution. The resulting equilibrium can be adjusted by adding a common ion instead of using pure water for the solvent. The key constraints will include ensuring that the solvent with the common ion can be produced economically. The common ion must not increase the toxicity of the product noticeably, and it has to be successful in reducing the amount of solid that dissolves. It might seem more practical simply to reduce the amount of solid used, but if the pack is to stay warm for some length of time, a minimum amount of salt may be needed initially.

SUMMARY

Although we often talk about chemical reactions in terms of reactants and products, all reactions can proceed in either direction. The exact mixture of reactants and products resulting from a reaction in a closed system represents equilibrium, a state where the rates of the forward and reverse reactions are equal.

We can describe the equilibrium mixture mathematically using the expression for the equilibrium constant, K. This expression is useful in many ways. It provides an easy way to assess the likely extent of a reaction: a large value of K indicates a reaction that favors product formation, and a small value of K indicates that reactants are favored. The equilibrium expression also provides a way to predict the amount of each species present at equilibrium, given a set of initial concentrations.

The concept of dynamic equilibrium is also an important way to organize our thinking about reactions. If we want to achieve a particular outcome from a reaction that is at equilibrium, we may need to adjust the environment to achieve that goal. LeChatelier's principle, which says that an equilibrium will shift to offset any applied stress, guides us in such choices. Stresses that can be applied include adding or removing chemicals, changes in temperature, and changes in pressure.

Specific examples of important equilibrium systems include sparingly soluble salts, weak acids, and weak bases. For salts, the solubility product constant, K_{sp}, provides a way to calculate equilibrium concentrations and molar solubilities. Weak acids and bases also have specifically named equilibrium constants: the acid ionization constant, K_a, and the base ionization constant, K_b.

The equilibrium position is dictated by the thermodynamics of the reaction, and the equilibrium constant is directly related to the standard free energy change for the reaction. So we can determine equilibrium constants from tabulated thermodynamic information. Alternatively, experimental measurements of K can also provide a route to thermodynamic data.

KEY TERMS

acid ionization constant (K_a) (12-7)
admixture (12-1)
amphoteric (12-7)
base ionization constant (K_b) (12-7)
Brønsted–Lowry acid (12-7)
common ion effect (12-6)
conjugate acid (12-7)
conjugate acid–base pair (12-7)
conjugate base (12-7)
dynamic equilibrium (12-2)
equilibrium constant (K) (12-3)
equilibrium expression (12-3)
heterogeneous equilibrium (12-3)
homogeneous equilibrium (12-3)
LeChatelier's principle (12-5)
molar solubility (12-6)
Portland cement (12-1)
reaction quotient (Q) (12-3)
solubility product constant (K_{sp}) (12-6)
sparingly soluble (12-6)

PROBLEMS AND EXERCISES

INSIGHT INTO Concrete Production and Weathering

12.1 Identify the first chemical step in the production of Portland cement. How is this reaction related to the chemistry that takes place in the carbonation of concrete?

12.2 Explain why the hydration process for concrete is exothermic by considering the chemical bonds in the reactants and products.

12.3 What fraction of the annual release of $CO_2(g)$ into the atmosphere is the result of concrete production? What is the main chemical step that leads to the production of CO_2?

12.4 In what geographical region of the country would a civil engineer be most likely to use concrete with air entraining admixtures in a design? Explain your answer.

12.5 In Chapter 11, we discussed several factors that can influence chemical kinetics. Use your understanding of those factors to offer an explanation as to why concrete hydration takes days to complete.

12.6 With what chemical in concrete does CO_2 actually react during carbonation?

12.7 A student finds a piece of old concrete that has recently broken off from the curb alongside a road. Describe what she would expect to observe if she sprayed a solution containing phenolphthalein on all the surfaces of this piece of concrete.

Chemical Equilibrium

12.8 On your desk is a glass half-filled with water and a square glass plate to use as a cover for the glass.

(a) How would you set up a dynamic equilibrium using the glass, water, and glass plate?

(b) Once equilibrium is established in (a), describe the processes that are occurring at the molecular level.

12.9 In the figure, orange fish are placed in one aquarium and green fish in an adjoining aquarium. The two tanks are separated by a removable partition that is initially closed.

(a) Describe what happens in the first few minutes after the partition is opened.

(b) What would you expect to see several hours later?

(c) How is this system analogous to dynamic chemical equilibrium?

12.10 At a particular temperature, iodine vapor, I_2, is added to a bulb containing $H_2(g)$. Describe what is happening to the concentrations of reactants on the molecular level:

(a) at the moment I_2 enters the reaction bulb

(b) as reaction proceeds

(c) at equilibrium

12.11 For the system in the preceding problem, show the equilibrium condition in terms of the rates of the forward and reverse reactions.

12.12 An equilibrium involving the carbonate and bicarbonate ions exists in natural waters:

$$HCO_3^-(aq) \rightleftarrows H^+(aq) + CO_3^{2-}(aq)$$

Assuming that the reactions in both directions are elementary processes:

(a) Write rate expressions for the forward and reverse reactions.

(b) Write a constant expression for the equilibrium, based on the rates of the forward and reverse reactions.

12.13 The graph represents the progress of the reaction, $A(g) \rightleftarrows B(g)$, which starts with only the reactant A present. Label the two lines to show which represents A and which represents B. Discuss what is occurring at each of the times indicated ((a), (b), and (c)).

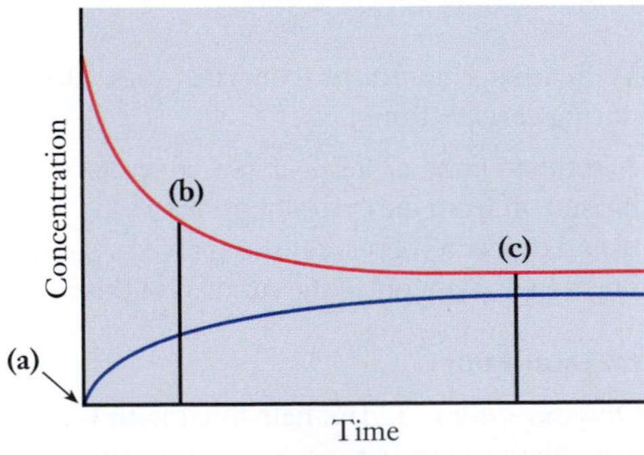

12.14 A small quantity of a soluble salt is placed in water. Equilibrium between dissolved and undissolved salt may or may not be attained. Explain.

12.15 In the following equilibrium in lake water, which of the inferences below can be drawn from the equation alone?

$$HCO_3^-(aq) \rightleftarrows H^+(aq) + CO_3^{2-}(aq)$$

(a) The rate of the forward reaction equals the rate of the reverse reaction.

(b) The equilibrium concentrations of all species are equal.

(c) Equilibrium was attained by starting with only $HCO_3^-(aq)$ in solution.

Equilibrium Constants

12.16 Write equilibrium (mass action) expressions for each of the following reactions:

(a) $H_2(g) + I_2(g) \rightleftarrows 2\ HI(g)$

(b) $2\ NO(g) + O_2(g) \rightleftarrows 2\ NO_2(g)$

(c) $N_2(g) + 3\ H_2(g) \rightleftarrows 2\ NH_3(g)$

(d) $CO(g) + NO_2(g) \rightleftarrows CO_2(g) + NO(g)$

(e) $2\ CO(g) + O_2(g) \rightleftarrows 2\ CO_2(g)$

12.17 Write equilibrium (mass action) expressions for each of the following reactions:

(a) $2\ NOBr(g) \rightleftarrows 2\ NO(g) + Br_2(g)$

(b) $4\ HCl(g) + O_2(g) \rightleftarrows 2\ H_2O(g) + 2\ Cl_2(g)$

(c) $SO_2(g) + \frac{1}{2} O_2(g) \rightleftarrows SO_3(g)$

(d) $CH_4(g) + 2\ O_2(g) \rightleftarrows CO_2(g) + 2\ H_2O(g)$

(e) $C_2H_5OH(g) + 3\ O_2(g) \rightleftarrows 2\ CO_2(g) + 3\ H_2O(g)$

12.18 What is the difference between *homogeneous equilibrium* and *heterogeneous equilibrium*?

12.19 Write equilibrium expressions for each of the following heterogeneous equilibria:

(a) $CaCO_3(s) \rightleftarrows Ca^{2+}(aq) + CO_3^{2-}(aq)$

(b) $AgCl(s) \rightleftarrows Ag^+(aq) + Cl^-(aq)$

(c) $Mg_3(PO_4)_2(s) \rightleftarrows 3\ Mg^{2+}(aq) + 2\ PO_4^{3-}(aq)$

(d) $Zn(s) + Cu^{2+}(aq) \rightleftarrows Cu(s) + Zn^{2+}(aq)$

12.20 Write equilibrium expressions for each of the following equilibria:

(a) $2\ C(s) + O_2(g) \rightleftarrows 2\ CO(g)$

(b) $Zn^{2+}(aq) + H_2S(g) \rightleftarrows ZnS(s) + 2\ H^+(aq)$

(c) $HCl(g) + H_2O(\ell) \rightleftarrows H_3O^+(aq) + Cl^-(aq)$

(d) $H_2(g) + \frac{1}{2} O_2(g) \rightleftarrows H_2O(g)$

12.21 For each of the following, are products or reactants favored?

(a) $AgCl(s) \rightleftarrows Ag^+(aq) + Cl^-(aq)$ $\quad K = 1.8 \times 10^{-10}$

(b) $Ca^{2+}(aq) + CO_3^{2-}(aq) \rightleftarrows CaCO_3(s)$ $\quad K = 2.1 \times 10^8$

(c) $N_2O_4 \rightleftarrows 2\ NO_2$ $\quad K = 1.10$

12.22 Which of the following is more likely to precipitate the hydroxide ion?

(a) $Cu(OH)_2(s) \rightleftarrows Cu^{2+}(aq) + 2\ OH^-(aq)$
$K = 1.6 \times 10^{-19}$

(b) $Ca(OH)_2(s) \rightleftarrows Ca^{2+}(aq) + 2\ OH^-(aq)$
$K = 7.9 \times 10^{-6}$

12.23 Which of the following is more likely to precipitate the sulfate ion?

(a) $PbSO_4(s) \rightleftarrows Pb^{2+}(aq) + SO_4^{2-}(aq) \quad K = 1.8 \times 10^{-8}$

(b) $CaSO_4(s) \rightleftarrows Ca^{2+}(aq) + SO_4^{2-}(aq) \quad K = 9.1 \times 10^{-6}$

12.24 The reaction, $3\ H_2(g) + N_2(g) \rightleftarrows 2\ NH_3(g)$, has the following equilibrium constants at the temperatures given:

at $T = 25°C$, $K = 2.8 \times 10^4$

at $T = 500°C$, $K = 2.4 \times 10^{-7}$

(a) At which temperature are reactants favored?

(b) At which temperature are products favored?

(c) What can you say about the reaction if the equilibrium constant is 1.2 at 127°C?

12.25 For each of the following equations, write the equilibrium expression for the reverse reaction:

(a) $2\ C(s) + O_2(g) \rightleftarrows 2\ CO(g)$

(b) $AgCl(s) \rightleftarrows Ag^+(aq) + Cl^-(aq)$

12.26 Consider the following equilibria involving $SO_2(g)$ and their corresponding equilibrium constants:

$$SO_2(g) + \frac{1}{2}O_2(g) \rightleftarrows SO_3(g) \qquad K_1$$

$$2\ SO_3(g) \rightleftarrows 2\ SO_2(g) + O_2(g) \qquad K_2$$

Which of the following expressions relates K_1 to K_2?

(a) $K_2 = K_1^2$

(b) $K_2^2 = K_1$

(c) $K_2 = K_1$

(d) $K_2 = 1/K_1$

(e) $K_2 = 1/K_1^2$

12.27 Using the following equations, determine each of the following:

(a) equilibrium expressions, K_1 and K_2

(b) the equation for the reaction that is the sum of the two equations

(c) the equilibrium expression, K_3, for the sum of the two equations

$$CO_3^{2-}(aq) + H^+(aq) \rightleftarrows HCO_3^-(aq) \qquad K_1 = ?$$

$$HCO_3^-(aq) + H^+(aq) \rightleftarrows H_2CO_3(aq) \qquad K_2 = ?$$

12.28 (a) In Exercise 12.27, if $K_1 = 2.1 \times 10^{10}$ and $K_2 = 2.3 \times 10^6$, what is the value of K_3?

(b) What is the equilibrium constant, K_3' for the reverse reaction?

12.29 An engineer is considering the use of bacteria called methanotrophs to remediate the production of small amounts of methane at a mine site. The reaction that takes place can be summarized as $CH_4(g) + 2\ O_2(g) \rightleftarrows CO_2(g) + H_2O(\ell)$. What is the equilibrium constant expression for this reaction?

12.30 Using the equations

$$HAsO_4^{2-}(aq) \rightleftarrows AsO_4^{3-}(aq) + H^+(aq) \qquad K_1 = 3.0 \times 10^{-13}$$

$$HAsO_4^{2-}(aq) + H^+(aq) \rightleftarrows H_2AsO_4^-(aq) \qquad K_2 = 1.8 \times 10^7$$

determine the equilibrium constant for the following reaction:

$$H_2AsO_4^-(aq) \rightleftarrows AsO_4^{3-}(aq) + 2\ H^+(aq)$$

12.31 In Exercise 12.30, which reaction has the greater tendency to go to completion as written, reaction 1 or reaction 2?

12.32 The following data were collected for the reaction, $H_2(g) + I_2(g) \rightleftarrows 2\ HI(g)$, at equilibrium at 25°C:

$[H_2] = 0.10\ mol\ L^{-1}$, $[I_2] = 0.20\ mol\ L^{-1}$, $[HI] = 4.0\ mol\ L^{-1}$

Calculate the equilibrium constant for the reaction at this temperature.

12.33 The following data were collected for a system at equilibrium at 140°C. Calculate the equilibrium constant for the reaction, $3\ H_2(g) + N_2(g) \rightleftarrows 2\ NH_3(g)$ at this temperature.

$[H_2] = 0.10\ mol\ L^{-1}$, $[N_2] = 1.1\ mol\ L^{-1}$, $[NH_3] = 3.6 \times 10^{-2}\ mol\ L^{-1}$

12.34 The equilibrium constant for the reaction, $3\ H_2(g) + N_2(g) \rightleftarrows 2\ NH_3(g)$, at a given temperature is 1.4×10^{-7}. Calculate the equilibrium concentration of ammonia, if $[H_2] = 1.2 \times 10^{-2}\ mol\ L^{-1}$ and $[N_2] = 3.2 \times 10^{-3}\ mol\ L^{-1}$.

Equilibrium Concentrations

12.35 Nitrosyl chloride, NOCl, decomposes to NO and Cl_2 at high temperatures:

$$2\ NOCl(g) \rightleftarrows 2\ NO(g) + Cl_2(g)$$

Suppose you place 2.00 mol NOCl in a 1.00-L flask and raise the temperature to 462°C. When equilibrium has been established, 0.66 mol NO is present. Calculate the equilibrium constant K_c for the decomposition reaction from these data.

12.36 Hydrogen gas and iodine gas react via the following equation:

$$H_2(g) + I_2(g) \rightleftarrows 2\ HI(g) \qquad K_c = 76 \text{ (at 600 K)}$$

If 0.050 mol HI is placed in an empty 1.0-L flask at 600 K, what are the equilibrium concentrations of HI, I_2, and H_2?

12.37 A system consisting of 0.100 mole of oxygen gas, O_2, is placed in a closed 1.00-L container and is brought to equilibrium at 600 K:

$$O_2(g) \rightleftarrows 2\ O(g) \qquad K = 2.8 \times 10^{-39}$$

What are the equilibrium concentrations of O and O_2?

12.38 The equilibrium constant of the reaction of $Cl_2(g)$ with $PCl_3(g)$ to produce $PCl_5(g)$ is 51 at a particular temperature. If the initial concentration of PCl_3 is $0.012\ mol\ L^{-1}$ and the initial concentration of Cl_2 is $0.125\ mol\ L^{-1}$, what are the equilibrium concentrations of all species?

12.39 The following reaction establishes equilibrium at 2000 K:

$$N_2(g) + O_2(g) \rightleftarrows 2\ NO \qquad K = 4.1 \times 10^{-4}$$

If the reaction began with $0.100\ mol\ L^{-1}$ of N_2 and $0.100\ mol\ L^{-1}$ of O_2, what were the equilibrium concentrations of all species?

12.40 The reaction in Exercise 12.39 was repeated. This time, the reaction began when only NO was injected into the reaction container. If $0.200\ mol\ L^{-1}$ NO was injected, what were the equilibrium concentrations of all species?

12.41 In the reaction in Exercise 12.39, another trial was carried out. The reaction began with an initial concentration of N_2 equal to the initial concentration of NO. Each had a concentration of 0.100 mol L^{-1}. What were the equilibrium concentrations of all species?

12.42 The experiment in Exercise 12.39 was redesigned so that the reaction started with 0.15 mol each of N_2 and O_2 being injected into a 1.0-L container at 2500 K. The equilibrium constant at 2500 K is 3.6×10^{-3}. What was the composition of the reaction mixture after equilibrium was attained?

12.43 Again the experiment in Exercise 12.39 was redesigned. This time, 0.15 mol each of N_2 and O_2 was injected into a 5.0-L container at 2500 K, at which the equilibrium constant is 3.6×10^{-3}. What was the composition of the reaction mixture at equilibrium?

12.44 At a particular temperature, the equilibrium constant, *K*, for the dissociation of N_2O_4 into NO_2 is 133. If the initial concentration of N_2O_4 is 0.100 mol L^{-1}, what are the concentrations of all species at equilibrium?

$$N_2O_4 \rightleftarrows 2\ NO_2$$

12.45 A student is simulating the carbonic acid–hydrogen carbonate equilibrium in a lake:

$$H_2CO_3(aq) \rightleftarrows H^+(aq) + HCO_3^-(aq) \qquad K = 4.4 \times 10^{-7}$$

She starts with 0.1000 M carbonic acid. What are the concentrations of all species at equilibrium?

12.46 Because carbonic acid undergoes a second ionization, the student in Exercise 12.45 is concerned that the hydrogen ion concentration she calculated is not correct. She looks up the equilibrium constant for the reaction

$$HCO_3^-(aq) \rightleftarrows H^+(aq) + CO_3^{2-}(aq)$$

Upon finding that the equilibrium constant for this reaction is 4.8×10^{-11}, she decides that her answer in 12.45 is correct. Explain her reasoning.

12.47 Because calcium carbonate is a sink for CO_3^{2-} in a lake, the student in Exercise 12.45 decides to go a step further and examine the equilibrium between carbonate ion and $CaCO_3$. The reaction is

$$Ca^{2+}(aq) + CO_3^{2-}(aq) \rightleftarrows CaCO_3(s)$$

The equilibrium constant for this reaction is 2.1×10^8. If the initial calcium ion concentration is 0.02 M and the carbonate concentration is 0.03 M, what are the equilibrium concentrations of the ions?

LeChatelier's Principle

12.48 The following reaction is in equilibrium in lake water:

$$HCO_3^-(aq) + H^+(aq) \rightleftarrows H_2CO_3(aq)$$

Predict the change in the reaction quotient, *Q*, for each disturbance below and use that prediction to explain how the equilibrium is shifted by the stress.

(a) $NaHCO_3$ is added to the lake.

(b) H_2CO_3 is added.

(c) NaOH is added.

12.49 In the following equilibrium in a closed system, indicate how the equilibrium is shifted by the indicated stress:

$$HF(g) + H_2O(\ell) \rightleftarrows H_3O^+(aq) + F^-(aq)$$

(a) Additional HF(g) is added to the system.

(b) Water is added.

(c) $Ca(NO_3)_2$ solution is added, and CaF_2 precipitates.

(d) The volume is reduced.

(e) KOH is added.

(f) A catalyst is added.

12.50 In each of the reactions, how does the equilibrium respond to an increase in pressure?

(a) $2\ SO_2(g) + O_2(g) \rightleftarrows 2\ SO_3(g)$

(b) $H_2(g) + I_2(g) \rightleftarrows 2\ HI(g)$

(c) $3\ H_2(g) + N_2(g) \rightleftarrows 2\ NH_3(g)$

12.51 In each of the reactions below, tell how the equilibrium responds to an increase in pressure.

(a) $CaCO_3(s) \rightleftarrows CaO(s) + CO_2(g)$

(b) $N_2O_4(g) \rightleftarrows 2\ NO_2(g)$

(c) $HCO_3^-(aq) + H^+(aq) \rightleftarrows H_2CO_3(aq)$

12.52 How would a temperature increase affect each of the equilibria below?

(a) $AgNO_3(s) \rightleftarrows Ag^+(aq) + NO_3^-(aq)$, $\quad \Delta H° = 22.6$ kJ mol^{-1}

(b) $C(s) + O_2(g) \rightleftarrows 2\ CO(g)$, $\quad \Delta H° = -209$ kJ mol^{-1}

12.53 The decomposition of NH_4HS,

$$NH_4HS(s) \rightleftarrows NH_3(g) + H_2S(g)$$

is an endothermic process. Using Le Chatelier's principle, explain how increasing the temperature would affect the equilibrium. If more NH_4HS is added to a flask in which this equilibrium exists, how is the equilibrium affected? What if some additional NH_3 is placed in the flask? What will happen to the pressure of NH_3 if some H_2S is removed from the flask?

12.54 Consider the following system:

$$4\ NH_3(g) + 3\ O_2(g) \rightleftarrows 2\ N_2(g) + 6\ H_2O(\ell) \qquad \Delta H = -1530.4 \text{ kJ}$$

(a) How will the amount of ammonia at equilibrium be affected by

(1) removing $O_2(g)$?

(2) adding $N_2(g)$?

(3) adding water?

(4) expanding the container at constant pressure?

(5) increasing the temperature?

(b) Which of the above factors will increase the value of *K*? Which will decrease it?

12.55 The following equilibrium is established in a closed container:

$$C(s) + O_2(g) \rightleftarrows CO_2(g) \qquad \Delta H° = -393 \text{ kJ mol}^{-1}$$

How does the equilibrium shift in response to each of the following stresses?

(a) The quantity of solid carbon is increased.

(b) A small quantity of water is added, and CO_2 dissolves in it.

(c) The system is cooled.

(d) The volume of the container is increased.

Solubility Equilibria

12.56 Write the K_{sp} expression for each of the following equilibria for dissolution of the salt in water:

(a) $AgI(s) \rightleftarrows Ag^+(aq) + I^-(aq)$

(b) $PbI_2(s) \rightleftarrows Pb^{2+}(aq) + 2\ I^-(aq)$

(c) $Hg_2I_2 \rightleftarrows Hg_2^{2+}(aq) + 2\ I^-(aq)$

(d) $Cu(IO_3)_2 \rightleftarrows Cu^{2+}(aq) + 2\ IO_3^-(aq)$

12.57 Write an equilibrium equation and a K_{sp} expression for the dissolution of each of the following in water. **(a)** AgBr, **(b)** $Cu(OH)_2$, **(c)** $BaSO_4$, **(d)** $PbCrO_4$, **(e)** $Mg_3(PO_4)_2$

12.58 Calculate the molar solubility of the following compounds. **(a)** MgF_2, **(b)** $Fe(OH)_3$, **(c)** $Mg_3(PO_4)_2$

12.59 The Safe Drinking Water Act of 1974 established the maximum permitted concentration of silver ion at 0.05 ppm. What is the concentration of Ag^+ in parts per million in a saturated solution of AgCl? (NOTE: 1 ppm = 1 mg of solute/L of solution.)

12.60 In Exercise 12.59, what is the allowed concentration of AgCl in g per 100 g of water?

12.61 Calculate the solubility of $ZnCO_3$ in **(a)** water, **(b)** 0.050 M $Zn(NO_3)_2$, and **(c)** 0.050 M K_2CO_3. K_{sp} of $ZnCO_3 = 3 \times 10^{-8}$

12.62 Because barium sulfate is opaque to X-rays, it is suspended in water and taken internally to make the gastrointestinal tract visible in an X-ray photograph. Although barium ion is quite toxic, barium sulfate's K_{sp} of 1.1×10^{-10} gives it such low solubility that it can be safely consumed.

(a) What is the molar solubility of $BaSO_4$.

(b) What is its solubility in grams per 100 g of water?

12.63 The ore cinnabar (HgS) is an important source of mercury. Cinnabar is a red solid whose solubility in water is 5.5×10^{-27} mol L^{-1}.

(a) Calculate its K_{sp}.

(b) What is its solubility in grams per 100 g of water?

12.64 Calculate the K_{sp} of the following compounds, given their molar solubilities. **(a)** AgCN, 7.73×10^{-9} M, **(b)** $Ni(OH)_2$, 5.16×10^{-6} M, **(c)** $Cu_3(PO_4)_2$, 1.67×10^{-8} M

12.65 From the solubility data given for the following compounds, calculate their solubility product constants.

(a) CuBr, copper(I) bromide, 1.0×10^{-3} g/L

(b) AgI, silver iodide, 2.8×10^{-8} g/10 mL

(c) $Pb_3(PO_4)_2$, lead(II) phosphate, 6.2×10^{-7} g/L

(d) Ag_2SO_4, silver sulfate, 5.0 mg/mL

12.66 The solubility of magnesium fluoride, MgF_2, in water is 0.016 g/L. What is the solubility, in grams per liter, of magnesium fluoride in 0.020 M sodium fluoride, NaF?

12.67 Solid Na_2SO_4 is added slowly to a solution that is 0.10 M in $Pb(NO_3)_2$ and 0.10 M in $Ba(NO_3)_2$. In what order will solid $PbSO_4$ and $BaSO_4$ form? Calculate the percentage of Ba^{2+} that precipitates just before $PbSO_4$ begins to precipitate.

12.68 Will a precipitate of $Mg(OH)_2$ form when 25.0 mL of 0.010 M NaOH is combined with 75.0 mL of a 0.10 M solution of magnesium chloride?

12.69 Use the web to look up boiler scale and explain chemically why it is a problem in equipment where water is heated (such as in boilers).

12.70 A student is studying a solution of the sparingly soluble salt Ag_2SO_4, for which the K_{sp} is 1.2×10^{-5}. The solubility of Ag_2SO_4 is high enough that silver ions often encounter sulfate ions in solution. What would be the effect on the measurement of the solubility of Ag_2SO_4 if a large number of ion pairs exist in the solution? Defend your answer.

Acids and Bases

12.71 Write the formula of the conjugate base of each of the following acids. **(a)** HNO_3, **(b)** H_2O, **(c)** HSO_4^-, **(d)** H_2CO_3, **(e)** H_3O^+

12.72 Write the formula of the conjugate acid of each of the following bases. **(a)** OH^-, **(b)** NH_3, **(c)** CH_3NH_2, **(d)** HPO_4^{2-}, **(e)** CO_3^{2-}

12.73 For each of the following reactions, indicate the Brønsted-Lowry acids and bases. What are the conjugate acid–base pairs?

(a) $CN^-(aq) + H_2O(\ell) \rightleftarrows HCN(aq) + OH^-(aq)$

(b) $HCO_3^-(aq) + H_3O^+(aq) \rightleftarrows H_2CO_3(aq) + H_2O(\ell)$

(c) $CH_3COOH(aq) + HS^-(aq) \rightleftarrows CH_3COO^-(aq) + H_2S(aq)$

12.74 What are the products of each of the following acid–base reactions? Indicate the acid and its conjugate base and the base and its conjugate acid.

(a) $HClO_4 + H_2O \rightarrow$

(b) $NH_4^+ + H_2O \rightarrow$

(c) $HCO_3^- + OH^- \rightarrow$

12.75 Write chemical equations and equilibrium expressions for the reactions of each of the following weak acids with water. **(a)** acetic acid, CH_3COOH, **(b)** propanoic acid, C_2H_5COOH, **(c)** hydrofluoric acid, HF, **(d)** hypochlorous acid, HClO, **(e)** carbonic acid, H_2CO_3

12.76 Write chemical equations and equilibrium expressions for the reactions of each of the following weak bases with water. **(a)** ammonia, NH_3, **(b)** methylamine, CH_3NH_2, **(c)** acetate ion, CH_3COO^-, **(d)** hydrogen carbonate ion, HCO_3^-, **(e)** aniline, $C_6H_5NH_2$

12.77 Hydrofluoric acid is a weak acid used in the building industry to etch patterns into glass for elegant windows. Because it dissolves glass, it is the only inorganic acid that must be stored in plastic containers. A 0.1 M solution of HF has a pH of 2.1. Calculate $[H_3O^+]$ in this solution.

12.78 The pH of a 0.129 M solution of a weak acid, HB, is 2.34. What is K_a for the weak acid?

12.79 Calculate the pH of a 0.10 M solution of propanoic acid and determine its percent ionization.

12.80 Find the pH of a 0.115 M solution of $NH_3(aq)$.

12.81 Acrylic acid is used in the polymer industry in the production of acrylates. Its K_a is 5.6×10^{-5}. What is the pH of a 0.11 M solution of acrylic acid, $CH_2CHCOOH$?

12.82 The ionization constant of a very weak acid, HA, is 4.0×10^{-9}. Calculate the equilibrium concentrations of H_3O^+, A^-, and HA in a 0.040 M solution of the acid.

12.83 Morphine, an opiate derived from the opium poppy (genus *Papaver*), has the molecular formula $C_7H_{19}NO_3$. It is a weakly basic amine, with a K_b of 1.6×10^{-6}. What is the pH of a 0.0045 M solution of morphine?

Free Energy and Chemical Equilibrium

12.84 In a particular experiment, the equilibrium constant measured for the reaction, $Cl_2(g) + NO_2(g) \rightleftarrows Cl_2NO_2(g)$, is 2.8.

(a) Based on this measurement, calculate $\Delta G°$ for this reaction.

(b) Calculate $\Delta G°$ using data from Appendix E at the back of the book and discuss the agreement between your two calculations.

12.85 For the reaction, $PCl_5(g) \rightleftarrows PCl_3(g) + Cl_2(g)$, the measured equilibrium constant is 7.7×10^{-3} at 25°C. **(a)** Based on this information, predict if $\Delta G°$ is positive or negative. Explain your answer. **(b)** Calculate $\Delta G°$ from this information.

12.86 Use values from Appendix E at the back of the book to calculate the equilibrium constant, K_p, for the following gaseous reactions:

(a) $H_2(g) + Cl_2(g) \rightleftarrows 2\ HCl(g)$

(b) $CH_4(g) + H_2O(g) \rightleftarrows CO(g) + 3\ H_2(g)$

(c) $SO_2(g) + Cl_2(g) \rightleftarrows SO_2Cl_2(g)$

(d) $2\ HCl(g) + F_2(g) \rightleftarrows 2\ HF(g) + Cl_2(g)$

12.87 A group of students working on chemistry homework calculates the equilibrium constants for three reactions. Reaction 1 has $\Delta G° = -50$ kJ, reaction 2 has $\Delta G° = -30$ kJ, and reaction 3 has $\Delta G° = -10$ kJ. The students are surprised when their answers for each reaction are very close to $K = 1$. Suggest what error the students must have made and calculate the actual values of K for these reactions.

INSIGHT INTO Bendable Concrete

12.88 What is meant by the term *crack propagation*? How does crack propagation in a brittle material differ from that in a less brittle material?

12.89 Ductile materials, like metal wires, are quite different from brittle materials like concrete. If you have a piece of wire, and no wire cutters, how could you break it into two pieces? How does this differ from breaking a piece of brittle material like glass?

12.90 If bendable concrete were widely available, would it be ideal for all engineering designs that currently use concrete? What types of structures might benefit most from having bendable concrete? What types of designs would seem unlikely to benefit from replacing traditional concrete with an ECC material?

12.91 The monomer of polyvinyl alcohol was described in Example Problem 7.6. (a) Depict a small section of the resulting polymer as a line diagram. (b) Based on your understanding of intermolecular forces, how might different polymer molecules that are adjacent to each other in a PVA fiber interact with each other?

Additional Problems

12.92 Consider the equilibrium

$$N_2(g) + O_2(g) \rightleftarrows 2\ NO(g)$$

At 2300 K the equilibrium constant $K_c = 1.7 \times 10^{-3}$. Suppose that 0.015 mol NO(g), 0.25 mol $N_2(g)$, and 0.25 mol $O_2(g)$ are placed into a 10.0-L flask and heated to 2300 K.

(a) Is the system at equilibrium?

(b) If not, in which direction must the reaction proceed to reach equilibrium?

(c) Calculate the equilibrium concentrations of all three substances.

12.93 Solid $CaCO_3$ is placed in a closed container and heated to 800°C. What is the equilibrium concentration of CO_2 in the following equilibrium, for which $K = 2.5 \times 10^{-3}$?

$$CaCO_3(s) \rightleftarrows CaO(s) + CO_2(g)$$

12.94 A reaction important in smog formation is

$$O_3(g) + NO(g) \rightleftarrows O_2(g) + NO_2(g) \qquad K = 6.0 \times 10^{34}$$

(a) If the initial concentrations are $[O_3] = 1.0 \times 10^{-6}$ M, $[NO] = 1.0 \times 10^{-5}$ M, $[NO_2] = 2.5 \times 10^{-4}$ M, and $[O_2] = 8.2 \times 10^{-3}$ M, is the system at equilibrium? If not, in which direction does the reaction proceed?

(b) If the temperature is increased, as on a very warm day, will the concentrations of the products increase or decrease? (HINT: You may have to calculate the enthalpy change for the reaction to find out if it is exothermic or endothermic.)

12.95 An engineer working on a design to extract petroleum from a deep thermal reservoir wishes to capture toxic hydrogen sulfide gases present by reaction with aqueous iron(II)nitrate to form solid iron(II)sulfide. **(a)** Write the chemical equation for this process, assuming that it reaches equilibrium. **(b)** What is the equilibrium constant expression for this system? **(c)** How can the process be manipulated so that it does not reach equilibrium, allowing the continuous removal of hydrogen sulfide?

12.96 A chemical engineer is working to optimize the production of acrylonitrile to be used in the manufacture of carbon fibers. The reaction being used is the combination of propene gas, ammonia, and oxygen. The reaction is normally carried out at moderately high temperatures so all species are in the gas phase.

$$2\ CH_3{-}CH{=}CH_2(g) + 2\ NH_3(g) + 3\ O_2(g) \rightleftarrows 2\ CH_2{=}CH{-}C{\equiv}N(g) + 6\ H_2O(g)$$

(a) Write the equilibrium constant expression for this reaction.

(b) The boiling point of acrylonitrile is 77°C, and that of propene is −48°C. What would the equilibrium expression be if this reaction were carried out at room temperature?

(c) What characteristic of this reaction might cause the engineer to desire carrying out this reaction at room temperature?

12.97 Methanol, CH_3OH, can be produced by the reaction of CO with H_2, with the liberation of heat. All species in the reaction are gaseous. What effect will each of the following have on the equilibrium concentration of CO?

(a) Pressure is increased, **(b)** volume of the reaction container is decreased, **(c)** heat is added, **(d)** the concentration of CO is increased, **(e)** some methanol is removed from the container, and **(f)** H_2 is added.

12.98 The following equilibrium is established in a closed container:

$$C(s) + O_2(g) \rightleftarrows CO_2(g) \qquad \Delta H^\circ = -393 \text{ kJ}$$

Set up a table like the one below and indicate whether equilibrium concentrations increase, decrease, or remain the same.

Change Applied to System	C(s)	O_2(g)	CO_2(g)
Pressure is decreased			
Temperature is increased			
Carbon is added			
Oxygen is added			
Carbon dioxide is added			
Volume of the system is increased			
Catalyst is added			
Heat source is removed			

12.99 Using the kinetic-molecular theory, explain why an increase in pressure produces more N_2O_4 in the following system:

$$N_2O_4(g) \rightleftarrows 2\ NO_2(g)$$

12.100 The solubility of KCl is 34.7 g per 100 g of water at 20°C. Calculate its molar solubility in **(a)** pure water and **(b)** 0.10 M NaCl.

12.101 To prepare community wastewater to be released to a lake, phosphate ions (among others) must be removed. $Al_2(SO_4)_3$ can be used for this purpose, precipitating $AlPO_4$, for which K_{sp} at 25°C is 1.3×10^{-20}. What mass of $Al_2(SO_4)_3$ must be added per gallon of water to assure that the concentration of PO_4^{3-} in the water released to the lake will be no higher than 1.0×10^{-12} M?

12.102 A nuclear engineer is considering the effect of discharging waste heat from a power plant into a lake and estimates that this may warm the water locally to 25°C. One question to be considered is the effect of this temperature change on the uptake of CO_2 by the water. The equilibrium constant for the reaction $CO_2 + H_2O \rightleftarrows H_2CO_3$ is $K = 1.7 \times 10^{-3}$ at 25°C. Because bonds form, the reaction is exothermic.

(a) Will this reaction progress further toward products at higher temperatures near the water discharge with its warmer water than it would in the cooler lake water? Explain your reasoning.

(b) Carbonic acid has a K_a of 2.5×10^{-4} at 25°C. What is the equilibrium constant for the reaction $CO_2 + 2\ H_2O \rightleftarrows HCO_3^- + H_3O^+$?

(c) What additional factor should the engineer be considering about CO_2 gas, probably before considering this reaction chemistry?

12.103 Copper(II) iodate has a solubility of 0.136 g per 100 g of water. Calculate its molar solubility in water and its K_{sp}.

12.104 In Exercise 12.103, what do you predict would happen to an aqueous copper(II) iodate equilibrium if sulfide ion, S^{2-}, were added? The K_{sp} of CuS is 8.7×10^{-36}.

FOCUS ON PROBLEM SOLVING EXERCISES

12.105 You have three white solids. What experiment could you carry out to rank them in order of increasing solubility?

12.106 You find a bottle in the laboratory. You know it is an acid and the label says it is a 1.05 M solution but doesn't say what acid it is. Assuming the label is correct for the concentration, how could you determine what acid it is? What would have to be measured and what would you have to look up?

12.107 You find a bottle of acid in the laboratory and it is labeled $H_2C_2O_4$. There is no concentration given and you don't have the reagents to carry out a titration. What other measurement could be made and what information would you need to look up to determine the concentration of the acid?

12.108 You need to find the equilibrium constant for a reaction in an industrial process. Describe how you can estimate this value using only thermodynamic data.

Cumulative Problems

12.109 A mixture of 1.00 mole of PCl_3 and 2.00 moles of PBr_3 is placed in a 1.00-L reaction vessel. At the temperature of the experiment, $K = 2.0$ for the following reaction:

$$PCl_3 + PBr_3 \rightleftarrows PCl_2Br + PClBr_2$$

(a) What is the equilibrium concentration of each species in the vessel?

(b) Because the number of bonds is the same on both sides of the equation, this reaction should be close to thermoneutrality, meaning that it is neither exothermic nor endothermic. What drives this reaction?

12.110 Sulfur dioxide can be generated by the reaction of sodium hydrogen sulfite with hydrochloric acid.**(a)** Write the balanced chemical equation for the reaction, given that the other products are sodium chloride and water. **(b)** If 1.9 g of sodium hydrogen sulfite is reacted with excess HCl, what mass of SO_2 is produced? **(c)** If the gas generated is in a 100-mL vessel at 25°C, and 0.05 mol of O_2 is introduced, equilibrium between SO_2, O_2, and SO_3 will be established.

$$SO_2 + \frac{1}{2}O_2 \rightleftarrows SO_3 \qquad K = 2.6 \times 10^{12}$$

Calculate the equilibrium concentration of each species.

12.111 The vapor pressure of water at 80.0°C is 0.467 atm. Find the value of K_c for the process

$$H_2O(\ell) \rightleftarrows H_2O(g)$$

at this temperature.

12.112 Although ammonia is made in enormous quantities by the Haber-Bosch process, sulfuric acid is made in even greater quantities by the *contact process*. A simplified version of this process can be represented by these three reactions:

$$S(s) + O_2(g) \rightleftarrows SO_2(g)$$
$$2\ SO_2(g) + O_2(g) \rightleftarrows 2\ SO_3(g)$$
$$SO_3(g) + H_2O(\ell) \rightleftarrows H_2SO_4(\ell)$$

(a) Use tabulated data to calculate ΔH° for each reaction.

(b) Which reactions are exothermic? Which are endothermic?

(c) In which of the reactions does entropy increase? In which does it decrease? In which does it stay about the same?

(d) For which reaction(s) do low temperatures favor formation of products?

13 Electrochemistry

CHAPTER OUTLINE

Rust is clearly part of the design for this sculpture in downtown Milwaukee. But in many engineering designs, the extensive corrosion seen here could prove disastrous. To prevent corrosion, engineers need to understand the chemistry of oxidation–reduction reactions. *Thomas A. Holme*

Electrochemistry deals directly with electrons and their movement. Whether the electrons move under the driving forces of spontaneous chemical processes or respond to an applied external electric potential, their movement brings about a whole new set of opportunities and challenges. We introduce the chapter with the topic of corrosion, one result of electrochemistry that costs the U.S. economy an estimated $300 billion per year.

Chapter Objectives

After mastering this chapter, you should be able to

- describe at least three types of corrosion and identify chemical reactions responsible for corrosion.
- define oxidation and reduction.

- write and balance half-reactions for simple redox processes.
- describe the differences between galvanic and electrolytic cells.
- use standard reduction potentials to calculate cell potentials under both standard and nonstandard conditions.
- use standard reduction potentials to predict the spontaneous direction of a redox reaction.
- calculate the amount of metal plated, the amount of current needed, or the time required for an electrolysis process.
- distinguish between primary and secondary batteries.
- describe the chemistry of some common battery types, and explain why each type of battery is suitable for a particular application.
- describe at least three common techniques for preventing corrosion.

13-1 Corrosion

Throughout the history of the U.S. space program, NASA has used the Kennedy Space Center (KSC) on the coast of Florida as its primary launch site. In stark contrast, the site of the space center used by Russian cosmonauts was constructed in the middle of the Moyunqum Desert. Although many factors are considered when choosing a site for such facilities, the contrast between the environments in these two places is striking. The placement of the KSC so close to the ocean means that rust has long been an important issue for the American space program. Rust is such a significant concern to NASA that they founded the Beach Corrosion Test Site in the 1960s to study the effect of the atmosphere on structures used in launching rockets. When rust forms on steel, it weakens the metal, and weaker materials are more prone to failure. Therefore, NASA studies rust and **corrosion**, the degradation of metals by chemical reactions with the environment. Corrosion generally involves a slow combination of oxygen with metals to form oxides (**Figure 13.1**).

Engineers realize that many metals corrode. The formation of rust—iron oxide—is just the most familiar example.

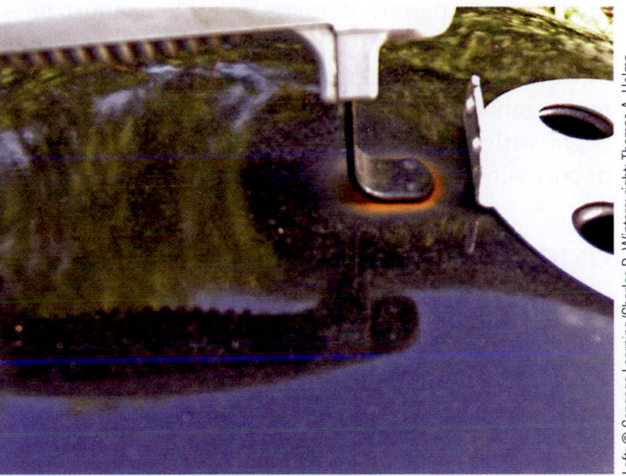

Left: © Cengage Learning/Charles D. Winters; right: Thomas A. Holme

Figure 13.1 ■ Corrosion occurs in a variety of forms. The chain on the left shows uniform corrosion, whereas the grill cover on the right shows corrosion only where the handle is attached. This is called crevice corrosion and can occur whenever small gaps exist between metal parts.

Researchers have identified many different forms of corrosion. The rusting of automobile bodies is an example of **uniform corrosion** and is one of the most visible forms of corrosion. What conditions are necessary for this process to occur? Another important form of corrosion is **galvanic corrosion**, which occurs only when two different metals contact each other in the presence of an appropriate electrolyte. What is so special about the contact of two different metals? Other forms of corrosion tend to require specific conditions, yet many of these situations are common in designs for machinery. **Crevice corrosion** is a major problem in many large machines. When two pieces of metal touch each other, they tend to leave a small gap (unless the joint is covered with a coating, such as paint). At that gap, or crevice, the metals are more likely to corrode. How can we understand the tendency to corrode using basic principles of electrochemistry?

A puzzling aspect of corrosion is the difference between iron (or steel) and other metals in their behavior. Aluminum, for example, has a greater tendency to corrode than iron, but the corrosion of aluminum is not problematic. Why not? As we will see, the answer lies in the nature of the products of the corrosion reaction. What differences are there between the corrosive formation of aluminum oxide and iron(III) oxide? How can we picture corrosion at an atomic level to account for these differences?

Although corrosion is an important example of electrochemical processes, we need to realize that it alone does not define electrochemistry. There are many important applications of electrochemistry that provide useful products for society rather than degrading them. The refining of some metal ores into useable materials and the production of batteries are two obvious examples of the practical utility of electrochemistry. But even in the production of batteries, we need to consider additional manufacturing steps to reduce the impact of corrosion on the useful lifetime of the battery. Where does corrosion occur in a battery, and what can be done to combat it? The study of electrochemical principles will help us answer these questions.

13-2 Oxidation–Reduction Reactions and Galvanic Cells

What happens to a piece of steel that sits outside, unprotected? In most locations, it rusts. Would you expect to observe the same thing if that piece of steel were inside a house or in a desert? Perhaps not. There must be some special conditions that promote the reaction of iron with oxygen to form iron(III) oxide. We could design a set of experiments to study the formation of rust, but from a laboratory perspective, rust formation is rather slow. To find out more about the basics of electrochemistry, let's begin with more easily observed reactions and then apply what we learn to examples of corrosion.

Oxidation–Reduction and Half-Reactions

Reactions involving the transfer of electrons are known as **oxidation–reduction reactions.** (The term is often inverted and shortened to **redox reactions.**) To delve more deeply into electrochemistry, we will first need to understand this class of chemical reactions and define a few relevant terms. **Oxidation** is the loss of electrons from some chemical species, whereas **reduction** is the gain of electrons. By now you will recognize that electrons can't simply be lost. They have to go somewhere, in keeping with conservation of charge and mass. This leads directly to one of the most important principles of redox chemistry: the electrons lost in oxidation must always be gained in the simultaneous reduction of some other species. In other words, we can't have oxidation unless we also have reduction. To illustrate this idea, let's consider the reaction

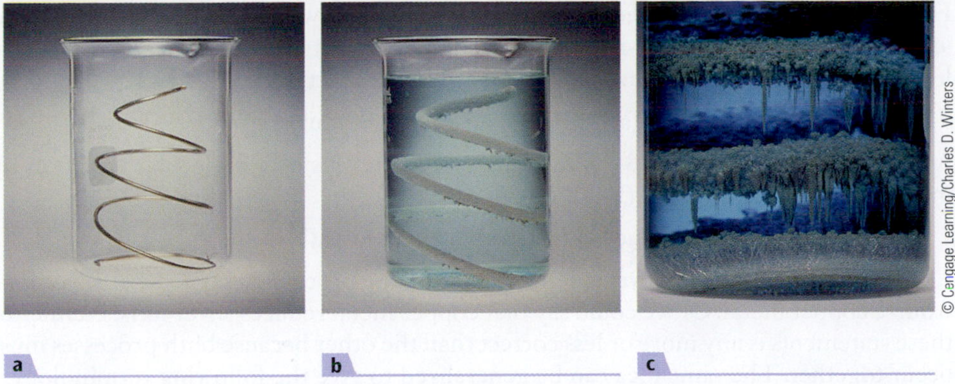

Figure 13.2 ■ When a clean copper wire is placed into a colorless solution of silver nitrate, it is quickly apparent that a chemical reaction takes place. Crystals of silver metal form on the wire, and the solution takes on a blue color. In this reaction, copper metal is oxidized to Cu^{2+}, and Ag^+ ions are reduced to silver metal.

that takes place when we place a copper wire in a solution of silver nitrate—a popular classroom demonstration (**Figure 13.2**).

As you can see in the photographs, the original solution of silver nitrate is clear and colorless, and the copper wire is smooth (Figure 13.2a). A short time after the wire is placed in the solution, the copper wire looks fuzzy, and the solution has become light blue (Figure 13.2b). These observations become more apparent over time. In the final photo, there is an obvious buildup of slivers of silver metal on the wire, and the solution is distinctly blue (Figure 13.2c). What has happened chemically? The blue color of the solution is indicative of the presence of aqueous Cu^{2+} ions. These ions must originate from the copper wire because no other source of copper is available. So in forming the cations, copper atoms must have lost electrons. We say that copper has been oxidized, and we could write an equation to describe this change:

$$Cu(s) \rightarrow Cu^{2+}(aq) + 2\ e^-$$

But if cations are produced at the wire, where did their electrons go? The accumulation of silver on the copper wire provides us with the answer. The only apparent way to form metallic silver in this system is from the silver ions in the original solution. So we must conclude that silver cations in solution have accepted the electrons lost by the copper. We say that silver has been reduced, and again we can write an equation to show the change:

$$Ag^+(aq) + e^- \rightarrow Ag(s)$$

The two equations we have written describe what are called **half-reactions** for the oxidation of copper and the reduction of silver. Neither one can occur on its own, because oxidation and reduction must take place in concert with one another.

As you examine the half-reactions with this in mind, you may notice a small discrepancy. Copper loses two electrons in the oxidation half-reaction, whereas silver gains only one electron in the reduction half-reaction. This tells us that in order to conserve electrons, two silver ions must be reduced for every copper atom that is oxidized. We can multiply the reduction by two to make this explicit, giving us the following pair of half-reactions:

$$Cu(s) \rightarrow Cu^{2+}(aq) + 2\ e^-$$
$$2\ Ag^+(aq) + 2\ e^- \rightarrow 2\ Ag(s)$$

Here we can more readily see that the silver gains the two electrons that the copper loses. If we add the two half-reactions together, the electrons will cancel and we'll be left with the net ionic equation for the overall redox reaction:

$$2\ Ag^{+}(aq) + Cu(s) \rightarrow 2\ Ag(s) + Cu^{2+}(aq)$$

As we discussed in Section 3-3, we could also write this as a molecular equation by including the spectator ions (NO_3^- in this case):

$$2\ AgNO_3(aq) + Cu(s) \rightarrow 2\ Ag(s) + Cu(NO_3)_2(aq)$$

Looking at our example above, we could describe the process by saying that silver ions oxidize copper metal. Or we could say that copper metal reduces silver ions. Neither of these statements is any more or less correct than the other because both processes must occur together. The same idea can be generalized to give the following terminology:

The species undergoing oxidation is referred to as a reducing agent.

The species undergoing reduction is referred to as an oxidizing agent.

An agent in this sense is something that allows the process to occur. In everyday language, for example, an agent of change is someone or something that helps facilitate change.

Many students find this confusing, but it should make sense if you consider the wording carefully.

Building a Galvanic Cell

The half-reactions we've just introduced provide us with a way of seeing the reaction as oxidation and reduction but seem like a rather artificial idea. Both half-reactions took place in the same beaker, after all. But what would happen if we could set up these half-reactions in two separate containers? Consider the following experiment.

Place a piece of copper metal in a solution of copper(II) ions. (**Figure 13.3a**; the spectator ion is not important—sulfate might be used because copper(II) sulfate is

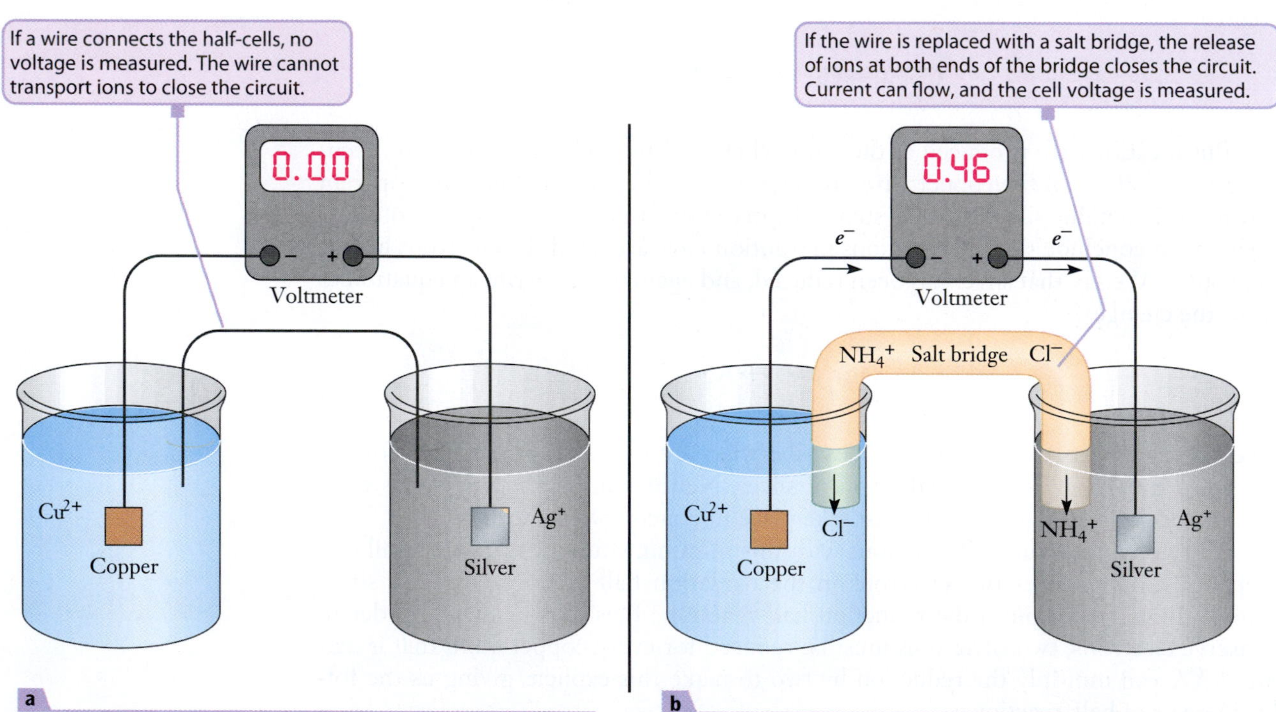

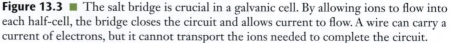

Figure 13.3 ■ The salt bridge is crucial in a galvanic cell. By allowing ions to flow into each half-cell, the bridge closes the circuit and allows current to flow. A wire can carry a current of electrons, but it cannot transport the ions needed to complete the circuit.

a common laboratory reagent.) In a separate beaker, place a piece of silver metal or a silver wire in a solution of silver ions. (Again, the spectator ion is not critical. We might use nitrate ions since silver nitrate is another common reagent.) We observe no reactions in the individual beakers. Now connect the solutions by a conducting wire and connect a voltmeter between the copper and silver, as shown. If you have a good measuring device, you might notice a brief electrical current, but there is otherwise no observable change in the system. Why? The wire cannot allow a continuous flow of charge between the solutions because the charged particles in the solutions are ions, not free electrons. Any initial flow of electrons from the copper-containing solution toward the silver-containing solution is quickly stopped by the buildup of negative charge in the beaker of silver solution. Connecting the two beakers in a second way, with a salt bridge, can alleviate the charge buildup (**Figure 13.3b**). A **salt bridge** contains a strong electrolyte that allows either cations or anions to migrate into the solution where they are needed to maintain charge neutrality. In this case, suppose that the salt bridge is filled with NH_4Cl. Ammonium ions will flow into the beaker of silver solution to offset the *removal* of Ag^+(aq), and chloride ions will flow into the beaker of copper solution to offset the *production* of Cu^{2+}(aq) in the copper solution beaker. In Figure 13.3b, we can see the various moving charges in this arrangement. The apparatus we have constructed in this way is an example of a *galvanic cell.*

Nothing flows through the salt bridge. It merely dispenses ions at each end.

The word galvanic *stems from the 18th-century Italian physicist, Luigi Galvani, who used frogs to conduct early experiments on the role of electricity in living systems.*

To define a galvanic cell formally, we should first note its most important characteristic. By separating the two half-reactions, inserting a salt bridge between the solutions, and allowing electrons to flow along a wire connecting the metals, we have generated a sustainable electric current. A **galvanic cell** is any electrochemical cell in which a spontaneous chemical reaction can be used to generate an electric current. The observation of an electric current in such a cell led to the name **electrochemistry**. This connection between electricity and chemistry was vitally important in the development of modern chemical principles.

Terminology for Galvanic Cells

The number of possible galvanic cells that could be built is almost unlimited. All of them have common features, and terminology has been devised to describe these features. The electrically conducting sites at which either oxidation or reduction take place are called **electrodes**. Oxidation occurs at the **anode** and reduction occurs at the **cathode**. These terms are also used for other types of electrochemical cells, too, not just for galvanic cells.

Because there are so many possible electrochemical cells, a shorthand notation for representing their specific chemistry has been devised. This **cell notation** lists the metals and ions involved in the reaction. A vertical line, |, denotes a phase boundary, and a double line, ||, represents the salt bridge. The anode is always written on the left and the cathode on the right:

Anode | electrolyte of anode || electrolyte of cathode | cathode

The previous example of copper and silver would be written as follows:

$$Cu(s) \mid Cu^{2+}(aq)(1\ M) \parallel Ag^{+}(aq)(1\ M) \mid Ag(s)$$

Once again, we see that the spectator ions are not identified in this notation. The concentration of the electrolyte is generally included, for reasons we will soon see. In the example above, we include concentrations of 1 M. These values have special meaning because they are the assigned concentrations for the **standard state** of an electrochemical cell. If the electrochemical half-reaction includes the production or consumption of a gas, the standard state is a pressure of 1 atm. A standard state also implies that the electrode material is in its thermodynamic standard state, which is commonly the case because most electrode materials are solids at room temperature.

Atomic Perspective on Galvanic Cells

To see the pervasive nature of corrosion or to understand how electrochemistry can be harnessed in a battery, we need to understand some fundamental laws. We have noted that electrons flow in a properly constructed cell, but what causes this flow to occur? To understand this question, we will build a galvanic cell again, but this time we will focus on what happens at the atomic level. If we construct the anode and cathode separately, what does the resulting system look like at each of the electrodes *before* we connect them with a salt bridge? Although it may conflict with your intuition, there is a buildup of charge at the interface between the solid electrode and the solution surrounding it. We are accustomed to thinking that chemical systems are electrically neutral, but this constraint is true only within a single phase, not necessarily between phases. So, a system that does not yet have a complete circuit undergoes partial reaction. At the anode, some oxidation occurs and cations dissolve into solution, leaving behind a negative charge on the anode. At the cathode, reduction brings positively charged ions to the electrode, and the result is a buildup of positive charge on that electrode. Thus before the circuit is complete, a form of equilibrium is quickly achieved where charge builds up on the individual electrodes, and it is locally offset by the ions in the solution, as shown in **Figure 13.4**. We could consider this equilibrium state at each electrode *the half-reaction equilibrium*—but it is *not* an oxidation–reduction equilibrium.

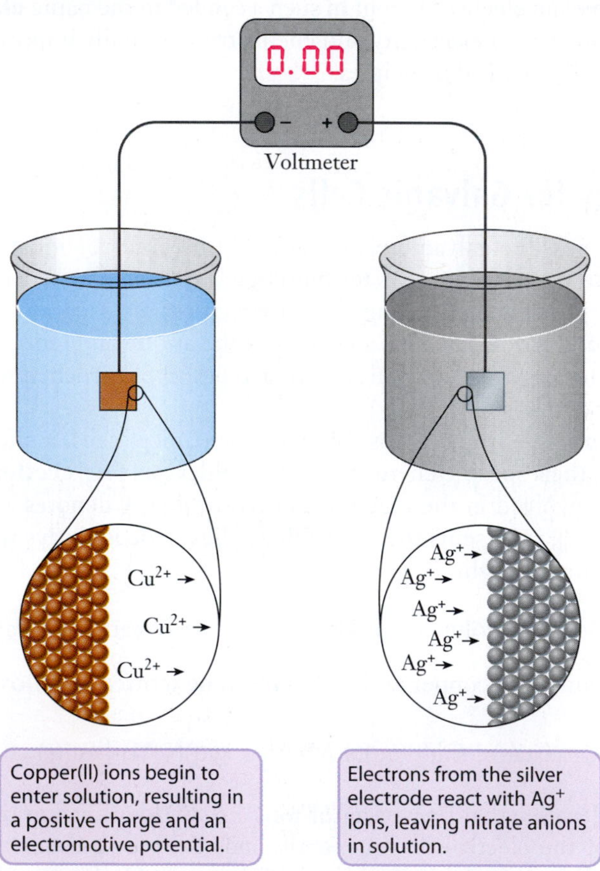
0.00
–
+
Voltmeter
Cu^{2+} →
Cu^{2+} →
Cu^{2+} →
Ag^+ →
Ag^+ →
Ag^+ →
Ag^+ →
Ag^+ →
Ag^+ →
Copper(II) ions begin to enter solution, resulting in a positive charge and an electromotive potential.
Electrons from the silver electrode react with Ag^+ ions, leaving nitrate anions in solution.

Figure 13.4 ■ Without a salt bridge to close the circuit, local charges will build up around both electrodes. Neither electrode reaction can proceed to any significant extent, so no cell voltage can be measured.

Still, the buildup of charge on the electrodes is important because it means that there is a potential for electrical work. This potential is called the **cell potential**, or **electromotive force (EMF)**. EMF is important because it is related to the maximum electrical work that can be obtained from an electrochemical cell:

$$w_{max} = qE \tag{13.1}$$

Here q is the charge that has moved (through the flow of electrons) and E is the cell potential. This equation fits the scientific concept of work because something (charge) is moved by the action of some force (the EMF).

How does the existence of EMF explain the tendency of metals to corrode? To answer this question, let's examine two types of corrosion.

Galvanic Corrosion and Uniform Corrosion

Although common instances of corrosion are often subtler than laboratory cells, metals in contact with a solution can establish the sort of oxidation half-reaction equilibrium we've just described. If the solution contains a substance that can undergo reduction, a redox reaction may result. Under some circumstances, when two metals are in contact, there is the possibility of galvanic corrosion. The process is similar to a galvanic cell in that oxidation and reduction take place in (at least slightly) different locations, but there is a connection of the metals that completes the circuit. The separation of oxidation and reduction may be quite subtle, however, as noted in **Figure 13.5**. The everyday "tin" can is mostly steel, though in some cases the steel is plated with tin. If there is a scratch in the tin and the underlying steel (mostly iron) is exposed to air and water, corrosion will occur rapidly. Why? Because the half-reaction equilibrium of the tin assists the process by which iron is oxidized, an example of galvanic corrosion. Pure steel will corrode more slowly, and zinc-plated steel will corrode even more slowly.

Experience tells us that a pile of nails left out in the yard rusts without contact with another metal (**Figure 13.6**). This implies that the second half-cell in this backyard redox reaction must involve a nonmetal. In uniform corrosion, rust can cover the surface of iron or steel. The second electrode is a second region of the iron itself, located some distance away from the first spot. Ions that can conduct current help facilitate this process, so when the chloride ions of salt are present the rate of rusting is enhanced.

Both galvanic and uniform corrosion involve the concept of electromotive force or cell potential, but to understand them more thoroughly, we need to quantify the cell potential.

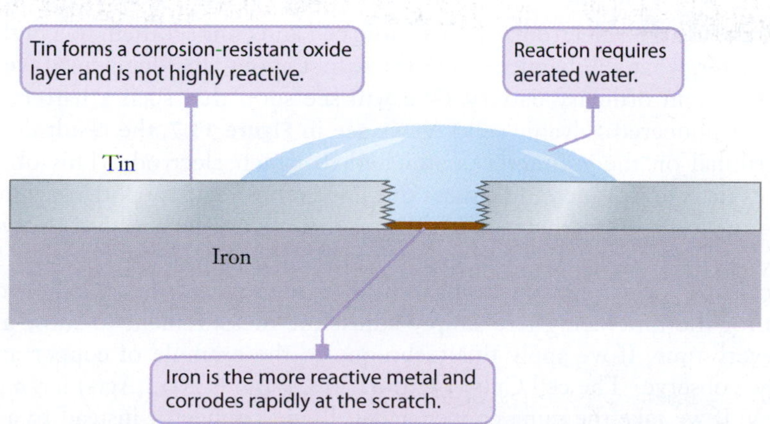
Tin forms a corrosion-resistant oxide layer and is not highly reactive.
Reaction requires aerated water.
Tin
Iron
Iron is the more reactive metal and corrodes rapidly at the scratch.

Figure 13.5 ■ A tin can is usually tin-plated steel. If the tin coating is scratched to expose the underlying steel, iron in the steel will corrode rapidly.

Figure 13.6 ■ The rusting of iron exposed to the weather—even support iron underneath concrete—is an example of uniform corrosion. Iron is oxidized, and oxygen from the air is reduced. Water is needed for ion mobility between the anodic and cathodic regions, and the presence of ionic salts speeds the reaction considerably.

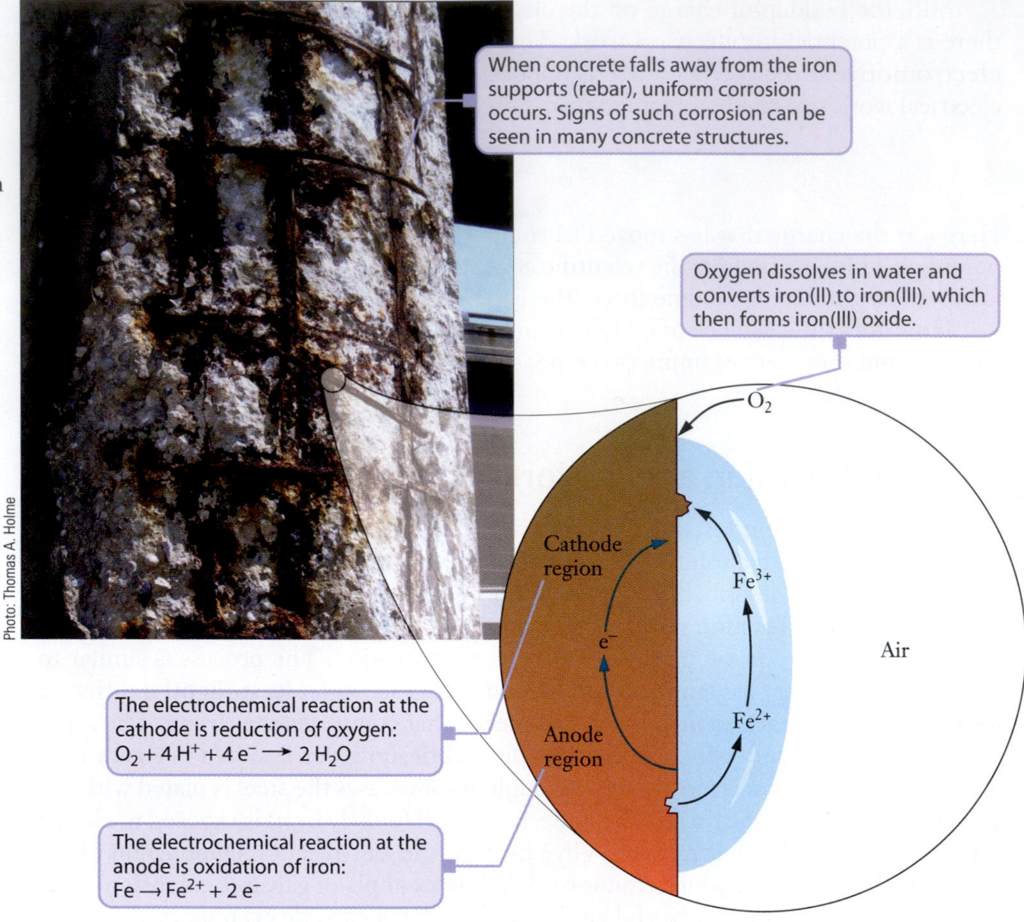

When concrete falls away from the iron supports (rebar), uniform corrosion occurs. Signs of such corrosion can be seen in many concrete structures.
Oxygen dissolves in water and converts iron(II) to iron(III), which then forms iron(III) oxide.
O_2
Cathode region
Fe^{3+}
e^-
Air
Fe^{2+}
Anode region
The electrochemical reaction at the cathode is reduction of oxygen: $O_2 + 4\,H^+ + 4\,e^- \longrightarrow 2\,H_2O$
The electrochemical reaction at the anode is oxidation of iron: $Fe \longrightarrow Fe^{2+} + 2\,e^-$
Photo: Thomas A. Holme

The positive terminal of the meter is connected with a red wire and the negative with a black wire in these photos.

09.22

-09.22
Photos: Thomas A. Holme

Figure 13.7 ■ An electric potential has a fixed polarity and voltage. Reversing the poles of a battery with respect to a voltmeter changes the sign on the measured voltage but does not actually influence the electrochemical reaction in the battery.

13-3 Cell Potentials

Our observation about the relative corrosivities of various plated steels suggests that we need to express the cell potential numerically. Just how strongly is steel driven toward corrosion when connected to tin? Can the tendency of a metal such as zinc to protect steel be of practical use? For answers to these questions, we return to the concept of cell potential.

Measuring Cell Potential

If we take the charged electrodes of a galvanic cell and connect them to a high impedance voltmeter, what will happen? Let's think first about what happens if we connect a voltmeter to an ordinary battery. (We will see soon that such a battery is really just a well-engineered galvanic cell.) As we see in **Figure 13.7**, the result depends on which terminal on the voltmeter is attached to which electrode. This observation reminds us that the voltmeter measures the size of the electrical potential and also its polarity—the locations of the negative charge (negative pole) and the positive charge (positive pole).

We can always change the terminal of the voltmeter to which we connect an electrode, so for the time being let's simply adjust the measurement to show a positive voltage every time. If we apply this technique to the example of copper and silver, what do we observe? The cell Cu(s) | Cu^{2+}(1 M) ‖ Ag^+(1 M) | Ag(s) has a potential of 0.462 V. If we take the same copper half-cell and connect it instead to a half-cell

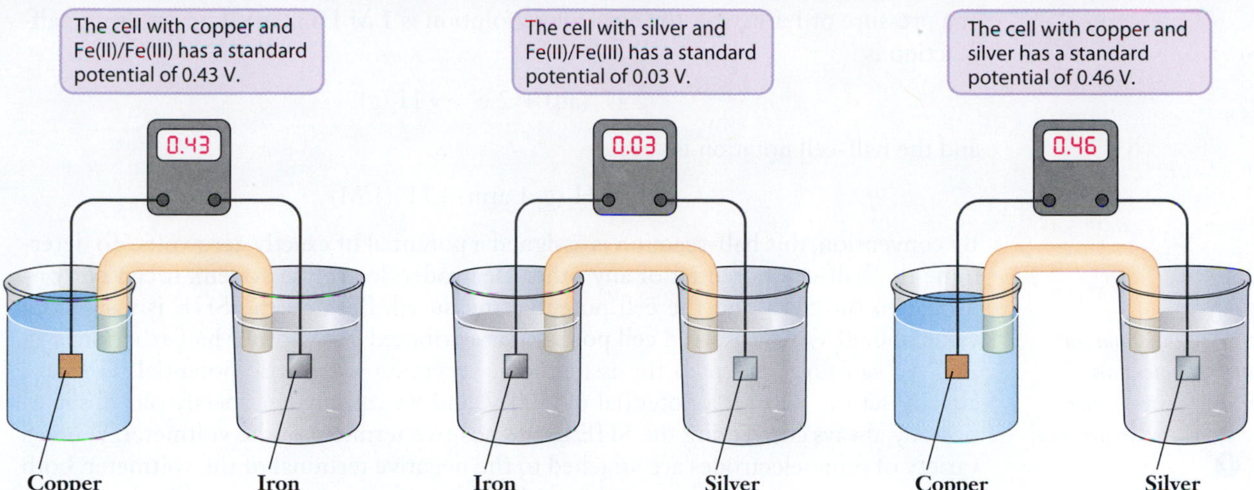

Figure 13.8 ■ Measurement of standard cell voltages for various combinations of the same half-reactions suggests that a characteristic potential can be associated with a particular half-reaction. (Voltages are shown here with fewer significant figures than in the text.)

that reduces iron(III) to iron(II), we find that the measured cell potential is 0.434 V. If this iron electrode is connected to the silver one, the resulting cell potential is 0.028 V. These numbers are clearly related: 0.462 = 0.434 + 0.028 (**Figure 13.8**). This observation is important for two reasons. First, it shows a behavior of cell potentials that is akin to that of a state function. Second, it suggests that if we choose a specific standard electrode to which we compare all other electrodes, we can devise a practical system for determining cell potential.

The choice for the standard component in cell potential measurements is the **standard hydrogen electrode (SHE)**, illustrated in **Figure 13.9.** A platinum wire or foil is the conducting source of electrons. Hydrogen gas is bubbled over the electrode

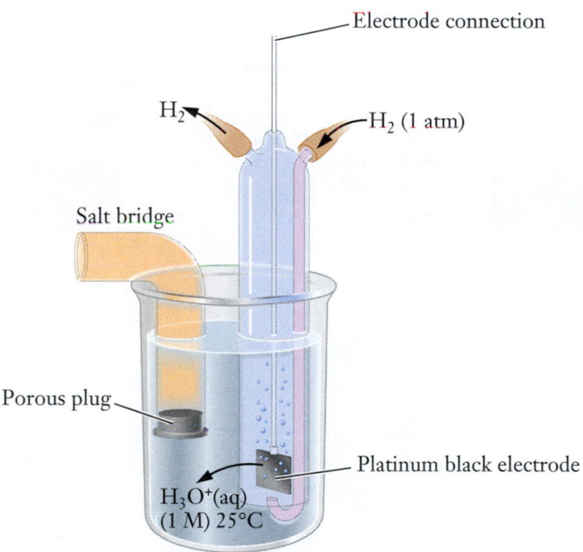

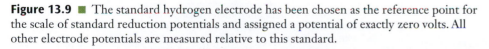

Figure 13.9 ■ The standard hydrogen electrode has been chosen as the reference point for the scale of standard reduction potentials and assigned a potential of exactly zero volts. All other electrode potentials are measured relative to this standard.

at a pressure of 1 atm, and the electrolyte solution is 1 M HCl(aq). The resulting half-reaction is

$$2\ H^+(aq) + 2\ e^- \rightarrow H_2(g)$$

and the half-cell notation is

$$Pt(s)\ |\ H_2(g, 1\ atm)\ |\ H^+(1\ M)$$

By convention, this half-reaction is assigned a potential of exactly zero volts. To determine the half-cell potential of any other electrode/electrolyte system, it can be connected to the SHE and the cell potential measured. Because the SHE is assigned a value of 0.00 V, the observed cell potential is attributed to the other half-reaction.

Although the SHE is the standard for establishing cell potentials, it is a cumbersome device. Other standards are often used in practical applications.

We can now return to the issue of the direction of electric potential. We have already set the SHE at a potential of 0.00 V, and we can further specify our observations by always connecting the SHE to the positive terminal of the voltmeter. When a variety of other electrodes are attached to the negative terminal of the voltmeter, both positive and negative values are observed. What is the connection of this observation to the chemistry that is taking place? The anode is the site of oxidation, so electrons are released there. The anode has a negative charge in a galvanic cell. In an electric circuit containing several components, the negative pole of one component is connected to the positive pole of the next component. So, if the anode is connected to the positive terminal of the voltmeter, the voltmeter shows a positive reading. By contrast electrons are consumed in the reduction reaction, at the cathode of a galvanic cell. The cathode carries a positive charge. If a cathode is connected to the positive terminal, we have violated the wiring convention of an electric circuit, and the voltmeter displays a negative voltage (**Figure 13.10**).

If we apply this reasoning to the measurement of standard potentials, where the standard hydrogen electrode is always connected to the positive terminal, the sign of the measured potential tells us the direction of the redox reaction. When the measured cell potential is positive, the SHE is the anode, and we know that H_2 is oxidized to $H^+(aq)$:

$$H_2(g) \rightarrow 2\ H^+(aq) + 2\ e^-$$

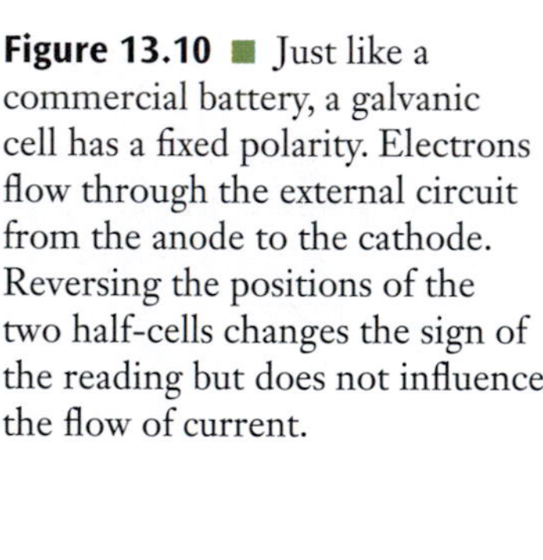

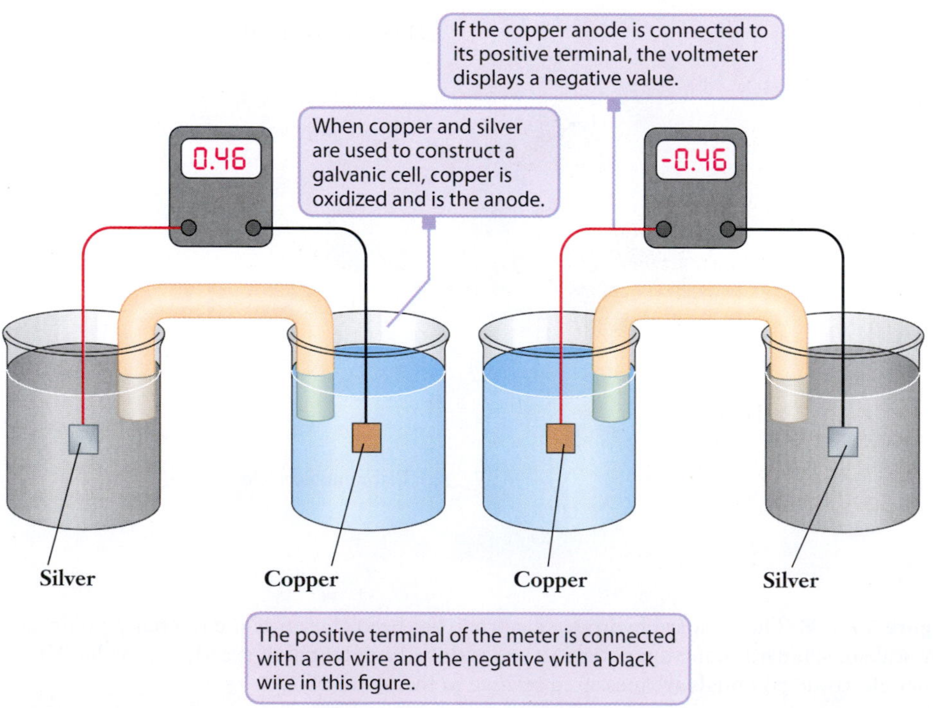

If the copper anode is connected to its positive terminal, the voltmeter displays a negative value.
When copper and silver are used to construct a galvanic cell, copper is oxidized and is the anode.
0.46
-0.46
Silver
Copper
Copper
Silver
The positive terminal of the meter is connected with a red wire and the negative with a black wire in this figure.

Figure 13.10 ■ Just like a commercial battery, a galvanic cell has a fixed polarity. Electrons flow through the external circuit from the anode to the cathode. Reversing the positions of the two half-cells changes the sign of the reading but does not influence the flow of current.

When the cell potential is negative, the SHE is the cathode, and H^+ is reduced to H_2. This reasoning allows us to use electrochemical measurements to assess the tendency of materials to undergo oxidation or reduction.

Standard Reduction Potentials

Just as some acids are stronger than others, the oxidizing and reducing strengths of various chemicals also have a range of values. To help organize the oxidation–reduction trends of species used in electrochemistry, all half-cell potentials are tabulated as reductions. Thus a table of **standard reduction potentials** shows the potential of any half-reaction when connected to an SHE. Note that all electrodes must be constructed to meet the conditions of standard states, i.e., all materials are in thermodynamic standard states, solutions have a concentration of 1 M, and gases have a pressure of 1 atm. A partial list of standard reduction potentials is given in **Table 13.1**, and a more extensive one appears in Appendix I. Several features of this table bear mentioning. First, although the half-reactions are all listed as reductions, one half-reaction in any electrochemical cell must be an oxidation. Second, some half-reactions have positive potentials, whereas others have negative potentials. Recall that these measurements are taken with the SHE connected to the positive terminal. If the voltage is positive, the SHE is the anode—the oxidation site. So, a positive voltage in the standard reduction potential means that the half-reaction proceeds as written (reduction occurs). If the value is negative, the half-reaction proceeds as an oxidation when connected to the SHE because the SHE is serving as the cathode. (See Figure 13.10.)

These observations have implications for the tendency for oxidation and reduction of the chemicals involved in a half-reaction. A large, positive value for the standard reduction potential implies that the substance is reduced readily and is therefore a good oxidizing agent. Conversely, if a reduction potential has a relatively large negative value, the reaction proceeds readily in reverse—the substance is oxidized. Thus negative standard reduction potentials identify good reducing agents. The product of the half-reaction as written in the table is the reducing agent because the reverse reaction is the oxidation. In either case, the magnitude of the value is important in deciding what is oxidized and what is reduced. *In any galvanic cell, the half-reaction with the more positive reduction potential will be the cathode.* The relationship between standard reduction potential and the direction of the redox reaction can be pictured by thinking of the reduction potentials along an axis, as shown in **Figure 13.11.**

Such a sequence of reactivities based on standard reduction potentials is sometimes called the electrochemical series.

Table ■ 13.1

Standard reduction potentials for several of the half-reactions involved in the cells discussed in the text. A more extensive table of potentials appears in Appendix I.

Half-Reaction	Standard Reduction Potential (V)
$Zn^{2+} + 2\ e^- \rightarrow Zn$	−0.763
$Fe^{2+} + 2\ e^- \rightarrow Fe$	−0.44
$2\ H^+ + 2\ e^- \rightarrow H_2$	0.000
$Cu^{2+} + 2\ e^- \rightarrow Cu$	+0.337
$Fe^{3+} + e^- \rightarrow Fe^{2+}$	+0.771
$Ag^+ + e^- \rightarrow Ag$	+0.7994

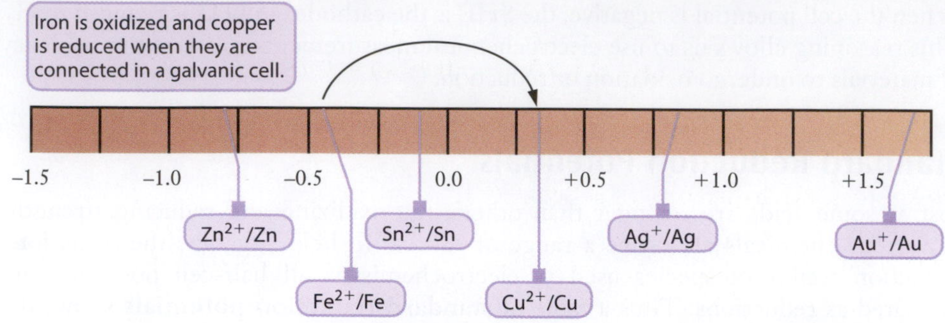

Figure 13.11 ■ If we think of standard reduction potentials arranged horizontally, as shown here, we can easily identify the anode and cathode in a galvanic cell. For any pair of electrodes, the one appearing farther to the left on such a scale will be the anode, and the one farther to the right will be the cathode in a galvanic cell.

With this information, we can determine the standard cell potential for any pair of half-reactions by using the equation:

$$E°_{cell} = E°_{red} - E°_{ox} \tag{13.2}$$

$E°_{red}$ is the standard reduction potential of the cathode, and $E°_{ox}$ is the standard reduction potential of the anode. The minus sign is needed because oxidation is the opposite of the tabulated reduction. Example Problem 13.1 examines this concept in the context of materials that are important in galvanic corrosion.

EXAMPLE PROBLEM 13.1

Copper and iron (generally in the form of steel) are two of the many metals used in designing machines. (a) Using standard reduction potentials, identify the anode and the cathode and determine the cell potential for a galvanic cell composed of copper and iron. Assume standard conditions. (b) We can also construct a galvanic cell using copper and silver. Confirm that the potential of the following galvanic cell is 0.462 V:

$$Cu(s) \mid Cu^{2+}(1\ M) \parallel Ag^{+}(1\ M) \mid Ag(s)$$

Strategy We must interpret the nature of an electrochemical system based on the information available in a table of standard reduction potentials. With two half-reactions there are only two possible outcomes—and one outcome yields a negative value for the cell potential. Because we know that a galvanic cell cannot have a negative $E°$ value, we must determine the combination of half-reactions that provides a positive value for $E°$.

Solution

(a) Using Table 13.1, we find the following two half-reactions:

$$Fe^{2+}(aq) + 2\ e^- \rightarrow Fe(s) \qquad E° = -0.44\ V$$

$$Cu^{2+}(aq) + 2\ e^- \rightarrow Cu(s) \qquad E° = 0.337\ V$$

Iron must be oxidized for a combination of these two half-reactions to yield a positive cell potential:

$$Fe(s) + Cu^{2+}(aq) \rightarrow Fe^{2+}(aq) + Cu(s) \qquad E°_{cell} = ?$$

Using Equation 13.2,

$$E^\circ_{cell} = 0.337\ \text{V} - (-0.44\ \text{V}) = 0.78\ \text{V}$$

Copper is the cathode in this cell because copper is reduced; iron is the anode because it is oxidized.

(b) Find the two half-reactions in Table 13.1:

$$Ag^+(aq) + e^- \rightarrow Ag(s) \qquad E^\circ = 0.7994\ \text{V}$$

$$Cu^{2+}(aq) + 2\ e^- \rightarrow Cu(s) \qquad E^\circ = 0.337\ \text{V}$$

From the values of the standard reduction potentials, we can see that copper is oxidized, and silver is reduced. We then use equation 13.2 to find the cell potential:

$$Cu(s) + 2\ Ag^+(aq) \rightarrow Cu^{2+}(aq) + 2\ Ag(s) \qquad E^\circ_{cell} = 0.7994\ \text{V} - 0.337\ \text{V} = 0.462\ \text{V}$$

Analyze Your Answer When calculating E° for a galvanic cell, the simplest check on our answer is to make sure it is positive. If we get a positive value, then as long as we have started from the correct reduction potentials, we can be confident that we have done the arithmetic correctly.

Discussion The calculation of cell potential might seem more complicated when both reduction potentials are negative. If we keep in mind that the overall cell potential must be positive for a galvanic cell, then only one of the two choices will be correct, no matter what values are given.

Check Your Understanding Suppose you wanted to make a 1.00-V cell using copper as one of the electrode materials. Of the electrodes listed in Table 13.1, which would provide a cell potential closest to 1.00 V when coupled with copper? What is the anode and what is the cathode in your cell?

Example Problem 13.1a in which copper and iron are used to construct an electrochemical cell points to the origin of galvanic corrosion. Thus far we have implied that the separation of the electrodes is something that is carefully constructed and achieved through a salt bridge. Deciding which electrode is the anode and which is the cathode and identifying the cell reactions are relatively simple tasks. But in real-world applications, the separation may be much more subtle.

As a starting point, let's consider the galvanic corrosion that can occur at a joint between copper and iron materials. The resulting degradation of the iron makes it obvious that iron acts as the anode. This means that the copper must be the cathode. But what is being reduced? It turns out that two reactions are involved at the cathode, and both involve the reduction of water. Which one will occur depends on the amount of oxygen present in the water and the cell potentials. If copper is not reacting, then is it even important that the material is copper? If it is water that is actually reacting, what would happen if some other metal such as zinc were present? Here is where the standard reduction potentials become useful.

Let's add the reduction of zinc to the list of half-reactions from Example Problem 13.1:

$$Fe^{2+}(aq) + 2\ e^- \rightarrow Fe(s) \qquad E^\circ = -0.44\ \text{V}$$

$$Cu^{2+}(aq) + 2\ e^- \rightarrow Cu(s) \qquad E^\circ = \ \ 0.34\ \text{V}$$

$$Zn^{2+}(aq) + 2\ e^- \rightarrow Zn(s) \qquad E^\circ = -0.763\ \text{V}$$

If iron is paired with zinc in an electrochemical cell, which electrode will be the anode? The one that is the more negative is always the anode, and that's zinc in this case. So if

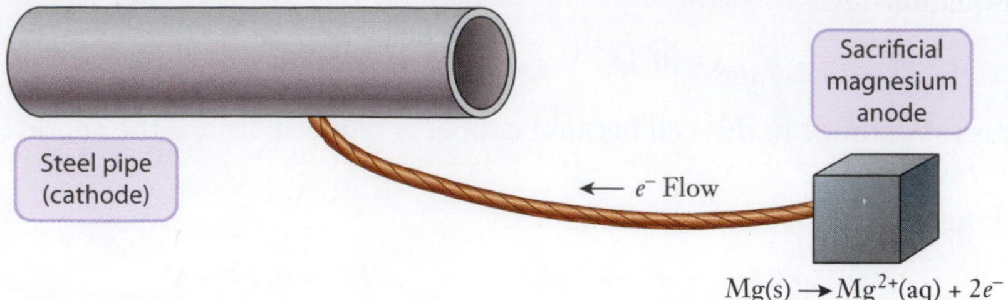

Figure 13.12 ■ Sacrificial anodes are one effective method of corrosion prevention. An unprotected iron or steel pipe buried in the ground would be at high risk for corrosion. By connecting the buried pipe to a metal such as magnesium, which is more easily oxidized, a galvanic cell is created with the pipe as the cathode. In the case of a buried pipe, the soil itself serves as the electrolyte. The anode is called "sacrificial" because it will be eaten away over time by oxidation. But replacing the anode—which may be nothing more than a metal block or stake—is much easier than replacing the buried pipe.

zinc and iron are both present, iron does not corrode because zinc serves as the anode. Copper, though, is a better oxidizing agent than iron, and so iron in contact with copper will corrode via galvanic corrosion. The fact that zinc can prevent corrosion provides a common method of protecting iron. **Galvanized steel** includes a thin layer of zinc as a coating, making it notably less prone to rust.

The electroplating industry is built upon the idea that metal coatings on objects can provide useful properties, including corrosion resistance.

Cathodic Protection

The fact that some materials are more easily oxidized than iron provides a way to construct galvanic corrosion conditions intentionally to protect the iron. If we choose a metal, such as magnesium, whose reduction potential is more negative than that of iron, the magnesium is oxidized, and iron is reduced:

$$Fe^{2+} + 2\,e^- \rightarrow Fe \qquad E° = -0.41\text{ V}$$

$$Mg^{2+} + 2\,e^- \rightarrow Mg \qquad E° = -2.39\text{ V}$$

If we wish to prevent iron from corroding, we can use a piece of magnesium as a **sacrificial anode**. Connecting the piece of magnesium to the iron forces the iron to be the cathode, thereby preventing the oxidation of iron. This process is called **cathodic protection**. By making the iron a cathode, we ensure that it cannot be an anode, and so it will not corrode. To be effective, the sacrificial anode must be replaced periodically, but this method finds many uses, including the protection of iron or steel pipelines (**Figure 13.12**).

Nonstandard Conditions

One thought you may have about corrosion is that we cannot expect standard conditions in real world applications. What must we do to account for the differences that arise when standard conditions are not present? This is a very important question with clear ties to thermodynamics. The equation that describes cell potentials under nonstandard conditions is called the **Nernst equation**:

$$E = E° - \frac{RT}{nF}\ln Q \tag{13.3}$$

Here Q is the reaction quotient we worked with in Chapter 12: a ratio of concentrations of reactants and products, each raised to the power of its stoichiometric coefficients. Thus for the generic reaction

$$a\,\text{A} + b\,B \rightleftarrows c\,\text{C} + d\,\text{D}$$

$$Q = \frac{[\text{C}]^c[\text{D}]^d}{[\text{A}]^a[\text{B}]^b} \tag{13.4}$$

In Equation 13.3, n is the number of electrons transferred in the redox reaction. The **Faraday constant (F)**, named after Michael Faraday, has a value of 96,485 J V^{-1} mol^{-1} or 96,485 C mol^{-1}. (Remember that 1 J = 1 C V.) Equation 13.3 bears a resemblance to Equation 12.7, which we used to derive the relationship between $\Delta G°$ and the equilibrium constant. We can apply the Nernst equation to estimate the potential of the electrochemical system in the corrosion of steel at more realistic concentrations. Example Problem 13.2 explores the use of the Nernst equation in an electrochemical cell that models corrosion.

Faraday is responsible for much of the terminology of electrochemistry. Taking the advice of a linguist friend (William Whewell), he abandoned ideas such as eastode and westode and settled on cathode and anode.

EXAMPLE PROBLEM 13.2

Suppose that you work for a company that designs the drive mechanisms for large ships. The materials in this mechanism will obviously come into contact with environments that enhance corrosion. To estimate the difficulties that corrosion might cause, you decide to build a model electrochemical cell using electrolyte concentrations that might be present in your system when it is in service. Assume that you have a cell that has an iron(II) concentration of 0.015 M and an H^+ concentration of 1.0×10^{-3} M. The cell temperature is 38°C, and the pressure of hydrogen gas is maintained at 0.04 atm. What would the cell potential be under these conditions?

Strategy This problem defines nonstandard conditions that must be addressed using the Nernst equation. Virtually anytime you are given concentrations of electrolytes present in a cell (other than 1 M), you need this equation. This problem also presents the challenge of identifying the reactions involved. Iron will be the anode, but we will need to scan the table of standard reduction potentials to identify a possible cathode reaction. The most likely suspect is the reduction of H^+ to H_2. Once we know both half-reactions, we can calculate the standard cell potential and fill in the appropriate values in the Nernst equation.

Solution

Anode: $Fe^{2+}(aq) + 2\,e^- \rightarrow Fe(s) \qquad E° = -0.44\text{ V}$

Cathode: $2\,H^+(aq) + 2\,e^- \rightarrow H_2(g) \qquad E° = 0.00\text{ V}$

These reactions tell us two things: First, the standard cell potential will be

$$E° = 0.00\text{ V} - (-0.44\text{ V}) = 0.44\text{ V}$$

Second, there are two electrons transferred in the overall redox reaction:

$$Fe(s) + 2\,H^+(aq) \rightarrow H_2(g) + Fe^{2+}(aq)$$

These facts, plus the values given in the problem and those of the constants allow us to use the Nernst equation to find the cell potential:

$$E = 0.44\text{ V} - \frac{(8.314\text{ J mol}^{-1}\text{K}^{-1} \times 311\text{ K})}{(2 \times 96{,}485\text{ J V}^{-1}\text{mol}^{-1})} \ln\left[\frac{(0.015)(0.04)}{(0.0010)^2}\right] = 0.35\text{ V}$$

Discussion The correction made in the cell potential is rather small, even though we are far from standard conditions. But what happens as the reaction proceeds? The concentration of H^+ decreases, and that of Fe^{2+} increases. So the potential is not a constant—ln Q becomes larger (a smaller number in the denominator, larger in the numerator), until eventually the potential falls to zero (at least in principle). So we can see that it can be vital to account for nonstandard conditions in many applications.

Check Your Understanding Suppose that you want to construct a galvanic cell from nickel and cadmium for a chemistry project. You take all the precautions necessary to deal with a toxic heavy metal like cadmium, and you decide to use a 0.01 M solution of cadmium nitrate as one of the electrolytes. If you need a cell potential of 0.17 V for this system, what concentration of $NiCl_2$ should you use? (For a follow-up web research project, compare this nickel-cadmium system with the one that is used in a commercial Ni-Cd battery.)

13-4 Cell Potentials and Equilibrium

Many metals are found in nature as oxide ores. Two common iron ores are hematite, Fe_2O_3, and magnetite, Fe_3O_4. Winning iron metal from its oxides in the refining process requires a great deal of energy. In the process of converting an oxide to the pure element, the entropy of the system decreases. The result is that refining is *not* a spontaneous process; the free energy change in the refining process is positive.

Corrosion can be thought of as the reverse of the process of refining a metal. When a metal corrodes, it is returning to its natural (oxidized) state, and it does so spontaneously. If we place an iron nail in a solution of aerated salt water, we know we will observe the formation of rust. Based on our experience from Chapter 10, the change in free energy for this process must be negative. Can we determine the free energy change in corrosion or in the operation of a galvanic cell?

Cell Potentials and Free Energy

Recall that free energy is related to the maximum possible amount of work that can be done by the system. In the case of a galvanic cell, the work done is electrical work, so we can state the relationship between free energy and the cell potential as

$$\Delta G^\circ = -nFE^\circ \qquad (13.5)$$

Here we have replaced the charge, q, from Equation 13.1 with nF because often we can determine moles more easily than charge in a chemistry problem. The minus sign is required due to our sign conventions. A galvanic cell always has a positive cell potential, so all three terms on the right are positive. The cell reaction is spontaneous, so ΔG° must be negative, as required for any spontaneous process. This equation allows us to calculate the standard free energy change for an electrochemical reaction, provided we can look up the relevant standard reduction potentials. Example Problem 13.3 shows how this works.

EXAMPLE PROBLEM 13.3

Suppose that we wish to study the possible galvanic corrosion between zinc and chromium, so we set up the following cell:

$$Cr(s) \mid Cr^{2+}(aq) \parallel Zn^{2+}(aq) \mid Zn(s)$$

What is the chemical reaction that takes place, and what is the standard free energy change for that reaction?

Strategy To calculate the free energy change, we must know two things: the cell potential and the number of electrons transferred in the reaction. Then we can simply use these values in Equation 13.5 to obtain the free energy change.

Solution First, we need the balanced chemical equation, which in this case can be written immediately because two electrons are transferred in each half-reaction ($n = 2$):

$$Zn^{2+}(aq) + Cr(s) \rightarrow Cr^{2+}(aq) + Zn(s)$$

Now if we look up the standard reduction potentials, we find

$$Zn^{2+}(aq) + 2\,e^- \rightarrow Zn(s) \qquad E° = -0.763\ V$$

$$Cr^{2+}(aq) + 2\,e^- \rightarrow Cr(s) \qquad E° = -0.910\ V$$

According to Equation 13.2, the cell potential is

$$E°_{cell} = -0.763\ V - (-0.910\ V) = 0.147\ V$$

Inserting this in Equation 13.5,

$$\Delta G° = -nFE° = -2\ mol \times 96{,}485\ J\ V^{-1}mol^{-1} \times 0.147\ V$$
$$= -2.84 \times 10^4\ J = -28.4\ kJ$$

Discussion The most common error students make in this type of problem is forgetting to include n, the number of electrons transferred. Remember that even if you must multiply a chemical equation that represents the reaction given in the standard reduction potential by some value to match the number of electrons between oxidation and reduction, you do *not* multiply the standard reduction potential. Reduction potentials are not like thermochemical quantities in this regard.

Check Your Understanding If this reaction could be reversed, what would be the value of the free energy change? What does this value tell you about the spontaneity of the reverse process?

How can we understand the spontaneity of corrosion using this type of calculation? What we must do is identify the half-reactions involved and imagine them as a galvanic cell. Then the potential of this imaginary cell tells us the standard free energy change of the corrosion reaction. Let's return to iron rusting in aerated water, which is surely the most common example of corrosion. Iron is initially oxidized to its 2+ state, and oxygen dissolved in the water undergoes the accompanying reduction. So we have the following half-reactions:

The process of rusting is more complicated than these half-reactions imply, so we should consider our result as a rough approximation of the free energy change.

$$2\,H_2O + O_2 + 4\,e^- \rightarrow 4\,OH^- \qquad E° = 0.40\ V$$

$$Fe^{2+} + 2\,e^- \rightarrow Fe \qquad E° = -0.44\ V$$

The cell potential of a galvanic cell comprising these two electrode reactions would be 0.84 V, so the free energy change $\Delta G° = -320$ kJ. This large negative value tells us that equilibrium strongly favors the oxidation of iron metal.

Equilibrium Constants

We also learned in Chapter 12 that $\Delta G°$ is related to the equilibrium constant, K, by Equation 12.7: $\Delta G° = -RT \ln K$. By substituting the right-hand side of this equation for $\Delta G°$ as given in Equation 13.5, we can obtain a relationship between cell potential and the equilibrium constant:

$$E° = \frac{RT}{nF} \ln K \qquad (13.6)$$

Here we have divided both sides of the equation by nF to isolate the cell potential in the equation. This equation also resembles the Nernst equation (Equation 13.4), and it is easy to see the connection. At equilibrium, the free energy change is zero and the reaction quotient, Q, is equal to the equilibrium constant, K.

We can gain some important insight into electrochemical reactions in general, and corrosion in particular, by replacing the natural logarithm with the common (base 10) logarithm to obtain a slightly different form of Equation 13.6:

$$E° = \frac{2.303RT}{nF} \log K \qquad (13.7)$$

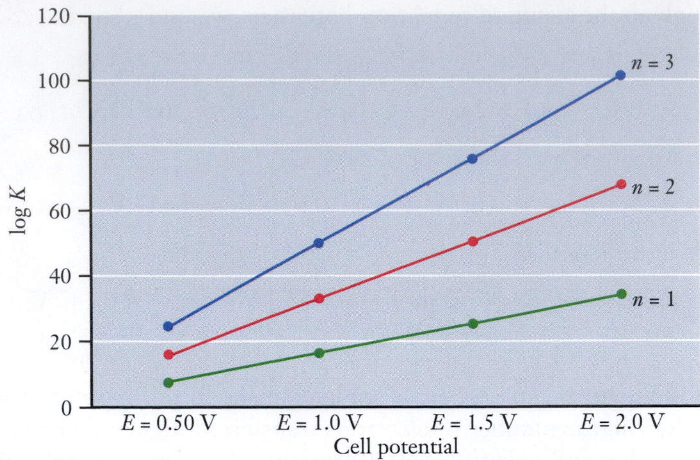

Figure 13.13 ■ The variation of equilibrium constant with cell potential is shown. The different lines correspond to reactions involving the transfer of one, two, or three electrons, as indicated.

The factor of 2.303 is the result of changing from the natural log to the common log function: ln (10) = 2.303. This change to the base 10 log function allows us to think in orders of magnitude. Both R and F are constants, so we can simplify Equation 13.7 by absorbing their values into the numerical term. Because we are most often interested in experiments carried out at the standard temperature of 25°C (298 K), we can insert that value as well. (Of course, the resulting version of the equation will be correct *only* at this temperature.)

This form of Equation 13.7 is valid only at 25°C.

$$E° = \frac{0.0592 \text{ V}}{n} \log K$$

If we choose an electrochemical reaction that transfers only one electron at a potential of 1.0 V and determine the value of K, we obtain

$$K = 10^{\frac{(1)(1.0 \text{ V})}{0.0592 \text{ V}}} = 7.8 \times 10^{16}$$

Thus a modest cell potential of only 1 V results in a very large value for K, even when only one electron is transferred. Recall that large values of K indicate that equilibrium strongly favors products. In corrosion, it is quite common to have reactions transferring two, three, or four electrons, which will tend to make the value of K even larger. This is illustrated in **Figure 13.13.** The vertical axis on this graph is the log of the equilibrium constant, (log K), so an increase of 20, for example, is actually an increase of 10^{20} in K. Notice how either a greater cell potential or a larger number of electrons transferred quickly pushes the equilibrium constant to larger values, indicating reactions that tend to proceed strongly toward products. From an equilibrium perspective, corrosion strongly favors the formation of metal oxide products.

13-5 Batteries

The tendency of many electrochemical reactions to move toward complete formation of products presents a significant challenge in the case of corrosion, but it also provides opportunities for using these reactions constructively. The most familiar example of this is the **battery**, a cell or series of cells that generates an electric current. Batteries are composed of many different materials and find many uses, but they all share one common property—they provide a means by which we harness the electrical work of a galvanic cell and use it productively.

Alessandro Volta, for whom the volt is named, was the first person to realize that the differences in chemical reactivity of metals could be used to generate electricity. His work was a precursor to the development of the battery.

Batteries are sometimes classified as either "primary" or "secondary." A **primary cell** such as the typical alkaline battery becomes useless once the underlying chemical reaction has run its course. The lifetime of the battery is determined by the amounts of reactants present, so a relatively large D-cell would last longer than a smaller AA-cell in the same application. The battery "dies" when the reactants have been converted into products, bringing the reaction to a halt. In practice, the voltage output of a battery usually begins to decrease near the end of its lifetime, and the cell will generally fail before the reactants are completely consumed.

A **secondary cell** is one that can be recharged, allowing for a much longer life cycle. To make a rechargeable battery, we must be able to reverse the redox reaction, converting the products back into reactants. Because we know that the cell reaction must be exothermic to supply energy, we can also see that the reverse reaction must be endothermic. So some external energy source will be needed to "push" the reaction back toward the reactants. This is the role of a battery charger, which typically uses electrical energy from some other source to drive the redox reaction in the energetically "uphill" direction. The question of whether or not a given reaction can be made reversible in this way determines whether a particular type of battery can be recharged easily.

In order to develop these ideas a little further, let's look more closely at the chemistry and engineering involved in some common types of batteries.

Primary Cells

The most prevalent type of primary battery in use today is the **alkaline battery**. Flashlights almost always rely on alkaline batteries, and many toys, radios, and other devices also use them. The widespread use of these batteries leads to stiff competition among manufacturers, as evidenced by the number of television commercials appearing for specific brands. Still, aside from some fairly minor engineering details, these alkaline batteries all rely on the same chemistry.

The anode in an alkaline battery is a zinc electrode, and the oxidation half-reaction can be written

$$Zn(s) + 2\ OH^-(aq) \rightarrow Zn(OH)_2(s) + 2\ e^-$$

The cathode is derived from manganese(IV) oxide, and the half-reaction is

$$2\ MnO_2(s) + H_2O(\ell) + 2\ e^- \rightarrow Mn_2O_3(s) + 2\ OH^-(aq)$$

We can combine these two half-reactions to yield a net equation that represents the chemistry of an alkaline dry cell battery:

$$Zn(s) + 2\ MnO_2(s) + H_2O(\ell) \rightarrow Zn(OH)_2(s) + Mn_2O_3(s)$$

The essential design features of an alkaline battery are shown in **Figure 13.14**. The electrolyte used is KOH, but rather than dissolving the electrolyte in liquid water, it is in the form of a paste or a gel—hence the term *dry cell*. The MnO_2 for the cathode is mixed with graphite to increase conductivity. The anode is a paste containing powdered zinc. (Powdering the zinc increases the surface area and improves performance.) The battery case is also important in the design. Electrons generated by oxidation are collected by a piece of tin-coated brass connected to the bottom of the battery case. The remainder of the battery case is in contact with the cathode, but a protrusion on top makes it easier for a consumer to identify the positive terminal. All of these design elements are used to facilitate the underlying chemistry that powers the battery.

Other primary cells are also used as batteries for certain applications. In some instances, the battery must be quite small. For a medical device, such as a heart pacemaker,

Figure 13.14 ■ Construction of a typical alkaline battery is illustrated.

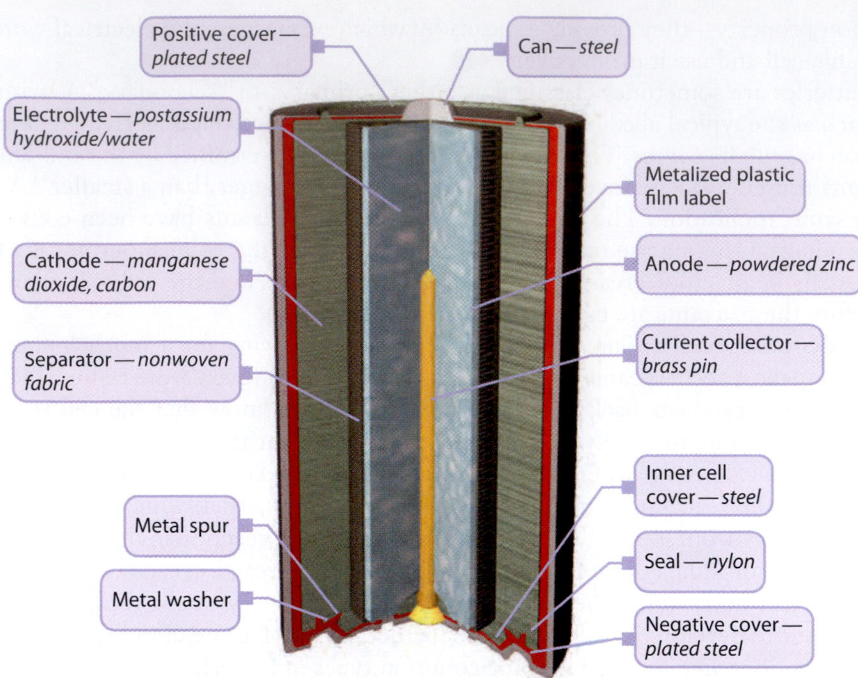

a battery should not only be small but also be long lasting. **Lithium batteries** have come to fill this role. In the most common type of lithium battery, shown in **Figure 13.15**, lithium metal is the anode:

These lithium batteries are distinct from rechargeable lithium-ion batteries, which we will discuss in Section 13-8.

$$Li(s) \rightarrow Li^{+} + e^{-}$$

As in the alkaline battery, the cathode uses manganese(IV) oxide. But in this case the MnO_2 reacts with lithium ions:

$$MnO_2(s) + Li^{+} + e^{-} \rightarrow LiMnO_2$$

In this half-reaction, manganese(IV) is reduced to manganese(III). The overall cell reaction is therefore

$$Li(s) + MnO_2(s) \rightarrow LiMnO_2(s)$$

The electrolyte is a lithium salt dissolved in an organic solvent, which can diffuse through the polypropylene separator. The entire battery is enclosed in a button-style case. The important feature of this battery is that it provides a stable current and electrical potential for long periods, despite its small size.

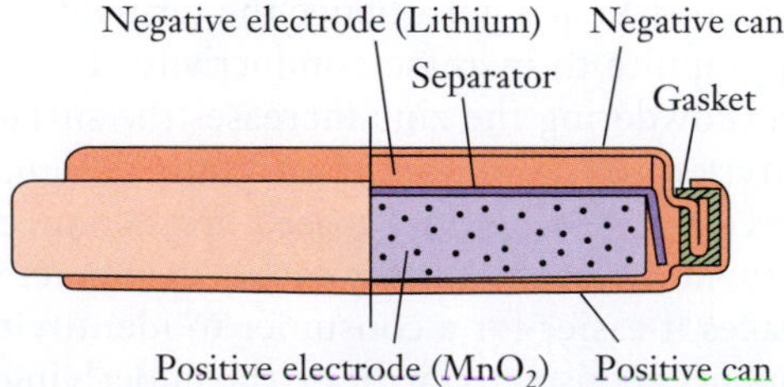

Figure 13.15 ■ The lithium battery shown here has a voltage output that is extremely stable over long times. These batteries are commonly used in devices where frequent battery changes would be a nuisance or hazard.

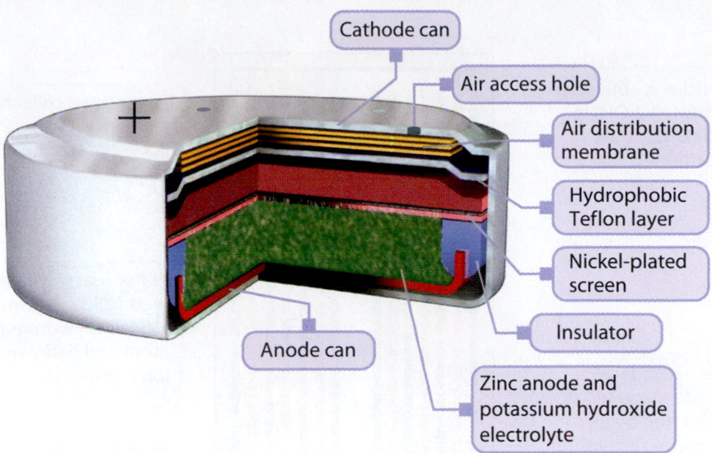

Figure 13.16 ■ In a zinc-air battery, one of the reactants is oxygen from the surrounding air. As a result, these batteries can offer a very attractive energy density. They can also be kept on hand in sealed bags and charged when needed by opening the bag to expose the battery to air.

Another interesting cell is the zinc-air battery, depicted in **Figure 13.16.** These batteries are sold as single-use, long lasting products for emergency use in cellular phones. The anode reaction involves the oxidation of zinc:

$$Zn(s) + 2\ OH^{-}(aq) \rightarrow Zn(OH)_2(s) + 2\ e^{-}$$

At the cathode, oxygen from the air is reduced:

$$\frac{1}{2} O_2\ (g) + H_2O(\ell) + 2\ e^{-} \rightarrow 2\ OH^{-}$$

With a significant portion of the working material obtained from the surrounding air, the prospect for lightweight zinc-air alkaline batteries is intriguing. There are, however, obstacles to the widespread use of this technology because environmental factors, such as humidity, affect its performance.

Secondary Cells

Rechargeable batteries are increasingly common in consumer products.

Nickel-metal-hydride batteries (**Figure 13.17**) are available in many standard sizes and can be used to replace alkaline batteries in most applications. Much larger versions serve as the main batteries in hybrid cars like the Toyota Prius. In this battery, the cathode reaction is complex, but is best represented by the equation

$$NiO(OH)(s) + H_2O(\ell) + e^{-} \rightarrow Ni(OH)_2(s) + OH^{-}(aq)$$

The anode reaction is

$$MH(s) + OH^{-}(aq) \rightarrow M + H_2O(\ell) + e^{-}$$

where M stands for some metal or metal alloy. Some of the alloys used in commercially available nickel-metal-hydride batteries contain as many as seven different metals. The reason for the strong performance of these alloys is not yet well understood and is an active research area.

Although these batteries are important for their utility in personal electronics, they are not the most widely selling rechargeable batteries. That distinction still belongs to

Figure 13.17 ■ Nickel-metal-hydride batteries have become popular as rechargeable cells and can be used in place of alkaline batteries in most applications.

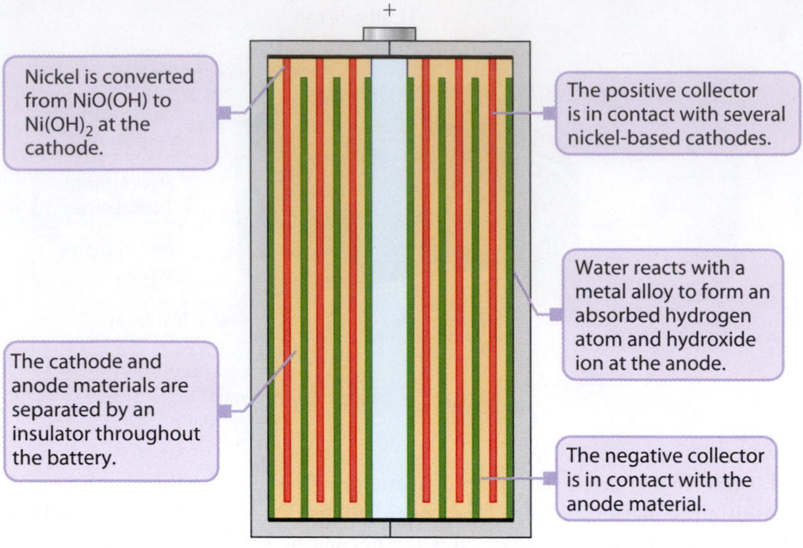

+
Nickel is converted from NiO(OH) to Ni(OH)$_2$ at the cathode.
The positive collector is in contact with several nickel-based cathodes.
Water reacts with a metal alloy to form an absorbed hydrogen atom and hydroxide ion at the anode.
The cathode and anode materials are separated by an insulator throughout the battery.
The negative collector is in contact with the anode material.
–

Lead-acid storage batteries account for 88% of all lead consumed annually in the United States.

the lead storage battery in automobiles, which have been using the same technology for roughly 100 years. The anode reaction in this battery is

$$Pb(s) + HSO_4^-(aq) \rightarrow PbSO_4(s) + H^+(aq) + 2\ e^-$$

and the cathode reaction is

$$PbO_2(s) + 3\ H^+(aq) + HSO_4^-(aq) + 2\ e^- \rightarrow PbSO_4(s) + 2\ H_2O(\ell)$$

Although these equations do not show it explicitly, the electrolyte for this battery is sulfuric acid. We can infer this from the presence of hydrogen sulfate ions (HSO_4^-) in the equations; sulfate ion is converted to HSO_4^- under the highly acidic conditions of the battery. Because of the importance of sulfuric acid in these batteries, they are sometimes referred to as **lead-acid storage batteries**.

The cell designated by the reactions given above generates a potential of nearly 2.0 V. Thus to obtain a standard 12-V automobile battery, we need six cells wired in series, so that their voltages are additive. **Figure 13.18** shows a diagram of a lead storage battery. These batteries are rechargeable because the lead sulfate product tends to

Figure 13.18 ■ The lead-acid storage battery consists of Pb anodes alternating with PbO_2 cathodes, all immersed in sulfuric acid. Although the underlying chemistry has remained the same for many years, engineering advances have drastically improved the lifetime and reliability of the modern lead-acid battery.

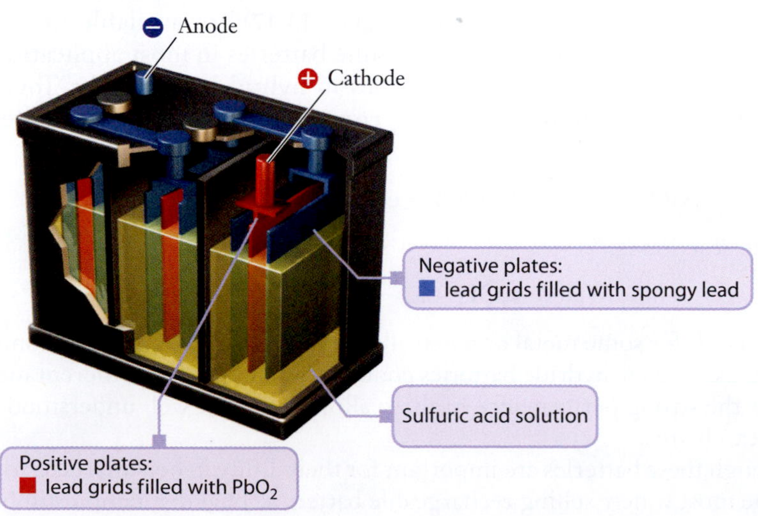

Anode
Cathode
Negative plates:
lead grids filled with spongy lead
Sulfuric acid solution
Positive plates:
lead grids filled with PbO$_2$

Table ■ 13.2

Comparison of some characteristics of common primary and rechargeable battery types

Primary Battery Comparison			
Attribute	**Zinc-Air**	**Alkaline**	**Lithium**
Energy density	High	Medium	High
Energy storage	High	Medium	Medium
Cost	Low	Low	High
Safety	High	High	Medium
Environment	High	High	Medium
Rechargeable Battery Comparison			
Attribute	**Lead-Acid**	**Nickel-Metal-Hydride**	**Lithium-Ion**
Energy density	Low	Medium	High
Energy storage	Medium	Medium	Medium
Cycle life	High	High	High
Cost	Low	Medium	High
Safety	High	High	Medium
Environment	Low	Medium	Medium

adhere to the electrode surfaces. Passing electric current (obtained from the alternator of the automobile) to drive the redox reactions in the nonspontaneous direction recharges the battery. Driving the reactions in a battery in the reverse direction to recharge it sometimes requires a larger voltage than the discharge voltage to ensure that all chemical species are returned to their original states. Over time, as the battery experiences mechanical shocks from bumping and jarring during the normal use of a car, the lead sulfate falls away from the electrodes. Eventually the battery can no longer be recharged and must be replaced.

Many new types of batteries have been introduced in recent years, as the power demands of portable electronic devices such as cell phones, music players, and laptop computers have soared. What factors influence the choice of battery types for a particular application? One of the most important considerations is energy density, just as we discussed earlier for fuels. **Table 13.2** summarizes typical characteristics and uses of several types of batteries.

Battery manufacturers usually report energy densities in units of W h/kg. Because a watt is 1 J/s, and an hour is 3600 s, 1 W h = 3600 J. Cost also plays a key role in the selection of a battery type. Although some people might be willing to pay hundreds of dollars for a good battery for an expensive laptop computer, you are not likely to want that same battery for your iPod® shuffle. Battery lifetime is also a factor. Digital cameras tend to require more power than traditional film cameras, for example. So consumers are much more willing to pay more for rechargeable batteries for digital cameras because they anticipate cost savings over time.

Fuel Cells

A **fuel cell** is a voltaic cell in which the reactants can be supplied continuously and the products of the cell reaction are continuously removed. Like a battery, it uses a chemical reaction to produce electrical energy. But unlike a battery, it can be refueled on an ongoing basis. The most common fuel cells are based on the reaction of hydrogen and oxygen to produce water. Hydrogen gas flows into the anode compartment, and oxygen gas flows into the cathode compartment. The electrodes in each compartment are usually porous carbon, which has been impregnated with platinum catalysts. Oxygen is reduced at the cathode:

$$O_2 + 4\,H^+ + 4\,e^- \rightarrow 2\,H_2O$$

Hydrogen is oxidized at the anode:

$$H_2 \rightarrow 2\,H^+ + 2\,e^-$$

Just as in other voltaic cells, the two half-reactions are physically separated. Electrons flow from the anode to the cathode through an external circuit, while protons pass through a special proton exchange membrane separating the two electrode compartments.

The overall cell reaction is simply the combination of hydrogen and oxygen to form water:

$$2\,H_2 + O_2 \rightarrow 2\,H_2O$$

We noted back in Chapter 3 that this reaction can produce an explosion, so clearly it can release significant amounts of energy. The fact that its reactants and products are common environmentally benign substances is also advantageous. Fuel cells using alternative fuels such as methanol or methane rather than hydrogen are also being explored.

Fuel cells are used in a variety of specialized applications, including powering instrumentation aboard spacecraft. Considerable research efforts aimed at producing fuel cells for consumer electronic devices such as laptop computers is ongoing. Significant engineering challenges in the design of fuel cells arise from the fact that the reactants are typically gases and therefore have much lower energy densities than the solids and liquids found in traditional batteries. The need for precious metal catalysts in the electrodes is also a drawback because such catalysts drive up the cost of the fuel cell.

Battery research has long been important in U.S. manufacturing. Thomas Edison held 147 patents in battery technology.

Limitations of Batteries

Battery manufacturing is a major industry, and the companies in this business carry out considerable research into the relevant technology. Longer lasting, better performing, or lighter batteries can be sold at a premium. Aside from expenditure of the materials that compose the cathodes or anodes, the next most common chemical cause for the loss of performance of a battery is corrosion. Perhaps once, you forgot to remove the batteries from a radio or flashlight that you did not use for many months. When you looked into the battery compartment, it is likely that you found that the batteries had corroded extensively, possibly even allowing fluids to leak out of the cells. Optimum conditions for corrosion are present in many batteries. In fact, older dry cell batteries were even more susceptible to corrosion than the common alkaline battery of today. How do manufacturers attempt to limit the performance-diminishing effects of corrosion on batteries? One method is the protective plating of materials used in the battery by another electrochemical process called electrolysis.

13-6 Electrolysis

From our consideration of equilibrium, we know that chemical reactions may proceed in the directions of either reactants or products. In our previous discussions, however, we found ourselves merely observing the direction that nature would use to establish equilibrium. Perhaps, if the mathematical signs on ΔH and ΔS were the same, we could manipulate the direction of equilibrium by raising or lowering the temperature. In acid–base chemistry or in the formation of precipitates, however, it was not apparent that we could overcome free energy and make the reaction go the way we would like. In contrast, redox reactions can be manipulated in a fairly simple way. The spontaneous direction of a redox reaction can be exploited in a galvanic cell to provide an electrical current, or we can use an external electric current to force a redox reaction in the nonspontaneous direction. **Electrolysis** is the process of passing an electric current through an ionic solution or molten salt to produce a chemical reaction.

A device used to carry out electrolysis is called an electrolytic cell.

Electrolytic cells can be divided into two categories based on the nature of the electrodes used. If the electrodes are chemically inert materials that simply provide a path for electrons, the process is called **passive electrolysis**. When the electrodes are part of the electrolytic reaction, we have **active electrolysis**. Passive electrolysis is used in industry to purify metals that corrode easily. Active electrolysis is used to plate materials to provide resistance to corrosion. We will consider one example of each form of electrolysis, but first we need to address the issue of electrical polarity in electrolytic cells.

Electrolysis and Polarity

Electrolysis changes the polarity of the electrodes in a system. Instead of the reactions that would occur spontaneously at the electrodes, the external power supply causes the reverse reactions to occur. Thus for reduction, electrons are forced to the cathode in an electrolytic reaction. The cathode is still the site of reduction, but in electrolysis it is negatively charged because the external power supply forces electrons (and their negative charges) to the location of the reaction. The anode is still the site of oxidation, but in electrolysis, it is positively charged. In forcing oxidation of the material, the external power supply pulls the electrons away from the region of the oxidation reaction. **Figure 13.19** illustrates the electrode charges in electrolysis.

Passive Electrolysis in Refining Aluminum

Aluminum metal finds widespread use as a structural material because it is resistant to catastrophic failure from corrosion. This does not imply, however, that aluminum does not corrode. Pure aluminum metal reacts with oxygen very quickly, as we pointed out in Chapter 1. However, when corrosion occurs, a thin layer of Al_2O_3 is formed at the surface of the metal. The oxide surface that is strongly bonded to

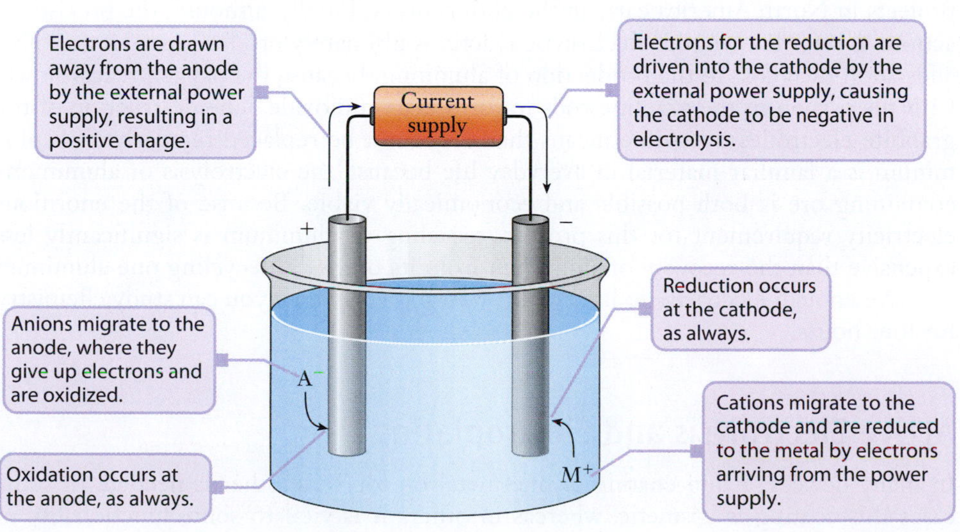
Electrons are drawn away from the anode by the external power supply, resulting in a positive charge.
Current supply
Electrons for the reduction are driven into the cathode by the external power supply, causing the cathode to be negative in electrolysis.
+
–
Reduction occurs at the cathode, as always.
Anions migrate to the anode, where they give up electrons and are oxidized.
A⁻
Cations migrate to the cathode and are reduced to the metal by electrons arriving from the power supply.
M⁺
Oxidation occurs at the anode, as always.

Figure 13.19 ■ In electrolysis, an external source of current drives a redox reaction that would otherwise not be spontaneous. The flow of ions through the solution completes the circuit. Electrolysis is useful for electroplating as well as for purifying several metals.

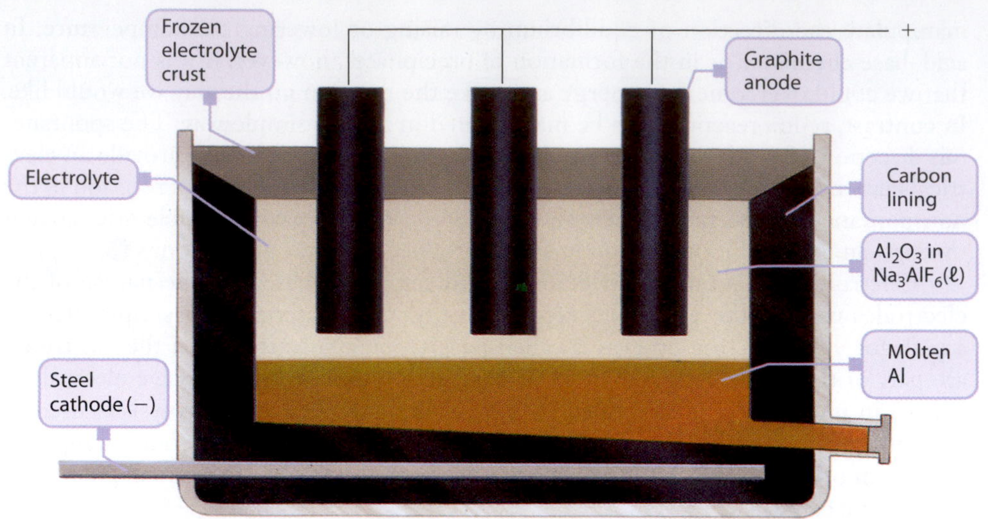

Figure 13.20 ■ The Hall–Heroult process involves the electrolytic refining of aluminum from Al_2O_3. Carrying out the electrolysis in a bath of cryolite (Na_3AlF_6) lowers the required temperature significantly, making the process economically feasible.

the pure aluminum underneath protects it and makes it useful in a variety of applications. The ready formation of Al_2O_3 can be predicted from the standard reduction potential of the Al(s) | Al^{3+} half-reaction. With $E° = -1.66$ V, aluminum is a considerably stronger reducing agent than iron, whose Fe(s) | Fe^{2+} half-reaction has a potential $E° = -0.44$ V. If the reaction to form aluminum ions is thermodynamically favorable, the reverse reaction will not be spontaneous. (Recall from Chapter 1 that the refining of aluminum from its ore is so difficult that aluminum was considered a precious metal until the mid-19th century, even though large sources of Al_2O_3-containing ore were known.)

Electrolysis provides the means to overcome the nonspontaneous reaction to separate aluminum from oxygen. The industrial process used to accomplish this task was discovered simultaneously in the United States by Charles Martin Hall and in France by Paul Heroult in 1886. The Hall–Heroult process uses carbon electrodes as inert sites for passive electrolysis, as illustrated in **Figure 13.20.** Three components of this process are important to its usefulness. The first is the inclusion of Na_3AlF_6, or cryolite, in the reactor, which lowers the melting point from 2045°C for Al_2O_3 to a temperature closer to 1000°C for the mixture. Lower temperatures can be achieved more readily and therefore make refining aluminum economically viable. Second, the process requires large amounts of electric current to produce substantial amounts of aluminum. Such vast amounts of electricity became available from large hydroelectric projects in North America early in the 20th century. Finally, although the precise reactions are not known, the electrolytic reactions ultimately produce oxygen gas. This adds additional cost to the production of aluminum because the oxygen reacts slowly with the carbon in the graphite rods to form carbon dioxide. The destruction of the graphite electrodes over time means that they must be replaced regularly. Still, aluminum is a familiar material in everyday life because the electrolysis of aluminum-containing ore is both possible and economically viable. Because of the enormous electricity requirement for this process, recycling of aluminum is significantly less expensive than the recovery of aluminum from its ore. Recycling one aluminum can save enough electricity to light a 100-watt bulb by which you can study chemistry for four hours!

Cryolite is referred to as a flux—a substance that lowers the melting point of a material.

We discussed several aspects of aluminum recycling in Section 10-8.

Active Electrolysis and Electroplating

In many devices, a thin coating of one metal on top of another is needed. In some cases this coating is cosmetic, whereas in others it is vital to some functionality of the coated piece. The process of depositing a thin coat of metal by using electricity is called **electroplating**. But where do we find electroplated products? One place is in the batteries we discussed in the last section. The electron collector that provides the external connection for the anode is electroplated. The metal coating gives this piece of a battery both corrosion protection and desirable conducting properties. Small parts such as this are plated in bulk by placing them in a barrel and then placing the barrel in the electrolyte solution. How does this process work? Let's consider the plating of silver to make parts for electrical devices.

The solution from which silver is plated normally contains cyanide ions, $CN^-(aq)$, which form complexes with silver. The need for uniform coatings makes this step important. The anode and cathode reactions are as follows:

Anode: $Ag(s) + 2\ CN^-(aq) \rightarrow Ag(CN)_2^-(aq) + e^-$

Cathode: $Ag(CN)_2^-(aq) + e^- \rightarrow Ag(s) + 2\ CN^-(aq)$

That the reactions are the reverse of each other is common in electroplating operations. Although a galvanic cell cannot be constructed from such a combination—it would have a cell potential of zero—the fact that an external current drives electrolysis means that the cell potential is not critical. Most electroplating processes are designed to run at low voltages however, so a zero cell potential is advantageous. The basic design of a barrel system for silver plating is illustrated in **Figure 13.21**. The barrel is rotated during the plating process so that all the small parts it contains obtain electrical contact with the cathode, thereby becoming part of the cathode themselves during the contact time. The overall process normally involves several preliminary steps, not shown here, to prepare the parts for the silver plating. For many applications, an undercoating of copper is plated first, and then silver is plated over the copper. Silver plating is industrially important because silver is a good conductor, is resistant to corrosion, and is generally considered attractive.

The use of plating to prevent corrosion in an electrical apparatus shows how important an understanding of corrosion can be. Because many modern devices are built from a multitude of parts, there are inevitably places where different metals will come into contact. As we have already seen, this can lead to galvanic corrosion. To prevent galvanic corrosion but not hinder the electrical performance of the device, parts are often plated. Sometimes, a limited area of a part is plated. For example, nuts might be plated with silver only on the inside of the threads. Doing so hinders galvanic corrosion between the nut and the bolt because the metals that are in contact are corrosion resistant. Because silver coatings are relatively expensive, plating only the inside reduces the cost. The external area of the nut does not make contact with other metals and therefore does not need plating.

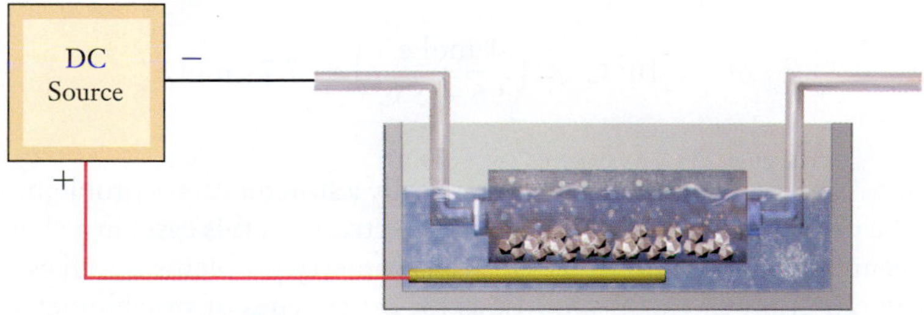

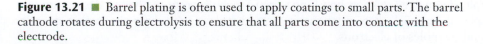

Figure 13.21 ■ Barrel plating is often used to apply coatings to small parts. The barrel cathode rotates during electrolysis to ensure that all parts come into contact with the electrode.

13-7 Electrolysis and Stoichiometry

Silver plating the inside of a nut suggests the importance of knowing how much material is being deposited on the surface. If the silver coating were to become too thick, the nut might not fit the bolt—the threads might become too narrow. So in electroplating, it can be vitally important to use carefully controlled amounts of materials. Fortunately, the flow of electrons in an electric current provides a ready source of information. Let's review some facts about electricity and then apply those facts to electrolytic cells.

Current and Charge

When the current is measured in an electric circuit, the observation is the flow of charge for a period of time. The base unit of current, the ampere (A), is a combined unit defined as one coulomb per second: $1\ A = 1\ C\ s^{-1}$. Devices called amp-meters (or ammeters) measure current. If a known current passes through a circuit for a known time, the charge can be easily calculated:

$$\text{Charge} = \text{current} \times \text{time}$$

$$Q = I \times t \qquad (13.8)$$

Typically, Q is in coulombs, I in amperes (coulombs/second), and t in seconds. We have already introduced the Faraday constant, which is the charge of 1 mole of electrons: $F = 96{,}485\ C\ mol^{-1}$. So if we can calculate the charge that passes through an electrolytic cell, we will know the number of moles of electrons that pass. Assuming that we know how many electrons were required to reduce each metal cation, it is simple to calculate the number of moles of material plated, as shown in Example Problem 13.4.

EXAMPLE PROBLEM 13.4

In a process called flash electroplating, a current of 2.50×10^3 A passes through an electrolytic cell for 5.00 minutes. How many moles of electrons are driven through the cell?

Strategy Because we know both the current and the time for which it was applied, we can use the relationship between charge and current in Equation 13.8 to obtain the charge. Then, Faraday's constant will let us convert that charge into moles of electrons. We must be careful with units, however, because the ampere is charge per second and time is given here in minutes.

Solution

$$Q = 2500\ A \times 300\ s = 7.50 \times 10^5\ C$$

Now use Faraday's constant:

$$7.50 \times 10^5\ C \times \left(\frac{1\ mol\ e^-}{96{,}485\ C}\right) = 7.77\ mol\ e^-$$

Discussion This two-step manipulation is really a stoichiometry problem, as it allows us to find the number of moles of something (electrons, in this case) in a chemical reaction. The central importance of moles in stoichiometry problems is well established at this point in our study of chemistry. Just as for other types of stoichiometry problems that we've considered, we sometimes need to use these relationships in reverse. For example, if you need to know how long to run a given current to obtain a desired amount of plated material, these relationships would still be used, only the known quantity would be moles of electrons.

Check Your Understanding You need 0.56 mol of electrons to deposit a thin covering of silver on a part that you are using for a prototype machine. How long would you need to run a current of 5.0 A to obtain this number of moles?

Although current can be measured readily, electricity use is often considered in terms of power. Electrical power is the rate of electricity consumption, and utility charges are based on consumption. The SI unit for power is the watt:

$$1 \text{ watt} = 1 \text{ J s}^{-1}$$

Thus wattage is the rate of energy use in J s^{-1}. To determine the amount of energy used, we multiply this rate by the time during which it occurs. To obtain numbers of a convenient magnitude, electrical utilities normally determine energy consumption in kilowatt-hours (kWh). One kilowatt-hour is equal to 3.60×10^6 J. In Chapter 2, we noted the relationship between electrical potential and energy:

$$1 \text{ J} = 1 \text{ C V}$$

So a second perspective on electrolysis, one that is very important for an electroplating company because of its relationship to cost of production, is in terms of energy expenditure. From this relationship, we can see why electroplating companies prefer low voltage applications. Example Problem 13.5 investigates the energy cost of electrolysis.

Waste treatment is another source of costs that we will not address, although the removal of wastes clearly involves stoichiometry.

EXAMPLE PROBLEM 13.5

Suppose that a batch of parts is plated with copper in an electrolytic bath running at 0.15 V and 15.0 A for exactly 2 hours. What is the energy cost of this process if the electric utility charges the company \$0.0500 per kWh?

Strategy We can determine the energy expended because we know the current, time, and voltage. The current multiplied by the time gives us the charge, which when multiplied by the voltage yields the energy. Once we know the energy expenditure we can convert the value we calculate (in J) to kWh to obtain the cost of the electricity.

Solution

$$Q = I \times t = 15.0 \text{ C s}^{-1} \times 7200 \text{ s} = 1.08 \times 10^5 \text{ C}$$

$$\text{Energy} = \text{charge} \times \text{voltage} = 1.08 \times 10^5 \text{ C} \times 0.15 \text{ V} = 1.6 \times 10^4 \text{ J}$$

Now convert to kWh and determine the cost:

$$(1.6 \times 10^4 \text{ J}) \times \left(\frac{1 \text{ kWh}}{3.60 \times 10^6 \text{ J}}\right) \times \left(\frac{\$0.0500}{1 \text{ kWh}}\right) = \$0.00023$$

Analyze Your Answer This value seems small—but does it make sense? Two factors contribute to the small result: a kilowatt-hour is a rather large energy unit, and the cost per kWh is low. Thus a very small cost is not problematic. This low cost points to the importance of running electroplating at low voltage: increasing voltage would lead directly to higher cost. This example is not representative of costs in the electroplating industry, but it does help us to get a handle on some of the variables that industry must consider when implementing electrochemical techniques.

Check Your Understanding If you need 25,000 A of current to carry out electrolysis for 15 minutes, what voltage would you need if you are to use 1.7 kWh for the process?

Calculations Using Masses of Substances in Electrolysis

Now that we have reviewed the important electrical concepts in quantitative applications of electrolysis, we can put these factors together with our customary approach to stoichiometry problems. In problems such as these, we generally encounter one of two questions. First, how much material is plated, given a specified current or electrical energy expenditure? And, second, how long must a given current pass through the cell to yield a desired mass of plated material? Example Problems 13.6 and 13.7 illustrate these two types of problems.

EXAMPLE PROBLEM 13.6

An electrolysis cell that deposits gold (from $Au^+(aq)$) operates for 15.0 minutes at a current of 2.30 A. What mass of gold is deposited?

Strategy As in any stoichiometry problem we need a balanced chemical equation, so we will start by writing the half-reaction for gold reduction. To determine the mass of gold deposited, we must calculate the number of moles of electrons used from the current and the time. We can use the half-reaction to obtain a mole ratio and convert moles of electrons into moles of gold. Once we have moles of gold, we convert to mass using the molar mass, as we have done many times in stoichiometry problems.

Solution First, write the balanced half-reaction:

$$\mathrm{Au^+(aq) + e^- \rightarrow Au(s)}$$

Next, calculate moles of electrons based on current and time:

$$Q = I \times t = (2.30\ \mathrm{C\ s^{-1}})(900\ \mathrm{s}) = 2.07 \times 10^3\ \mathrm{C}$$

$$(2.07 \times 10^3\ \mathrm{C}) \times \left(\frac{1\ \mathrm{mol\ e^-}}{96{,}485\ \mathrm{C}}\right) = 2.15 \times 10^{-2}\ \mathrm{mol\ e^-}$$

Now we note that the mole ratio of electrons to gold is 1:1, which means that we also have 2.15×10^{-2} mol of Au. Finally, convert from moles to mass:

$$(2.15 \times 10^{-2}\ \mathrm{mol\ Au}) \times (197\ \mathrm{g\ mol^{-1}}) = 4.23\ \mathrm{g\ Au}$$

Discussion The latter part of this problem looks familiar, as we have been carrying out stoichiometry problems similar to this since Chapter 3. The only changes from our previous experience were the use of a half-reaction (so we can see the electrons explicitly) and the use of some physical laws that govern electricity to obtain moles of electrons in the first place.

Check Your Understanding Suppose that a machine part requires 400 mg of copper to provide a coating that is thick enough for a certain electrical function. If you run an electrolysis using copper(II) ions for 10 minutes at a current of 0.5 A, will you meet the requirement?

Now we can consider an example where we approach the problem from the other direction.

EXAMPLE PROBLEM 13.7

Suppose that you have a part that requires a tin coating. You've calculated that you need to deposit 3.60 g of tin to achieve an adequate coating. If your electrolysis cell (using Sn^{2+}) runs at 2.00 A, how long must you operate the cell to obtain the desired coating?

Strategy This problem is different from the last one because it requires that we use the balanced half-reaction to relate moles of electrons and moles of a desired substance, in this case tin. Because we are given a mass of tin required, we can obtain the moles of tin, and that will tell us the moles of electrons required. We then determine the charge of those electrons with the Faraday constant, and we can use our understanding of electricity to get the time needed at the given current.

Solution

$$Sn^{2+}(aq) + 2\ e^- \rightarrow Sn(s)$$

$$3.60\ \text{g Sn} \times \left(\frac{1\ \text{mol Sn}}{118.7\ \text{g Sn}}\right) \times \left(\frac{2\ \text{mol e}^-}{1\ \text{mol Sn}}\right) \times \left(\frac{96{,}485\ \text{C}}{1\ \text{mol e}^-}\right) = 5.85 \times 10^3\ \text{C}$$

Now recall that $Q = I \times t$, so $t = Q/I$.

$$t = \frac{5.85 \times 10^3\ \text{C}}{2.00\ \text{C s}^{-1}} = 2930\ \text{s} = 48.8\ \text{min}$$

Analyze Your Answer This may seem like a long time to run an experiment. So does this answer make sense? Often, plating a few grams of material at low currents requires on the order of thousands of seconds (as in the last problem for gold). The fact that tin requires two electrons in this process largely accounts for more than double the time compared to the last problem, so the answer is not unreasonable.

Discussion Perhaps this problem seems more complicated than some other stoichiometry problems because we divided it into two steps. Still, the fundamental components of the problem are the same as always. We used a balanced chemical equation to relate moles of a substance (tin) to moles of electrons. Then we used the relationship, $Q = I \times t$, to find the time.

Check Your Understanding Suppose that you need a coating of rhodium (a precious metal) on a part that is to be used as a mirror in a laser. Because rhodium is so expensive, you wish to deposit only 4.5 mg on the surface of the part. If you run an electrolytic apparatus for 5.0 min, what current do you need to deposit the desired amount from a solution containing Rh^{3+} ions?

An additional type of calculation that engineers and chemists in electroplating companies sometimes need to make is to determine the amount of material based on coating thickness. In such a case, the surface area of the part must also be known. Multiplying the coating thickness by the surface area provides a volume, and then multiplying by the density of the material provides a mass. When considering the plating of parts of a battery to protect the battery from corrosion, these calculations are made more complicated by the fact that in barrel plating, hundreds or even thousands of parts are plated at the same time. Still, many electroplating companies establish a significant portion of their business from the need to protect parts from the effects of corrosion, so this type of quantitative information is quite important.

Additional factors such as the rate of coating affect the quality of the product, so plating does not always use the simplest reactions. Cyanide solutions, for example, are used to control the rate of deposition in silver plating.

13-8 Batteries in Engineering Design

As we've noted several times in this chapter, the development of ever smaller and more powerful portable electronic devices has spurred significant advances in battery technology. One relatively new type of battery that has been widely adopted is

Figure 13.22 ■ The battery on the left suffered an on-board fire during a flight. The battery on the right shows an undamaged version of the same system.

the **lithium-ion battery**, which is found in products ranging from the iPhone to the Chevrolet Volt electric car. Recently, however, the lithium-ion battery has come under scrutiny because of its failure in a high-profile application. After many years in design and development, the Boeing 787 Dreamliner was grounded, at least temporarily, within months of its first commercial flight after lithium-ion batteries overheated and caught fire in flight **(Figure 13.22)**. What factors led to the Dreamliner's battery problems, and what implications might they have for lithium-ion batteries in other applications?

All modern planes have electrical systems that rely on battery power, so a reasonable question might be why is the 787 different? The answer is twofold and ultimately stems from specific design factors. First, the Dreamliner requires higher levels of electrical power than more conventional aircraft. This is due in large part to a design decision to replace the traditional hydraulic system for controlling the flight of the plane with an all-electronic alternative. Second, increased fuel efficiency was a major goal in the Dreamliner design, so it was important to minimize weight. The competing demands for high energy capacity and low weight led to the decision to use lithium-ion batteries.

The most attractive feature of the lithium-ion battery is its high energy density when compared to other batteries. Even other advanced battery technologies like the nickel-metal-hydride batteries that are used in many hybrid cars have roughly half the energy density of lithium-ion batteries. Nonetheless, packing a great deal of electrical potential energy into a small volume carries certain risks, as Boeing's experience vividly illustrates. How do lithium-ion batteries compare to those described earlier in this chapter?

As is true in any galvanic cell, the anode and cathode in a lithium-ion battery are kept physically separate. In this case a semi-permeable separator allows the flow of lithium ions between the anode and cathode compartments, as shown in **Figure 13.23**. The anode is a form of graphite into which lithium atoms have been incorporated, or "intercalated." The cathode is cobalt oxide, and it also has lithium incorporated into it. (Other designs use different metal oxides for the cathode.) Because the electrodes both consist of solids with lithium embedded in them, it is difficult to write a simple

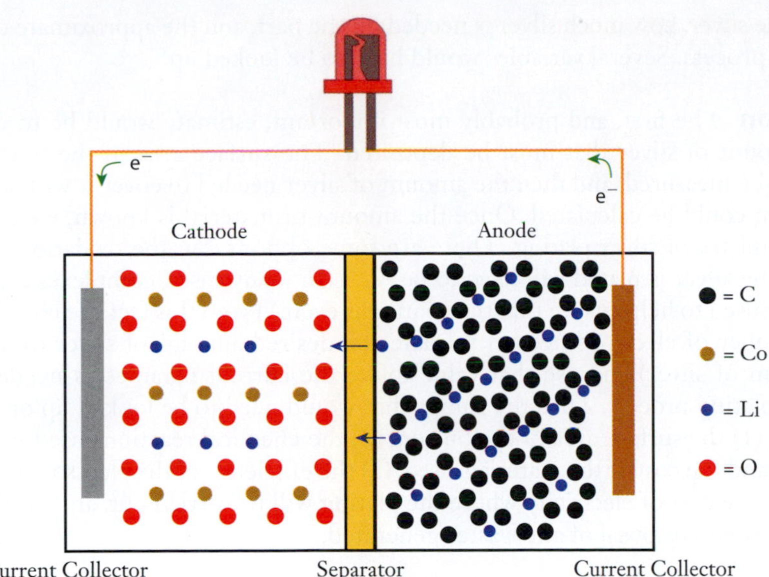

Figure 13.23 ■ A schematic diagram of the chemistry of a lithium-ion battery is shown. Lithium ions move through the separator from the anode to the cathode during the spontaneous cell reaction. When the battery is recharged, an external power source forces the ions to flow in the opposite direction, from cathode to anode.

chemical equation for the cell reaction. But the chemistry is summarized reasonably by the following reactions:

$$\text{Anode:} \quad \text{Li}_n\text{C}_6 \rightarrow n\ \text{Li}^+ + n\ \text{e}^- + 6\ \text{C}$$

$$\text{Cathode:} \quad \text{Li}_{1-n}\text{CoO}_2 + n\ \text{Li}^+ + n\ \text{e}^- \rightarrow \text{LiCoO}_2$$

There are two reasons why the energy density for a battery based on these half-reactions will be high. First, the standard reduction potential for Li^+/Li is among the largest observed, as seen in Appendix I. Second, both lithium and carbon, two of the important materials in the battery, are relatively light. So the combination of a high cell potential and a low mass leads to a desirable energy density.

In part because of these desirable energy characteristics, lithium-ion batteries are also susceptible to their environment a bit more than other advanced batteries. In particular, high temperatures will cause lithium-ion batteries to degrade fairly quickly, so engineering designs that use them generally must account for this. And, in very rare cases, as seen in the Dreamliner incidents, the batteries can catch fire. Estimates suggest that about 1 battery out of 10,000,000 catches fire. In addition to their use in aircraft, lithium-ion batteries are widely used in cell phones, laptops, and other portable devices, so it is important to recognize that the danger of fire is one that can be largely handled with appropriate design. Of course, one key concern is that human error can affect the safe operation of batteries. For example, the rechargeable nature of lithium-ion batteries means that under some circumstances they may be overcharged. When this happens, too much lithium moves to the anode side of the battery and small collections of lithium called dendrites can form, and these structures may cause short circuits that can contribute to the types of failures seen in the Boeing 787.

FOCUS ON PROBLEM SOLVING

Question You are working on the design of a battery and need to have a small button plated with silver (because it has useful conductive properties and is corrosion resistant). You determine that a coating of 1 mm will suffice for the design. What would you have to look up to devise an initial cost estimate for this process?

Strategy This is a question about electrolysis from a practical perspective. Cost versus coating thickness can be determined by finding out which materials are used to plate the silver, how much silver is needed on the part, and the approximate cost of the plating process. Several variables would have to be looked up.

Solution The first, and probably most important, estimate would be to determine the amount of silver that must be deposited. The surface area of the button would have to be measured, and then the amount of silver needed to cover it with a thickness of 1 mm could be calculated. Once the amount of material is known, we can look at the chemistry of silver plating. There are some options, but the oxidation state from which the silver is plated will have to be Ag^+. (In many cases, complexes that contain Ag^+ are used to help ensure that the coating is even.) From this fact, we could calculate the number of electrons needed to plate the desired amount of silver (one electron per atom of silver) and use that value to set the current parameters needed for the electroplating process. Thus the things that would need to be looked up or measured include (1) the surface area of the button, (2) the chemical reactions used in the silver plating and the cost of the materials used, (3) the efficiency of the electroplating operation, (4) the cost of electricity where the plating will be carried out, and (5) the cost of treatment and disposal of any wastes generated.

SUMMARY

Many chemical reactions can be understood in terms of the transfer of electrons between the species involved. These reactions are called oxidation–reduction, or redox reactions. Because electrons also move in generating and delivering electricity, oxidation–reduction reactions provide a connection to electrical systems.

If the oxidation process (or half-reaction) is physically separated from the reduction half-reaction, electrons can be made to traverse a circuit. A chemical reaction used to generate an electric current is called a galvanic cell, and a commercially important example of such a cell is the battery. A number of factors influence the nature and function of a galvanic cell. Most importantly, the chemicals involved have different propensities to undergo redox reactions, and the cell potential (or voltage) is a practical measure of these different reactivities. Because the concentration of materials also affects the voltage, standard conditions have been defined, and standard cell potentials can be calculated for many possible pairs of half-reactions from tabulated data. Nonstandard conditions can also be accounted for by using the Nernst equation.

Like any chemical system, oxidation–reduction reactions will progress toward equilibrium. Because of this fact, the cell potential provides a way to measure equilibrium constants or free energy changes.

Important manifestations of electrochemistry include battery technology and corrosion. Different types of batteries can be identified by the half-reactions involved. Batteries are classified as either primary cells, which cannot be recharged, or secondary batteries, which are rechargeable. The chemical origin of these differences is that the reaction in a primary cell results in products that cannot be reversed to regenerate the original reactants, whereas in secondary cells, the reaction can be driven in reverse by an applied voltage. Corrosion is another example of electrochemistry with practical implications, albeit not productive ones. In corrosion, the electrochemical reaction may be similar to those of a battery. But the flow of electrons is not captured in a circuit, and the reaction products are generally not as useful as the original material.

Electrochemical reactions can also be driven by external voltage sources in a process called electrolysis. Electrolysis has important industrial applications for coating materials with metals. The amount of material in a coating can be determined accurately through a stoichiometric calculation in which the current and time tell us the number of electrons involved in the process.

KEY TERMS

active electrolysis (13-6)
alkaline battery (13-5)
anode (13-2)
battery (13-5)
cathode (13-2)
cathodic protection (13-3)
cell notation (13-2)
cell potential (13-2)
corrosion (13-1)
crevice corrosion (13-1)
electrochemistry (13-2)
electrode (13-2)
electrolysis (13-6)
electromotive force (EMF) (13-2)
electroplating (13-6)
Faraday constant (*F*) (13-3)
fuel cell (13-5)
galvanic cell (13-2)
galvanic corrosion (13-1)
galvanized steel (13-3)
half-reaction (13-2)
lead-acid storage battery (13-5)
lithium battery (13-5)
lithium-ion battery (13-8)

Nernst equation (13-3)
nickel-metal-hydride battery (13-5)
oxidation (13-2)
oxidation–reduction reaction (13-2)
passive electrolysis (13-6)
primary cell (13-5)
redox reaction (13-2)
reduction (13-2)
sacrificial anode (13-3)
salt bridge (13-2)
secondary cell (13-5)
standard hydrogen electrode (SHE) (13-3)
standard reduction potential (13-3)
standard state (cell) (13-2)
uniform corrosion (13-1)
zinc-air battery (13-5)

PROBLEMS AND EXERCISES

INSIGHT INTO Corrosion

13.1 When you look at several older cars that are showing initial signs of rust formation, where do you expect to find the most rust? What does this observation imply about conditions that lead to corrosion?

13.2 What type of corrosion is associated with each of the following circumstances?

(a) Two metals are touching each other.

(b) A small crack separates two metals.

(c) A metal automobile body rusts in the atmosphere.

13.3 Using the Internet, find three cases where corrosion was the cause of some sort of failure or malfunction of a device or a structure.

13.4 Using the Internet, find three cases where concern about corrosion is used to help advertise a product or to convince people to buy a product.

Oxidation–Reduction Reactions and Galvanic Cells

13.5 For the following oxidation–reduction reactions, identify the half-reactions and label them as oxidation or reduction.

(a) $Cu(s) + Ni^{2+}(aq) \rightarrow Ni(s) + Cu^{2+}(aq)$

(b) $2\ Fe^{3+}(aq) + 3\ Ba(s) \rightarrow 3\ Ba^{2+}(aq) + 2\ Fe(s)$

13.6 Explain why a substance that is oxidized is called a reducing agent.

13.7 What is the role of a salt bridge in the construction of a galvanic cell?

13.8 Which half-reaction takes place at the anode of an electrochemical cell? Which half-reaction takes place at the cathode?

13.9 If a salt bridge contains KNO_3 as its electrolyte, which ions diffuse into solution in the anode compartment of the galvanic cell? Explain your answer.

13.10 If a salt bridge contains KNO_3 as its electrolyte, which ions diffuse into solution in the cathode compartment of the galvanic cell? Explain your answer.

13.11 The following oxidation–reduction reactions are used in electrochemical cells. Write them using cell notation.

(a) $2\ Ag^+(aq)(0.50\ M) + Ni(s) \rightarrow 2\ Ag(s) + Ni^{2+}(aq)(0.20\ M)$

(b) $Cu(s) + PtCl_6^{2-}(aq)(0.10\ M) \rightarrow Cu^{2+}(aq)(0.20\ M) + PtCl_4^{2-}(aq)(0.10\ M) + 2\ Cl^-(aq)(0.40\ M)$

(c) $Pb(s) + SO_4^{2-}(aq)(0.30\ M) + 2\ AgCl(s) \rightarrow PbSO_4(s) + 2\ Ag(s) + 2\ Cl^-(aq)(0.20\ M)$

(d) In a galvanic cell, one half-cell contains 0.010 M HCl and a platinum electrode, over which H_2 is bubbled at a pressure of 1.0 atm. The other half-cell is composed of a zinc electrode in a 0.125 M solution of $Zn(NO_3)_2$.

13.12 Write a balanced chemical equation for the overall reaction in each of the following galvanic cells.

(a) $Ag(s)\ |\ Ag^+(aq)\ ||\ Sn^{4+}(aq), Sn^{2+}(aq)\ |\ Pt(s)$

(b) $Al(s)\ |\ Al^{3+}(aq)\ ||\ Cu^{2+}(aq)\ |\ Cu(s)$

(c) $Pt(s)\ |\ Fe^{2+}(aq), Fe^{3+}(aq)\ ||\ MnO_4^-(aq), Mn^{2+}(aq)\ |\ Pt(s)$

13.13 For the reactions in (a) and (b) in the preceding problem, no anions at all are shown in the cell notation. Explain why this is not a concern.

13.14 Explain why the terms *cell potential* and *electromotive force* mean the same thing in electrochemical cells.

13.15 How does galvanic corrosion differ from uniform corrosion of iron?

13.16 How can each of the following serve as both the cathode and the anode of an electrochemical cell? **(a)** a single piece of metal, **(b)** two connected pieces of metal

13.17 A student who has mercury amalgam fillings in some of her teeth is eating a piece of candy. She accidentally bites down on a piece of the aluminum foil wrapper and experiences a sharp sensation in her mouth. Explain what has happened in terms of electrochemistry.

Cell Potentials

13.18 Based on the cell potential measured for the cells

$$Co(s)\ |\ Co^{2+}(aq)\ ||\ Cu^{2+}(aq)\ |\ Cu(s) \qquad E° = 0.614\ V$$
$$Fe(s)\ |\ Fe^{2+}(aq)\ ||\ Cu^{2+}(aq)\ |\ Cu(s) \qquad E° = 0.777\ V$$

what potential should you expect to find for the following cell?

$$Fe(s)\ |\ Fe^{2+}(aq)\ ||\ Co^{2+}(aq)\ |\ Co(s)$$

13.19 If the reaction at a standard hydrogen electrode is $2\ H^+ + 2\ e^- \rightarrow H_2$, why is there a platinum foil in the system?

13.20 Four voltaic cells are set up. In each, one half-cell contains a standard hydrogen electrode. The second half-cell is one of the following: Cr^{3+}(aq, 1.0 M) | Cr(s), Fe^{2+}(aq, 1.0 M) | Fe(s), Cu^{2+} (aq, 1.0 M) | Cu(s), or Mg^{2+}(aq, 1.0 M) | Mg(s).

(a) In which of the voltaic cells does the hydrogen electrode serve as the cathode?

(b) Which voltaic cell produces the highest voltage? Which produces the lowest voltage?

13.21 If the SHE was assigned a value of 3.00 V rather than 0.00 V, what would happen to all of the values listed in the table of standard reduction potentials?

13.22 Explain why a large negative value for the standard reduction potential indicates a half-cell that is likely to undergo oxidation.

13.23 In tables of standard reduction potentials that start from large positive values at the top and proceed through 0.0 V to negative values at the bottom, the alkali metals are normally at the bottom of the table. Use your chemical understanding of alkali metals and how they behave in bonding to explain why this is so.

13.24 In the table of standard reduction potentials, locate the half-reactions for the reductions of the following metal ions to the metal: $Sn^{2+}(aq)$, $Au^{+}(aq)$, $Zn^{2+}(aq)$, $Co^{2+}(aq)$, $Ag^{+}(aq)$, and $Cu^{2+}(aq)$. Among the metal ions and metals that make up these half-reactions:

(a) Which metal ion is the weakest oxidizing agent?

(b) Which metal ion is the strongest oxidizing agent?

(c) Which metal is the strongest reducing agent?

(d) Which metal is the weakest reducing agent?

(e) Will Sn(s) reduce $Cu^{2+}(aq)$ to Cu(s)?

(f) Will Ag(s) reduce $Co^{2+}(aq)$ to Co(s)?

(g) Which metal ions on the list can be reduced by Sn(s)?

(h) What metals can be oxidized by $Ag^{+}(aq)$?

13.25 Using values from the table of standard reduction potentials, calculate the cell potentials of the following cells.

(a) $Ga(s) \mid Ga^{3+}(aq) \parallel Ag^{+}(aq) \mid Ag(s)$

(b) $Zn(s) \mid Zn^{2+}(aq) \parallel Cr^{3+}(aq) \mid Cr(s)$

(c) $Fe(s), FeS(s) \mid S^{2-}(aq) \parallel Sn^{2+}(aq) \mid Sn(s)$

13.26 Using values from the table of standard reduction potentials, calculate the cell potentials of the following cells.

(a) $Fe(s) \mid Fe^{2+}(aq) \parallel Hg^{2+}(aq) \mid Hg(\ell)$

(b) $Pt(s) \mid Fe^{2+}(aq), Fe^{3+}(aq) \parallel MnO_4^{-}(aq), H^{+}(aq), Mn^{2+}(aq) \mid Pt(s)$

(c) $Pt(s) \mid Cl_2(g) \mid Cl^{-}(aq) \parallel Au^{+}(aq) \mid Au(s)$

13.27 One half-cell in a voltaic cell is constructed from a copper wire dipped into a 4.8×10^{-3} M solution of $Cu(NO_3)_2$. The other half-cell consists of a zinc electrode in a 0.40 M solution of $Zn(NO_3)_2$. Calculate the cell potential.

13.28 Four metals, A, B, C, and D, exhibit the following properties.

(i) Only A and C react with 1.0 M hydrochloric acid to give $H_2(g)$.

(ii) When C is added to solutions of the ions of the other metals, metallic B, D, and A are formed.

(iii) Metal D reduces B^{n+} to give metallic B and D^{n+}.

Based on this information, arrange the four metals in order of increasing ability to act as reducing agents.

13.29 Use the Nernst equation to calculate the cell potentials of the following cells at 298 K.

(a) $2\,Ag^{+}(aq)(0.50\ M) + Ni(s) \rightarrow 2\,Ag(s) + Ni^{2+}(aq)(0.20\ M)$

(b) $Cu(s) + PtCl_6^{2-}(aq)(0.10\ M) \rightarrow Cu^{2+}(aq)(0.20\ M) + PtCl_4^{2-}(aq)(0.10\ M) + 2\,Cl^{-}(aq)(0.40\ M)$

(c) $Pb(s) + SO_4^{2-}(aq)(0.30\ M) + 2\,AgCl(s) \rightarrow PbSO_4(s) + 2\,Ag(s) + 2\,Cl^{-}(aq)(0.20\ M)$

13.30 One half-cell in a voltaic cell is constructed from a silver wire dipped into a $AgNO_3$ solution of unknown concentration. The other half-cell consists of a zinc electrode in a 1.0 M solution of $Zn(NO_3)_2$. A potential of 1.48 V is measured for this cell. Use this information to calculate the concentration of $Ag^{+}(aq)$.

13.31 We noted that a tin-plated steel can corrodes more quickly than an unplated steel can. In cases of galvanic corrosion, one cannot expect standard conditions. Suppose that you want to study the galvanic corrosion of tin-plated steel by constructing a cell with low concentrations of the ions. You have pieces of tin and iron. You also have a solution of tin(II) chloride that is 0.05 M and one of iron(II) nitrate that is 0.01 M.

(a) Describe the half-reactions you construct for this experiment.

(b) Which half-reaction will be the anode and which the cathode?

(c) Based on the solutions you have, calculate the cell potential for your experiment.

13.32 The following half-cells are available: $Ag^{+}(aq, 1.0\ M) \mid Ag(s)$; $Zn^{2+}(aq, 1.0\ M) \mid Zn(s)$; $Cu^{2+}(aq, 1.0\ M) \mid Cu(s)$; and $Co^{2+}(aq, 1.0\ M) \mid Co(s)$. Linking any two half-cells makes a voltaic cell. Given four different half-cells, six voltaic cells are possible. These are labeled, for simplicity, Ag-Zn, Ag-Cu, Ag-Co, Zn-Cu, Zn-Co, and Cu-Co.

(a) In which of the voltaic cells does the copper electrode serve as the cathode? In which of the voltaic cells does the cobalt electrode serve as the anode?

(b) Which combination of half-cells generates the highest potential? Which combination generates the lowest potential?

13.33 You are a chemical engineer in a firm that manufactures computer chips. A new accountant, who understands budgets but not chemistry, sends out the following memo:

> "Because of the high price of gold, the company should change from gold-plated pins on our computer chips to pins made of copper, which is much cheaper."

Write a memo to your boss explaining why the accountant's choice of copper is a poor one and why gold should continue to be used in pin manufacturing.

13.34 Immediately after a person has his or her ears (or other body parts) pierced, a solid 14K gold post should be used to keep the hole open as it heals.

(a) Explain why solid gold is preferred over other metals, such as steel.

(b) Why do you think that a gold-plated steel post might not offer the same protection as the solid gold post?

(c) Surgical steel is often used as the metal post in a newly pierced ear because it is less expensive than gold. Using the Internet or other resources, find the composition of surgical steel.

13.35 In May 2000, a concrete pedestrian walkway collapsed in North Carolina, injuring more than 100 people. Investigation revealed that $CaCl_2$ had been mixed into the grout that filled the holes around the steel reinforcing cables inside the concrete, resulting in corrosion of the cables, thus weakening the structure. Based on your understanding of corrosion, explain why the use of chloride compounds in steel-reinforced concrete is discouraged by the concrete industry.

13.36 A chemical engineering student is studying the effect of pH on the corrosion of iron. The following data are collected:

Corrosion Rate (mm/year)	pH
0.03	13.00
0.06	12.00
0.16	11.00
0.22	10.00
0.25	9.00
0.25	8.00
0.25	7.00
0.25	6.00
0.25	5.00
0.27	4.00
0.37	3.00
0.38	2.80
0.63	2.80
1.00	2.80

Plot corrosion rate vs. pH. (HINT: Label the x axis with *decreasing* pH.)

(a) How would you describe the dependence of corrosion on pH?

(b) When the pH goes below 3, bubbles appear in the solution. What change takes place in the reduction half-reaction at low pH?

(c) How do you explain the shape of the curve at the lowest pH?

(Based on a graph in the *Kirk-Othmer Encyclopedia of Chemical Technology*, 4th ed., vol 7. New York: John Wiley & Sons, p. 559.)

Cell Potentials and Equilibrium

13.37 How is the relationship between maximum electrical work and cell potential used to determine the free energy change for electrochemical cells?

13.38 Calculate the standard free energy change for the following cells.

(a) $Ga(s) \mid Ga^{3+}(aq) \parallel Ag^{+}(aq) \mid Ag(s)$

(b) $Zn(s) \mid Zn^{2+}(aq) \parallel Cr^{3+}(aq) \mid Cr(s)$

(c) $Fe(s), FeS(s) \mid S^{2-}(aq) \parallel Sn^{2+}(aq) \mid Sn(s)$

13.39 Calculate the standard free energy change for the following reactions using the standard cell potentials for the half-reactions that are involved.

(a) $Fe(s) + Hg_2{}^{2+}(aq) \rightarrow Fe^{2+}(aq) + 2\ Hg(\ell)$

(b) $Fe^{3+}(aq) + Ag(s) + Cl^{-}(aq) \rightarrow Fe^{2+}(aq) + AgCl(s)$

(c) $2\ MnO_4{}^{-}(aq) + 5\ Zn(s) + 16\ H_3O^{+}(aq) \rightarrow 2\ Mn^{2+}(aq) + 5\ Zn^{2+}(aq) + 24\ H_2O(\ell)$

13.40 Suppose that you cannot find a table of standard reduction potentials. You remember that the standard reduction potential of $Cu^{2+} + 2\ e^{-} \rightarrow Cu(s)$ is 0.337 V. Given that $\Delta G_f^\circ(Cu^{2+}) = 65.49$ kJ mol^{-1} and that $\Delta G_f^\circ(Ni^{2+}) = -45.6$ kJ mol^{-1}, determine the standard reduction potential of $Ni \mid Ni^{2+}$ from these data.

13.41 Use the potential of the galvanic cell, $Co(s) \mid Co^{2+} \parallel Pb^{2+} \mid Pb(s)$, to determine $\Delta G_f^\circ(Pb^{2+})$, given that $\Delta G_f^\circ(Co^{2+}) = -54.4$ kJ mol^{-1}.

13.42 Which of the following reactions is (are) spontaneous at standard conditions?

(a) $Zn(s) + 2\ Fe^{3+}(aq) \rightarrow Zn^{2+}(aq) + 2\ Fe^{2+}(aq)$

(b) $Cu(s) + 2\ H^{+}(aq) \rightarrow Cu^{2+}(aq) + H_2(g)$

(c) $2\ Br^{-}(aq) + I_2(s) \rightarrow Br_2(\ell) + 2\ I^{-}(aq)$

13.43 Consult a table of standard reduction potentials and determine which of the following reactions are spontaneous under standard electrochemical conditions.

(a) $Mn(s) + 2\ H^{+}(aq) \rightarrow H_2(g) + Mn^{2+}(aq)$

(b) $2\ Al^{3+}(aq) + 3\ H_2(g) \rightarrow 2\ Al(s) + 6\ H^{+}(aq)$

(c) $2\ Cr(OH)_3(s) + 6\ F^{-}(aq) \rightarrow 2\ Cr(s) + 6\ OH^{-}(aq) + 3\ F_2(g)$

(d) $Cl_2(g) + 2\ Br^{-}(aq) + \rightarrow Br_2(\ell) + 2\ Cl^{-}(aq)$

13.44 The equilibrium constant for a reaction is 3×10^{-15}. **(a)** Without carrying out any calculation, discuss whether ΔG° for the reaction is positive or negative. **(b)** Calculate ΔG° for this reaction.

13.45 Some calculators cannot display results of an antilog calculation if the power of 10 is greater than 99. This shortcoming can come into play for determining equilibrium constants of redox reactions, which are sometimes quite large. Solve the following expressions for K: **(a)** $\log K = 45.63$, **(b)** $\log K = 25.00$, **(c)** $\log K = 20.63$. What is the relationship among the three expressions and the three answers? How can you use this relationship to solve problems that exceed 10^{99}, even if your calculator will not carry out the calculation directly?

13.46 Calculate the equilibrium constant for the following reactions using data from the standard reduction potential tables.

(a) $Cl_2(g) + 2\ Br^{-}(aq) \rightarrow Br_2(g) + 2\ Cl^{-}(aq)$

(b) $Ni(s) + 2\ Ag^{+}(aq) \rightarrow 2\ Ag(s) + Ni^{2+}(aq)$

(c) $I_2(s) + Sn^{2+}(aq) \rightarrow 2\ I^{-}(aq) + Sn^{4+}(aq)$

13.47 Use the standard reduction potentials for the reactions:

$AgCl(s) + e^{-} \rightarrow Ag(s) + Cl^{-}(aq)$ and $Ag^{+}(aq) + e^{-} \rightarrow Ag(s)$

to calculate the value of K_{sp} for silver chloride at 298 K. How does your answer compare with the value listed in Table 12.4?

13.48 Hydrogen peroxide is often stored in the refrigerator to help keep it from decomposing according to the reaction:

$$2\ H_2O_2(\ell) \rightarrow 2\ H_2O(\ell) + O_2(g)$$

Use information from the standard reduction potentials to determine the equilibrium constant of this reaction.

13.49 Calculate the equilibrium constant for the redox reactions that could occur in the following situations and use that value to explain whether or not any reaction will be observed.

(a) A piece of iron is placed in a 1.0 M solution of $NiCl_2(aq)$.

(b) A copper wire is placed in a 1.0 M solution of $Pb(NO_3)_2(aq)$.

13.50 If a logarithmic scale had not been used for the graph of Figure 13.13, what would the plots look like?

13.51 An engineer is assigned to design an electrochemical cell that will deliver a potential of exactly 1.52 V. Design and sketch a cell to provide this voltage, detailing the solutions, their concentrations, and the electrodes you will need. Write equations for all relevant reactions.

13.52 A magnesium bar with a mass of 6.0 kg is attached to a buried iron pipe to serve as a sacrificial anode. An average current of 0.020 A flows between the bar and the pipe. **(a)** What reaction takes place at the surface of the magnesium bar? **(b)** What reaction takes place at the surface of the iron pipe? **(c)** In which direction do electrons flow between the two surfaces? **(d)** How many years would it take for the entire magnesium bar to be consumed?

13.53 Would nickel make an acceptable sacrificial anode to protect steel? Explain your answer.

13.54 Zinc is used as a coating for galvanized steel, where it helps prevent corrosion. Explain why zinc would also make an acceptable sacrificial anode for a steel pipeline. Are these two uses related in terms of the science that gives rise to their utility?

13.55 Sacrificial anodes are sometimes connected to steel using a copper wire. If the anode is completely corroded away and not replaced, so that only the copper wire remains, what could happen where the copper and iron meet? Explain your answer.

Batteries

13.56 What is the principal difference between a primary and a secondary cell?

13.57 Based on the chemistry that takes place, explain why an alkaline battery is called "alkaline."

13.58 If you put a 9-volt battery in a smoke detector in your home or apartment, you are not installing a single galvanic cell. Explain how and why this is so.

13.59 If alkaline batteries were not alkaline but rather acidic (as in the older dry cell batteries), what extra difficulties could you envision with corrosion, based on reactions that are part of the table of standard reduction potentials?

13.60 What would happen to the voltage of an alkaline battery if the zinc were replaced by steel? Assume that the zinc reaction is simply $Zn \mid Zn^{2+}$ and that steel is iron.

13.61 Battery manufacturers often assess batteries in terms of their specific energy (or energy capacity). The weight capacity of a battery is defined as $q \times V$/mass. Why would a battery manufacturer be interested in this quantity?

13.62 Suppose you had a fresh battery and wanted to measure its weight capacity, as described in the previous problem. Devise and describe an experiment that you could carry out to make this measurement. Do you think performing your experiment on a single battery would yield reliable results?

13.63 What product forms from the lead components of a lead storage battery? Why does mechanical shock (bumps) ultimately degrade the performance of the lead storage battery?

13.64 The electrolyte solution in a zinc-air battery (Figure 13.16) is aqueous KOH. For a zinc-air battery to obtain O_2 from the air, there are tiny openings in the battery. However, these openings also allow water vapor to escape and enter the battery as the humidity changes. The optimum relative humidity for efficient zinc-air battery function, at which the KOH electrolyte is in equilibrium with water in the surrounding air, is 60% at 25°C. On the molecular level, describe what will happen to a zinc-air battery if used consistently in **(a)** an arid environment and **(b)** an extremely humid environment.

13.65 You are an electrical engineer investigating different button batteries for possible use in a design. You obtain the discharge curves shown below for four different batteries under the same conditions and discharge load.

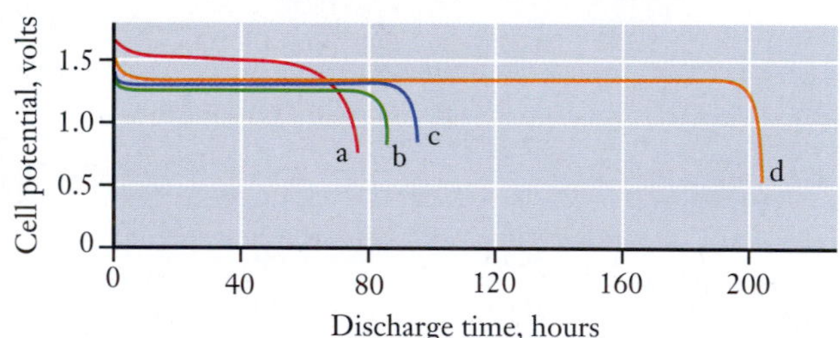

(a) Write a paragraph in which you interpret curve (a) and curve (d) to a customer seeking a battery for an electronic device.

(b) Both (b) and (d) represent zinc-air cells. Explain why there is a difference in these two curves. (HINTS: Which electrode in a zinc-air cell has an unlimited life? Compare the approximate area of the other electrode for the two cells.)

13.66 Assume the specifications of a Ni-Cd voltaic cell include delivery of 0.25 A of current for 1.00 h. What is the minimum mass of the cadmium that must be used to make the anode in this cell?

13.67 On the Internet, access the website of a major battery company, such as Rayovac, Duracell, or Eveready (Energizer). Search the site for a battery type not covered in this chapter. Find out all you can about the battery from engineering data or specification (spec) sheets. Print spec sheets that include graphs of discharge characteristics, operating temperature data, etc. Summarize and interpret your findings in a one-page report to which you attach the spec sheets.

Electrolysis

13.68 What is the difference between active and passive electrolysis? Based on the common meanings of the words *active* and *passive*, what part of electrolysis is the focus of the name?

13.69 Why is it easier to force an oxidation–reduction reaction to proceed in the nonspontaneous direction than it is to force an acid–base reaction in the nonspontaneous direction?

13.70 In any electrochemical cell, reduction occurs at the cathode. Why does the cathode carry a negative charge, when electrolysis occurs, rather than the positive charge it carries in a galvanic cell?

13.71 When aluminum is refined by electrolysis from its oxide ores, is the process used active or passive electrolysis? Explain your answer.

13.72 What complication arises in the electrolytic refining of aluminum because of the production of oxygen?

13.73 In an electroplating operation, the cell potential is sometimes 0 V. Why is a zero potential possible in electrolysis but not in a galvanic cell?

13.74 When copper wire is placed in a solution of silver nitrate, we say that the silver plates out on the copper, but we don't call the process electroplating. Explain the difference.

13.75 In barrel plating, why is the barrel containing the small parts rotated?

13.76 When a coating is plated onto a metal, two different metals are in contact with each other. Why doesn't galvanic corrosion occur at this interface?

13.77 Use the Internet to find electroplating companies that carry out silver plating. Popular impressions are that silver plating is mostly done for cosmetic reasons, making objects (like silverware) more attractive. Based on your Internet research, do you believe this popular impression accurately reflects the plating industry?

13.78 Based on the reaction used for silver plating, suggest why there are environmental challenges associated with industrial plating companies. Use the Internet to find out how some electroplating companies address their responsibilities to protect the environment.

Electrolysis and Stoichiometry

13.79 If a current of 15 A is run through an electrolysis cell for 2.0 hours, how many moles of electrons have moved?

13.80 Suppose somebody in a laboratory doesn't quite turn off the current in an electrolysis cell so that 4.2 mA of current continues to flow. If the cell is then left untouched for exactly 3 weeks, how many moles of electrons have flowed through the cell in that time?

13.81 If a barrel plating run uses 200.0 A for exactly 6 hours for an electroplating application at 0.30 V, how many kilowatt-hours have been used in the run? If the voltage is 0.90V, what is the power usage (in kWh)?

13.82 An electrical engineer is analyzing an electroplating run and wants to calculate the charge that has been used. She knows that the cell voltage is 0.25 V and that 10.3 kWh was expended. What was the charge? If the current of the apparatus is 3.0 A, how long was it running?

13.83 In a copper plating experiment in which copper metal is deposited from a copper(II) ion solution, the system is run for 2.6 hours at a current of 12.0 A. What mass of copper is deposited?

13.84 A metallurgist wants to gold-plate a thin sheet with the following dimensions: 1.5 in × 8.5 in × 0.0012 in. The gold plating must be 0.0020 in thick.

(a) How many grams of gold (d = 19.3 g/cm^3) are required?

(b) How long will it take to plate the sheet from AuCN using a current of 7.00 A? (Assume 100% efficiency.)

13.85 Tin-plated steel is used for "tin" cans. Suppose that in the production of sheets of tin-plated steel, a line at a factory operates at a current of 100.0 A for exactly 8 hours on a continuously fed sheet of unplated steel. If the electrolyte contains tin(II) ions, what is the total mass of tin that has plated out in this operation?

13.86 An electrolysis cell for aluminum production operates at 5.0 V and a current of 1.0×10^5 A. Calculate the number of kilowatt-hours of energy required to produce 1 metric ton (1.0×10^3 kg) of aluminum. (1 kWh = 3.6×10^6 J and 1 J = 1 C V.)

13.87 If a plating line that deposits nickel (from $NiCl_2$ solutions) operates at a voltage of 0.40 V with a current of 400.0 A and a total mass of 49.0 kg of nickel is deposited, what is the minimum number of kWH consumed in this process?

13.88 Suppose that a student mistakenly thinks that copper is plated from solutions that contain copper(I) ions rather than copper(II) ions. How would this mistake affect the answer that student would get for the mass of copper plated out for a given time and current? (Is it too high, too low, and by what factor?)

13.89 When a lead storage battery is recharged by a current of 12.0 A for 15 minutes, what mass of $PbSO_4$ is consumed?

13.90 A small part with a surface area of 2.62 cm^2 is plated with a gold coating that is 5.00×10^{-4} mm thick. The density of gold is 19.32 g cm^{-3}. What mass of gold is in this coating? If a barrel plating run has 5000 of these parts in a single barrel, how long would the run take if the current was 15.0 A? (Gold is effectively plated from Au^+(aq).)

13.91 An engineer is designing a mirror for an optical system. A piece of metal that measures 1.3 cm by 0.83 cm will have a coating of rhodium plated on its surface to serve as the mirror. The rhodium thickness will be 0.00030 mm, and the electrolyte contains Rh^{3+} ions. If the operating current of the electrolysis is 1.3 A, how long must it be operated to obtain the desired coating? What mass of rhodium is deposited? (The density of rhodium is 12.4 g cm^{-3}.)

INSIGHT INTO Batteries in Engineering Design

13.92 What two characteristics of lithium-ion batteries are responsible for their desirability in aerospace applications?

13.93 Explain why lithium-ion batteries tend to be relatively lightweight.

13.94 What characteristic of lithium ions makes it practical to build a semipermeable separator that allows them to flow through in a lithium-ion battery? Why would a sodium-ion battery be harder to design even if the oxidation–reduction reactions were similar?

13.95 Looking at Figure 13.23, describe how the operation of a lithium-ion battery does not lead to a charge build up on one side of the battery or the other?

13.96 Use the web to research how Boeing addressed the issue of battery fires in the Dreamliner, and write a short paragraph summarizing the company's solution to the problem.

FOCUS ON PROBLEM SOLVING EXERCISES

13.97 For a voltage-sensitive application, you are working on a battery that must have a working voltage of 0.85 V. The materials to be used have a standard cell potential of 0.97 V. What must be done to achieve the correct voltage? What information would you need to look up?

13.98 For a battery application, you need to get the largest possible voltage per cell. If voltage were the only concern, what chemical reactions would you choose? Use the Internet or your own background to speculate on the likely practical consequences of a choice driven by voltage alone.

13.99 An oxidation–reduction reaction using Sn(s) to remove N_2O(g) from a reaction vessel has been proposed, and you need to find its equilibrium constant. You cannot find any thermodynamic information on one product of the reaction, NH_3OH^+(aq). How could you estimate the equilibrium constant of the reaction?

13.100 You are designing a system for a rather corrosive environment and are using an expensive alloy composed primarily of titanium. How would you choose a sacrificial anode to help protect this expensive component in your design? What information would you have to look up?

13.101 You need to use a gold-plated connector in a design to take advantage of the conductivity and corrosion resistance of gold. To justify this choice to the project director, you must devise a cost projection for several levels of plating. What parameters could be varied? What information would you have to look up to provide the cost information requested?

Cumulative Problems

13.102 **(a)** What happens when a current is passed through a solution of dilute sulfuric acid to carry out electrolysis? **(b)** A 5.00-A current is passed through a dilute solution of sulfuric acid for 30.0 min. What mass of oxygen is produced?

13.103 A current is passed through a solution of copper(II) sulfate long enough to deposit 14.5 g of copper. What volume of oxygen is also produced if the gas is measured at 24°C and 0.958 atm of pressure?

13.104 Hydrazine, N_2H_4, has been proposed as the fuel in a fuel cell in which oxygen is the oxidizing agent. The reactions are

$$N_2H_4(aq) + 4\,OH^-(aq) \rightarrow N_2(g) + 4\,H_2O(\ell) + 4\,e^-$$

$$O_2(g) + 2\,H_2O(\ell) + 4\,e^- \rightarrow 4\,OH^-(aq)$$

(a) Which reaction occurs at the anode and which at the cathode?

(b) What is the net cell reaction?

(c) If the cell is to produce 0.50 A of current for 50.0 h, what mass in grams of hydrazine must be present?

(d) What mass in grams of O_2 must be available to react with the mass of N_2H_4 determined in part (c)?

13.105 A button case for a small battery must be silver coated. The button is a perfect cylinder with a radius of 3.0 mm and a height of 2.0 mm. For simplicity, assume that the silver solution used for plating is silver nitrate. (Industrial processes often use other solutions.) Assume that the silver plating is perfectly uniform and is carried out for 3.0 min at a current of 1.5 A. **(a)** What mass of silver is plated on the part? **(b)** How many atoms of silver have plated on the part? **(c)** Calculate an estimate of the thickness (in atoms) of the silver coating. (Silver has a density of 10.49 g/cm^3 and an atomic radius of 160 pm.)

Nuclear Chemistry

14

The Spitzer Space Telescope, which is sensitive to infrared radiation, is shown here against an infrared image of the sky. Because the intensity of cosmic rays increases with altitude, electronic equipment on satellites such as Spitzer is especially susceptible to damage from ionizing radiation. *Courtesy of NASA, JPL/Caltech*

CHAPTER OUTLINE

Throughout this text, we have focused on the ways that atoms interact with one another to form molecules. Although we have discussed the fact that atoms are made up of smaller subatomic particles, we have regarded atoms as stable particles that cannot be converted from one element to another. As we've seen, this approach is quite useful for understanding a broad array of phenomena in chemistry and technology.

But when we think of atoms as immutable, we also preclude considering other important phenomena and technologies. We cannot attempt to understand radioactivity or nuclear reactions without acknowledging that atoms of one element *can* be converted into atoms of another element. In this chapter, we will explore the realm of nuclear chemistry, which is at the heart of such applications as nuclear energy, medical radiation therapy, and carbon dating. We begin with a glimpse into one result of nuclear reactions in the universe: cosmic rays.

Chapter Objectives

After mastering this chapter, you should be able to

- describe cosmic rays and some of the ways that they influence Earth and its atmosphere.
- write, balance, and interpret equations for simple nuclear reactions.

- define and distinguish among various modes of nuclear decay, including alpha decay, beta decay, positron emission, and electron capture.
- interpret the kinetics of radioactive decay using first-order rate equations.
- use the chart of the nuclides to understand and explain how radioactive decay processes increase nuclear stability.
- use Einstein's equation to calculate the binding energies of nuclei and the energy changes of nuclear reactions.
- describe nuclear fission and fusion and explain how both processes can be highly exothermic.
- discuss the potential of both fission and fusion as energy sources and identify the pros and cons of the two technologies.
- explain how penetrating power and ionizing power combine to determine the effect of radiation on materials, including living tissues.
- describe how radioisotopes can be used in medical imaging techniques to monitor organ function.

INSIGHT INTO

14-1 Cosmic Rays and Carbon Dating

The usual definition of cosmic rays restricts the term to particles and excludes photons.

Earth is constantly bombarded by **cosmic rays**—particles traveling at high speeds. The vast majority of cosmic rays are atomic nuclei. In space, roughly 87% of cosmic rays are hydrogen nuclei (i.e., protons), and about 12% are helium nuclei. The rest are heavier nuclei. Where do these energetic particles come from, and what are their effects?

Some cosmic rays emanate from the sun, where solar flares can accelerate highly charged cations until they approach the speed of light. The distribution of elements in these cosmic rays reflects the composition of the sun itself. Although hydrogen and helium are the most prevalent, isotopes of such elements as carbon, nitrogen, oxygen, neon, magnesium, silicon, and iron are also present. Other cosmic rays originate outside the solar system.

The energies of cosmic rays are far higher than those in any of the other areas of chemistry that we've studied. Chemical energies are usually expressed in kilojoules per mole, but those of cosmic rays are typically expressed in **electron volts (eV)**, with 1 eV = 96.5853 kJ/mol. A typical chemical bond or the energy released in a chemical reaction is on the order of a few electron volts, whereas the energy of a cosmic ray is typically in the megaelectron volt or gigaelectron volt range. Particles with energies as high as 10^{20} eV have been reported. These energies are many times the ionization energy of atoms, and this helps explain why cosmic rays consist of bare nuclei rather than atoms. But with energies this large, what new reactive processes become possible? Are there engineering problems whose design solutions must account for the possible impact of cosmic rays?

As cosmic rays enter the atmosphere, the probability of collisions with gas molecules increases greatly. These collisions can induce **nuclear reactions** in the atmosphere, so that the particles actually reaching the surface of Earth might be quite different from those entering the atmosphere. One such nuclear reaction leads to the formation of the radioactive isotope ^{14}C, which we will see is used in carbon dating of archaeological artifacts.

Where does carbon-14 originate? In Earth's upper atmosphere, high-energy cosmic rays strike nuclei and induce nuclear reactions. In one such process, for example, a free neutron is absorbed by a nitrogen nucleus. The result is a carbon nucleus and a proton.

Naturally occurring terrestrial carbon is 98.9% carbon-12, $^{12}_{6}C$, and 1.11% carbon-13, $^{13}_{6}C$. Both are stable isotopes. Carbon-14, $^{14}_{6}C$, is unstable; it undergoes spontaneous **radioactive decay** or disintegration, ejecting particles from the nucleus and forming an atom of nitrogen. How does this process enable radiocarbon dating? To address the questions raised in this section, we'll need to understand some of the fundamentals of nuclear chemistry.

Radioactive decay *is any process by which an unstable atom or nucleus spontaneously emits subatomic particles.*

14-2 Radioactivity and Nuclear Reactions

Radioactive Decay

We've noted that the extremely high energies of cosmic rays can induce nuclear reactions in the atmosphere and that one such reaction is responsible for the formation of ^{14}C. In describing that process, we noted that *a free neutron is absorbed by a nitrogen nucleus and the result is a carbon nucleus and a proton.* The language used here is very similar to that we might use to describe a chemical reaction: the neutron and the nitrogen nucleus are reactants, and the carbon nucleus and the proton are products. It is convenient to summarize this information in an appropriate nuclear equation. How might we write such an equation? As for an ordinary chemical reaction, we place the starting materials on the left-hand side and the final products on the right, with an arrow separating the two. In this case, though, our reactants and products will be atoms or subatomic particles instead of molecules.

In Chapter 2, we saw that the symbol of a **nuclide**, E, with mass number A and atomic number Z can be written in the following form:

The term **nuclide** *includes atoms, ions, and nuclei. Nuclear reactions often involve highly ionized species.*

$$^{\text{mass number}}_{\text{atomic number}}\text{E} \text{ or } ^{A}_{Z}\text{E}$$

The most common isotope of nitrogen is nitrogen-14, for which the symbol is $^{14}_{7}N$. If we recognize that the atomic number is really just the charge on the nucleus, we can also write similar symbols for subatomic particles, including neutrons ($^{1}_{0}n$), protons ($^{1}_{1}p$), and electrons ($^{0}_{-1}e$). Using this idea, we can assemble the following equation for the nuclear reaction described above:

$$^{14}_{7}N + ^{1}_{0}n \rightarrow ^{14}_{6}C + ^{1}_{1}p$$

It is easy to see that this equation is not balanced in the same way as an ordinary chemical reaction. We have nitrogen on the reactant side and carbon on the product side. Because nuclear reactions can convert atoms of one element into atoms of another, the usual rules for balancing cannot apply. But nuclear equations must also obey their own set of conservation rules. Specifically, notice that this equation is balanced with respect to both charge and mass number. The sum of the mass numbers on each side is 15, whereas the sum of the charges on each side is 7. Every nuclear equation should be balanced with respect to both mass number and charge. Although the appearance of protons and neutrons in this equation may seem strange at first, at least we can say that the equation contains only familiar species: atoms, neutrons, and protons. For other nuclear reactions, we will need to account for different types of radioactivity as well.

Soon after radioactivity was discovered emanating from uranium, physicist Ernest Rutherford demonstrated that two distinct types of radiation could be distinguished, as shown in **Figure 14.1(a)**. One type was stopped by thin pieces of aluminum, whereas the other penetrated the metal sheets. Those that were stopped by the metal he called **alpha rays,** and those that passed through were called **beta rays**. In a magnetic or electric field, the two types of radiation were deflected in different directions, indicating that they had opposite charges (**Figure 14.1(b)**). One of the particles was deflected more than the other, indicating that their charge to mass ratios were different. This experiment also revealed a third type of radiation, which passed through the field undeflected. Rutherford called it a **gamma ray**.

Radiation *refers to particles or photons emitted in nuclear decay.*

In the years since these observations were made, nuclear scientists have characterized these forms of radiation in detail. The more massive and positively charged particle is an **alpha particle** (α), which is actually a helium nucleus, $^{4}_{2}He$. The negatively

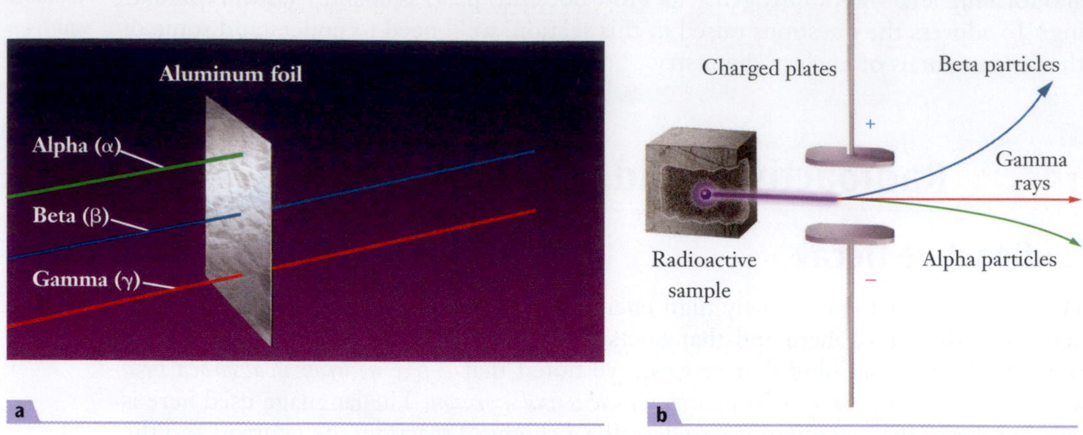

Figure 14.1 ■ (**a**) In early studies of radioactivity, Rutherford observed two types of radiation, which he called alpha and beta rays. A thin sheet of aluminum foil blocked the alpha rays, but beta rays passed through the foil. (**b**) Rutherford also studied the effect of an electric field on radiation and saw that alpha and beta rays were deflected in opposite directions. This experiment also revealed a third type of radiation, gamma rays, that passed through the field undeflected. (Like beta rays, gamma rays pass through the foil in the experiment shown in (**a**).)

charged particle is an electron, ${}_{-1}^{0}e$, but electrons emitted from the nucleus are usually called **beta particles** (β^- or ${}_{-1}^{0}\beta$). (There is also a positive beta particle, called a **positron** (${}_{1}^{0}\beta$), which we'll examine later in this section.) The particles unaffected by the magnetic field are **gamma rays** (γ), high-energy photons of electromagnetic radiation.

Alpha Decay

When a nucleus undergoes **alpha decay**, it ejects an alpha particle so that its mass number decreases by 4 and its atomic number decreases by 2. Uranium-238 is one example of a nuclide that decays by alpha emission:

$${}_{92}^{238}U \rightarrow {}_{90}^{234}Th + {}_{2}^{4}He$$

The atomic number of the new nucleus identifies it as thorium, and its mass number is 234. In radioactive decay, it is common to refer to the reactant nucleus as the *parent* and the product nucleus as the *daughter*. So here we would say that ^{238}U is the parent and ^{234}Th is the daughter. To be sure that we have accounted for every particle, compare the mass numbers on the left and right sides of the equation (238 = 234 + 4). Repeat the process for the atomic numbers (92 = 90 + 2). Example Problem 14.1 provides additional practice with this concept.

EXAMPLE PROBLEM 14.1

Complete the equations for each of the following nuclear decay processes:

$${}_{84}^{210}Po \rightarrow {}_{82}^{206}Pb + ?$$

$${}_{90}^{230}Th \rightarrow ? + {}_{2}^{4}He$$

Strategy Nuclear equations must be balanced with respect to both total mass and charge. As long as only one particle is missing from each equation, we can use these two criteria to determine its identity.

Solution Consider the first reaction. Looking at the two isotopes shown, the difference in atomic number is 84 − 82 = 2, and the difference in mass number is

210 − 206 = 4. That means that the missing particle must have a mass number of 4 and an atomic number of 2, making it an alpha particle. The completed equation is

$$^{210}_{84}\text{Po} \rightarrow {}^{206}_{82}\text{Pb} + {}^{4}_{2}\text{He}$$

Next, consider the second reaction. Again, looking at the two species shown, the difference in atomic number is 90 − 2 = 88 and the difference in mass number is 230 − 4 = 226. The fact that the atomic number is 88 tells us the missing isotope is radium, Ra, and the mass number tells us it must be radium-226. This lets us complete the equation:

$$^{230}_{90}\text{Th} \rightarrow {}^{226}_{88}\text{Ra} + {}^{4}_{2}\text{He}$$

Discussion This type of problem for alpha decay is normally quite straightforward. Later in the chapter, we will find nuclear reactions where, for example, several neutrons are produced as products. That type of problem is slightly more complicated because the stoichiometry changes from the simple one-to-one ratios seen here.

Check Your Understanding Identify the element and isotope of the missing reactant in the following nuclear equation:

$$? \rightarrow {}^{205}_{82}\text{Pb} + {}^{4}_{2}\text{He}$$

Beta Decay

A ^{14}C nucleus undergoes spontaneous decay by emitting a beta particle, β^-, or $_{-1}^{0}\beta$, an electron ejected from the nucleus. But how can an electron be ejected *from the nucleus*? The answer is that in beta decay, a neutron must decay into a proton and an electron. Detailed study of the energetics of **beta decay** shows that an additional particle, with no charge and virtually no mass, must also be emitted. This particle, called an *antineutrino*, is written as $\bar{\nu}$.

A neutrino and its antineutrino have identical masses, estimated to be on the order of a million times less than the mass of an electron.

$$^{1}_{0}\text{n} \rightarrow {}^{1}_{1}\text{p} + {}_{-1}^{0}\beta + \bar{\nu}$$

The proton remains in the nucleus and increases the atomic number by 1.

Recall that in the *Insight* we said that ^{14}C is produced by neutron absorption by a nitrogen-14 nucleus:

$$^{14}_{7}\text{N} + {}^{1}_{0}\text{n} \rightarrow {}^{14}_{6}\text{C} + {}^{1}_{1}\text{p}$$

Carbon-14 is radioactive and eventually undergoes beta decay:

$$^{14}_{6}\text{C} \rightarrow {}^{14}_{7}\text{N} + {}_{-1}^{0}\beta + \bar{\nu}$$

In this process—and in any β^- decay—*the atomic number increases by 1* because a proton has taken the place of the neutron that decayed. Because the emitted beta particle is released, we can detect it fairly easily. Carbon dating schemes are based on the fact that this decay proceeds at a known rate, as we will see in Section 14-3. Example Problem 14.2 provides practice with nuclear equations for β^- decay.

EXAMPLE PROBLEM 14.2

We have seen the nuclear equation for the beta decay involved in radiocarbon dating. Now complete the equations for each of the following β^- decay reactions, using $_{-1}^{0}\beta$ to represent the beta particle:

$$^{234}_{90}\text{Th} \rightarrow {}^{234}_{91}\text{Pa} + \quad ?$$

$$^{234}_{91}\text{Pa} \rightarrow \; ? \; + {}_{-1}^{0}\beta + \bar{\nu}$$

Strategy We can use both the mass numbers and the charges given to determine the missing particles in the equations, much as in the previous example for alpha decay.

Solution Consider each equation separately, starting with the thorium-234 decay:

$$^{234}_{90}\text{Th} \rightarrow {}^{234}_{91}\text{Pa} + ?$$

A mass number of zero does not mean that a particle has zero mass!

The fact that both Th and Pa have mass numbers of 234 indicates that the other particle in the equation must have a mass number of zero. This is consistent with beta decay. Balancing the charges on the reactant and product sides requires that the unknown has a charge of 1− (so that 91 − 1 = 90). Because this is beta decay, an antineutrino is also emitted. The equation is

$$^{234}_{90}\text{Th} \rightarrow {}^{234}_{91}\text{Pa} + {}^{0}_{-1}\beta + \bar{\nu}$$

Next, turn to the decay of protactinium-234:

$$^{234}_{91}\text{Pa} \rightarrow ? + {}^{0}_{-1}\beta + \bar{\nu}$$

Because the beta particle given as a product has no mass number, we know that the unknown particle has a mass of 234. The beta particle has a charge of 1−, so the unknown particle must have an atomic number of 92 (so 92 − 1 = 91). The missing particle is ^{234}U, so the equation is

$$^{234}_{91}\text{Pa} \rightarrow {}^{234}_{92}\text{U} + {}^{0}_{-1}\beta + \bar{\nu}$$

Check Your Understanding The only remaining way to look at this type of question results when the unknown is the reactant. Suppose you observe beta decay and then determine that the product nuclide is ^{218}At. What was the reactant nuclide?

Gamma Decay

Recall from Chapter 6 that a photon is a quantum of electromagnetic radiation.

Gamma decay is the emission of a high-energy photon and tends to accompany other types of decay. In the nucleus, protons and neutrons occupy energy levels, analogous to the energy levels occupied by electrons in an atom. As we discussed in Section 6-3, when an atom is in an excited state, it can emit a photon as an electron moves from a higher energy orbital to a lower energy orbital. Similarly, when alpha and beta particles leave the nucleus, some energy levels in the nucleus are no longer occupied. The nucleus is in an excited state, and to return to its ground state, it emits a photon. Because the spacing between nuclear energy levels is very large, this photon will take the form of very high energy gamma radiation. The wavelength of gamma radiation is on the order of 10^{-12} m and the frequency is about $3 \times 10^{20}\ \text{s}^{-1}$. This corresponds to energy of 10^{8} kJ/mol, which is several orders of magnitude larger than the energies of ordinary chemical reactions.

Gamma radiation, which changes neither the mass number nor the atomic number of a nuclide, accompanies the beta decay of most nuclei, including carbon-14. We can rewrite the decay of carbon-14 as follows, explicitly showing the emission of gamma radiation:

$$^{14}_{6}\text{C} \rightarrow {}^{14}_{7}\text{N} + {}^{0}_{-1}\beta + \bar{\nu} + {}^{0}_{0}\gamma$$

Note that the balance of the equation is unchanged by the emission of gamma radiation: gamma rays are electromagnetic radiation, and so they have neither mass nor charge.

Electron Capture

In **electron capture**, the nucleus captures an electron from the first ($n = 1$) shell in the atom. Because that first level is also called the "K shell," electron capture is often referred to as **K capture**. The result is that a proton in the nucleus is converted to a neutron. In effect, electron capture is the reverse of beta emission. As in beta decay, an additional particle is needed to conserve energy; in this case, that particle is a neutrino, ν:

$$^{1}_{1}\text{p} + {}^{0}_{-1}\text{e} \rightarrow {}^{1}_{0}\text{n} + \nu$$

The result of electron capture is that the nuclear charge decreases by 1, as illustrated by the following equation for electron capture by aluminum-26:

$$^{26}_{13}\text{Al} + {}^{0}_{-1}\text{e} \rightarrow {}^{26}_{12}\text{Mg} + \nu$$

Positron Emission

A positron is a positively charged electron, β^+ or $^{0}_{1}\beta$. A positron and an electron form a *matter–antimatter* pair; they are identical in mass and spin but opposite in charge. Collisions of particles and their antiparticles, such as the electron and the positron, result in the annihilation of both particles and the conversion of their combined masses to energy (**Figure 14.2**). The collision of a positron with an electron produces two 511-keV gamma-ray photons, traveling in opposite directions. (These gamma rays are the basis for positron emission tomography, which we will discuss in Section 14-8.) In β^+ decay, a proton decays into a neutron and a positron:

The neutrino and antineutrino are a matter–antimatter pair, like the positron and the electron.

$$^{1}_{1}\text{p} \rightarrow {}^{1}_{0}\text{n} + {}^{0}_{1}\beta + \nu$$

Positron decay has the same effect as electron capture: *The nuclear charge decreases by 1.*

Positron decay has an important application in medicine, as we'll see in Section 14-8. Example Problem 14.3 provides some practice in writing nuclear equations that involve positron emission or electron capture.

EXAMPLE PROBLEM 14.3

Complete the following equations with the correct particles and identify the mode of decay.

(a) $^{15}_{8}\text{O} \rightarrow {}^{15}_{7}\text{N} + \ ?$

(b) $^{40}_{19}\text{K} \rightarrow \ ? \ + {}^{0}_{-1}\beta + \bar{\nu}$

(c) $^{40}_{19}\text{K} + \ ? \rightarrow {}^{40}_{18}\text{Ar} + \bar{\nu}$

Strategy As before, we rely on both mass numbers and charges to identify the missing species in the equations. Once each species has been identified, we can use our definitions of the various nuclear reactions to identify the decay mode.

Solution Consider each equation in turn.

(a) $^{15}_{8}\text{O} \rightarrow {}^{15}_{7}\text{N} + \ ?$

There is no change in the mass number from oxygen-15 to nitrogen-15, so the unknown particle has a mass number of zero. For the sum of the charges on the product side to equal eight, the unknown particle must have a charge of 1+. These two facts tell us that the unknown particle is a positron, and the event is positron emission. A neutrino is also needed to complete the equation:

$$^{15}_{8}\text{O} \rightarrow {}^{15}_{7}\text{N} + {}^{0}_{1}\beta + \nu$$

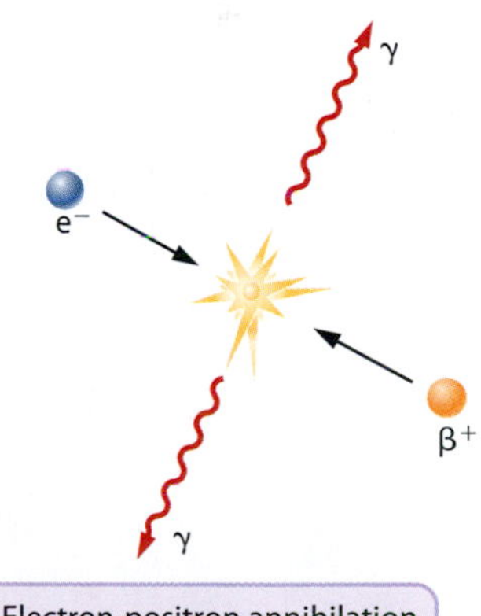

Figure 14.2 ■ Collision of any particle with its antiparticle results in the annihilation of both particles and the complete conversion of their masses to energy. The annihilation of a positron and an electron produces two 511-keV gamma-ray photons.

(b) ${}^{40}_{19}\text{K} \rightarrow\ ?\ + {}^{0}_{-1}\beta + \bar{\nu}$

We are looking at beta decay, so the mass number of the decaying isotope does not change. Thus the mass number of the product must be 40. The sum of the charges of the products must be 19, so the unknown must have a value of 20 (19 = 20 − 1). The unknown is ${}^{40}\text{Ca}$, and the process is beta decay:

$$ {}^{40}_{19}\text{K} \rightarrow {}^{40}_{20}\text{Ca} + {}^{0}_{-1}\beta + \bar{\nu} $$

(c) ${}^{40}_{19}\text{K} +\ ?\ \rightarrow {}^{40}_{18}\text{Ar} + \nu$

There is no change in mass number from potassium to argon in this reaction, so the mass number of the missing particle must be zero. For the sum of the charges of the reactants to equal 18 requires a charge of 1−. The missing particle is an electron, and the event is electron capture:

$$ {}^{40}_{19}\text{K} + {}^{0}_{-1}\text{e} \rightarrow {}^{40}_{18}\text{Ar} + \nu $$

Discussion The examples here point out that it is not always easy to distinguish between various possible decay processes. We wrote the last equation as electron capture, for example, but ${}^{40}_{19}\text{K}$ could also decay to ${}^{40}_{18}\text{Ar}$ by positron emission:

$$ {}^{40}_{19}\text{K} \rightarrow {}^{40}_{18}\text{Ar} + {}^{0}_{1}\beta + \nu $$

Also notice that in the last two equations, we have the same parent nucleus (${}^{40}_{19}\text{K}$) decaying to two different daughter nuclei (${}^{40}_{20}\text{Ca}$ and ${}^{40}_{18}\text{Ar}$). We've seen similar behavior in ordinary chemical reactions. Complete and incomplete combustion are one example in which the same reactants can produce different sets of products.

Check Your Understanding Carbon-11 is another unstable isotope of carbon. It decays by positron emission. Write the equation for this nuclear reaction.

14-3 Kinetics of Radioactive Decay

In Chapter 11, we learned that the rate of a chemical reaction can be expressed in terms of either the rate at which a reactant is consumed or the rate at which the product is formed. So we study the rates of chemical reactions by monitoring the concentration of one or more of the substances involved. In studying radioactive decay, however, we can generally measure the rate directly. Each decay produces a high-energy particle or photon, which allows us to count the decays in a given time period. The rate at which a sample decays is called the *activity* of the sample. For a sample of N nuclei, the rate of disintegration is given by $\Delta N/\Delta t$. The SI unit of nuclear activity is the **becquerel (Bq)**, defined as one nuclear disintegration per second. An older term, the **curie (Ci)**, is a much larger unit, originally defined as the number of disintegrations per second in 1 gram of radium-226. According to the currently accepted definition, 1 Ci is exactly 3.7×10^{10} Bq.

To explore the kinetics of radioactive decay, let's consider the decay of ${}^{131}\text{I}$, an isotope used in diagnosing thyroid irregularities. It decays by beta emission:

$$ {}^{131}_{53}\text{I} \rightarrow {}^{131}_{54}\text{Xe} + {}^{0}_{-1}\beta + \bar{\nu} $$

If we start with exactly 100 μg of ${}^{131}\text{I}$ and monitor its radioactivity, the initial reading will be 4.60×10^{11} Bq. But because each decay reduces the number of ${}^{131}\text{I}$ atoms remaining in the sample, the activity will decrease over time. If we monitor the activity for 40 days, we will obtain the data shown in **Figure 14.3**.

Looking at the graph in Figure 14.3, it should be apparent that the activity decreases exponentially with time. Because the activity is proportional to the number of nuclei present, N also decreases exponentially. Every 8 days, the value of N

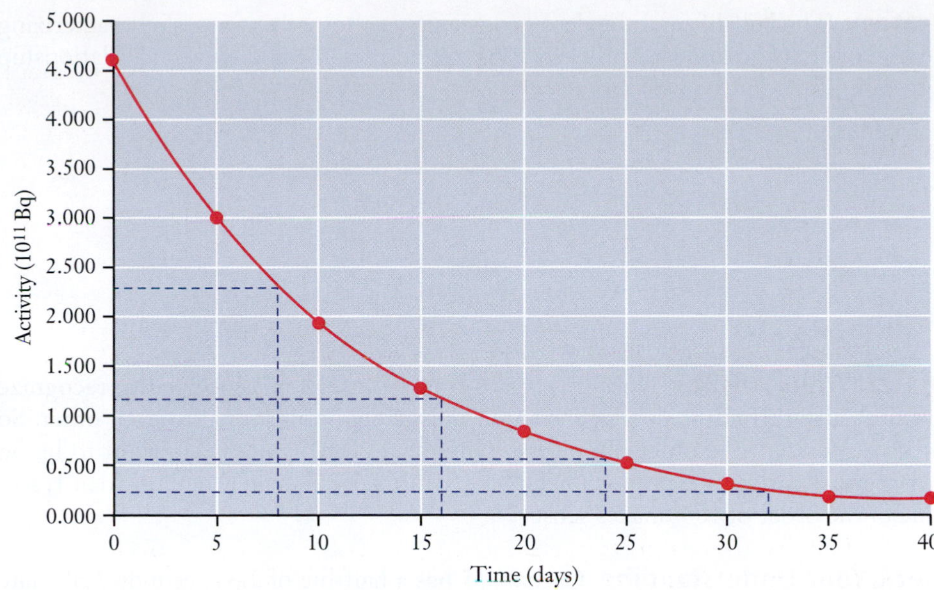

Figure 14.3 ■ Radioactive decay always follows first-order kinetics. This means that the half-life is constant for any given radioisotope. The beta decay of ^{131}I has a half-life of 8 days, as shown here.

drops by one-half, as shown by the dashed blue lines in the graph. The equation for this curve is

$$N = N_0 e^{-kt} \tag{14.1}$$

where N_0 is the initial number of nuclei and k is the **decay constant.**

The decay constant for nuclear decay is often designated as λ. We use k here to emphasize the similarity to other rate constants.

Taking the natural logarithm of each side and rearranging, the equation becomes

$$\ln N = \ln N_0 - kt$$

$$\ln N_0 - \ln N = kt$$

$$\ln \frac{N_0}{N} = kt \tag{14.2}$$

Recalling our treatment of kinetics in Chapter 11, we see that radioactive decay is a first-order process (Section 11-4), and the **half-life** of ^{131}I can be found by finding the time, $t_{1/2}$, it takes for one-half of a sample to disintegrate.

The time it takes for N_0 to decay to one-half its original value is given by

$$\ln\left(\frac{N_0}{\frac{1}{2}N_0}\right) = \ln\left(\frac{2N_0}{N_0}\right) = \ln(2) = kt_{1/2}$$

Solving this equation for the half-life gives

$$t_{1/2} = \frac{\ln 2}{k} = \frac{0.693}{k} \tag{14.3}$$

These equations for first-order kinetics are widely used in nuclear chemistry, as illustrated in Example Problem 14.4.

EXAMPLE PROBLEM 14.4

The half-life of carbon-14, used in radiocarbon dating, is 5730 years. What is the decay constant for carbon-14?

Strategy The half-life and the decay constant are alternative ways of characterizing the rate of decay, so knowing one allows us to find the other. The needed relationship is given in Equation 14.3.

Solution

$$t_{1/2} = \frac{0.693}{k}$$

$$k = \frac{0.693}{t_{1/2}} = \frac{0.693}{5730 \text{ yr}} = 1.21 \times 10^{-4} \text{ yr}^{-1}$$

Analyze Your Answer Does this answer make sense? First, we need to recognize that the half-life and the decay constant are inversely proportional to each other. So with a half-life on the order of 10^3 yr, we might expect the decay constant to be on the order of 10^{-3} yr. The proportionality constant, 0.693, is somewhat less than 1, so a value of the order of 10^{-4} makes sense.

Check Your Understanding Carbon-15 has a half-life of 2.45 seconds. Calculate the decay constant of carbon-15 and compare it to that of carbon-14.

Radiocarbon Dating

With these kinetic equations in hand, we can now see how radiocarbon dating allows us to determine the age of artifacts. As we mentioned in the chapter opening *Insight*, ^{14}C is constantly formed through the interaction of cosmic rays with the atmosphere. This ^{14}C is incorporated into plants and animals, so that the $^{14}C/^{12}C$ ratio in *living* organisms remains relatively constant over time. Once a plant or animal dies, decay of ^{14}C continues. So the $^{14}C/^{12}C$ ratio in the remains of plant or animal tissues decreases as time passes. The amount of ^{14}C formed in the atmosphere depends on the composition of the atmosphere itself, as well as on the incoming flux of cosmic rays. If we look at the elemental composition of the atmosphere for the past 70,000 years, we see little change. So it is reasonable to expect that the production and availability of ^{14}C to plants (and thus the $^{14}C/^{12}C$ ratio) has remained nearly constant during that time. When the plant or animal dies, this ratio becomes the initial point for decay because that is when the $^{14}C/^{12}C$ ratio starts to decrease due to ^{14}C decay. If we can determine the current value of this ratio, therefore, we have all the information we need to use the kinetic equations for radioactive decay and determine the age of artifacts. Example Problem 14.5 provides practice with this type of exercise.

Dendrochronology, which is based on counting growth rings in long-lived trees, has been used to calibrate carbon dating.

EXAMPLE PROBLEM 14.5

A piece of cloth is discovered in a burial pit in the southwestern United States. A tiny sample of the cloth is burned to CO_2, and the $^{14}C/^{12}C$ ratio is 0.250 times the ratio in today's atmosphere. How old is the cloth?

Strategy We have the decay constant, k, for the relevant process from Example Problem 14.4. This will allow us to determine t, the time that has passed since the fiber in the cloth was harvested.

Solution Recall the following from Example Problem 14.4:

$$t_{1/2} = \frac{0.693}{k}$$

$$k = \frac{0.693}{t_{1/2}} = \frac{0.693}{5730 \text{ yr}} = 1.21 \times 10^{-4} \text{ yr}^{-1}$$

Equation 14.2 relates the amount of ^{14}C to the time, which in this case is the age of the artifact:

$$\ln \frac{N_0}{N} = kt$$

Because the amount of ^{12}C does not change over time, we can use the $^{14}C/^{12}C$ ratio to find N_0/N. We are told that the measured $^{14}C/^{12}C$ ratio is 25.0% of the current ratio in the atmosphere, so we can set $N_0/N = 1/0.25$. Solving for t gives

$$t = \frac{1}{k} \ln \frac{N_0}{N}$$
$$= \frac{1}{1.21 \times 10^{-4}\ \text{yr}^{-1}} \ln \frac{1}{0.250}$$
$$= 11{,}500\ \text{yr}$$

Analyze Your Answer We can consider this answer with the same order of magnitude reasoning we used in Example Problem 14.4. The time is inversely proportional to the decay constant. (Units provide a good hint if you forget this.) With the decay constant on the order of 10^{-4}, our answer should be on the order of 10^4, which it is. We can also refer to the context of this problem: most carbon dating is carried out on archaeological artifacts, so a time frame on the order of 10,000 years seems reasonable.

Check Your Understanding We can also use the activity of a particular sample to solve a radiocarbon dating problem. The activity of ^{14}C in the atmosphere (and therefore in all living things) today is 0.255 Bq/g of *total* carbon. Suppose that archaeologists found carbonized wheat grains in a fire pit at a dig site in the plains of northeastern Colorado. Measurement of the decay rate showed 0.070 Bq/g of carbon. How long ago was the wheat harvested?

Although radiocarbon dating does not produce an exact age (its margin of error is about ± 40 to 100 years), the technique is an invaluable method of determining the ages of objects less than 60,000 years old. Longer-lived nuclei, such as uranium, can be used to date much older minerals and geological formations. **Table 14.1** contains half-lives of some common nuclei.

Table ■ 14.1

Half-lives of some radioactive isotopes

Isotope	Decay Process	Half-life
^{3}H	$^{3}_{1}H \rightarrow ^{3}_{2}He + ^{0}_{-1}\beta + \bar{\nu}$	12.33 y
^{8}Be	$^{8}_{4}Be \rightarrow 2\ ^{4}_{2}He$	~10^{-16} s
^{14}C	$^{14}_{6}C \rightarrow ^{14}_{7}N + ^{0}_{-1}\beta + \bar{\nu}$	5730 y
^{15}O	$^{15}_{8}O \rightarrow ^{15}_{7}N + ^{0}_{1}\beta + \nu$	122.24 s
^{18}F	$^{18}_{9}F \rightarrow ^{18}_{8}O + ^{0}_{1}\beta + \nu$	1.83 h
^{131}I	$^{131}_{53}I \rightarrow ^{131}_{54}Xe + ^{0}_{-1}\beta + \bar{\nu}$	8.02 d
^{235}U	$^{235}_{92}U \rightarrow ^{231}_{90}Th + ^{4}_{2}He$	7.04×10^{8} y
^{238}U	$^{238}_{92}U \rightarrow ^{234}_{90}Th + ^{4}_{2}He$	4.47×10^{9} y
^{259}Sg	$^{259}_{106}Sg \rightarrow ^{255}_{104}Rf + ^{4}_{2}He$	0.9 s

14-4 Nuclear Stability

Although carbon dating provides an interesting and useful tool for archaeology, it also raises some important questions. For example, why does ^{14}C decay, whereas ^{12}C and ^{13}C do not? To address this question in detail requires more knowledge of nuclear science than can be included here, but we can search for patterns of behavior that will allow us to predict some aspects of nuclear chemistry and radioactive decay.

We have relied on the periodic table to help us to understand many trends and patterns in the chemical behavior of the elements. To understand nuclear stability, we might try looking for the same sort of trends and patterns in nuclear behavior. One way to visualize these patterns is through the **chart of the nuclides**, a plot of the number of protons versus the number of neutrons in all known, stable nuclei, as shown in **Figure 14.4**.

It is clear that virtually all stable nuclides fall in a central region in the chart of the nuclides. This area (shown as the blue dots in Figure 14.4) is often called the **band of stability**. The region outside the band is referred to as the **sea of instability**. For low atomic numbers, the stable nuclides lie along a line with Z and N approximately equal. But looking at the figure, we can see that as we move toward higher atomic numbers

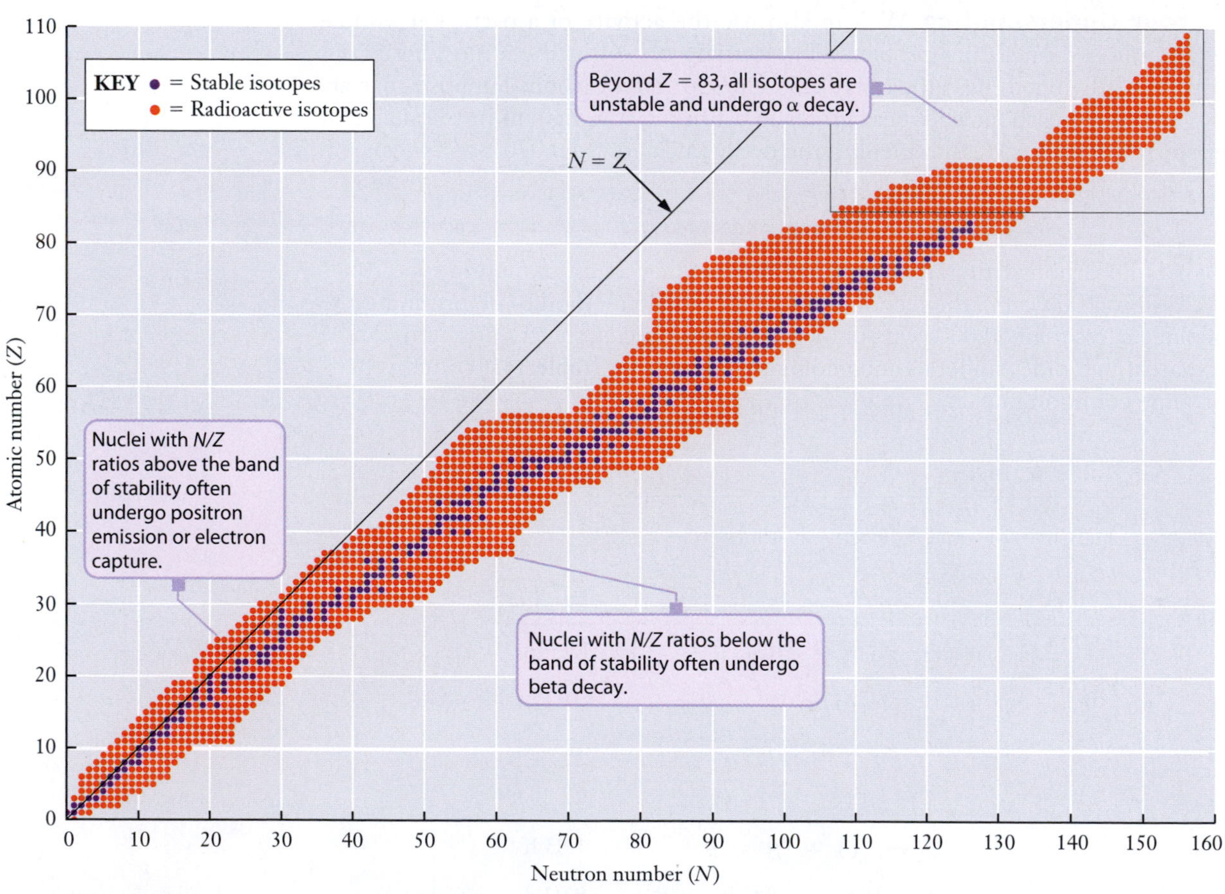

Figure 14.4 ■ The chart of the nuclides is a plot of atomic number (Z) versus neutron number (N) for all known nuclides. All stable isotopes are shown as blue dots, and lie in a region that is referred to as the band of stability. Many general chemistry textbooks reverse the axes in this chart so that N is on the y axis and Z is on the x axis. But the form shown here is used throughout the nuclear science field, and we choose to follow that convention.

(beginning at about $Z = 20$), the band of stability begins to deviate from that line. A nucleus with more protons appears to require additional neutrons to maintain its stability. Eventually, when $Z > 83$, no number of neutrons can stabilize the nucleus and its large collection of positive charges.

This is consistent with examples we have already seen. For a light element such as carbon, we can expect isotopes with roughly equal numbers of protons and neutrons to be stable. So ^{12}C, with six protons and six neutrons is stable, whereas ^{14}C, which has eight neutrons, is unstable and undergoes beta decay, producing ^{14}N with seven protons and seven neutrons. If we examine many more isotopes and their modes of emission, we find that isotopes *below* or to the *right* of the band of stability tend to emit beta particles to gain stability, whereas those *above* or to the *left* undergo positron emission or electron capture. Isotopes to the right of the band have more neutrons than necessary, so beta emission occurs to produce an additional proton (Section 14-2). Isotopes to the left of the band, in contrast, have more protons than needed, so positron emission and electron capture occur to produce an additional neutron and increase stability.

The stable neutron–proton ratio increases to about 1.5:1 by the time we reach $Z = 83$ (bismuth), at the end of the band of stability. Heavier nuclei tend to emit alpha particles, decaying successively, often with beta particle emissions as well, until a stable nucleus is formed. Alpha emission is a quick way to lower both the proton and neutron numbers. In **Figure 14.5**, the **decay series** for ^{238}U illustrates the role of alpha emission in reaching a stable isotope of lead.

From the form of the chart of the nuclides, it seems clear that neutrons play an important role in nuclear stability. But why is this? To explore this issue, we'll need to examine the forces in the nucleus.

In Section 7-2, we discussed Coulomb's law, which states that the electrostatic force between two ions is a function of the inverse square of the distance between them. In the nucleus, the distances between nucleons are so small that the repulsive force between a pair of protons is enormous. So our intuition tells us that the nucleus should not exist—it should fly apart. But it doesn't. Clearly, some attractive force must be at work to outweigh coulombic repulsion.

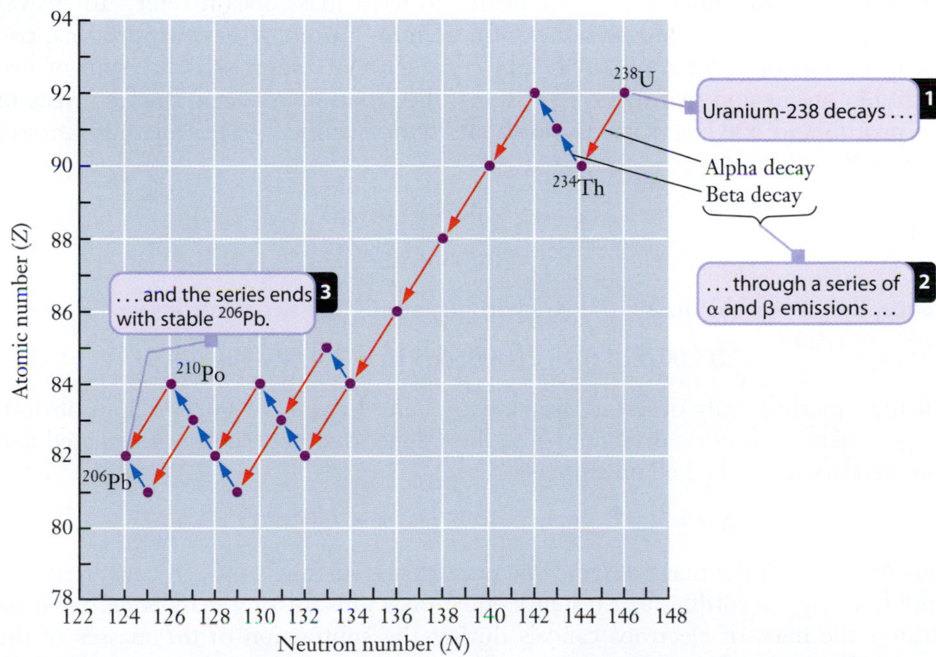

Figure 14.5 ■ The decay series starting with ^{238}U involves a series of alpha and beta emissions before it eventually produces a stable ^{206}Pb product.

Nuclear scientists know that one of the forces that holds the nucleus together is the **strong force**, which acts over very short (i.e., nuclear) distances between any two nucleons. Attraction due to the strong force overcomes the repulsive coulombic forces between protons and binds the nucleus together.

The strong force arises from interactions between fundamental particles called quarks *and* gluons.

And what is the role of the neutron? The strong force is not dependent on charge, so it acts between *any* pair of nucleons: proton–proton, proton–neutron, or neutron–neutron. So neutrons help hold the nucleus together. Another function of the neutron may be to "dilute" the protons, keeping them somewhat farther apart from one another and diminishing repulsions. These two neutron functions may help explain the need for a larger ratio of neutrons to protons in heavier nuclei.

14-5 Energetics of Nuclear Reactions

We introduced nuclear reactions by considering the effects of high-energy cosmic rays entering the atmosphere. You are undoubtedly familiar with the notion that nuclear technology, in the forms of both weapons and power plants, can be used to release tremendous amounts of energy. But where does this energy come from, and why are the energies of nuclear reactions so much larger than those in ordinary chemical reactions? To consider these questions, we'll need to turn our attention to nuclei and the forces that hold them together.

We know that a stable nucleus doesn't fly apart, so how much energy might be required to take it apart? The **nuclear binding energy** is the energy that would be released if the nucleus were formed from a collection of free nucleons. The greater the binding energy, the more stable the nucleus. Where does this energy come from?

Binding Energy

The answer is contained in Einstein's famous equation, which describes the interconversion of mass and energy:

$$E = mc^2 \tag{14.4}$$

The speed of light, c, is 2.9979×10^8 m s^{-1}. This equation may be familiar, but it is not generally understood. Specifically, to what mass does it refer? To answer this question, let's calculate the mass of a helium-4 atom. It has two protons, two neutrons, and two electrons, so we can consider that the mass is the sum of two hydrogen atoms and two neutrons. (Remember that a hydrogen atom consists of one proton and one electron.) Experimentally determined values for those masses are as follows:

$$^{1}\text{H} = 1.007825 \text{ u}$$

$$^{1}n = 1.008665 \text{ u}$$

Hence, the sum of the masses of two hydrogen atoms and two neutrons is

$$2(1.007825 \text{ u}) + 2(1.008665 \text{ u}) = 4.032980 \text{ u}$$

But the experimentally observed mass of a helium-4 atom is 4.002603 u, in obvious disagreement with our calculation. The difference, Δm, between the calculated and observed masses of the helium atom is

$$\Delta m = 4.032980 \text{ u} - 4.002603 \text{ u} = 0.030377 \text{ u}$$

This Δm is called the **mass defect**. The mass that is missing has been converted into binding energy according to Einstein's equation. (This missing mass is *nuclear* mass because the mass of electrons cancels during the subtraction of the masses of the atoms.) Thus using Einstein's equation, we can calculate the binding energy, E_b, of the nucleus. If we first convert the mass defect from atomic mass units to kilograms, then the resulting value for mc^2 will be in joules (1 J = 1 kg m^2 s^{-2}):

$$
\begin{aligned}
E_b &= (\Delta m)c^2 \\
&= 0.030377\ \text{u}\left(\frac{1.66054 \times 10^{-27}\ \text{kg}}{\text{u}}\right)\left(\frac{2.99792 \times 10^8\ \text{m}}{\text{s}}\right)^2 \\
&= 4.5335 \times 10^{-12}\ \text{kg m}^2\ \text{s}^{-2} \\
&= 4.5335 \times 10^{-12}\ \text{J}
\end{aligned}
$$

We've found the energy released when one ^{4}He nucleus forms from its constituent nucleons. For a mole of ^{4}He nuclei, the energy released is

$$
\begin{aligned}
E_b &= (4.5335 \times 10^{-12}\ \text{J})\,(6.02214 \times 10^{23}\ \text{mol}^{-1}) \\
&= 2.7301 \times 10^{12}\ \text{J/mol} \\
&= 2.7301 \times 10^{9}\ \text{kJ/mol}
\end{aligned}
$$

The binding energy of 1 mole (or just 4 g) of helium is equivalent to the energy required to drive an average automobile about 30 times around the earth at the equator! Obviously, this is much more energy than could be derived chemically by burning a comparable amount of any conceivable fuel. (Recall that the energy released in an exothermic chemical reaction is typically on the order of hundreds of kJ per mole.)

Calculating and comparing the binding energies of various nuclei can provide us with some added insight into nuclear stability. A convenient way to visualize this is through a plot of binding energy per nucleon versus mass number for the most stable isotope of each element (**Figure 14.6**). (The energies in this figure are in megaelectron volts, or MeV. 1 MeV = 1.60×10^{-13} J.) The plot shows that the binding energy per nucleon goes through a maximum, indicating that there is a region of maximum nuclear stability. This region is centered around ^{56}Fe, indicating that this is the most stable of all known nuclei.

The plot suggests that two processes might produce greater stability. If two lighter nuclei can be combined, the result would be a nucleus of greater stability. Such a process is called **nuclear fusion**. On the other hand, some of the heaviest nuclei can split, resulting in two or more lighter nuclei that might fall within the region of greatest stability. This is called **nuclear fission**. We'll look at both processes in more detail later.

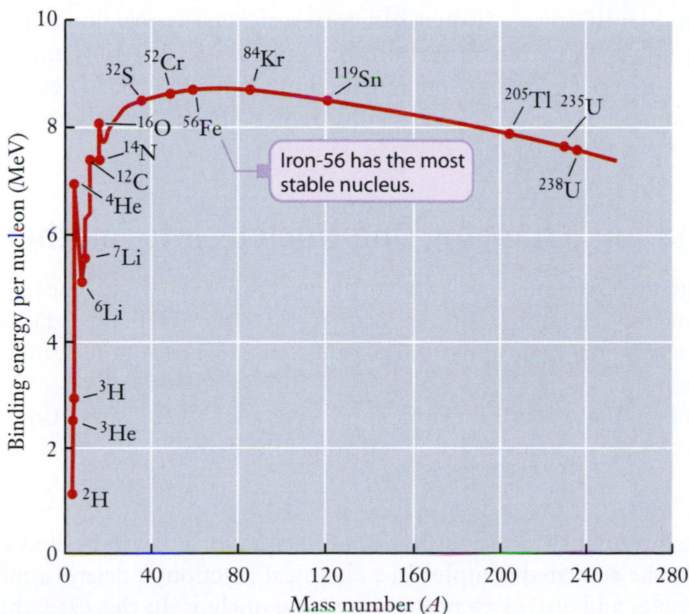

Figure 14.6 ■ The binding energy per nucleon is plotted as a function of mass number for the elements from hydrogen through uranium. The curve goes through a maximum at ^{56}Fe, which means that ^{56}Fe is the most stable of all nuclei.

Magic Numbers and Nuclear Shells

In Chapter 6, we discussed the relative stability of filled shells or subshells of atomic orbitals. The noble gases, with atomic numbers of 2, 10, 18, 36, 54, and 86, provide the clearest examples of this behavior. These elements, with their characteristic np^6 electron configurations, are so stable that they are almost completely unreactive.

Much like the periodic table reveals patterns in the stability of elements, an examination of a list of stable isotopes also shows various patterns. Of more than 260 stable nuclei, most have even numbers of both protons and neutrons. Only a handful, including ^{14}N, have odd numbers of both. The rest have either even numbers of protons and odd numbers of neutrons, or vice versa.

It is also possible to identify a set of isotopes that show special stability. These unusually stable isotopes have atomic numbers, Z, or neutron numbers, N, of 2, 8, 20, 28, 50, 82, 126, or 184. Nuclear scientists refer to these values as **magic numbers**. Any isotope in which Z or N is a magic number can be expected to be especially stable. If both Z and N have magic number values, the effect can be even greater and the nucleus is said to be doubly magic.

These observations, along with more detailed studies of nuclear binding energies, point toward the idea that nucleons occupy shells or energy levels, much like those we've seen for electrons in the quantum mechanical model of the atom.

Like electrons, many nuclei also have spin. Nuclear spins are the key to magnetic resonance imaging (MRI) techniques.

Although it has not yet been observed, element number 126 is of particular interest to nuclear chemists and physicists. Its magic atomic number should indicate an unusually stable nucleus for its size. But getting to element 126 presents significant experimental challenges. The so-called superheavy elements (those after fermium, Fm, in the periodic table) pose unusual problems for the researcher because of their extremely short half-lives and because only a few atoms can be produced at a time.

14-6 Transmutation, Fission, and Fusion

The fact that nitrogen can be converted into carbon by cosmic rays in the atmosphere points to an important parallel between nuclear reactions and chemical reactions. In chemical reactions, atoms are rearranged to form new molecules. In nuclear reactions, nucleons are rearranged to form new nuclei. As we've seen in the nuclear equations we've written, this means that the identities of atoms actually change in nuclear reactions. We can identify three distinct categories of nuclear reactions that change the identity of a nucleus. In **transmutation**, one nucleus changes to another, either by natural decay or in response to some outside intervention, such as neutron bombardment. In fission, a heavy nucleus splits into lighter nuclei, and in fusion, light nuclei merge into a heavier nucleus.

Transmutation: Changing One Nucleus into Another

One of the goals of the medieval alchemists was to use chemical means to change a base metal, such as lead, into gold, a precious metal. The alchemists failed, of course. As we now know, such a transmutation of one element into another can occur only by a nuclear reaction.

The production of ^{14}C in the atmosphere is a natural transmutation. Another interesting nuclear reaction arises from neutron capture by ^{10}B:

$$^{10}_{5}B + ^{1}_{0}n \rightarrow ^{11}_{5}B^* \rightarrow ^{7}_{3}Li + ^{4}_{2}He$$

The unstable intermediate nucleus ($^{11}_{5}B^*$ in the equation above) is called a **compound nucleus**. Like the activated complex in a chemical reaction, it decays almost instantly, emitting particles and energy to produce a stable nucleus. In this case, the compound

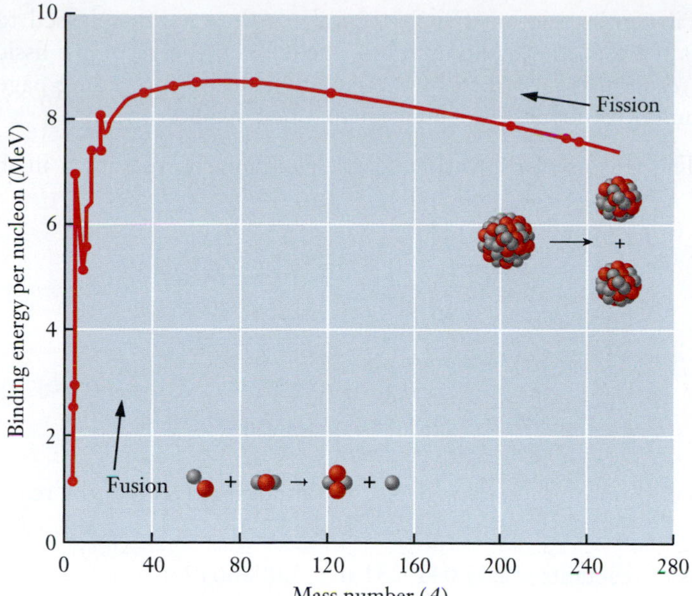

Figure 14.7 ■ Fission and fusion both lead to more stable nuclei. In fusion, light nuclei combine to form a heavier product, moving up toward the maximum in the nuclear binding energy plot. In fission, a very heavy nucleus splits into smaller pieces, moving back toward the maximum in the binding energy plot.

nucleus decays by alpha particle emission. This reaction is especially interesting because ^{11}B is normally a stable isotope. The nuclear reaction, however, produces a compound nucleus with the same mass number as a stable isotope, but which nonetheless decays because it is formed in a highly excited state.

Transmutation reactions are used to produce a number of medically useful radioisotopes. We'll look at applications of some of these isotopes in the closing *Insight* section.

Fission

Earlier, we discussed the idea that if a large, unstable nucleus could split into two smaller ones, the resulting nuclei might fall in the region of greater stability (**Figure 14.7**). In fission, this splitting of a nucleus is exactly what happens. Not all nuclei can undergo fission, however, and those that do are said to be *fissionable* or *fissile*. Some fission reactions are spontaneous, meaning that a large nucleus simply breaks into smaller pieces. Other fission reactions must be induced by neutron bombardment. (Neutron bombardment can also be used to increase the rate of decay for what might already be a spontaneously fissile nucleus.)

In induced fission, a neutron is absorbed by a large fissile nucleus, such as ^{235}U, producing a highly unstable intermediate compound nucleus, such as ^{236}U. This compound nucleus separates into two smaller parts, emitting neutrons in the process. One of the many possible fission pairs that could be formed from ^{236}U is barium and krypton:

$$^{235}_{92}U + {}^{1}_{0}n \rightarrow {}^{236}_{92}U^{*} \rightarrow {}^{141}_{56}Ba + {}^{92}_{36}Kr + 3\,{}^{1}_{0}n$$

We have already noted that nuclear reactions convert mass into energy. The fission of ^{235}U is at the heart of the nuclear power industry, so we might ask how much energy is released by this reaction. Example Problem 14.6 answers this question and provides practice in determining the energy release from fission.

EXAMPLE PROBLEM 14.6

Calculate the energy released by a nucleus of uranium-235 if it splits into a barium-141 nucleus and a krypton-92 nucleus according to the equation above.

Strategy Einstein's equation ($E = mc^2$) relates the energy released to the difference in mass between the fissile uranium nuclide and the resulting fission products. The fission reaction is described by the equation on the preceding page, so we just need to account for the masses of all participating particles.

Solution The masses of the various particles involved are shown in the following table:

Particle	Mass, u
^{235}U	235.0439231
^{141}Ba	140.9144064
^{92}Kr	91.9261528
Neutron	1.0086649

We can sum the appropriate particle masses to find the masses of the reactants and the products:

$$\text{Mass of reactants} = 235.0439231 \text{ u} + 1.0086649 \text{ u}$$
$$= 236.0525880 \text{ u}$$

$$\text{Mass of products} = 140.9144064 \text{ u} + 91.9261528 \text{ u} + 3(1.0086649 \text{ u})$$
$$= 235.8665539 \text{ u}$$

Now we subtract to find the mass defect.

$$\Delta m = 235.8665539 \text{ u} - 236.0525880 \text{ u} = -0.1860341 \text{ u}$$

$$E = (\Delta m)c^2$$
$$= -0.1860341 \text{ u}\left(\frac{1.66053886 \times 10^{-27} \text{ kg}}{\text{u}}\right)\left(\frac{2.99792458 \times 10^{8} \text{ m}}{\text{s}}\right)^2$$
$$= -2.776406 \times 10^{-11} \text{ kg m}^2/\text{s}^2$$
$$= -2.776406 \times 10^{-11} \text{ J}$$

Analyze Your Answer The first thing we notice about our answer is that it is negative. Does this make sense? Notice that we have calculated Δm as the mass of the products minus the mass of the reactants, in keeping with the conventions we used in our discussions of thermodynamics. This means that the resulting value, and the corresponding energy, will also follow the thermodynamic sign conventions. So the fact that we have a negative energy confirms that this is the amount of energy *released* in the fission reaction, as expected. How can we assess the magnitude of our answer? First, keep in mind that this is the energy released by the fission of a single atom of ^{235}U. Then, remember that we expect a nuclear reaction to release much more energy than an ordinary chemical reaction. We know that the energy released in an exothermic chemical reaction is typically on the order of hundreds or perhaps thousands of kJ/mol. So we might convert our answer to kJ/mol to make a comparison. The result above corresponds to 10^{10} kJ/mol. This is clearly much larger than what would be seen in any chemical reaction, so it seems plausible for the nuclear reaction considered here.

Check Your Understanding A single neutron can also induce the fission of ^{235}U to produce ^{90}Sr and ^{143}Xe. Calculate the energy released when 1.00 kg of ^{235}U undergoes this reaction. The experimentally determined masses are m_{Sr} = 89.9077376 u, m_{Xe} = 142.9348900 u.

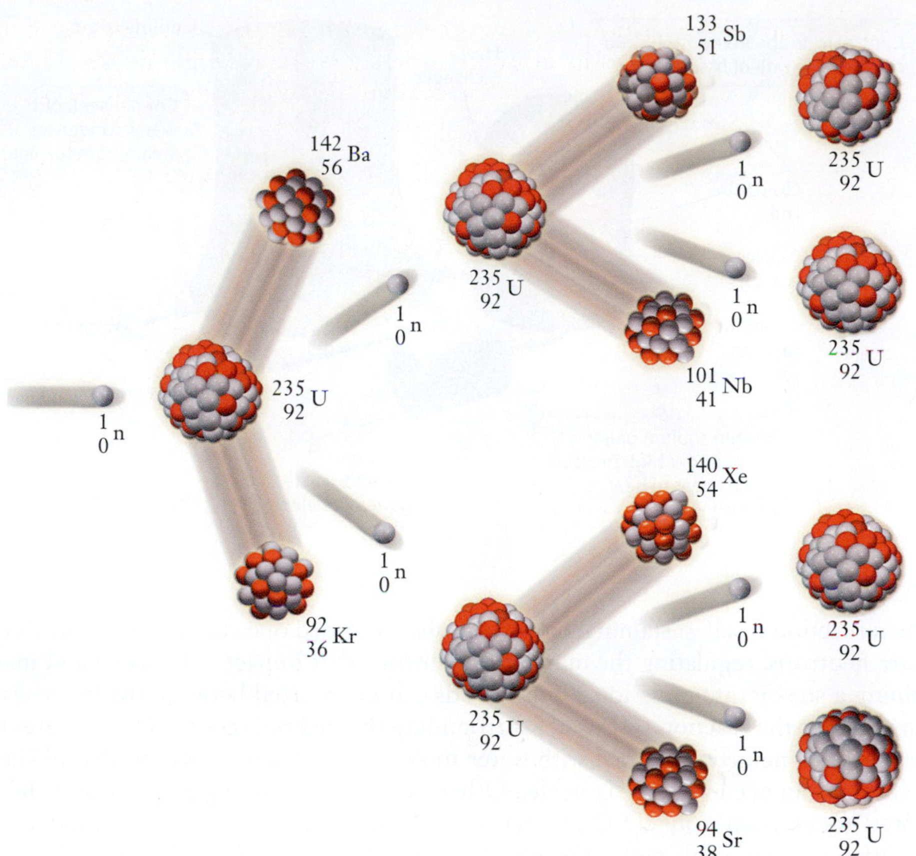

Figure 14.8 ■ Bombarding enriched uranium with neutrons induces the fission of ^{235}U, and each decay produces additional neutrons. Because each of those neutrons can induce further fission, a self-sustaining chain reaction is possible.

If you look at the equation we've been using for the fission of ^{235}U, you may notice one unusual feature: neutrons appear on both sides of the equation. The original neutron bombardment initiates the process, but the subsequent fission reaction releases three neutrons. Because each of these neutrons can induce further fission, a **chain reaction** is possible (**Figure 14.8**). For such a chain reaction to be sustainable, there must be enough fissile atoms to ensure that the neutrons induce fission before they escape from the sample. The amount of material required for a sustainable chain reaction is called the **critical mass**, and it depends on the particular fissile fuel being used. The critical mass for ^{235}U is only a few kilograms. A reaction initiated with a critical mass of ^{235}U results in an uncontrolled chain reaction in which an enormous amount of energy is released almost instantaneously. But if the reaction can be controlled so that only a limited number of the neutrons released are allowed to induce additional decays, controlled fission can be used to generate electricity.

Most strategies for controlling the proliferation of nuclear weapons rely on preventing the attainment of a critical mass of fissionable material.

Nuclear Reactors

Commercial nuclear reactors rely on the controlled fission of ^{235}U as their source of energy. Only about 0.72% of naturally occurring uranium atoms are fissile ^{235}U. So to be used as reactor fuel, uranium must be enriched first to increase this percentage. Enriching uranium requires separating the ^{235}U isotope from ^{238}U, which makes up more than 99% of naturally occurring uranium. This isotope separation is technologically challenging but can produce uranium in which the percentage of ^{235}U is enhanced significantly. Weapons-grade uranium typically contains >90% ^{235}U, whereas that used in nuclear reactors contains 3–5% ^{235}U.

Pellets of enriched uranium oxide are embedded in **fuel rods** and covered with water (**Figure 14.9**) in the **reactor core**. To initiate fission, a source of neutrons must be incorporated into the reactor design. Once this neutron source initiates fission, the

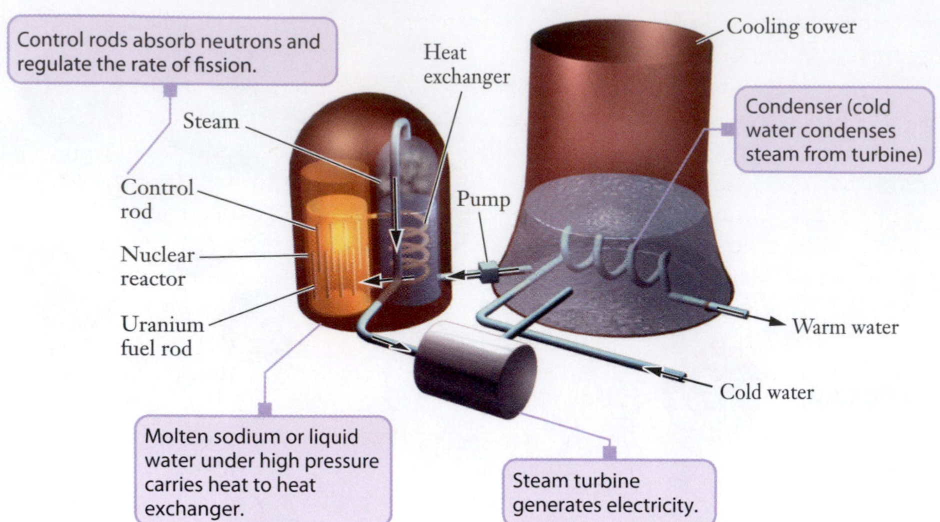

Figure 14.9 ■ The general design shown here is used in all U.S. nuclear power plants.

chain reaction is self-sustaining. **Control rods** composed of cadmium or boron absorb extra neutrons, regulating the number of neutrons that impact ^{235}U nuclei and maintaining a steady rate of fission. Control rods can be inserted between the fuel rods to slow or stop the reaction. The water surrounding the fuel rods cools and moderates the reactor; fast neutrons interact with water molecules and slow down, so that effective collisions can occur with ^{235}U nuclei. Other materials, including graphite and "heavy water" (deuterium oxide, D_2O, also written ^2H_2O), are sometimes used as moderators in different types of reactors. The cooling water carries the heat released in the reaction to the steam turbine. Steam turns the turbine, and electricity is generated.

As you'd probably imagine, nuclear reactors pose significant engineering design challenges. Structural integrity, containment of radioactive materials, and effective cooling are all essential to the safe operation of a nuclear reactor. The disaster at the Japanese Fukushima Daiichi power plant in 2011 provided a vivid illustration of the importance of planning for highly unlikely events. When an earthquake and tsunami struck the region, the plant's reactors first had to withstand the initial physical shock. Automated safety systems quickly inserted control rods into the three operational reactors, stopping the fission process. But shutting down the chain reaction does not immediately stop a reactor from producing heat. Many of the fission products are themselves radioactive, and continue to decay after the reactor has been shut down. Consider the fission pair we examined earlier in Example Problem 14.6:

$$^{235}_{92}U + ^{1}_{0}n \rightarrow ^{236}_{92}U^* \rightarrow ^{141}_{56}Ba + ^{92}_{36}Kr + 3\,^{1}_{0}n$$

Both ^{141}Ba and ^{92}Kr lie below the band of stability and so will undergo beta decay. The emitted beta particles will interact with surrounding materials, and their energy will be converted to heat. The resulting **decay heat** will continue to be produced even after the fission reaction has been "turned off." In the case of the Fukushima disaster, the ability to cool the reactors was compromised due to damage from the earthquake and flooding from the tsunami, ultimately resulting in the melting of the fuel rods.

Nuclear Waste

Several of the products of fission in a nuclear reactor are themselves radioactive isotopes. Much of this radioactive material is concentrated in used or "spent" fuel rods. Because of the large amounts of radioactivity in these rods, they are referred to as *high-level nuclear waste*. Some of the radioisotopes involved have very long half-lives, so the storage or disposal of this high-level waste presents enormous challenges for both engineering and public policy.

It is possible to reprocess spent rods to produce new fuel pellets. But this process has not been carried out in the United States since the 1970s because of regulatory concerns and nuclear nonproliferation treaties. All high-level waste produced in U.S. nuclear reactors is currently stored on-site at the reactor. But clearly this practice cannot continue indefinitely and raises its own safety and security concerns. In 1982, Congress established the Nuclear Waste Policy Act mandating that the U.S. Department of Energy find and develop an appropriate underground disposal facility. In 1987, Yucca Mountain in southwest Nevada was chosen as the site for an extensive feasibility study (**Figure 14.10**). Yucca Mountain has several characteristics that make it a possible site for burying high-level waste. It is extremely remote, the climate is dry, and the water table is about 1000 feet below the potential burial vault. The project faced strong opposition from Utah residents and their congressional representatives, however, and in 2011 funding for the development of the Yucca Mountain site was terminated. Thus the United States currently has no plan in place for the long-term storage of nuclear waste.

Beyond the political and policy aspects of this issue are also tremendous engineering challenges. An acceptable storage facility for high-level nuclear waste will have to be designed so that it can remain intact for thousands of years. Even predicting

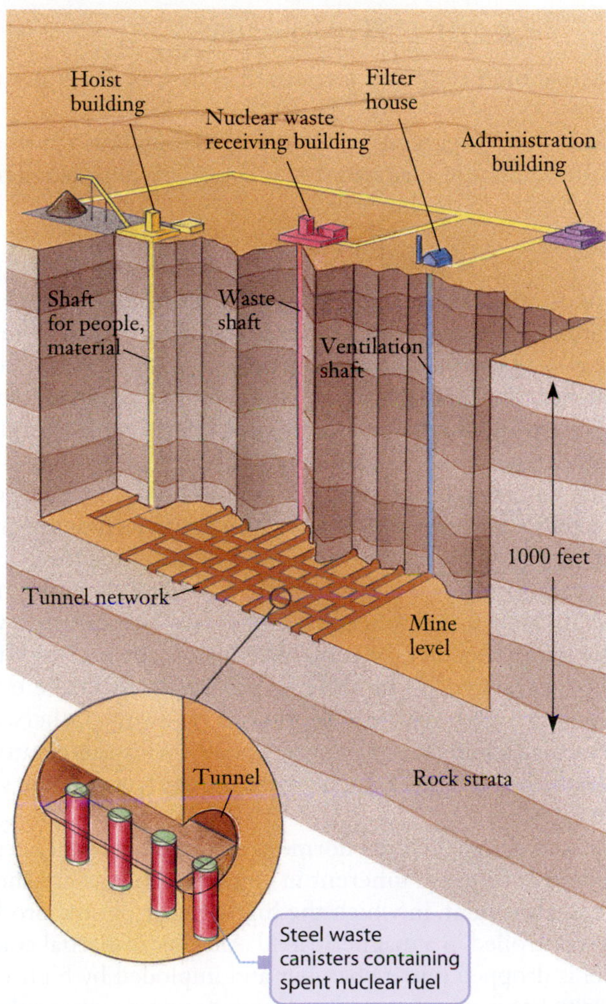

Figure 14.10 ■ The proposed Yucca Mountain storage facility for high-level nuclear waste is shown in this artist's drawing. Waste would be stored in a 3-square-mile network of tunnels 1000 feet below ground level.

environmental conditions over such a timescale is obviously highly speculative. Moreover, the construction materials will need to withstand the effects of high levels of radiation exposure over their long lifetime. As we'll see in Section 14-7, radiation can have significant effects on many materials.

Fusion

Nuclear fission has been harnessed to produce electricity for decades, but nuclear fusion has yet to realize its potential as a relatively safe source of energy. In a fusion reaction, small nuclei combine to form larger, more stable nuclei. The energy of the sun originates in a fusion reaction in which four hydrogen nuclei combine to form a helium nucleus:

$$4\,{}^{1}_{1}\text{H} \rightarrow {}^{4}_{2}\text{He} + 2\,{}^{0}_{1}\beta + 2\nu + \text{energy}$$

However, as we saw in our discussion of chemical kinetics (Section 11-6), the simultaneous collision of four particles is very unlikely. Instead, the reaction proceeds in a stepwise fashion, one hydrogen nucleus colliding with another to produce an intermediate that collides with a third proton, and so on. The reaction of four protons is too slow to be used in a fusion reactor.

A more practical approach is to fuse two heavier isotopes of hydrogen, deuterium (^{2}H) and tritium (^{3}H):

$${}^{2}_{1}\text{H} + {}^{3}_{1}\text{H} \rightarrow {}^{4}_{2}\text{He} + {}^{1}_{0}\text{n}$$

This reaction produces more energy per nucleus than fission, and it is attractive for a number of reasons.

Deuterium is a naturally occurring isotope; it makes up 0.015% of all hydrogen atoms. So the available supply of deuterium is practically unlimited. Tritium can be produced from ^{6}Li:

$${}^{6}_{3}\text{Li} + {}^{1}_{0}\text{n} \rightarrow {}^{4}_{2}\text{He} + {}^{3}_{1}\text{H}$$

Fusion does not produce the high-level radioactive waste that fission generates. The primary product is ordinary helium, which poses no radiation threat. The fact that energetic neutrons are also produced does pose some problems. Neutron bombardment could induce nuclear reactions in materials surrounding the fusion reactor, producing some level of radioactivity. But this also exists in fission reactors, and careful choices of engineering materials can minimize the risks involved.

The use of fusion to generate electricity is complicated by a number of factors, one of which is the repulsion of positively charged nuclei. Enormous energy is needed to force nuclei close enough together to overcome the coulombic forces between protons, so the reaction can be initiated only at temperatures on the order of 10^{6} K. In the hydrogen bomb—the first application of fusion—the initiator was a fission bomb that forced the nuclei together. In controlled fusion, the means of initiating the reaction must not be destructive. To produce an economically viable energy source, it must require less energy input than will be released by fusion. Otherwise, energy is consumed by the process rather than produced. Finally, with temperatures reaching 10^{6} K, the reaction has to be confined somehow despite the fact that any solid material would melt.

In kinetic terms, fusion has an extremely high activation energy.

Taken together, these factors present enormous engineering challenges. Two means of solving the confinement problem inherent in controlled fusion are showing promise. One is magnetic confinement, in which the high-energy plasma produced at very high temperatures is controlled in a magnetic field. Another is inertial confinement, in which a pellet of fuel is dropped into a chamber and imploded by high-energy lasers. So far, although energy has been produced in fusion reactors, getting the reaction to go forward has required more energy than it produced. Fusion research continues, and an international collaboration is finalizing plans to build the next generation reactor, known as ITER, in Cadarache, France. (The name ITER comes from the Latin word *iter,* which means "the way.") It is estimated that fusion may become a viable source of energy late in the 21st century.

14-7 The Interaction of Radiation and Matter

The effects of radiation on matter are governed by three main factors. The first is simply the amount of radiation to which matter is exposed. As you would probably expect, higher doses of radiation usually have more serious effects. But the impact of any radiation exposure also depends critically on two characteristics of the radiation itself, commonly referred to as **penetrating power** and **ionizing power**.

Ionizing and Penetrating Power of Radiation

Any radiation can be classified as either *ionizing* or *nonionizing*; the distinction is based on the energy carried by a photon or particle. If that energy is greater than the ionization energy of typical atoms or molecules, then we should expect that the radiation could induce ionization in whatever material it encounters. Nonionizing radiation includes visible light, radio waves, and microwaves; all have photon energies smaller than typical ionization energies. X-rays and gamma rays, on the other hand, have much larger photon energies, and so they are ionizing radiation. Alpha and beta particles are also ionizing radiation. Ionizing radiation is much more likely to cause significant damage to any materials that it encounters, including living tissue.

When ionizing radiation passes through living tissue, it can eject electrons from atoms or molecules it encounters, producing free radicals. ◀ⓘ These highly reactive particles, either atoms or molecular fragments, can then scavenge electrons from other molecules. When free radicals react in living tissue, results can range from surface burns to serious genetic damage, as might occur if part of a DNA molecule is changed. One commonly formed molecular fragment is the extremely reactive hydroxyl radical, $:\dot{\underset{\cdot\cdot}{O}}—H$. Once formed in a cell, the hydroxyl radical will react with nearby molecules, and the cell may be irreparably damaged. Thus ionizing radiation can lead directly to cell death.

We first considered the reactivity of free radicals in Section 2-8 when we discussed polymerization reactions. ⓘ

Ionizing power alone cannot predict the impact of radiation on matter, however. We must also consider penetrating power—how far a particle penetrates into a material before its energy is absorbed or dissipated. The high energies of alpha particles, for example, give them large ionizing power. But because an alpha particle is relatively large, it does not tend to penetrate deeply into matter. Alpha particles can be stopped by cells in the outer layer of skin, where they pick up electrons to become helium atoms. Because the energy of the alpha particle is dissipated primarily in the skin, it can cause surface burns. But alpha particles originating outside of the body usually do not cause more serious harm because they do not penetrate sufficiently to reach internal organs.

Alpha particles produced *inside* the body are a much greater danger because their energy will be deposited in internal organs. This is the primary risk from radon gas. Radon is a naturally occurring gas, and the amount in any area varies based on the types of soil and rocks present. As a member of Group 18—the noble gases—radon is chemically inert. But it is radioactive and decays by alpha emission:

$$^{222}_{86}\mathrm{Rn} \rightarrow {}^{4}_{2}\mathrm{He} + {}^{218}_{84}\mathrm{Po}$$

$$^{218}_{84}\mathrm{Po} \rightarrow {}^{4}_{2}\mathrm{He} + {}^{214}_{82}\mathrm{Pb}$$

Because radon is a gas, it can be inhaled. If an inhaled radon atom undergoes decay, alpha particles will be produced in the lungs. The dissipation of the energy of those alpha particles can cause serious tissue damage, leading to increased risk of lung cancer.

Beta particles typically have lower energy than alpha particles, and so they might seem safer. But because beta particles are smaller, they can pass several centimeters

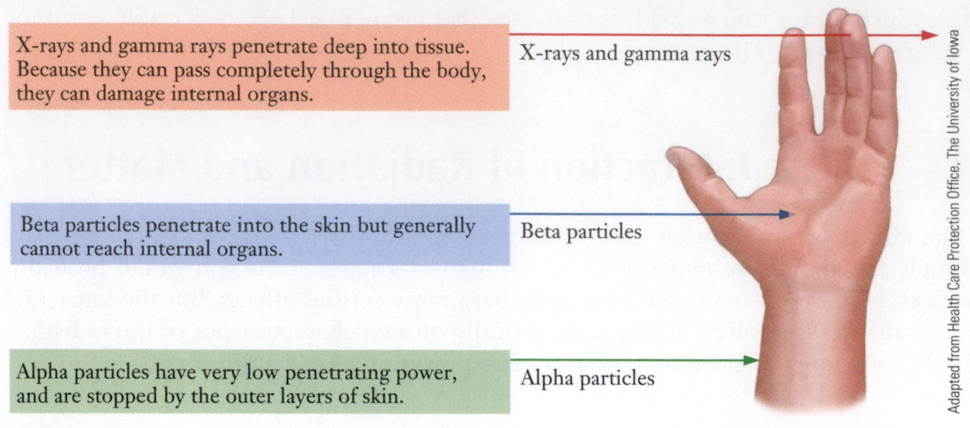

X-rays and gamma rays penetrate deep into tissue. Because they can pass completely through the body, they can damage internal organs.
X-rays and gamma rays
Beta particles penetrate into the skin but generally cannot reach internal organs.
Beta particles
Alpha particles have very low penetrating power, and are stopped by the outer layers of skin.
Alpha particles
Adapted from Health Care Protection Office, The University of Iowa

Figure 14.11 ■ The possible health hazards from exposure to ionizing radiation depend on the penetrating power of the radiation. Alpha particles are stopped in outer layers of the skin, whereas beta particles penetrate several centimeters into the body. Gamma rays can pass completely through the body.

Note the similarities between this figure and Figure 14.1.

into the body. Because of this greater penetrating power, beta radiation is often more dangerous than alpha radiation. Gamma rays can pass entirely through the body, and their interactions with atoms and molecules in vital organs can do a great deal of damage. **Figure 14.11** illustrates the penetrating power of various forms of ionizing radiation.

As our discussion thus far implies, ideas such as ionizing and penetrating power are important in assessing the possible health effects of radiation exposure. But these same ideas also govern the effect of radiation on materials or electronic devices and can have important engineering implications. In the opening section of this chapter, we learned that the distribution of cosmic rays entering the atmosphere is very different from that reaching Earth's surface. So electronic devices in space are exposed to a much higher level of radiation than they would be at ground level. As modern communications have become more dependent on satellite-based electronics, the need to study and protect against the possible effects of cosmic rays has grown.

Computer chips and other solid-state electronic devices rely on a carefully controlled distribution of electrons and holes in semiconductor materials. So the production of ions within those materials can cause catastrophic failures. The cosmic rays most likely to induce ionization are heavy nuclei at high energy, which are much more likely to be encountered in space than at ground level. Even a single such particle can produce a large number of ions, leading to what is known as a *single event effect.* In some cases, these effects can be temporary and perhaps require resetting a device. But other single event effects can be much more serious, causing permanent hardware damage. As a result, chips for use in satellites are frequently packaged in special hardened materials, designed to protect them from the cosmic rays they will encounter. Dangers from exposure to cosmic rays also represent a significant obstacle for long-term manned space missions, such as travel to other planets.

Methods of Detecting Radiation

To assess radiation doses, we must have suitable methods for measuring the type and amount of the radiation. Rutherford's early experiments used a fluorescent screen coated with a zinc sulfide (ZnS) phosphor. The fluorescent screen produced a pinpoint of light each time it was struck by an alpha particle. The tiny flashes of light had to be observed and counted by Rutherford and his assistants, an arduous and potentially

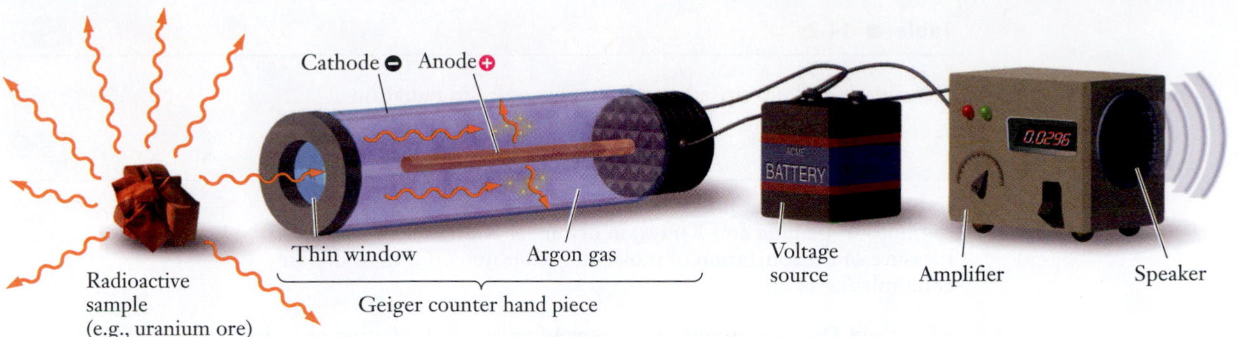

Figure 14.12 ■ In a Geiger-Mueller tube, or Geiger counter, radiation passes through a thin window into a gas-filled tube. Energy from the radiation produces ions in the gas, releasing electrons. The resulting ions and electrons are then attracted to oppositely charged electrodes, producing a pulse of electric current. In a typical Geiger counter, the current pulse is converted to an audible clicking sound.

inaccurate task. Such a method is clearly not suitable for modern applications in medical imaging or carbon dating.

The scintillation counter is a modern adaptation of the visually recorded light pulse on a screen. Such a counter uses a fluorescent material to detect radiation, which produces photons. Those photons can then be counted using a photomultiplier tube and appropriate electronics.

The portable **Geiger counter** is commonly used in laboratories and by radiation safety teams to measure radioactivity (**Figure 14.12**). A glass tube containing a gas at a low pressure (about 0.1 atm) is coated on the inside with a metal that acts as a cathode. A wire anode runs down the center of the tube. A high voltage is applied across the electrodes. Alpha and beta particles enter a window in the tube and ionize atoms of gas. Electrons released from the gas atoms are attracted to the anode, and, as they travel to the anode, they cause the ionization of more gas atoms, releasing more electrons. An avalanche of electrons reaches the anode, and a current pulse is recorded. If it is connected to a speaker, the pulse is registered as an audible click.

A film-badge **dosimeter**, commonly worn by people who work with radioactive isotopes, takes advantage of Becquerel's discovery that radiation darkens photographic plates (**Figure 14.13**). The darkened badge, coupled with a record of each exposure to a radioactive species, provides a warning mechanism if safe exposure levels are exceeded.

Methods of detecting radiation must properly account for **background radiation**. Cosmic rays are a constant source of radioactivity, for example, as are natural radioactive isotopes in soil, air, and water. This radiation is always around us and must be subtracted from measurements of radioactive sources.

Measuring Radiation Dose

Radiation detectors such as those described above give us the ability to quantify the level of radiation in a particular environment. Because of the interplay between ionizing power and penetrating power, there are a number of different ways to express radiation dose. The **exposure** measures the number of ions produced in air. The **absorbed dose**, as the name implies, measures the amount of radiation actually absorbed by a particular material. The **equivalent dose** attempts to quantify the resulting damage to human tissue. To further complicate things, two different sets of units are commonly used to account for the differences in ionizing power and penetrating power in assessing the extent of effective damage in human tissue. These quantities and their respective units are summarized in **Table 14.2**.

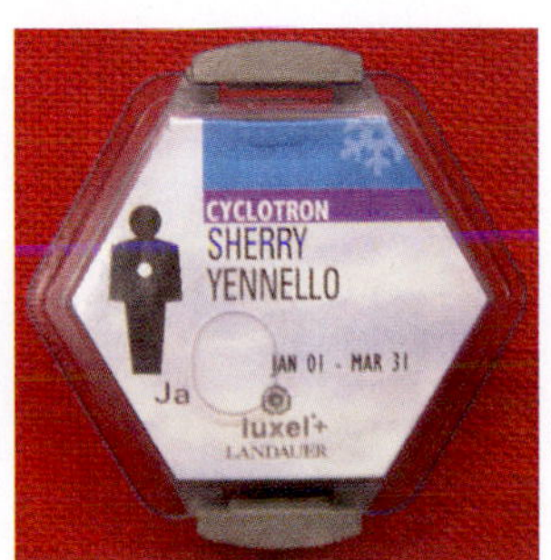

Figure 14.13 ■ People who work around radioactive materials wear film badges (like the one shown here) to monitor their exposure to radiation.

Table ■ 14.2

Definitions and units used to quantify exposure to radiation

	US	SI
Exposure—an older unit for measuring exposure to photons (gamma and X-rays) in dry air. It is a measure of the ionization of molecules of air in coulombs/kg of air.	Roentgen (R) 1 R = 2.58 × 10^4 C/kg of dry air	
Absorbed Dose—amount of any type of radiation absorbed in any kind of material. Doesn't describe biological effects of different types of radiation.	Radiation absorbed dose (rad) 1 rad = 1 erg/g (1 erg = 10^{-2} J)	Gray (Gy) 1 Gy = 100 rads
Equivalent Dose—relates the absorbed dose to the damage produced by different types of radiation. A quality factor, Q, is multiplied by the absorbed dose to get the equivalent dose.	Roentgen equivalent man (rem) 1 rem = Q × absorbed dose in rad	Sievert (Sv) 1 Sv = 100 rem 1 Sv = Q × absorbed dose in Gy

The **quality factor (*Q*)**, used in calculating the equivalent dose is also known as the **relative biological effectiveness (RBE)**. It varies from a value of 1 for high-energy photons (gamma and X-rays) to about 20 for alpha particles. The definition of equivalent dose attempts to incorporate information about exposure and toxicity simultaneously.

INSIGHT INTO

14-8 Modern Medical Imaging Methods

Radiation has been used as a tool for medical diagnosis since the introduction of X-rays as a way to see into the body. More recent developments include the use of various radioactive isotopes to obtain images of specific organs, as well as more elaborate techniques such as **positron emission tomography (PET)**.

Radiation is also used to treat various cancers.

In a traditional X-ray, radiation passes through the body and a photographic image is produced based on the extent to which the radiation is absorbed. Because bones absorb X-rays much more strongly than organs or other tissues, X-ray images are excellent tools for orthopedic diagnosis. X-ray images can also be used to examine the structure of some organs, as in a chest X-ray to examine the heart or lungs. But they are not generally suitable for studying the function of organs. Instead, a range of techniques has been developed around the idea that if a small amount of an appropriate radioisotope can be selectively deposited in the target organ, then monitoring the radiation from that isotope can produce a detailed image of the organ. If such images can be taken with sufficient time resolution, it becomes possible to monitor processes such as blood flow or oxygen uptake, providing a powerful and relatively noninvasive medical tool.

How can radioactive isotopes be introduced into specific organs? The answer relies on the fact that biochemistry tells us that certain elements and compounds are taken up specifically by particular organs. The thyroid gland uses iodine to produce thyroid hormone, for example. So, radioactive ^{131}I is often used to diagnose thyroid problems. The patient is given an injection of a small amount of the radioisotope, and then natural biochemical pathways carry the ^{131}I to the thyroid. ^{131}I undergoes beta decay, so detection of radiation emanating from the thyroid can be used to produce an

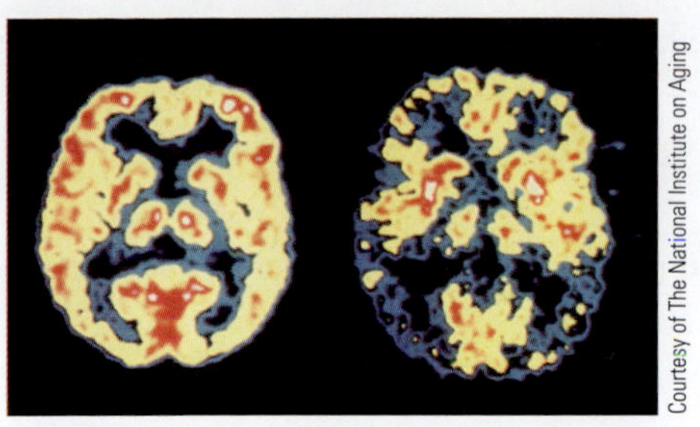

Figure 14.14 ■ Positron emission tomography (PET) produces high-quality images of the brain and other organs. The technique is based on the simultaneous detection of pairs of gamma ray photons produced when a positron and an electron collide. This makes it possible to measure even low signal levels very accurately.

image of the gland. Here the organ itself becomes the light source, so only an external radiation detection scheme is needed to record the image. Several other isotopes are used in similar ways to image different organs. So long as the radiation dose is fairly small and the half-life of the isotope being used is not too long, these procedures are extremely safe.

As the name implies, PET images (**Figure 14.14**) are based on isotopes that emit positrons. Recall from Section 14-4 that positron emission tends to occur in neutron-deficient isotopes. Among the available positron emitters are ^{11}C, ^{18}F, ^{13}N, and ^{15}O. Because all of these elements are found in common organic molecules, it is relatively easy to incorporate them into appropriate biological molecules. ^{15}O might be incorporated into glucose, for example. Because our bodies rely on glucose as a source of energy, it will be taken up by most organs.

Each decay of the radioisotope releases a positron, but the lifetime of a positron in the body is extremely short. In most cases, a positron will travel no more than a couple of millimeters before it encounters an electron. The positron and electron then undergo matter–antimatter annihilation, which releases a pair of gamma-ray photons traveling in opposite directions. Detectors register the gamma rays, and computers map out the path taken by the tagged compound. The result is a map of a slice through the body. The half-life of ^{15}O is just 122 s, so decay to stable (and harmless) ^{15}N is quick. This short half-life is beneficial because the radiation is only in the body for a very short time. But it also provides some challenges in carrying out the procedure because the short-lived radioisotope cannot be stored. Because of this, PET scans can be performed only in facilities that can produce the needed isotopes on-site.

FOCUS ON PROBLEM SOLVING

Question Suppose that you are working for a company that prepares radioisotopes for medical imaging. You receive an order from a researcher requesting 15 mg of ^{45}Ti for an experiment. The customer's laboratory is located roughly 1.5 hours from your production source. How would you determine what mass of the isotope to produce to fill this order?

Strategy The key issue here is that the isotope you produce will decay during the time that it is in transit to the research lab. Because the experiment requires a set mass of the isotope, you must consider the kinetics and produce more than is required so that there will be enough ^{45}Ti left upon delivery. The most important piece of information to look up is the half-life of ^{45}Ti. Once that value is known, we can use the integrated rate law for first-order kinetics to carry out the needed calculations.

Solution Once we know the half-life, we can determine the rate constant, k:

$$k = \frac{0.693}{t_{1/2}}$$

Next, we turn to the integrated rate law for a first-order reaction. In Equation 14.1, we wrote this as

$$N = N_0 e^{-kt}$$

Here we can use masses in place of the N and N_0 terms, giving us

$$m = m_0 e^{-kt}$$

Our goal is to find m_0, the initial mass of ^{45}Ti needed. We have to make an assumption about the delivery time and the amount of time it would take to package the isotope for delivery. If we assume 30 minutes from production to getting the isotope aboard the delivery vehicle, we might use 120 minutes as our time:

$$m = m_0 e^{-kt}$$

$$m_0 = \frac{m}{e^{-kt}} = \frac{15 \text{ mg}}{e^{-k(120 \text{ min})}}$$

SUMMARY

Nuclear reactions involve the conversion of one chemical element into another and are typically accompanied by the emission of radiation in the form of high-energy particles and photons. Some nuclear reactions occur naturally, as in radioactive decay, whereas others can be induced by human intervention. Nuclear decay processes can be categorized in terms of the type of radiation emitted.

We can write equations for nuclear reactions similar to those for ordinary chemical reactions. But in a nuclear reaction, the number of atoms of each element is not conserved, so we must use different ideas to balance the equations. In a properly balanced nuclear equation, both mass number and charge number should be conserved.

All radioactive decay processes follow first-order kinetics, and their rates are commonly described in terms of their half-lives. The standard equations for first-order reactions can thus be used to predict the activity of a radioactive sample as a function of time. Rates of decay are usually reported in units of becquerels or curies. Carbon dating relies on the fact that the kinetics of the decay of ^{14}C are well understood.

Nuclear chemists rely on the chart of the nuclides to summarize their knowledge of the stability of various isotopes. All known stable isotopes fall into a region known as the band of stability, and the position of unstable isotopes with respect to this band can be used to predict the type of decay expected.

Compared to ordinary chemical reactions, nuclear reactions release tremendous amounts of energy. The energy change of a nuclear process can be calculated from the mass defect by using the Einstein equation, $E = mc^2$. The same type of calculation allows us to find the binding energy for any given nucleus and reveals that nuclear binding energy goes through a maximum for ^{56}Fe. Thus very heavy nuclei can undergo fission, splitting into two or more lighter nuclei with increased stability. Lighter nuclei, in contrast, may gain stability by undergoing fusion, in which two or more nuclei combine to form a larger and more stable nucleus. Current nuclear reactor technology is based on the fission of ^{235}U. Fusion reactors also hold great promise as future energy sources but still face significant engineering challenges.

The effects of radiation on matter—including living tissues—depend on both the type of radiation and the extent of exposure. In assessing its effect on matter, radiation is characterized in terms of ionizing power and penetrating power.

KEY TERMS

absorbed dose (14-7)
alpha decay (14-2)
alpha particle (14-2)
alpha ray (14-2)
background radiation (14-7)
band of stability (14-4)
becquerel (Bq) (14-3)
beta decay (14-2)
beta particle (14-2)
beta ray (14-2)
chain reaction (14-6)
chart of the nuclides (14-4)
compound nucleus (14-6)
control rod (14-6)
cosmic ray (14-1)

critical mass (14-6)
curie (Ci) (14-3)
decay constant (14-3)
decay series (14-4)
dosimeter (14-7)
electron capture (14-2)
electron volt (eV) (14-1)
equivalent dose (14-7)
exposure (14-7)
fuel rod (14-6)
gamma ray (14-2)
Geiger counter (14-7)
half-life (14-3)
ionizing power (14-7)
K capture (14-2)
magic number (14-5)
mass defect (14-5)
nuclear binding energy (14-5)
nuclear fission (14-5)
nuclear fusion (14-5)
nuclear reaction (14-1)
nuclide (14-2)
penetrating power (14-7)
positron (14-2)
positron emission tomography (PET) (14-8)
quality factor (Q) (14-7)
radioactive decay (14-1)
reactor core (14-6)
relative biological effectiveness (RBE) (14-7)
sea of instability (14-4)
strong force (14-4)
transmutation (14-6)

PROBLEMS AND EXERCISES

INSIGHT INTO Cosmic Rays and Carbon Dating

14.1 Cosmic rays are sometimes referred to as *corpuscular rays.* How does this term distinguish them from other rays of the sun that reach the planet?

14.2 Who would have greater exposure to cosmic rays: a commercial pilot or an engineer who helps build the runways the pilot uses? Explain your answer.

14.3 Use the web to find information on the composition of cosmic rays beyond hydrogen and helium.

14.4 A fastball thrown by a professional baseball pitcher travels at about 100 mph. How does the kinetic energy of the baseball thrown at this speed compare to cosmic radiation at 15 MeV? (The mass of a baseball is about 145 g.)

14.5 (a) How does carbon-14 enter a living plant?
(b) Write the equation for this reaction.

14.6 Explain the following:
(a) The ratio of ^{14}C to ^{12}C in a living plant is constant.
(b) The constant ratio cited in (a) is a steady-state condition, not an equilibrium.

Radioactivity and Nuclear Reactions

14.7 Refer to Figure 14.1 and describe the experiment in which Ernest Rutherford demonstrated that different types of radiation emanate from uranium.

14.8 List two ways in which a gamma ray differs from an alpha or beta particle.

14.9 Match the following forms of radioactive decay with the appropriate result.

(a) Alpha	1. No change in mass number or atomic number
(b) Positron	2. Atomic number decreases by 1
(c) Gamma	3. Atomic number decreases by 2, mass number decreases by 4
(d) K capture	4. Atomic number increases by 1
(e) Beta	

14.10 Why does positron decay have the same effect on the nucleus as electron capture?

14.11 Complete each equation and name the particle ejected from the nucleus.
(a) ${}^{20}_{8}\text{O} \rightarrow ? + {}^{\ 0}_{-1}\beta + \bar{\nu}$
(b) $? \rightarrow {}^{232}_{92}\text{U} + {}^{4}_{2}\text{He}$
(c) ${}^{201}_{82}\text{Pb} \rightarrow {}^{201}_{83}\text{Bi} + ?$

14.12 Complete the following nuclear equations. Write the mass number, atomic number, and symbol for the remaining particle.
(a) ${}^{9}_{4}\text{Be} + ? \rightarrow {}^{6}_{3}\text{Li} + {}^{4}_{2}\text{He}$
(b) $? + {}^{1}_{0}\text{n} \rightarrow {}^{24}_{11}\text{Na} + {}^{4}_{2}\text{He}$
(c) ${}^{40}_{20}\text{Ca} + ? \rightarrow {}^{40}_{19}\text{K} + {}^{1}_{1}\text{H}$
(d) ${}^{241}_{95}\text{Am} + {}^{4}_{2}\text{He} \rightarrow {}^{243}_{97}\text{Bk} + ?$
(e) ${}^{246}_{96}\text{Cm} + {}^{12}_{6}\text{C} \rightarrow 4\,{}^{1}_{0}\text{n} + ?$
(f) ${}^{238}_{92}\text{U} + ? \rightarrow {}^{249}_{100}\text{Fm} + 5\,{}^{1}_{0}\text{n}$

14.13 Write equations for the following nuclear reactions.
(a) Alpha decay by ^{188}Bi
(b) Beta emission by ^{87}Rb
(c) Positron emission by ^{40}K
(d) Electron capture by ^{138}La

14.14 Write equations for the following nuclear reactions.
(a) ^{94}Tc undergoes electron capture.
(b) ^{89}Nb ejects a positron.
(c) ^{217}Fr emits an alpha particle.
(d) ^{220}Ra is formed by alpha emission.

14.15 (a) Explain why tritium, ^{3}H, cannot undergo alpha decay.
(b) What type of decay would you expect tritium to undergo?
(c) Write the nuclear equation for the expected decay.

14.16 By what processes do these transformations occur?
(a) Thorium-230 to radium-226
(b) Cesium-137 to barium-137

(c) Potassium-38 to argon-38

(d) Zirconium-97 to niobium-97

14.17 One way to convert lead into gold is to irradiate ^{206}Pb with neutrons. There are several steps in the process, and three neutrons are required for each lead atom. In addition to gold, three beta particles and three alpha particles are produced. (This process requires a particle accelerator, and the cost is much higher than the value of any gold that could be produced.)

(a) What isotope of gold will this process produce?

(b) Write the overall nuclear reaction for the process.

Radioactive Decay Rates

14.18 What is the half-life of a radioisotope if it decays to 12.5% of its radioactivity in 12 years?

14.19 ^{137}Cs has a half-life of 30.2 years. How many years will it take for a 100.0-g sample to decay to 0.01 g?

14.20 Cobalt-60 is used extensively in medicine as a source of γ-rays. Its half-life is 5.27 years.

(a) How long will it take a Co-60 source to decrease to 18% of its original activity?

(b) What percent of its activity remains after 29 months?

14.21 The half-life of Sb-110 is 23.0 s.

(a) Determine its decay constant in s^{-1}.

(b) Compute the activity of a 1.000-g sample of ^{110}Sb in Bq and Ci.

14.22 Gold-198 is used in the diagnosis of liver problems. The half-life of ^{198}Au is 2.69 days. If you begin with 2.8 μg of this gold isotope, what mass remains after 10.8 days?

14.23 The half-life of ^{19}O is 29 s. Suppose that a scientist wishes to incorporate ^{19}O into a molecule and then use its radioactivity to trace the fate of that molecule in a reaction. To make an accurate measurement, there must be at least 1.50 mg of ^{19}O left by the end of the study, which will take 2.5 min each time the experiment is run. What is the minimum mass of ^{19}O that must be used each time the experiment is conducted?

14.24 Archaeologists use the year 1950 as the date of the present. So 1000 B.C. is 2950 B.P. (Before the Present). An archaeologist finds that the carbon-14 content of a wooden needle is 11% of that of trees living today. At what date B.P. was the tree from which the needle was made cut down?

14.25 A wooden artifact is burned and found to contain 21 g of carbon. The ^{14}C activity of the sample is 105 disintegrations/min. What is the age of the artifact?

14.26 Some mammoth bones found in Arizona were found by carbon-14 dating to be 1.13×10^4 years old. What must have been the activity of the carbon-14 in the bones in disintegrations per minute per gram? Assume the original activity was 15.3 disintegrations per minute per gram.

14.27 Tritium, ^3H, is a radioactive isotope of hydrogen produced by cosmic rays in the atmosphere, where it oxidizes and becomes part of the water cycle. Its half-life is 12.33 yr. Which of the following could be dated using tritium? Explain your answer.

(a) 4000-year-old ice from Antarctica

(b) Modern French wine

(c) Ancient Egyptian beer from a tomb in the Valley of the Kings

14.28 Explain why uranium can be used for geological dating but not for dating artifacts from human activity on Earth.

14.29 A piece of a spear handle is found in an archaeological dig in Central America. It contains 12.5% as much ^{14}C as a tree living today. Based only on the half-life of carbon-14, how old is the spear handle?

14.30 Beginning in the late 19th century and continuing into the 20th century, the Industrial Revolution was fueled by coal and spewed enormous volumes of ^{14}C-depleted CO_2 into the atmosphere.

(a) Why is coal ^{14}C-depleted?

(b) How would the increase in ^{14}C-depleted CO_2 affect the date of a very old object in which its $^{14}C/^{12}C$ ratio was compared to the 20th-century ratio? Explain.

Nuclear Stability

14.31 What is the N/Z ratio for each of the following nuclides?
(a) $^{14}_{7}N$ (b) $^{114}_{50}Sn$ (c) $^{234}_{90}Th$

14.32 Write a nuclear equation for the type of decay each of these unstable isotopes is most likely to undergo.

(a) Neon-19

(b) Thorium-230

(c) Bromine-82

(d) Polonium-212

14.33 Based on their positions relative to the band of stability, predict the type of decay that each of the following will undergo. Write equations for the expected decays.
(a) $^{73}_{36}Kr$ (b) $^{13}_{8}O$ (c) $^{126}_{50}Sn$

14.34 The following successive decays occur, starting with ^{24}F, which has a half-life of 0.3 s, and ending at a stable isotope:

$$^{24}_{9}F \xrightarrow{\beta^-, t_{1/2} = 0.3\text{ s}} ? \xrightarrow{\beta^-, t_{1/2} = 3.4\text{ m}} ? \xrightarrow{\beta^-, t_{1/2} = 15\text{ h}} ?$$

(a) Identify each of the decay products.

(b) Using the chart of the nuclides, explain why successive beta emissions make more sense than (i) alpha emission; (ii) positron emission; or (iii) electron capture in producing a stable nucleus from ^{24}F.

14.35 The thorium-232 radioactive decay series, beginning with ^{232}Th and ending with ^{208}Pb, occurs in the following sequence: $\alpha, \beta^-, \beta^-, \alpha, \alpha, \alpha, \alpha, \beta^-, \beta^-, \alpha$. Write an equation for each step in this series.

14.36 Starting with ^{238}U and ending with an isotope of Pb, write a decay scheme in which every nuclide emits only an α particle. Compare your scheme to the decay series in Figure 14.5 and suggest a reason why β^- emissions must occur in the ^{238}U decay series.

14.37 The ^{238}U series can be used to date minerals containing uranium by measuring the ratio of ^{238}U to ^{206}Pb, but all of the decay products have to remain in the mineral through several of their half-lives so that essentially all the ^{238}U that decays becomes ^{206}Pb. One of the decay products in this series is the noble gas, ^{222}Rn. How would the experimentally determined age of the mineral be affected if some of the radon diffused from the rock?

14.38 Both ^{238}U and ^{232}Th decay to stable isotopes of lead. ^{222}Rn ($t_{1/2}$ = 3.8 d) is an intermediate nuclide in the ^{238}U series. ^{218}Rn ($t_{1/2}$ = 56 s) is an intermediate in the ^{232}Th decay series. In which series would diffusion of gaseous radon cause the greater error in the computed age of a rock (see Exercise 14.37)?

Energetics of Nuclear Reactions

14.39 Calculate the binding energy of 1 mole of ^{14}C nuclei. The experimentally determined mass of a carbon-14 atom (including its six electrons) is 14.003242 u.

14.40 Collision of an electron and a positron results in formation of two γ-rays. In the process, their masses are converted completely into energy.

(a) Calculate the energy evolved from the annihilation of an electron and a positron in kilojoules per mole.

(b) Using Planck's equation (Equation 6.2), determine the frequency of the γ-rays emitted in this process.

14.41 Compute the binding energy of the ^{7}Li nucleus, whose experimentally determined mass is 7.016004 u.

14.42 Find the mass defect in the reaction in which a neutron decays to a proton and an electron.

14.43 Compute the binding energy of the ^{14}N nucleus, which has an experimentally determined mass of 14.003074 u.

14.44 The actual mass of a ^{108}Pd atom is 107.90389 amu. **(a)** Calculate the mass defect in amu per atom and in g per mole for this isotope. **(b)** What is the nuclear binding energy in kilojoules per mole for this isotope?

14.45 It takes 360 kJ to keep a 100-watt lightbulb burning for 1 hour. Assuming that the lightbulb won't burn out, for how many years could the energy of the mass defect of 1 mole of ^{14}C keep a 100-W lightbulb burning?

14.46 It has been suggested that element 114 has unusual stability, but 114 is not a magic number. What is a possible stable neutron number for this element?

14.47 Lead has a magic number of protons. A doubly magic nuclide should be exceptionally stable. Why isn't lead-164 stable?

14.48 Tin has a magic number of protons. One of its isotopes, ^{132}Sn, has a magic number of neutrons, as well, but it is unstable. Explain.

14.49 Use the web to identify the heaviest element that has been observed to form a chemical compound. What element is it and what compound does it form? Are heavier elements incapable of forming compounds?

14.50 Glenn Seaborg was a leader in the search for the heaviest elements, and element 106 is named Seaborgium (Sg) in his honor. In an article in the *Journal of Chemical Education*, Seaborg summarized the history of the search for the transuranium elements up to the mid-1980s (Glenn T. Seaborg, "The Transuranium Elements," *Journal of Chemical Education*, vol. 62, 1985, pp. 463–467). Read the article and answer the following questions.

(a) Who was responsible for the position of the lanthanides in the periodic table?

(b) What did Enrico Fermi think was being produced when he bombarded uranium with neutrons?

(c) How did Ida Noddack's paper straighten out the confusion about the identity of the products of Fermi's nuclear reaction?

(d) Compare Figures 1 and 2, the 1930 periodic table and Seaborg's 1945 version of the table. What was Seaborg's insight that gave us the modern periodic table?

(e) State the two possible reasons Seaborg gives for the fact that the "island of stability" has not been reached.

(f) Why does fusion of two elements with heavy nuclei produce only a few atoms of a new, superheavy nucleus?

(g) In answer to a student's question, Seaborg announced to the world the discovery of two new elements in 1945 on "The Quiz Kids," a radio program in Chicago. Seaborg wrote, "This was the first time in the history of the world that the announcement of the discovery of chemical elements was sponsored by Alka Seltzer." What are the atomic numbers and names of the two elements?

Transmutation, Fission, and Fusion

14.51 Ernest Rutherford carried out the first artificial transmutation in 1919 when he bombarded ^{14}N with alpha particles. The result was an oxygen nucleus:

$$^{14}_{7}N + ^{4}_{2}He \rightarrow ^{18}_{9}F^{*} \rightarrow ^{17}_{8}O + ^{1}_{1}p$$

Radiochemists use a shorthand notation to represent transmutations. Omitting the intermediate $^{18}_{9}F^{*}$ nucleus, the reaction above can be represented as ^{14}N (α, p) ^{17}O. Write balanced equations for each of the indicated nuclear bombardments.

(a) ^{27}Al (p, γ) ^{28}Si

(b) ^{40}Ar (n, β^{-}) ^{41}K

14.52 **(a)** Write the equation for the reaction that occurs when a ^{200}Hg atom is bombarded by a proton. One of the products is an alpha particle. (Include the compound nucleus.)

(b) Represent the reaction above in the shorthand notation introduced in the previous problem, omitting the compound nucleus.

(c) The reaction above must take place in a particle accelerator so that the protons strike the mercury nuclei at very high energies. Why isn't this method of producing the precious metal product in **(a)** the answer to the alchemist's dreams?

14.53 Slow neutrons can be absorbed by some nuclei to produce new elements. Why does a neutron make a better projectile for a transmutation reaction than an alpha or beta particle?

14.54 What is the missing product in each of these fission equations?

(a) $^{235}_{92}U + ^{1}_{0}n \rightarrow ? + ^{93}_{38}Sr + 3\ ^{1}_{0}n$

(b) $^{235}_{92}U + ^{1}_{0}n \rightarrow ? + ^{132}_{51}Sb + 3\ ^{1}_{0}n$

(c) $^{235}_{92}U + ^{1}_{0}n \rightarrow ? + ^{141}_{56}Ba + 3\ ^{1}_{0}n$

14.55 How much energy is released in the fission of 1 kg of ^{235}U according to the equation below? The experimentally determined masses of the products are 136.92532 u and 96.91095 u, respectively. (The masses of ^{235}U and the neutron are given in Example Problem 14.6.)

$$^{235}_{92}U + ^{1}_{0}n \rightarrow ^{137}_{52}Te + ^{97}_{40}Zr + 2\ ^{1}_{0}n$$

14.56 Identify each of the following and indicate its purpose in a fission reactor.

(a) moderator **(c)** control rods

(b) fuel **(d)** coolant

14.57 Because the fissionable nucleus of uranium is ^{235}U, the large quantity of ^{238}U in a nuclear reactor isn't used. However, it is possible to produce fissionable ^{239}Pu from ^{238}U by the reaction ^{238}U (n, β^-) ^{239}Np. ^{239}Np decays by β^- emission to ^{239}Pu. A reactor used for this purpose is called a *breeder reactor*.

(a) What is the compound nucleus formed by ^{238}U?

(b) Write balanced equations for production of ^{239}Pu.

14.58 **(a)** What mass of octane, C_8H_{18}, would have to undergo combustion to provide the same quantity of energy as that released by the fission of one kilogram of ^{235}U to produce ^{90}Sr and ^{143}Xe?

(b) How many gallons of octane is this? (The density of octane is 0.702 g/cm^3.)

14.59 According to the U.S. Nuclear Regulatory Commission, in 2011 there were 104 commercial nuclear reactors in the United States. In 2011, 4.1×10^6 gigawatt-hours of electricity were generated for use, and 19.43% of that was generated by nuclear power.

(a) A nuclear technician assumed that the products of the fission of ^{235}U by one neutron were ^{97}Zr and ^{137}Te (see Problem 14.55). Ignoring decay products, the technician estimated the total mass of zirconium and tellurium in metric tons that would have to be disposed of in 2011 as nuclear waste (1 watt-hour = 3600 joules; 1 metric ton = 1000 kg). What was the technician's result?

(b) Look up the half-lives of ^{97}Zr and ^{137}Te. When the waste was disposed of, do you think there was a measurable quantity of either isotope? Explain.

(c) Look up the decay mode of each isotope and write the decay series to a stable isotope for each one. Which decay series has a radioactive isotope that will remain in the waste for many years?

(d) Using the stable isotope products in your calculation, how far off was the technician's estimate of total waste? Explain.

14.60 **(a)** Write the equation for the reaction of deuterium and tritium that could be used in a fusion reactor.

(b) What are two obstacles to this reaction as a viable source of energy?

14.61 A fusion reaction that has been proposed involves bombarding boron-11 with a proton and producing a single, stable product.

(a) Write the equation for the reaction.

(b) Which of the obstacles in the previous problem is overcome by this reaction?

14.62 Use the Internet and other resources to find an estimate of the amount of electrical energy that may be required worldwide by the year 2025. Write a 2- to 3-page essay in which you briefly discuss the pros and cons of each of the following methods of electrical energy production to meet the estimated demand. **(a)** solar energy, **(b)** fossil fuels, **(c)** fission, **(d)** fusion

14.63 In 1998, the Nuclear Regulatory Commission suggested that residents who live near a nuclear power plant should have a supply of potassium iodide on hand in the event of a nuclear plant emergency. Explain why the suggestion makes sense.

14.64 The Yucca Mountain nuclear waste burial site was touted as a potential solution to the problem of storing high-level nuclear waste. Using the Internet, find information about the site and, in a well-reasoned essay, formulate arguments for and against using Yucca Mountain to store nuclear waste. Based on your arguments, if building the site were your decision, what would you do?

14.65 A nuclear waste storage facility like the Yucca Mountain site must last for roughly 10,000 years. What engineering uncertainties are introduced for the design of such a facility compared to a design for a tunnel for transportation services, for example?

14.66 What is the critical feature of nuclear fusion that requires special containment ideas, such as magnetic or inertial containment?

The Interaction of Radiation and Matter

14.67 Use information from Section 6-7 to estimate which form of electromagnetic radiation is the lowest energy ionizing radiation.

14.68 Is it the production of ions that makes ionizing radiation dangerous to living tissue or some other effect? Explain your answer.

14.69 Rank the most common forms of radioactivity in order of increasing penetrating power.

14.70 Both penetrating power and ionizing power are important in determining the danger of radiation to living tissue. The goal of understanding damage to living tissues is to limit that damage. If you were an engineer working for a firm that designs safety clothes and apparatus for working in radioactive environments,

(a) Under what conditions is ionizing power the more important factor to consider?

(b) Under what conditions is penetrating power the more important factor?

14.71 Polonium-210 is an alpha-emitting nuclide found in cigarette smoke. How is the fact that it is found in such smoke related to its health hazard?

14.72 Why is ionizing cosmic radiation of greater concern for engineering designs on satellites than for ground-based electronics designs?

14.73 How is the damage from the possible impact of ionizing cosmic rays on an electronic device in a satellite related to the materials from which electronic devices are constructed?

14.74 A radioactive sample used in medicine registers an activity of 10 μCi. What is its activity in Bq?

14.75 A Geiger counter is set to click once for every 100 disintegrations that it detects. The counter registers 20 clicks per second for a sample of an unknown isotope. What is the activity of the isotope in Ci? In μCi?

14.76 You are a biologist studying Cu^{2+} absorption from water in certain legumes. You have a small quantity of $^{67}CuSO_4$ and a Geiger counter. (^{67}Cu is a β^- emitter with a half-life of 61 h.) Design and describe an experiment that demonstrates whether the legume takes up Cu^{2+} from water.

14.77 An average person is exposed to about 360 millirem of background radiation per year from a variety of natural and man-made sources. About two-thirds of that typically comes from inhalation of ^{222}Rn produced in soil by the decay of ^{238}U. Given that ^{222}Rn decays by alpha emission, estimate the absorbed dose in **(a)** joules and **(b)** grays for a 60-kg person from one year's inhalation of radon.

14.78 In the early part of the 20th century, watches were produced with hand-painted dials that glowed in the dark. The paint contained ZnS, phosphor, and radium, which decayed to produce alpha particles that excited the ZnS to glow when struck by a charged particle. The people (mostly women) who painted the dials often put a fine point on their paintbrushes by rolling the brush on their tongues. Many workers in this industry subsequently died of bone cancer and leukemia when their bone marrow was destroyed. Imagine that you are writing a letter from the future to a family member working in the early-20th-century watch industry and explain to your relative what the hazards are and what safety equipment your relative should use.

14.79 Concern over radiation exposure has led some local, state, and federal agencies to propose a zero dose limit above background radiation (see Problem 14.77). A zero dose limit means that industry could not allow anyone to be exposed to radiation above the average background exposure per year. Why does background radiation change with altitude? Suggest a reason why the airline industry might oppose a zero dose limit.

14.80 Plutonium-239 is produced in breeder reactors by successive reactions in which ^{238}U is bombarded by neutrons and the product decays by two successive beta emissions.

(a) Write the equations for the three nuclear reactions.

(b) Inhalation of ^{239}Pu, which decays by alpha emission, can cause lung cancer. Explain why inhalation of ^{239}Pu may be more dangerous than getting an equal quantity of the isotope on the skin.

INSIGHT INTO Modern Medical Imaging

14.81 Write the equation for the nuclear decay of each of the following isotopes used in positron emission tomography (PET). **(a)** ^{18}F **(b)** ^{11}C **(c)** ^{13}N

14.82 In nuclear medicine, radioactive isotopes are often ingested or otherwise introduced into the human body. Discuss the properties you would expect medical radioisotopes to exhibit.

14.83 Write the nuclear equation for the antimatter annihilation of an electron by a positron to produce two gamma rays.

14.84 In boron neutron capture therapy (BNCT), ^{10}B absorbs a neutron, producing a nucleus that decays by alpha emission.

(a) Write the equations for these two successive events, including the compound nucleus.

(b) Alpha emission by inhaled ^{239}Pu causes lung cancer, but alpha emission in BNCT destroys cancerous tumors. Suggest a reason for this difference.

14.85 Osteoporosis (bone loss) is a growing medical problem as the average age of the U.S. population increases. It can be diagnosed using the isotope gadolinium-153. Using the Internet and other resources, research and write a paragraph about using ^{153}Gd to determine bone density.

Additional Problems

14.86 Fission reactions in the three operating reactors at the Fukushima Daiichi power plant were shut down immediately when the earthquake struck on March 11, 2011. Three days later, on March 14, decay heat was being produced at a rate of roughly 16 MW. Estimate the mass of water that would be vaporized per minute by this heat. (Recall that 1 W = 1 J/s, so 1 MW is 10^6 J/s. Assume that the cooling water is initially at 10°C and that once the water is converted to steam at 100°C it is vented from the reactor.)

14.87 Recall that the definition of the curie is 1 Ci = 3.70×10^{10} disintegrations per second and that this value was originally chosen so that the activity of 1.0 g of ^{226}Ra would be equal to 1 Ci. Given that ^{226}Ra has a molar mass of 226.025 g/mol and a half-life of 1600 years, calculate Avogadro's number.

14.88 Several radioactive isotopes of cesium are produced in the fission of ^{235}U, including those listed in the table below. These cesium isotopes can pose significant health concerns, in large part because the fact that cesium's chemical properties are similar to those of sodium or potassium means that cesium can be retained by the body over fairly long times. In the aftermath of the Fukushima earthquake and the ensuing nuclear disaster, many news stories focused on the dangers posed by the release of ^{137}Cs. Explain why the ^{137}Cs isotope is considered a bigger health threat than any of the other isotopes.

Isotope	Decay Mode	Half-life
^{135}Cs	β^- decay	2.3×10^6 years
^{136}Cs	β^- decay	13.16 days
^{137}Cs	β^- decay	30.17 years

14.89 Tellurium-128 undergoes an extremely rare form of radioactive decay know as double beta decay:

$$^{128}_{52}Te \rightarrow {}^{128}_{54}Xe + 2\ {}^{0}_{-1}\beta + 2\ \bar{\nu}$$

This is an extremely low probability event; its half-life of 2.2×10^{24} years is among the longest half-lives that have been determined. Estimate the activity in a 128-g (1 mol) sample of ^{128}Te. How many ^{128}Te atoms in such a 1-mol sample would decay in a day?

14.90 Radioactivity is commonly measured by counting events, and the activity of a particular sample can be reported as a count rate (counts per second, etc.). Suppose a sample of gallium contains ^{70}Ga with a half-life of 21.1 minutes. If the sample initially produces 455 counts per second, after what time would the sample produce 115 counts per second?

14.91 Two radioactive isotopes of iron, ^{52}Fe and ^{53}Fe, are both positron emitters with half-lives of 8.28 hours and 8.51 minutes, respectively. A particular sample of iron initially emits positrons at a rate of 210 counts per minute. What would be the activity of the sample after 30 minutes if the original sample contained the following?

(a) only ^{52}Fe, **(b)** only ^{53}Fe, **(c)** a mixture of 50% of each isotope

14.92 Both ^{192}Ir and ^{137}Cs are used in brachytherapy, the use of radioactive isotopes to treat various cancers. Both are beta emitters.

(a) Explain how this observation is consistent with the information presented in Figure 14.4.

(b) How do the *N/Z* ratios for these two medical isotopes compare to each other? How do they compare to their nearest stable isotopes?

14.93 Tailings from uranium mining contain radium, which decays to produce radon. The radon is also radioactive and undergoes further decay. Assume the radium present is all ^{226}Ra, and write the pair of reactions for these two radioactive events. Use Figure 14.4 to help choose the most plausible pathway for the decay of radon.

14.94 One decay series for radon is α, α, β, α, β, α, β, β. **(a)** What is the final isotope produced by this decay series? **(b)** Another possible decay series for radon is α, β, β, α, α, β, β, α. Does this lead to the same isotope as in (a)?

14.95 Both ^{192}Ir and ^{137}Cs are used in cancer treatments. The half-life of ^{192}Ir is 74 days and that of ^{137}Cs is 137 days. Suppose you work for a company that produces medical isotopes. If the radiation dosage needed from either of these isotopes is the same for a given treatment, which product would require a larger sample to be produced prior to shipment to a hospital that is 2 days away from your plant? Explain your reasoning.

14.96 Uranium-233 is another fissile isotope of uranium. It can be produced through neutron absorption by thorium-232. Protactinium-233 is formed along the way to the production of ^{233}U. Propose a set of nuclear reactions for this route to ^{233}U.

14.97 An interesting application of nuclear decay is the radioisotope thermoelectric generator, or RTG. In such a device, energy from radioactive decay is initially converted to heat in a manner similar to the decay heat discussed in Section 14-6. The heat is then converted to electrical energy using thermocouples in what is known as the Seebeck effect. The RTG is a long-lived source of low levels of electrical energy and has been used to power various space probes, including *Cassini*, which is currently orbiting Saturn. The most commonly used isotope in RTGs is ^{238}Pu. **(a)** Write a nuclear equation for the alpha decay of ^{238}Pu. **(b)** The energy output from the decay of ^{238}Pu is about 0.54 kW per kilogram of plutonium. Suppose the fuel for a particular RTG is 7.8 kg of $^{238}PuO_2$, and that the conversion of heat to electrical energy is about 6.5% efficient. Estimate the electrical power output in Watts for such an RTG.

14.98 One pathway for the decay of ^{26}Al is positron emission to form ^{26}Mg. The half-life of ^{26}Al is 7.17×10^5 years. **(a)** How many ^{26}Al nuclei would decay per second in a 1-mol sample of ^{26}Al? **(b)** The mass of ^{26}Al is 25.986892 u, the mass of ^{26}Mg is 25.982593 u, and the mass of a positron is 5.486×10^{-4} u. What is the energy released in a single ^{26}Al decay event? **(c)** If all the energy released is assumed to be heat, how much heat is evolved in one minute by the decay of the 1-mol sample of ^{26}Al from part (a)?

14.99 The half-life of ^{237}Np is 2.14×10^6 years, and Earth is 4.54×10^9 years old. If there was some amount of ^{237}Np present when Earth was formed, and no ^{237}Np has been formed since, what fraction of the original ^{237}Np remains today?

FOCUS ON PROBLEM SOLVING EXERCISES

14.100 A researcher wants to use radioactive metals to label an enzyme that is found in high concentration in cancer cells. Either ^{64}Cu or ^{68}Ga will bind with the enzyme. Suppose that you are the engineer assigned to this order by your company. What would you need to know about these isotopes and what questions would you ask the researcher to help with the selection?

14.101 Thallium-201 has a long history of use in nuclear medicine. Suppose that a customer requires 140 mg of the isotope, but delivery will take 18 hours by an express courier service. You are assigned to determine how much isotope you need to produce. What information would you need to look up to answer the question?

14.102 A customer has ordered 70 mg of ^{18}F radioisotope from your company. The delivery truck that normally would have made the trip from your factory to the customer in 15 minutes is delayed by a major traffic jam and will now take slightly more than an hour to arrive. If you packaged 100 mg of the isotope when it left the factory, will enough be left upon delivery? What information would you have to look up?

14.103 Suppose that you want to use radioactive iodine to run an experiment to treat a tumor. You can choose either ^{123}I or ^{125}I for this work. What do you need to know about these isotopes? What design parameters for the experiment factor into your decision?

Appendix A

INTERNATIONAL TABLE OF ATOMIC WEIGHTS

Atomic Number	Symbol	Name	Atomic Weight	Atomic Number	Symbol	Name	Atomic Weight
89	Ac	Actinium	[227]	72	Hf	Hafnium	178.49
13	Al	Aluminum	26.9815386	108	Hs	Hassium	[270]
95	Am	Americium	[243]	2	He	Helium	4.002602
51	Sb	Antimony	121.760	67	Ho	Holmium	164.93032
18	Ar	Argon	39.948	1	H	Hydrogen*	1.008
33	As	Arsenic	74.92160	49	In	Indium	114.818
85	At	Astatine	[210]	53	I	Iodine	126.90447
56	Ba	Barium	137.327	77	Ir	Iridium	192.217
97	Bk	Berkelium	[247]	26	Fe	Iron	55.845
4	Be	Beryllium	9.012182	36	Kr	Krypton	83.798
83	Bi	Bismuth	208.98040	57	La	Lanthanum	138.90547
107	Bh	Bohrium	[272]	103	Lr	Lawrencium	[262]
5	B	Boron*	10.81	82	Pb	Lead	207.2
35	Br	Bromine	79.904	116	Lv	Livermorium	[293]
48	Cd	Cadmium	112.411	3	Li	Lithium*	6.94
55	Cs	Cesium	132.9054519	71	Lu	Lutetium	174.9668
20	Ca	Calcium	40.078	12	Mg	Magnesium	24.3050
98	Cf	Californium	[251]	25	Mn	Manganese	54.938045
6	C	Carbon*	12.011	109	Mt	Meitnerium	[276]
58	Ce	Cerium	140.116	101	Md	Mendelevium	[258]
17	Cl	Chlorine*	35.45	80	Hg	Mercury	200.59
24	Cr	Chromium	51.9961	42	Mo	Molybdenum	95.96
27	Co	Cobalt	58.933195	60	Nd	Neodymium	144.242
112	Cn	Copernicium	[285]	10	Ne	Neon	20.1797
29	Cu	Copper	63.546	93	Np	Neptunium	[237]
96	Cm	Curium	[247]	28	Ni	Nickel	58.6934
110	Ds	Darmstadtium	[281]	41	Nb	Niobium	92.90638
105	Db	Dubnium	[268]	7	N	Nitrogen*	14.007
66	Dy	Dysprosium	162.500	102	No	Nobelium	[259]
99	Es	Einsteinium	[252]	76	Os	Osmium	190.23
68	Er	Erbium	167.259	8	O	Oxygen*	15.999
63	Eu	Europium	151.964	46	Pd	Palladium	106.42
100	Fm	Fermium	[257]	15	P	Phosphorus	30.973762
114	Fl	Flerovium	[289]	78	Pt	Platinum	195.084
9	F	Fluorine	18.9984032	94	Pu	Plutonium	[244]
87	Fr	Francium	[223]	84	Po	Polonium	[209]
64	Gd	Gadolinium	157.25	19	K	Potassium	39.0983
31	Ga	Gallium	69.723	59	Pr	Praseodymium	140.90765
32	Ge	Germanium	72.63	61	Pm	Promethium	[145]
79	Au	Gold	196.966569	91	Pa	Protactinium	231.03588

(Continued)

Atomic Number	Symbol	Name	Atomic Weight	Atomic Number	Symbol	Name	Atomic Weight
88	Ra	Radium	[226]	52	Te	Tellurium	127.60
86	Rn	Radon	[222]	65	Tb	Terbium	158.92535
75	Re	Rhenium	186.207	81	Tl	Thallium*	204.38
45	Rh	Rhodium	102.90550	90	Th	Thorium	232.03806
111	Rg	Roentgenium	[280]	69	Tm	Thulium	168.93421
37	Rb	Rubidium	85.4678	50	Sn	Tin	118.710
44	Ru	Ruthenium	101.07	22	Ti	Titanium	47.867
104	Rf	Rutherfordium	[267]	74	W	Tungsten	183.84
62	Sm	Samarium	150.36	118	Uuo	Ununoctium	[294]
21	Sc	Scandium	44.955912	115	Uup	Ununpentium	[288]
106	Sg	Seaborgium	[271]	113	Uut	Ununtrium	[284]
34	Se	Selenium	78.96	92	U	Uranium	238.02891
14	Si	Silicon*	28.0855	23	V	Vanadium	50.9415
47	Ag	Silver	107.8682	54	Xe	Xenon	131.293
11	Na	Sodium	22.98976928	70	Yb	Ytterbium	173.054
38	Sr	Strontium	87.62	39	Y	Yttrium	88.90585
16	S	Sulfur*	32.06	30	Zn	Zinc	65.38
73	Ta	Tantalum	180.94788	40	Zr	Zirconium	91.224
43	Tc	Technetium	[98]				

Values shown are from *Atomic weights of the elements 2009 (IUPAC Technical Report)*, M.E. Wiesser and T. B. Coplen, Pure Appl. Chem. **83**, 359 (2011). For those elements marked with an asterisk (*), IUPAC has defined the atomic weight as an interval. In such cases the values shown here are conventional atomic weights for use in calculations.

Appendix B

PHYSICAL CONSTANTS

Quantity	Symbol or Abbreviation	Value
Acceleration of gravity	g	9.80665 m/s
Atomic mass unit	amu or u	$1.660538921 \times 10^{-27}$ kg
Avogadro's number	N_A	$6.02214129 \times 10^{23}$ particles/mol
Boltzmann's constant	k	$1.3806488 \times 10^{-23}$ J/K
Charge to mass ratio of electron	e/m	$1.758820088 \times 10^{11}$ C/kg
Electronic charge	e	$1.602176565 \times 10^{-19}$ C
Electron rest mass	m_e	$9.10938291 \times 10^{-31}$ kg
Faraday constant	F	96,485.3365 C/mol e^-
Molar volume (STP)	V_m	22.414 L/mol
Neutron rest mass	m_n	$1.674927351 \times 10^{-27}$ kg
Planck's constant	h	$6.62606957 \times 10^{-34}$ J s
Proton rest mass	m_p	$1.672621777 \times 10^{-27}$ kg
Speed of light	c	2.99792458×10^{8} m s^{-1}
Universal gas constant	R	8.3144621 J mol^{-1} K^{-1}
		0.082057 L atm mol^{-1} K^{-1}
		62.364 L torr mol^{-}1 K^{-}1

Appendix C

ELECTRON CONFIGURATIONS OF ATOMS IN THE GROUND STATE

Z	Element	Configuration
1	H	$1s^1$
2	He	$1s^2$
3	Li	$[He]2s^1$
4	Be	$[He]2s^2$
5	B	$[He]2s^22p^1$
6	C	$[He]2s^22p^2$
7	N	$[He]2s^22p^3$
8	O	$[He]2s^22p^4$
9	F	$[He]2s^22p^5$
10	Ne	$[He]2s^22p^6$
11	Na	$[Ne]3s^1$
12	Mg	$[Ne]3s^2$
13	Al	$[Ne]3s^23p^1$
14	Si	$[Ne]3s^23p^2$
15	P	$[Ne]3s^23p^3$
16	S	$[Ne]3s^23p^4$
17	Cl	$[Ne]3s^23p^5$
18	Ar	$[Ne]3s^23p^6$
19	K	$[Ar]4s^1$
20	Ca	$[Ar]4s^2$
21	Sc	$[Ar]3d^14s^2$
22	Ti	$[Ar]3d^24s^2$
23	V	$[Ar]3d^34s^2$
24	Cr	$[Ar]3d^54s^1$
25	Mn	$[Ar]3d^54s^2$
26	Fe	$[Ar]3d^64s^2$
27	Co	$[Ar]3d^74s^2$
28	Ni	$[Ar]3d^84s^2$
29	Cu	$[Ar]3d^{10}4s^1$
30	Zn	$[Ar]3d^{10}4s^2$
31	Ga	$[Ar]3d^{10}4s^24p^1$
32	Ge	$[Ar]3d^{10}4s^24p^2$
33	As	$[Ar]3d^{10}4s^24p^3$
34	Se	$[Ar]3d^{10}4s^24p^4$
35	Br	$[Ar]3d^{10}4s^24p^5$
36	Kr	$[Ar]3d^{10}4s^24p^6$
37	Rb	$[Kr]5s^1$
38	Sr	$[Kr]5s^2$
39	Y	$[Kr]4d^15s^2$
40	Zr	$[Kr]4d^25s^2$
41	Nb	$[Kr]4d^45s^1$
42	Mo	$[Kr]4d^55s^1$
43	Tc	$[Kr]4d^65s^2$
44	Ru	$[Kr]4d^75s^1$
45	Rh	$[Kr]4d^85s^1$
46	Pd	$[Kr]4d^{10}$
47	Ag	$[Kr]4d^{10}5s^1$
48	Cd	$[Kr]4d^{10}5s^2$
49	In	$[Kr]4d^{10}5s^25p^1$
50	Sn	$[Kr]4d^{10}5s^25p^2$
51	Sb	$[Kr]4d^{10}5s^25p^3$
52	Te	$[Kr]4d^{10}5s^25p^4$
53	I	$[Kr]4d^{10}5s^25p^5$
54	Xe	$[Kr]4d^{10}5s^25p^6$
55	Cs	$[Xe]6s^1$
56	Ba	$[Xe]6s^2$
57	La	$[Xe]5d^16s^2$
58	Ce	$[Xe]4f^15d^16s^2$
59	Pr	$[Xe]4f^36s^2$
60	Nd	$[Xe]4f^46s^2$
61	Pm	$[Xe]4f^56s^2$
62	Sm	$[Xe]4f^66s^2$
63	Eu	$[Xe]4f^76s^2$
64	Gd	$[Xe]4f^75d^16s^2$
65	Tb	$[Xe]4f^96s^2$
66	Dy	$[Xe]4f^{10}6s^2$
67	Ho	$[Xe]4f^{11}6s^2$
68	Er	$[Xe]4f^{12}6s^2$
69	Tm	$[Xe]4f^{13}6s^2$
70	Yb	$[Xe]4f^{14}6s^2$
71	Lu	$[Xe]4^{14}5d^16s^2$
72	Hf	$[Xe]4f^{14}5d^26s^2$
73	Ta	$[Xe]4f^{14}5d^36s^2$
74	W	$[Xe]4f^{14}5d^46s^2$
75	Re	$[Xe]4f^{14}5d^56s^2$
76	Os	$[Xe]4f^{14}5d^66s^2$
77	Ir	$[Xe]4f^{14}5d^76s^2$
78	Pt	$[Xe]4f^{14}5d^96s^1$
79	Au	$[Xe]4f^{14}5d^{10}6s^1$
80	Hg	$[Xe]4f^{14}5d^{10}6s^2$
81	TI	$[Xe]4f^{14}5d^{10}6s^26p^1$
82	Pb	$[Xe]4f^{14}5d^{10}6s^26p^2$
83	Bi	$[Xe]4f^{14}5d^{10}6s^26p^3$
84	Po	$[Xe]4f^{14}5d^{10}6s^26p^4$
85	At	$[Xe]4f^{14}5d^{10}6s^26p^5$
86	Rn	$[Xe]4f^{14}5d^{10}6s^26p^6$
87	Fr	$[Rn]7s^1$
88	Ra	$[Rn]7s^2$
89	Ac	$[Rn]6d^17s^2$
90	Th	$[Rn]6d^27s^2$
91	Pa	$[Rn]5f^26d^17s^2$
92	U	$[Rn]5f^36d^17s^2$
93	Np	$[Rn]5f^46d^17s^2$
94	Pu	$[Rn]5f^67s^2$
95	Am	$[Rn]5f^77s^2$
96	Cm	$[Rn]5f^76d^17s^2$
97	Bk	$[Rn]5f^97s^2$
98	Cf	$[Rn]5f^{10}7s^2$
99	Es	$[Rn]5f^{11}7s^2$
100	Fm	$[Rn]5f^{12}7s^2$
101	Md	$[Rn]5f^{13}7s^2$
102	No	$[Rn]5f^{14}7s^2$
103	Lr	$[Rn]5f^{14}6d^17s^2$
104	Rf	$[Rn]5f^{14}6d^27s^2$
105	Db	$[Rn]5f^{14}6d^37s^2$
106	Sg	$[Rn]5f^{14}6d^47s^2$
107	Bh	$[Rn]5f^{14}6d^57s^2$
108	Hs	$[Rn]5f^{14}6d^67s^2$
109	Mt	$[Rn]5f^{14}6d^77s^2$

Appendix D

PHYSICAL CONSTANTS OF SOME COMMON SUBSTANCES

SPECIFIC HEATS AND HEAT CAPACITIES

Substance	Specific Heat, c ($J\ g^{-1}\ K^{-1}$)	Molar Heat Capacity, C_p ($J\ mol^{-1}\ K^{-1}$)
Al(s)	0.900	24.3
Ca(s)	0.653	26.2
Cu(s)	0.385	24.5
Fe(s)	0.444	24.8
Hg(ℓ)	0.138	27.7
H_2O(s), ice	2.09	37.7
H_2O(ℓ), water	4.184	75.3
H_2O(g), steam	2.03	36.4
CO_2(g)	0.843	37.1
C_6H_6(ℓ), benzene	1.74	136
C_6H_6(g), benzene	1.04	81.6
C_2H_5OH(ℓ), ethanol	2.46	113
C_2H_5OH(g), ethanol	0.954	420
$(C_2H_5)_2O$(ℓ), diethyl ether	3.74	172
$(C_2H_5)_2O$(g), diethyl ether	2.35	108

HEATS AND TEMPERATURES FOR PHASE CHANGES

Substance	Melting Point (°C)	Heat of Fusion ($J\ g^{-1}$)	ΔH_{fus} ($kJ\ mol^{-1}$)	Boiling Point (°C)	Heat of Vaporization ($J\ g^{-1}$)	ΔH_{fus} ($kJ\ mol^{-1}$)
Al	658	395	10.6	2467	10520	284
Ca	851	233	9.33	1487	4030	162
Cu	2083	205	13.0	2595	4790	305
H_2O	0	334	6.01	100	2260	40.7
Fe	1530	267	14.9	2735	6340	354
Hg	−39	11	23.3	357	292	58.6
CH_4	−182	58.6	0.92	−164	510	8.18
C_2H_5OH	−117	109	5.02	78.3	855	39.3
C_6H_6	5.48	127	9.92	80.1	395	30.8
$(C_2H_5)_2O$	−116	97.9	7.66	35	351	26.0

Appendix E

SELECTED THERMODYNAMIC DATA AT 298.15 K

Species	$\Delta H_f°$ (kJ mol^{-1})	$S°$ (J mol^{-1} K^{-1})	$\Delta G_f°$ (kJ mol^{-1})
Aluminum			
Al(s)	0	28.3	0
$AlCl_3$(s)	−704.2	110.7	−628.9
Al_2O_3(s)	−1676	50.92	−1582
$AlPO_4$(s)	−1733.8	90.76	−1617.9
$Al_2(SO_4)_3$(s)	−3440.84	239.3	−3099.94
Antimony			
Sb(s)	0	45.69	0
Sb_4O_6(s)	−1417.1	246.0	−1253.0
Sb_2O_5(s)	−971.9	125.1	−829.2
$SbCl_5$(ℓ)	−440.2	301	−350.1
$SbCl_3$(s)	−382.17	184.1	−259.4
Argon			
Ar(g)	0	154.843	0
Arsenic			
As(s)	0	35.1	0
H_3As(g)	66.44	222.78	68.93
As_2O_5(s)	−924.87	105.4	−782.3
$AsCl_3$(ℓ)	−305.0	216.3	−259.4
Barium			
Ba(s)	0	62.8	0
$BaCl_2$(s)	−860.1	126	−810.9
$BaCO_3$(s)	−1216.3	112.1	−1137.6
BaF_2(s)	−1207.1	96.36	−1156.8
$Ba(NO_3)_2$(s)	−992.07	213.8	−796.59
BaO(s)	−553.5	70.42	−525.1
$BaSO_4$(s)	−1465	132	−1353
Beryllium			
Be(s)	0	9.54	0
$BeCl_2$(s)	−490.4	82.68	−445.6
BeF_2(s)	−1026.8	53.35	−979.4
BeO(s)	−609.6	14.14	−580.3
$Be(OH)_2$(s)	−907.1	50.2	−817.6
$BeSO_4$(s)	−1205.20	77.91	−1093.80
Bismuth			
Bi(s)	0	56.74	0
$BiCl_3$(s)	−379.1	177.0	−315.0
Bi_2O_3(s)	−573.88	151.5	−493.7
Bi_2S_3(s)	−143.1	200.4	−140.6
Boron			
B(s)	0	5.86	0
BCl_3(ℓ)	−427.2	206.3	−387.4
B_2H_6(g)	35.6	232.11	86.7
BF_3(g)	−1137.00	254.12	−1120.33
H_3BO_3(s)	−1094.33	88.83	−968.92
$NaBH_4$(s)	−183.34	104.68	−119.54
Bromine			
Br(g)	111.8	174.9	82.4
Br_2(ℓ)	0	152.23	0
Br_2(g)	30.91	245.4	3.14
BrF_3(g)	−255.6	292.4	−229.5
HBr(g)	−36.4	198.59	−53.43
Cadmium			
Cd(s)	0	51.76	0
$CdCl_2$(s)	−391.50	115.27	−343.93
CdO(s)	−258.2	54.8	−228.4
CdS(s)	−161.9	64.9	−156.5
Calcium			
Ca(s)	0	41.6	0
Ca(g)	192.6	154.8	158.9
Ca^{2+}(aq)	−542.8	−53.1	−553.5
$CaBr_2$(s)	682.8	130	−663.6
CaC_2(s)	−62.8	70.3	−67.8
$CaCO_3$(s)	−1207	92.9	−1129
$CaCl_2$(s)	−795.0	114	−750.2
CaF_2(s)	−1215	68.87	−1162
CaH_2(s)	−189	42	−150
CaO(s)	−635.5	40	−604.2

Species	$\Delta H_f°$ (kJ mol^{-1})	$S°$ (J mol^{-1} K^{-1})	$\Delta G_f°$ (kJ mol^{-1})
$CaS(s)$	−482.4	56.5	−477.4
$Ca(NO_3)_2(s)$	−938.39	193.3	−743.07
$Ca(OH)_2(s)$	−986.6	76.1	−896.8
$Ca(OH)_2(aq)$	−1002.8	76.15	−867.6
$Ca_3(PO_4)_2(s)$	−4120.8	236.0	−3884.7
$CaSO_4(s)$	−1433	107	−1320
Carbon			
C(s, graphite)	0	5.740	0
C(s, diamond)	1.897	2.38	2.900
C(g)	716.7	158.0	671.3
$CCl_4(\ell)$	−135.4	216.4	−65.27
$CCl_4(g)$	−103	309.7	−60.63
$CHCl_3(\ell)$	−134.5	202	−73.72
$CHCl_3(g)$	−103.1	295.6	−70.37
$CH_2Cl_2(g)$	−121.46	177.8	−67.26
$CH_3Cl(g)$	−80.83	234.58	−57.37
$CF_4(g)$	−925	261.61	−879
$CF_2Cl_2(g)$	−477	301	−440
$CH_3CF_3(\ell)$	−737	280	−668
$CH_4(g)$	−74.81	186.2	−50.75
$C_2H_2(g)$	226.7	200.8	209.2
$C_2H_4(g)$	52.26	219.5	68.12
$C_2H_6(g)$	−84.86	229.5	−32.9
$C_3H_6(g)$	20.41	266.9	62.75
$C_3H_8(g)$	−103.8	269.9	−23.49
$C_4H_{10}(g)$ (n-butane)	−126.5	310.03	−15.71
$C_6H_6(\ell)$	49.03	172.8	124.5
$C_8H_{18}(\ell)$ (n-octane)	−250.3	361.2	16.32
$CH_3OH(\ell)$	−238.66	126.8	−166.27
$C_2H_5OH(\ell)$	−277.7	161	−174.9
$C_2H_5OH(g)$	−235.1	282.6	−168.6
HCOOH(aq)	−425.6	92.0	−351
$HCOOH(\ell)$	−424.72	128.95	−361.35
HCHO(g)	−108.57	218.77	−102.53
$CH_3CHO(g)$	−166.19	250.3	−128.86
$CH_3COOH(\ell)$	−484.5	159.8	−389.9
$H_2C_2O_4(s)$	−828.93	115.6	−697.2
$H_2C_2O_4(aq)$	−825.1	45.61	−673.9
HCN(g)	135.1	201.78	124.7
$CH_3NH_2(g)$	23.0	243.3	32.16
$CO(NH_2)_2(s)$	−334	105	−198

Species	$\Delta H_f°$ (kJ mol^{-1})	$S°$ (J mol^{-1} K^{-1})	$\Delta G_f°$ (kJ mol^{-1})
$CH_3CN(\ell)$	54	150	99
$C_3H_3N(\ell)$	172.9	188	208.6
CO(g)	−110.5	197.6	−137.2
$CO_2(g)$	−393.5	213.6	−394.4
$CO_2(aq)$	−413.80	117.6	−385.98
$COCl_2(g)$	−223.0	289.2	−210.5
$CS_2(g)$	117.4	237.7	67.15
$(CH_3)_2SO(\ell)$	−203	188	−99
$C_6H_{12}O_6(s)$ (glucose)	−1274.5	212.1	−910.56
$C_{12}H_{22}O_{11}(s)$ (sucrose)	−2221.7	360.24	−1544.3
Cesium			
Cs(s)	0	85.23	0
$Cs^+(aq)$	−248	133	−282
CsF(aq)	−568.6	123	−558.5
Chlorine			
$Cl_2(g)$	0	223.0	0
Cl(g)	121.7	165.1	105.7
$Cl^-(g)$	−226	—	—
$Cl^-(aq)$	−167.29	56.48	−131.26
ClO(g)	101.22		
$ClO_2(g)$	104.60	249.4	123.4
$Cl_2O(g)$	80	266	98
ClF(g)	−54.48	217.89	−55.94
$ClF_3(g)$	−163.2	281.61	−123.0
HCl(g)	−92.31	186.8	−95.30
HCl(aq)	−167.4	55.10	−131.2
Chromium			
Cr(s)	0	23.8	0
$Cr_2O_3(s)$	−1139.7	81.2	−1058.1
$Cr_2Cl_3(s)$	−556.5	123.0	−486.1
$(NH_4)_2Cr_2O_7(s)$	−1807	—	—
Cobalt			
Co(s)	0	30.04	0
CoO(s)	−237.94	52.97	−214.20
$Co_3O_4(s)$	−891	102.5	−774
$CoCl_2(s)$	−312.5	109.16	−269.8
$CoSO_4(s)$	−888.3	118.0	−782.3

(Continued)

Species	$\Delta H_f°$ (kJ mol^{-1})	$S°$ (J mol^{-1} K^{-1})	$\Delta G_f°$ (kJ mol^{-1})
Copper			
Cu(s)	0	33.15	0
CuBr(s)	−104.6	96.11	−100.8
CuCl(s)	−137.2	86.2	−119.86
$CuCl_2$(s)	−220.1	108.07	−175.7
CuI(s)	−67.8	96.7	−69.5
CuO(s)	−157	42.63	−130
Cu_2O(s)	−168.6	93.14	−146.0
CuS(s)	−53.1	66.5	−53.6
Cu_2S(s)	−79.5	120.9	−86.2
$CuSO_4$(s)	−771.36	109	−661.8
Fluorine			
F^-(g)	−322	—	—
F^-(aq)	−332.6	—	−278.8
F(g)	78.99	158.6	61.92
F_2(g)	0	202.7	0
HF(g)	−271	173.7	−273
HF(aq)	−320.08	88.7	−296.8
Germanium			
Ge(s)	0	3	0
GeH_4(g)	91	217	113
$GeCl_4$(g)	−496	348	−457
GeO_2(s)	−551	55	−497
Gold			
Au(s)	0	48	0
Helium			
He(g)	0	126.150	0
Hydrogen			
H(g)	218.0	114.6	203.3
H_2(g)	0	130.6	0
$H_2O(\ell)$	−285.8	69.91	−237.2
H_2O(g)	−241.8	188.7	−228.6
$H_2O_2(\ell)$	−187.8	109.6	−120.4
Iodine			
I(g)	106.6	180.66	70.16
I_2(s)	0	116.1	0
I_2(g)	62.44	260.6	19.36
ICl(g)	17.78	247.4	−5.52
HI(g)	26.5	206.5	1.72

Species	$\Delta H_f°$ (kJ mol^{-1})	$S°$ (J mol^{-1} K^{-1})	$\Delta G_f°$ (kJ mol^{-1})
Iron			
Fe(s)	0	27.3	0
$FeCl_2$(s)	−340.67	117.9	−302.3
$FeCl_3$(s)	−399.49	142.3	−334.00
$FeCO_3$(s)	−741	93	−667
$Fe(CO)_5(\ell)$	−774.0	338	−705.4
$Fe(CO)_5$(g)	−733.8	445.2	−697.3
FeO(s)	−272	—	—
Fe_2O_3 (s, hematite)	−824.2	87.40	−742.2
Fe_3O_4 (s, magnetite)	−1118	146	−1015
$Fe(OH)_2$(s)	−569.0	88	−486.5
$Fe(OH)_3$(s)	−823.0	106.7	−696.5
FeS_2(s)	−177.5	122.2	−166.7
$FeSO_4$(s)	−928.4	107.5	−820.8
Lead			
Pb(s)	0	64.81	0
$PbCl_2$(s)	−359.4	136	−314.1
PbO(s, yellow)	−217.3	68.70	−187.9
PbO_2(s)	−277.4	68.6	−217.33
$Pb(OH)_2$(s)	−515.9	88	−420.9
PbS(s)	−100.4	91.2	−98.7
$PbSO_4$(s)	−919.94	148.57	−813.14
Lithium			
Li(s)	0	28.0	0
$LiAlH_4$(s)	−116.3	78.74	−44.7
LiCl(s)	−408.61	59.33	−384.37
LiF(s)	−615.97	35.65	−587.71
LiH(s)	−90.54	20.008	−68.35
$LiNO_3$(s)	−483.13	90.0	−381.1
LiOH(s)	−487.23	50	−443.9
LiOH(aq)	−508.4	4	−451.1
Magnesium			
Mg(s)	0	32.5	0
Mg^{2+}(aq)	−454.668	−138.1	−455.57
$MgBr_2$(s)	−524.3	117.2	−503.8
$MgCl_2$(s)	−641.8	89.5	−592.3
$MgCO_3$(s)	−1095.8	65.7	−1012.1
MgF_2(s)	−1123.4	57.24	−1070.2
MgI_2(s)	−364.0	129.7	−358.2

Species	$\Delta H_f°$ (kJ mol^{-1})	$S°$ (J mol^{-1} K^{-1})	$\Delta G_f°$ (kJ mol^{-1})
$Mg_3N_2(s)$	−461.1	87.86	−401.1
$Mg(NO_3)_2(s)$	−790.65	164.0	−589.4
$MgO(s)$	−601.8	27	−569.6
$Mg(OH)_2(s)$	−924.7	63.14	−833.7
$Mg_3(PO_4)_2(s)$	−3780.7	189.20	−3538.7
$MgS(s)$	−347	—	—
$MgSO_4(s)$	−1284.9	91.6	−1170.6
Manganese			
$Mn(s)$	0	32.01	0
$MnCl_2(s)$	−481.29	118.24	−440.50
$MnO(s)$	−385.22	59.71	−362.90
$MnO_2(s)$	−520.03	53.05	−465.14
$Mn_2O_3(s)$	−959.0	110.5	−881.1
$Mn_3O_4(s)$	−1387.8	155.6	−1283.2
$MnSO_4(s)$	−1065.25	112.1	−957.36
Mercury			
$Hg(\ell)$	0	76.02	0
$Hg(g)$	61.32	174.96	31.82
$HgCl_2(s)$	−224.3	146	−178.6
$Hg_2Cl_2(s)$	−224	146	−179
HgO(s, red)	−90.83	70.29	−58.56
HgS(s, red)	−8.2	82.4	−50.6
Neon			
$Ne(g)$	0	146.328	0
Nickel			
$Ni(s)$	0	30.1	0
$NiCl_2(s)$	−82.0	52.97	−79.5
$Ni(CO)_4(g)$	−602.9	410.4	−587.3
$NiO(s)$	−244	38.6	−216
$NiS(g)$	−82.0	52.97	−79.5
$NiSO_4(s)$	−872.91	92	−759.7
Nitrogen			
$N_2(g)$	0	191.5	0
$N(g)$	472.704	153.19	455.579
$NH_3(g)$	−46.11	192.3	−16.5
$NH_3(aq)$	−80.29	111.3	−26.50
$NH_4^+(aq)$	−132.51	113.4	−79.31
$N_2H_4(\ell)$	50.63	121.2	149.2
$(NH_4)_3AsO_4(aq)$	−1268	—	—
$NH_4Br(s)$	−270.83	113	−175.2
$NH_4Cl(s)$	−314.4	94.6	−201.5
$NH_4Cl(aq)$	−300.2	—	—
$NH_4HCO_3(s)$	−847	12.1	−666
$NH_4I(s)$	−201.4	117	−113
$NH_4NO_3(s)$	−365.6	151.1	−184.0
$(NH_4)_2SO_4(s)$	−1180.85	220.1	−901.67
$NF_3(g)$	−125	260.6	−83.3
$NO(g)$	90.25	210.7	86.57
$NO_2(g)$	33.2	240.0	51.30
$N_2O(g)$	82.05	219.7	104.2
$N_2O_3(g)$	83.72	321.28	139.46
$N_2O_4(g)$	9.16	304.2	97.82
$N_2O_5(g)$	11	356	115
$N_2O_5(s)$	−43.1	178	114
$NOCl(g)$	52.59	264	66.36
$HNO_3(\ell)$	−174.1	155.6	−80.79
$HNO_3(g)$	−135.1	266.2	−74.77
HNO_3 (aq)	−206.6	146	−110.5
Oxygen			
$O_2(g)$	0	205.0	0
$O(g)$	249.2	161.0	231.8
$O_3(g)$	143	238.8	163
$OF_2(g)$	23	246.6	41
Phosphorus			
$P(g)$	314.6	163.1	278.3
P_4(s, white)	0	177	0
P(s, red)	−18.4	22.8	−12.1
$P_4(g)$	58.91	279.98	24.44
$PCl_3(g)$	−306.4	311.7	−286.3
$PCl_5(g)$	−398.9	353	−324.6
$PF_3(g)$	−918.8	273.24	−897.5
$PH_3(g)$	5.4	210.1	13
$P_4O_{10}(s)$	−2984	228.9	−2698
$H_3PO_4(s)$	−1281	110.5	−1119
$H_3PO_4(aq)$	−1288.34	158.2	−1142.54
Potassium			
$K(s)$	0	63.6	0
$KBr(s)$	−393.798	95.90	−380.66
$KCl(s)$	−436.5	82.6	−408.8
$KClO_3(s)$	−391.2	143.1	−289.9
$KClO_4(s)$	−432.747	82.59	−409.14

(Continued)

Species	$\Delta H_f°$ (kJ mol^{-1})	$S°$ (J mol^{-1} K^{-1})	$\Delta G_f°$ (kJ mol^{-1})
$K_2CO_3(s)$	−1151.02	155.52	−1063.5
$K_2Cr_2O_4(s)$	−1403.7	200.12	−1295.7
$K_2Cr_2O_7(s)$	−2061.5	291.2	−1881.8
KF(s)	−567.27	66.57	−537.75
KI(s)	−327.9	106.4	−323.0
$KMnO_4(s)$	−837.2	171.71	−737.6
$KNO_3(s)$	−494.63	133.05	−394.86
$KO_2(s)$	284.93	116.7	−239.4
$K_2O_2(s)$	−494.1	102.1	−425.1
KOH(s)	−424.7	78.91	−378.9
KOH(aq)	−481.2	92.0	−439.6
$K_2SO_4(s)$	−1437.79	175.56	−1321.37
Rubidium			
Rb(s)	0	76.78	0
RbOH(aq)	−481.16	110.75	−441.24
Selenium			
Se(s)	0	42.442	0
$H_2Se(g)$	−9.7	219.02	15.9
Silicon			
Si(s)	0	18.8	0
$SiBr_4(\ell)$	−457.3	277.8	−443.9
SiC(s)	−65.3	16.6	−62.8
$SiCl_4(g)$	−657.0	330.6	−617.0
$SiCl_4(\ell)$	−687.0	239.7	−619.84
$SiH_4(g)$	34.3	204.5	56.9
$SiF_4(g)$	−1615	282.4	−1573
$SiI_4(g)$	−132	—	—
$SiO_2(s)$	−910.9	41.84	−856.7
$H_2SiO_3(s)$	−1189	134	−1092
$H_4SiO_4(s)$	−1481.1	192	−1332.9
$Na_2SiO_3(s)$	−1079	—	—
$H_2SiF_6(aq)$	−2331	—	—
Silver			
Ag(s)	0	42.55	0
$Ag^+(aq)$	105.79	73.86	77.12
AgBr(s)	−100.37	107.1	−96.90
AgCl(s)	−126.904	96.085	−109.8
AgI(s)	−61.84	115.5	−66.19
$AgNO_3(s)$	−124.39	140.92	−33.41
$Ag_2O(s)$	−31.05	121.3	−11.20
AgS(s)	−32.59	144.01	−40.67
AgSCN(s)	87.9	131.0	101.36
$Ag_2SO_4(s)$	−715.88	200.4	−618.41
Sodium			
Na(s)	0	51.0	0
Na(g)	108.7	153.6	78.11
$Na^+(g)$	601	—	—
$Na^+(aq)$	−240.2	59.0	−261.9
NaBr(s)	−359.9	86.82	−348.98
NaCl(s)	−411.0	72.38	−384
NaCl(aq)	−407.1	115.5	−393.0
$NaClO_4(s)$	−383.30	142.3	−254.85
NaCN(s)	−87.49	118.49	−76.4
$NaCH_3COO(s)$	−708.80	123.0	−607.18
$Na_2CO_3(s)$	−1131	136	−1048
NaF(s)	−573.647	51.46	−543.494
NaH(s)	−56.275	40.016	−33.46
$NaHCO_3(s)$	−950.81	101.7	−851.0
$NaHSO_4(s)$	−1125.5	113.0	−992.8
$NaH_2PO_4(s)$	−1538	128	−1387
$Na_2HPO_4(s)$	−1749	151	−1609
NaI(s)	−287.78	98.53	−286.06
$NaNO_3(s)$	−467.85	116.52	−367.00
$Na_2O(s)$	−414.22	75.06	−375.46
$Na_2O_2(s)$	−510.87	95.0	−447.7
NaOH(s)	−426.7	64.45	−379.49
NaOH(aq)	−469.6	49.8	−419.2
$Na_3PO_4(s)$	−1917.40	173.80	−1788.80
$Na_2S(s)$	−364.8	83.7	−349.8
$Na_2SO_4(s)$	−1387.08	149.58	−1270.16
$Na_2S_2O_3(s)$	−1123.0	155	−1028.0
Strontium			
Sr(s)	0	52.3	0
$SrCl_2(s)$	−828.9	114.85	−781.1
$SrCO_3(s)$	−1220.1	97.1	−1140.1
SrO(s)	−592.0	54.4	−561.9
Sulfur			
S(s, rhombic)	0	31.8	0
S(g)	278.8	167.8	−38.3
$S_8(g)$	102.30	430.98	49.63

Species	$\Delta H_f°$ (kJ mol^{-1})	$S°$ (J mol^{-1} K^{-1})	$\Delta G_f°$ (kJ mol^{-1})
S_2Cl_2(g)	−18	331	−31.8
SF_6(g)	−1209	291.7	−1105
H_2S(g)	−20.6	205.7	−33.6
H_2SO_4(ℓ)	−814.0	156.9	−690.1
H_2SO_4(aq)	−907.5	17	−742.0
SO_2(g)	−296.8	248.1	−300.2
SO_3(g)	−395.6	256.6	−371.1
$SOCl_2$(ℓ)	−206	—	—
SO_2Cl_2(ℓ)	−389	—	—
Tellurium			
Te(s)	0	49.71	0
TeO_2(s)	−322.6	79.5	−270.3
Tin			
Sn(s, white)	0	51.55	0
Sn(s, gray)	−2.09	44.1	0.13
$SnCl_2$(s)	−350	—	—
$SnCl_4$(ℓ)	−511.3	258.6	−440.2
$SnCl_4$(g)	−471.5	366	−432.2
SnO(s)	−285.8	56.5	−256.9
SnO_2(s)	−580.7	52.3	−519.7
Titanium			
Ti(s)	0	30.6	0
$TiCl_4$(ℓ)	−804.2	252.3	−737.2
$TiCl_4$(g)	−763.2	354.8	−726.8
Tungsten			
W(s)	0	32.6	0
WO_3(s)	−842.9	75.90	−764.1
Uranium			
U(s)	0	50.21	0
UF_6(s)	−2147.4	377.9	−2063.7
UO_2(s)	−1084.9	77.03	−1031.7
UO_3(s)	−1223.8	96.11	−1145.9
Xenon			
Xe(g)	0	169.683	0
XeF_2(g)	−130	260	−96
XeF_4(g)	−215	316	−138
XeO_3(g)	502	287	561
Zinc			
Zn(s)	0	41.63	0
$ZnCl_2$(s)	−415.05	111.46	−369.398
$ZnCO_3$(s)	−812.78	82.4	−731.52
ZnO(s)	−348.3	43.64	−318.3
$Zn(OH)_2$(s)	−643.25	81.6	−555.07
ZnS(s)	−205.6	57.7	−201.3
$ZnSO_4$(s)	−982.8	110.5	−871.5

Appendix F

IONIZATION CONSTANTS OF WEAK ACIDS AT 25°C

Acid	Formula and Ionization Equation	K_a
Acetic	$CH_3COOH + H_2O \rightleftarrows CH_3COO^- + H_3O^+$	1.8×10^{-5}
Arsenic	$H_3AsO_4 + H_2O \rightleftarrows H_2AsO_4^- + H_3O^+$	$K_{a1} = 2.5 \times 10^{-4}$
	$H_2AsO_4^- + H_2O \rightleftarrows HAsO_4^{2-} + H_3O^+$	$K_{a2} = 5.6 \times 10^{-8}$
	$HAsO_4^{2-} + H_2O \rightleftarrows AsO_4^{3-} + H_3O^+$	$K_{a3} = 3.0 \times 10^{-13}$
Arsenous	$H_3AsO_3 + H_2O \rightleftarrows H_2AsO_3^- + H_3O^+$	$K_{a1} = 6.0 \times 10^{-10}$
	$H_2AsO_3^- + H_2O \rightleftarrows HAsO_3^{2-} + H_3O^+$	$K_{a2} = 3.0 \times 10^{-14}$
Benzoic	$C_6H_5COOH + H_2O \rightleftarrows C_6H_5COO^- + H_3O^+$	6.3×10^{-5}
Boric	$B(OH)_3 + H_2O \rightleftarrows BO(OH)_2^- + H_3O^+$	$K_{a1} = 7.3 \times 10^{-10}$
	$BO(OH)_2^- + H_2O \rightleftarrows BO_2(OH)^{2-} + H_3O^+$	$K_{a2} = 1.8 \times 10^{-13}$
	$BO_2(OH)^{2-} + H_2O \rightleftarrows BO_3^{3-} + H_3O^+$	$K_{a3} = 1.6 \times 10^{-14}$
Butanoic	$CH_3CH_2CH_2COOH + H_2O \rightleftarrows CH_3CH_2CH_2COO^- + H_3O^+$	1.5×10^{-5}
Carbonic	$H_2CO_3 + H_2O \rightleftarrows HCO_3^- + H_3O^+$	$K_{a1} = 4.4 \times 10^{-7}$
	$HCO_3^- + H_2O \rightleftarrows CO_3^{2-} + H_3O^+$	$K_{a2} = 4.8 \times 10^{-11}$
Citric	$C_3H_5O(COOH)_3 + H_2O \rightleftarrows C_4H_5O_3(COOH)_2^- + H_3O^+$	$K_{a1} = 7.4 \times 10^{-3}$
	$C_4H_5O_3(COOH)_2^- + H_2O \rightleftarrows C_5H_5O_5COOH^{2-} + H_3O^+$	$K_{a2} = 1.7 \times 10^{-5}$
	$C_5H_5O_5COOH^{2-} + H_2O \rightleftarrows C_6H_5O_7^{3-} + H_3O^+$	$K_{a3} = 7.4 \times 10^{-7}$
Cyanic	$HOCN + H_2O \rightleftarrows OCN^- + H_3O^+$	3.5×10^{-4}
Formic	$HCOOH + H_2O \rightleftarrows HCOO^- + H_3O^+$	1.8×10^{-4}
Gluconic	$HOCH_2(CHOH)_4COOH + H_2O \rightleftarrows HOCH_2(CHOH)_4COO^- + H_3O^+$	2.4×10^{-4}
Heptonic	$HOCH_2(CHOH)_5COOH + H_2O \rightleftarrows HOCH_2(CHOH)_5COO^- + H_3O^+$	1.3×10^{-5}
Hydrazoic	$HN_3 + H_2O \rightleftarrows N_3^- + H_3O^+$	1.9×10^{-5}
Hydrocyanic	$HCN + H_2O \rightleftarrows CN^- + H_3O^+$	6.2×10^{-10}
Hydrofluoric	$HF + H_2O \rightleftarrows F^- + H_3O^+$	6.3×10^{-4}
Hydrogen peroxide	$H_2O_2 + H_2O \rightleftarrows HO_2^- + H_3O^+$	2.4×10^{-12}
Hydrosulfuric	$H_2S + H_2O \rightleftarrows HS^- + H_3O^+$	$K_{a1} = 1.0 \times 10^{-7}$
	$HS^- + H_2O \rightleftarrows S^{2-} + H_3O^+$	$K_{a2} = 1.0 \times 10^{-19}$
Hypobromous	$HOBr + H_2O \rightleftarrows OBr^- + H_3O^+$	2.5×10^{-9}
Hypochlorous	$HOCl + H_2O \rightleftarrows OCl^- + H_3O^+$	2.9×10^{-8}
Hypoiodous	$HOI + H_2O \rightleftarrows OI^- + H_3O^+$	2.3×10^{-11}
Nitrous	$HNO_2 + H_2O \rightleftarrows NO_2^- + H_3O^+$	4.5×10^{-4}
Oxalic	$(COOH)_2 + H_2O \rightleftarrows COOCOOH^- + H_3O^+$	$K_{a1} = 5.9 \times 10^{-2}$
	$COOCOOH^- + H_2O \rightleftarrows (COO)_2^{2-} + H_3O^+$	$K_{a2} = 6.4 \times 10^{-5}$
Phenol	$HC_6H_5O + H_2O \rightleftarrows C_6H_5O^- + H_3O^+$	1.3×10^{-10}
Phosphoric	$H_3PO_4 + H_2O \rightleftarrows H_2PO_4^- + H_3O^+$	$K_{a1} = 7.5 \times 10^{-3}$
	$H_2PO_4^- + H_2O \rightleftarrows HPO_4^{2-} + H_3O^+$	$K_{a2} = 6.2 \times 10^{-8}$
	$HPO_4^{2-} + H_2O \rightleftarrows PO_4^{3-} + H_3O^+$	$K_{a3} = 3.6 \times 10^{-13}$

Acid	Formula and Ionization Equation	K_a
Phosphorous	$H_3PO_3 + H_2O \rightleftarrows H_2PO_3^- + H_3O^+$	$K_{a1} = 1.6 \times 10^{-2}$
	$H_2PO_3^- + H_2O \rightleftarrows HPO_3^{2-} + H_3O^+$	$K_{a2} = 7.0 \times 10^{-7}$
Propanoic	$CH_3CH_2COOH + H_2O \rightleftarrows CH_3CH_2COO^- + H_3O^+$	1.3×10^{-5}
Salicylic	$C_6H_4(OH)COOH + H_2O \rightleftarrows C_6H_4(OH)COO^- + H_3O^+$	1.1×10^{-3}
Selenic	$H_2SeO_4 + H_2O \rightleftarrows HSeO_4^- + H_3O^+$	K_{a1} = Very large
	$HSeO_4^- + H_2O \rightleftarrows SeO_4^{2-} + H_3O^+$	$K_{a2} = 1.2 \times 10^{-2}$
Selenous	$H_2SeO_3 + H_2O \rightleftarrows HSeO_3^- + H_3O^+$	$K_{a1} = 2.7 \times 10^{-3}$
	$HSeO_3^- + H_2O \rightleftarrows SeO_3^{2-} + H_3O^+$	$K_{a2} = 2.5 \times 10^{-7}$
Sulfuric	$H_2SO_4 + H_2O \rightleftarrows HSO_4^- + H_3O^+$	K_{a1} = Very large
	$HSO_4^- + H_2O \rightleftarrows SO_4^{2-} + H_3O^+$	$K_{a2} = 1.2 \times 10^{-2}$
Sulfurous	$H_2SO_3 + H_2O \rightleftarrows HSO_3^- + H_3O^+$	$K_{a1} = 1.2 \times 10^{-2}$
	$HSO_3^- + H_2O \rightleftarrows SO_3^{2-} + H_3O^+$	$K_{a2} = 6.2 \times 10^{-8}$
Tellurous	$H_2TeO_3 + H_2O \rightleftarrows HTeO_3^- + H_3O^+$	$K_{a1} = 2 \times 10^{-3}$
	$HTeO_3^- + H_2O \rightleftarrows TeO_3^{2-} + H_3O^+$	$K_{a2} = 1 \times 10^{-8}$

Appendix G

IONIZATION CONSTANTS OF WEAK BASES AT 25°C

Base	Formula and Ionization Equation	K_b
Ammonia	$NH_3 + H_2O \rightleftarrows NH_4^+ + OH^-$	1.8×10^{-5}
Aniline	$C_6H_5NH_2 + H_2O \rightleftarrows C_6H_5NH_3^+ + OH^-$	4.2×10^{-10}
Dimethylamine	$(CH_3)_2NH + H_2O \rightleftarrows (CH_3)_2NH_2^+ + OH^-$	7.4×10^{-4}
Ethylenediamine	$(CH_2)_2(NH_2)_2 + H_2O \rightleftarrows (CH_2)_2(NH_2)_2H^+ + OH^-$	$K_{b1} = 8.5 \times 10^{-5}$
	$(CH_2)_2(NH_2)_2H^+ + H_2O \rightleftarrows (CH_2)_2(NH_2)_2H_2^{2+} + OH^-$	$K_{b2} = 2.7 \times 10^{-8}$
Hydrazine	$N_2H_4 + H_2O \rightleftarrows N_2H_5^+ + OH^-$	$K_{b1} = 8.5 \times 10^{-7}$
	$N_2H_5^+ + H_2O \rightleftarrows N_2H_6^{2+} + OH^-$	$K_{b2} = 8.9 \times 10^{-16}$
Hydroxylamine	$NH_2OH + H_2O \rightleftarrows NH_3OH^+ + OH^-$	6.6×10^{-9}
Methylamine	$CH_3NH_2 + H_2O \rightleftarrows CH_3NH_3^+ + OH^-$	5.0×10^{-4}
Pyridine	$C_5H_5N + H_2O \rightleftarrows C_5H_5NH^+ + OH^-$	1.5×10^{-9}
Trimethylamine	$(CH_3)_3N + H_2O \rightleftarrows (CH_3)_3NH^+ + OH^-$	7.4×10^{-5}

Appendix H

SOLUBILITY PRODUCT CONSTANTS OF SOME INORGANIC COMPOUNDS AT 25°C

Substance	K_{sp}
Aluminum compounds	
$AlAsO_4$	1.6×10^{-16}
$Al(OH)_3$	1.9×10^{-33}
$AlPO_4$	1.3×10^{-20}
Antimony compounds	
Sb_2S_3	1.6×10^{-93}
Barium compounds	
$Ba_3(AsO_4)_2$	1.1×10^{-13}
$BaCO_3$	8.1×10^{-9}
$BaCrO_4$	2.0×10^{-10}
BaF_2	1.7×10^{-6}
$Ba_3(PO_4)_2$	1.3×10^{-29}
$BaSeO_4$	2.8×10^{-11}
$BaSO_3$	8.0×10^{-7}
$BaSO_4$	1.1×10^{-10}
Bismuth compounds	
$BiOCl$	7.0×10^{-9}
$BiO(OH)$	1.0×10^{-12}
BiI_3	8.1×10^{-19}
$BiPO_4$	1.3×10^{-23}
Bi_2S_3	1.6×10^{-72}
Cadmium compounds	
$Cd_3(AsO_4)_2$	2.2×10^{-32}
$CdCO_3$	2.5×10^{-14}
$Cd(CN)_2$	1.0×10^{-8}
$Cd_2[Fe(CN)_6]$	3.2×10^{-17}
$Cd(OH)_2$	1.2×10^{-14}
CdS	3.6×10^{-29}
Calcium compounds	
$Ca_3(AsO_4)_2$	6.8×10^{-19}
$CaCO_3$	4.8×10^{-9}
$CaCrO_4$	7.1×10^{-4}
CaF_2	1.7×10^{-10}
$Ca(OH)_2$	7.9×10^{-6}
$CaHPO_4$	2.7×10^{-7}
$Ca(H_2PO_4)_2$	1.0×10^{-3}
$Ca_3(PO_4)_2$	2.0×10^{-33}
Chromium compounds	
$CrAsO_4$	7.8×10^{-21}
$Cr(OH)_3$	6.7×10^{-31}
$CrPO_4$	2.4×10^{-23}
Cobalt compounds	
$Co_3(AsO_4)_2$	7.6×10^{-29}
$CoCO_3$	8.0×10^{-13}
$Co(OH)_2$	2.5×10^{-16}
CoS (α)	5.9×10^{-21}
CoS (β)	8.7×10^{-23}
$Co(OH)_3$	4.0×10^{-45}
Co_2S_3	2.6×10^{-124}
Copper compounds	
$CuBr$	5.3×10^{-9}
$CuCl$	1.9×10^{-7}
$CuCN$	3.2×10^{-20}
CuI	5.1×10^{-12}
Cu_2S	1.6×10^{-48}
$CuSCN$	1.6×10^{-11}
$Cu_3(AsO_4)_2$	7.6×10^{-36}
$CuCO_3$	2.5×10^{-10}
$Cu_2[Fe(CN)_6]$	1.3×10^{-16}
$Cu(OH)_2$	1.6×10^{-19}
CuS	8.7×10^{-36}
Gold compounds	
$AuBr$	5.0×10^{-17}
$AuCl$	2.0×10^{-13}
AuI	1.6×10^{-23}
$AuBr_3$	4.0×10^{-36}
$AuCl_3$	3.2×10^{-25}
$Au(OH)_3$	1.0×10^{-53}
AuI_3	1.0×10^{-46}
Iron compounds	
$FeCO_3$	3.5×10^{-11}
$Fe(OH)_2$	7.9×10^{-15}
FeS	4.9×10^{-18}
$Fe_4[Fe(CN)_6]_3$	3.0×10^{-41}
$Fe(OH)_3$	6.3×10^{-38}
Fe_2S_3	1.4×10^{-88}
Lead compounds	
$Pb_3(AsO_4)_2$	4.1×10^{-36}
$PbBr_2$	6.3×10^{-6}
$PbCO_3$	1.5×10^{-13}
$PbCl_2$	1.7×10^{-5}
$PbCrO_4$	1.8×10^{-14}
PbF_2	3.7×10^{-8}
$Pb(OH)_2$	2.8×10^{-16}
PbI_2	8.7×10^{-9}
$Pb_3(PO_4)_2$	3.0×10^{-44}
$PbSeO_4$	1.5×10^{-7}
$PbSO_4$	1.8×10^{-8}
PbS	8.4×10^{-28}
Magnesium compounds	
$Mg_3(AsO_4)_2$	2.1×10^{-20}
MgC_2O_4	8.6×10^{-5}
MgF_2	6.4×10^{-9}
$Mg(OH)_2$	1.5×10^{-11}
$MgNH_4PO_4$	2.5×10^{-12}
Manganese compounds	
$Mn_3(AsO_4)_2$	2.1×10^{-20}
$MnCO_3$	1.8×10^{-11}
$Mn(OH)_2$	4.6×10^{-14}
MnS	5.1×10^{-15}
$Mn(OH)_3$	1.0×10^{-36}
Mercury compounds	
Hg_2Br_2	1.3×10^{-22}
Hg_2CO_3	8.9×10^{-17}

(Continued)

Substance	K_{sp}	Substance	K_{sp}	Substance	K_{sp}
Hg_2Cl_2	1.1×10^{-18}	*Silver compounds*		$Sr_3(PO_4)_2$	1.0×10^{-31}
Hg_2CrO_4	5.0×10^{-9}	Ag_3AsO_4	1.1×10^{-20}	$SrSO_3$	4.0×10^{-8}
Hg_2I_2	4.5×10^{-29}	$AgBr$	5.3×10^{-13}	$SrSO_4$	2.8×10^{-7}
Hg_2SO_4	6.8×10^{-7}	Ag_2CO_3	8.1×10^{-12}		
Hg_2S	5.8×10^{-44}	$AgCl$	1.8×10^{-10}	*Tin compounds*	
$Hg(CN)_2$	3.0×10^{-23}	Ag_2CrO_4	9.0×10^{-12}	$Sn(OH)_2$	2.0×10^{-26}
$Hg(OH)_2$	2.5×10^{-26}	$AgCN$	6.0×10^{-17}	SnI_2	1.0×10^{-4}
HgI_2	4.0×10^{-29}	$Ag_4[Fe(CN)_6]$	1.6×10^{-41}	SnS	1.0×10^{-28}
HgS	3.0×10^{-53}	AgI	1.5×10^{-16}	$Sn(OH)_4$	1.0×10^{-57}
		Ag_3PO_4	1.3×10^{-20}	SnS_2	1.0×10^{-70}
Nickel compounds		Ag_2SO_3	1.5×10^{-14}		
$Ni_3(AsO_4)_2$	1.9×10^{-26}	Ag_2SO_4	1.2×10^{-5}	*Zinc compounds*	
$NiCO_3$	6.6×10^{-9}	Ag_2S	1.0×10^{-49}	$Zn_3(AsO_4)_2$	1.1×10^{-27}
$Ni(CN)_2$	3.0×10^{-23}	$AgSCN$	1.0×10^{-12}	$ZnCO_3$	1.5×10^{-11}
$Ni(OH)_2$	2.8×10^{-16}			$Zn(CN)_2$	8.0×10^{-12}
NiS (α)	3.0×10^{-21}	*Strontium compounds*		$Zn_2[Fe(CN)_6]$	4.1×10^{-16}
NiS (β)	1.0×10^{-26}	$Sr_3(AsO_4)_2$	1.3×10^{-18}	$Zn(OH)_2$	4.5×10^{-17}
NiS (γ)	2.0×10^{-28}	$SrCO_3$	9.4×10^{-10}	$Zn_3(PO_4)_2$	9.1×10^{-33}
		$SrCrO_4$	3.6×10^{-5}	ZnS	1.1×10^{-21}

Appendix I

STANDARD REDUCTION POTENTIALS IN AQUEOUS SOLUTION AT 25°C

Half-Reaction, Acidic Solution	Standard Reduction Potential, $E°$ (volts)
$Li^+(aq) + e^- \rightarrow Li(s)$	−3.045
$K^+(aq) + e^- \rightarrow K(s)$	−2.925
$Rb^+(aq) + e^- \rightarrow Rb(s)$	−2.925
$Ba^{2+}(aq) + 2\ e^- \rightarrow Ba(s)$	−2.90
$Sr^{2+}(aq) + 2\ e^- \rightarrow Sr(s)$	−2.89
$Ca^{2+}(aq) + 2\ e^- \rightarrow Ca(s)$	−2.87
$Na^+(aq) + e^- \rightarrow Na(s)$	−2.714
$Mg^{2+}(aq) + 2\ e^- \rightarrow Mg(s)$	−2.37
$H_2(g) + 2\ e^- \rightarrow 2\ H^-(aq)$	−2.25
$Al^{3+}(aq) + 3\ e^- \rightarrow Al(s)$	−1.66
$Zr^{4+}(aq) + 4\ e^- \rightarrow Zr(s)$	−1.53
$ZnS(s) + 2\ e^- \rightarrow Zn(s) + S^{2-}(aq)$	−1.44
$CdS(s) + 2\ e^- \rightarrow Cd(s) + S^{2-}(aq)$	−1.21
$V^{2+}(aq) + 2\ e^- \rightarrow V(s)$	−1.18
$Mn^{2+}(aq) + 2\ e^- \rightarrow Mn(s)$	−1.18
$FeS(s) + 2\ e^- \rightarrow Fe(s) + S^{2-}(aq)$	−1.01
$Cr^{2+}(aq) + 2\ e^- \rightarrow Cr(s)$	−0.91
$Zn^{2+}(aq) + 2\ e^- \rightarrow Zn(s)$	−0.763
$Cr^{3+}(aq) + 3\ e^- \rightarrow Cr(s)$	−0.74
$HgS(s) + 2\ H^+(aq) + 2\ e^- \rightarrow Hg(\ell) + H_2S(g)$	−0.72
$Ga^{3+}(aq) + 3\ e^- \rightarrow Ga(s)$	−0.53
$2\ CO_2(g) + 2\ H^+(aq) + 2\ e^- \rightarrow (COOH)_2(aq)$	−0.49
$Fe^{2+}(aq) + 2\ e^- \rightarrow Fe(s)$	−0.44
$Cr^{3+}(aq) + e^- \rightarrow Cr^{2+}(aq)$	−0.41
$Cd^{2+}(aq) + 2\ e^- \rightarrow Cd(s)$	−0.403
$Se(s) + 2\ H^+(aq) + 2\ e^- \rightarrow H_2Se(aq)$	−0.40
$PbSO_4(s) + 2\ e^- \rightarrow Pb(s) + SO_4^{\ 2-}(aq)$	−0.356
$Tl^+(aq) + e^- \rightarrow Tl(s)$	−0.34
$Co^{2+}(aq) + 2\ e^- \rightarrow Co(s)$	−0.28
$Ni^{2+}(aq) + 2\ e^- \rightarrow Ni(s)$	−0.25
$[SnF_6]^{2-}(aq) + 4\ e^- \rightarrow Sn(s) + 6\ F^-(aq)$	−0.25
$AgI(s) + e^- \rightarrow Ag(s) + I^-(aq)$	−0.15
$Sn^{2+}(aq) + 2\ e^- \rightarrow Sn(s)$	−0.14
$Pb^{2+}(aq) + 2\ e^- \rightarrow Pb(s)$	−0.126
$N_2O(g) + 6\ H^+(aq) + H_2O(\ell) + 4\ e^- \rightarrow 2\ NH_3OH^+(aq)$	−0.05
$2\ H^+(aq) + 2\ e^- \rightarrow H_2(g)$	0.000
$AgBr(s) + e^- \rightarrow Ag(s) + Br^-(aq)$	0.10

(Continued)

Half-Reaction, Acidic Solution	Standard Reduction Potential, $E°$ (volts)
$S(s) + 2\ H^+(aq) + 2\ e^- \rightarrow H_2S(aq)$	0.14
$Sn^{4+}(aq) + 2\ e^- \rightarrow Sn^{2+}(aq)$	0.15
$Cu^{2+}(aq) + e^- \rightarrow Cu^+(aq)$	0.153
$SO_4^{2-}(aq) + 4\ H^+(aq) + 2\ e^- \rightarrow H_2SO_3(aq) + H_2O(\ell)$	0.17
$SO_4^{2-}(aq) + 4\ H^+(aq) + 2\ e^- \rightarrow SO_2(g) + 2\ H_2O(\ell)$	0.20
$AgCl(s) + e^- \rightarrow Ag(s) + Cl^-(aq)$	0.222
$Hg_2Cl_2(s) + 2\ e^- \rightarrow 2\ Hg(\ell) + 2\ Cl^-(aq)$	0.27
$Cu^{2+}(aq) + 2\ e^- \rightarrow Cu(s)$	0.337
$[RhCl_6]^{3-}(aq) + 3\ e^- \rightarrow Rh(s) + 6\ Cl^-(aq)$	0.44
$Cu^+(aq) + e^- \rightarrow Cu(s)$	0.521
$TeO_2(s) + 4\ H^+(aq) + 4\ e^- \rightarrow Te(s) + 2\ H_2O(\ell)$	0.529
$I_2(s) + 2\ e^- \rightarrow 2\ I^-(aq)$	0.535
$H_3AsO_4(aq) + 2\ H^+(aq) + 2\ e^- \rightarrow H_3AsO_3(aq) + H_2O(\ell)$	0.58
$[PtCl_6]^{2-}(aq) + 2\ e^- \rightarrow [PtCl_4]^{2-}(aq) + 2\ Cl^-(aq)$	0.68
$O_2(g) + 2\ H^+(aq) + 2\ e^- \rightarrow H_2O_2(aq)$	0.682
$[PtCl_4]^{2-}(aq) + 2\ e^- \rightarrow Pt(s) + 4\ Cl^-(aq)$	0.73
$SbCl_6^-(aq) + 2\ e^- \rightarrow SbCl_4^-(aq) + 2\ Cl^-(aq)$	0.75
$Fe^{3+}(aq) + e^- \rightarrow Fe^{2+}(aq)$	0.771
$Hg_2^{2+}(aq) + 2\ e^- \rightarrow 2\ Hg(\ell)$	0.789
$Ag^+(aq) + e^- \rightarrow Ag(s)$	0.7994
$Hg^{2+}(aq) + 2\ e^- \rightarrow Hg(\ell)$	0.855
$2\ Hg^{2+}(aq) + 2\ e^- \rightarrow Hg_2^{2+}(aq)$	0.920
$NO_3^-(aq) + 3\ H^+(aq) + 2\ e^- \rightarrow HNO_2(aq) + H_2O(\ell)$	0.94
$NO_3^-(aq) + 4\ H^+(aq) + 3\ e^- \rightarrow NO(g) + 2\ H_2O(\ell)$	0.96
$Pd^{2+}(aq) + 2\ e^- \rightarrow Pd(s)$	0.987
$AuCl_4^-(aq) + 3\ e^- \rightarrow Au(s) + 4\ Cl^-(aq)$	1.00
$Br_2(\ell) + 2\ e^- \rightarrow 2\ Br^-(aq)$	1.08
$ClO_4^-(aq) + 2\ H^+(aq) + 2\ e^- \rightarrow ClO_3^-(aq) + H_2O(\ell)$	1.19
$IO_3^-(aq) + 6\ H^+(aq) + 5\ e^- \rightarrow \frac{1}{2}I_2(aq) + 3\ H_2O(\ell)$	1.195
$Pt^{2+}(aq) + 2\ e^- \rightarrow Pt(s)$	1.2
$O_2(g) + 4\ H^+(aq) + 4\ e^- \rightarrow 2\ H_2O(\ell)$	1.229
$MnO_2(s) + 4\ H^+(aq) + 2\ e^- \rightarrow Mn^{2+}(aq) + 2\ H_2O(\ell)$	1.23
$N_2H_5^+(aq) + 3\ H^+(aq) + 2\ e^- \rightarrow 2\ NH_4^+(aq)$	1.24
$Cr_2O_7^{2-}(aq) + 14\ H^+(aq) + 6\ e^- \rightarrow 2\ Cr^{3+}(aq) + 7\ H_2O(\ell)$	1.33
$Cl_2(g) + 2\ e^- \rightarrow 2\ Cl^-(aq)$	1.360
$BrO_3^-(aq) + 6\ H^+(aq) + 6\ e^- \rightarrow Br^-(aq) + 3\ H_2O(\ell)$	1.44
$ClO_3^-(aq) + 6\ H^+(aq) + 5\ e^- \rightarrow \frac{1}{2}Cl_2(aq) + 3\ H_2O(\ell)$	1.47
$Au^{3+}(aq) + 3\ e^- \rightarrow Au(s)$	1.50
$MnO_4^-(aq) + 8\ H^+(aq) + 5\ e^- \rightarrow Mn^{2+}(aq) + 4\ H_2O(\ell)$	1.507
$NaBiO_3(s) + 6\ H^+(aq) + 2\ e^- \rightarrow Bi^{3+}(aq) + Na^+(aq) + 3\ H_2O(\ell)$	1.6
$Ce^{4+}(aq) + e^- \rightarrow Ce^{3+}(aq)$	1.61
$2\ HOCl(aq) + 2\ H^+(aq) + 2\ e^- \rightarrow Cl_2(g) + 2\ H_2O(\ell)$	1.63
$Au^+(aq) + e^- \rightarrow Au(s)$	1.68
$PbO_2(s) + SO_4^{2-}(aq) + 4\ H^+(aq) + 2\ e^- \rightarrow PbSO_4(s) + 2\ H_2O(\ell)$	1.685
$NiO_2(s) + 4\ H^+(aq) + 2\ e^- \rightarrow Ni^{2+}(aq) + 2\ H_2O(\ell)$	1.7

Half-Reaction, Acidic Solution	Standard Reduction Potential, $E°$ (volts)
$H_2O_2(aq) + 2\ H^+(aq) + 2\ e^- \rightarrow 2\ H_2O(\ell)$	1.77
$Pb^{4+}(aq) + 2\ e^- \rightarrow Pb^{2+}(aq)$	1.8
$Co^{3+}(aq) + e^- \rightarrow Co^{2+}(aq)$	1.82
$F_2(g) + 2\ e^- \rightarrow 2\ F^-(aq)$	2.87

STANDARD REDUCTION POTENTIALS IN AQUEOUS SOLUTION AT 25°C

Half-Reaction, Basic Solution	Standard Reduction Potential, $E°$ (volts)
$SiO_3^{2-}(aq) + 3\ H_2O(\ell) + 4\ e^- \rightarrow Si(s) + 6\ OH^-(aq)$	−1.70
$Cr(OH)_3(s) + 3\ e^- \rightarrow Cr(s) + 3\ OH^-(aq)$	−1.30
$[Zn(CN)_4]^{2-}(aq) + 2\ e^- \rightarrow Zn(s) + 4\ CN^-(aq)$	−1.26
$Zn(OH)_2(s) + 2\ e^- \rightarrow Zn(s) + 2\ OH^-(aq)$	−1.245
$[Zn(OH)_4]^{2-}(aq) + 2\ e^- \rightarrow Zn(s) + 4\ OH^-(aq)$	−1.22
$N_2(g) + 4\ H_2O(\ell) + 4\ e^- \rightarrow N_2H_4(aq) + 4\ OH^-(aq)$	−1.15
$SO_4^{2-}(aq) + H_2O(\ell) + 2\ e^- \rightarrow SO_3^{2-}(aq) + 2\ OH^-(aq)$	−0.93
$Fe(OH)_2(s) + 2\ e^- \rightarrow Fe(s) + 2\ OH^-(aq)$	−0.877
$2\ NO_3^-(aq) + 2\ H_2O(\ell) + 2\ e^- \rightarrow N_2O_4(g) + 4\ OH^-(aq)$	−0.85
$2\ H_2O(\ell) + 2\ e^- \rightarrow H_2(g) + 2\ OH^-(aq)$	−0.828
$Fe(OH)_3(s) + e^- \rightarrow Fe(OH)_2(s) + OH^-(aq)$	−0.56
$S(s) + 2\ e^- \rightarrow S^{2-}(aq)$	−0.48
$Cu(OH)_2(s) + 2\ e^- \rightarrow Cu(s) + 2\ OH^-(aq)$	−0.36
$CrO_4^{2-}(aq) + 4\ H_2O(\ell) + 3\ e^- \rightarrow Cr(OH)_3(s) + 5\ OH^-(aq)$	−0.12
$MnO_2(s) + 2\ H_2O(\ell) + 2\ e^- \rightarrow Mn(OH)_2(s) + 2\ OH^-(aq)$	−0.05
$NO_3^-(aq) + H_2O(\ell) + 2\ e^- \rightarrow NO_2^-(aq) + 2\ OH^-(aq)$	0.01
$O_2(g) + H_2O(\ell) + 2\ e^- \rightarrow OOH^-(aq) + OH^-(aq)$	0.076
$HgO(s) + H_2O(\ell) + 2\ e^- \rightarrow Hg(\ell) + 2\ OH^-(aq)$	0.0984
$[Co(NH_3)_6]^{3+}(aq) + e^- \rightarrow [Co(NH_3)_6]^{2+}(aq)$	0.10
$N_2H_4(aq) + 2\ H_2O(\ell) + 2\ e^- \rightarrow 2\ NH_3(aq) + 2\ OH^-(aq)$	0.10
$2\ NO_2^-(aq) + 3\ H_2O(\ell) + 4\ e^- \rightarrow N_2O(g) + 6\ OH^-(aq)$	0.15
$Ag_2O(s) + H_2O(\ell) + 2\ e^- \rightarrow 2\ Ag(s) + 2\ OH^-(aq)$	0.34
$ClO_4^-(aq) + H_2O(\ell) + 2\ e^- \rightarrow ClO_3^-(aq) + 2\ OH^-(aq)$	0.36
$O_2(g) + 2\ H_2O(\ell) + 4\ e^- \rightarrow 4\ OH^-(aq)$	0.40
$Ag_2CrO_4(s) + 2\ e^- \rightarrow 2\ Ag(s) + CrO_4^{2-}(aq)$	0.446
$NiO_2(s) + 2\ H_2O(\ell) + 2\ e^- \rightarrow Ni(OH)_2(s) + 2\ OH^-(aq)$	0.49
$MnO_4^-(aq) + e^- \rightarrow MnO_4^{2-}(aq)$	0.564
$MnO_4^-(aq) + 2\ H_2O(\ell) + 3\ e^- \rightarrow MnO_2(s) + 4\ OH^-(aq)$	0.588
$ClO_3^-(aq) + 3\ H_2O(\ell) + 6\ e^- \rightarrow Cl^-(aq) + 6\ OH^-(aq)$	0.62
$2\ NH_2OH(aq) + 2\ e^- \rightarrow N_2H_4(aq) + 2\ OH^-(aq)$	0.74
$OOH^-(aq) + H_2O(\ell) + 2\ e^- \rightarrow 3\ OH^-(aq)$	0.88
$ClO^-(aq) + H_2O(\ell) + 2\ e^- \rightarrow Cl^-(aq) + 2\ OH^-(aq)$	0.89

Appendix J

ANSWERS TO *CHECK YOUR UNDERSTANDING EXERCISES*

Chapter 1

1.1 See Figure 1.4.

1.2 **(a)** 3, **(b)** 1

1.3 **(a)** 16.10 m, **(b)** 0.005 g

1.4 \$299. (We ignore significant figures here since they would not usually be taken into account in calculating a price.) The amount purchased might still have to be a multiple of 5 gallons, so it might exceed the actual amount needed.

1.5 4.3×10^9 W

1.6 3.34×10^4 g, or 33.4 kg

1.7

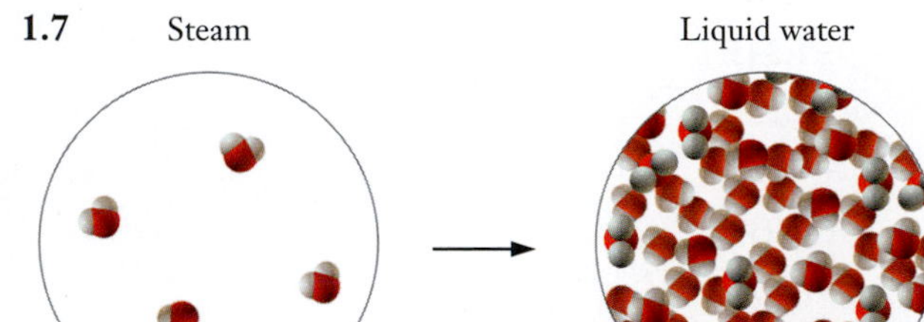

Chapter 2

2.1 28.1 amu

2.2 8 carbon atoms, 12 hydrogen atoms, 4 nitrogen atoms; C_2H_3N

2.3

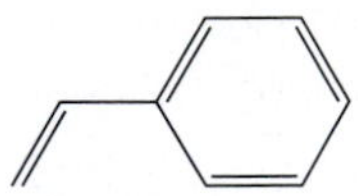

2.4 C_6H_9NO

2.5 **(a)** carbon disulfide, **(b)** sulfur hexafluoride, **(c)** dichlorine heptoxide

2.6 **(a)** copper(II) sulfate, **(b)** silver phosphate, **(c)** vanadium(V) oxide

Chapter 3

3.1 $C_3H_8 + 2\ O_2 \rightarrow 3\ CH_2O + H_2O$

3.2 Compounds (a), (b), and (c) are all soluble; (d) is insoluble.

3.3 Molecular $\rightarrow$ $2\ HCl(aq) + Ca(OH)_2(aq) \rightarrow CaCl_2(aq) + 2\ H_2O(\ell)$

Total ionic $\rightarrow$ $2\ H^+(aq) + 2\ Cl^-(aq) + Ca^{2+}(aq) + 2\ OH^-(aq) \rightarrow Ca^{2+}(aq) + 2\ Cl^-(aq) + 2\ H_2O(\ell)$

Net ionic $\rightarrow$ $H^+(aq) + OH^-(aq) \rightarrow H_2O(\ell)$

(These could also be written with H_3O^+ instead of H^+.)

3.4 Molecular: $Na_2SO_4(aq) + Pb(NO_3)_2(aq) \rightarrow PbSO_4(s) + 2\ NaNO_3(aq)$

Total ionic: $2\ Na^+(aq) + SO_4^{2-}(aq) + Pb^{2+}(aq) + 2\ NO_3^-(aq) \rightarrow PbSO_4(s) + 2\ Na^+(aq) + 2\ NO_3^-(aq)$

Net ionic: $SO_4^{2-}(aq) + Pb^{2+}(aq) \rightarrow PbSO_4(s)$

3.5 **(a)** 180.112 g mol^{-1}, **(b)** 286.327 g mol^{-1} **(c)** 133.103 g mol^{-1}

3.6 41.4 mol

3.7 7×10^6 tons

3.8 $C_{15}H_{28}O_2$

3.9 57 mol % Au, 22 mol % Ag, and 21 mol % Cu

3.10 596.75 moles

3.11 0.216 M

Chapter 4

4.1 13.5 mol H_2O

4.2 3.27 g H_2

4.3 9.6 g SO_2

4.4 $>4.5 \times 10^3$ moles H_2

4.5 Same result as in the example: S_8 is the limiting reactant.

4.6 20.8 g B_2H_6

4.7 The remaining fuel is 262.2 g of aluminum.

4.8 80.9% yield

4.9 63.3 mL

4.10 0.828 M NaOH

Chapter 5

5.1 82.1 torr

5.2 310 K

5.3 0.77 ft^3

5.4 $V = 29.6$ L, $P_{O_2} = 0.349$ atm, $P_{N_2} = 1.78$ atm

5.5 6.4 g SO_2, 3.4 g SO_3; $P_{SO_2} = 0.98$ atm, $P_{SO_3} = 0.42$ atm

5.6 0.96 g $NaHCO_3$

5.7 9.67 L SO_2

5.8 $P_{ideal} = 0.817$ atm, $P_{VdW} = 0.816$ atm

Chapter 6

6.1 $5.600 \times 10^{16}\ s^{-1}$

6.2 1.5405×10^{-10} m, or 154.05 pm

6.3 2.6×10^{-20} J

6.4 97.27 nm

6.5 $4d$

6.6 $1s^22s^22p^63s^23p^2$

6.7 [Ar] $4s^23d^{10}4p^3$

6.8 [Ar] $4s^23d^8$

6.9 F (smallest) < Si < Cr < Sr < Cs (largest)

6.10 Rb (lowest IE) < Mg < Si < N < He (highest IE)

Chapter 7

7.1 The first ionization energy will be small, the second ionization energy will be slightly larger, and the third ionization energy will be extremely large.

7.2 RbBr will have the smallest lattice energy. The lattice includes only singly charged ions, and the Rb^+ and Br^- ions are larger than those in KCl.

7.3 N—Cl

7.4

```
   H   H
   |   |
H— C — C —H
   |   |
   H   H
```

7.5

```
⎡    H    ⎤+
⎢    |    ⎥      ⎡  ..   ..   .. ⎤−
⎢ H— N —H ⎥      ⎢ :O — Cl — O:  ⎥
⎢    |    ⎥      ⎣  ..   ..   .. ⎦
⎣    H    ⎦
```

7.6 :O=C=O:

7.7 Octahedral

7.8 **(a)** T-shape, **(b)** linear

7.9 **(a)** Trigonal planar, **(b)** linear

Chapter 8

8.1 Values are given in the text: 68% for body-centered cubic and 52% for simple cubic.

8.2 The dopant should be an element with more than four valence electrons. Phosphorus is the most likely choice. The band diagram will look like that in Figure 8.12.

8.3 Ne (lowest boiling point) < CH_4 < CO < NH_3 (highest boiling point)

8.4

```
     H           O
     |           ‖
     N           
   /   \   /\   / \
  H         |       OH
            |
```

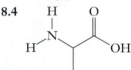

Chapter 9

9.1 +223 J

9.2 16.3°C (or 16.3 K)

9.3 24.4 J mol^{-1} $°C^{-1}$

9.4 64.0°C

9.5 29.3 kJ/g

9.6 -3.28×10^4 J

9.7 1.9 kJ mol^{-1}

9.8 −252.6 kJ

9.9 −1215 kJ (for reaction of 2 moles of CH_4)

9.10 41.2 g NO, 19.2 g N_2

Chapter 10

10.1 266.8 J mol^{-1} K^{-1}

10.2 Gases condense at low temperatures. The process is enthalpy driven because it is exothermic, but it decreases the entropy of the system. Enthalpy dominates at low temperatures.

10.3 8.80 J mol^{-1} K^{-1}

10.4 185.9 kJ mol^{-1}

Chapter 11

11.1 NO_2: 8.0×10^{-6} mol L^{-1} s^{-1}; O_2: 2.0×10^{-6} mol L^{-1} s^{-1}

11.2 **(a)** Reaction is first order in H_2, first order in Br_2, and second order overall. **(b)** Reaction is first order in N_2O_5 and first order overall.

11.3 No, as long as we do the calculation correctly the results will agree within any experimental error.

11.4 Units for rate constants depend on the overall reaction order.

11.5 870 seconds

11.6 First-order reaction, $k = 0.0023\ s^{-1}$

11.7 $1.6 \times 10^{-5}\ s^{-1}$

11.8 42 kJ mol^{-1}

11.9 **(a)** Yes, it does match the actually stoichiometry and does not involve any steps in which three or more molecules collide. **(b)** N_2O_3 and NO are both intermediates in this mechanism. **(c)** The first two steps are unimolecular, and the third step is bimolecular.

Chapter 12

12.1 $K = \dfrac{[HNO_2]^2}{[NO][NO_2][H_2O]}$

12.2 $3\ Cu^{2+}(aq) + 2\ PO_4^{3-}(aq) \rightarrow Cu_3(PO_4)_2(s)$;

$K = \dfrac{1}{[Cu^{2+}]^3[PO_4^{3-}]^2}$

12.3 Because the value of K for Mn^{2+} is larger than that for either Ca^{2+} or Mg^{2+}, Mn^{2+} has a higher ability to precipitate OH^- than either of the other cations.

12.4 **(a)** $K = \dfrac{[NO]^2[H_2O]^3}{[NH_3]^2[O_2]^{5/2}}$; **(b)** $K = \dfrac{[NH_3]^2[O_2]^{5/2}}{[NO]^2[H_2O]^3}$; **(c)** $K = \dfrac{[NO]^4[H_2O]^6}{[NH_3]^4[O_2]^5}$

12.5 $K = \dfrac{[NO]^2}{[N_2][O_2]}$ and the reaction is $N_2(g) + O_2(g) \rightleftarrows 2\ NO(g)$.

12.6 $[H_2] = [I_2] = 0.002$ M, [HI] = 0.016 M

12.7 $[Cl_2] = [PCl_3] = 0.036$ M, $[PCl_5] = 0.004$ M

12.8 Possibilities include removing HCN from the system or adding an acid (other than HCN), which would increase $[H^+]$.

12.9 **(a)** Shift toward reactants, **(b)** no change

12.10 **(a)** $K_{sp} = [Mn^{2+}][OH^-]^2$, **(b)** $K_{sp} = [Cu^{2+}][S^{2-}]$, **(c)** $K_{sp} = [Cu^+][I^-]$, **(d)** $K_{sp} = [Al^{3+}]^2[SO_4^{2-}]^3$

12.11 7.7×10^{-9} M; 1.0×10^{-7} g/100 g H_2O

12.12 3.5×10^{-4} M in H_2O; 7.6×10^{-9} M in 0.15 M NaF

12.13 Hydrogen carbonate ion (HCO_3^-) is the conjugate acid of carbonate ion (CO_3^{2-}), and hydroxide ion (OH^-) is the conjugate base of water.

12.14 10.5% ionized. In this case, the assumption is introducing an error of roughly 0.6% so that whether or not it is acceptable would depend on the required accuracy.

12.15 ln K = 236, which means that K for this reaction would be on the order of 10^{102}. This extremely large number indicates that the complete oxidation is the very strongly preferred pathway.

Chapter 13

13.1 Using zinc for the other electrode will give a cell potential of 1.100 V. Copper is the cathode and zinc is the anode.

13.2 $[Ni^{2+}] = 5 \times 10^{-2}$ M

13.3 Reversing the reaction would change the sign of ΔG, giving +28.4 kJ. The positive value indicates that the reverse process is not spontaneous.

13.4 The 5-A current must flow for 1.1×10^4 s, which is about 3 hours.

13.5 0.27 V

13.6 No. The stated conditions would deposit only 100 mg of copper, so the coating will reach only about one-fourth of the desired thickness.

13.7 0.042A

Chapter 14

14.1 $^{209}_{84}Po \rightarrow {}^{205}_{82}Pb + {}^{4}_{2}He$

14.2 ^{218}Po

14.3 $^{11}_{6}C \rightarrow {}^{11}_{5}B + {}^{0}_{1}\beta + \nu$

14.4 0.283 s^{-1}; this is roughly 11 orders of magnitude larger than the decay constant for ^{14}C.

14.5 1.1×10^4 years

14.6 7.034435×10^{10} kJ

Appendix K

ANSWERS TO ODD-NUMBERED END-OF-CHAPTER EXERCISES

Chapter 1

1.1 Approximately 200 g heavier. (This web exercise will have varying answers depending on the websites found.)

1.3 The first step in the scientific method is observation. Everything begins with observing something in nature and then trying to understand it.

1.5 This web exercise will have varying answers depending on the websites found.

1.7 Observations in the laboratory are normally made in the macroscopic regime of chemistry.

1.9 At the macroscopic level, distinguishing phases involves (1) determining their ability to maintain their shapes (only solids do) and (2) determining if the system fills the entire volume available to it, or only part of the volume—this observation helps distinguish liquids from gases.

1.11 A picture such as this would be appropriate:

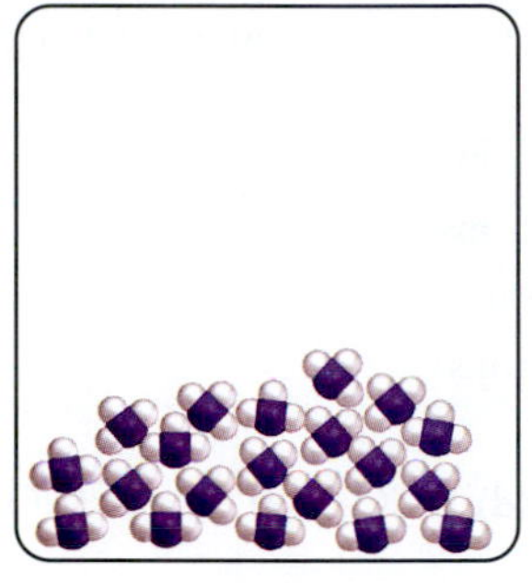

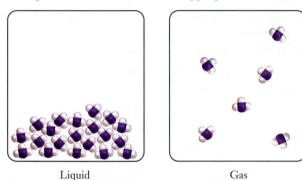

Liquid

Gas

1.13 **(a)** Chemical change; **(b)** physical change; **(c)** chemical change; **(d)** chemical change; and **(e)** physical change.

1.15 When steel rusts (a chemical change), it becomes more brittle and changes color. (Other answers possible.)

1.17 Looking at the picture in Exercise 1.11, the depiction of the gas shows a significant amount of empty space between particles. The empty space contributes nothing to the mass of the system even though the entire volume is occupied, so the density of the gas is lower than that of the liquid (which occupies less than the complete volume and has less space between particles).

1.19 If you wanted to know how many cities were in a particular state, you would need to define what constitutes a city. Is population the basis, or is the specific nature of incorporation, for example? (Other answers possible.)

1.21 Precise. The measurements might well be close to each other even though they are not close to the actual answer.

1.23 In deductive reasoning, you consider all the information (the facts) and put them together. In inductive reasoning, you infer what seems accurate and then devise ways to determine if subsequent observations fit the inferred conclusion.

1.25 Inductive reasoning. Scientists make predictions, inferences, and then seek to prove or refute the predictions by making further observations. They don't start with a number of facts and then draw conclusions by putting them together.

1.27 Probably not. All observations that could be explained by the model conform to the predictions it makes. When this occurs, a model may inspire sufficient confidence to be labeled a theory.

1.29 Somebody asks to meet with you at 9:00 and you go in the evening when they expected you in the morning (but didn't specify a.m.). (Other answers possible.)

1.31 Meters and amperes are base units. Grams are not (because kg is the base unit for mass). Joules and liters are both derived units.

1.33 Smallest to largest: nano-, centi-, kilo-, and giga-

1.35 The erg was an energy unit associated with work being done, and the calorie was associated with heat—at the time, scientists thought these two forms of energy were different things.

1.37 One million, 1×10^6

1.39 The upper temperature, 100, was set by body temperature. The lower temperature was set by the lowest temperature at which water would freeze (with added salt) and it was set to 0. This shows some limitations of measurement because we know that body temperature can vary over time or from one person to another.

1.41 **(a)** 263 K **(b)** 273.15 K **(c)** 553 K **(d)** 1.7×10^3 K

1.43 **(a)** 6.213×10^1 **(b)** 4.14×10^{-4} **(c)** 5.1×10^{-6} **(d)** 8.71×10^8 **(e)** 9.1×10^3

1.45 **(a)** 4 **(b)** 3 **(c)** 4 **(d)** 4 **(e)** 3

1.47 **(a)** 80.0 **(b)** 0.7615 **(c)** 14.712 **(d)** 0.029

1.49 0.58 g/mL

1.51 5.663 g

1.53 125 g

1.55 1.73 m, 5 feet 8.1 inches

1.57 **(a)** 3.47 μg **(b)** 0.273 mL **(c)** 7.25×10^{-7} s **(d)** 1.3×10^{-3} km

1.59 **(a)** 0.0223 L **(b)** 13.6 in^3 **(c)** 0.0236 qt

1.61 523 g/L

1.63 4.1 L

1.65 8.97×10^{14} L

1.67 2.35 cm

1.69 Each bullet should have a mass of 9700 mg. (Calculation gives a range of 9696–9744 mg, but the data given have just two significant figures, and therefore both ends of the range round to 9700 mg.)

1.71 In a crystal, atoms or molecules are close together and arranged in an orderly lattice. In a liquid, atoms or molecules are still fairly close together, but the arrangement is much more disordered. A picture similar to the first two panels of Figure 1.3 would illustrate the difference.

1.73 4900 cans

1.75 3.84 g/cm^3

1.77 Aluminum has an elastic modulus of 10.0×10^6 which is less than the given number. Therefore, a frame made with the new material will seem stiffer than one made with 1aluminum.

1.79 Compared to steel and titanium, aluminum has lower yield strength; so if strength is important to a design, aluminum would *not* be a good choice.

1.81 The primary difference appears to be the size of the tubes used. Aluminum frames use larger tubing than steel-frame bikes.

1.83 This web exercise will have varying answers depending on the websites found. In addition to the material costs, the cost of welding, etc., vary from one material to another and contribute to the overall price differences.

1.85 If we knew the mass of the acid solution in the battery, we could multiply that by 0.380 to find the mass of sulfuric acid present in the solution. Since we are given the density of the battery acid, looking up or measuring either the mass or the volume of the battery acid would enable us to complete the calculation.

1.87 We could estimate the percentage of ethanol if we knew the densities of both water and ethanol. The information given allows us to find the density of the solution, and then we can set that equal to a weighted average of the densities of the two components. If we let x equal the fraction of ethanol, then we would have:

$$\rho_{\text{solution}} = x(\rho_{\text{ethanol}}) + (1 - x)(\rho_{\text{water}})$$

Note that such a calculation assumes that the volume of the solution is the sum of the volumes of the individual liquids, but this is not necessarily true.

1.89 An object will float in water if its mass is less than the mass of water that would be displaced if the entire object were submerged. We are given dimensions for the cork and the lead cube but not masses. So we might start by looking up densities for cork and for lead. (Since the lead will be much heavier, we might neglect the mass of the cork.) If the combined mass of the cork and the lead is less than the mass of water that would occupy the combined volume of the cork and the lead, then they should float.

Chapter 2

2.1 A polymer is a large molecule made of small repeating units of atoms called monomers that are bonded together. A monomer is a small group of atoms bonded together.

2.3 Responses will vary. Carpet fibers, sofa fabric, drinking cups, and calculator cases are examples.

2.5 Different atoms convey different physical properties. By making polymers that contain different types of atoms, one can design polymers that have certain desirable traits.

2.7 Annual production is roughly 45 billion pounds. Depending on how usage is grouped, the top three uses are most likely (1) product containers such as shampoo bottles, (2) thin polymer films, and (3) grocery bags and trash bags.

2.9 Because the number of protons distinguishes one type of atom from another. It dictates how many electrons are present in the neutral atom and this is what makes atoms of different elements unique.

2.11 Isotopes are different atomic forms of an element that differ only in the number of neutrons in the nucleus.

2.13 **(a)** magnesium-24: 12 protons, 12 neutrons, 12 electrons

(b) tin-119: 50 protons, 69 neutrons, 50 electrons

(c) thorium-232: 90 protons, 142 neutrons, 90 electrons

(d) carbon-13: 6 protons, 7 neutrons, 6 electrons

(e) copper-63: 29 protons, 34 neutrons, 29 electrons

(f) bismuth-205: 83 protons, 122 neutrons, 83 electrons

2.15 200.59 amu

2.17 It is the average of several values that also takes into account the frequency of occurrence of each value.

2.19 The percent abundance of ^{63}Cu is 69.0%, and the percent abundance of ^{65}Cu is 31.0%.

2.21 78.72%

2.23 (b)

2.25 Cations have positive charges, whereas anions have negative charges.

2.27 **(a)** Na^+: 11 protons, 10 electrons

(b) Al^{3+}: 13 protons, 10 electrons

(c) S^{2-}:16 protons, 18 electrons

(d) Br^-: 35 protons, 36 electrons

2.29 $^{126}_{52}Te^{2-}$

2.31 **(a)** Ba^{2+} **(b)** Ti^{4+} **(c)** $PO_4{}^{3-}$ **(d)** $HCO_3{}^-$ **(e)** S^{2-} **(f)** $ClO_4{}^-$ **(g)** Co^{2+} **(h)** $SO_4{}^{2-}$

2.33 One common chemical used for ionic polymerization is butyl-lithium. This molecule decomposes into a lithium ion (positive charge) and a butyl ion (negative charge). An advantage of ionic polymerization is that the resulting polymer is easier to process further to induce desirable properties.

2.35 In the formula $Ba(OH)_2$, there are one Ba atom, two O atoms, and two H atoms.

2.37 Electrons are either shared between atoms to form covalent chemical bonds or are transferred from one atom to another, creating ions that are held by ionic chemical bonds.

2.39 Molecules are groups of atoms bonded to one another with covalent bonds. Ionic compounds consist of arrays of positive and negative ions held by electrostatic attraction (ionic bonds). Therefore, the term *molecule* does not apply, and the formula of an ionic compound is represented by the smallest whole-number combination of ions that has no net charge, the formula unit.

2.41 Both involve lattice systems, but metals do not have alternating positive and negative ions. Instead, metals have positively charged nuclei (with some electrons) occupying the lattice points surrounded by a "sea" of mobile, negatively charged electrons.

2.43 Molecules consist of atoms that share pairs of electrons rather than having electrons donated from one atom to another (ionic compounds) or mobile across an entire lattice (metals). Such a shared pair of electrons between two atoms creates an attractive force known as a covalent bond.

2.45 Polymers exist as large numbers (hundreds, even thousands) of monomer units bonded together. It is more convenient to represent a polymer by its empirical or monomer formula.

2.47 The empirical formula for butadiene, C_4H_6, is C_2H_3.

2.49 The periodic table is an organization of the elements with respect to size and structure. It allows correlating chemical and physical properties and predicting behavior trends.

2.51 It is observed that the number of hydrogen atoms with which an individual atom of various elements will combine varies cyclically from one side of the periodic table to another. This demonstrates periodicity—regular variations in the behavior of the elements.

2.53 **(a)** K: Group 1 **(b)** Mg: Group 2 **(c)** Ar: Group 18 **(d)** Br: Group 17

2.55 Sulfur and selenium are both in Group 16, so they should exhibit similar properties.

2.57 A metalloid is an element that exhibits some properties of metals and some of nonmetals. They are said to be intermediate in character and occupy the region on the periodic table between metals and nonmetals.

2.59 **(a)** Si: metalloid **(b)** Zn: metal **(c)** B: metalloidv **(d)**N: nonmetal **(e)** K: metal **(f)** S: nonmetal

2.61 The chemistry of transition metals is more complicated because most transition metals can form multiple different cations (whose charges vary).

2.63 *Organic chemistry* is a field of chemistry focused on molecules that consist primarily of carbon atoms bonded to each other.

2.65 A *functional group* is a specific group of atoms bonded to each other that display certain properties and influences the properties of organic molecules when they are present. Organic molecules are classified and named according to which functional groups are present.

2.67 The molecular and empirical formula for the molecule on the left is $C_{15}H_{10}N_2O_2$. The molecular formula for the one on the right is $C_2H_6O_2$; the empirical formula is CH_3O.

2.69 The molecular formula for GABA is $C_4H_9NO_2$.

2.71 In molecules, atoms can combine in many different ratios and the names must reflect those particular formulas. In ionic compounds, two specific ions can combine in only one way, thus the different rules for naming ionic versus molecular compounds.

2.73 **(a)** dinitrogen pentoxide **(b)** disulfur dichloride **(c)** nitrogen tribromide **(d)** tetraphosphorus decoxide.

2.75 **(a)** SF_6 **(b)** BrF_5 **(c)** S_2Cl_2 **(d)** S_4N_4

2.77 **(a)** magnesium chloride **(b)** iron(II) nitrate **(c)** sodium sulfate **(d)** calcium hydroxide **(e)** iron(II) sulfate

2.79 **(a)** phosphorus pentachloride **(b)** sodium sulfate **(c)** calcium nitride **(d)** iron(III) nitrate **(e)** sulfur dioxide **(f)** dibromine pentoxide

2.81 Low-density polyethylene (LDPE) is made of branched molecules that cannot pack very closely together, resulting in lower density for the polymer. High-density polyethylene (HDPE) consists of straight-chain molecules that can pack together very closely, giving a higher density. LDPE has lower density, is softer and weaker; HDPE has higher density, is harder, and much stronger.

2.83 According to data from the American Plastics Council, 7.8 billion pounds of LDPE and 15.7 billion pounds of HDPE were produced in the United States in 2003. As noted earlier, HDPE is used primarily for consumer product bottles, whereas LDPE is used in industrial liners and garbage bags.

2.85 You need to use the fact that oxygen nearly always forms an oxide ion with a 2− charge in compounds with metals. Find compounds with the metal, look at their formulas, and calculate the charge on the metal that would offset the charge that is present from the oxygen atoms. In the case of iron, FeO must have Fe^{2+} and Fe_2O_3 must have Fe^{3+} for the compounds to be neutral.

2.87 The average mass is closest to the mass number of the heaviest isotope, ^{88}Sr. This can only be true if ^{88}Sr is the most abundant isotope.

2.89 Because oxide ions are always 2−, these two compounds tell us the charge of iron ions in the compounds. FeO tells us we have a 2+ ion and Fe_2O_3 tells us we have a 3+ ion. Next we need to recall that chloride ions are always 1−, so we can tell that the two common compounds must be $FeCl_2$ and $FeCl_3$.

2.91 Steel is mostly iron, so we will compare the densities of aluminum, iron, and titanium. The shading in Figure 2.12 shows aluminum with a lower density than the other two elements, and this is consistent with the data given. Figure 2.12 does not differentiate between the densities of iron and titanium. If more levels of shading were used, we would expect the figure to show that iron is actually denser than titanium.

2.93 The density is 0.95 g/cm^3, so it is HDPE.

Chapter 3

3.1 **(a)** 0.10% **(b)** 72% land, 28% aquatic **(c)** 83%

3.3 $6\ CO_2 + 6\ H_2O \rightarrow C_6H_{12}O_6 + 6\ O_2$

3.5 Answers will vary depending on the websites visited. In addition to ethanol, products of biofuel refining include other fuels, such as bio-LPG, as well as chemical feedstocks.

3.7 s: solid; ℓ: liquid; g: gas; and aq: aqueous

3.9 Law of conservation of matter—matter is neither created nor destroyed.

3.11 **(a)** $4\ Al(s) + 3\ O_2(g) \rightarrow 2\ Al_2O_3(s)$
(b) $N_2(g) + 3\ H_2(g) \rightarrow 2\ NH_3(g)$
(c) $2\ C_6H_6(\ell) + 15\ O_2(g) \rightarrow 6\ H_2O(\ell) + 12\ CO_2(g)$

3.13 $2\ C_3H_6N_6O_6 + 3\ O_2 \rightarrow 6\ H_2O + 6\ CO_2 + 6\ N_2$

3.15 $C_2H_5OH + 3\ O_2 \rightarrow 2\ CO_2 + 3\ H_2O$

3.17 **(a)** $N_2(g) + 3\ H_2(g) \rightarrow 2\ NH_3(g)$
(b) $2\ H_2(g) + CO(g) \rightarrow CH_3OH(\ell)$
(c) $2\ S(s) + 3\ O_2(g) + 2\ H_2O(\ell) \rightarrow 2\ H_2SO_4(\ell)$

3.19 $3\ SiCl_4 + 16\ NH_3 \rightarrow Si_3N_4 + 12\ NH_4Cl$

3.21 Solution: a homogenous mixture of two or more substances.
Solute: the minor component(s) of a solution, the component(s) in lesser amount.
Solvent: the major component of a solution, the component found in the greatest amount.

3.23 Concentrated solution: a solution in which the number of solute particles is high.
Dilute solution: solution in which the number of solute particles is low.

3.25 An electrolyte is any solute that dissolves to produce a solution that conducts electrical current. Strong electrolytes dissociate completely, whereas weak electrolytes dissociate only partially. At similar concentrations, a solution of a strong electrolyte such as HCl will conduct electricity better than a solution of a weak electrolyte such as CH_3COOH.

3.27 **(a)** K^+, OH^- **(b)** K^+, SO_4^{2-} **(c)** Li^+, NO_3^- **(d)** NH_4^+, SO_4^{2-}

3.29 1.40×10^3 mL

3.31 **(a)** strong base; K^+, OH^-
(b) strong base (but only slightly soluble); Mg^{2+}, OH^-
(c) weak acid; H^+, OH^-
(d) strong acid; H^+, Br^-
(e) strong base; Li^+, OH^-
(f) weak acid; H^+, HSO_3^-, SO_3^{2-}

3.33 The total ionic equation shows all species present. The net ionic equation shows only those species that undergo reaction—the spectator ions are removed.

3.35 **(a)** $Zn(s) + 2\ HCl(aq) \rightarrow H_2(g) + ZnCl_2(aq)$
$Zn(s) + 2\ H^+(aq) \rightarrow H_2(g) + Zn^{2+}(aq)$
(b) $Mg(OH)_2(s) + 2\ HCl(aq) \rightarrow MgCl_2(aq) + 2\ H_2O(\ell)$
$Mg(OH)_2(s) + 2\ H^+(aq) \rightarrow Mg^{2+}(aq) + 2\ H_2O(\ell)$
(c) $2\ HNO_3(aq) + CaCO_3(s) \rightarrow Ca(NO_3)_2(aq) + H_2O(\ell) + CO_2(g)$
$2\ H^+(aq) + CaCO_3(s) \rightarrow Ca^{2+}(aq) + H_2O(\ell) + CO_2(g)$
(d) $3\ (NH_4)_2S(aq) + 2\ FeCl_3(aq) \rightarrow 6\ NH_4Cl(aq) + Fe_2S_3(s)$
$3\ S^{2-}(aq) + 2\ Fe^{3+}(aq) \rightarrow Fe_2S_3(s)$

3.37 The mole is the number of particles in exactly 12 grams of carbon-12. The mole is a convenient way for chemists to relate large numbers of molecules/atoms to one another. 1 mole $= 6.022 \times 10^{23}$.

3.39 7.8×10^{19} cm^3; 7.8×10^{16} L

3.41 **(a)** Fe_2O_3: 159.70 g/mol
(b) BCl_3: 117.170 g/mol
(c) $C_6H_8O_6$: 176.126 g/mol

3.43 **(a)** PbS: 239.2 g/mol, 86.60% Pb, 13.40% S
(b) C_2H_6: 30.079 g/mol, 79.887% C, 20.11% H
(c) CH_3COOH: 60.052 g/mol, 40.002% C, 6.7135% H, 53.285% O
(d) NH_4NO_3: 80.0432 g/mol, 34.9979% N, 5.0368% H, 59.9654% O

3.45 **(a)** $Mg_3(PO_4)_2$: 262.8578 g/mol
(b) Na_2S: 78.046 g/mol
(c) N_2O_4: 92.0110 g/mol

3.47 NaOH: 39.9971 g/mol, so 79.9942 g NaOH are needed

3.49 **(a)** 67 g **(b)** 0.0698 g **(c)** 0.060 g **(d)** 1.32×10^4 g

3.51 **(a)** $C_7H_5N_3O_6$: 1.577 moles
(b) CH_3NO_2: 1.35 moles
(c) $C_3H_6N_6O_6$: 7.56 moles

3.53 6.8×10^{13} molecules CO_2

3.55 4.29×10^{24} O atoms

3.57 **(a)** 4.7×10^{22} molecules/min
(b) 2.8×10^{24} molecules/hr
(c) 6.8×10^{25} molecules/day

3.59 $C_8H_8O_3$

3.61 **(a)** 40 at % Cu, 60 at % Al **(b)** 38 wt % NiO, 62 wt % MgO **(c)** 50 mol % MgO, 50 mol % FeO

3.63 **(a)** 5.80 M **(b)** 4.2 M **(c)** 3.41 M **(d)** 0.71 M

3.65 **(a)** 0.16 mol **(b)** 0.0083 mol **(c)** 0.11 mol **(d)** 0.027 mol

3.67 **(a)** 0.74 M **(b)** 0.0646 M **(c)** 4.9×10^{-3} M **(d)** 0.294 M

3.69 0.208 L

3.71 Greenhouse gases absorb infrared light and trap energy in the atmosphere.

3.73 A carbon reservoir is a natural feature of the earth in which carbon can accumulate. The two largest carbon reservoirs on earth are fossil fuels and the oceanic pool.

3.75 Although the burning of fossil fuels contributes a relatively small amount of carbon to the atmosphere, that additional carbon is enough to tip the balance toward accumulation of carbon in the atmosphere. The large amount of carbon fixed by photosynthesis is already offset by plant and soil respiration.

3.77 $3\ NO_2 + H_2O \rightarrow 2\ HNO_3 + NO$

3.79 $4\ C_5H_5N + 25\ O_2 \rightarrow 20\ CO_2 + 10\ H_2O + 2\ N_2$

3.81 $2\ NH_3 + 2\ CH_4 + 3\ O_2 \rightarrow 2\ HCN + 6\ H_2O$

3.83 2 [cyclohexane] $+ 5\ O_2 \rightarrow$ 2 HO(C=O)–(CH$_2$)$_4$–(C=O)OH $+ 2\ H_2O$

3.85 **(a)** 6.022×10^{23} octane molecules, **(b)** 24 moles F, **(c)** ∽6 g C

3.87 Molecular formula is C_9H_{12}.

3.89 37%

3.91 Atom percentage and mole percentage must be the same, so the range would be 34.0–81.8 mol % Pt.

3.93 Only MgS is an electrolyte.

3.95 0.494 g $KMnO_4$

3.97 1.7×10^6 gallons

3.99 4×10^{10} Cu atoms

3.101 **(a)** 42.2 at % Zn, 7.85 at % Cd, and 50.00 at % O
(b) 42.2 mol % Zn, 7.85 mol % Cd, and 50.00 mol % O
(c) 62.1 wt % Zn, 19.9 wt % Cd, and 18.0 wt % O

3.103 We know that 1 mole of chlorophyll contains 1 mole of magnesium. We can look up the molar mass of magnesium. Then we can say that the mass of one mole of magnesium must equal 2.72% of the molar mass of chlorophyll. This will let us solve the problem.

3.105 We can calculate the mass percentage of chlorine in both NaCl and $MgCl_2$. The percentage of chlorine in the sample will be a weighted average of those two values, with each one multiplied by the fraction of its component that is actually present. As long as NaCl and $MgCl_2$ are the only compounds present, we will be able to solve for the fraction of each.

3.107 We have the total mass of CO_2 exhaled per day, and we can convert that into molecules. So all we need is an estimate of the number of exhalations per day. It should be easy to find an estimate of the average human respiration rate, which would enable us to complete the calculation.

3.109 The atomic masses of iron and oxygen are fixed, so the compound with the highest ratio of oxygen atoms to iron atoms will have the greatest mass percentage of oxygen. Looking at the formulas, it should be easy to see that this will be Fe_2O_3, and no further information is necessary.

3.111 Ionic bonding is expected in precipitates.

3.113 0.0762 M $Fe(NO_3)_2$ and 0.152 M NO_3^-

3.115 If the molar masses to be used have four or five significant figures, they will limit the accuracy of a calculation only if the other data have five or more significant figures. For a mass on the order of 100 g, this will be true if we measure masses with a resolution beyond 0.01 g. Many laboratory balances can easily do so.

Chapter 4

4.1 Combustion is not always complete, and products are not always carbon dioxide and water. Moreover, gasoline is not a single compound. So describing the process with a single chemical equation is also an approximation.

4.3 In complete combustion, all carbon present is converted to carbon dioxide, whereas in incomplete combustion, some fraction of the carbon is converted only to carbon monoxide.

4.5 They are gases at room temperature, so they would evaporate from liquid gasoline.

4.7 **(a)** $\left(\frac{2\text{ mol H}_2}{1\text{ mol O}_2}\right)$ or $\left(\frac{2\text{ mol H}_2}{2\text{ mol H}_2\text{O}}\right)$ or $\left(\frac{1\text{ mol O}_2}{2\text{ mol H}_2\text{O}}\right)$

or $\left(\frac{1\text{ mol O}_2}{2\text{ mol H}_2}\right)$ or $\left(\frac{2\text{ mol H}_2\text{O}}{2\text{ mol H}_2}\right)$ or $\left(\frac{2\text{ mol H}_2\text{O}}{1\text{ mol O}_2}\right)$

(b) $\left(\frac{2\text{ mol H}_2\text{O}_2}{2\text{ mol H}_2\text{O}}\right)$ or $\left(\frac{2\text{ mol H}_2\text{O}_2}{1\text{ mol O}_2}\right)$ or $\left(\frac{2\text{ mol H}_2\text{O}}{2\text{ mol H}_2\text{O}_2}\right)$

or $\left(\frac{1\text{ mol O}_2}{2\text{ mol H}_2\text{O}_2}\right)$ or $\left(\frac{1\text{ mol O}_2}{2\text{ mol H}_2\text{O}}\right)$ or $\left(\frac{2\text{ mol H}_2\text{O}}{1\text{ mol O}_2}\right)$

(c) $\left(\frac{1\text{ mol P}_4}{5\text{ mol O}_2}\right)$ or $\left(\frac{1\text{ mol P}_4}{1\text{ mol P}_4\text{O}_{10}}\right)$ or $\left(\frac{1\text{ mol P}_4\text{O}_{10}}{1\text{ mol P}_4}\right)$

or $\left(\frac{5\text{ mol O}_2}{1\text{ mol P}_4}\right)$ or $\left(\frac{5\text{ mol O}_2}{1\text{ mol P}_4\text{O}_{10}}\right)$ or $\left(\frac{1\text{ mol P}_4\text{O}_{10}}{5\text{ mol O}_2}\right)$

(d) $\left(\frac{2\text{ mol KClO}_3}{2\text{ mol KCl}}\right)$ or $\left(\frac{2\text{ mol KClO}_3}{3\text{ mol O}_2}\right)$.

or $\left(\frac{2\text{ mol KCl}}{2\text{ mol KClO}_3}\right)$ or $\left(\frac{3\text{ mol O}_2}{2\text{ mol KClO}_3}\right)$.

or $\left(\frac{3\text{ mol O}_2}{2\text{ mol KCl}}\right)$ or $\left(\frac{2\text{ mol KCl}}{3\text{ mol O}_2}\right)$.

4.9 **(a)** $S_8 + 8\ O_2 \rightarrow 8\ SO_2$ **(b)** 25 S_8 molecules **(c)** 200 SO_2 molecules

4.11 11.3 moles

4.13 **(a)** 13 moles, **(b)** 3.6 moles, **(c)** 4.9 moles, **(d)** 2.6×10^{-3} mol.

4.15 **(a)** 3.6 g, **(b)** 3.67 g, **(c)** 2.5 mg, **(d)** 12 kg.

4.17 3.2 g Cl_2

4.19 0.883 metric tons C; 5.48 metric tons Fe

4.21 910 g N_2

4.23 O_2 is limiting.

4.25 We must have more than 24 moles of Br_2.

4.27 Conditions ARE conducive for the formation of carbon monoxide.

4.29 Sulfuric acid is the limiting reactant and forms 5.88 g SO_2.

4.31 Sulfuric acid is limiting and 790 g $Al_2(SO_4)_3$ forms.

4.33 Nitrogen dioxide is limiting and 3.3 kg of HNO_3 forms.

4.35 SiO_2 is limiting, and 11.5 g of C will remain.

4.37 Side reactions produce something other than the desired product. Sometimes the reaction itself doesn't proceed to completion.

4.39 Reaction 1: 36.6%; Reaction 2: 1.3% Reaction 3: 34% Reaction 4: 103%

4.41 88%

4.43 150 g

4.45 66.5%

4.47 H_2S is limiting and the percentage yield is 79%.

4.49 **(a)** 6.0×10^6 g SiO_2 and 3.6×10^6 g C **(b)** 0.78 metric ton of coal per metric ton of sand

4.51 0.302 g H_2

4.53 It changes color when the chemical being added has consumed the chemical of interest in the solution.

4.55 22.9 mL

4.57 3.61 g

4.59 **(a)** 7.62×10^{-3} M H_2SO_4 **(b)** 50.1 g $CuFeS_2$ **(c)** 2.2%

4.61 2.27 g H_2

4.63 Octane ratings measure the quality of a gasoline in terms of the extent to which it can be compressed before it ignites without a spark. Octane itself is rather compressible, so if you had pure octane the rating would be 100. It is possible to have an octane rating greater than 100 for a fuel that is more compressible than octane.

4.65 Web question that will vary depending on location.

4.67 Lead reacts with metals such as rhodium, palladium, and platinum, effectively covering these metals and eliminating their catalytic activity.

4.69 104 g/mol

4.71 2.0×10^{13} g $Ca_5(PO_4)_3F$

4.73 0.031 M

4.75 5×10^8 kg NH_3

4.77 Yes, CO was a limiting reactant. $CO + 2\ H_2 \rightarrow CH_3OH$

4.79 Sulfur is limiting and 67.4 g of S_2Cl_2 forms.

4.81 HF is limiting and 1049 kg of Na_3AlF_6 forms.

4.83 HCl is limiting and 1.5 g CO_2 forms. The concentration of Ca^{2+} is 0.050 M.

4.85 588.2 g

4.87 13.48 g CH_4

4.89 0.041 M Ca^{2+}

4.91 $CaCO_3$ and $Ca(OH)_2$ could each neutralize two moles of HCl for each mole of base, whereas NH_3 and NaOH could only neutralize 1 mole of HCl for each mole of base. Since the costs are given in terms of mass, we also need to consider the molar masses. Combining the reaction stoichiometry with the molar masses allows us to calculate the cost per mole of acid to be neutralized, and the results indicate that NH_3 offers the lowest cost.

4.93 We are given the mass of Cr_2O_3, and we can convert that into moles. We can see from the balanced equation that we need 2 moles of Al for each mole of Cr_2O_3, and this information will enable us to find the number of moles of Al needed. That can be converted to mass if desired; no additional information should be needed.

4.95 The increase in mass is due to oxygen, so we know that 0.5386 g of the unknown element combines with 0.1725 g of oxygen. We can convert that mass of oxygen into moles and then use the 3:2 mole ratio from the formula to find the number of moles of element X in the original sample. Dividing the mass by the number of moles will give the desired molar mass.

4.97 There are several ways we could approach this problem. One would be to calculate the mass percentage of selenium in Bi_2Se_3. That would allow us to find the mass of selenium in 5 kg of Bi_2Se_3. Comparing that mass to the 560 g available would then tell us whether or not we have enough selenium.

4.99 We can convert 500 g O_2 into moles and then use the 2:1 mole ratio from the balanced equation to find the number of moles of $KMnO_4$ needed.

4.101 Convert 150 tons of Freon-12 to grams and then to moles. Use the 2:1 mole ratio from the balanced equation to find the number of moles of sodium oxalate needed to react with all of the Freon-12. The number of moles of sodium oxalate needed could then be converted to mass if desired.

4.103 C_5H_3

4.105 91.6% Al

Chapter 5

5.1 Nitrogen oxides and VOCs (volatile organic chemicals). Nitrogen oxides are emitted from automobile exhaust as are small amounts of VOCs. Gasoline vapors and spills and chemical plants also release VOCs.

5.3 Because NO is being consumed by reactions that are involved in the formation of photochemical smog. The rates of the reactions that consume NO exceed those that form it during daylight hours.

5.5 No, it would not contribute significantly to the formation of smog. The organic chemicals present in asphalt are not volatile, so they don't fulfill the same role as VOCs in the formation of smog.

5.7 Gases become less dense as their temperature increases. The air in a hot-air balloon is at a higher temperature than the surrounding air. The result is that it is less dense and therefore floats on the denser, cooler air.

5.9 Atmospheric pressure is created when the mass of the gases attracted to the surface of the earth by gravity generates a force that produces pressure (force/unit area).

5.11 For a liquid with half the density of mercury, atmospheric pressure will generate a column twice as high as that generated for mercury. The column will be 750 mm × 2 = 1500 mm.

5.13 1.5×10^2 atm

5.15 **(a)** 0.9474 atm **(b)** 9.50×10^2 mm Hg **(c)** 542 torr **(d)** 98.7 kPa **(e)** 6.91 atm

5.17 The increased pressure due to the weight of the water compresses the air in your ear, which can cause a "pop" in your ear upon expansion.

5.19 113 mL

5.21 A balloon will return to its shape because volume is proportional to absolute temperature. At constant pressure and moles of gas, lowering *T* will decrease the volume of the balloon, but returning to room temperature will restore the shape.

5.23 Jacques Charles first observed that when pressure and moles of gas are held constant, volume and temperature vary inversely. This relationship is a straight line when *T* is plotted vs. *V*. Although it is not possible to lower the temperature of all gases to absolute zero, if the lines that are generated by different gases are extrapolated, all lead to the same temperature and it is absolute zero (or −273°C).

5.25 At constant volume, the pressure of a gas is directly related to absolute temperature. Increasing temperature could cause dangerously high pressure and possible tank rupture.

5.27 6.9 mL

5.29 1.38×10^3 L

5.31 61.7 g/mol

5.33 **(a)** 5.39 g/L **(b)** 1.96 g/L **(c)** 1.63 g/L

5.35 670 K

5.37 Partial pressure is the pressure exerted by a gas in a mixture of gases, behaving as if it were alone in the container.

5.39 $P_i = P_{total} X_i$

5.41 0.72 atm

5.43 1.9×10^{-3} mol NO_2 and 3.3×10^{-3} mol SO_2

5.45 $P_{CO_2} = 640$ torr, $P_{NO} = 37$ torr, $P_{SO_2} = 5.7$ torr, and $P_{H_2O} = 70.$ torr

5.47 Total pressure is 20.0 atm: $P_{H_2O} = 6.28$ atm, $P_{O_2} = 2.91$ atm, and $P_{N_2} = 10.8$ atm.

5.49 Web question—answers will vary depending on the sites found by students.

5.51 0.10 g NH_4Cl

5.53 1.07 g Zn

5.55 160 L O_2

5.57 1.00 atm

5.59 Ammonia is the limiting reactant and 8.0 g urea form.

5.61 0.52 L

5.63 74 g NaN_3

5.65 187 torr

5.67 3×10^{-17} atm

5.69 High pressure and/or low temperature

5.71 CH_4, N_2, Kr, CH_2Cl_2

5.73 At higher temperature, the distribution shifts toward higher molecular speeds, so the blue curve represents 1000 K.

5.75 Mean free path is the average distance a particle travels between collisions with other particles.

5.77 **(a)** 16.3 atm **(b)** 15.8 atm

5.79 Statements (a), (b), (c), and (d) are true. Statement (e) is false.

5.81 An electric current is the target of the measurement. Because the number of electrons is related to the number of molecules of gas, the current can be converted into pressure by counting the number of molecules (via the current they produce upon ionization).

5.83 Responses will vary depending on the types of gauges found.

5.85 2.19 g/L

5.87 34.8 torr

5.89 79% Zn

5.91 3×10^4 L air

5.93 **(a)** $4\ Fe^{2+}(aq) + 4\ H^+(aq) + O_2(g) \rightarrow 4\ Fe^{3+}(aq) + 2\ H_2O(\ell)$

(b) 1.4×10^3 g H_2O

(c) 5.1 atm

5.95 539 hours

5.97 444 torr

5.99 0.314

5.101 **(a)** The only gas commonly encountered in aerospace applications is air. **(b)** 29 g/mol **(c)** 290 $m^2\ s^{-2}\ K^{-1}$

5.103 **(a)** The final pressure will be greater. The first, second, and fourth reactions do not change the number of moles of gas, but the third reaction generates gaseous products from solid a reactant. The number of moles will increase, and the pressure will rise. **(b)** 2.3 atm

5.105 For this problem, you need to determine whether air is present by measuring various densities and determining if there are differences. First, the unopened can must be weighed and its volume determined (by displacement if possible). Next, the soda must have its density determined separately, perhaps by measuring the mass of a volumetric container such as a graduated cylinder when empty and when filled with a known volume of soda. There's some chance that the density of the aluminum can be ignored in this problem, but if necessary, it could be measured. If the density measured for the whole system varies from that measured for the soda alone, the most logical source for the difference would be gas in the "head space."

5.107 To carry out this problem requires first determining the amount of carbonate present (by multiplying the amount of ore by the fraction of carbonate). A stoichiometry problem can then be set up using the chemical equation to convert between moles of carbonate and moles of carbon dioxide. Finally, the volume can be determined using the ideal gas law, as long as the temperature and pressure are known or measurable.

5.109 N_2: 75.518% O_2: 23.15% Ar: 1.29% CO_2: 0.048%

5.111 **(a)** 2×10^{27} molecules **(b)** 4×10^{21} molecules

5.113 $C_2H_2F_4$

Chapter 6

6.1 Nondestructive testing can be carried out without damaging the part or sample being analyzed, whereas other forms of analysis may require that the sample be digested (or consumed). Nondestructive testing of a part or device can often be carried out while the part or device is in use.

6.3 The amounts of light absorbed at different wavelengths provide information on the relative amounts of each element present.

6.5 Because of conservation of energy, the energy of the emitted photon can never be larger than the energy of the photon originally absorbed to excite the emitting atom.

6.7 When the term *light* is used, normally it refers to visible light. Like other forms of radiation, visible light is composed of an electrical field that lies perpendicularly to a magnetic field—and thus it is called electromagnetic radiation.

6.9 Radio, IR, visible, UV

6.11 480 nm: 6.2×10^{14} Hz green light
530 nm: 5.7×10^{14} Hz yellow light
580 nm: 5.2×10^{14} Hz orange light
700 nm: 4×10^{14} Hz red light

6.13 *Photon* is the term used to refer to a discrete amount of energy imparted by light. It is sometimes thought of as a packet of electromagnetic radiation that carries energy.

6.15 **(a)** 1.01×10^{-20} J **(b)** 1.1×10^{-15} J **(c)** 4.12×10^{-23} J

6.17 **(a)** 5.7×10^{-6} m (IR) **(b)** 2.3 m (radio) **(c)** 2.8×10^{-9} m (X-ray) **(d)** 36 m (radio)

6.19 **(a)** 2.5×10^{-19} J **(b)** 3.1×10^{-19} J **(c)** 4.90×10^{-19} J

6.21 **(a)** 3.00×10^{-3} m **(b)** 6.63×10^{-23} J/photon **(c)** 39.9 J/mol

6.23 7.1×10^{-19} J

6.25 A continuous spectrum has no missing frequencies of light, similar to a rainbow or the light given off by the sun. A discrete spectrum consists of specific, well-separated frequencies, like the emission spectra of elements.

6.27 The Bohr model depicted electrons circling the nucleus in orbits that could only be at certain discrete distances from the nucleus. Because energy is related to the distance between charged particles, these orbits corresponded to energy levels. For the excited electron in the Bohr atom to return to the original orbit, the atom releases a photon of energy equal to the energy difference between the two orbits. The spectra of hydrogen atoms therefore correspond to the specific energies released when electrons drop from higher energy orbits to lower energy ones.

6.29 The ground state occurs when all electrons in an atom are in the lowest energy states possible.

6.31 493.1 nm

6.33 **(a)** 4.558×10^{-19} J for the 435.8 nm transition and 3.638×10^{-19} J for the 546.1 nm transition. **(b)** 1.389×10^{14} Hz.

6.35 The wave function is used to describe electrons around the nucleus of an atom because electrons can be diffracted. Diffraction is usually associated with electromagnetic radiation (light), but electrons also produce diffraction patterns. This means that electrons also exhibit particle–wave duality. In some instances, electrons are treated as particles; in other instances, they are treated as waves.

6.37 The principal quantum number, n, has integral values starting at 1 and continuing to infinity. The secondary quantum number, ℓ, is limited in value by the principal quantum number and has integral values from 0 through $n - 1$.

6.39 4d

6.41 **(a)** 8 **(b)** 6 **(c)** 2 **(d)** 1

6.43 An s orbital is a sphere, which means it can be rotated by any number of degrees and its appearance does not change. There are three p orbitals in a subshell, and these orbitals must be arranged so that the electrostatic repulsion from the electrons in the orbitals is minimized. This occurs when the p orbitals are arranged along the x, y, and z axes—or in specific directions. The same type of reasoning is true for the 5d orbitals, though they don't point along the x, y, and z axes.

6.45

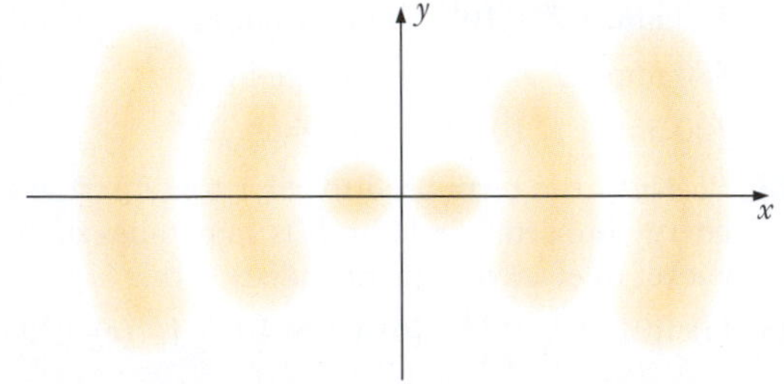

6.47 Spin paired means that the two paired electrons in the same orbital have opposites spins, +1/2 and −1/2, or ⇅.

6.49 Shielding describes the effect that the inner core electrons have on the nuclear charge that the outer, valence electrons experience. The inner core electrons partially mask the nuclear charge on the nucleus from the valence electrons.

6.51 The effective nuclear charge is the apparent nuclear charge that the valence electrons experience as the result of shielding by the inner core electrons. As the effective nuclear charge increases, the electrons contained within an orbital feel a stronger attractive force toward the nucleus. The electrons are pulled toward the nucleus, reducing the size of the atomic orbital.

6.53 Hund's rule states that for a set of equivalent atomic orbitals, each orbital is filled with one electron before pairing is allowed to occur. When two electrons are placed in the same atomic orbital, electron repulsion occurs. The atomic orbital is now at a higher energy than before (repulsion increases energy). The lowest energy state is the most stable state for an atom, so each atomic orbital in a set of equivalent orbitals receives one electron until all orbitals contain one electron, so that the atom remains at the lowest possible energy level.

6.55 Ground state configurations: (a) and (b)

Excited state configurations: (c), (e), and (f)

Impossible configuration: (d)

6.57 The electronic configurations for F, Cl, and Br are listed below:

F = $1s^22s^22p^5$ or [He] $2s^22p^5$

Cl = $1s^22s^22p^63s^23p^5$ or [Ne] $3s^23p^5$

Br = $1s^22s^22p^63s^23p^64s^23d^{10}4p^5$ or [Ar] $4s^23d^{10}4p^5$

The valence electron configuration of each is identical: $ns^2 np^5$ (the d electrons in Br are not valence electrons because only the outermost s and p electrons and electrons in unfilled d orbitals are considered valence electrons). Because these three halogens have similar valence electron electronic configurations, they will have similar chemistries (the valence electrons are where chemistry occurs). For example, each halogen will accept one electron to form an anion, F^-, Cl^-, and, Br^-. When paired with Na^+, they form sodium salts, NaF, NaCl, and NaBr.

6.59 All elements in a group of the periodic table have the same basic electron configuration for their valence electrons. These valence electrons are responsible for the chemistries of the elements. Because the group elements have similar configurations, they will have similar chemical properties. For example, oxygen, sulfur, selenium, and tellurium have the same valence electronic configuration, $ns^2 np^4$. They react with hydrogen to form H_2O, H_2S, H_2Se, and H_2Te.

6.61 The 7p block would be completely filled; there would be an eighth s row, and a g block (8g) would be added to the periodic table. The f and d blocks would increase by one row, as would the eighth row in the p block. This assumes that the last current element is Uuq, Z = 114.

6.63 A halogen lamp contains halogen vapor, which reacts with the vaporized tungsten from the filament. The tungsten halogen compound then reacts with the filament, depositing the evaporated tungsten back on the filament. The result is that the filament lasts longer and can be heated to a higher temperature, generating more light.

6.65 As the effective nuclear charge increases across the periodic table, the size of the atomic orbitals decreases. The decrease in the size of the atomic orbitals results in a smaller atomic size.

6.67 **(a)** Be, Li, Na
(b) F, N, P
(c) O, I, Sn

6.69 The first large jump in ionization energies in Cl would occur between the fifth and sixth ionization energy. The fifth ionization energy would remove the last $3p$ electron. The sixth ionization would remove the first $3s$ electron. It is harder to remove electrons from a completely filled set of orbitals than it is from a set of partially filled orbitals. The next large jump in ionization energy would occur between removing the last $3s$ electron and the first $2p$ electron (the seventh and eighth ionization energies).

6.71 The electron configuration for oxygen is $1s^2 2s^2 2p^4$. The electron configuration for nitrogen is $1s^2 2s^2 2p^3$. For oxygen, there is one set of paired electrons in the $2p$ orbitals. There are no paired electrons in the $2p$ orbitals for nitrogen. The paired electrons in oxygen repel each other. The electron repulsion raises the energy of these paired $2p$ electrons in oxygen, reducing the amount of energy needed to remove one of the paired electrons to form a cation, compared to nitrogen.

6.73 **(a)** Se **(b)** Br^- **(c)** Na **(d)** N **(e)** N^{3-}

6.75 **(a)** Na **(b)** C **(c)** Na, Al, B, C

6.77 778.0 nm

6.79 7.6×10^8 J

6.81

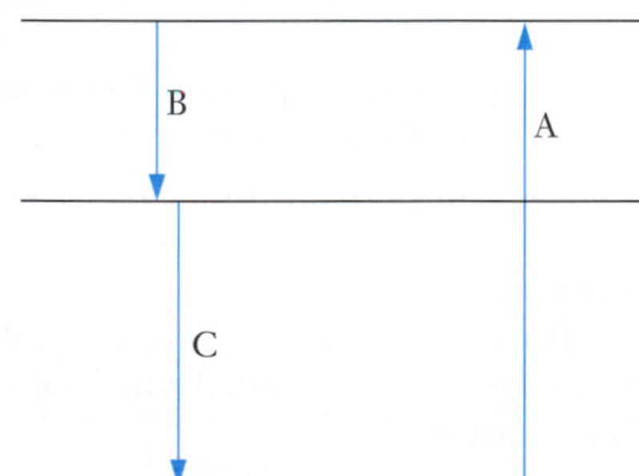

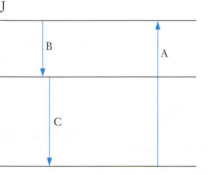

A is the absorption event at 185 nm, and B is the emission event at 4.924×10^{14} Hz. C is the energy difference between the ground state and the final excited state. The ground state is assumed to have an energy value of zero.

$E_C = 7.5 \times 10^{-19}$J

6.83 The unknown metal must be lithium.

6.85 **(a)** Ga^{3+}, Ca^{2+}, K^+
(b) Be^{2+}, Mg^{2+}, Ca^{2+}, Ba^{2+}
(c) Al^{3+}, K^+, Sr^{2+}, Rb^+
(d) Ca^{2+}, K^+, Rb^+

6.87 $\nu = 5.3 \times 10^{13}\ s^{-1}$, $\lambda = 5.7 \times 10^{-6}$ m

6.89 3×10^3 W, 1×10^{22} photons/s

6.91 The key features of the experiment is that the manganese atoms absorb light near 403 nm. The amount of light absorbed will be related to the number of manganese atoms present. So, if the observed absorption can be calibrated against some standard for concentration, the comparison will tell us the amount of manganese in the sample.

6.93 We are using inductive reasoning. If we assume that any energy level is possible, we would infer that the spectrum would be continuous, which disagrees with observations. Therefore, we have used inductive reasoning to say that the energy levels must be specific.

6.95 **(a)** 1.91×10^{-7} m **(b)** 6.4×10^{18} photons

Chapter 7

7.1 Biocompatibility is the ability of a material to coexist with a natural biological system into which it has been implanted.

7.3 PMMA bone cement works somewhat like a grout, filling space between bone and a surgical implant. A putty containing liquid methyl methacrylate and other materials is applied by the surgeon and then hardens in place as the MMA polymerizes. The composition of the cement must be chosen such that polymerization will proceed at a reasonable rate at body temperature.

7.5 Coatings can be used to improve biocompatibility. Reactants used to generate such coatings are called precursors.

7.7 Because removing a second electron from sodium ion, Na^+, involves a core electron. The energy needed to accomplish this is quite large, so it is not observed in nature.

7.9 **(a)** N **(b)** Ba^{2+} **(c)** Se **(d)** Co^{3+}

7.11 **(a)** Cl^-, S^{2-}, P^{3-}
(b) O^{2-}, S^{2-}, Se^{2-}
(c) N^{3-}, S^{2-}, Br^-
(d) Cl^-, Br^-, I^-

7.13 Because the ions are closely packed in an ionic solid, the next nearest neighbors are close enough to contribute their repulsive interactions to the overall lattice energy (and so on).

7.15 **(a)** Repulsive **(b)** According to Equation 7.2, the Coulomb potential varies as $1/r$. So as we move from first nearest neighbors to second, third, or fourth nearest neighbors, the magnitude of V must decrease, and $V_{1st} > V_{2nd} > V_{3rd} > V_{4th}$.

7.17 In a covalent bond, the electrons involved are shared between the nuclei, whereas in ionic compounds, an electron is transferred from one atom to another. Ultimately, the difference lies in the extent to which an electron is transferred during bond formation.

7.19

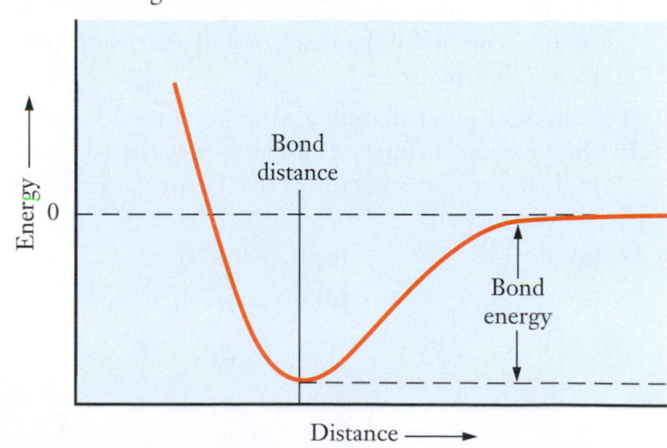

7.21 The negatively charged electrons are attracted to both positively charged nuclei, so the forces involved are still coulombic.

7.23 The number of bonds that can be formed is limited by the number of openings each element has in the valence shell. It's possible to get slightly more bonding than dictated by this number (commonly called the valence) in some cases, but it's not possible to keep going forever. From a physical perspective, the repulsion that would arise from packing too much electron density together forbids the formation of too many bonds.

7.25 :L̤̇t·

7.27 A lone pair is a pair of electrons located only on a single atom, not shared between atoms.

7.29 A double bond is stronger than two single bonds but not twice as strong. This leads to some reactivity trends where the "second" bond is broken and replaced with a new single bond (going from a double bond to two single bonds).

7.31 Electron affinity is defined for an isolated atom and measures the energy change for adding an electron. Electronegativity is defined within chemical compounds and it estimates the propensity of an atom to attract electrons to itself.

7.33

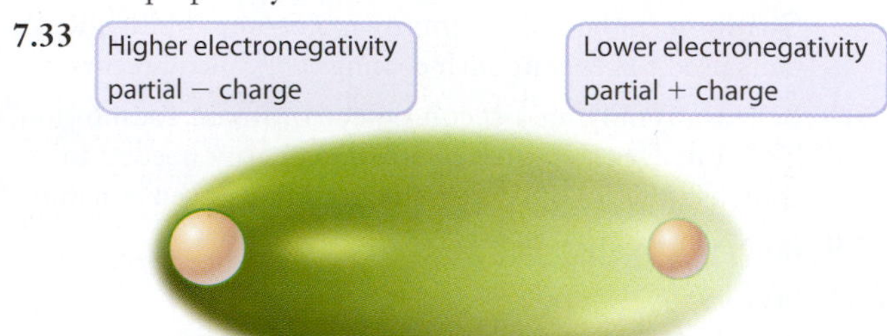

If we allow the density of electrons to be shown by shading, there is greater electron density near the nucleus on the left and that portion of the bond would have a negative charge.

7.35 Such a bond is called "polar" because the asymmetrical distribution of electrons results in a small electric dipole (or field). The existence of poles in this field is the origin of the terminology calling it a polar bond.

7.37 **(a)** O—H is the most polar; C—H and S—H have very similar polarity because they have the same difference in electronegativity. **(b)** H—Cl is the most polar bond and Cl—Cl is the least (nonpolar). **(c)** C—F is the most polar and F—F is the least (nonpolar). **(d)** N—H is the most polar and N—Cl is the least polar.

7.39 No, they cannot. Since fluorine is the most electronegative element, no other element can pull electron density away from a fluorine atom.

7.41 The most polar diatomic would be At—F because it has the greatest difference in electronegativity. (Astatine is radioactive. For nonradioactive elements, I—F would be the most polar.)

7.43 **(a)** :C≡O: **(b)** H—S̤̈—H

(c) :F̤̈—S(F)(F)(F)(F)—F̤̈: (S bonded to six F atoms, each F with three lone pairs)

(d) :C̤̈l—N̈—C̤̈l: with :C̤̈l: bonded below N

7.45 **(a)** :F̤̈:(F)C=C(F):F̤̈: (each C bonded to two :F̤̈: atoms)

(b) H(H)C=C(H)—C≡N: (H and H on one carbon; H and C≡N on the other)

7.47 In Lewis structures, resonance occurs when there are two or more equivalent ways to draw a double bond. The actual structure is an average of all of the equivalent structures.

7.49 It is an average of all of the equivalent structures. In plants, crossbred varieties are referred to as hybrids and they are an average of the two original plants.

7.51 When resonance structures exist so that there is a hybrid, the electrons involved in that component of the molecule are not localized to the same small space commonly associated with a chemical bond. Electrons with "more space" will have lower energy than more localized electrons.

7.53 PH_3 is the one with a single lone pair. F_2 has three lone pairs on each fluorine atom, CH_4 has no lone pairs, and the carbonate ion has lone pairs on each oxygen atom.

7.55 **(a)** There are too many valence electrons shown. O_3 should have only 18 valence electrons. **(b)** There are too few valence electrons shown. There should be 36 and only 34 are shown

7.57 Wave interference is the buildup or diminishment of the amplitude of a wave because of the interaction of two or more waves. Buildup results from constructive interference, and diminishment results from destructive interference.

7.59 Overlap is what is needed for constructive or destructive interference of electron waves. If electrons didn't behave as waves, they wouldn't interfere. So the wave behavior of electrons implies the importance of overlap.

7.61 Both sigma and pi bonds build electron density in the space between nuclei (so that is how they are similar). They differ based on the precise location of the electron density. For a sigma bond, it builds up on the line between nuclei, whereas for a pi bond, the built-up density is above and below that line.

7.63 **(a)** S̤̈=C=S̤̈ **(b)** H—C(H)(H)—C̤̈l: (C bonded to three H and one :C̤̈l:)

(c) :Ö̤—N̈=Ö̤ **(d)** Ö̤=S̈—Ö̤:

7.65 Molecular shapes provide a motivation to consider an extension to the concept of atomic orbitals combining to form molecules. For a molecule such as water, if only *p* orbitals are available, the bond angle would have to be 90°. But that is not the observed angle, so we must have additional flexibility, and hybrid orbitals provide that.

7.67 This type of hybrid is called an sp^3 hybrid. Because four atomic orbitals are used, there must be four hybrid orbitals as well.

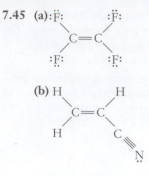

7.69 (a) sp^3 (b) sp^3 (c) sp (d) sp^2

7.71 Like charges repel each other. Clouds of electron density will all be negatively charged, so they will tend to repel each other.

7.73 (a) T-shape (b) Trigonal pyramid (c) Seesaw (d) Tetrahedral

7.75 XeF_2, CO_2, and BeF_2 are linear. XeF_2 and OF_2 have lone pairs on the central atom.

7.77

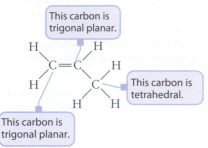

This carbon is trigonal planar.

$H_2C{=}CH{-}CH_3$

This carbon is tetrahedral.

This carbon is trigonal planar.

7.79 The geometry about the C═C double bond is trigonal planar, but when the reaction occurs, the geometry becomes tetrahedral.

7.81 The honeycomb structure shown in Figure 7.18 means that nearly all of the SiO_4 groups are at a surface.

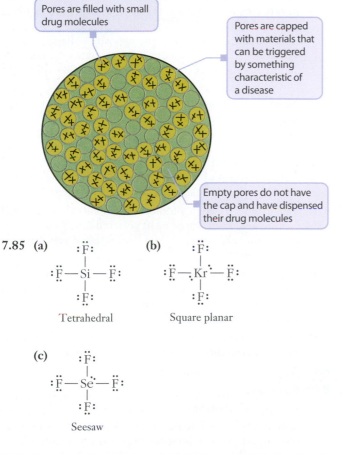

7.83

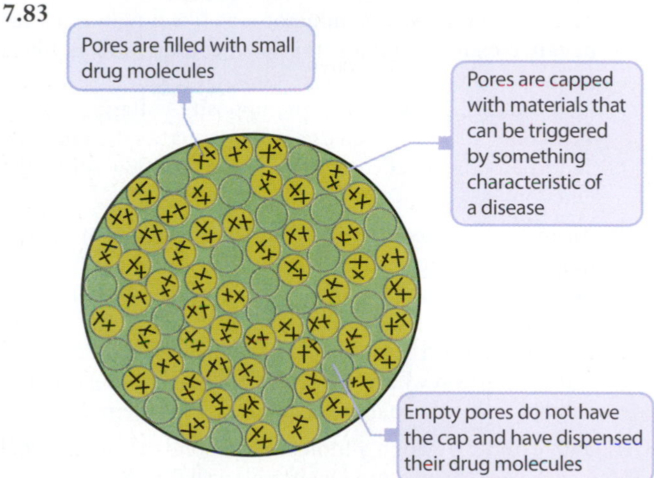

7.85 (a) :F̤: — Si — with four F atoms (Lewis structure)

Tetrahedral

(b) Kr with four F atoms and two lone pairs (Lewis structure)

Square planar

(c) Se with four F atoms and one lone pair (Lewis structure)

Seesaw

7.87 Based on the valence of each atom, a neutral molecule would have 32 electrons. There are 34 electrons shown in the structure, so it must have a 2− charge. Each carbon has one double bond, so the hybridization must be sp^2. The structure has five sigma bonds and two pi bonds.

7.89 :C̤l̤ — M̤ — C̤l̤: with :C̤l̤: below M

One metal that would make sense as M is bismuth, Bi. It is in the same group as N, which would produce a similar structure. Since bismuth is the only metal in that group, it is the best choice.

7.91 In order of decreasing ionic character, the four compounds would rank as follows: $SnF_4 > SnCl_4 > SnBr_4 > SnI_4$. This order is based on the difference in electronegativity between tin and each of the four halogens.

7.93 (a) ~120° (b) There will be one shorter bond. (The double bond on the left hand side of the diagram is the shorter bond. The two N—O bonds on the right would be equivalent, due to resonance, so both would be longer than the double bond.)

7.95 Because it is linear, a CO_2 molecule is nonpolar. Even though each C═O bond is somewhat polar due to the electronegativity difference, the dipole vectors from the two bonds cancel one another.

7.97 No, lead selenide should not be considered ionic. According to Figure 7.8, the electronegativities are 2.4 for selenium and 1.7 for lead. A difference of just 0.7 is not large enough to give rise to ionic bonding.

7.99 The Lewis structure of nitrogen gas shows that it has a triple bond, :N≡N: Such a bond is quite strong, so it is relatively difficult to break. This makes nitrogen a relatively inert gas at normal temperatures. The triple bond can be broken, however, at higher temperatures such as those in internal combustion engines.

7.101 The key feature that would need to be in any picture is that for a current to flow, a charge must be able to move. In an ionic solid, oppositely charged ions are held in place and cannot move to conduct electricity. In pure water, there are no charges that could move. When an ionic solid dissolves, however, the charges are now free to move and electricity can be conducted. Any molecular-level picture has to show these features.

7.103 Because all have two *ns* and two *np* electrons in their valence shells, they show four dots.

7.105

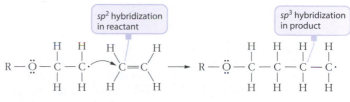

sp^2 hybridization in reactant

sp^3 hybridization in product

R—Ö—CH_2—CH_2· + $H_2C{=}CH_2$ ⟶ R—Ö—CH_2—CH_2—CH_2—CH_2·

Chapter 8

8.1 There are three solid forms of carbon: diamond, graphite, and buckminsterfullerene (and other forms of fullerenes and nanotubes.)

8.3 For engineering applications, the hardness of diamonds leads to important applications. If you consider a material in which you wish to drill a hole, the material you use in your drill bit must be harder than the substance. Because diamond is so hard, it can drill through most substances—even rather hard ones—so it has useful applications

8.5 Carbon nanotubes have "caps" on their ends. These ends close off the cylindrical nanotube. The caps are one hemisphere of a molecule of buckminsterfullerene.

8.7 Answers will vary slightly based on websites found, but they should note that the atmosphere needs to be inert. If an arc is conducted in air, the oxygen present will react with the carbon and buckminsterfullerene will not form.

8.9 The packing efficiency tells us the amount of space occupied by particles in a solid. The more space filled, the more efficient the packing.

8.11 Lower efficiency

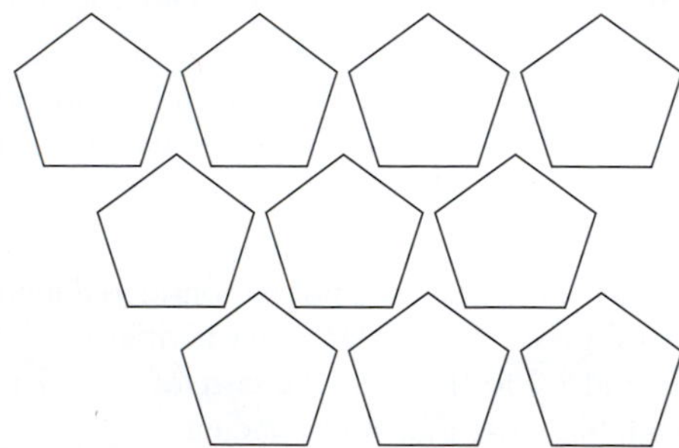

Higher efficiency

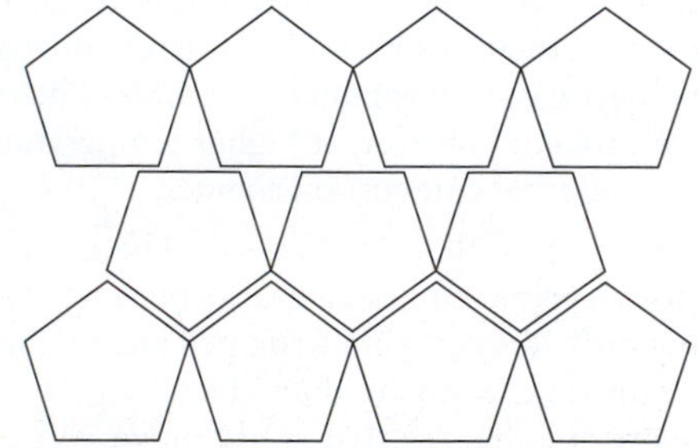

8.13 For the square, it is 78.5%. For the hexagonal, it is 90.7%.

8.15 The coordination number is 4. Each carbon atom is bonded to four other carbon atoms.

8.17 $r = 1.67 \times 10^{-8}$ cm

8.19 $a = 4.76 \times 10^{-8}$ cm

8.21 From the data given, Avogadro's number must be 6.023×10^{23}, a value fairly close to the accepted value.

8.23 Malleability requires that the metal remain bonded despite being deformed. The sea of electrons model accounts for this property because the electrons can move rapidly to adjust to changes in the positions of the positively charged cores. Changing the shape of the metal changes the positions of these cores, but it doesn't affect the bonding significantly.

8.25 To conduct electricity, the material must allow charge to move. In the sea of electrons model, the electrons in the "sea" can respond to an electric field by moving. So metals conduct electricity.

8.27 In a bonding orbital, there is a buildup of electron density between the nuclei being bonded. In an antibonding orbital there is a node, a region where the electrons cannot be, between the nuclei involved.

8.29 The picture looks like this:

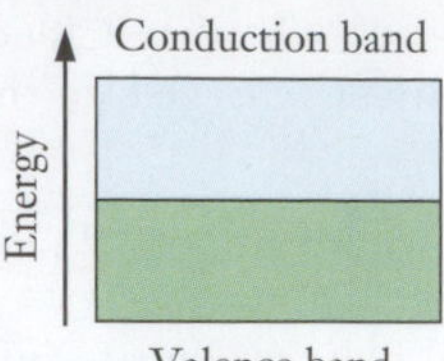

8.31 Semiconductors have filled valence bands, but the conduction band is nearby in energy, so that some electrons occupy it by thermal excitation. As temperature is increased, thermal excitation increases, so a higher temperature will tend to increase the conductivity of the semiconductor.

8.33 Doping either provides additional electrons that can be part of the conduction band, or it creates "holes" that also can be used to increase the conductivity of the material.

8.35 A semiconductor is named p-type because a dopant creates "positive" holes in the valence band. To do this, the dopant must have fewer valence electrons than the atom that is replaced.

8.37 For a voltage-gated device, the flow of charge across some plane (in this case a junction between the two types of semiconductors) is regulated by a change in voltage. Because an n-type semiconductor acts like a material with negative charges and a p-type semiconductor acts like a material with positive charges, the charge carriers of these materials will flow toward the opposite voltage polarity. With one polarity, the charge will flow across the junction, and when it is switched, the charge will not flow across the junction.

8.39 An instantaneous dipole is a small separation of charge (positive and negative) that happens because of a fluctuation in the positions of the electrons of an atom or molecule. If the electron density is momentarily in one region, that portion of the molecule will be negatively charged and the portion where the electrons "would otherwise be" will be positively charged, leading to a transient dipole.

8.41 The extent to which a molecule responds to an external electric field is measured by its polarizability. So a material that is not very polarizable will not respond very strongly to such a field.

8.43 For a substance to be a gas, its particles must not interact very strongly. Small molecules often have relatively small interactions. Large polymer molecules experience strong attractive forces and exist only in condensed phases.

8.45 When a polar solvent is present, the dipole on the solvent molecules can interact with the electric charge on an ion.

8.47 All have high electronegativity.

8.49 **(a)** dispersion

(b) hydrogen bonding

(c) Dispersion and dipole–dipole forces are both present. Dispersion may be more important due to the large size of the chlorine atoms.

(d) dispersion

(e) hydrogen bonding

(f) dipole–dipole

8.51 Carbon tetrachloride is rather large for a molecule with only five atoms because the chlorine atoms are fairly large. Large molecules tend to be polarizable, so the dispersion forces in CCl_4 are large enough that it is a liquid at room temperature. Ammonia, with only nitrogen and hydrogen has small dispersion forces, so its interactions are dominated by dipole–dipole and hydrogen bonding interactions. The fact that CCl_4 is a liquid and NH_3 is a gas implies that the large dispersion forces in the former are stronger than the hydrogen bonding and dipole forces in the latter.

8.53 For molecules to leave a liquid and enter the gas phase, they must overcome interactions with other molecules in the liquid. The stronger these interactions, the less likely it is that molecules will escape the liquid. So, the vapor pressure will be lower.

8.55 (a) increases

(b) increases

(c) does not change

(d) increases

8.57 (a) O_2 (b) SO_2 (c) HF (d) GeH_4

8.59 The vapor pressure will be low and the boiling point will be high.

8.61 Substance A has the highest vapor pressure. Substance C will have the next highest vapor pressure and finally, Substance B will have the lowest vapor pressure.

8.63 The liquid would look like this drawing:

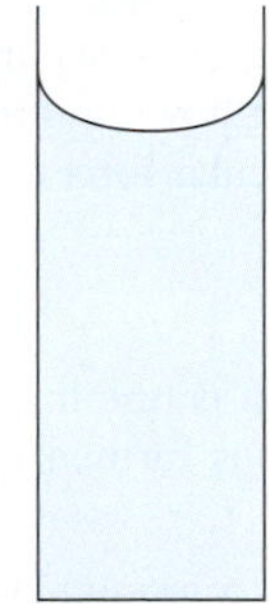

8.65 For addition polymerization to occur, there must be a carbon–carbon double bond present, so this would be the thing to look for in a monomer.

8.67 Polymerization reactions can be viewed as the growing of a molecular chain. The reactions that increase the length of the chain and leave it capable of continuing to grow are called propagation. Eventually a reaction occurs that doesn't allow the chain to keep growing, and that is the chain termination reaction.

8.69 Because the atoms connected to the polymer backbone are all the same. They cannot be distinguished, so there are no isotactic or syndiotactic forms.

8.71 There are always two products—the polymer that forms and some small molecule, often water.

8.73 In a block copolymer, the different monomers present are in groups that are still part of the backbone. In graft copolymers, the different groupings of monomers are attached to the backbone as side chains.

8.75 Thermoplastics soften (and eventually melt) at high temperatures.

8.77 When a thermosetting polymer is heated, it forms bridges or chemical links between chains. These links are called cross-links, and they tend to prevent the resulting polymer from softening when heated.

8.79 Requires web research—answers will vary though it is unlikely that students will report that melting an eraser will allow it to be remolded and still work.

8.81 The substrate is the original solid on which the MEMS device is to be fabricated. The mask is a pattern used to control the deposition of material onto the substrate.

8.83 The rate at which the silicon dioxide mask is etched must be much slower than the rate at which the underlying silicon substrate is etched.

8.85 $\sim 5 \times 10^{13}$ Si atoms

8.87 (a) about 350 mmHg

(b) Ethanol has stronger intermolecular forces.

(c) about 84°C

(d) 46°C (CS_2), 78°C (C_2H_5OH), 98°C (C_7H_{16})

(e) Heptane is a liquid. Carbon disulfide and ethanol are gases.

8.89 (a)

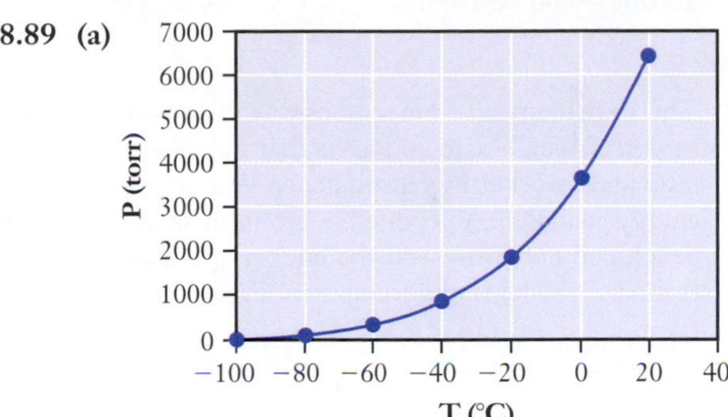

From the graph, we can estimate the normal boiling point to be about −45°C.

(b) The temperature is roughly 35°C, so we can extrapolate from the graph and estimate the pressure to be 12,000 torr, or roughly 16 atm.

(c) The storage tanks will need to be strong enough to withstand high internal pressure.

8.91 The pressure from evaporation of the full amount in a closed room of this size is roughly 17 torr. This is smaller than the vapor pressure, so the entire volume will evaporate.

8.93 Answers will vary somewhat depending on the websites found. Many solar panels are based on p-n junctions with large areas.

8.95 (a) The alternating chain for syndiotactic polystyrene allows better packing of the polymers so that the melting point is higher. (b) Designs that require higher performance, including a higher melting point for the plastic, might require the syndiotactic polymer. One possible example would be medical devices that must be exposed to high temperatures for sterilization.

8.97 To avoid creating an n-type or a p-type semiconductor, we must make sure the dopant has the same number of electrons as silicon. To lower conductivity, we must increase the size of the band gap, and so the dopant has to be smaller than silicon. The only candidate is carbon, whose atoms have four valence electrons and are smaller than silicon atoms.

8.99 The magnet is providing an external field that is inducing an alignment of electrons in the pin. Polarization is similar in that an external field is used to alter the electron distribution of a molecule.

8.101 To measure the degree of polymerization, the engineer would have to measure the molecular mass of the polymer and divide by the molecular mass of the monomer.

8.103 Cryolite is a very useful flux for separating the Al and O in aluminum oxide.

8.105 Answers will vary depending on the websites found. But a typical doping level might be on the order of one dopant atom per 1,000,000 silicon atoms. To produce an n-type semiconductor, we might use phosphorous as the dopant. We will calculate our answers using these choices. **(a)** Doping 1 metric ton of silicon with phosphorous at a level of one part per million would require just 1.1 g of phosphorous. **(b)** The mole fraction of the phosphorous dopant would be 10^{-6}.

Chapter 9

9.1 There are many possible reasons why these two items could be correlated. The main idea is that the productivity of an individual worker increases, depending on the amount of energy available. A productive economy is usually a well-developed economy—so the energy spent per worker is related to the level of development in an economy.

9.3 **(a)** Europe would probably be relatively similar. **(b)** The main difference between the United States and a developing country would be the total size of the graph. Chances are also good that the percentage of waste in the developing country would be larger.

9.5 The energy imports shown in Figure 9.2 are equivalent to 5.3 billion barrels of oil a year. This means we must import nearly 15 million barrels a day. (Note that this number is high because some of our energy imports are in forms other than oil.)

9.7 The internal energy is determined by the total of the kinetic and potential energy of the atoms in the system.

9.9 6.0×10^{-20} J

9.11 $P = \dfrac{\text{Force}}{\text{Area}} = \dfrac{\text{kg m}^1\text{ s}^{-2}}{\text{m}^2} = \text{kg m}^{-1}\text{ s}^{-2}$ so $P = \text{kg m}^2\text{ s}^{-2}$, which is an energy unit.

9.13 A hydrocarbon is a compound composed only of hydrogen and carbon. Many fuels are hydrocarbons.

9.15 **(a)** 3.61×10^3 cal **(b)** 5.9×10^5 J **(c)** 3.00×10^{-5} Btu

9.17 $+6.83 \times 10^4$ kJ

9.19 A negative sign on ΔE means that internal energy has been lost by (or has left) the system.

9.21 Work plays a more important role in society and therefore is more valued. Consider how fuel efficiency is measured for a car. It is in miles per gallon—miles driven represent work. There is little value placed on how warm the engine has gotten.

9.23 **(a)** $w = -836$ J **(b)** $w = -30$ J. So more work is done in the system in (a).

9.25 Increased efficiency does not always imply lower household energy demand. In some cases, people may choose to own more appliances such as televisions or computers. Even though each appliance is more efficient, because more are present there is no decrease in total energy demand.

9.27 According to one site, only 8% of energy use in the United States is attributable to lighting, and of that small percentage, only 27% is from residential lighting. So only slightly more than 2% of energy use in the United States can be attributed to the lighting in residences. Although it might make sense to conserve energy in this way, it's not likely to have a major impact on total energy use.

9.29 Calorimetry is the science of measuring heat flow. It is the primary way that we determine changes in the energy of systems.

9.31 3.50×10^5 J

9.33 5.2 °C

9.35 0.52 J g^{-1} $°C^{-1}$

9.37 383 J $°C^{-1}$

9.39 700 MJ

9.41 Under constant pressure

9.43 An exothermic process releases heat to the surroundings, and an endothermic one absorbs heat from the surroundings.

9.45 Because a phase change is occurring, there is no change in temperature. Heat flowing into the substance is used to overcome intermolecular forces.

9.47 -3.4×10^2 kJ/mol

9.49 5.87 kg acetone

9.51 A formation reaction is one in which exactly 1 mole of a chemical compound is formed from its elements in their standard states.

9.53 **(a)** Two moles of the compound are formed. **(b)** The equation is not balanced. **(c)** The standard state of oxygen is O_2.

9.55 –217.3 kJ/mol; 262 kJ is released.

9.57 $C_2H_2(g) + H_2(g) \rightarrow C_2H_4$ $\Delta H° = -175$ kJ

9.59 -126 kJ/mol

9.61 The sum of the bond energies in the products must be greater than those in the reactants.

9.63 3.2×10^3 kJ

9.65 699 kJ

9.67 480 kJ

9.69 31.9 g/mol

9.71 Because in transportation, not only are products (and their mass) moved around, the fuel in the vehicles must also be moved around. Keeping the mass of the fuel as low as feasible means that as little energy as possible is spent moving fuel (and a greater percentage is spent moving products).

9.73 Transmission and distribution substations both move electricity from one point to another in the grid. The main difference is that transmission substations step up the voltage for transmission over long distances, whereas distribution substations step down the voltage to levels more suitable for transmission to customers.

9.75 A smart grid would use technology to gather information and make informed decisions regarding the production and distribution of electricity. In contrast with the traditional grid, the smart grid would be able to respond to various situations in an automated fashion.

9.77 9°C

9.79 **(a)** Within experimental tolerances, the temperature change should be the same. Twice as much heat is released, but twice as much solution is present to absorb it. **(b)** In this case, the temperature rise will be greater. Twice as much heat is released, and here the solution volume is the same as in the original trial.

9.81 33.0°C

9.83 54.7 kJ is released.

9.85 106 g CH_4

9.87 **(a)** 23°C

(b) 360 cm

(c) 0.038%

(d) Over the length of the wire, there is an expansion of 0.139 cm. This is not likely to be significant except in the most exacting of designs.

9.89 **(a)** Graph A **(b)** 14.2 kJ

9.91 First, we would have to assume that the heat transfer process is 100% efficient. Then, we must calculate the amount of heat it takes to cool the 500 kg of fruit (using some reasonable value for the initial temperature to calculate ΔT) using the equation, Heat = mass × specific heat × ΔT. Then this amount of heat would go into evaporating the refrigerant, whose mass can be calculated because we know that 680 J is absorbed per gram evaporated.

9.93 Upon drawing the Lewis structures, it becomes apparent that the total number of chemical bonds for reactants and products in this reaction remains the same—that is, there are 3 B—Br and 3 B—Cl bonds in both reactants and products. Although there may be slight changes because of the specific molecular environment in which the bonds are found, the net change is going to be small.

9.95 For any reaction, the heat of the reaction is given (according to Hess's law) by the sum of the heats of formation multiplied by their "signed stoichiometric coefficients." In this case, the enthalpy change for the reaction has been measured. If every other heat of formation other than silicon nitride is known (or can be looked up), we can solve the Hess's law problem for the heat of formation of silicon nitride.

9.97 This thermochemical equation refers to the energy of combustion of 2 moles of octane. To obtain the energy released per gram we must determine the molar mass of octane and convert from moles to grams.

9.99 Generally, it would be best to have a material with high heat capacity. Such a material would be capable of absorbing a large amount of heat without having a large change in temperature according to the equation, Heat = mass × specific heat × ΔT. Having the temperature stay relatively constant would have an advantage because the material would not get too hot for safe use when people might touch it.

9.101 6.21×10^{16} kJ

9.103 1.26×10^{5} kJ per gallon. Roughly 69% efficient. Assumptions, such as no loss to friction, etc., are major and would generally tend to push this number toward lower efficiency.

Chapter 10

10.1 From an environmental perspective, the recycling process involves the input of energy and often other resources that may have an environmental cost.

10.3 Most uses are associated with the production of fibers for things like durable carpeting or insulating fibers such as fiberfill.

10.5 Web research required—answer depends on site found.

10.7 **(a)** Spontaneous: this is the sublimation of dry ice.

(b) Not spontaneous: table salt is solid at room temperature.

(c) Not spontaneous: table salt does not dissociate at room temperature

(d) Not spontaneous: CO_2 is stable and does not dissociate into its elements.

10.9 Combustion of a fuel is spontaneous, for example, once it has been initiated by a spark. Spontaneous does not imply instantaneous upon mixing.

10.11 **(a)** Nonspontaneous. Opposite: Formation of O_2 from oxygen atoms. **(b)** Nonspontaneous. Opposite: A tray of ice melts on a warm day. **(c)** Nonspontaneous. Opposite: An acid and base form a salt solution after the reaction is complete. **(d)** Spontaneous. **(e)** Nonspontaneous. Opposite: SO_3 and water react to form sulfuric acid.

10.13 **(a)** This process is not spontaneous at high temperatures.

(b) This process is not spontaneous at very high temperatures.

(c) This process is always spontaneous.

(d) This process may not be spontaneous at high temperatures.

10.15 No, at the very least the system must include oxygen (often from air).

10.17 Alice said, "With a wall so narrow your balance must sometime fail you—chances are you will soon fall. The ground is not so narrow, so you cannot fall from it. And as for the king—he may mean well but I doubt he is a magician. If you break into 1000 pieces what will become of you if the king finds only 995 of them? I don't think you should sit on that wall waiting to find out."

10.19 He found that when he calculated the heat around the whole path divided by temperature (q/T), the result was zero.

10.21 Nature spontaneously tends toward a more probable state.

10.23 **(a)** CO_2 vapor has higher entropy because gases are more randomly arranged than solids and thereby have more microstates. **(b)** Sugar dissolved in water will have higher entropy because the molecules are dispersed, and more dispersed systems have more microstates. **(c)** The mixed liquids have more entropy because they have dispersed into each other.

10.25 $\Delta S < 0$. This process is the opposite of mixing, so the sign on ΔS must be negative.

10.27 In each case the side of the equation with more moles of gas will have the higher entropy. So the signs would be **(a)** positive, **(b)** positive, **(c)** negative, **(d)** negative, and **(e)** positive.

10.29 The reduction of entropy in the leaf itself as it grows must be accompanied by a larger increase in entropy elsewhere. Decomposition of the leaf after it falls from the tree leads to a large increase in entropy.

10.31 The thermodynamic entropy increase occurs mainly in heating the surroundings.

10.33 Once in a deck, playing cards would not become a disarrayed pile, whereas atoms and molecules will constantly move and mix, leading to the more probable state without any intervention.

10.35 It increases.

10.37 To convert heat into work requires a decrease in randomness, which would lower entropy.

10.39 The exothermic nature of the change means that heat is released to the surroundings. This leads to an entropy increase in the surroundings which is greater than the decrease in the entropy of the system.

10.41 It generally has no influence on the ΔS of the system. The exothermic part has a major effect on the entropy of the surroundings.

10.43 The reaction, $Br_2(g) + Cl_2(g) \rightarrow 2\ BrCl(g)$, has the greater entropy change. There are more possible "pairings" for the heteronuclear molecules, so in each case those have the higher entropy $\Delta S°$.

10.45 The process is endothermic.

10.47 **(a)** $\Delta S° = -152.4$ J/K **(b)** $\Delta S° = -268.0$ J/K **(c)** $\Delta S° = -185.0$ J/K **(d)** $\Delta S° = -116$ J/K **(e)** $\Delta S° = +134.2$ J/K

10.49 $\Delta S° = -114.6$ J/K. This negative value for entropy change arises from the fact that water molecules solvate ions in solution, thereby creating order.

10.51 **(a)** $\Delta S° = 164.0$ J/K **(b)** Plants carry out this type of reaction in conjunction with other reactions that increases entropy. As long as the overall entropy change of the universe is negative, the coupled reactions can proceed spontaneously.

10.53 $H_2C_2O_4(s) < C_2H_5OH(\ell) < CH_4(g)$. The entropy of a gas is generally much greater than that of a liquid, which in turn is generally greater than that of a solid.

10.55 **(a)** At the molecular level, the water molecules in the solid (ice) are becoming less ordered as it melts. The ice water is still at 0°C because all of the heat that flows into it melts the ice. **(b)** Inside the container, the entropy must be increasing because the system is isolated.

10.57 When the process takes place at constant pressure and temperature.

10.59 **(a)** 514 kJ **(b)** −1295 kJ **(c)** 35 kJ

10.61 Both the enthalpy change and the entropy change are positive. At lower temperatures, the enthalpy term dominates. But at higher temperatures, the $T\Delta S$ term becomes more important, leading to a negative ΔG.

10.63 A reversible change is a small change where the system can be restored to its original state. An irreversible change places the system in a state from which the original state cannot be restored.

10.65 **(a)** $\Delta S° = -97.4$ J/K **(b)** $\Delta G° = -1226.5$ kJ **(c)** At high enough temperatures, the $T\Delta S$ term could dominate this equation. **(d)** Using the equation $\Delta G° = \Delta H° - T\Delta S°$ might be questionable for determining this range because it requires that $\Delta H \neq f(T)$.

10.67 **(a)** $SiH_4(g) + 2\ O_2(g) \rightarrow SiO_2(s) + 2\ H_2O(g)$ **(b)** $\Delta S° = -195.3$ J/K **(c)** $\Delta G° = -1370.8$ kJ, so the process is spontaneous. **(d)** For methane combustion, $\Delta G° = -800.9$ kJ, which is not nearly as negative as for silane. Both reactions are enthalpy driven.

10.69 **(a)** $\Delta G° = 111.7$ kJ **(b)** $\Delta G° = +8.3$ kJ **(c)** $\Delta G° = -817.6$ kJ **(d)** $\Delta G° = -40.42$ kJ **(e)** $\Delta G° = +617.0$ kJ

10.71 For dissolving, $\Delta G°(NaCl) = -9$ kJ and $\Delta G°(AgCl) = 55.7$ kJ. The negative value of $\Delta G°$ for NaCl suggests that it dissolves, whereas the positive value of $\Delta G°$ for AgCl suggests that it doesn't.

10.73 **(a)** positive **(b)** negative **(c)** zero **(d)** positive **(e)** negative

10.75 **(a)** spontaneous at lower temperature ($\Delta H < 0, \Delta S < 0$)
(b) spontaneous at higher temperature ($\Delta H > 0, \Delta S > 0$)
(c) not spontaneous at any temperature ($\Delta H > 0, \Delta S < 0$)
(d) spontaneous at higher temperature ($\Delta H > 0, \Delta S > 0$)

10.77 No. An endothermic reaction cannot be spontaneous if it also has $\Delta S < 0$.

10.79 If the long chain breaks into a shorter chain and a small piece, randomness has increased.

10.81 $\Delta S = 84.4$ J/K/mol

10.83 $T_{bp} = 2153$°C.

10.85 **(a)** $\Delta G° = -16.2$ kJ **(b)** $\Delta H° = -17$ kJ

10.87 **(a)** False. An exothermic reaction is often, but not always, spontaneous.

(b) False. When $\Delta G°$ is positive, the reaction cannot occur unless it is driven by an external system.

(c) False. $\Delta S°$ is positive for a reaction in which there is an increase in the number of moles of gas in going from reactants to products.

(d) False. If $\Delta H°$ and $\Delta S°$ are both negative, $\Delta G°$ will be negative at low temperatures.

10.89 **(a)** If we assume an initial temperature of 25°C, then the maximum temperature will be ~48°C. **(b)** Probably not, because heat will be dissipated to the surroundings and whatever products arise from the exothermic process will also have some heat capacity. **(c)** Assuming the process occurs at constant pressure, ΔH is –1500 J. **(d)** $\Delta S = q/T = 1500/298 = 5.0$ J/K

10.91 **(a)** 467.9 kJ **(b)** 87.4 J/K **(c)** 301.2 kJ **(d)** $T > 838.6$ K

10.93 **(a)** –210 J/K **(b)** –95 kJ

10.95 Use the sample to carry out a combustion reaction in a calorimeter. From the heat of combustion, along with heats of formation of the products (carbon dioxide and water), you can calculate the heat of formation of cyclohexane.

10.97 The equation $\Delta G° = \Delta H° - T\Delta S°$ provides a relationship between free energy, enthalpy, and entropy. If we can obtain values for any two of these variables (in this case, the entropy and free energy changes) for a given reaction, we can calculate the third. So if we calculate $\Delta S°$ for the formation reaction of the substance in question, we can use that value along with the known value of $\Delta G_f°$ to find $\Delta H_f°$. (We would use the standard temperature of 298 K.)

10.99 The key feature lies in the sign of the enthalpy and entropy change. If they have the same signs (both positive or both negative), the spontaneity of the process will be temperature dependent. Melting snow is an endothermic process ($\Delta H > 0$), but it increases entropy ($\Delta S > 0$), so it will tend to occur at warmer temperatures. Snow formation is exothermic but lowers entropy, so it will tend to occur at lower temperatures.

10.101 **(a)** $2\ F_2 + 2\ H_2O \rightarrow 4\ HF + O_2$ **(b)** $\Delta G° = -618$ kJ, so the reaction is spontaneous. **(c)** $\Delta H° = -512$ kJ **(d)** So, 233 kJ of heat is released when this much F_2 is bubbled through.

10.103 **(a)** MnO: manganese(II) oxide; MnO_2: manganese(IV) oxide; Mn_2O_3: manganese(III) oxide

(b) $3\ MnO + 1/2\ O_2 \rightarrow Mn_3O_4$; $3\ MnO_2 \rightarrow Mn_3O_4 + O_2$; $3\ Mn_2O_3 \rightarrow 2\ Mn_3O_4 + 1/2\ O_2$

(c) for the MnO reaction, $\Delta G° = -194$ kJ; for the MnO_2 reaction, $\Delta G° = +113$ kJ; and for the Mn_2O_3 reaction, $\Delta G° = +78$ kJ. These numbers suggest that MnO_2 is the most stable form.

Chapter 11

11.1 An allotrope is a single physical and chemical form of an element. Both diatomic oxygen and ozone are allotropes of the element oxygen.

11.3 Ozone is considered a pollutant when it forms near the surface of the earth in the troposphere. When it forms in the stratosphere as part of the Chapman cycle, however, it has the beneficial attribute of protecting the planet from harmful UV radiation.

11.5 There is no net reaction—and therein lies the reason it is called a cycle.

11.7 The breaking of the double bond in diatomic oxygen requires the most energy and the shortest wavelength light.

11.9 No—it is a seasonal phenomenon. The depletion of stratospheric ozone that is referred to as the ozone hole occurs in the springtime over polar regions.

11.11 **(a)** beats per minute **(b)** feet per second **(c)** rotations per second **(d)** volume per millisecond.

11.13 An instantaneous rate is measured at one moment in a reaction, whereas an average rate is measured over some finite period of time. The hot dog eating rate of 5 per minute over the duration of the contest is an example of an average rate. The reported rates of 6 hot dogs per minute at 1.0 minute and 3 hot dogs per minute at 3.0 minutes are examples of instantaneous rates at those points in the contest.

11.15 The combustion of hydrocarbon fuels produces carbon dioxide and water, so the production of these chemicals could also be measured to determine the combustion rate of the candle.

11.17 The rate of oxygen consumption is 5.29×10^{-4} mol L^{-1} s^{-1}. NO is produced at the rate of 4.23×10^{-4} mol L^{-1} s^{-1}. Water is produced at the rate of 6.35×10^{-4} mol L^{-1} s^{-1}.

11.19 For the first 8.3 minutes, the rate is 0.51 L/min. Over the entire 19-minute experiment, the rate is 0.605 L/min.

11.21 Estimating the slope from a graph, we obtain a rate of 0.37 torr/min.

11.23 The rate will increase by a factor of 9 if [A] is tripled. The rate will decrease by a factor of 4 if [A] is halved.

11.25 **(a)** 1st order in X, 1st order in Y, and 1st order in Z; 3rd order overall

(b) 2nd order in X, 0.5 order in Y, and 1st order in Z; 3.5 order overall

(c) 1.5 order in X and −1 order in Y; 0.5 order overall

(d) 1st order in X and −2 order in Y; −1 order overall

11.27 Solving the rate law for k, we have $k = \text{rate}/[H_2][Br_2]$. Inserting the given units for the rate and mol/L for each of the two concentrations gives us units of L mol^{-1} s^{-1} for k.

11.29 $x = 1$ and $y = 2$

11.31 First order with respect to A, second order with respect to B, zero order with respect to C, and third order overall.

11.33 Rate $= k[NO_2^-][NH_4^+]$. The value of the rate constant is 3.7×10^{-4} L mol^{-1} s^{-1}.

11.35 Rate $= k[NO]^2[H_2]$, and the rate constant is 23 L^2 mol^{-2} s^{-1}.

11.37 Looking at the figure of the cube,

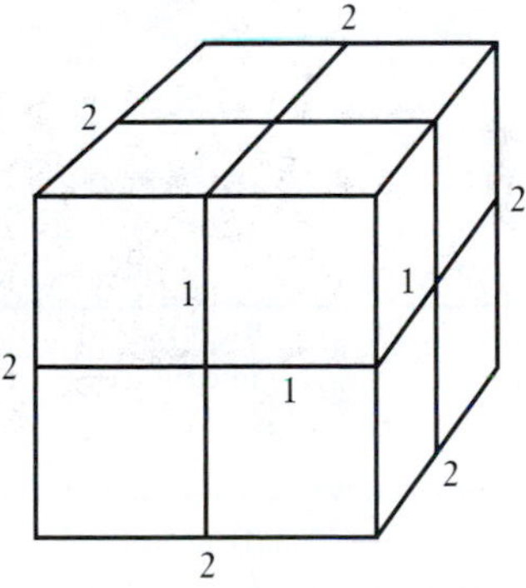

we can calculate the surface area of the whole cube as $6 \times (2 \times 2) = 24$ cm^2. The surface area of each cube in the group of eight smaller cubes is $6 \times (1 \times 1) = 6$ cm^2, and for eight such cubes, the surface area is $8 \times 6 = 48$ cm^2. Dividing the eight small cubes into even smaller cubes will increase the surface area again. Consider a log versus all the toothpicks that could be made from it. The toothpicks are analogous to the smaller cubes in this picture. They would have large surface area compared to the log, and it would be much easier to start a fire with the toothpicks than with a complete log.

11.39 $[N_2O_5] = 0.146$ M

11.41 0.4 mg

11.43 260 mg

11.45 **(a)** 0.16 mol/L **(b)** 90 s **(c)** 120 s

11.47 We have $\frac{1}{[X]_0} = kt_{1/2}$ and $t_{1/2} = \frac{1}{k[X]_0}$ The half-life is concentration dependent and will change during the course of the reaction.

11.49 Looking at the various possible plots, we find a straight line for the plot of ln[picloram] vs. time so the reaction is first order. $k = 1.7 \times 10^{-2}$ days^{-1}, and $t_{1/2} = 42$ days.

11.51 This problem requires plotting data to find straight lines. Test plots for first- and second-order kinetics are as follows:

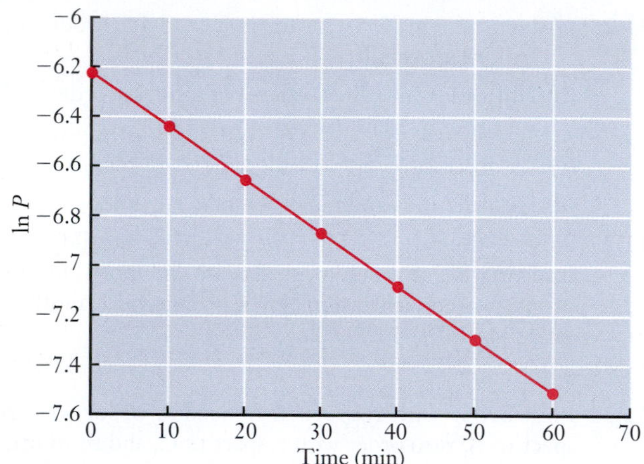

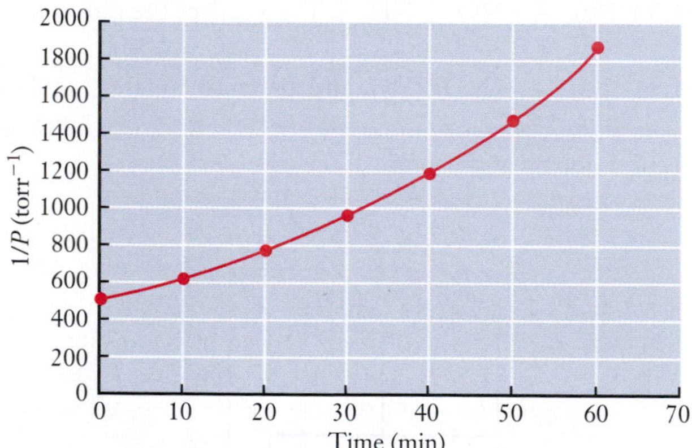

From these graphs, we can see that the plot of ln P versus time is linear, so the reaction is first order. The slope of that graph is equal to $-k$. A linear regression fit to the line gives a slope of -0.0217 min^{-1}, so the rate constant is $k = 0.0217$ min^{-1}.

11.53 **(a)** The reaction is first order. **(b)** $k = 3.3 \times 10^{-3}$ s^{-1}. **(c)** Initial rate = 3.3×10^{-3} M s^{-1}.

11.55 $k = 5 \times 10^{-70}$ s^{-1}

11.57 $E_a = 120$ kJ/mol

11.59 $E_a = 3.2 \times 10^5$ J/mol = 320 kJ/mol

11.61 This experiment would need some measurement to determine if the milk has soured—perhaps the smell of the milk could be noted at a given time during the day. The next step would be holding the temperature steady at several values. Then, assuming the only source of lactic acid is the reaction in the bacteria, the activation energy could be estimated using the Arrhenius equation.

11.63 A reaction mechanism cannot be proven correct. Even if a particular intermediate is observed, there is always the possibility that another mechanism could be proposed that also generates that intermediate. A mechanism can be much more readily proven incorrect. For example, any measurement that demonstrates the existence of an intermediate that is not part of the proposed mechanism proves that the mechanism is wrong.

11.65 Because the Chapman cycle includes species that are intermediates in the overall reaction, it must be a mechanism of the overall reaction. The molecularity of the three steps in order are

$O_2 \rightarrow 2\ O$ (unimolecular—light has been left out of this equation)

$O_2 + O \rightarrow O_3$ (bimolecular)

$O_3 \rightarrow O_2 + O$ (unimolecular—light left out in this equation)

$O_3 + O \rightarrow 2\ O_2$ (bimolecular)

11.67 No, this equation is clearly not a mechanism. If it were, it would imply a six-body collision.

11.69 **(a)** $2\ NO + Br_2 \rightarrow 2\ NOBr$ **(b)** the first step **(c)** $NOBr_2$ is the intermediate. **(d)** Both are bimolecular. **(e)** Equation #1: Rate = $k[NO][Br_2]$, and Equation #2: Rate = $k[NOBr_2][NO]$.

11.71 **(a)**

$$\begin{array}{l} NO_2(g) + NO_2(g) \rightarrow NO(g) + NO_3(g) \\ NO_3(g) + CO(g) \rightarrow NO_2(g) + CO_2(g) \\ \hline NO_2(g) + CO(g) \rightarrow NO(g) + CO_2(g) \end{array}$$

(b) Each step is bimolecular.

(c) Rate = $k[NO_2]^2$

(d) $NO_3(g)$ is an intermediate.

11.73 To the extent that the success of the sports team parallels the progress of a chemical reaction, the reserve that enhances the success of the sports team works the same way as a chemical that enhances the rate of a reaction.

11.75 Homogeneous catalysis takes place completely within a single phase, whereas heterogeneous catalysis takes place in two phases, often on the surface of a solid.

11.77 It would do the same thing—it would lower the activation energy of both the forward and reverse reactions.

11.79 Answers will vary depending on the websites visited.

11.81 The key initiating reaction for smog formation requires light. So this is a photochemical reaction.

11.83 VOC is an acronym for volatile organic chemical. VOCs are significant in photochemical smog formation because they destroy radicals that form in some of the initiating reactions. If these radicals were not destroyed, the reactions that formed them would tend to reverse, and the smog formation would be reversed almost as soon as it started.

11.85 The presence of nitrogen oxides and VOCs are the key factors. In places where there are many automobiles, such as urban centers, there are ample sources for these two compounds.

11.87 Because several reactions are required from the initial generation of the pollutants, such as nitrogen oxides, before ozone is formed. It takes some time for these reactions to take place—so there is a lag in time before ozone forms.

11.89 Answer varies depending on location. The Environmental Protection Agency website at www.epa.gov represents a good source to answer this question.

11.91 $C_6H_{14} + HO\cdot \rightarrow C_6H_{13}\cdot + H_2O$

$C_6H_{13}\cdot + O_2 + M \rightarrow C_6H_{13}O_2\cdot + M$

$C_6H_{13}O_2\cdot + NO \rightarrow C_6H_{13}O\cdot + NO_2$

The NO_2 can then be recycled into the ozone generation process by creating O. In addition, by consuming NO, the reaction prevents some of the destruction of ozone as it is created (because NO reacts with ozone, as noted in Equation 11.10.c).

11.93 **(a)** :F̤—Ẍ̤e—F̤: It's a linear molecule.

(b) In this case, the partial pressures can be determined from the volumes, so the partial pressure of Xe is 459 mm Hg, and the partial pressure of F_2 is 301 mm Hg.

(c) $n_{F_2} = 0.0107$ mol and $n_{Xe} = 0.0164$ mol, so the fluorine is the limiting reactant. There are 0.0107 mol of XeF_2 formed, so the theoretical yield = 1.81 g XeF_2.

(d) Precipitation reactions are usually very fast compared to this reaction. The need to dissociate the F_2 molecules in the gas phase reaction is one reason for this.

11.95 The value observed is 0.13 ppm. Assuming atmospheric pressure of 1 atm and a temperature of 298 K, we can calculate that there are 3.2×10^{15} molecules of O_3.

11.97 **(a)** Chances are that it slows as the process continues. **(b)** Eventually, the rate is the same in both directions.

11.99 The rate of ammonia formation is much greater at high temperatures. The reaction must have a high activation energy such that high temperatures are required to produce collisions with enough energy to overcome the activation barrier. This is consistent with the fact that the triple bond in N_2 is extremely strong.

11.101 The reaction should be 2 Be + Si $\rightarrow$ Be_2Si, and 401 g of Be are needed.

11.103 76 kJ/mol

11.105 First, it's important to realize that radioactivity is essentially a first-order kinetics problem. Knowing the percentage of counts as a fraction of that seen for living organisms allows us to set up the ratio, I/I_o, which is equal to e^{-kt}. To use this relationship, we need to know the value of k or of the half-life, and then we can solve for time, t, which will be the age of the artifact.

11.107 To solve this problem, we need to consider factors that change the rate of reaction. In some cases, the presence of products will slow the reaction. Another strategy would be to lower the concentration of the chemical or chemicals in the rate law of the reaction or process that is being used.

11.109 Two things. First, the bonds fluorine atoms form with other elements are stronger than F—F bonds, making the reactions highly exothermic. Second, the activation energies for the reactions are probably not very large.

11.111 Radicals are very reactive—so for them to be useful, they must be generated in the reaction mixture (it's nearly impossible to have a bottle of them). The high temperatures allow bonds to break and radicals to form.

Chapter 12

12.1 The first reaction is the decomposition of calcium carbonate, $CaCO_3 \rightarrow CaO + CO_2$, and this reaction is essentially the opposite reaction that takes place in the carbonation of concrete.

12.3 Different sources estimate this number differently, but roughly 5% of annual CO_2 production can be attributed to concrete production. This is largely due to the decomposition of calcium carbonate.

12.5 The reaction is slow, so the activation energy for it to occur must be relatively high.

12.7 The parts of the concrete that were exposed to air are likely to be carbonated and therefore show no color change upon exposure to phenolphthalein. The parts that had been internal were not exposed to air, so the concrete should retain its original chemical composition and turn pink when exposed to phenolphthalein.

12.9 **(a)** In the first few minutes, orange fish will tend to swim toward the right, and green fish will tend to swim to the left. **(b)** After several hours, the distribution should be uniform throughout. **(c)** This is analogous to a dynamic equilibrium because fish swim in all directions, but there is no net change in the distribution of the locations of the different colored fish.

12.11 $\text{Rate}_{H_2+I_2\rightarrow 2HI} = \text{rate}_{2HI\rightarrow H_2+I_2}$

12.13 The top line is the concentration of A, and the bottom line is the concentration of B. At **(a)**, only A is present, so the rate of disappearance of A is large, as is the rate of appearance of B. At **(b)**, the rate of the forward reaction has slowed somewhat, and the rate of the reverse reaction is becoming appreciable. Finally, at **(c)**, the rates of the forward and reverse reactions are the same, so there is no net change in the concentration of either chemical.

12.15 **(a)** This inference can be made because the equilibrium symbol implies this fact. **(b)** This is not implied by equilibrium. Although concentrations do not change at equilibrium, this does not imply that they are all the same. **(c)** This is not implied either. Chemical equilibrium can be approached from either "direction"—starting with all reactants or with all products are just two possible starting points.

12.17 **(a)** $K = \dfrac{[NO]^2[Br_2]}{[NOBr]^2}$ **(b)** $K = \dfrac{[H_2O]^2[Cl_2]^2}{[HCl]^4[O_2]}$

(c) $K = \dfrac{[SO_3]}{[SO_2][O_2]^{1/2}}$ **(d)** $K = \dfrac{[CO_2][H_2O]^2}{[CH_4][O_2]^2}$

(e) $K = \dfrac{[CO_2]^2[H_2O]^3}{[C_2H_5OH][O_2]^3}$

12.19 **(a)** $K = [Ca^{2+}][CO_3^{\,2-}]$ **(b)** $K = [Ag^+][Cl^-]$

(c) $K = [Mg^{2+}]^3[PO_4^{\,3-}]^2$ **(d)** $K = \dfrac{[Zn^{2+}]}{[Cu^{2+}]}$

12.21 **(a)** Reactants are favored. **(b)** Products are favored. **(c)** Reactants and products are present in roughly in equal concentrations.

12.23 **(a)** Lead is more likely to precipitate sulfate ions.

12.25 (a) For the reverse reaction, $K = \dfrac{[O_2]}{[CO]^2}$.

(b) For the reverse reaction, $K = \dfrac{1}{[Ag^+][Cl^-]}$.

12.27 (a) $K_1 = \dfrac{[HCO_3^-]}{[H^+][CO_3^{\,2-}]}$ and $K_2 = \dfrac{[H_2CO_3]}{[H^+][HCO_3^{\,-}]}$

(b) Summing the two reactions leads to $CO_3^{2-} + 2\,H^+ \rightleftarrows H_2CO_3$

(c) The equilibrium expression, is $K_3 = \dfrac{[H_2CO_3]}{[H^+][CO_3^{\,2-}]}$, which is the product of K_1 and K_2.

12.29 $K = \dfrac{[CO_2]}{[CH_4]\,[O_2]^2}$

12.31 Because K_2 is greater than K_1, reaction #2 (from problem 12.30) has a greater tendency to go toward completion.

12.33 $K = 1.2$

12.35 $K = 0.080$

12.37 The equilibrium concentrations are $[O_2] = 0.100$ M and $[O] = 1.7 \times 10^{-20}$ M.

12.39 $[NO] = 0.0020$ M and $[N_2] = [O_2] = 0.099$ M.

12.41 $[NO] = 0.002$ M, $[N_2] = 0.149$ M, and $[O_2] = 0.0491$ M.

12.43 $[NO] = 0.0017$ M, $[N_2] = 0.029$ M, and $[O_2] = 0.029$ M.

12.45 $[H_2CO_3] = 0.0998$ M, $[H^+] = 0.00021$ M, and $[HCO_3^{\,-}] = 0.00021$ M

12.47 $[Ca^{2+}] = 0.0$ M and $[CO_3^{\,2-}] = 0.010$ M. (Within the significant figures we have available, the concentration of calcium ion is zero.)

12.49 **(a)** Shifts the equilibrium toward products. **(b)** Shifts equilibrium toward products. (Adding water will reduce $[H_3O^+]$ and $[F^-]$, making $Q < K$.) **(c)** More products will be formed. **(d)** The equilibrium shift will be toward products. **(e)** The equilibrium shift will be toward products. **(f)** Addition of a catalyst has no effect on the equilibrium.

12.51 **(a)** Shift toward reactants. **(b)** Shift toward the reactant. **(c)** There is no gas involved in this equilibrium, so pressure changes do not affect it directly.

12.53 Because the process is endothermic, increasing the temperature will drive the reaction to the right. Adding more NH_4HS to the flask will also drive the reaction to the right. If additional NH_3 is added to the flask, the reaction will be driven to the left. If H_2S is removed from the flask, the reaction will be driven to the right, and the pressure of NH_3 will increase.

12.55 **(a)** No effect on equilibrium concentrations. **(b)** Shifts toward products. **(c)** Shifts toward products. **(d)** Does not affect the equilibrium.

12.57 **(a)** $AgBr(s) \rightleftarrows Ag^+(aq) + Br^-(aq)$; $K_{sp} = [Ag^+][Br^-]$

(b) $Cu(OH)_2(s) \rightleftarrows Cu^{2+}(aq) + 2\,OH^-(aq)$; $K_{sp} = [Cu^{2+}][OH^-]^2$

(c) $BaSO_4(s) \rightleftarrows Ba^{2+}(aq) + SO_4^{\,2-}(aq)$; $K_{sp} = [Ba^{2+}][SO_4^{\,2-}]$

(d) $PbCrO_4(s) \rightleftarrows Pb^{2+}(aq) + CrO_4^{\,2-}(aq)$; $K_{sp} = [Pb^{2+}][CrO_4^{\,2-}]$

(e) $Mg_3(PO_4)_2(s) \rightleftarrows 3\,Mg^{2+}(aq) + 2\,PO_4^{\,3-}(aq)$; $K_{sp} = [Mg^{2+}]^3[PO_4^{\,3-}]^2$

12.59 1.4 ppm

12.61 **(a)** 2×10^{-4} M **(b)** 6×10^{-7} M **(c)** 6×10^{-7} M

12.63 **(a)** $K_{sp} = 3.0 \times 10^{-53}$ **(b)** 1.3×10^{-25} g HgS /100 g H_2O

12.65 **(a)** 4.9×10^{-11} **(b)** 1.4×10^{-16} **(c)** 2.8×10^{-44} **(d)** 1.6×10^{-5}

12.67 $BaSO_4$ will precipitate first. A saturated $BaSO_4$ solution has a Ba^{2+} concentration of 1.05×10^{-5} M, which will be the concentration of Ba^{2+} when the lead starts to precipitate. This means that 99.99% of the Ba^{2+} has precipitated.

12.69 Web question—answer will vary depending on site found. The key idea is that calcium carbonate is less soluble in water at higher temperatures, leading to the formation of scale in equipment used to boil or heat water.

12.71 **(a)** $NO_3^{\,-}$ **(b)** OH^- **(c)** $SO_4^{\,2-}$ **(d)** $HCO_3^{\,-}$ **(e)** H_2O

12.73 For the purpose of this guide, we'll just list them rather than drawing brackets. (Conjugate acid written first in each case.)

(a) HCN & CN^- and H_2O & OH^-

(b) H_3O^+ & H_2O and H_2CO_3 & $HCO_3^{\,-}$

(c) H_2S & HS^- and CH_3COOH & CH_3COO^-

12.75 **(a)** $CH_3COOH(aq) + H_2O(\ell) \rightleftarrows H_3O^+(aq) + CH_3COO^-(aq)$

$$K_a = \frac{[H_3O^+][CH_3COO^-]}{[CH_3COOH]}$$

(b) $C_2H_5COOH(aq) + H_2O(\ell) \rightleftarrows H_3O^+(aq) + C_2H_5COO^-(aq)$

$$K_a = \frac{[H_3O^+][C_2H_5COO^-]}{[C_2H_5COOH]}$$

(c) $HF(aq) + H_2O(\ell) \rightleftarrows H_3O^+(aq) + F^-(aq)$

$$K_a = \frac{[H_3O^+][F^-]}{[HF]}$$

(d) $HClO(aq) + H_2O(\ell) \rightleftarrows H_3O^+(aq) + ClO^-(aq)$

$$K_a = \frac{[H_3O^+][ClO^-]}{[HClO]}$$

(e) $H_2CO_3(aq) + H_2O(\ell) \rightleftarrows H_3O^+(aq) + HCO_3^{\,-}(aq)$

$$K_a = \frac{[H_3O^+][HCO_3^{\,-}]}{[H_2CO_3]}$$

12.77 $[H_3O^+] = 0.008$ M

12.79 pH = 2.95, 1.1% ionized

12.81 pH = 2.61

12.83 pH = 9.93

12.85 **(a)** Because the value of K is less than 1, it suggests that equilibrium lies toward reactants, and therefore $\Delta G°$ should be a positive number. **(b)** Because we want the standard free energy change, we use $T = 298$ K, and $\Delta G° = 12.1$ kJ mol^{-1}.

12.87 This was probably an error in units. The value for R is in J and the $\Delta G°$ value is in kJ. If the numbers obtained for $\Delta G°$ are converted to J, the differences among the reactions become more obvious. The correct K values are 5.8×10^8, 1.8×10^5, and 57, respectively.

12.89 The wire could be broken by bending or twisting it until it eventually breaks. For a ductile material, the failure would be gradual, requiring the application of stress over some time. For a brittle material, the failure would take place more immediately.

12.91 **(a)** [structure: repeating unit with OH OH OH groups, subscript n]

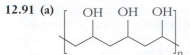

(b) Because of the –OH groups, hydrogen bonding is the most likely mode of interaction between adjacent PVA molecules.

12.93 $[CO_2] = 2.5 \times 10^{-3}$ M

12.95 **(a)** $H_2S(aq) + Fe(NO_3)_2(aq) \rightleftarrows FeS(s) + 2\ HNO_3(aq)$

(b) $K = \dfrac{[HNO_3]^2}{[H_2S][Fe(NO_3)_2]}$

(c) Removal of one or both of the products will prevent it from reaching equilibrium.

12.97 **(a)** Proceeds toward products. **(b)** Proceeds toward products. **(c)** Proceeds toward reactants. **(d)** Shifts equilibrium toward products. **(e)** Shifts equilibrium toward products. **(f)** Shifts equilibrium toward products.

12.99 An increase in pressure increases the collision frequency of the gaseous molecules. Because two NO_2 molecules react to form one N_2O_4, an increase in the number of collisions will tend to increase the probability that this reaction will occur until the collision frequency returns to its original value.

12.101 8.4×10^{-6} g $Al_2(SO_4)_3$ per gallon

12.103 Molar solubility = 3.29×10^{-3} M and $K_{sp} = 1.42 \times 10^{-7}$.

12.105 Measure out the same small mass of the three white solids. Add distilled water dropwise to each and count the number of drops it takes to dissolve each of them. The one that is most soluble will take the fewest drops; the one that is least soluble will take the most drops.

12.107 If you have a pH meter, by reading the pH of the solution you would be able to do an equilibrium calculation to determine the initial concentration (because you would be able to look up the K_a of this acid.)

12.109 **(a)** $[PCl_3] = 0.24$ M, $[PBr_3] = 1.24$ M, $[PCl_2Br] = [PClBr_2] = 0.76$ M. **(b)** This reaction is entropy driven.

12.111 $K_c = [H_2O(g)] = 0.0159$

Chapter 13

13.1 Normally, the first location where rust shows is toward the bottom of the car, particularly in the wheel-wells. These areas of the car encounter water (with salt as an electrolyte where snow removal is required), and the presence of water is necessary for the electrochemical reactions associated with corrosion to occur.

13.3 Examples will depend on Internet research.

13.5 **(a)** The oxidation half-reaction is $Cu(s) \rightarrow Cu^{2+}(aq) + 2\ e^-$, and the reduction half-reaction is $Ni^{2+}(aq) + 2\ e^- \rightarrow Ni(s)$.

(b) The oxidation half-reaction is $Ba(s) \rightarrow Ba^{2+}(aq) + 2\ e^-$, and the reduction half-reaction is $Fe^{3+}(aq) + 3e^- \rightarrow Fe(s)$.

13.7 It provides a source of charge so that the electrical circuit can be completed.

13.9 The anode is the site of oxidation. Generally, the site of oxidation in a galvanic cell will release positive ions into solution. So to maintain overall charge neutrality in the solution, nitrate ions from the salt bridge would be released into this compartment.

13.11 **(a)** $Ni(s) \mid Ni^{2+}(aq)(0.20\ M) \parallel Ag^+(aq)(0.50\ M) \mid Ag(s)$

(b) $Cu(s) \mid Cu^{2+}(aq)(0.20\ M) \parallel PtCl_6^{2-}(aq)(0.10\ M), PtCl_4^{2-}(aq)(0.10\ M), Cl^-(aq)\ (0.40\ M) \mid Pt(s)$

(c) $Pb(s) \mid PbSO_4(s) \mid SO_4^{2-}(aq)(0.30\ M) \parallel Cl^-(aq)\ (0.20\ M) \mid AgCl(s) \mid Ag(s)$

(d) $Zn(s) \mid Zn^{2+}(aq)(0.125\ M) \parallel H^+(aq)(0.010\ M) \mid H_2(g)(1.0\ atm) \mid Pt(s)$

13.13 The anions in this case are spectator ions, so they don't need to be included in the cell notation.

13.15 The electromotive potential in galvanic corrosion is supplied largely by the electrical contact of two metals with different electromotive activities. In uniform corrosion, the same metal (normally iron) undergoes both oxidation and reduction, so only one metal is required for it to happen.

13.17 The aluminum wrapper has an oxide coating on the aluminum, and aluminum oxide is not reactive. When a person bites the aluminum, however, the much more reactive aluminum is exposed, and an electrochemical reaction between the amalgam in the filling and the exposed aluminum occurs—producing a noticeable electric current.

13.19 There must be a conductive metal present for the electrons to move in a circuit and provide the electrons for the oxidation–reduction reaction.

13.21 As long as the SHE was still the standard, all values would shift by 3.00 V.

13.23 A chemical that appears at the bottom of the standard reduction potential table with a sizeable negative value means that the chemical being considered has a strong tendency to be oxidized (and very little tendency to be reduced). This observation is in line with expectations for alkali metals that tend to lose an electron easily to form mono-positive cations.

13.25 **(a)** 1.33 V **(b)** 0.02 V **(c)** 0.87 V

13.27 1.043 V

13.29 **(a)** $E = 1.05$ V **(b)** $E = 0.39$ V **(c)** $E = 0.604$ V

13.31 **(a)** $Sn^{2+}(aq) + 2\ e^- \rightarrow Sn(s)$ and $Fe(s) \rightarrow Fe^{2+}(aq) + 2\ e^-$

(b) Iron will be the anode and tin will be the cathode.

(c) $E = 0.32$ V

13.33 The memo would need to point out the chemical properties (such as corrosion resistance) of gold that make its use important. The physical properties of gold that are useful in microchip applications (electrical resistance, for example) could also be included.

13.35 Because corrosion is an example of an electrochemical reaction, an electrolyte in concrete assists the corrosion process. Chloride-containing salts such as $CaCl_2$ provide chloride ions as an electrolyte.

13.37 Use the equation, $\Delta G = nFE$.

13.39 **(a)** −237 kJ **(b)** −53.0 kJ **(c)** -2.190×10^3 kJ

13.41 $\Delta G_f°\ (Pb^{2+}) = -25\ kJ\ mol^{-1}$

13.43 **(a)** spontaneous

(b) not spontaneous

(c) not spontaneous

(d) spontaneous

13.45 (1) $K = 4.3 \times 10^{45}$, (2) $K = 1.0 \times 10^{25}$, (3) 4.3×10^{20} Looking at these values, we can see that the (1) = (2) × (3). This result suggests that if a calculator cannot handle the inverse log calculation, one need only break it into two smaller calculations and multiply the results.

13.47 $K_{sp} = 1.7 \times 10^{-10}$. This value is quite close to the tabulated value of 1.8×10^{-10}.

13.49 **(a)** $K = 3 \times 10^6$, so this reaction could be observed.

(b) $K = 2 \times 10^{-16}$, so this reaction would not be observed.

13.51 A cell based on the reaction, $3\ Ag^+(aq) + Cr(s) \rightarrow 3\ Ag(s) + Cr^{3+}(aq)$ has $E° = 1.54$ V, so it is a good candidate. One set of concentrations that would work is to let the silver ion concentration be 0.50 M and the concentration of chromium(III) be 1.2 M. The sketch should reflect a calculation using the Nernst equation. NOTE: Other combinations are certainly possible.

13.53 No. Steel is mostly iron, with a standard reduction potential more negative than that of nickel, so it would still be the iron that is oxidized if it is coupled with nickel.

13.55 The steel would be susceptible to galvanic corrosion because copper has a greater positive value for its standard reduction potential. This would mean that the iron in the steel would tend to corrode spontaneously when in contact with the relatively good oxidizing agent, copper.

13.57 Because the acidic electrolyte used in older batteries is replaced by a basic (or alkaline) electrolyte.

13.59 The electrolyte in alkaline batteries includes MnO_2. In acidic solution, the half-reaction $MnO_2(s) + 4\ H^+(aq) + 2\ e^- \rightarrow Mn^{2+}(aq) + 2\ H_2O$ has a positive standard reduction potential of 1.23 V. So this reaction would tend to proceed when coupled with other half-reactions. Also, because the half-reaction for $Zn^{2+} \mid Zn$ has a negative standard cell potential, Zn will tend to be oxidized in acidic solution.

13.61 Batteries are used to do work. If the battery itself is massive, the amount of work needed to move it (along with whatever else is part of the device) would be needlessly high. Moreover, in consumer products, lightweight devices (to allow them to be easily carried) are also important.

13.63 $PbSO_4(s)$ forms as a product. This material can be converted back to either Pb(s) or $PbO_2(s)$ as long as it remains at the electrode. If the battery is jarred, however, and the $PbSO_4$ is removed from the electrode mesh, it can't be converted back and the battery slowly degrades.

13.65 **(a)** The paragraph should emphasize that battery (a) has a higher voltage when fully charged (which could be beneficial). The long lifetime of battery (d) would be the most salient factor to be noted. **(b)** The zinc electrode is the component that limits the lifetime of the battery (because oxygen is derived from the air.) Battery (d) must have a larger surface area for the zinc electrode than that in battery (b).

13.67 Requires Internet research, and answers will vary depending on the site found.

13.69 Because oxidation–reduction reactions are already moving electrons between various reactants to form products, it is relatively easy to force electrons to move in the opposite direction by applying an external electrical potential. By contrast, in acid–base chemistry, the water formed as a product is very stable, and breaking the strong O—H bonds in the water molecule is difficult. Thus it is not easy to reverse this type of reaction.

13.71 It is passive electrolysis. The quickest way to see this is that the electrodes are carbon—they merely provide a way for electrons to move in the system (though oxygen produced in the process reacts with the carbon and slowly erodes it).

13.73 Because electrical energy is being forced into the electrolysis cell via an external potential, the cell reaction can have zero potential. A galvanic cell with zero potential is not possible because there would be no driving force for reaction.

13.75 Rotating the barrel shifts the small parts and therefore prevents individual parts from being "plated" for longer times than other individual parts. This would lead to uneven coatings.

13.77 Depends on Internet research. Some electroplating companies do emphasize cosmetic silver plating, so the paper could support the assertion. It is more likely that the site found will have more industrial applications for silver plating, and a paper would reflect that fact.

13.79 1.1 mol e^-

13.81 0.36 kWh at 0.30 V; 1.1 kWh at 0.90 V

13.83 37 g Cu

13.85 1.772 kg Sn

13.87 18 kWh

13.89 34 $PbSO_4$

13.91 Mass of coating = 4.0×10^{-4} g; plating time = 0.87 s

13.93 Lithium is the lightest metal in the periodic table, so it provides a very light anode. Carbon also plays an important role in the battery design, and it is light, too.

13.95 Li^+ ions flow through the separator from anode to cathode, but electrons also flow from anode to cathode through the external circuit. The numbers of electrons and Li^+ ions must be equal, so there is no net accumulation of charge.

13.97 To solve this problem, you need the Nernst equation. This means we must determine the number of electrons transferred and we must be able to calculate the standard cell potential from tabulated data. We can plug this information into the Nernst equation and adjust the concentration to obtain the desired voltage.

13.99 If you can build a galvanic cell with the tin as one half-cell and the N_2O as the other half-cell, the measurement of the standard cell potential would provide the best means to determine the equilibrium constant. You could also calculate this standard cell potential if the necessary reduction potentials for the relevant half-reactions are available.

13.101 The key variable is the thickness of the gold coating. A thinner coating will cost less. Other possibilities include using low voltage electroplating to reduce production costs and using barrel plating to increase the throughput of part production to lower production costs. The main pieces of information that you need to look up to make estimates of the electrolytic process are the standard reduction potentials of the half-reactions involved.

13.103 Oxygen is produced from the water solvent, $2\ H_2O \rightarrow O_2 + 4\ H^+ + 4\ e^-$. The volume produced is 2.9 L.

13.105 **(a)** 0.30 g Ag **(b)** 1.7×10^{21} atoms **(c)** ~10^6 layers of atoms

Chapter 14

14.1 Cosmic rays are composed of particles (which are discrete) rather than electromagnetic radiation (which is continuous).

14.3 Answer may vary based on site found and how many different particles the site lists.

14.5 The primary means is through photosynthesis, which can be represented by the equation, $6\ CO_2 + 6\ H_2O \rightarrow C_6H_{12}O_6 + 6\ O_2$.

14.7 Rutherford examined the ability of radiation emitted from uranium to pass through thin aluminum foil. His key observation was that some particles emitted by the uranium went through foil but others were blocked by it. Rutherford concluded that there must be at least two types of particles.

14.9 Alpha decay: Atomic number decreases by 2, mass number decreases by 4. Positron decay: Atomic number decreases by 1. Gamma decay: No change in mass number or atomic number. K capture: Atomic number decreases by 1. Beta decay: Atomic number increases by 1.

14.11 **(a)** $^{20}_{9}F$ **(b)** $^{236}_{94}Pu$ **(c)** $_{-1}^{0}\beta$

14.13 **(a)** $^{188}_{83}Bi \rightarrow\, ^{184}_{81}Tl + \,^{4}_{2}He$ **(b)** $^{87}_{37}Rb \rightarrow\, ^{87}_{38}Sr + \,_{-1}^{0}\beta + \bar{\nu}$
(c) $^{40}_{19}K \rightarrow\, ^{40}_{18}Ar + \,^{0}_{1}\beta + \nu$ **(d)** $^{138}_{57}Bi + \,_{-1}^{0}e \rightarrow\, ^{138}_{56}Ba + \nu$

14.15 **(a)** Tritium has only three nucleons, so it cannot emit an alpha particle, which itself must contain four nucleons.
(b) Beta emission **(c)** $^{3}_{1}H \rightarrow\, ^{3}_{2}He + \,_{-1}^{0}\beta + \bar{\nu}$

14.17 **(a)** $^{197}_{79}Au$ **(b)** $^{206}_{82}Pb + 3\,^{1}_{0}n \rightarrow\, ^{197}_{79}Au + 3\,^{4}_{2}He + 3\,_{-1}^{0}\beta$

14.19 4.0×10^2years

14.21 **(a)** $0.0301\ s^{-1}$ **(b)** 1.7×10^{20} Bq, or 4.5×10^9 Ci

14.23 54 mg

14.25 9200 years

14.27 **(a)** Not useful; too little tritium left
(b) Could be used; a modern wine is probably within a few half-lives (<60 or so years) old
(c) Not useful; tomb objects are too old.

14.29 17,200 years

14.31 **(a)** 1 **(b)** 1.28 **(c)** 1.6

14.33 **(a)** Electron capture: $^{73}_{36}Kr + \,_{-1}^{0}e \rightarrow\, ^{73}_{35}Br + \nu$
(b) Electron capture: $^{13}_{6}O + \,_{-1}^{0}e \rightarrow\, ^{13}_{5}N + \nu$
(c) Beta decay: $^{126}_{50}Sn \rightarrow\, ^{126}_{51}Sb + \,_{-1}^{0}\beta + \bar{\nu}$

14.35 $^{232}_{90}Th \rightarrow\, ^{228}_{88}Ra + \,^{4}_{2}He$
$^{228}_{88}Ra \rightarrow\, ^{228}_{89}Ac + \,_{-1}^{0}\beta + \bar{\nu}$
$^{228}_{89}Ac \rightarrow\, ^{228}_{90}Th + \,_{-1}^{0}\beta + \bar{\nu}$
$^{228}_{90}Th \rightarrow\, ^{224}_{88}Ra + \,^{4}_{2}He$
$^{224}_{88}Ra \rightarrow\, ^{220}_{86}Rn + \,^{4}_{2}He$
$^{220}_{86}Rn \rightarrow\, ^{216}_{84}Po + \,^{4}_{2}He$
$^{216}_{84}Po \rightarrow\, ^{212}_{82}Pb + \,^{4}_{2}He$
$^{212}_{82}Pb \rightarrow\, ^{212}_{83}Bi + \,_{-1}^{0}\beta + \bar{\nu}$
$^{212}_{83}Bi \rightarrow\, ^{212}_{84}Po + \,_{-1}^{0}\beta + \bar{\nu}$
$^{212}_{84}Po \rightarrow\, ^{208}_{82}Pb + \,^{4}_{2}He$

14.37 Loss of radon from the sample would increase the U/Pb ratio, so the measurement would underestimate the age of the rock.

14.39 1.01585×10^{10} kJ/mol

14.41 6.2877×10^{-12} J/nucleus, or 3.7865×10^9 kJ/mol

14.43 1.67682×10^{-11} J/nucleus, or 1.00981×10^{10} kJ/mol

14.45 3219 years

14.47 With an N/Z ratio of 1, ^{164}Pb would lie far above the band of stability, making it susceptible to positron emission or electron capture.

14.49 The heaviest element known to form a compound is fermium, which forms $FmCl_2$. Presumably, heavier elements could form chemical compounds, but the small amounts formed in experiments, combined with the short half-lives of many isotopes, make observation of such compounds very difficult.

14.51 **(a)** $^{27}_{13}Al + \,^{1}_{1}p \rightarrow\, ^{28}_{14}Si + \gamma$
(b) $^{40}_{18}Ar + \,^{1}_{0}n \rightarrow\, ^{41}_{19}K + \,_{-1}^{0}\beta + \bar{\nu}$

14.53 Because it does not have a charge, a neutron is not repelled by the charges on electrons or protons in the target atom.

14.55 7.6089×10^{13} J/kg of ^{235}U

14.57 **(a)** $^{239}_{92}U^*$ **(b)** $^{239}_{93}Np \rightarrow\, ^{239}_{94}Pu + \,_{-1}^{0}\beta + \bar{\nu}$

14.59 **(a)** 37 metric tons
(b) The half-lives are 16.9 hours for $^{97}_{40}Zr$ and 2.49 s for $^{137}_{52}Te$. So, virtually none of these isotopes would accumulate over the course of a year. Most of the daughter nuclei will decay either while still in the reactor core or while in short-term storage at the power plant.
(c) $^{97}Zr \xrightarrow{\beta-} \,^{97}Nb \xrightarrow{\beta-} \,^{97}Mo$
$^{137}Te \xrightarrow{\beta-} \,^{137}I \xrightarrow{\beta-} \,^{137}Xe \xrightarrow{\beta-} \,^{137}Cs \xrightarrow{\beta-} \,^{137}Ba$
The half-life of ^{137}Cs is about 30 years. All the other isotopes involved have half-lives less than 1 day.
(d) From the half-lives, we can conclude that virtually all of the ^{97}Zr will have decayed to stable ^{97}Mo before the waste would be disposed of. The ^{137}Te will have decayed to ^{137}Cs and will still be radioactive due to the longer half-life of that isotope. Thus we might

estimate that the amount of radioactive waste actually would be 137/(97 + 137) of the original estimate, or 22 metric tons. This assumes that there is an easy and affordable way to separate the stable ^{97}Mo from the radioactive ^{137}Cs.

14.61 **(a)** $^{11}_{5}\text{B} + ^{1}_{1}\text{p} \rightarrow ^{12}_{6}\text{C} + \gamma$ (The gamma ray carries away excess energy.)

(b) Boron-11 occurs naturally, whereas tritium would have to be produced artificially to carry out deuterium–tritium fusion. Also, the reaction considered here does not release neutrons, making it easier to contain radiation.

14.63 In the event of a nuclear plant emergency, there is a strong chance that radioactive iodine could be released. Because iodine accumulates in the thyroid, this could damage the thyroids of people who breathe the air nearby. Potassium iodide tablets act by saturating the thyroid with normal iodine and preventing the gland from taking up appreciable amounts of radioactive iodine.

14.65 Designing a structure that must last for thousands of years poses many challenges. An engineer working on such a project must attempt to anticipate conditions far into the future. Dramatic climate change could occur on this time-scale, for example, and might impact structural requirements. Predicting human activity far in the future is even more difficult. Roads built in the days of horse-drawn carriages could not be expected to withstand modern traffic, for example.

14.67 The smallest ionization energies are on the order of 500 kJ/mol. This corresponds to a photon energy of about 10^{-18} J. This means that ultraviolet light is likely to be the lowest energy ionizing radiation.

14.69 $\gamma > \beta > \alpha$

14.71 Inhaling the smoke brings ^{210}Po into the lungs, and alpha particles emitted from inside the body are much more dangerous than those coming from external sources.

14.73 Modern electronic devices rely on the carefully controlled distribution and flow of electric charges. Cosmic rays can produce ions in these devices, disrupting their normal function. Chips for use in satellites are often encased in materials that are less susceptible to penetration by radiation to minimize damage by cosmic rays.

14.75 5.4×10^{-7} Ci, or 0.54 μCi

14.77 **(a)** 7200 J **(b)** 0.12 Gy

14.79 Because cosmic rays interact with gases in the atmosphere, natural radiation levels tend to increase with altitude. Radiation levels at the altitudes where commercial airliners fly are noticeably higher than those at ground level, and shielding planes to block all such radiation would be difficult and expensive.

14.81 **(a)** $^{18}_{9}\text{F} \rightarrow ^{18}_{8}\text{O} + ^{0}_{1}\beta + \nu$ **(b)** $^{11}_{6}\text{C} \rightarrow ^{11}_{5}\text{B} + ^{0}_{1}\beta + \nu$
(c) $^{13}_{7}\text{N} \rightarrow ^{13}_{6}\text{C} + ^{0}_{1}\beta + \nu$

14.83 $^{0}_{1}\beta + ^{0}_{-1}\beta \rightarrow 2\,\gamma$

14.85 Responses will vary depending on the sources listed. The method is called dual photon absorptiometry, or DPA, and is based on the fact that when ^{153}Gd decays it emits gamma ray photons at multiple energies. Comparing the intensities of the emission at two of these energies allows calculating bone density because calcium absorbs one of these photon energies more strongly than the other.

14.87 6.1×10^{23}

14.89 5.2×10^{-4} disintegrations per day (or one atom every 5.3 years)

14.91 **(a)** 201 counts per minute **(b)** 18 counts per minute **(c)** 21 counts per minute

14.93 $^{226}_{88}\text{Ra} \rightarrow ^{222}_{86}\text{Rn} + ^{4}_{2}\text{He}$
$^{222}_{86}\text{Rn} \rightarrow ^{218}_{84}\text{Po} + ^{4}_{2}\text{He}$

14.95 A larger sample would be needed for ^{192}Ir, because more of the originally prepared sample would have decayed in transit. (If we are comparing samples by mass, then we should also realize that if the masses of the two samples are equal, there would be more nuclei in the sample of ^{137}Cs because of its smaller molar mass.)

14.97 **(a)** $^{238}_{94}\text{Pu} \rightarrow ^{234}_{92}\text{U} + ^{4}_{2}\text{He}$ **(b)** 240 W

14.99 $0.5^{2120} \cong 8 \times 10^{-605}$

14.101 The most important piece of information needed is the half-life of ^{201}Tl. It might also be useful to know how quickly the isotope was expected to be used once it had been delivered.

14.103 Relevant information that could be found easily includes the half-life and decay mode for each isotope. It would also be important to understand what type and energy of radiation was most likely to be effective in killing tumor cells. This information might be more difficult to look up.

Glossary

A

Absolute temperature (5-6): Temperature measured on a scale whose zero point is absolute zero. Kelvin is the most common absolute temperature scale, and Rankine is another example.

Absorbed dose (14-7): The amount of any type of radiation absorbed in any kind of material.

Acceptor level (8-3): Energy state associated with the dopant atoms and lying slightly above the valence band in a p-type semiconductor.

Accuracy (1-3): How close the observed value is to the "true" value.

Acid (3-3): Any substance that dissolves in water to produce H_3O^+ (or H^+) ions. See also *Brønsted–Lowry acid.*

Acid ionization constant (K_a) (12-7): Equilibrium constant for the reaction of a weak acid with water, producing H_3O^+ and the conjugate base of the weak acid.

Actinides (2-5): The elements thorium through lawrencium, with valence electrons in the 5*f* subshell. They appear as the second row of elements below the main body of the periodic table.

Activated complex (11-5): The arrangement of atoms corresponding to the maximum potential energy along the lowest energy path from reactants to products. Also called the transition state.

Activation barrier (11-5): See *activation energy*.

Activation energy (E_a) (11-5): The minimum energy that reactant molecules must possess to undergo a particular chemical reaction.

Active electrolysis (13-6): Electrolysis in which the electrode materials are reactants and products in the cell reaction.

Actual yield (4-4): The amount of product obtained in a chemical reaction.

Addition polymerization (8-6): Reaction in which monomers are added to one another to form a polymer, and no other products are formed.

Addition reaction (2-6): Reaction in which two or more smaller molecules combine to form a single larger product molecule.

Adhesion (8-5): Attractive forces between two dissimilar substances or surfaces.

Admixture (12-1): Chemical substances added to concrete to give it certain desirable characteristics or properties.

Alkali metals (2-5): Elements of Group 1, the first column of the periodic table, except hydrogen: Li, Na, K, Rb, Cs, and Fr.

Alkaline battery (13-5): The most common single-use (primary) battery, so named because it uses a paste of KOH as the electrolyte.

Alkaline earth metals (2-5): Elements of Group 2, the second column of the periodic table: Be, Mg, Ca, Sr, Ba, and Ra.

Alkanes (4-1): Hydrocarbons in which the carbon atoms are all linked together by single bonds.

Alpha decay (14-2): A radioactive decay process in which an atom's nucleus emits an alpha particle (^4_2He) and that reduces the atomic number by two and the mass number by four.

Alpha particle (14-2): An energetic particle with a charge of 2+ and a mass number of 4 emitted in some radioactive decays. A helium nucleus.

Alpha ray (14-2): Radiation consisting of alpha particles.

Alternating copolymer (8-6): Polymer in which molecules contain a regular alternating pattern of two different monomers.

Amorphous solid (8-2): Noncrystalline solid in which the arrangement of atoms or molecules has no long-range order.

Amphoteric (12-7): Able to react as either an acid or a base. (Also referred to as *amphiprotic.*)

Amplitude (6-2): The size or "height" of a wave.

Anion (2-3): A negatively charged atom or group of atoms.

Anode (13-2): Electrode at which oxidation takes place.

Antibonding molecular orbital (8-3): An orbital resulting from the combination of atomic orbitals from two or more atoms that has higher energy than any of the atomic orbitals from which it was formed.

Aqueous solution (3-3): Solution in which water is the solvent.

Arrhenius behavior (11-5): Variation of a rate constant with temperature in accordance with the Arrhenius equation.

Arrhenius equation (11-5): Mathematical description of the dependence of the rate constant on temperature: $k = Ae^{E_a/RT}$.

Atactic polymer (8-6): A polymer in which substituent groups attached to the polymer chain are not arranged in a regular or ordered fashion.

Atmosphere (atm) (5-2): Unit of pressure equal to 760 torr or 101,325 Pa. Normal atmospheric pressure at sea level is approximately 1 atm.

Atomic absorption spectroscopy (AAS) (6-1): Analytical technique that determines elemental composition from the specific wavelengths of light absorbed by a sample.

Atomic mass unit (amu or u) (2-2): Unit used to express the relative masses of atoms. The amu is defined as one-twelfth the mass of an atom of ^{12}C, and 1 amu = 1.6606×10^{-24} g.

Atomic number (*Z*) (2-2): The number of protons in the nucleus of an atom of a particular element.

Atomic spectrum (6-3): The pattern of wavelengths of light absorbed or emitted by a particular element.

Atom (1-2): An unimaginably small particle that could not be made any smaller and still behave as a chemical system. Atoms are the smallest particles that can retain an element's chemical identity.

Aufbau principle (6-5): Description of the process by which electrons fill orbitals in order of increasing energy.

Average rate (11-2): Rate of reaction measured over some period of time.

Average speed (5-6): The mean speed of molecules in a gas; can be found from the Maxwell-Boltzmann distribution of speeds.

Avogadro's law (5-3): At fixed temperature and pressure, the volume of a gas is proportional to the number of molecules (or moles) of gas present.

Avogadro's number (N_A) (3-4): The number of particles in a mole of any substance (6.022×10^{23}).

B

Background radiation (14-7): Radiation originating from both natural and man-made sources to which an entire population is routinely exposed.

Band gap (8-3): Difference in energy between the valence band and the conduction band in a solid.

Band of stability (4-4): A region in the chart of the nuclides in which all known stable nuclides are located. The band of stability lies near $N/Z = 1$ for light nuclei but curves toward higher N/Z ratios for heavier elements.

Band theory (8-3): Description of chemical bonding in solids in terms of bands of orbitals. Used to explain the electrical properties of metals, insulators, and semiconductors.

Barometer (5-2): A device that measures atmospheric pressure.

Base (3-3): Any substance that dissolves in water to produce OH^- ions.

Base ionization constant (K_b) (12-7): Equilibrium constant for the reaction of a weak base with water, producing OH^- and the conjugate acid of the weak base.

Battery (13-5): An electrochemical cell or series of cells in which a spontaneous redox reaction generates an electric current.

Becquerel (Bq) (14-3): The SI unit of nuclear activity, equal to one nuclear disintegration per second.

Beta decay (14-2): A radioactive decay process in which a neutron decays into a proton and an electron (or beta particle, ${}_{-1}^{0}\beta$). The beta particle and an antineutrino are emitted while the proton remains in the nucleus, thus increasing the atomic number by one and leaving the mass number unchanged.

Beta particle (14-2): An energetic particle with a charge of 1– and a mass number of 0 emitted in some radioactive decays. An electron emitted from the nucleus.

Beta ray (14-2): Radiation consisting of beta particles.

Bimolecular (11-6): Describes an elementary step involving a collision between two reactant molecules.

Binary compound (2-7): Any chemical substance containing two different elements.

Binding energy (6-2): Amount of energy required to remove an electron from a metal surface. Also called the *work function*. See also *nuclear binding energy*.

Biocompatibility (7-1): Ability of a material to coexist with a natural biological system into which it has been implanted.

Biofuel (3-1): Fuel whose energy content is derived from carbon fixation in plants.

Biomass (3-1): Biological material from living or recently living organisms, most commonly from plants.

Block copolymer (8-6): Polymer in which molecules contain long sections or blocks of two or more different monomers.

Body-centered cubic (bcc) (8-2): Crystal lattice in which particles are arranged at the corners of a cube and also at the center of the cube.

Bohr model (6-3): Early model of atomic structure that proposed that electrons are restricted to circular orbits of specific sizes. Although this model did account for some features of atomic spectra, it is no longer viewed as correct.

Bond energy (7-3): The amount of energy released when isolated atoms form a covalent bond. Also equal to the amount of energy that must be supplied to break that bond.

Bond length (7-3): The distance between the nuclei of two bonded atoms.

Bonding molecular orbital (8-3): An orbital resulting from the combination of atomic orbitals from two or more atoms that has lower energy than any of the atomic orbitals from which it was formed.

Bonding pair (7-3): A pair of electrons shared in a covalent bond between two atoms.

Boundary (9-3): The line or surface—often hypothetical—that separates the system and the surroundings in a thermodynamic problem.

Boyle's law (5-3): At constant temperature, the pressure and volume of a fixed amount of gas are inversely proportional.

British thermal unit (Btu) (9-1): A unit of energy defined as the amount of energy needed to raise the temperature of 1 lb of water by 1°F.

Brønsted–Lowry acid (12-7): Any species that donates a proton (H^+ ion) in a chemical reaction.

C

Calorie (cal) (9-2): A unit of energy originally defined as the amount of energy required to heat 1 g of water from 14.5 to 15.5°C. 1 cal = 4.184 J. (Note that a "food calorie," abbreviated "Cal," is actually a kilocalorie. 1 Cal = 4.184 kJ)

Calorimetry (9-4): Experimental technique for measuring the heat flow into or out of a thermodynamic system.

Carbon sequestration (3-6): The process of capturing atmospheric CO_2 for long-term storage.

Catalyst (11-7): Any substance that increases the rate of a chemical reaction but is neither created nor destroyed in the process.

Cathode (13-2): Electrode at which reduction takes place.

Cathodic protection (13-3): A method for preventing corrosion whereby the metal to be protected is connected electrically to a more reactive metal. The protected metal thus becomes the cathode, so it cannot be oxidized.

Cation (2-3): A positively charged atom or group of atoms.

Cell notation (13-2): Shorthand notation for an electrochemical cell that shows the electrodes, gases, and solutions in the cell reaction as well as the phase boundaries separating them.

Cell potential (13-2): The electromotive force of an electrochemical cell. The amount of work the cell can do per coulomb of charge.

Chain reaction (14-6): A self-sustaining series of reactions. In nuclear fission, a situation in which one or more neutrons released in one fission event induce further fission.

Chapman cycle (11-1): A series of chemical reactions responsible for the formation and destruction of ozone in the stratosphere.

Charles's law (5-3): The volume and temperature of a fixed amount of gas at constant pressure are directly proportional.

Chart of the nuclides (14-4): A plot of atomic number (Z) versus neutron number (N) for all known nuclei.

Chemical bond (2-4): Attractive force between two atoms holding them together to form a molecule or a part of a molecule.

Chemical compound (2-4): A pure substance made up of atoms of two or more elements joined together by chemical bonds.

Chemical energy (9-2): Potential energy stored in chemical bonds, which can be released through an exothermic reaction.

Chemical equation (3-2): Symbolic representation of a chemical reaction.

Chemical formula (2-4): Symbolic description of a molecule or compound showing its constituent elements and the relative number of each.

Chemical nomenclature (2-7): The systematic assignment of names to chemical compounds on the basis of their constituent elements or ions and their structures.

Chemical property (1-2): Property of a substance that is associated with the types of chemical changes that the substance undergoes.

Chemical reaction (3-2): Process in which substances (reactants) are transformed into other substances (products) by the combination, rearrangement, or separation of atoms.

Coherent light (6-8): Light in which all waves are in phase, such as the light produced by a laser.

Cohesion (8-5): Attractive forces between two similar substances or surfaces.

Combustion (1-2): Reaction in which an element or compound burns in oxygen or air. Complete combustion of hydrocarbons produces CO_2 and H_2O as products.

Common ion effect (12-6): Shift in equilibrium when one or more ions that are part of the equilibrium are introduced from an outside source.

Compound nucleus (14-6): A high-energy intermediate formed in a nuclear reaction, similar to an activated complex in a chemical reaction.

Concentration (3-3): A measure of the relative quantities of solute and solvent in a solution.

Condensation (9-5): The process by which a molecule from the gas phase enters the liquid phase. Also the process by which a bulk gas is converted to a liquid.

Condensation polymer (8-6): A polymer made from monomers containing two or more functional groups through a reaction in which water (or another small molecule) is formed as a byproduct.

Conduction band (8-3): A band of orbitals or energy states at higher energy than the valence band and typically unoccupied in the ground state of a solid.

Conjugate acid (12-7): Molecule or ion formed by the addition of a hydrogen ion to its conjugate base.

Conjugate acid–base pair (12-7): Two chemical species related to one another by the loss or gain of a single hydrogen ion.

Conjugate base (12-7): Molecule or ion formed by the removal of a hydrogen ion from its conjugate acid.

Control rod (14-6): A rod inserted into a nuclear reactor to regulate the rate of fission. Control rods are made from elements that can absorb neutrons without undergoing subsequent decay.

Coordination number (8-2): The number of lattice points immediately adjacent to any given lattice point in the structure of a solid.

Copolymer (8-6): A polymer made from more than one type of monomer.

Core electrons (6-5): Electrons in the filled inner shell(s) of an atom.

Corrosion (13-1): The degradation of metals exposed to the environment by chemical oxidation.

Cosmic ray (14-1): Energetic particles impinging upon the atmosphere from beyond Earth.

Coulomb's law (2-3): Mathematical description of the force between two charged particles.

Covalent bond (2.4, 7.3): Attractive force resulting from the sharing of electrons between pairs of atoms.

Crevice corrosion (13-1): Corrosion in a small gap between two pieces of metal.

Criteria pollutants (5-1): Chemical compounds found in urban air that have a variety of negative effects on health, the environment, and property. The six principal criteria pollutants identified by the EPA are CO, NO_2, O_3, SO_2, Pb, and particulate matter (PM).

Critical mass (14-6): The minimum mass of a particular fissile fuel that can sustain a chain reaction.

Crystalline solid (8-2): Solid featuring regular, repeating geometric arrangements of atoms, ions, or molecules.

Curie (Ci) (14-3): A unit of nuclear activity, exactly equal to 3.7×10^{10} Bq.

D

Dalton's law of partial pressures (5-4): The total pressure of a mixture of gases is the sum of the partial pressures of the component gases.

Decay constant (14-3): The first-order rate constant for a radioactive decay. We use the symbol k in this text, but λ is also commonly used.

Decay series (14-4): A series of nuclear decays starting with an unstable isotope and eventually producing a stable nuclide.

Deductive reasoning (1-3): A thought process that takes two or more statements or assertions and combines them so that a clear and irrefutable conclusion can be drawn.

Degree of polymerization (8-6): The average number of repeating units in a polymer.

Density (ρ) (1-2): The ratio of mass to volume.

Differential rate law (11-3): Mathematical expression for the rate of a chemical reaction defined in terms of the rate of change in the concentration of a reactant or product.

Diffraction (6-4): Interaction of a wave with a regular pattern of lines or slits.

Dilution (3-5): Process in which solvent is added to a solution to decrease the solute concentration.

Dimensional analysis (1-5): Problem-solving strategy in which one inspects the units on all quantities in a calculation to check for correctness.

Dipole (7-4): Two equal and opposite electric charges separated by a short distance.

Dipole–dipole forces (8-4): Attractive interaction between two polar molecules or between two polar regions of a single large molecule.

Dispersion forces (8-4): Attractive forces between molecules resulting from the interaction of transient (induced) dipoles. Also called London forces or van der Waals forces.

Dissociation reaction (3-3): Chemical reaction in which an ionic solid dissolves in water and dissociates into its constituent ions.

Distribution function (5-6): Mathematical function describing the variation of some property across a population.

Donor level (8-3): Energy state associated with dopant atoms and lying slightly below the conduction band in an n-type semiconductor.

Doping (8-3): The controlled addition of a very small concentration of one element (the dopant) to an intrinsic semiconductor such as silicon or germanium.

Dosimeter (14-7): Any device used to measure an individual's cumulative exposure to radiation.

Double bond (7-3): A covalent bond formed by the sharing of two pairs of electrons between the same pair of atoms.

Ductile (8-3): Easily drawn or pulled into a wire.

Dynamic equilibrium (8-5, 12-2): State achieved when the rates of the forward and backward processes in a chemical reaction become equal to one another.

E

Effective nuclear charge (Z_{eff}) (6-5): Net positive charge "experienced" by outer shell electrons in a many-electron atom.

Elastic modulus (1-6): A measure of the stiffness of a material, defined as the ratio of stress to strain.

Electrical grid (9-8): An interconnected system for delivering electricity from producers to consumers.

Electrochemistry (13-2): The study of the connection between oxidation–reduction reactions and the flow of electrons.

Electrode (13-2): An electrically conducting site at which either oxidation or reduction takes place.

Electrolysis (13-6): Process in which an electric current is used to drive a redox reaction in its nonspontaneous direction.

Electrolyte (3-3): A substance that ionizes or dissociates in water to produce an aqueous solution that conducts electricity.

Electromagnetic spectrum (6-2): The various forms of light, all of which consist of oscillating electric and magnetic fields traveling at the speed of light (2.998×10^8 m/s).

Electromotive force (EMF) (13-2): The difference in electric potential between the two electrodes in a galvanic cell. See also *cell potential*.

Electron affinity (6-7): The amount of energy needed to add an electron to a neutral atom to form an anion.

Electron capture (14-2): A radioactive decay process in which an atom's nucleus captures an inner shell electron, reducing the atomic number by one and leaving the mass number unchanged. Also called *K capture*.

Electron configuration (6-5): A complete description of the distribution of electrons among orbitals in an atom or ion.

Electronegativity (7-4): A relative measure of the tendency of an atom in a molecule to attract the shared electrons in a covalent bond.

Electrons (2-2): Negatively charged subatomic particles that occupy most of the volume of an atom.

Electron volt (eV) (14-1): A unit of energy equal to 96.5853 kJ mol^{-1}.

Electroplating (13-6): The process of depositing a thin coat of a metal on a surface by electrolysis.

Elemental analysis (3-5): Experimental measurement of the mass percentage of each element in a compound.

Elementary step (11-6): Chemical reaction taking place through a single collision between molecules, ions, or atoms. A series of elementary steps makes up a *reaction mechanism*.

Element (1-2): A substance that cannot be broken down into two or more new substances by chemical or physical processes.

Empirical formula (2-4): Chemical formula that shows the composition of a compound in terms of the simplest whole-number ratio of the atoms of each element.

Endothermic (9-5): A process in which heat flows into a thermodynamic system. For a chemical reaction, this means ΔH is positive.

Energy density (9-7): The amount of energy that can be released per gram of fuel burned.

Energy economy (9-1): The production and use of energy in the developed world.

Enthalpy (H) (9-5): A state function defined as $E + PV$ and equal to the heat flow at constant pressure.

Enthalpy diagram (9-6): Conceptual device for visualizing the enthalpy changes in a chemical reaction.

Enthalpy driven (10-6): Description of a process for which ΔH and ΔS are both negative and which is spontaneous only at low temperatures.

Entropy (S) (10-3): A state function that measures the dispersal of energy in a system. Entropy is correlated with the degree of randomness or disorder of a system on the molecular level.

Entropy driven (10-6): Description of a process for which ΔH and ΔS are both positive and which is spontaneous only at high temperatures.

Equilibrium constant (K_{eq}, K) (12-3): The ratio of concentrations of products to those of reactants; each concentration is raised to a power that matches its stoichiometric coefficient in the balanced chemical equation. Equilibrium constants vary with temperature.

Equilibrium expression (12-3): Equation defining the equilibrium constant in terms of the concentrations of the various substances involved in the reaction.

Equivalent dose (14-7): A measure of radiation exposure that attempts to measure the tissue damage likely to be done by a particular type of radiation. Equal to the absorbed dose multiplied by a *quality factor*.

Etching (8-7): Chemical process designed to selectively remove material from an object, as in the fabrication of MEMS devices.

Excited state (6-3): Any state formed when an atom or molecule absorbs energy, as when an electron is promoted to a higher energy orbital.

Exothermic (9-5): A process in which heat flows out of a thermodynamic system. For a chemical reaction, this means that ΔH is negative.

Exposure (14-7): A measure of radiation exposure based on the ionization of molecules of air, expressed in coulombs per kilogram of air.

F

Face-centered cubic (fcc) (8-2): Crystal lattice in which particles are arranged at the corners of a cube and also at the center of each face of the cube. Also called *cubic close packing*.

Factor-label method (1-5): See *dimensional analysis*.

Faraday constant (F) (13-3): The electric charge on 1 mole of electrons: 96,485 C mol^{-1}.

First law of thermodynamics (9-3): Energy can be transformed from one form to another but cannot be created or destroyed; the total energy of the universe is constant.

First-order reaction (11-3): Any reaction whose rate is given by a rate law of the form, Rate $= k$[A], where A represents a chemical species in the reaction (usually a reactant).

Fissile (14-6): Able to undergo nuclear fission. Also called *fissionable*.

Formation reaction (9-5): A chemical reaction by which 1 mole of a compound is formed from its constituent elements in their standard states.

Formula unit (2-4): The smallest whole number ratio of anions and cations in an ionic compound.

Free radicals (2-8): Extremely reactive chemical species containing one or more unpaired electrons.

Freezing (9-5): Phase change in which a liquid is converted to a solid.

Frequency (ν) (6-2): The number of complete cycles of a wave passing a given point per unit time.

Frequency factor (11-5): The preexponential term (A) in the Arrhenius equation. This term accounts for the frequency of collisions as well as any geometric or orientational constraints that influence the reaction rate. Also called the *preexponential factor*.

Fuel additives (4-6): Substances added to gasoline or other fuels, typically intended to improve performance or reduce exhaust emissions.

Fuel cell (13-5): An electrochemical cell in which reaction between a fuel and an oxidant produces electricity. Unlike a battery, a fuel cell uses a replenishable external supply of reactants.

Fuel rod (14-6): Component of a nuclear reactor containing fissionable uranium fuel.

Functional group (2-6): An atom or group of atoms that imparts characteristic chemical properties to any organic compound in which it is found.

Fusion (9-5): The phase change in which a solid is converted to a liquid; melting. See also *nuclear fusion*.

G

Galvanic cell (13-2): Any electrochemical cell in which a spontaneous chemical reaction generates an electric current.

Galvanic corrosion (13-1): Degradation of metal parts through oxidation that occurs only when two different metals are in contact with each other in the presence of an appropriate electrolyte.

Galvanized steel (13-3): Steel coated with a thin layer of zinc to prevent corrosion.

Gamma ray (14-2): Radiation consisting of high-energy photons.

Gas (1-2): State of matter in which a substance has no fixed shape and expands to occupy the entire volume of its container.

Geiger counter (14-7): A radiation detector generally sensitive only to alpha and beta radiation. Also called a *Geiger-Mueller counter*.

Gibbs free energy (G) (10-6): A thermodynamic state function that can be used to predict the direction of spontaneous change at constant temperature and pressure. $\Delta G = \Delta H - T\Delta S$, and $\Delta G < 0$ for any spontaneous process.

Graft copolymer (8-6): Material whose molecules contain large segments of one type of molecule "grafted" onto or attached to chains of a second type of polymer. The graft copolymer will display properties of each of the constituent polymers.

Ground state (6-3): The lowest energy state of an atom or molecule.

Group (2-5): Vertical column of the periodic table.

H

Half-life ($t_{1/2}$) (11-4, 14-3): The time it takes for the concentration of a reactant to fall to one-half of its original value. For a first-order reaction, the half-life is independent of the initial concentration.

Half-reaction (13-2): A reaction showing only the oxidation or reduction process in a redox reaction.

Halogens (2-5): The elements in Group 17 of the periodic table: F, Cl, Br, I, and At.

Heat (q) (9-2): The flow of thermal energy between two objects, from the warmer to the cooler because of the difference in their temperatures.

Heat capacity (9-4): The amount of energy that must be transferred to an object or substance to raise its temperature by a specified amount (usually 1 K).

Heat of formation ($\Delta H_f°$) (9-5): Enthalpy change for the formation of 1 mole of a substance from its constituent elements in their thermodynamic standard states.

Heat of reaction ($\Delta H°$) (9-5): Enthalpy change for a chemical reaction.

Heisenberg's uncertainty principle (6-4): It is impossible to determine both the position and momentum of an electron simultaneously and with arbitrary accuracy.

Hess's law (9-6): The enthalpy change for any process is independent of the way the process is carried out.

Heterogeneous catalyst (11-7): Catalyst in a phase different from the reacting substances.

Heterogeneous equilibrium (12-3): Chemical equilibrium in which not all reactants and products are in the same phase.

High-density polyethylene (HDPE) (2-8): Form of polyethylene featuring long, straight chain molecules. Because these chains pack together more tightly, HDPE is harder and stronger than low-density polyethylene.

Homogeneous catalyst (11-7): Catalyst in the same phase as the reacting substances.

Homogeneous equilibrium (12-3): Chemical equilibrium in which all reactants and products are in the same phase.

Hund's rule (6-5): Within a subshell of an atom, electrons occupy orbitals individually first before pairing.

Hybrid orbitals (7-7): Orbitals formed by combining two or more orbitals from the same atom.

Hybridization (7-7): The process of forming hybrid orbitals. The idea that when two or more atoms approach one another closely enough to form chemical bonds, the interactions can be strong enough to reshape the orbitals of those atoms.

Hydrates (2-4): Compounds in which one or more water molecules are incorporated into the crystal lattice.

Hydrocarbons (2-6, 4-1): Molecules containing only carbon and hydrogen atoms.

Hydrogen bonding (8-4): An unusually strong dipole–dipole interaction between a hydrogen atom bonded to F, N, or O and a lone pair of electrons on another highly electronegative atom such as F, N, or O.

I

Indicator (4-5): A dye added to a titration to show when a reaction is complete.

Inductive reasoning (1-3): Thought process that begins with a series of specific observations and attempts to generalize to a larger, more universal conclusion.

Inorganic chemistry (2-6): The study of all elements, besides carbon, and their compounds.

Insoluble (3-3): Description of a solid that does not dissolve in solution to any significant extent.

Instantaneous rate (11-2): The rate of a chemical reaction measured at a single instant.

Insulator (8-3): A material that does not conduct electricity, characterized by a large energy gap between its valence and conduction bands.

Integrated rate law (11-4): Mathematical equation that shows how the concentration of a reactant or product changes as a function of time.

Intermolecular forces (8-4): Attractive forces between molecules.

Internal energy (*E*) (9-2): The combined kinetic and potential energies of the atoms and molecules that make up a substance or object.

Ion (2-3): An atom or group of atoms with a net electrical charge.

Ionic bonds (7-2): Chemical bonds formed by the mutual electrostatic attraction of oppositely charged ions.

Ionic bonding (2-4): Bonding in which one or more electrons are transferred from one species to another.

Ionization energy (6-7): The amount of energy required to remove an electron from a free atom in the gas phase.

Ionization gauge (5-7): A pressure sensor in which gas molecules are first ionized and then collected and measured as a current. Suitable for measurement of low pressures, between 10^{-5} and 10^{-11} torr.

Ionizing power (14-7): A measure of the ability of a particular type of radiation to produce ions in matter with which it interacts.

Irreversible (10-6): A process, often far from equilibrium, which cannot be reversed by a small incremental change in any variable.

Isomers (4-1): Compounds having the same molecular formula but in which the atoms are arranged differently.

Isotactic polymer (8-6): A polymer in which substituent groups are all on the same side of the polymer chain.

Isotopes (2-2): Atoms of the same element that have different numbers of neutrons.

Isotopic abundance (2-2): The percentage of the atoms in a naturally occurring sample of an element that consists of a specified isotope.

J

Joule (J) (9-2): The SI unit of energy; $1\ \text{J} = 1\ \text{kg m}^2/\text{s}^2$

K

K capture (14-2): A radioactive decay process in which an atom's nucleus captures an inner-shell electron, reducing the atomic number by one and leaving the mass number unchanged. Also called *electron capture*.

Kinetic energy (9-2): Energy from the motion of an object; given by $\frac{1}{2}mv^2$, where m is the mass of the object and v is its velocity.

Kinetic–molecular theory (5-6): Description of a gas as a collection of a very large number of atoms or molecules in constant, random motion. The ideal gas law can be derived from the postulates of the kinetic theory.

L

Lanthanides (2-5): The elements cerium through lutetium, whose valence electrons are in the 4*f* subshell. They appear as the first row of elements below the main body of the periodic table.

Laser (6-8): A source of high-intensity, coherent, monochromatic light. Originally an acronym for "*l*ight *a*mplification by *s*timulated *e*mission of *r*adiation."

Lattice (2-4, 7-2): A regular, periodic arrangement of atoms, ions, or molecules in a crystalline solid.

Lattice energy (7-2): Enthalpy change from the formation of a crystal lattice from its separate constituent ions.

Law (1-3): A statement of an accepted scientific principle, based on evidence from a significant number of experimental observations.

Law of conservation of matter (3-2): Matter is neither created nor destroyed in ordinary chemical reactions.

Lead-acid storage battery (13-5): Secondary (rechargeable) battery used in the ignition system of automobiles, whose name derives from the use of lead in the electrodes and sulfuric acid as the electrolyte.

LeChatelier's principle (12-5): When a system at equilibrium is stressed, it responds by reestablishing equilibrium so that the applied stress is partially offset.

Lewis dot symbol (7-3): Visual representation of the valence electrons of an atom.

Lewis structure (7-3): Visual representation of the structure of a covalently bonded molecule or ion that shows the distribution of all valence electrons.

Light-emitting diode (LED) (6-8): Semiconductor device in which monochromatic light is produced as electrons undergo a transition across the band gap.

Limiting reactant (4-3): The reactant that is completely consumed in a reaction. The available amount of it determines the maximum possible reaction yield.

Line structure (2-6): Visual representation of organic compounds in which carbon atoms and the hydrogen atoms attached to them are not shown explicitly.

Liquid (1-2): State of matter in which a substance has a fixed volume but flows to take on the shape of its container.

Lithium battery (13-5): A commercially available primary battery with a lithium anode and a manganese dioxide cathode. Used in calculators, watches, and medical devices.

Lithium-ion battery (13-8): A common secondary battery in which lithium ions move from the anode to the cathode during use and in the opposite direction during recharging. Used in cell phones, laptop computers, and other portable electronic devices as well as some electric vehicles.

Lone pair (7-3): Paired valence electrons that are associated with a single atom and not involved in a covalent bond; also called *nonbonding electrons*.

Low-density polyethylene (LDPE) (2-8): Form of polyethylene featuring branched chain molecules. Because these chains do not pack together as tightly, LDPE is softer and more flexible than high-density polyethylene.

M

Macroscopic perspective (1-2): Viewpoint of chemistry focusing on samples of matter that are large enough to be seen, measured, or handled easily.

Magic number (14-5): A number of neutrons or protons associated with unusually stable nuclei. Known magic numbers are 2, 8, 20, 28, 50, 82, 126, and 184.

Magnetic quantum number (m_ℓ) (6-4): One of the three quantum numbers identifying an atomic orbital. Specifies the spatial orientation of an orbital. Allowed values of m_ℓ are related to the value of the secondary quantum number (ℓ).

Main group elements (2-5): Elements of Groups 1, 2, and 13 − 18, for which the highest lying valence electron occupies an *s* or *p* orbital. Also called *representative elements.*

Malleability (1-2): Ability of a material to be shaped or formed by the application of pressure, as through hammering or rolling.

Malleable (8-3): Able to be shaped or formed by the application of pressure, by hammering or rolling, for example.

Manometer (5-7): A device used to measure pressure.

Mass defect (14-5): The difference in mass between a nucleus and the sum of the masses of the neutrons and protons that make up the nucleus. Related to nuclear binding energy by Einstein's equation.

Mass density (1-3): The ratio of the mass of a substance to its volume. Also known simply as *density*.

Mass number (2-2): The combined total of protons and neutrons in an atom.

Matter (1-2): Anything that has mass and occupies space.

Maxwell-Boltzmann distribution (5-6): Mathematical function describing the fraction of molecules traveling at a particular speed in a gas. Depends on the mass of the molecules and the temperature of the gas.

Mean free path (5-6): The average distance a particle travels between collisions with other particles.

Metal (2-5): An element that is malleable, ductile, conducts electricity, and tends to form cations in its chemical compounds.

Metallic bonding (2-4): Bonding in a solid metal, where the nuclei and some fraction of their electrons comprise a positively charged "core" localized at these lattice points, and valence electrons form a surrounding "sea" of negative charge.

Metalloid (2-5): An element that displays some properties of metals and some properties of nonmetals. Sometimes called a *semimetal.*

Methyl tertiary-butyl ether (MTBE) (4-6): Compound with the formula $C_5H_{12}O$ that has been used as an oxygenating additive in gasoline.

Micro-electrical-mechanical systems (MEMS) (8-7): Miniaturized mechanical or electro-mechanical devices made using techniques of microfabrication, with typical dimensions on the micron to millimeter scale.

Microscopic perspective (1-2): Viewpoint of chemistry focusing on samples of matter at the atomic and molecular level, where samples cannot be seen, measured, or handled easily. Note that this scale is smaller than the resolution of a traditional microscope. Also called the *particulate perspective*.

Microstate (10-3): One possible way for a collection of particles to assume a given total energy.

Molar concentration (3-5): See *Molarity.*

Molar heat capacity (*C*) (9-4): The amount of heat required to raise the temperature of 1 mole of a substance by 1 degree (°C or K).

Molar mass (3-4): The average mass of 1 mole of any element or compound.

Molar solubility (12-6): The concentration of a solid in a saturated solution, expressed in moles per liter.

Molarity (M) (3-5): Unit of concentration defined as the number of moles of solute per liter of solution.

Mole (mol) (3-4): The particular quantity by which chemists usually choose to count atoms or molecules. The number of particles in 1 mole is defined as the number of atoms in *exactly* 12 grams of ^{12}C, which is also called *Avogadro's number.*

Mole fraction (*X*) (5-4): The ratio of the number of moles of one component to the total number of moles in a mixture.

Mole ratio (3-4, 4-2): A stoichiometric ratio relating the number of moles of one reactant or product in a chemical reaction to the number of moles of some other reactant or product in the same reaction. Obtained from a balanced chemical equation.

Molecular equation (3-3): Chemical equation in which all substances are shown as molecular formulas.

Molecular formula (2-4): Chemical formula showing the number of atoms of each element in one molecule of a compound.

Molecular shape (7-8): The geometry of a molecule or ion, defined by the positions of its constituent atoms.

Molecule (1-2, 2-4): Two or more atoms joined together by chemical bonds, forming the smallest particle that retains the chemical properties of a substance.

Monochromatic (6-8): Light consisting only of a single wavelength or color, such as the light emitted by a laser.

Monomer (2-1): The small repeating unit from which a polymer molecule is made.

Most probable speed (5-6): The speed at which the most molecules move in a gas; the maximum in the Maxwell-Boltzmann distribution of speeds.

N

n-type semiconductor (8-3): Material formed by doping silicon or germanium with an element such as phosphorus or arsenic that adds extra valence electrons. So named because the charge carriers in an n-type semiconductor are negatively charged electrons.

Nanotube (8-1): A long, cylindrical carbon molecule. Nanotubes hold promise for a wide range of future engineering applications.

Nernst equation (13-3): Mathematical equation relating the cell potential to the concentrations and pressures of chemical species in the cell reaction.

Net ionic equation (3-3): A chemical equation in which those ions and molecules actually involved in the reaction are shown, but any spectator ions are omitted.

Neutralization (3-3): A chemical reaction in which an acid and a base react to produce water and a salt.

Neutrons (2-2): Electrically neutral subatomic particles found in an atom's nucleus.

Nickel-metal-hydride (NiMH) battery (13-5): Secondary (rechargeable) battery used in many small electronic devices; involves the oxidation of hydrogen from a metal hydride and the reduction of nickel oxyhydroxide (NiO(OH)). Compared with nicad batteries, NiMH eliminates the use of toxic cadmium and is less prone to memory effects.

Noble gases (2-5): The elements in Group 18 of the periodic table: He, Ne, Ar, Kr, Xe, and Rn. Also called *rare gases*.

Node (6-4): A point or plane where the amplitude of a wave is equal to zero. For atomic orbitals, a point or plane where there is zero probability of finding an electron.

Nonattainment area (5-1): A region where detected concentrations of one or more criteria pollutants exceed the limits specified in the EPA's primary standards.

Nondestructive testing (6-1): Analytical techniques that can be used to determine the chemical composition of a sample or object without damaging it in the process.

Nonelectrolyte (3-3): A substance that dissolves in water to produce a solution that does not conduct electricity.

Nonmetal (2-5): An element that does not display the chemical and physical properties of metals.

Nuclear binding energy (14-5): The amount of energy released by the formation of an atomic nucleus from isolated nucleons. Also the amount of energy required to separate a nucleus into its constituent nucleons.

Normal boiling point (8-5): The temperature at which a substance's vapor pressure is equal to 1 atmosphere.

Nuclear fission (14-5): A process in which a single heavy nucleus splits into two or more smaller and more stable nuclei.

Nuclear fusion (14-5): A process in which two or more smaller nuclei combine to form a single heavier and more stable nucleus.

Nuclear reaction (14-1): Any process in which an atom or nucleus is converted into a different chemical element.

Nucleus (2-2): The small, dense core of an atom, containing neutrons and protons.

Nuclide (14-2): A specific nucleus with the numbers of protons and neutrons specified. The nucleus of a particular isotope.

O

Octet rule (7-3): In forming chemical bonds, main group elements gain, lose, or share electrons to achieve a configuration in which they are surrounded by eight valence electrons.

Operator (6-4): Notation used to indicate one or more operations to be carried out on a mathematical function.

Orbital (6-4): The wave representation of electrons in an atom; a mathematical or visual depiction of a region of space in which there is a high probability of finding an electron.

Orbital overlap (7-6): Interaction between orbitals—and hence electrons—from two different atoms responsible for the formation of a covalent bond.

Order of reaction (11-3): The way in which a reaction rate depends on the concentration(s) of the substance(s) involved, expressed as the exponents on the concentration term(s) in the rate law. Typically a small integer or a small rational fraction.

Organic chemistry (2-6): The study of compounds of the element carbon.

Oxidation (13-2): The loss of electrons by some chemical species in a chemical reaction.

Oxidation–reduction reaction (13-2): Chemical reaction involving the transfer of electrons from one species to another. Also called a *redox reaction*.

Oxyanions (2-7): Polyatomic anions that contain oxygen.

Oxygenated fuels (4-6): Blends of gasoline or other fuels to which oxygen-containing additives such as methanol, ethanol, or MTBE have been added to reduce undesirable exhaust emissions.

P

p-n junction (8-3): Point at which a piece of p-type material meets a piece of n-type material; an important building block for integrated circuits.

p-type semiconductor (8-3): Material formed by doping silicon or germanium with an element such as aluminum or gallium that reduces the number of valence electrons. So named because the charge carriers in a p-type semiconductor can be thought of as positively charged vacancies or "holes."

Packing efficiency (8-2): The percentage of the total volume that is actually occupied by the atoms in a given crystal lattice.

Partial pressure (5-4): The pressure that would be exerted by one component of a gas mixture if only that component were present at the same temperature and volume as the mixture.

Particulate perspective (1-2): Viewpoint of chemistry focusing on samples of matter at the atomic and molecular level, where samples cannot be seen, measured, or handled easily. Also called the *microscopic perspective*.

Parts per billion (ppb) (1-4): A unit of concentration defined as the number of particles of a particular component per every billion (10^9) molecules of a mixture. Often used to describe the levels of pollutants or other trace components in air.

Parts per million (ppm) (1-4, 5-1): A unit of concentration defined as the number of particles of a particular component per every million (10^6) molecules of a mixture. Often used to describe the levels of pollutants or other trace components in air.

Pascal (Pa) (5-2): The SI unit for pressure. $1 \text{ Pa} = 1 \text{ N m}^{-2}$

Passive electrolysis (13-6): Electrolysis in which the electrodes are chemically inert materials that simply provide a pathway for electrons to enter and leave the electrolytic cell.

Pauli exclusion principle (6-5): No two electrons in an atom may have the same set of four quantum numbers.

Penetrating power (14-7): A measure of the depth to which a particular type of radiation is expected to penetrate in matter with which it interacts.

Percentage yield (4-4): The ratio of the actual yield to the theoretical yield in a particular chemical reaction, expressed as a percentage.

Periodic law (2-5): When the elements are arranged by atomic number, they display regular and periodic variation in their chemical properties.

Periodicity (2-5): Regular patterns in the chemical and physical behavior of groups of elements.

Periods (2-5): Horizontal rows of the periodic table.

Phase diagram (8-1): A diagram showing what state of an element is most stable under different combinations of temperature and pressure.

Phases (1-2): One of the three states of matter: solid, liquid, or gas.

Photochemical reaction (3-2, 5-1): A chemical reaction in which light provides the energy needed to initiate the reaction.

Photochemical smog (11-8): Common urban air condition characterized by a brownish haze; produced by a complex series of reactions involving ozone (O_3), oxides of nitrogen (NO_x), and volatile organic compounds (VOCs).

Photoelectric effect (6-2): Emission of electrons from metals illuminated by light of an appropriate wavelength.

Photon (6-2): A massless "particle" of light, whose energy is equal to Planck's constant times the frequency of the light ($h\nu$).

Physical property (1-2): Property that can be observed or measured while the substance being observed retains its composition and identity.

Pi (π) bond (7-6): A chemical bond in which the electron density is not concentrated along the line of centers of the two atoms, as occurs through the "sideways" interaction of two *p* orbitals.

Polar bond (7-4): A chemical bond between two atoms that have different electronegativities, such that one end of the bond takes on a partial positive charge and the other end takes on a partial negative charge. Also called a polar covalent bond.

Polarizability (8-4): Susceptibility of a molecule to perturbations of its charge distribution in the presence of an external electric field.

Poly(ethylene terephthalate) (PET) (10-8): A recyclable polyester commonly used in bottles for soft drinks and other liquids.

Poly(vinyl chloride) (PVC) (10-8): A thermoplastic polymer made from vinyl chloride monomer; widely used in construction materials, pipes, and for wire and cable insulation.

Polyatomic ion (2-3): A group of atoms that carries an electrical charge.

Polymer backbone (2-1): The long chain of atoms—usually carbon atoms—running the length of a polymer molecule.

Polypropylene (10-8): A thermoplastic polymer made from propylene monomer; widely used in syringes and other medical supplies because it can be sterilized easily.

Polystyrene (10-8): A thermoplastic polymer made from styrene monomer; used in plastic model kits and plastic cutlery. Can be aerated to produce a foam (Styrofoam®) and made into coffee cups or insulated food containers.

Portland cement (12-1): Cement most commonly used in the production of concrete, consisting of a mixture of oxides of calcium, silicon, and aluminum.

Positron (14-2): The antiparticle of an electron, having the same mass as an electron but with a charge of 1+. Also called a β^+ particle.

Positron emission tomography (PET) (14-8): A medical imaging technique in which trace amounts of a positron-emitting isotope are injected into the body so that they accumulate in a particular organ. Detection of gamma rays produced in the annihilation of positrons is used to construct images of the organ being studied.

Potential energy (9-2): Energy an object possesses because of its position.

Precipitation reaction (3-3): Chemical reaction in which an insoluble solid (called the precipitate) is formed by the reaction of two or more solutions.

Precision (1-3): The spread in values obtained from a measurement. A precise measurement will produce similar results upon repeated observations.

Preexponential factor (11-5): The preexponential term (A) in the Arrhenius equation. This term accounts for the frequency of collisions as well as any geometric or orientational constraints that influence the reaction rate. Also called the *frequency factor*.

Pressure (*P*) (5-2): The force exerted on an object per unit area.

Primary cell (or primary battery) (13-5): A single-use battery that cannot be recharged.

Primary standard (5-1): Science-based limit established by the EPA for acceptable levels of *criteria pollutants* based solely on health considerations. (See also *secondary standards*.) (Also a substance of very high purity that can be used as a reference in determining the concentration of other substances.)

Principal quantum number (*n*) (6-4): One of the three quantum numbers identifying an atomic orbital. Specifies the shell in which an orbital is found. Allowed values of n are positive integers.

Products (3-2): Substances formed in a chemical reaction that appear on the right-hand side of a chemical equation.

Protons (2-2): Positively charged subatomic particles found in an atom's nucleus. Also H^+ ions.

PV-work (9-2): Work done by the expansion of a gas against an external pressure.

Q

Quality factor (14-7): A variable to quantify the ability of a particular type of radiation to cause tissue damage. Used in calculating the equivalent dose. Also called the *relative biological effectiveness*.

Quantized (6-3): Restricted to certain allowed values.

Quantum number (6-4): One of a number of indexes that identify a particular quantum state. An electron in an atom can be specified by a set of four quantum numbers.

R

Radioactive decay (14-1): The spontaneous emission of subatomic particles or photons by an unstable nucleus, resulting in the formation of a new nuclide.

Random error (1-3): Uncertainty associated with the limitations of the equipment with which a measurement is made; random error may cause the measurement to be either too high or too low, and so it can be minimized by averaging repeated measurements.

Rate constant (*k*) (11-3): A proportionality constant relating the rate of a chemical reaction to the appropriate product of concentration terms for the species involved. Rate constants vary with temperature according to the Arrhenius equation.

Rate law (11-3): A mathematical equation relating the rate of a chemical equation to the concentrations of reactants, products, or other species.

Rate-determining step (11-6): The slowest step in a reaction mechanism.

Reactant (3-2): Substance present at the outset of a chemical reaction that is consumed during the reaction; reactants appear on the left-hand side of a chemical equation.

Reaction mechanism (11-6): Series of one or more elementary steps that provide a pathway for converting reactants into products.

Reaction quotient (*Q*) (12-3): Expression identical in form to the equilibrium constant but in which the concentrations do not correspond to equilibrium values. Comparison of the reaction quotient to the equilibrium constant predicts the direction of spontaneous change.

Reaction rate (11-2): The change in the concentration of a reactant or product as a function of time, normalized for stoichiometry.

Reactive intermediate (11-6): A chemical species that is generated in one step in a reaction mechanism and consumed in a later step. Intermediates do not appear in the overall chemical equation for a reaction.

Reactor core (14-6): The central part of a nuclear reactor that contains the uranium fuel, moderator, coolant, and supporting structures.

Redox reaction (13-2): A chemical reaction involving the transfer of electrons from one species to another. See also *Oxidation–reduction reaction*.

Reduction (13-2): The gain of electrons by some chemical species in a chemical reaction.

Reformulated gasoline (RFG) (4-6): Gasoline containing additives to provide at least 2% oxygen by weight. Reformulated gasoline is currently required in some urban areas in California, Texas, Kentucky, and Virginia, as well as in Washington, D.C.

Refraction (6-2): The bending of light when it moves between two different media, such as air and the glass of a prism or lens.

Relative biological effectiveness (RBE) (14-7): A variable to quantify the ability of a particular type of radiation to cause tissue damage. Used in calculating the equivalent dose. Also called the *quality factor*.

Representative elements (2-5): Elements of Groups 1, 2, and 13–18, for which the highest lying valence electron occupies an *s* or *p* orbital. Also called *main group elements*.

Resonance hybrid (7-5): The actual structure of a molecule represented by a combination of two or more different Lewis structures.

Resonance structure (7-5): One of two or more individual Lewis structures that contribute to the resonance hybrid for a molecule. The resonance structures differ from one another only in the placement of electrons, but all show the same arrangement of the atoms themselves.

Reversible (10-6): A process near equilibrium, which can be reversed by a small incremental change of some variable.

Root-mean-square speed (5-6): The square root of the average of the squares of the velocities of a collection of particles.

S

Sacrificial anode (13-3): A piece of a more reactive metal used to protect another metal from corrosion. The presence of the sacrificial anode assures that the part to be protected will be the cathode and hence will not be oxidized.

Salt (3-3): Ionic compound that can be formed by the reaction of an acid with a base.

Salt bridge (13-2): Device used to allow the flow of ions into the individual half-cell compartments in an electrochemical cell and maintain charge neutrality.

Schrödinger equation (6-4): Mathematical equation for the allowed quantum mechanical wave functions for an atom or molecule.

Scientific method (1-1): An approach to understanding that begins with the observation of nature, continues to hypothesis or model building in response to that observation, and ultimately includes further experiments that either bolster or refute the hypothesis.

Scientific models (1-3): Empirical or mathematical descriptions that scientists create to make sense of a range of observations.

Scientific notation (1-4): Representation of a number by a value between 1 and 10 times a power of 10. (Engineers in some fields prefer to use a number between 1 and 100 times a power of 10.) Scientific notation is a convenient way of displaying very large or small numbers, as well as a way of making clear which digits are significant.

Sea-of-electrons model (8-3): Simplified description of metallic bonding in which the valence electrons of metal atoms are delocalized and move freely throughout the solid rather than being tied to any specific atom.

Sea of instability (14-4): The region surrounding the band of stability in the chart of the nuclides, in which unstable nuclides are found.

Second law of thermodynamics (10-4): The entropy of the universe is constantly increasing. For any spontaneous process, $\Delta S_u > 0$.

Second-order reaction (11-4): Any reaction whose rate is given by a rate law of the form, Rate = $k[A]^2$ or Rate = $k[A][B]$, where A and B represent chemical species in the reaction (usually reactants).

Secondary cell (or secondary battery) (13-5): A battery that can be recharged by applying an external electric current to reverse the cell reaction and regenerate the reactants.

Secondary quantum number (ℓ) (6-4): One of the three quantum numbers identifying an atomic orbital. Specifies the subshell in which an orbital is found. Allowed values of ℓ are related to the value of the principal quantum number (n).

Secondary standard (5-1): Science-based limit established by the EPA for acceptable levels of *criteria pollutants* based on considerations of our environment and property. (See also *primary standard*.)

Semimetals (2-5): Elements displaying some of the properties of metals and some of the properties of nonmetals.

Shell (6-4): Set of orbitals in an atom that have the same value of the principal quantum number (n).

Shielding (6-5): Reduction of the effective nuclear charge experienced by valence electrons due to the presence of inner shell electrons in an atom.

Side reactions (4-4): Any reactions that lead to products other than those desired in an industrial or laboratory synthesis.

Sigma (σ) bond (7-6): A chemical bond in which the electron density is concentrated along the line of centers of the two atoms, such as occurs through the interaction of two *s* orbitals or the "end-to-end" interaction of two *p* orbitals.

Significant figures (1-4): A method for expressing the degree of uncertainty in a measurement in which only those digits known accurately are reported.

Simple cubic (sc) (8-2): Crystal lattice in which particles are arranged at the corners of a cube. Also called *primitive cubic*.

Solid (1-2): State of matter in which a substance has a fixed shape and volume.

Solubility (3-3): The ability of a compound to dissolve in solution. It can be expressed in terms of the maximum amount of solute for a given amount of solvent at a particular temperature.

Solubility product constant (K_{sp}) (12-6): Equilibrium constant for the dissolution of a sparingly soluble compound, consisting of the concentrations of the resulting ions raised to appropriate stoichiometric powers.

Soluble (3-3): Description of a solid that dissolves in solution to a reasonably large extent.

Solute (3-3): The minor component(s) of a solution.

Solution (3-3): A homogeneous mixture of two or more substances in a single phase.

Solvent (3-3): The component present in the greatest amount in any solution.

Sparingly soluble (12-6): Dissolving in water only to a very small extent in accordance with an appropriate solubility product constant.

Specific heat (c) (9-4): The amount of energy required to raise the temperature of one gram of a substance by one degree (°C or K).

Spectator ions (3-3): Ions that are present in a solution but do not actively participate in a chemical reaction.

Spin paired (6-5): Used to describe two electrons, meaning they have opposite spins. To occupy the same orbital, two electrons must be spin paired.

Spin quantum number (m_s) (6-4): Quantum number specifying the spin of an electron. The two possible values (+1/2 and −1/2) are usually referred to as "spin up" and "spin down."

Spontaneous process (10-2): Any thermodynamic process that takes place without continuous intervention. Note that being spontaneous does not necessarily mean that a process will be fast.

Standard Gibbs free energy change ($\Delta G°$) (10-7): Free energy change for a reaction in which all participating substances are present in their thermodynamic standard state (gases at 1 atm, solutions at 1 M, T + 25°C). See also *Gibbs free energy*.

Standard hydrogen electrode (SHE) (13-3): Electrode used to define the scale of standard reduction potentials, consisting of a platinum electrode at which 1 M H_3O^+ ions are reduced to H_2 gas at 1 atm.

Standard molar entropy ($S°$) (10-5): Absolute entropy of 1 mole of a substance at 25°C.

Standard reduction potential (13-3): The potential of an electrochemical cell consisting of the given half-reaction and a standard hydrogen electrode; provides a convenient measure of the tendency of a species to undergo oxidation or reduction.

Standard state (9-5): The most stable form of a substance at 25°C and 1 atm.

Standard temperature and pressure (STP) (5-5): For gases, $T = 0°C$ and $P = 1$ atm.

State function (9-6): A thermodynamic variable whose value depends only on the state of the system, not on its history.

Statistical mechanics (10-3): A mathematical approach to thermodynamics which treats macroscopic systems by considering the statistical laws governing the interactions between their constituent particles.

Stoichiometric coefficients (3-2): The numbers placed in front of each species to balance a chemical equation.

Stoichiometry (3-2): The study of quantitative relationships between the amounts of reactants and products in a chemical reaction.

Strong electrolyte (3-3): A substance that ionizes or dissociates completely in water to produce an aqueous solution in which only the constituent ions and not intact molecules are present.

Strong force (14-4): A force that acts at very short distances and is responsible for holding atomic nuclei together.

Subshell (6-4): Set of orbitals in an atom that have the same values for the n and ℓ quantum numbers.

Surface tension (8-5): Energy required to overcome the attractive forces between molecules at the surface of a liquid.

Surroundings (9-3): The rest of the universe except for the thermodynamic system under consideration, with which the system can exchange energy.

Symbolic perspective (1-2): Viewpoint of chemistry focusing on symbolic representations of the substances involved through formulas, equations, etc.

Syndiotactic polymer (8-6): A polymer in which substituent groups are arranged on alternate sides of the polymer chain.

System (9-3): The part of the universe that is being considered in a thermochemical problem.

Systematic error (1-3): Uncertainty from some unknown bias or flaw in the equipment with which a measurement is made; a systematic error makes the measurement consistently either too high or too low and cannot be minimized by averaging repeated measurements.

T

Temperature scales (1-4): Systems for measuring temperature, defined by choosing two reference points and setting a fixed number of degrees between them. Common temperature scales include Fahrenheit, Celsius, Kelvin, and Rankine.

Termolecular (11-6): Description of an elementary step in which three reactant particles must collide with one another.

Theoretical yield (4-4): The hypothetical maximum amount of product that can be obtained in a chemical reaction under ideal conditions.

Thermochemical equation (9-5): A balanced chemical equation that includes the enthalpy change for the corresponding reaction.

Thermochemistry (9-1): The study of the energetic consequences of chemistry.

Thermocouple gauge (5-7): A sensor in which the conduction of heat away from a hot filament is used to determine pressure.

Thermoplastic polymer (8-6): A polymeric material or plastic that can be repeatedly softened by heating and hardened by cooling.

Thermosetting polymer (8-6): A polymeric material or plastic that forms covalent crosslinks such that it cannot be melted without decomposing.

Third law of thermodynamics (10-5): The entropy of a perfect crystal of any pure substance approaches zero as the temperature approaches absolute zero.

Titration (4-5): A process whereby a solution-phase reaction is carried out under controlled conditions using a known amount of one reactant so that the amount of the other reactant can be determined with high precision.

Torr (5-2): A unit of pressure equal to 1 mm Hg. 760 torr = 1 atm.

Total ionic equation (3-3): A chemical equation in which all ions and molecules present in solution, including spectator ions, are shown.

Trace analysis (6-1): Highly sensitive analytical techniques that can determine elemental composition at or below the parts per million level.

Transition metals (2-5): The elements in Groups 3 through 12 of the periodic table, in which valence electrons occupy *d* orbitals.

Transmutation (14-6): Reaction in which one nucleus changes to another, either by natural decay or in response to some outside intervention, such as neutron bombardment.

Triple bond (7-3): A covalent bond formed by the sharing of three pairs of electrons between the same pair of atoms.

Turnover number (11-7): The effectiveness of a catalyst, expressed as the number of molecules that can react per catalyst binding site per unit time.

U

Ultra-high molecular weight polyethylene (UHMWPE) (2-8): Polyethylene consisting of extremely long polymer chains and displaying unusually high strength.

Uncertainty principle (6-4): It is impossible to determine both the position and momentum of an electron simultaneously and with arbitrary accuracy. Also called *Heisenberg's uncertainty principle.*

Uniform corrosion (13-1): Degradation of a metal surface by oxidation from uniform exposure to environmental moisture.

Unimolecular (11-6): Description of an elementary step involving only a single reactant particle.

Unit cell (8-2): The smallest collection of atoms or molecules that can be repeated in three dimensions to produce the overall structure of a solid.

Units (1-4): Labels designating the type of quantity measured and the particular scale on which the measurement was made. Most measurements have no physical meaning if their units are omitted.

Universal gas constant (*R*) (5-1): The proportionality constant in the ideal gas law; its value in SI units is 8.314 J K^{-1} mol^{-1}, and a frequently useful value is 0.08206 L atm mol^{-1} K^{-1}.

Universe (9-3): In thermodynamics, the system under consideration plus its surroundings.

V

Valence band (8-3): A band of orbitals or energy states populated by the valence electrons in the ground state of a bulk solid.

Valence bond model (7-6): A description of chemical bonding in which all bonds result from overlap between atomic orbitals of the two atoms forming the bond.

Valence electrons (6-5): The electrons in an atom's highest occupied principal quantum shell, plus any electrons in partially filled subshells of lower principal quantum number. The electrons available for bond formation.

van der Waals equation (5-6): An empirical equation of state used to describe quantitatively the behavior of a real gas by taking into account the volume of molecules and the interaction between them.

Vapor pressure (8-5): The equilibrium gas phase pressure of a substance in equilibrium with its pure liquid or solid in a sealed container.

Vaporization (9-5): The phase change in which a liquid is converted to a gas.

Visible light (6-2): The portion of the electromagnetic spectrum that can be detected by human eyes, typically wavelengths between 400 and 700 nm.

Volatile (8-5): Having a high vapor pressure; able to vaporize easily.

Volatile organic chemical (VOC) (5-1): Carbon-based molecules that evaporate readily, are often emitted into the atmosphere, and are involved in the formation of smog.

VSEPR theory (7-8): Valence-shell electron-pair repulsion theory. A simple method for predicting the shapes of molecules; it states that molecules assume a shape that allows them to minimize the repulsions between bonds and lone pairs in the valence shell of the central atom.

W

Wave function (ψ) (6-4): A mathematical function providing the quantum mechanical wave representation of an electron in an atom; a solution to the Schrödinger wave equation.

Wavelength (λ) (6-2): The distance between adjacent corresponding points (peaks, valleys, etc.) of a wave.

Wave–particle duality (6-2): The notion that, in some situations, light or electrons are best described as waves, while in other cases a particle description works better.

Weak electrolyte (3-3): A substance that ionizes or dissociates only partially in water to produce an aqueous solution in which both individual ions and intact molecules are present.

Work (*w*) (9-2): A transfer of energy accomplished by a force moving a mass some distance against resistance.

X

X-ray fluorescence (XRF) (6-1): Highly sensitive nondestructive analytical technique that can determine elemental composition of a sample by detecting wavelengths of X-rays emitted after excitation with high-energy photons.

Y

Yield strength (1-6): A measure of the amount of force required to produce a specified deformation of a material.

Z

Zero-order reaction (11-4): Any reaction whose rate is independent of the concentrations of all participating substances.

Zinc-air battery (13-5): A primary cell in which zinc is oxidized at the anode and oxygen is reduced at the cathode. Because oxygen can be obtained on demand from the air, these batteries can have high energy density and are suitable for use as emergency power supplies for cell phones or other devices.

Index

Italicized page numbers indicate illustrations, page numbers followed by "t" indicate tables, those followed by "q" indicate end-of-chapter questions, and those followed by "k" indicate glossary entries.

H

I

Q

R

S

T